Artificial Intelligence and Biological Sciences

Editor
P.V. Mohanan
Sree Chitra Tirunal Institute for Medical Sciences and Technology
Department of Science and Technology, Government of India
Thiruvananthapuram, India

CRC Press
Taylor & Francis Group
Boca Raton London New York

CRC Press is an imprint of the
Taylor & Francis Group, an **informa** business
A SCIENCE PUBLISHERS BOOK

First edition published 2025
by CRC Press
2385 NW Executive Center Drive, Suite 320, Boca Raton FL 33431

and by CRC Press
4 Park Square, Milton Park, Abingdon, Oxon, OX14 4RN

CRC Press is an imprint of Taylor & Francis Group, LLC

Library of Congress Cataloging-in-Publication Data (applied for)

ISBN: 978-1-032-78120-4 (hbk)
ISBN: 978-1-032-79560-7 (pbk)
ISBN: 978-1-003-49272-6 (ebk)

DOI: 10.1201/9781003492726

Typeset in Times New Roman
by Prime Publishing Services

Preface

Artificial intelligence (AI) is the ability of machines to simulate the intelligence capability of higher organisms. The term was first coined by John McCarthy in 1956. AI has established its prevalence in many fields like mathematics, physiology, computing, biology and psychology. Ideally, an AI system should respond logically, be self-aware, have the ability to learn from the experience, and be able to discern and respond to external environments. With deep learning and machine learning algorithms, intellect systems that can perform the activities requiring human intellect, can be developed.

Advance in AI in medical and biological sciences have opened new ways for drug developments. Novel therapeutic molecules and its target action can be easily predicted and can be modified. AI helps in disease detection and makes diagnosis much faster. AI has made personalized precision medicines a reality. These approaches will make decision-making faster, with lesser errors. That can improve better risk analysis especially disease diagnosis and selecting treatment strategy. In agricultural practices, the exact combination of fertilizers, pesticides, herbicides, soil management, water requirement analysis, yield prediction and overall crop management can be achieved by implementing AI interventions. Thus AI could bring about improvement in agriculture, medical research, pharmaceuticals and bio-based industries for a sustainable life.

The book has 28 chapters consisting of the latest information on current status and methodologies of Artificial Intelligence in Biological Science and its applications. Chapters have been written by leading experts in the field. The book provides an in-depth understanding on the applications of AI in the biological sciences. It is useful for the pharma companies, CROs, product developers, students, researchers, academicians, policy makers and practitioners.

It has been a great pleasure to work with multidisciplinary researchers/scientists/professors, who have several years of experience in their chosen areas. Without their cooperation, it would not have been possible to bring out this book, focusing on Artificial Intelligence and its applications. Furthermore, I wish to express my sincere gratitude to the Director and Head, of the Biomedical Technology Wing, Sree Chitra Tirunal Institute for Medical Sciences and Technology (Department of Science and Technology, Government of India), Trivandrum, Kerala for their phenomenal support and advice throughout this work. The support and encouragement from family and friends is greatly acknowledged.

July, 2024

P.V. Mohanan
Editor

Contents

Section II: Applications of Artificial Intelligence in Biological Sciences

List of Contributors

AB Pant
Systems Toxicology Group, CSIR-Indian Institute of Toxicology Research Vishvigyan Bhavan, Mahatma Gandhi Marg, Lucknow, India., Academy of Scientific and Innovative Research (AcSIR), Ghaziabad, India.

Abhishek Sengupta
Systems Biology and Data Analytics Lab, Amity Institute of Biotechnology, Amity University, Noida, India.

Abhishek Subramanian
Department of Biotechnology, Indian Institute of Technology Hyderabad, Telangana 502284, India.

Aishath Shaheeda
Department of Cell and Molecular Biology, Manipal School of Life Sciences, Manipal Academy of Higher Education, Manipal, India.

Alakto Choudhury
Systems Biology and Data Analytics Lab, Amity Institute of Biotechnology, Amity University, Noida, India.

Alex C Varghese
Fertility Associates, Auckland, New Zealand

Alex Thomas
Toxway Research Solutions, Mumbai, Maharshtra, India.

Anjana Soorajkumar
Dubai Health, Mohammed Bin Rashid University of Medicine and Health Sciences, United Arab Emirates, Dubai.

Arasu Narayan
Northeastern University, Seattle, USA.

Aravind Kumar Rengan
Department of Biomedical Engineering, Indian Institute of Technology Hyderabad, India.

Aravind Sridhar
Auxillium Health Inc, Milwaukee, USA.

Arya Vinod
School of Biological Sciences, Central University of Kerala, India.

Ashtami J
Toxicology Division, Sree Chitra Tirunal Institute of Medical Science and Technology, Thiruvananthapuram, Kerala.

Ashwini S Galande
Medical Optics and Sensors Laboratory, Department of Biomedical Engineering, Indian Institute of Technology, Hyderabad, India.

Aswathy Vijay
Medical Optics and Sensors Laboratory, Department of Biomedical Engineering, Indian Institute of Technology, Hyderabad, India.

Athira SS
Division of Toxicology, Sree Chitra Tirunal Institute of Medical Science and Technology, Thiruvananthapuram, Kerala, India.

Ayush Khandelwal
Department of Cell and Molecular Biology, Manipal School of Life Sciences, Manipal Academy of Higher Education, Manipal, India.

Baiju G Nair
Department of Bioscience and Engineering, National Institue of Technology, Calicut, Kerala, India

Bala Pesala
Auxillium Health Inc, Milwaukee, USA.

Basu Bhattacharjee
Department of Biomedical Engineering, Indian Institute of Technology Hyderabad, India.

Deepak Modi
Department of Molecular and Cellular Biology, National Institute for Research in Reproductive Health and Child Health, Mumbai, India.

Dhananjay K. Pandey
Amity Institute of Biotechnology, Amity University Jharkhand, Ranchi, India.

Geethanjali Radhakrishnan
Auxillium Health Inc, Milwaukee, USA.

Grace R. Raji
School of Biological Sciences, Central University of Kerala, India.

Greeshma N
Department of Biomedical Engineering, Indian Institute of Technology Hyderabad, Hyderabad, India.

Gunpreet Kaur
University Centre of Excellence in Research, Baba Farid University of Health Sciences, Faridkot, India.

Jeffrey Niezgoda
Auxillium Health Inc, Milwaukee, USA.

Kamlesh Yadav
School of Engineering Medicine, Texas A&M University, Houston, USA.

Kannan Pakshirajan
Department of Biosciences and Bioengineering, Indian Institute of Technology Guwahati, Guwahati, India.

Katarzyna Swist-Szulik
Royal Manchester Children's Hospital, UK.

Kavati Pavan Kumar
Department of Biomedical Engineering, Indian Institute of Technology Hyderabad, India.

Kavipriya Murugaiyan
Department of Biomedical Engineering, Indian Institute of Technology Hyderabad, India.

Khader Shameer
Faculty of Medicine, School of Public Health, Imperial College of London.

Kritika Shukla
Systems Biology and Data Analytics Lab, Amity Institute of Biotechnology, Amity University, Noida, India.

Kunal Kumar
Amity Institute of Biotechnology, Amity University Jharkhand, Ranchi, India.

Kunal Ranjan
Amity Institute of Biotechnology, Amity University Jharkhand, Ranchi, India.

Manjoosha Ratna Yerrapragada
Department of Biomedical Engineering, Indian Institute of Technology Hyderabad, Hyderabad, India.

Manjoosha RY
Department of Biomedical Engineering, Indian Institute of Technology, Hyderabad, India.

Mannat Singh
Systems Biology and Data Analytics Lab, Amity Institute of Biotechnology, Amity University, Noida, India.

Medha Bairy
Department of Cell and Molecular Biology, Manipal School of Life Sciences, Manipal Academy of Higher Education, Manipal, India.

Megha KB
Division of Toxicology, Sree Chitra Tirunal Institute of Medical Science and Technology, Thiruvananthapuram, Kerala.

Miguel Pais-Vieira
iBiMED – Institute of Biomedicine, Department of Medical Sciences, University of Aveiro Portugal.

Mohanan PV
Division of Toxicology, Sree Chitra Tirunal Institute of Medical Science and Technology, Thiruvananthapuram, Kerala, India.

Mrudula Gore
Department of Cell and Molecular Biology, Manipal School of Life Sciences, Manipal Academy of Higher Education, Manipal, India.

Nagarajan Ganapathy
Department of Biomedical Engineering, Indian Institute of Technology Hyderabad, India.

Naorem Bela Devi
Department of Biosciences and Bioengineering, Indian Institute of Technology Guwahati, Guwahati, India.

Neha Pinky
Department of Toxicology, Jamia Hamdard, India.

Nishant Kumar
Department of Biotechnology, Indian Institute of Technology Hyderabad, Telangana 502284, India.

Pankhi Vatsa
Systems Toxicology Group, CSIR-Indian Institute of Toxicology Research Vishvigyan Bhavan, Mahatma Gandhi Marg, Lucknow, India.
Academy of Scientific and Innovative Research (AcSIR), Ghaziabad, India.

Parveen Bansal
University Centre of Excellence in Research, Baba Farid University of Health Sciences, Faridkot, India.

Parveen Kumar Goyal
Department of Pharmacy, Panipat Institute of Engineering and Technology (PIET), Samalkha, Panipat, India

Pratap Reddy Kuppasani
Medical Electronics, Tata Elxsi Limited, Pune, India.

Priyanka Narad
Division of Biomedical Informatics (BMI), Indian Council of Medical Research, Ansari Nagar, New Delhi, India.

Rahul Raghavan Thekke Veedu
Hospital for Special Surgery-Weill Cornell Medicine, New York, U.S.A.

Rajani Sharma
Amity Institute of Biotechnology, Amity University Jharkhand, Ranchi, India.

Ranjini Kudva
Department of Pathology, Kasturba Medical College, Manipal Academy of Higher Education, Manipal, India

Ravinder Sharma
Faculty of Pharmaceutical Sciences, ICFAI University, Baddi, India.

Renu John
Department of Biomedical Engineering, Indian Institute of Technology, Hyderabad, India.

Rishi Kumar
Indian Pharmacopoeia Commission, Ghaziabad, India.

Romy Roy
Type 3 Civilization Technologies Pvt Ltd, Kerala, India.

Roopal Behera
Department of Cell and Molecular Biology, Manipal School of Life Sciences, Manipal Academy of Higher Education, Manipal, India.

Sachin Shetty
Department of Cell and Molecular Biology, Manipal School of Life Sciences, Manipal Academy of Higher Education, Manipal, India.

Sandeep Gopalakrishnan
Auxillium Health Inc, Milwaukee, USA., College of Health Professions and Sciences, University of Wisconsin-Milwaukee, USA.

Sandeep Karunakaran
Medical Optics and Sensors Laboratory, Department of Biomedical Engineering, Indian Institute of Technology, Hyderabad, India.

Shajil UK
Department of Bioscience and Engineering, National Institute of Technology, Calicut, Kerala India

Shama Prasada K
Department of Cell and Molecular Biology, Manipal School of Life Sciences, Manipal Academy of Higher Education, Manipal, India.

Shanthi PB
Department of Computer Science and Engg., Manipal Institute of Technology, Manipal Academy of Higher Education, Manipal, India.

Soniya Shende
Type 3 Civilization Technologies Pvt Ltd, Kerala, India.

Subh Naman
Faculty of Pharmaceutical Sciences, ICFAI University, Baddi, India.

Suhel Parvez
Department of Toxicology, Jamia Hamdard, India.

Sujith A
Department of Chemistry, National Institute of Technology, Calicut, India.

Suparna Sen
Department of Biosciences and Bioengineering, Indian Institute of Technology Guwahati, Guwahati, India.

Swarubini PJ
Department of Biomedical Engineering, Indian Institute of Technology Hyderabad, India.

V.B. Sameer Kumar
School of Biological Sciences, Central University of Kerala, India.

V.P Thejashwini
Type 3 Civilization Technologies Pvt Ltd, Kerala, India.

Vikas Gupta
University Centre of Excellence in Research, Baba Farid University of Health Sciences, Faridkot, India.

Vikas Thapa
Medical Optics and Sensors Laboratory, Department of Biomedical Engineering, Indian Institute of Technology, Hyderabad, India.

Zitin Wali
Department of Toxicology, Jamia Hamdard, India.

CHAPTER

1

Overview of Artificial Intelligence

Sachin Shetty,[1] *Ayush Khandelwal,*[1] *Mrudula Gore,*[1] *Roopal Behera,*[1] *Aishath Shaheeda,*[1] *Medha Bairy* and *Shama Prasada K*[1*]

1. Introduction

Artificial intelligence (AI) is a rapidly expanding field that offers greater insights into how to simplify human life. John McCarthy first used the term "artificial intelligence" in 1956. He defined "the science and engineering of making intelligent machines". Artificial intelligence (AI) and machine learning (ML) have had significant impacts on humanity over the past few years, and they have been applied in a wide range of industries, including engineering, communications, manufacturing, biotechnology and healthcare (Holzinger et al., 2023). Biotechnology already uses AI extensively to address a range of issues. In this chapter, we have discussed the role of AI in industrial biotechnology, healthcare, stem cell and regenerative medicine, biomedical research and bioinformatics.

2. AI in Industrial Biotechnology

Industrial biotechnology is a promising field in biotechnology involving the application of biological systems and processes to produce materials, chemicals, and pharmaceuticals. Industrial biotechnology is a field at the intersection of biology, chemistry, and engineering and has experienced remarkable advancements over the decade. AI has emerged as a transformative technology in the realm of industrial biotechnology. The incorporation of AI into industrial biotechnology processes has ushered in a new era of efficiency, innovation, and sustainability. By harnessing the power of machine learning, deep learning, data analytics, and natural language processing, AI has the potential to enhance productivity, reduce costs, and accelerate the development of biotechnological products. This approach has unlocked novel avenues for innovation and efficiency, paving the way for sustainable solutions to address global challenges in healthcare, agriculture, and environmental

[1] Department of Cell and Molecular Biology, Manipal School of Life Sciences, Manipal Academy of Higher Education, Manipal, India.

* Corresponding author: shama.prasada@manipal.edu

sustainability. By harnessing the power of AI, biotechnologists are revolutionizing the way we develop and manufacture products in the biotech industry (Nielsen et al., 2022).

2.1 AI-driven Bioprocessing Optimization

Bioprocess optimization is an important field where major progress has been made in industrial biotechnology. Bioprocessing and manufacturing in biotechnology involve complex, large-scale production of biopharmaceuticals, vaccines, and other bioproducts. Traditional trial-and-error approaches in fermentation, enzyme production, and bioreactor operation are being replaced by AI-driven solutions. Machine learning algorithms for microalgae identification, wastewater treatment, fungal bioprocessing, composting, thermochemical conversion, and anaerobic digestion, etc., can analyse vast datasets and determine the best conditions for maximum yield, reduced resource consumption, and product quality improvement. AI algorithms, such as fuzzy intelligence (FL), genetic algorithm (GA), particle swarm optimization (PSO), Gaussian process (GP), support vector machine (SVM), gradient boosting machine (GBM), artificial neural network (ANN), principal component analysis (PCA), partial least squares regression (PLSR), and response surface methodology (RSM), can analyse real-time bioprocess data from bioprocesses while optimizing conditions such as temperature, pH, nutrient levels, and fermentation times. This not only improves yield but also reduces costs and minimizes waste and environmental impacts. This approach also minimizes human intervention and ensures consistent product quality. AI-powered bioprocessing promises to increase the availability of life-saving medications, vaccines, and biologics (Mondal et al., 2023).

2.2 AI Predictive Modelling for Strain Engineering

In the realm of strain development, AI plays a pivotal role in predictive modelling. By analysing genetic information, metabolic pathways, and growth kinetics, AI algorithms can predict how genetic modifications will affect the production of valuable compounds. This has streamlined the design and construction of microbial strains for bioproduction, leading to faster development cycles and more efficient bioprocesses. AI-driven tools can predict the behaviour of genetically modified microorganisms and help design strains that are better suited for bioproduction. These tools can optimize metabolic pathways to increase the production of desired compounds, such as biofuels, enzymes, or pharmaceuticals (Radamson and Simeon, 2018).

2.3 AI in Protein Engineering

AI-driven models, such as the Alpha Fold model, can predict protein structures with remarkable accuracy, which is vital for rational drug design, vaccine development, understanding protein–protein interactions, understanding protein functions and developing targeted therapies. In industrial biotechnology, enzymes are critical catalysts for various processes. AI algorithms, coupled with high-throughput screening, enable the design of novel enzymes with tailored properties. This approach has vast implications for biofuel production, and bioplastic synthesis among other processes. (Lim et al., 2022).

2.4 AI in Drug Discovery and Development

The drug discovery procedure is conventionally time-consuming and costly. However, AI has emerged as a game changer, drastically reducing the time and resources required for drug development. AI is accelerating drug discovery by analysing vast datasets of biological and chemical information. Machine learning models can predict the potential of compounds, identify drug candidates, and simulate drug interactions, significantly reducing the time and cost of bringing new drugs to the market. In industrial biotechnology, AI-driven drug discovery platforms have enabled the identification of novel enzymes, microbial strains, and biocatalysts for various applications. AI-driven molecular modelling techniques enable the rapid design and optimization of drug molecules. Machine learning models can analyse vast datasets of genomic and proteomic data to identify potential drug targets, validate their relevance in disease pathways, predict their efficacy, optimize drug candidates and assess potential side effects. This approach significantly speeds up the drug discovery process and reduces costs. This approach accelerates the process of selecting promising candidates for further research. AI can assist in designing drug formulations, optimizing drug delivery methods, and predicting how different formulations will behave in the body. Machine learning models can be used to analyse disease mechanisms and identify potential drug targets. AI algorithms can process and interpret genomic data to identify disease-associated genetic variations, predict disease risk, and improve custom-made treatment strategies. Moreover, AI can analyse existing drug databases and biological data to identify new therapeutic uses for existing drugs. This approach can save time and resources in drug development (Çelik et al., 2021).

2.5 AI in Biomarker Discovery, Personalized Medicine and Genomic Analysis

AI plays a pivotal role in this endeavour by analysing vast genomic datasets to detect genetic markers linked with diseases or treatment responses. AI can identify and validate biomarkers for disease diagnosis, prognosis, and monitoring, which is crucial for early disease detection and precision medication. In biopharmaceuticals, AI helps tailor treatments to individual patients based on their genetic makeup and health history, optimizing therapeutic outcomes. AI can assist in designing clinical trials by optimizing patient selection criteria, predicting patient responses, and identifying suitable biomarkers for monitoring drug efficacy. AI algorithms can provide physicians with valuable insights into which treatment options are most likely to succeed for a particular patient, reducing adverse effects and improving overall treatment outcomes. The use of AI can help patients analyse their genetic, molecular, and clinical data to tailor treatments to their individual characteristics, improve treatment outcomes and minimize side effects. AI can analyse real-world patient data and adverse event reports to detect potential safety concerns and adverse reactions to drugs, enabling quicker regulatory responses (Mann et al., 2021; Radamson, 2018).

2.6 AI in Gene Editing

CRISPR-based gene editing techniques can be enhanced with AI to predict the outcomes of genetic modifications more accurately. These methods rely on algorithms that leverage the continually increasing number of genes editing datasets, aiding in the prediction of CRISPR gRNA activity and specificity scores. Consequently, recent research has introduced several algorithms aimed at forecasting both the desired and unintended effects of CRISPR gRNAs, enhancing the performance and precision of the CRISPR–Cas9 system. This has accelerated the development of genetically modified organisms for industrial applications (Bhat et al., 2022).

2.7 AI in Quality Control and Assurance and Regulatory Compliance

Computer vision and AI can be used to automate quality control processes and inspect products for defects, contaminants, or inconsistencies in real time. AI-driven sensors and monitoring systems have revolutionized quality control in industrial biotechnology. Real-time data collection and analysis enable early detection of deviations and process irregularities, ensuring product consistency and compliance with quality standards. Automated systems can make immediate adjustments, minimize human error and improve overall product quality. AI-powered image recognition and data analytics are transforming quality control in biotech manufacturing. Automated systems can detect deviations from desired product specifications, reducing waste and ensuring consistency in production. AI can assist in ensuring regulatory compliance by analysing data and documentation to meet stringent industry standards and quality control requirements (Zanca et al, 2022).

2.8 AI in Laboratory Automation, Natural Language Processing and Supply Chain Optimization

AI-powered robotics and automation systems can perform repetitive tasks in laboratories, improving efficiency and reducing errors. NLP techniques can help researchers sift through vast amounts of scientific literature, extract valuable insights, and aid in the discovery process. AI can optimize supply chains by predicting demand, managing inventory, and optimizing transportation routes. This approach is crucial for ensuring that bioproducts reach their destinations efficiently (Çelik et al., 2022). The use of AI in biotechnology is a rapidly evolving field, and its applications are continually expanding. It has the potential to accelerate research, reduce costs, improve patient outcomes, and drive innovation in the biotechnology industry. However, it also presents challenges associated with data confidentiality, ethics, and regulatory deliberations that need to be spoken as it becomes more integrated into the industry (Çelik et al., 2022).

2.9 Challenges and Ethical Considerations

Despite the tremendous benefits of AI in industrial biotechnology, there are challenges and ethical considerations that need to be addressed. Guaranteeing data privacy and safety, avoiding bias in AI algorithms, and addressing concerns about job displacement are important issues. Moreover, the ethical use of AI in areas such as genetic engineering and synthetic biology raises questions about potential risks and regulations that need careful consideration. Additionally, the potential for bias in AI algorithms and its impact on healthcare disparities must be carefully addressed. It is crucial for scientists, policymakers, and industry leaders to collaborate in establishing ethical guidelines and regulations to harness the full potential of AI in biotechnology while mitigating potential risks (Çelik et al., 2022).

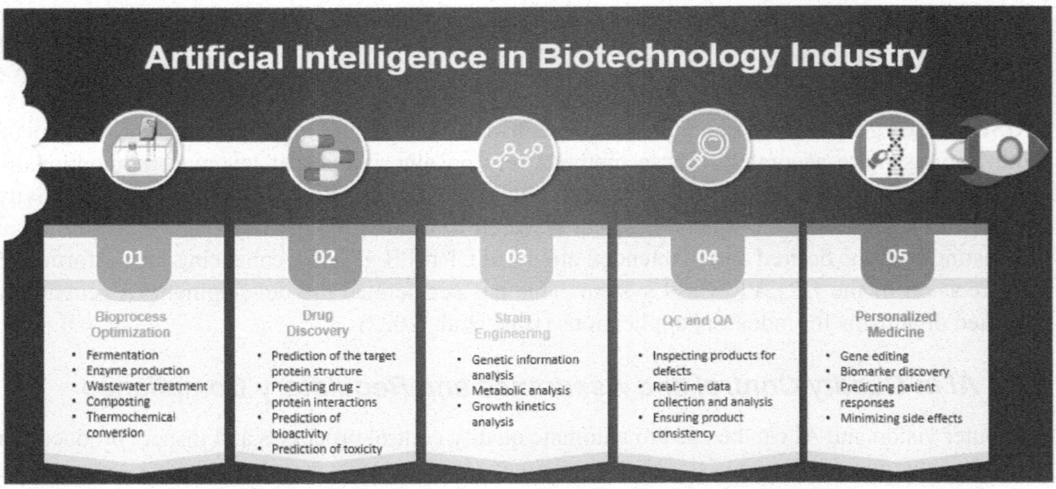

Fig. 1 Role of artificial intelligence in the biotechnology industry

2.10 Future Prospects in Industrial Biotechnology

The integration of AI in industrial biotechnology is still evolving, with many uncharted territories and challenges. The future of AI in industrial biotechnology is promising. We can expect further integration of AI with other emerging technologies, such as CRISPR–Cas9, for precision genome editing, advanced robotics for automated laboratory workflows, and blockchain for enhanced traceability and transparency in supply chains. Additionally, AI-powered biotechnology is likely to play a vital role in addressing universal challenges such as climate change and sustainable agriculture by optimizing processes and developing eco-friendly solutions. Privacy concerns, data security, and the need for interpretability in AI models remain paramount. However, the potential benefits in terms of efficiency, innovation, and sustainability are too compelling to ignore. In conclusion, artificial intelligence has emerged as a transformative force in industrial biotechnology. This approach has accelerated bioprocess optimization, revolutionized drug discovery, and enabled efficient bioproduction. Moreover, AI has facilitated sustainable practices and addressed global challenges in healthcare, agriculture, and environmental sustainability. However, challenges associated with data privacy, bias, and ethical considerations must be carefully navigated. As AI continues to advance, its integration with biotechnology promises a future of innovation and progress, benefiting humanity and the planet (Fig. 1).

3. AI in Medicine and Health Care

Do intelligent machines pose challenges or augment our vision in various sectors where further upgrading is imperative? We are witnessing an era where intelligent machines play a significant role in all sectors by helping people interpret the world, understand and analyse language and learn from real-world examples, as humans do. Like an experienced man who easily deals with problems, an experienced machine trained on an artificial neural network (ANN) is used to analyse and interpret new data from similar data. Intelligence is the ability to collect knowledge, find the reason behind it, and solve various problems we encounter. In that sense, artificial intelligence involves the study and development of machines and/or software that can learn and gather knowledge and can reason, communicate and manipulate objects. It is an amalgamation of computation, reasoning, perception and action (Mustapha et al., 2020). In this era of silicon chip and technology, interactions with various artificial intelligence applications have become invaluable, as each of us has become familiar or unknowingly involved in financial affairs, education, e-commerce and healthcare. Among these, the role of AI in healthcare and medical fields is appreciable and has already played an appreciable role in enhancing the quality of life through various applications, such as drug design, precision medicine, personalized treatment, documentation, and medical imaging (Schork, 2019).

Since 2012, medical and healthcare have developed as the most dominant field in the field of artificial intelligence. The convergence of two important events, the development of graphics processing units (GPUs) and big data, was pivotal for the development and broad application of AI. In the early part of the 21st century, the development of graphics processing units (GPUs) over central processing units (CPUs) aided in the analysis of voluminous data, and the development of big data helped in the resurgence of artificial neural networks under the new title "Machine learning". Machine learning relies entirely upon the enormous amount of data fed on it and its analysis, which helps it analyse a new set of data. Indeed, this new development in machine learning has expanded its application from basic digit learning in computer science to deep learning from image data, especially MRIs, X-rays and CT scans, to identify the minute details. This helps in diagnosis and disease prognosis as well as predictive analysis of patients and thereby optimized treatment (Jones et al., 2018). DNA analysis through recurrent neural networks has helped in drug design and genomic studies, enhanced utilization of patient data and course collaboration in various fields.

3.1 AI in Drug Design

Drug design is part of the virtual component of artificial intelligence in which mathematical algorithms are used, which we discuss as machine learning. The major hurdles we faced early in drug design were, lack of efficiency and off-target delivery, as well as the challenges in managing comparatively large amounts of proteomic, genomic and microarray data. Medication for a disease was not an effortless deal prior to the age of AI evolution. Because the steps in drug discovery are tedious and include identifying appropriate targets for drug action, screening with multiple drugs is needed to obtain promising results and optimization, preclinical and clinical studies are needed for safety purposes, and large-scale manufacturing practices and these procedures are also resource intensive. Unlike traditional computational methods, which are tedious and time consuming, the advent of artificial intelligence marked the beginning of computer-aided drug design. Drug discovery begins with the identification of target genes. To identify the targets that may be proteins or the genes underlying them, multiple methodologies, viz., gene expression data analysis and genome-wide association analysis, can be used to determine the interrelations between genomic variants and associated disorders, and genome and exome sequencing data can be used to identify mutations. The National Center for Biotechnology Information (NCBI)-Gene Expression Omnibus (GEO), The Cancer Genome Atlas (TCGA), and Array Express are the data repositories used to find gene expression patterns, whereas the GWAS Central and NHGRI-EBL GWASs are the data repositories for genome-wide sequence analysis. To obtain sequence data, the Sequence Read Archive and National Cancer Institute Genomic Data Commons were used. In addition, the published data in PubMed were also available to identify the target molecule based on previously published data (Gupta et al., 2021).

Once we find a target molecule, we must screen for a competent drug against that gene or protein. Through high-throughput output screening based on several factors, we screened out the "lead molecules" among the pool of drugs we screened. The factors we usually consider in drug screening include drug likeness, PAINS calculations for removing misleading drugs, and ADMET analysis (absorption, distribution, metabolism, excretion and toxicity) properties of the drug. These drugs were subsequently subjected to machine learning to identify potential therapeutic targets. These AI models include deep learning, which utilizes an artificial neural network to analyse the data based on previous experience; random forest, which is another machine learning algorithm that uses flowchart methods to predict the outcome; and so on. Following the analysis, the chosen drugs were selected for further analysis using QSAR. The quantitative structure-activity relationship predicts the behaviour of a molecule from its structure through pharmacophore models, setting a standard for the selection of the most appropriate drug. The chosen drug was subjected to molecular docking with its target, after which the binding energy was calculated to determine the stability of the drug. Once the drug was found to be successful in all these aspects, we then performed clinical trials both in vitro and in vivo.

The AI tools and algorithms, such as support vector machine (SVM), decision tree, random forest (RF), and artificial neural network (ANN), have helped in the identification of lead drugs from millions of candidates using various filters, such as QSAR and ADMET analysis. Compared with the previous computational and quantum methods used in drug design, these tools have significantly reduced the time consumption of drug design, increased efficacy and reduced toxicity due to off-target delivery. The AI revolution has significantly impacted drug design and discovery (Figure 2).

3.2 AI in Precision Medicine

In precision medicine, the pathological condition and therapeutic options are determined by medical history, behavioral and environmental factors combined with multiomics data. Rather than providing a universal medication for a particular pathophysiological condition, the significance of precision medicine is to diagnose and provide medications based on the unique genomic and health status of that candidate. This helps in early accurate diagnosis and tailored treatment. By combining genomic data with electronic health records (EHRs), doctors can compare new disease phenotypes and target treatment. According to the scheme provided by the CTSA (Central Translational Science Award), general drug development for treating a particular disease involves clinical trials from phase 0 to phase IV, starting from the T0 phase, in which the drug and target are found; T1, T2, and T3 clinical trials, in which population sizes increase in size compared to previous trials; and the T4 phase,

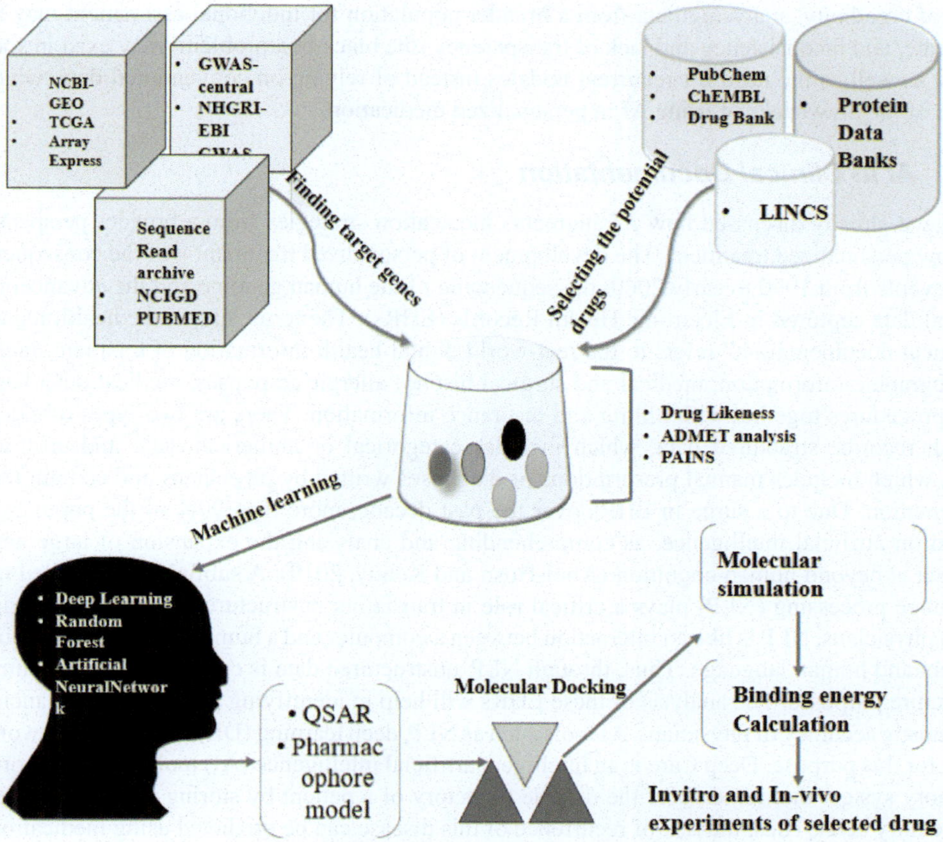

Fig. 2 Illustration describing multiple steps in drug design.

involving re-evaluation and implementation. This workflow develops the medication regimen considering the large population rather than considering each individual separately.

Instead, an analogous process of drug development is practised for the case of personalized medicine, and the workflow begins with the P0 stage, when the disease is diagnosed or the possibility of being affected by a particular disease is determined. AI aids this by analysing the DNA sequence and cross-reference with the latest research related to mutations or phenotypic changes. AI-based analyses of blood analytes have helped in diagnosing different cancers with an accuracy of 69-98%, depending on the cancer type (Cohen et al., 2018). The transition from P0 to P1 helps in understanding the pathophysiological conditions that are responsible for a particular disease or might result in the disease condition in the future. The P2 stage involves finding appropriate medications for the disease condition identified in the previous stages. The AI can scour through databases for novel mutations in diseases, analyse the underlying mechanism and identify the potential of drugs targeting similar pathways, allowing us to analyse the potential of off-label use of preexisting drugs. Of course, manual database analysis of a large dataset of novel mutations will be resource intensive and impossible. The tested medication was applied to the candidate in the P3 stage. Based on previous clinical data from electronic medical records, the AI can predict the response of a candidate to a particular medication. AI tools such as IBM Watson for Oncology help in personalized treatment based on previous clinical data from the real world. The P4 stage is meant for appropriate storage of data from a particular candidate in databases for further application, which helps in identifying similar patterns in future. The UK Biobank initiative in the United Kingdom and All of Us study in the USA are government-sponsored national initiatives in data warehousing. Even though artificial intelligence represents a breakthrough in personalized medicine, owing to the lack of ergodicity, analysing data from a broader population for individual assessment may not be accurate, and inconsistency and lack of transparency (the black box problem) may exist in some AI tools, as well as the need for recurrent updates instead of relying on conventional data, which are some of the drawbacks of using AI in personalized medication.

3.3 AI in Clinical Documentation

We have already discussed how AI augments medication strategies from a broader perspective to narrow personalized treatment. The development of personalized treatment was the consequence of two events from 1990 to early 2000: the sequencing of the human genome and the advancement of digital data captured in Electronic Health Records (EHRs). The term "electronic health record" or "clinical documentation" refers to the real-world digital health information of a patient, including demographic information, medical and surgical history, allergic responses, medications, diagnosis and procedures together with billing and insurance information. There are two types of electronic health records: structured data, which includes categorical or numerical data; and unstructured data, which includes manual prescriptions or diagnoses written by physicians and contain medical information. Due to a surge in EHRs over the past decade, more than 90% of the population has relied on artificial intelligence, as comprehending and analysing the expansion of large amounts of data is beyond human cognition (Abul-Husn and Kenny, 2019). A subfield of AI called natural language processing (NLP) plays a critical role in translating unstructured diagnostic descriptions from physicians. NLP is like an interaction between a computer and a human, where the machine can understand human language. Thus, through NLP, unstructured data is categorized and documented for future purposes. The analysis of these EHRs will help in identifying patterns and enhancing the diagnostic accuracy of physicians. AI tools such as NLP, deep learning (DL), and neural networks are used for this purpose. Deep Care is an intelligent artificial intelligence (AI) tool that has a short-term memory system that remembers the disease trajectory of a patient by storing past medical records in memory cells. Thus, the risk of recurrence of this disease can be predicted using medication, and the future trajectories of that condition can be anticipated. The disease progression of diabetic and mental health patients was demonstrated and found to be supportive using the Deep Care platform (Pham et al., 2017).

3.4 The Physical Component of AI in Healthcare

The physical component of artificial intelligence encompasses robotic assistance in multiple aspects, such as surgical assistance, rehabilitation and patient assistance, and assistive robots for differently abled and aged candidates. Robotic assistance has become an important part of augmenting surgical procedures. Like RIBA, assistive robots help patients move from one position to another, and caring service robots such as MARIO (managing active and healthy ageing with use of caring service robots) are used to address the emotional aspects of problems such as dementia among older people. Cognitive assistance platforms such as Virtrael are also among the advancements of AI in healthcare and are useful for rehabilitating people after brain injuries (Bohr and Memarzadeh, 2020). The application of AI and its subtypes in health care is numerous and is still a rapidly growing field. We have witnessed a revolution in artificial intelligence over the past decade, where people have accepted its application in multiple sectors, including healthcare and medicine. The purpose of AI is only to augment human capabilities, not to replace them completely. The application of AI in medicine and healthcare is magnificent from virtual components such as machine learning - used to identify patterns and minute details in medical imaging to providing guidance in physicians for better learning - used to identify patterns health assistants enhance patient care. On the physical side AI contributes to robotic assistance in surgery, and nanorobotic drug delivery. In this era of nanochips, human and intelligent machines have augmented their ability to serve society better, especially in the medical and health fields, which is promising for addressing currently intractable problems.

4. Artificial Intelligence (AI): The New Frontier in Biomedical Research

The 21st century, known as the century of technical advancements, is marked by advancements in informatics that have fundamentally altered how we view data analysis. Although access to information was a major issue in the past, today's fast-expanding datasets are accessible to the scientific community at any time and can be easily shared, examined, and kept in databases. The expansion of convoluted biomedical systems, such as innovative clinical and dashboard-based decision support systems, insight acquisition and management, medical imaging, computational intelligence in bio clinical medicine, molecular medicine, and managerial aspects of healthcare, has been made possible by recent technological advancements in medical informatics and biomedicine. Artificial intelligence (AI) has an enormous impact on the fields of biology, biotechnology, and medicine in general. Machine learning (ML) techniques, brain/neural computing, expert systems, fuzzy logic, evolutionary algorithms, and Bayesian modelling can all be used to implement AI in practical applications (Lu, 2019).

Artificial intelligence (AI) in biomedicine refers to the use of sophisticated algorithmic frameworks and computer programs to emulate human intelligence. Vision, robotics, data analytics, problem solving, natural language processing (NLP), decision science, bias, and linguistics are just a few of the numerous uses of AI. It is important to understand that there are two different methods available for artificial intelligence: the generative or white-box approach and the discriminative or black-box approach. White-box models are decipherable because they include simple, rule-based and pattern-based algorithms that demonstrate how data inputs lead to outputs. In contrast, black-box models illustrate algorithms and techniques that include extremely complex mathematical functions and hidden inputs and actions, making them difficult for observers to understand (Zednik, 2021).

4.1 AI in Life Sciences

The use of AI techniques with tech computers has significantly changed the extent of biological data available and has the potential to completely change the life sciences sector. Modern biology is increasingly dependent on machine learning (ML), particularly because it creates conjecturing models from biological datasets. Specifically, structural biology can be utilized to explore complex systems by creating models and biological networks from multidimensional datasets. Additionally, an increasing variety of artificial neural network (ANN) models, including genetic algorithms, DL, random forests, SVMs, and other techniques, are being applied to advance genomics research, address medication development procedures, and support clinical trials. AI is expected to lower the exorbitant cost of healthcare and create quicker, more effective therapies. The biomedical industry is experiencing advances in AI across the board, from primary research and extensive data to target discovery and tailored medication (Leite et al., 2021).

4.2 AI in Medical Imaging

It used to appear unrealistic that surgeons could look inside the human body without making any incisions. However, medical imaging in radiology has made progress, and the most recent AI-driven approaches are going much further. Imaging is evolving from its original purpose of diagnosing medical diseases to becoming an essential component of treatment, particularly in the case of cancer. To monitor tumors and the escalation of cancer cells and to determine the effectiveness of treatments, physicians are starting to rely increasingly more on medical imaging. Medical imaging's new function will lead to changes in the types of therapies patients receive. Additionally, the data physicians receive about treatment effectiveness will improve, enabling better decisions regarding the necessary treatments (Tadiboina, 2022). The primary objective of AI in radiomics is organ and lesion segmentation, which automatically segments organs at risk for treatment planning. Deep learning algorithms have been used to segment the brain, lung, prostate, kidney, pelvis, and other head and neck organs. On thoracic CT scans, Dong et al. trained deep neural networks for the segmentation of numerous organs using U-Net-GAN, a generative adversarial network. A group of U-Nets are trained as generators via U-Net-GAN, while fully convolutional networks (FCNs) are trained as discriminators. To create the ideal segmentation map of several organs, the generator and discriminator engage in an adversarial learning process. The feasibility and dependability of the suggested algorithm were proven when segmenting five different organs (Dong et al., 2019). Another example is the deep learning algorithm developed by Ardila et al., which analyses a patient's present and previous CT volumes to forecast the probability of lung cancer. On 6716 patients from the National Lung Cancer Screening Study, the model performed at the state-of-the-art level (94.4% area under the curve) and performed well on a self-sufficient clinical validation set of 1139 patients.

There are obstacles to be overcome before AI can become more dependable and widely used in the clinic, despite the excitement it has caused in the field of medical imaging research. The lack of good quality, large-volume, and long-term outcome data is a limitation of AI. Each clinic operates in a unique manner. A significant obstacle to AI-based medical imaging research is how to standardize the organization of the data produced by various procedures. The organization of medical imaging data may merit its own research area.

4.3 Predictive Analytics

Modern AI algorithms can find clinically important information hidden in the enormous volume of healthcare data by using applicable clinical queries created in partnership with healthcare practitioners. Most clinical trials for medical research incorporate specific hypotheses or questions that have been well defined. Given a set of features, artificial intelligence algorithms are trained to predict specific events, and insights can be gained through the prediction. Diabetes must be managed with long-term medication because it is a chronic illness. Krieg suggested high-order networks et al. to model the intricate relationships between diseases. The suggested approach is utilized to forecast type 2 diabetes clinical stages and replicate disease trajectories. The outcomes demonstrated that the proposed network outperforms the first-order network in terms of performance. Another example is the dilated recurrent neural network (DRNN), which was proposed by Zhu et al. to predict the upcoming glucose levels in patients with type 1 diabetes. In addition, several fields of biomedical research employ the use of predictive analysis, which includes predetermining sepsis, MRI scans, small abnormalities in the ECG, physicians, and more. Below is an illustration of how AI performs predictive analysis.

Deploying AI in the medical industry, faces several challenges including data protection, societal issues, ethical concerns, hacker risks, and developer considerations. Today's scientists aim to create a fully functional universal AI with cutting-edge, reliable algorithms. The tasks performed by this diverse range of AI are also more complex than those performed by the current AI. To address these issues adopting an advanced systems approach strategy in the health is crucial, treating the use of AI in healthcare as a dynamic learning process.

4.4 Virtual Health Assistant

An AI-powered application called a virtual health assistant helps users with diverse healthcare-related chores, provides medical information, and provides individualized health assistance. These virtual assistants employ ML techniques and natural language processing (NLP) to comprehend and reply to user requests. Digital transformation can help healthcare organizations become more consumer friendly. Ninety-two percent of health system representatives, according to Deloitte's research, want to improve the patient experience through digital transformation. They use a consumer-centered, outside-in approach to design processes and experiences from the standpoint of the consumer to foster trust and loyalty, increase patient experience, and improve more advanced ways of providing care using digital technologies. Technology can simultaneously enhance internal operations, expedite procedures, and boost output. As a result, we may anticipate more creative applications of conversational AI in healthcare that will change the way we approach patient care.

The global market for AI in healthcare is anticipated to reach $31.3 billion (approximately $96 per person in the U.S.) by 2025, with an exponential annual growth rate of 41.5% from 2020 to 2025, according to a Grand View Research analysis.

NLP is frequently employed in smart healthcare and has substantially enhanced the number of applications. An NLP-driven application consists of two primary components: the user interface (UI) and the back-end. The backend receives text or speech input from the user through the user interface (UI), processes it using NLP models, and then returns results by providing specific services via the same UI. The adoption of electronic health records, the proliferation of smartphones, and the widespread use of activity trackers have collectively enabled unprecedented access to digital data. Additionally, these advancements allow for the effective utilzation of AI in monitor patients. In ways we've never experienced before, we now have access to information on sleep patterns, blood pressure, heart rate, and other vital data. These advancements have led to improvements not only in these specific areas but also in others. For example, waveform pattern information can enhance the monitoring and analysis of hospital electrocardiograms, electroencephalograms, electromyograms, and Doppler ultrasounds. In intensive care units, the heart and lungs can be monitored using AI-enabled software that interprets vital signals (Reddy, 2023). Supporting elderly and disabled people: NLP techniques can enhance their quality of life and social integration. Voice-controlled home automation systems and robots offer daily support to the elderly and disabled. Androids and other speaking robots can even encourage social interactions. Additionally, NLP approaches aid people with disabilities, including dyslexia, hearing loss, speech difficulties, and neurological conditions (Chahboun et al., 2021). Depression is increasingly challenging in the medical field. Effective communication is plays a crucial role in improving mental wellbeing. However, patients often lack someone to confide in when needed. Medical chatbots are gaining popularity, as a potential solution, further advancements are necessary.

AI is a paradigm that is constantly expanding, with new developments and algorithms being created daily to serve various spheres and sectors of society that use it to advance findings and studies that were previously impractical or challenging to complete. AI has provided additional information on issues in biomedical research that require this clever engineering to provide sophisticated answers to challenging inquiries. Further research into this paradigm is made possible by computers' capacity to produce computations of previously unheard-of size and scope. This also makes it possible to assess the status of AI and the ways in which the creation of this intelligent tool is assisting researchers in their respective fields. From its applications in biomedical image analysis and radiology to treatment response and result prediction and many other biomedical fields that require computational advances and high-throughput processing beyond the capabilities of the human brain—ironically built on the very principles of our human cortex-AI has the potential to unlock some of the greatest mysteries and opportunities in biomedical science- a field on which humanity has only just begun (Figure 3).

5. AI in Stem Cell Therapy and Regenerative Medicine

The way doctors have "reacted" to disease thus far in medicine has involved treating symptoms. Stem cell therapy is one example whereby modern medicine is shifting away from reactive treatment toward regenerative medicine. Stem cells are a population of cells that can self-renew and develop into any cell type. The mechanisms and processes that underlie normal human developmental physiology and disease pathology could be better comprehended through the study of stem cells. The key characteristic of stem cells that distinguishes them from terminally differentiated cells is self-renewal. Stem cell therapy utilizes stem cells or their derivatives to promote the repair of infected, malfunctioning, or wounded tissue. Many disorders, including autoimmune, inflammatory, neurological, orthopaedic, and others, can be treated using stem cell therapy. It could also be the key to preventing aging and ensuring a disease-free state. The history of stem cell therapy started in the late 1950s when scientists who were accidentally exposed to radiation received successful allogeneic

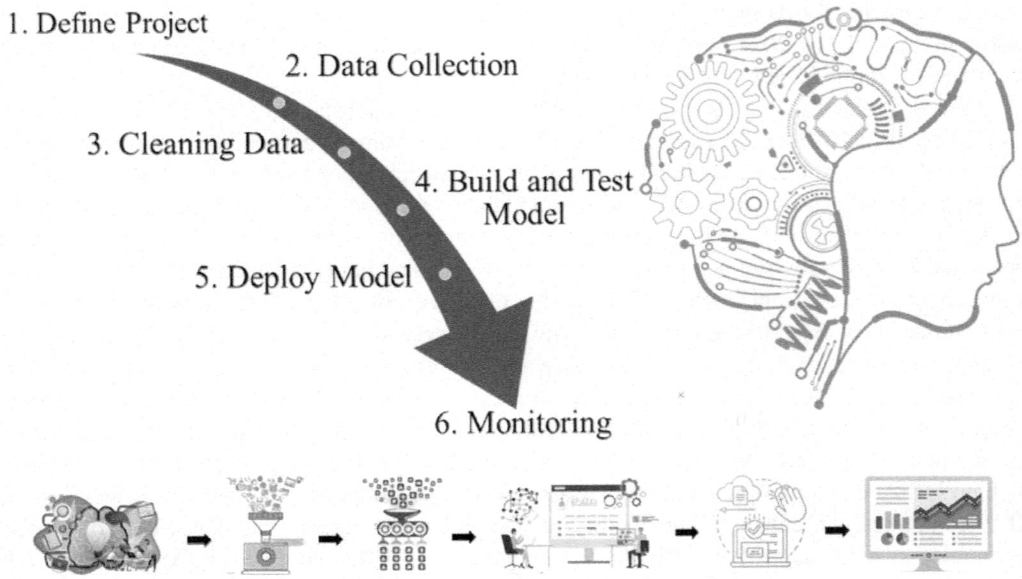

Fig. 3 Illustration depicting the methodology of predictive analysis performed by AI.

bone marrow transplants. Following this event, major advancements in the administration of stem cells for treatments were developed. The scientist John Gurdon made a ground-breaking discovery in the 1960s after he successfully cloned frogs from somatic cells. This work disproves the notion that only pluripotent cells can differentiate and proposes that somatic cells can also acquire a pluripotent state.

Typically, stem cells can be categorized into three types: induced pluripotent stem cells (iPSCs), adult stem cells, and embryonic stem cells (ESCs). When human ESCs were originally obtained in 1998, their usage was constrained because of moral reservations about the destruction of the embryo involved in their collection. When Yamanaka and Takahashi effectively converted multipotent ASCs into pluripotent ASCs, this problem was resolved (Takahashi and Yamanaka, 2006). In comparison to ESCs, iPSCs have been shown to be better candidates for research on regenerative medicine, disease remodelling, disease onset, drug development, and other topics given that they share many of the same features as ESCs—such as morphology, growth kinetics, specific proteins, and gene expression—while raising fewer ethical concerns.

Despite the wide range of uses for iPSCs, there are still certain limitations, such as the complexity of the cells, human handling errors, and difficulties in measuring iPSC-derived colony-related characteristics. Resolving these issues is necessary before attempting additional experimental or therapeutic methods, and this demands time. It is challenging to obtain the correct computations, and training is recommended. An automated approach is necessary to overcome these constraints. To solve this problem, artificial intelligence (AI) has been developed as an emerging technology. AI enables an automated platform to review iPSC-related data owing to deep learning-based strategies. By employing a convolutional neural network (CNN), deep learning modifies the data. CNNs are a form of multilayered neural network that can precisely distinguish between cells based on morphological and textural changes and mimic the organization of the human cerebral circuitry. Moreover, CNNs facilitate lineage tracking via microscopic imaging, detection of iPSC-derived endothelial cells without any need for immunostaining, grouping iPSC colonies or many other methods (Kommeri, R. and Thomas, V, 2023). The application of AI in SC biology involves finding the most efficient stem cells, each of which has distinct properties and potential for guiding clinicians in optimizing delivery methods by analysing patient data, such as medical imaging scans; assisting clinicians in tracking stem cell differentiation; analysing SCs using mathematical models; and predicting the mortality risk associated with SC transplantation.

5.1 AI in Cell Culture

Cells differ from one another in many ways making it difficult to predict what the cells will do at any given time. There are millions of parameters, which means millions of ways that research or therapy could go wrong. The variability associated with cell culture and difficulty in reproducing the results has led to research relating to the automation of the process. Genki Kanda and Taku Tsuzuki, worked on a robotics and artificial intelligence-based cell processing station that reliably carries out complex cell culture procedures performed by trained scientists. Kanda et al. provided an example of how to achieve advanced optimization using a combined robotics and artificial intelligence system. Characterization of cultured cells and establishment of an optimal cell culture protocol eliminate human intervention and reduce variability. Furthermore, the lessons learned by one robot can be copied to other robots. Microscopes are vital instruments in the medical industry. A key aspect of cytology is the identification of grown cells using a microscope, but this process is tedious, prone to error, and time-consuming because each cell has distinct morphological features. Kusumoto and Yuasa (2019) suggested that molecular biology is an important topic where deep learning has potential use. Over the years, several image analysis methods have been created; however, implementing each approach for every cell marker and cell type usually requires reprogramming the software and altering the parameters. Machine learning has an array of applications in these domains. Computational modelling and simulation models make it feasible to analyse high-throughput data on biological complexes, molecular-level interactions, DNA variations, and changes at the gene level as well as to determine the function of biomolecules.

However, few studies have applied CNNs to identify cells. One of the most frequently employed deep neural networks is the In-silico labelling method developed by Christiansen et al., which utilizes images from bright field microscopy as opposed to immunolabelling approaches to predict the location and texture of cell nuclei, the state and type of a cell, and the type of subcellular structure. For instance, using images from microscopes as input, Buggenthin et al. utilized deep learning systems to distinguish hematopoietic cells based on their specific shape and biological properties. Ugawa et al. employed "ghost cytometry" to group cells without the need for molecular labels. In this approach, morphological traits are converted to wave information through a barcode system. Fan et al. evaluated the morphological modifications of the mesenchymal-to-epithelial transition during cellular reprogramming by performing time-lapse-based imaging studies utilizing bright-field microscopy. They developed an ML model for the categorization, differentiation, and statistical evaluation of colony selection. Waisman et al. programmed a CNN algorithm to distinguish pluripotent stem cells from early differentiating cells by employing light microscopy images of pluripotent stem cells. With the goal of testing the quality of stem cells used to treat age-related macular degeneration (AMD), Schaub et al. partnered with the National Eye Institute and used a DNN. Their program learned a way to foresee how cells would act in various situations and environments from cell images. It was able to quickly assess photos of laboratory-grown eye tissues and categorize them as healthy or unhealthy. Of the thirty-six predictions that were generated to test this system, the AI method yielded only one inaccurate prediction. This non-invasive technique for identifying viable stem cells is called quantitative brightfield absorbance microscopy (QBAM).

Gene expression profiles and image processing are examples of areas where machine learning has been applied. Using a predetermined set of transcription factors for reprogramming, Danter et al. developed the DeepNEU deep machine learning approach to mimic artificial iPSC systems (Figure 4).

5.2 AI in Tissue Engineering

The discipline of biomedical engineering, referred to as "tissue engineering", blends biology and engineering to generate tissues or biological products outside the body or to use the knowledge obtained to regulate the repair of tissues more effectively inside the body. Tissue engineering has been defined as "an interdisciplinary and multidisciplinary field that aims at the development of

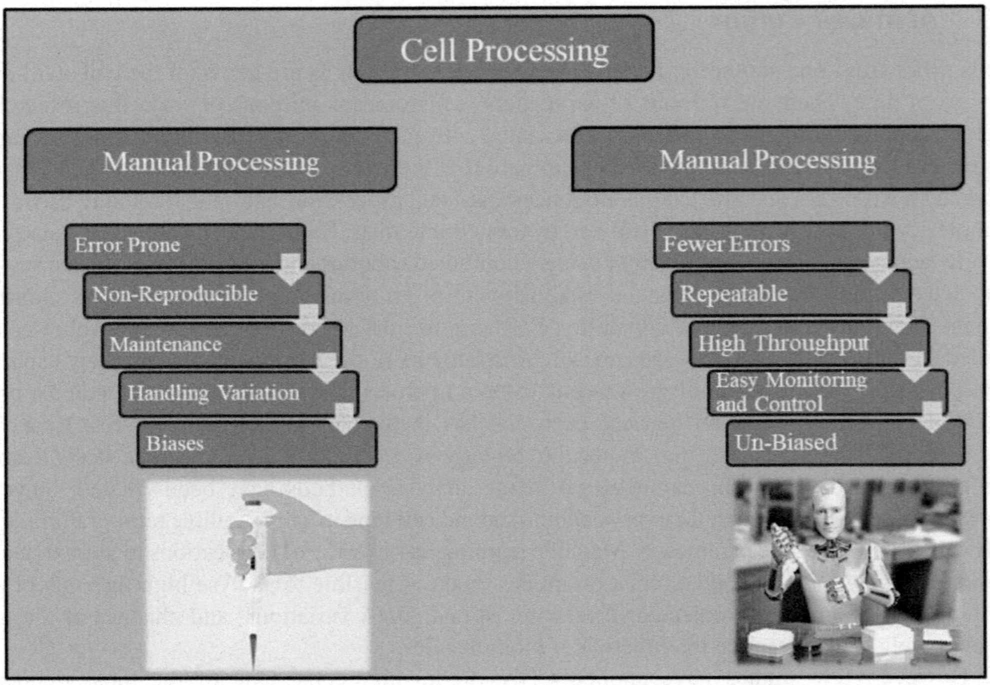

Fig. 4 Implementing robotics and artificial intelligence in cell culture

biological substitutes that restore, maintain, or improve tissue function" by the National Institute of Biomedical Imaging and Bioengineering (NIBIB). Manufacturing tissues and organs for implantation and creating tissue models for study are both addressed in tissue engineering. Challenges in tissue engineering include the need for appropriate biomaterial designs that are yet to be produced and a lack of knowledge of stem cells.

5.3 AI in Bioprinting

Additive manufacturing (AM) or 3D printing technology involves printing an object by laying up raw materials one layer at a time while being directed by a computer-aided design (CAD) model. It is an emerging technology that is currently used in the healthcare, agricultural, automotive, and aviation industries, among others. Furthermore, 3D printing is employed to create tissue systems and organs, tools for intricate surgery, prosthetic limbs, disease models, mimics of the human body, and numerous other things. One of the most significant approaches to tissue engineering and regenerative medicine is 3D printing using artificial materials, constructs, or "scaffolds", which emulate the properties of human tissues and organs and provide the appropriate mechanical stimulation to damaged tissues. Information on the materials, porosity, and lattice structure is essential for designing scaffolds. Due to the intricate geometrical features and complex mechanical properties of living tissues, conventional design and simulation methods cannot be utilized. In a world where organ shortage is an agonizing reality, implementing AI to pioneer organ 3D printing would certainly prove useful. AI can be used to determine the best materials for artificial organs in regenerative medicine, comprehend anatomical problems during treatment, and improve design with algorithms to increase the efficiency of 3D printing. 3D printing with the help of AI can also be utilized in developing technologies, such as three-dimensional organoids, organ-on-a-chip technology, the construction of multicellular tissues, the construction of functional organs from hPSCs, the construction of better models for understanding complex pathways at the tissue and cell levels, and intra- and interorgan interactions.

Bioinks are the major requirement in the manufacture of tissue structures. Bioinks are solutions of the desired cell type, and some biocompatible biomaterials in hydrogels usually encapsulated cells. Bioinks should have certain specifications, such as mechanical strength, adjustable gel consistency, biocompatibility, biodegradability and compatibility for chemical modifications, according to the cells used. The application of bioinks is limited because of the lack of appropriate bioinks that meet the above mentioned requirements. Machine learning can help foresee a large information base of materials and plans and thus can aid in the selection of an appropriate bioink in tissue engineering.

5.4 AI in Biofabrication

Biofabrication is cutting-edge research involving the production of complex biological systems, and therapeutic products using cells, proteins, biological material, and matrices as building blocks through bioprinting. Among the various applications of biofabrication, tissue engineering and regenerative medicine demand strict quality. It is vital to achieve biomimetic fidelity in biofabricated products, which depends on accurate functional and structural physiological properties. In accordance with a review by R Bardini and S di Carlo, visualizations from MRI or CT scans of patients can be utilized to generate computer-aided design (CAD) models, which can then be utilized to create 3D reconstructions of tissues or organs. In brief, the process starts from the creation of 3D CAD models from CT scan slices, followed by 3D bioprinting utilizing either cell-free or cell-laden biomaterials, development, and remodelling of tissue. This method is useful for creating 3D heart valves, regenerating lost teeth in braces, and creating artificial tissues. Machine learning assists in many other ways, including predicting material properties; mixing compositions of materials used for 3D printing; and developing high-resolution models of intracellular, cell–cell, and cell-biomaterial interactions. 4D bioprinting' is a recent development arising from integrating the concept of 'time' with 3D bioprinting. In this case, printed materials can alter their shape or functionality over time when exposed to an external stimulus. The use of 4D bioprinting in numerous biomedical sectors, including tissue engineering, has increased because of this printing innovation.

5.5 Synthetic Biology

Synthetic biology is the multidisciplinary field of science that involves designing and modifying biological parts, devices, and organisms to perform specific, well-defined functions. It integrates chemistry, physics, biological sciences, material sciences, computer sciences and engineering concepts to design or redesign biological systems. In this field, AI-dependent approaches assist in mining complex, large amounts of data relating to biological properties and designing optimal bioparts. Various reviews on the application of AI and synthetic biology in metabolic engineering, drug discovery, gene therapy and so on are available. AlphaFold is a popular and well-known example of a CNN-based model application. For the prediction of protein structure, Senior employed the tool to predict distance and torsion distributions.

To illustrate how human brain stem cells grown on a microchip could be taught to solve problems, scientists recently partnered on a project called The Neu-ChiP project. In this study, brain stem cells were utilized to create artificial intelligence (AI) devices by leveraging the brain's information processing. Electronic methodologies for machine learning currently have drawbacks because they demand an abundance of energy and computer resources. The goal of the prevailing "neuromorphic computing" strategy, which electrically mimics human neural activity, is to utilize the processing power of the brain to enhance computer capacity. The Neu-ChiP study investigated how complex live neural circuits specifically developed for this reason might transform the way humans analyse information. Networks of stem cells simulating the human brain were stacked onto microchips in this experiment and activated by the firing of varying light beam patterns. The goal of this experiment was to clarify how the brain processes information and solves problems. This could help with the development of creative and captivating human–machine interfaces. It is amazing to see how computation in live cells is developing and how it is bringing together interdisciplinary experts in the framework of synthetic biology.

AI, machine learning, deep learning, and various algorithms could improve the accuracy of various aspects of stem cell therapy, tissue engineering, bioprinting and synthetic biology. The optimization of variables, such as protocols for guided differentiation of stem cells, maintenance of cells, and engineering of materials, poses significant difficulties in the field of stem cells and regenerative medicine. Such challenges arise from the vast design space for biomaterials, the need for determining suitable chemical and biochemical substances, insufficient knowledge about complex biological interactions and numerous additional factors. AI can be used to predict cell behaviour across different environments; handle and analyse large amounts of genomic and proteomic data; help with the selection of compatible materials that satisfy the required parameters and model and replicate cellular behaviour; and analyse and process images for cell tracking and tissue engineering. Scientists believe that integrating AI will accelerate the translation of regenerative medicine into clinical practice. The integration of AI in stem cell therapy and regenerative medicine will soon help develop individualized and efficient treatments. The long-term prospects for AI, stem cells, and regenerative medicine in the healthcare sector are optimistic.

6. AI in Bioinformatics

Bioinformatics has been an integral part of modern biology; as the field of molecular biology has progressed, there has been a need to incorporate computational tools for quantitative analysis in addition to mathematical and statistical tools that are already indispensable to the field of modern biology. Along with advancements in biological science techniques and methods, there has been a need to store, manage, refer and analyse the copious amounts of data that was being produced, where bioinformatics came to be an interdisciplinary quantitative analysis field that was an amalgamation of biological sciences and computer science. Bioinformatics in life sciences has two functions, namely, developing quantitative computational tools and databases and interpreting and applying these tools for better quantitative analysis of biological research. It has been applied in three broad fields, structure, sequence and function analysis. In structure analysis, bioinformatics can be applied to the structure prediction of nucleic acids and different proteins, along with protein structure classification and composition. Structural analysis helps in identifying and visualizing targets, identifying target binding sites for various proteins, revealing information on protein functionality, and revealing interactions with neighbouring proteins and molecules (PPIN), which are mostly applied in in silico expression (Figure 5) analysis, drug discovery and protein engineering. Sequence analysis holds importance in every field, genomics, molecular evolution, identification of genetic defects, variations, mutations, identification of promoters and other regulatory elements, etc., in the fields of evolutionary genetics, forensics, disease identification, early diagnosis and therapy. The functional analysis includes metabolic pathway prediction and analysis, protein–protein interaction (PPI) analysis, protein subcellular localization, gene expression profiling, and gene and metabolite identification in different biological processes, especially in the fields of drug discovery, biomarker identification, disease diagnosis and targeted therapies. Hence, bioinformatics holds importance in the field of biological sciences at a very foundational level, and over the past 40 years, it has evolved along with advances in molecular biology and genomics. Moreover, significant progress has been made due to the incorporation of AI. Since the invention of AI, technology has become more precise, convenient and compact. Technology has transformed from mechanical to digital. For example, we started with the industrial automation of the mainframe and PCs in the 1960s and then developed the broadband internet in the 2000s. After the invention of the internet, the technology skyrocketed in the following years, leading to massive digital transformation with smartphones, cloud computing, social networks and big data. Finally, at present, technological transformation is at its peak with intelligent technologies such as artificial intelligence, machine learning, deep learning, neural networks and cloud computing.

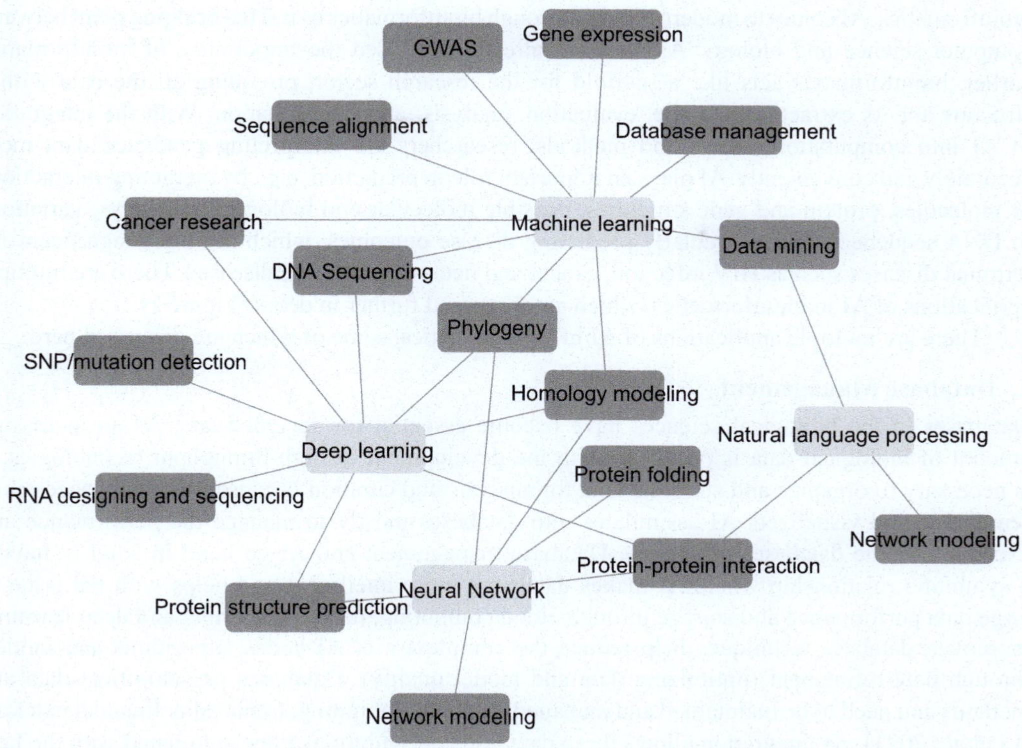

Fig. 5 The interaction of AI tools with various bioinformatics research tools in the form of a network was assessed using the interactive tool CYTOSCAPE, which uses machine learning to visualize and predict protein–protein interactions.

Talking about AI, what is artificial intelligence? It is a machine's ability to duplicate or simulate human intelligence as well as other higher organisms based on its ability to learn, reason and solve problems. The basic explanation of AI was given by Alan M Turing through a test, which he called the imitation game, also known as Turing's test, where an individual unknown to the sources is incapable of differentiating which of the sources is human or a machine (Turning, 2009). The goal of AI could be interpreted in two ways, as a scientific goal—AI—is mostly implemented by scientists, psychologists and philosophers, where AI is expected to emulate human behaviour and cognition through machine learning and deep learning, integrating the problem solving and decision-making attributes of humans. This approach finds application in research, healthcare and medicine, drug discovery, etc. As an engineering goal, AI aims at providing human convenience and comfort, mostly solving real-life problems. Through automation, robotics and natural language processing (NLP), AI is currently prevalent mostly in transportation, finance, education, entertainment and media (Kothari, 2023). Current applications revolve mostly around fulfilling engineering purposes, for example, humanoid robots such as Sophia and, recently, iRobot-developed Roomba automated vacuum cleaners that are powered by artificial intelligence (AI). In addition, there is avid use of AI in smart assistants such as Siri, Alexa and Cortana (Daley, 2023). Self-driving cars, especially social media and entertainment involving the film and gaming industries, have redefined the user experience entirely through the help of AI. The discipline of biological sciences has also been able to upgrade and transform with the advancements of technology and the invention of AI. AI holds great potential in healthcare and may even revolutionize the healthcare and research sector entirely.

Coming to the biological sciences, AI has possible important applications in healthcare, biomedical engineering, agriculture, regenerative medicine, and cancer and precision medicine. However, bioinformatics has been incorporated into the backbone of life sciences in one discipline, bioinformatics. AI comes to modern biology through bioinformatics as it is the bridging point between computer science and biology. As we have already discussed the importance of bioinformatics earlier, bioinformatics acts like a scaffold for the research sector, providing all the data with a structure for its extraction, storage, evaluation, analysis, and interpretation. With the integration of AI into computational tools and methods, researchers are interpreting generated data more accurately and conveniently. AI plays an important role in prediction, e.g., by predicting interactions of molecules, protein and gene structures, possible molecular and biological pathways, variations in DNA sequences, and particularly predicting disease outcomes, which are more significant for terminal diseases such as HIV infection, cancer and neurodegenerative diseases. There are multiple applications of AI in bioinformatics, which are discussed further in detail (Figure 1).

There are multiple applications of AI in bioinformatics, some of which are discussed here:

1. Database Management

Databases in the biological sciences have become essential for several years. As an increasing amount of biological data is produced with the development of high-throughput technologies, it is necessary to organize and store the data for analysis and curation because it is easily accessible, searchable and visualized. AI assimilates into databases mostly to manage the performance and execution of the databases efficiently. Database management and AI go hand in hand in having a symbiotic relationship where AI makes databases more intelligent and helps with the issue of large data performance and storage through cloud computing, machine learning and deep learning; moreover, database techniques help reduce the complexity of AI codes, algorithms and models through data refinement (optimizing data and model quality). Databases use empirical data and methods and need to be maintained and managed manually for tuning, troubleshooting and patching (Li et al., 2021). AI integration allows these database configurations to be automated with the help of machine learning. Database configurations such as tuning that controls memory allocation, input/output control, and big data that usually require human involvement for maintenance have become automated with the help of machine learning. Learning-based AI can recommend indexes and views, explore different types of partition keys with the implementation of neural networks, and remove redundancy through automatic query rewriting, enhancing database performance for query execution (Kunjir and Babu, 2020).

For database optimization, with AI, the optimizer is better able to estimate the cost of a query; in this sense, any query has a particular amount of data, and the amount of data that can be processed at a time is called the cardinality of that query; hence, the greater the amount of data, the greater the cardinality and the work to be done; hence, the query uses more resources and time and, hence, the higher the cost. With the help of AI, the cost estimation of a query by an optimizer is more accurate and efficient; hence, the optimizer chooses low-cost queries that can be processed in less time. Through learning-based AI, an optimizer is also able to efficiently, based on low join cost, select good plans for joining ordering of the queries in the sense of joining multiple data in a particular order. AI helps design an end-to-end optimizer that can generate a full plan for database optimization (Li et al., 2021). In database design, databases are usually designed by database architects based on their knowledge and experience; however, with AI learning-based tools, multiple novel designs can be predicted based on combinations of new data structures, automatically considering better indexes and views and workload management. Database monitoring and security: AI machine learning techniques automate monitoring database health-related data, diagnosis query speed, errors and failures. However, AI can detect but not troubleshoot. It monitors and records all database activities. In the case of database security, learning-based AI automatically detects and protects sensitive data and database vulnerability by imposing access control (Li et al., 2021). Apart from all the aid that AI puts into DBs, DBs provide support for AI to function by providing high-quality data, AI model training and inference. Hence, it is a collaborative operation on both sides.

2. Functional Genomics

Since the Human Genome Project (1990-2003), the amount of biological data generated has increased, suggesting that biology is a major producer of big data. Genomics has revolutionized the field of molecular and data sciences with advancements in high-throughput technologies. Currently, AI is a catch-all phrase that includes neural networks, deep learning, and machine learning. Machine learning is a subset of AI that can learn on its own. Machine learning differs from artificial intelligence (AI) in that it develops an understanding of data without having to be programmed. However, deep learning goes a little further and imitates human traits such as learning from examples. It is a subset of machine learning; however, it eliminates much of the human intervention that ML still relies on. It is more useful for large datasets, as it uses the concept of developing neural networks such as those of the human brain. The data is filtered through layers of networks, improving its quality and accuracy. Finally, neural networks form the innermost intricate layer of the AI tract, which consists of several nodes that represent artificial neurons that connect to each other and form multiple layers through which the data is processed. In the field of genomics, the application of AI is needed. Currently, genomics has gained recognition and pace through HGP and has now transformed into a whole separate discipline under "omics technology". Functional genomics examines how the entire genome functions in a biological system to determine whether a certain gene or intergenic region results in a specific phenotype. There are several subdisciplines, such as genomics (related to the genome, i.e., the DNA content of the organism), epigenomics (related to the regulation and modifications of the DNA content), transcriptomics (related to the RNA content of the genome), epitranscriptomics (regulation and modifications of the RNA content of the genome), proteomics (protein translation and interactions), and metabolomics (metabolite functioning, interactions and regulation) (Caudai et al., 2021). AI plays a role in prediction and visualization. Many applications in research areas, e.g., biomarker prediction, cancer diagnosis, DNA methylation, evolutionary biology, protein structure and interactions, have been implemented in combination with artificial intelligence (AI) machine learning algorithms since the 1980s, such as linear regression, support vector machines (SVMs), random decision forests, principal component analysis (PCA), clustering and Bayesian networks (Caudai et al., 2021). However, more recently, deep learning has been applied in the past few years due to advancements in computer sciences, large genome-scale "omics" data production and simultaneous decreases in sequencing costs. Additionally, deep learning has advantages over ML in terms of direct prediction based on raw data and a greater ability to interpret the data, reduce background noise and handle large genomic datasets. However, with automation and higher abstraction levels comes the downside of reduced detail and transparency of the process, which raises the risk of bias in the results. However, few AI algorithms that are integrated into different "omics" technologies are discussed here.

Learning-based AI may be supervised or unsupervised. In supervised learning, ML/DL-based AI interprets data based on previous experience; here, labelled data/training datasets are fed into the algorithm to collect and produce results for a particular dataset. Additionally, this approach is not very efficient when using big data. However, unsupervised learning requires no such training or labelling; rather, unsupervised learning involves direct interpretation from raw data and is less costly and time-consuming. However, the accuracy or traceability of the data processing methods have become slightly ambiguous. Several ML methods that are traditionally used include the following:

1. **Linear regression** Linear regression is a supervised machine learning technique that predicts a linear relationship between one dependent variable and one or more independent variables. Multiple linear regression methods have been implemented in functional genomics. For example, univariate/multivariate linear regression is mostly used for survival analysis to determine the correlation between the time and occurrence of an expected event and is also used to find genotype–phenotype associations (quantitative trait locus linkage analysis) (Anderson et al., 2006).

2. **Support vector machines (SVMs)** SVMs are supervised ML methods that are used for classification, regression and outlier detection, especially for high-dimensional data. For example, it is used for gene, protein and cancer classification and for the diagnosis and classification of various other diseases (Liao and Li, 2007).
3. **Random Decision Forests** Based on the idea of ensemble learning, a supervised machine learning technique can be used for classification. Several classifiers (decision trees) are integrated to solve a problem and enhance model performance. A greater number of trees in a forest is associated with greater accuracy. For example, it is used for the analysis of microarray expression data, GWASs, nucleic acid sequence analysis, and computational drug screening (Chen and Ishwaran, 2012).
4. **Principal component analysis (PCA)** PCA is an unsupervised machine learning approach used to predict correlations between a group of variables. The amount of data needed to produce a statistically significant result grows exponentially as the number of features or dimensions in a dataset increases. As a result, there may be problems with overfitting, longer computation times, and less accurate machine learning models. Hence, PCA reduces the high dimensionality of the dataset without altering the patterns or relationships between the variables. It can be used for feature extraction, visualization, and data preprocessing steps such as classification and clustering. For example, it is used to efficiently explain variations in gene expression (Ma and Dai, 2011).
5. **A dynamic Bayesian** This network is an unsupervised ML method that relates variables to each other over a period. For example, it is used for predicting gene regulatory networks and signaling pathways and is also used in GWASs to relate SNPs to various diseases (Dojer et al., 2006).
6. **k-Means clustering** This approach is also an unsupervised ML method that creates clusters of unlabelled datasets. For example, it is used to cluster gene expression data (Handhayani and Hiryanto, 2015).

However, few examples of the use of deep learning methods in genomics exist.

1. **Deep Neural Networks** A supervised technique that uses artificial neurons in many hidden layers, with the output of one layer serving as the input for the layer above it.
2. **Multilayer perceptron** It is a supervised DL method of neural networks consisting of multiple layers. A layer's neurons learn optimal weights for each input signal and produce distinct outputs.
3. **Convolutional Neural Networks** These are supervised DL algorithms that draw inspiration from biological pathways that use both convolutions and pooling layers to detect complex global and local patterns by scanning multidimensional arrays. To be efficient, multiple layers and large datasets are needed.
4. **Generative adversarial networks (GANs)** GANs are unsupervised DL neural networks that consist of neural networks that are set against each other. The first generates new instances, while the second evaluates whether the instances belong to the training dataset. GANs produce new information that can be distinguished from observed data.

Several of the applications of DL algorithms in functional genomics include one of the earliest implementations of AI in sequencing, dating back to 1992, when the Human Genome Project used neural networks to find protein-coding regions and splice junctions in DNA sequences. This AI approach takes examples from sensor-based robotic systems where the information comes from multiple sensors that produce overlapping data; these data are further optimized through deep learning. DNA sequence recognition through learning-based AI uses the same approach to locate protein-coding exons. The input for the neural networks is the DNA of interest (DOI) sequence. The neural network tries to learn genetic code as a triplet base code from the data processed by the sensors that use mathematical algorithms to produce statistical data for the sequence (Mural et al., 1992).

With the emergence of massive parallel sequencing, NGS has been applied in multiple fields, such as whole-genome sequencing, DNA sequencing data analysis (SNV/mutation/SNP detection), DNA methylation, RNA sequencing analysis, quantification of mRNA, discovery of RNA variants, gene expression analysis, identification of genetic aberrations, cancer research-biomarker identification, tumor heterogeneity analysis, identification of drug targets (DNA–protein interactions) and discovery of new pathogens. However, NGS can generate large amounts of data that need to be analysed. The use of AI in combination with NGS helps in the management of the large amount of sequencing data produced through AI techniques, such as pattern recognition and correlation. Moreover, NGS uses AI for accurate read alignment, variant calling and annotation. DL algorithms can distinguish patterns to classify genotypes. Moreover, they can predict splice sites, clinical relevance and severity of genetic variations and mutations (Dlamini et al., 2020). One of the major contributions of AI to sequencing is COVID-19 genome sequencing. For the SARS-CoV-2 virus, a deep learning method was used for DNA sequencing through neural networks. The AI model was trained to distinguish different viral genome strains of the coronavirus family through multiple experimental data integrated and labelled into DL algorithms (Bagabir et al., 2022).

DeepBind and DeepSEA are two deep learning architectures in which DeepSEA estimates the effects of sequence changes at the base pair level on chromatin, and DeepBind can accurately predict the sequence requirements of DNA/RNA-binding proteins (Zhou and Troyanskaya, 2015).

3. Protein structure design and prediction

Even though proteomics is based on functional genomics, the necessity of this field demands a separate research. Proteins are the major components of any biological system, and proteomics is more complex than genomics and forms the core of biological research. These proteins are essential for biochemical pathways because they are drivers and important for the structure, function and regulation of the whole system. It changes from cell to cell and over time, impacts an entire cascade of molecular functions. Most regulatory molecules are proteinaceous in nature; hence, understanding their functionality, which depends strongly on their structure, is imperative for analysing their function, regulation, PTMs, movement, interactions and expression. Proteomics covers multiple aspects of proteins, and how AI plays a role in each aspect is discussed further. The protein structure dictates protein function; hence, its stability, binding, affinity, interaction and activity largely depend on its structural conformation. However, in research, especially in drug discovery, to study protein regulation and expression, small molecules need to be targeted, and designing these small molecules is pertinent to the protein structure. There are traditional experimental methods for analysing protein structure, such as Y2H, mass spectrometry, DNA and protein microarray, synthetic lethality, phage display, NMR spectroscopy, FRET, X-ray crystallography, AFM, and electron microscopy. These methods provide information on protein physical interactions and functional associations through identification or characterization of specific biochemical and physicochemical properties of a protein complex. However, for protein design, computational approaches are used that are mostly based on experimental findings. For large-scale studies, computational tools are necessary.

Multiple levels of protein structure were analyzed; information on protein homology and families, motif identification, multiple sequence alignment, and sequence classification are all parts of the primary structure. The identification of substructures is aided by secondary structure, and 3D protein structures provide details on the chemical and topological characteristics of the protein as well as 3D patterns in the structure, folding, structure comparison, and domain classification. However, the issue prevails in predicting protein structure from its sequence since the 3D structures of most proteins have not been experimentally resolved, as experimental methods such as X-ray crystallography, NMR and cryogenic electron microscopy are expensive and time consuming. Additionally, different protein sequences can adopt different conformations, thereby acquiring different functionalities, which is very difficult to track. Hence, computational databases and tools provide support for time and cost conservation by providing data and tools for the protein binding site, protein–ligand docking, and free energy calculations. Homology modelling is a template-based method that is currently extensively used for structure prediction and is based on sequence comparison. Tools such as SWISS-MODEL, Modeller, and JPred are 3D modelling tools. Even with these databases and tools, protein structure prediction still lags since most of the tools depend on experimental data. AI integration into protein structure prediction improves the quality of protein models, which has a great impact on drug discovery. With the help of prediction databases, it is now possible to map entire proteomics datasets and address larger datasets. Secondary structure prediction provides insight into protein folding. Alpha helices, beta sheets, and other secondary structure motifs are used as the basis for secondary structure prediction, which employs the ensemble learning principle, which combines multiple classifiers via multivariate linear regression. Prediction tools such as JPred, Distill and PSIPRED are based on neural networks that classify these structures using a hidden Markov model. In addition to these tools, the PDB uses a bidirectional recurrent neural network, which processes sequential data in both the forward and backwards directions.

The data collected from these hidden layers are subsequently input to a predictive layer. It is more accurate since it can process information in both directions regarding both past and future contexts and can take up variable length sequences. Apart from these, deep learning is implemented in DeepCNF, a tool that uses a deep convolution neural network, where it makes it possible to consider secondary structures at different positions and their correlations over a longer range of sequences (Wang et al. 2016). The critical assessment of protein structure prediction (CASP) is a neural network-based program that uses machine learning to predict tertiary structure and successfully predicts the overall folding of proteins and the distance between residues. AlphaFold is a popular prediction tool used. This self-supervised learning model is based on a two-tower architecture in which information is collected from public databases and incorporates types of information for a given protein sequence that is multiple sequence alignment and a list of predicted similar structures. Protein function classification was based on structural characterization (Wei, 2019). Hence, DL methods such as recurrent neural networks, for example, DeepFunc and DeepPred, are implemented for primary sequences. Percolator is a different approach used in proteomics that maximizes the number of true hits for a variety of protein sequences for a given FDR and is mostly utilized in peptide identification with enzyme and modification specificity. The MS2PIP is another ML tool that is used for spectrum prediction via ensemble learning (random forest). pDEEP is a DL algorithm that can accurately predict the spectrum from the peptide sequence (Zhou et al., 2017). In Prosit, DL models are trained on fragmentation data of different modes acquired from different collision energies. Because of these advancements, our understanding of protein structure and function is expanding, and much of the related knowledge is available in drug discovery and personalized medicine through virtual screening of potential drug candidates through information on protein interactions. AI models are also used for repurposing existing drugs toward new therapies.

4. Evolutionary biology

One of the major prediction roles of ML in bioinformatics is in evolutionary studies. Since the emergence of sequencing techniques, bioinformatics has played an essential role in understanding evolution. In evolutionary biology, there are extensive studies, including sequencing and phylogenetic studies. However, in phylogeny, sequence alignment poses an issue since it aims to align sequences to exhibit maximum similarity, which is difficult to achieve due to years and years of accumulation of mutations and indels. AlexSys is a machine learning model that can determine whether a method is appropriate for multiple sequence alignment for the sequences being aligned (Aniba et al. 2010). The supertree construction problem, which entails creating an evolutionary tree that is consistent with a collection of trees on overlapping species, has been resolved using Answer Set Programming (ASP) (Koponen et al. 2015). The Relational database TreeBASE offers a web service interface API for taxonomy alignment, consistency verification, and inconsistency repair. The alignment of trees can improve search efficiency by splitting characters into subsets and using anchor trees to constrain the tree's search (Ford et al. 2015).

Bioinformatics is a field in which incomplete data extraction, prediction and storage problems occur, offering opportunities for artificial intelligence studies. However, coping with rapidly advancing technology is challenging. The use of weighted sequence data to encompass all omics observations and the availability of both qualitative and quantitative material are two notable trends. Chromosomal 3D structures, epigenetic factors, coding and noncoding RNA types, translation efficiency of proteins, and interactions between DNA and proteins are all now being identified via high-throughput sequencing. A molecular biology discovery that makes it possible to access individual observations, beginning with the DNA of individual species, enables personalized medicine and elaborate population representation at all levels. Successful research demands a thorough comprehension of the characteristics of a certain biological subject, ensuring that a biologist is always informed. Understanding biological systems is the goal of systems biology and has led to advancements in bioengineering, drug development, and sustainability. Despite their complexity, principles govern these evolvable systems. Systems biology can benefit from the application of AI technologies such as text mining, qualitative physics, and machine learning, especially in drug discovery and coral reef systems biology. Another potential field for collaboration between computer science and biology is synthetic biology, a subject delving into life engineering that focuses on creating living components with organized, fully controllable behaviors. Many areas of biological research, including genomics, proteomics, drug discovery, and personalized medicine, have been transformed by artificial intelligence (AI). With the aid of artificial intelligence (AI) methods such as machine learning and deep learning, the accuracy and efficiency of gene prediction have increased, making it possible to identify genetic variants and determine their effects on function. However, issues, including data privacy concerns, interpretability of AI models, and ethical issues in healthcare decision-making, continue to be problematic. The development of more interpretable AI models, the improvement of algorithms, and the promotion of interdisciplinary cooperation should be the main goals of future research.

References

Abul-Husn, N.S. and Kenny, E.E. (2019). Personalized medicine and the power of electronic health records. *Cell*, *177*(1), 58–69.

Anderson, C.A., McRae, A.F. and Visscher, P.M. (2006). A simple linear regression method for quantitative trait loci linkage analysis with censored observations. *Genetics*, *173*(3), 1735–1745.

Aniba, M.R., Poch, O., Marchler-Bauer, A. and Thompson, J.D. (2010). AlexSys: a knowledge-based expert system for multiple sequence alignment construction and analysis. Nucleic acids research, 38(19), 6338–6349.

Bagabir, S.A., Ibrahim, N.K., Bagabir, H.A. and Ateeq, R.H et al. (2022). COVID-19 and Artificial Intelligence: Genome sequencing, drug development and vaccine discovery. *Journal of Infection and Public Health*, *15*(2), 289–296.

Bardini, R. and Di Carlo, S. (2023). Computational modeling and optimization of biofabrication in Tissue Engineering and Regenerative Medicine—A literature review. bioRxiv, 2023–03.

Bohr, A. and Memarzadeh, K. (2020). The rise of artificial intelligence in healthcare applications. In *Artificial Intelligence in healthcare* (pp. 25–60). Academic Press.

Buggenthin, F., Marr, C., Schwarzfischer, M., Hoppe, P.S., Hilsenbeck, O., Schroeder, T. and Theis, F.J. (2013). An automatic method for robust and fast cell detection in bright field images from high-throughput microscopy. *BMC bioinformatics*, *14*(1), 1–12.

Caudai, C., Galizia, A., Geraci, F., Le Pera, L., Morea, V., Salerno, E.,... and Colombo, T. (2021). AI applications in functional genomics. *Computational and Structural Biotechnology Journal*, *19*, 5762–5790.

Çelik, İ., Arslan, F., Tunç, R. and YILDIZ, İ. (2021). Artificial intelligence on drug discovery and development İlaç keşfİ ve gelİştİrİlmesİnde yapay zekâ. Ankara Universitesi Eczacilik Fakultesi Dergisi, 45(2).

Chahboun, S., Kvello, Ø. and Page, A.G. (2021). Extending the field of extended language: a literature review on figurative language processing in neurodevelopmental disorders. Frontiers in Communication, 6, 661528.

Chen, X. and Ishwaran, H. (2012). Random forests for genomic data analysis. *Genomics*, *99*(6), 323–329.

Christiansen, E.M., Yang, S.J., Ando, D.M., Javaherian, A., Skibinski, G., et al. (2018). In silico labeling: predicting fluorescent labels in unlabeled images. *Cell*, *173*(3), 792–803.

Cohen, J.D., Li, L., Wang, Y., Thoburn, C., Afsari, B., Danilova, L.,... and Papadopoulos, N. (2018). Detection and localization of surgically resectable cancers with a multi-analyte blood test. *Science*, *359*(6378), 926–930.
Daley, S. (36). Artificial Intelligence Examples Shaking Up Business Across Industries. *Preuzeto*, *20*, 2023.
Dlamini, Z., Francies, F.Z., Hull, R. and Marima, R. (2020). Artificial intelligence (AI) and big data in cancer and precision oncology. *Computational and structural biotechnology journal*, *18*, 2300–2311.
Dojer, N., Gambin, A., Mizera, A., Wilczyński, B. and Tiuryn, J. et al. (2006). Applying dynamic Bayesian networks to perturbed gene expression data. *BMC bioinformatics*, *7*, 1–11.
Dong, X., Lei, Y., Tian, S., Wang, T., Patel, P., Curran, W.J. et al. (2019). Synthetic MRI-aided multi-organ segmentation on male pelvic CT using cycle consistent deep attention network. *Radiotherapy and Oncology*, *141*, 192–199.
Fan, K., Zhang, S., Zhang, Y., Lu, J., Holcombe, M. et al. (2017). A machine learning assisted, label-free, non-invasive approach for somatic reprogramming in induced pluripotent stem cell colony formation detection and prediction. *Scientific reports*, *7*(1), 13496.
Ford, E., St. John, K. and Wheeler, W. C. (2015). Towards improving searches for optimal phylogenies. Systematic Biology, 64(1), 56–65.
Gupta, R., Srivastava, D., Sahu, M., Tiwari, S., Ambasta, R.K. and Kumar, P. et al. (2021). Artificial intelligence to deep learning: machine intelligence approach for drug discovery. *Molecular diversity*, *25*, 1315–1360.
Handhayani, T. and Hiryanto, L. (2015). Intelligent kernel k-means for clustering gene expression. *Procedia Computer Science*, *59*, 171–177.
Holzinger, A., Keiblinger, K., Holub, P., Zatloukal, K. and Müller, H. et al. (2023). AI for life: Trends in artificial intelligence for biotechnology. *New Biotechnology*, *74*, 16–24.
Jones, L.D., Golan, D., Hanna, S.A. and Ramachandran, M. (2018). Artificial intelligence, machine learning and the evolution of healthcare: A bright future or cause for concern? *Bone & joint research*, *7*(3), 223–225.
Kanda, G.N., Tsuzuki, T., Terada, M., Sakai, N., Motozawa, N. et al. (2022). Robotic search for optimal cell culture in regenerative medicine. *Elife*, *11*, e77007.
Kommeri, R. and Thomas, V. (2023). Prospects of artificial intelligence in regeneration and repair of organs. In *Artificial Intelligence in Tissue and Organ Regeneration* (pp. 117-132). Academic Press.
Koponen, L., Oikarinen, E., Janhunen, T. and Säilä, L. (2015). Optimizing phylogenetic supertrees using answer set programming. Theory and Practice of Logic Programming, 15(4-5), 604–619.
Kothari, S. (2023). What are the Major Goals of Artificial Intelligence? Simplilearn.com. https://www.simplilearn.com/tutorials/artificial-intelligence-tutorial/major-goals-of-ai
Krieg, S.J., Robertson, D.H., Pradhan, M.P. and Chawla, N.V. (2020, November). Higher-order networks of diabetes comorbidities: disease trajectories that matter. In *2020 IEEE International Conference on Healthcare Informatics (ICHI)* (pp. 1–11). IEEE.
Kunjir, M. and Babu, S. (2020, June). Black or white? how to develop an autotuner for memory-based analytics. In *Proceedings of the 2020 ACM SIGMOD International Conference on Management of Data* (pp. 1667–1683).
Kusumoto, D., Yuasa, S. and Fukuda, K. (2022). Induced pluripotent stem cell-based drug screening by use of artificial intelligence. *Pharmaceuticals*, *15*(5), 562.
Leite, M.L., de Loiola Costa, L.S., Cunha, V.A., Kreniski, V., de Oliveira Braga Filho et al. (2021). Artificial intelligence and the future of life sciences. *Drug Discovery Today*, *26*(11), 2515–2526.
Li, G., Zhou, X. and Cao, L. (2021, June). AI meets database: AI4DB and DB4AI. In *Proceedings of the 2021 International Conference on Management of Data* (pp. 2859–2866).
Liao, C. and Li, S. (2007). A support vector machine ensemble for cancer classification using gene expression data. In *Bioinformatics Research and Applications: Third International Symposium, ISBRA 2007, Atlanta, GA, USA, May 7-10, 2007. Proceedings 3* (pp. 488–495). Springer Berlin Heidelberg.
Lim, H., Cankara, F., Tsai, C.J., Keskin, O., Nussinov, R. et al. (2022). Artificial intelligence approaches to human-microbiome protein–protein interactions. Current Opinion in Structural Biology, 73, 102328.
Lu, Y. (2019). Artificial intelligence: a survey on evolution, models, applications and future trends. *Journal of Management Analytics*, *6*(1), 1–29.
Ma, S. and Dai, Y. (2011). Principal component analysis-based methods in bioinformatics studies. *Briefings in bioinformatics*, *12*(6), 714–722.
Mann, M., Kumar, C., Zeng, W.F. and Strauss, M.T. (2021). Artificial intelligence for proteomics and biomarker discovery. Cell systems, 12(8), 759–770.
Mondal, Partha Pratim, Abhinav Galodha, Vishal Kumar Verma, Vijai Singh, et al. "Review on machine learning-based bioprocess optimization, monitoring, and control systems." Bioresource technology 370 (2023): 128523.
Mural, R.J., Einstein, J.R., Guan, X., Mann, R.C. and Uberbacher, E.C. et al. (1992). An artificial intelligence approach to DNA sequence feature recognition. *Trends in biotechnology*, *10*, 66–69.

Mustapha, F.Z., Haruna, A.A. and Muhammad, U.T. (2020). An Overview of Artificial Intelligence. Journal of Applied Sciences & Environmental Sustainability, 6(12), 60–74.

Nielsen, J., Tillegreen, C.B. and Petranovic, D. (2022). Innovation trends in industrial biotechnology. *Trends in Biotechnology*, *40*(10), 1160–1172.

Pham, T., Tran, T., Phung, D. and Venkatesh, S. (2017). Predicting healthcare trajectories from medical records: A deep learning approach. *Journal of biomedical informatics*, *69*, 218–229.

Radamson, H.H. and Simeon, E. (2018). * Institute of Microelectronics of Chinese Academy of Sciences (IMECAS), Beijing, PR China, University of Chinese Academy of Sciences (UCAS), Beijing, PR China. CMOS Past, Present and Future, 105.

Reddy, S. (Ed.). (2023). Translational Application of Artificial Intelligence in Healthcare: -A Textbook. CRC Press.

Riahi, Y., Saikouk, T., Gunasekaran, A. and Badraoui, I. (2021). Artificial intelligence applications in supply chain: A descriptive bibliometric analysis and future research directions. Expert Systems with Applications, 173, 114702.

Schaub, N.J., Hotaling, N.A., Manescu, P., Padi, S., Wan, Q. et al. (2020). Deep learning predicts function of live retinal pigment epithelium from quantitative microscopy. *The Journal of clinical investigation*, *130*(2), 1010–1023.

Schork, N.J. (2019). Artificial intelligence and personalized medicine. *Precision medicine in Cancer therapy*, 265–283.

Senior, Abrahan Bechara. *Modeling Fiber Damage during Processing of Long Fiber-Reinforced Thermoplastic Composites*. The University of Wisconsin-Madison, 2021.

Tadiboina, S.N. (2022). The use of AI in advanced medical imaging. *Journal of Positive School Psychology*, *6*(11), 1939–1946.

Takahashi, K. and Yamanaka, S. (2006). Induction of pluripotent stem cells from mouse embryonic and adult fibroblast cultures by defined factors. *cell*, *126*(4), 663–676.

Turing, A.M. (2009). *Computing machinery and intelligence* (pp. 23–65). Springer Netherlands.

Ugawa, Masashi, Yoko Kawamura, Keisuke Toda, Kazuki Teranishi et al. "In silico-labeled ghost cytometry." *Elife* 10 (2021): e67660.

Waisman, A., La Greca, A., Möbbs, A.M., Scarafía, M.A., Velazque, N.L.S. et al. (2019). Deep learning neural networks highly predict very early onset of pluripotent stem cell differentiation. *Stem cell reports*, *12*(4), 845–859.

Wang, S., Li, W., Liu, S. and Xu, J. (2016). RaptorX-Property: a web server for protein structure property prediction. *Nucleic acids research*, *44*(W1), W430–W435.

Wei, G.W. (2019). Protein structure prediction beyond AlphaFold. *Nature Machine Intelligence*, *1*(8), 336–337.

Xiong, J. (2006). Essential bioinformatics. https://doi.org/10.1017/cbo9780511806087

Zanca, F., Brusasco, C., Pesapane, F., Kwade, Z., Beckers, R. and Avanzo, M. (2022, October). Regulatory aspects of the use of artificial intelligence medical software. In *Seminars in Radiation Oncology* (Vol. 32, No. 4, pp. 432–441). WB Saunders.

Zednik, C. (2021). Solving the black box problem: A normative framework for explainable artificial intelligence. *Philosophy & technology*, *34*(2), 265–288.

Zhou, J. and Troyanskaya, O.G. (2015). Predicting effects of noncoding variants with deep learning–based sequence model. *Nature methods*, *12*(10), 931–934.

Zhou, X.X., Zeng, W.F., Chi, H., Luo, C., Liu, C., Zhan, J. et al. (2017). pDeep: predicting MS/MS spectra of peptides with deep learning. *Analytical chemistry*, 89(23), 12690–12697.

Zhu, T., Li, K., Chen, J., Herrero, P. and Georgiou, P. (2020). Dilated recurrent neural networks for glucose forecasting in type 1 diabetes. *Journal of Healthcare Informatics Research*, *4*, 308–324.

CHAPTER

2

Evolution of Artificial Intelligence

Aravind Sridhar,[1] *Bala Pesala*,[1] *Geethanjali Radhakrishnan*,[1] *Jeffrey Niezgoda*[1] and *Sandeep Gopalakrishnan*[1,2*]

1. Introduction

The history of artificial intelligence (AI) is one of the most fascinating and influential threads in the enduring fabric of human knowledge. Over many decades, AI has surpassed its status as a mere technological advancement and has become a formidable influence, fundamentally reshaping healthcare and the biological sciences. This chapter explores the historical development of AI and its significant contributions to healthcare and biosciences. The narrative around AI is characterised by intrigue, unwavering determination, and innovative thinking. This narrative traces the evolution of the subject matter from its origins in the 1950s to its contemporary revitalisation, propelled by state-of-the-art technology and unwavering human resolve. The chapter presents a cohesive storyline encompassing the initial stages of AI research, followed by the problematic period known as the AI winter and the following revival that has given rise to machine learning, deep learning, and the era characterised by the abundance of large-scale data. This discourse expounds upon AI's crucial role in healthcare, expanding the frontiers of medical diagnoses, treatment modalities, and comprehension of the human anatomy.

During its nascent stage in the 1950s, the area of AI was characterised by a limited number of pioneering individuals. The Dartmouth Workshop, held in 1956 and led by prominent figures such as John McCarthy, Marvin Minsky, Nathaniel Rochester, and Claude Shannon, is widely regarded as the seminal event that gave rise to the field of AI (Artificial Intelligence [AI] Coined at Dartmouth | Dartmouth. [n.d.]). At that moment, the prestigious corridors of academia were reverberating with a profound inquiry: Was it possible to endow machines with the cognitive capabilities of humans? Symbolic reasoning was the fundamental core of early AI investigations.

[1] Auxillium Health Inc, Milwaukee, USA.
[2] College of Health Professions and Sciences, University of Wisconsin-Milwaukee, USA.
* Corresponding author: sandeep@uwm.edu

The prevailing notion posited that robots may attain human-like thinking and reasoning capacity by organising knowledge and employing symbolic manipulation. The commencement of a novel epoch has arrived. The Logic Theorist, a programme developed by Allen Newell and Herbert A Simon in 1955, emerged during this period. The Logic Theorist demonstrated its pioneering nature by successfully proving mathematical theorems using symbolic reasoning. This accomplishment has significant implications for the advancement of automated problem-solving techniques and the subsequent creation of future AI systems (Newell, A. et al., 1956). The narrative around artificial intelligence would only be complete by acknowledging the pivotal role played by expert systems, which served as a vital component within the field throughout this era. The systems mentioned above, which were developed to encapsulate human expertise in a structured and rule-based manner, introduced a groundbreaking notion within the realm of healthcare (Swartout, 1985). An exemplary instance is Dendral, created during the late 1960s by (Edward Feigenbaum, Joshua Lederberg, and Bruce G. Buchanan 1978). Dendral was an expert system that found use in chemical analysis within the realm of mass spectrometry. The ability to discern chemical compounds from spectral data had the potential to advance chemistry and its implications for healthcare substantially.

Despite the extraordinary early accomplishments in AI, there were limitations and obstacles to overcome. AI system capabilities were severely constrained by the exceedingly limited computational capacity of the period. The task of encoding human knowledge in a manner that machines can understand and modify has proven to be a significant and challenging undertaking. Furthermore, the pragmatic aspect of managing uncertainty in real-world scenarios presented a substantial challenge for early AI investigations. During this arduous period, we shift into a distinct era sometimes referred to as the "AI winter," which encompasses the timeframe between the 1970s and the 1980s. During a certain period, AI development saw a notable deceleration, marked by a decline in financial support and waning enthusiasm. The discrepancy between exaggerated promises and disappointing outcomes resulted in a prevailing scepticism and disillusionment. The consequences of this winter extended beyond academic settings and impacted other fields, such as healthcare and biological sciences. In these sectors, efforts involving artificial intelligence faced setbacks and met challenges.

Resilience and flexibility are fundamental characteristics that contribute to advancements in several domains, including the science of artificial intelligence. The 1990s marked the revival of AI research. Machine learning has evolved as a very influential and transformational method, demonstrating its ability to extract valuable knowledge from extensive datasets and effectively adapt to ever-changing contexts. Deep learning, a subdiscipline within the broader subject of machine learning, has garnered significant attention because of its notable contributions to image analysis and natural language processing (LeCun, Y. et al., 2015). The utilisation of big data, which refers to vast repositories of information, has revitalised the field of artificial intelligence (AI), notably in healthcare. Critical advances in AI research underscore the significance of this revival. The healthcare industry has undergone significant transformation due to the development of AI algorithms and the increasing attention given to ethical and regulatory concerns. The accomplishments mentioned, including AlphaFold's significant advancements in protein structure predictions and the regulatory approvals of AI-based diagnostic tools like IDx-DR for detecting diabetic retinopathy, exemplify the profound impact of artificial intelligence (Jumper, J. et al., 2021; Abràmoff, M.D. et al., 2018).

The essential essence of the theory is realised through its application in practical contexts. Clinical decision support systems (CDSS) play a crucial role in aiding healthcare professionals in the process of illness diagnosis and treatment (Kawamoto, K. et al., 2005). AI expedites the drug discovery process, reducing the lengthy journey from identifying molecules to developing market-ready pharmaceuticals. AI also plays a crucial role in medical imaging by detecting and diagnosing abnormalities and medical conditions in radiology and pathology pictures (Litjens, G. et al., 2017). The prominence of personalised medicine has been increasingly evident with the integration of AI in customising medical interventions for specific patients. This development provides a look into a prospective healthcare landscape where the provision of medical care is as distinctive as the individuals it caters to (Topol, E., 2019).

As we approach the climax of the evolutionary trajectory of AI in healthcare and biological sciences, we direct our gaze towards the future. We possess a forward-looking perspective on the forthcoming patterns and obstacles that will influence the AI domain inside the healthcare sector. The proposition to incorporate a distinct chapter focused on AI within wound care signifies the continuous expansion of AI's utilisation within the healthcare field. The significance of multidisciplinary cooperation is emphasised, highlighting the necessity for integrating expertise from AI specialists and healthcare professionals. The ethical implications and legal constraints associated with implementing AI in healthcare emphasise the importance of responsible AI usage within a rapidly evolving digital healthcare environment. Within the broader context of AI's progression in healthcare and biological sciences, this chapter represents a significant phase characterised by initial developments, obstacles, achievements, and future ambitions. This text explores the progression of AI from its first challenges to its present resurgence, providing insight into a potential future where AI significantly contributes to the advancement of human health. As we traverse these elements, we extend an invitation to delve into the intricate realm of AI within the biological sciences, uncovering its limitless capabilities and steadfast dedication to improving the welfare of humankind.

Early AI Research (1950s-1970s)

The 1950s is a significant period in the annals of technical history. During this specific era, the human race endeavoured to investigate a profound inquiry: the feasibility of constructing robots endowed with cognitive abilities and logical thinking, presently referred to as artificial intelligence (AI). The Dartmouth Workshop, held in 1956 and organised by John McCarthy, Marvin Minsky, Nathaniel Rochester, and Claude Shannon, was a significant turning point (Artificial Intelligence [AI] Coined at Dartmouth | Dartmouth. [n.d.]) (Figure 1 and 2). It initiated an exploration into the unexplored domains of machine intelligence and bestowed the field with its name.

A bold objective characterised the AI research conducted throughout the 1950s-1970s: emulating human-like cognitive abilities in artificial systems. Symbolic thinking arose as a fundamental cornerstone of AI during this particular era. The notion posited that with the organisation of information into an extended framework and the subsequent manipulation of these symbols, robots might replicate the cognitive processes exhibited by humans. The era was characterised by a profound optimism fuelled by the belief that reasoning and addressing challenges might be mechanised (McCarthy, J., 1958).

Fig. 1 Marvin Minsky, Claude Shannon, Ray Solomonoff and other scientists at the Dartmouth Summer Research Project on Artificial Intelligence
(**Photo:** Margaret Minsky)

Fig.2 Dartmouth Hall Commemorative Plaque.
Photo: Joe Mehling

The Logic Theorist was a computer program developed in the 1950s by Allen Newell and Herbert A Simon at the RAND Corporation and was one of the seminal advancements in the nascent field of artificial intelligence (A. Newell and H. Simon, 1956). The Logic Theorist was a significant advancement in automated problem-solving and symbolic reasoning in proving mathematical theorems, which signified an important milestone in developing automated logical thinking. This initial AI system established the basis for subsequent research endeavours, showcasing the feasibility of mechanising human thinking processes (Newell, A. et al., 1959). Chess and checkers emerged as the arenas for initial AI testing. Arthur Samuel, a notable contributor to the development of artificial intelligence, created a program for playing the game of Checkers in 1952 (Samuel, A.L., 1960). What distinguished Samuel's work was utilising a self-play approach in which his program acquired the ability to play checkers. Through self-play and iterative strategy refinement, the system exhibited a manifestation of machine learning. This seminal development established the groundwork for the more expansive notion of contemporary machine learning (A.L. Samuel, 1959).

Expert systems emerged as a logical progression during this period. These systems intended to capture human expertise into organised, rule-based frameworks (Clancey, W.J., 1983). The development constituted a noteworthy advancement in artificial intelligence, explicitly emphasising its use in healthcare. Expert systems were developed to provide diagnostic and decision help comparable to the expertise of professionals in a specific field (Swartout, W.R., 1985). Developed by Edward Shortliffe in the 1970s, the MYCIN system represented a notable advancement within the medical field (Swartout, W.R., 1985). The system served as a virtual specialist in infectious diseases, employing a series of inquiries about the patient's symptoms, medical background, and laboratory findings. Based on the available information, MYCIN could produce diagnostic suggestions. Despite being considered unsophisticated by contemporary standards, MYCIN demonstrated the significant potential of AI in supporting medical practitioners in intricate decision-making procedures. (Clancey, W.J., 1983; Computer-Based Medical Consultations: Mycin, 1976). This era of AI research was pivotal in shaping the future of artificial intelligence in healthcare and biological sciences. The theoretical principles and early applications birthed during this period have paved the way for the remarkable progress and innovations we witness today. The forthcoming discussion will shift our focus to the AI Winter (1970s-1980s), a period that was less sanguine but pivotal in the annals of AI.

The AI Winter (1970s-1980s)

The decade of the 1970s marked the commencement of a tumultuous era within the realm of artificial intelligence (AI). This period, also known as the 'AI winter' was marked by a decline in excitement, decreased financial support, and a sceptical outlook on the possibilities of artificial intelligence. During this period, several significant events and difficulties emerged, resulting in a brief stagnation of progress in artificial intelligence. During the 1960s, there was a notable sense of optimism over the capacity of AI to overcome language barriers through machine translation. The primary driving force behind this endeavour was the publication of the Automatic Language Processing Advisory Committee (ALPAC) report in 1966 (ALPAC, 1966) Figure 3. The study evaluated the advancements made in machine translation research that the United States government has financially supported. The ALPAC report marked a significant shift in the trajectory of the subject matter at hand. The findings indicated that the introductory machine translation systems did not meet the anticipated standards. The research included a critical assessment of the calibre of machine-translated output, emphasising the frequent necessity for substantial post-editing by human professionals. The dissatisfaction with machine translation prompted a realignment of financial resources, diverting them from language translation endeavours to alternative uses of artificial intelligence (ALPAC, 1966).

The lack of success in machine translation posed a significant setback for the emerging field of artificial intelligence and served as an early indication of the obstacles ahead.

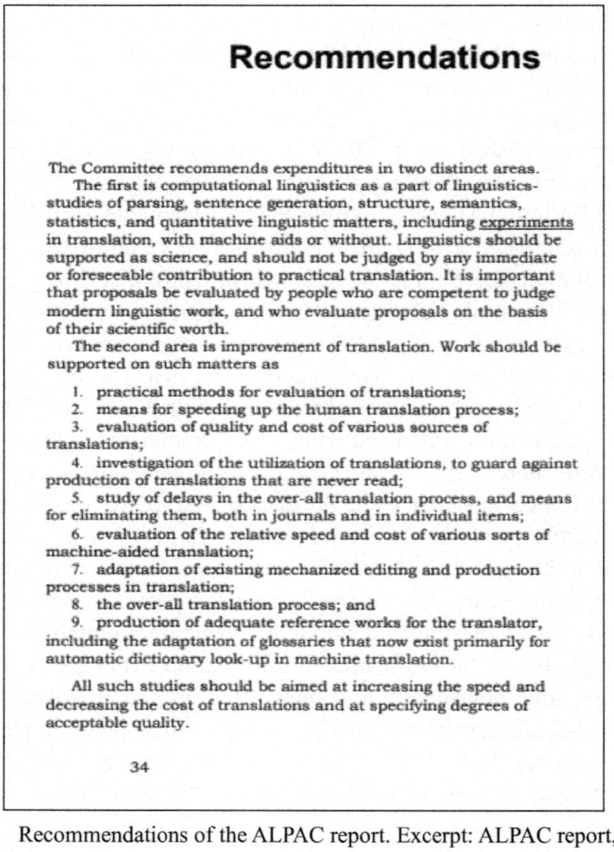

Recommendations

The Committee recommends expenditures in two distinct areas.

The first is computational linguistics as a part of linguistics-studies of parsing, sentence generation, structure, semantics, statistics, and quantitative linguistic matters, including <u>experiments</u> in translation, with machine aids or without. Linguistics should be supported as science, and should not be judged by any immediate or foreseeable contribution to practical translation. It is important that proposals be evaluated by people who are competent to judge modern linguistic work, and who evaluate proposals on the basis of their scientific worth.

The second area is improvement of translation. Work should be supported on such matters as

1. practical methods for evaluation of translations;
2. means for speeding up the human translation process;
3. evaluation of quality and cost of various sources of translations;
4. investigation of the utilization of translations, to guard against production of translations that are never read;
5. study of delays in the over-all translation process, and means for eliminating them, both in journals and in individual items;
6. evaluation of the relative speed and cost of various sorts of machine-aided translation;
7. adaptation of existing mechanized editing and production processes in translation;
8. the over-all translation process; and
9. production of adequate reference works for the translator, including the adaptation of glossaries that now exist primarily for automatic dictionary look-up in machine translation.

All such studies should be aimed at increasing the speed and decreasing the cost of translations and at specifying degrees of acceptable quality.

34

Fig. 3 Recommendations of the ALPAC report. Excerpt: ALPAC report, 1966

During the latter half of the 1950s and the early 1960s, there was a notable surge of enthusiasm surrounding the notion of perceptrons, considered early iterations of artificial neural networks. The perceptrons, which Frank Rosenblatt initially developed, showed considerable potential in the domain of pattern recognition tasks (Rosenblatt, F., 1958). Nevertheless, the inherent constraints of single-layer perceptrons became evident since they could only resolve linearly separable issues. When endeavouring to tackle intricate, non-linear issues, perceptrons showed their insufficiency. The publication of the book "Perceptrons" by Marvin Minsky and Seymour Papert in 1969 was a significant milestone in the field of neural networks since it included a critical analysis of the constraints inherent in these initial models (Minsky, M. and Papert, S., 1969). The research shed light on perceptrons' limitations and provided evidence of their inability to address specific foundational challenges effectively. The critique above resulted in a substantial decline in interest and financial support for artificial neural networks, laying the foundation for the onset of the AI winter.

From 1971 to 1975, the Defence Advanced Research Projects Agency (DARPA) experienced a sense of dissatisfaction with the progress made in speech understanding research. In the early 1970s, the Speech Understanding Research (SUR) programme was initiated by the Defence Advanced Research Projects Agency (DARPA) of the U.S. Department of Defence (Walker, 1976). The objective was to further the development of voice recognition systems capable of comprehending and transcribing oral communication. Carnegie Mellon University assumed a pivotal role in this endeavour. Nonetheless, the SUR program encountered significant obstacles, such as sluggish advancement and lofty objectives beyond the existing technological capabilities. DARPA was dissatisfied with the program's limited progress, resulting in decreased financial support and diminished enthusiasm for AI research on speech comprehension (Walker, 1976). This failure exacerbated the already prevailing coldness of the AI winter.

The year 1973 witnessed a notable setback in the domain of AI inside the United Kingdom, which was characterised by the release of the Lighthill report. Professor Sir James Lighthill, an esteemed scholar in aerodynamics, was assigned to assess the current status of AI development inside the nation. The paper "Artificial Intelligence: A Comprehensive Examination" critically analysed the accomplishments and prospects of AI research (James Lighthill, 1973). According to the Lighthill research, AI did not fulfil its ambitious claims. Lighthill's evaluation posited that a significant portion of the study exhibited excessive ambition and a dearth of concrete outcomes (James Lighthill, 1973). The outcome of this critical assessment led to a substantial reduction in financial resources allocated to AI research within the United Kingdom. Consequently, some researchers were disheartened by this development, abandoning their pursuits in the subject.

In essence, the artificial intelligence discipline underwent a period of introspection and austerity during the AI winter of the 1970s and 1980s. Inadequacies were observed in machine translations; criticism directed towards perceptrons (Minsky, M., and Papert, S., 1969), DARPA's dissatisfaction with speech understanding research (Walker, 1976), the Lighthill report's critical assessment of AI in the United Kingdom (James Lighthill, 1973), and DARPA's reduction in funding for academic AI research have collectively raised concerns and prompted a reassessment of the objectives and potential of AI research. The AI winter, characterised by a decline in financial support and excitement, was crucial in setting the stage for the subsequent revival of AI research throughout the 1990s. This resurgence was fuelled by advancements in machine learning, deep learning, and big data, which propelled the discipline to unprecedented achievements in healthcare and biological sciences.

Moving forward, we shall consider an alternative epoch that reignited the potential of artificial intelligence. The 1990s witnessed the emergence of a renaissance that brought about a significant transformation in the field of AI research, leading to its profound impact on the domains of healthcare and biological sciences. The subsequent part, "The Resurgence of AI (1990s-present)," will examine how AI transformed and adopted machine learning, deep learning, and big data. The earlier achievements, previously considered hypothetical in the 1970s, have significantly transformed the healthcare sector, resulting in novel opportunities and prospects.

The Resurgence of AI (1990s-present)

As the sun set on the AI winter of the 1970s and 1980s, a new dawn emerged in the 1990s, heralding the resurgence of artificial intelligence. During this renaissance, AI underwent significant advancements that converted it from a domain of theoretical potential to a tangible and influential force, particularly in healthcare and biological sciences. This section will examine the narrative surrounding the resurgence of AI, focusing on significant events that have shaped its trajectory from the 1990s until now.

1997: IBM's Deep Blue Defeats Garry Kasparov

The resurgence of AI can be attributed to a significant event in 1997, namely the victory of IBM's Deep Blue (Campbell, M. et al., 2002; Allen Newell et al., 1958), a computer designed for playing chess, over the reigning world chess champion Garry Kasparov (Figure 4). This pivotal event proved that artificial intelligence was not solely a theoretical endeavour but a pragmatic power that could surpass human intelligence in intricate domains. Deep Blue's triumph demonstrated machine intelligence's capacity, captivating the interest of researchers and the general public.

Fig. 4 World Chess Champion Garry Kasparov (L) makes a move during his fourth game against the IBM Deep Blue chess computer. Credit: Stan Honda Getty Images

2012: Google Brain's Deep Learning Breakthrough

In 2012, Google Brain achieved a notable advancement in deep learning. Researchers at Google's X lab developed a neural network of unprecedented magnitude, consisting of 16,000 computer processors interlinked by one billion connections (Dean, J. et al., 2012). The computational architecture of significant magnitude underwent a pioneering training regimen, employing a vast collection of images from YouTube. Significantly, the training procedure deliberately avoided providing explicit guidance, enabling the neural network to explore and comprehend patterns independently. An intriguing phenomenon was observed during the training process of the neural network. Over time, it demonstrated a developing capacity to recognise attributes that resemble those found in felines. The pinnacle of this self-directed learning process was evident in the network's exceptional ability to precisely identify cat videos on YouTube. This significant milestone represents a critical turning point in utilising deep neural networks, specifically facial and image recognition. These tasks have gained widespread prevalence in modern artificial intelligence frameworks. This significant advancement exemplified the integration of immense computational power and unsupervised learning at an unparalleled magnitude. The neural network demonstrated exceptional performance and accuracy, signifying a significant advancement in the capabilities of artificial intelligence systems. The implications of this accomplishment resonate within the contemporary AI field, exerting influence on subsequent advancements and accelerating a phase of rapid expansion.

2015: DeepMind's AlphaGo Defeats a Human Go Champion

AlphaGo, developed by DeepMind, made history in artificial intelligence in 2015 when it achieved a revolutionary victory over a human Go champion. The emergence of AlphaGo, a creation of Google's DeepMind, signified a significant turning point, shedding light on the vast capabilities of artificial intelligence in achieving mastery over complex strategic games (Figure 5). The journey towards mastery for AlphaGo commenced with a rigorous training regimen (Silver, D. et al., 2016). The AI system assimilated a vast collection of novice-level Go games, fully engaging with the intricacies of human gameplay. Following that, the programme embarked on an unwavering process of self-improvement through engaging in numerous games against various versions of itself. This was achieved by employing reinforcement learning techniques to correct errors and enhance strategic approaches (Silver, D. et al., 2016; Maddison, C.J. et al., 2014).

Fig. 5 AlphaGo in training. Credits-Google DeepMind

AlphaGo emerged victorious by a score of 5-0 against Fan Hui, its first professional human opponent and a three-time European Champion, in October 2015. This event represented a significant shift in the prevailing paradigm, indicating the emergence of a new era in which artificial intelligence not only engaged in competitive activities but also exhibited exceptional competence when pitted against experienced human players. In March 2016, AlphaGo peaked when it faced the esteemed Go player Lee Sedol, who had amassed 18 world titles. AlphaGo, boasting a worldwide viewership surpassing 200 million, attained a resounding victory of 4-1, thus marking a significant milestone that materialised a decade earlier than anticipated (Metz, C., 2016). The underscored advancements in artificial intelligence have demonstrated remarkable progress in understanding and manoeuvring complex strategic domains.

Throughout these momentous matches, AlphaGo demonstrated its remarkable ability to employ innovative strategies that defied conventional wisdom. In the second game, Move 37 played a crucial role in shifting the outcome significantly in favour of AlphaGo, with a probability of occurrence as low as 1 in 10,000. In contrast, Lee Sedol exhibited a noteworthy response in the fourth game by executing Move 78, referred to as "God's Touch" due to its remarkable and innovative characteristics (Metz, C, 2016). This significant occurrence served as a symbolic victory for AlphaGo and marked a noteworthy milestone for artificial intelligence, showcasing its capability to surmount intricate strategic obstacles. AlphaGo's triumph resonated throughout scholarly communities, shedding light on the unexplored capabilities of artificial intelligence in excelling in complex and traditionally human-centric endeavours.

2016: IBM Watson for Oncology

In the pivotal year of 2016, a landmark collaboration between IBM Watson Health and Pfizer Inc. took place, indicating a significant step forward in integrating AI within healthcare, particularly in the challenging domain of immuno-oncology (IBM Watson Health and Pfizer Inc. Collaboration, 2016). To accelerate Pfizer's immuno-oncology research efforts, this collaborative effort leveraged IBM Watson for Drug Discovery, highlighting the growing role of AI in advancing pharmaceutical exploration. IBM Watson for Oncology has emerged as a pivotal component in this transformative change, providing medical practitioners and individuals having cancer with vital information and tailored, evidence-based perspectives to guide decisions regarding its treatment. By utilising machine learning, natural language processing, and cognitive reasoning technologies, this artificial intelligence platform enabled the discovery of new drug targets, the investigation of combination therapies, and the development of patient selection strategies in immuno-oncology. Concurrently, IBM Watson was implemented in six hospitals of the Manipal Hospitals Group, a well-known healthcare organisation in India, which marked the beginning of a nationwide campaign to improve cancer care. Using Watson for Oncology by oncologists at Manipal Hospitals in India has allowed for personalised and evidence-based care to patients with breast, colorectal, and lung cancers. This technology has been employed to address the diverse healthcare needs within the expansive healthcare landscape of India.

2019: FDA Approves AI-Based Diagnostic System

The U.S. Food and Drug Administration (FDA) approved several noteworthy AI-based diagnostic systems in 2019, significantly advancing AI into medical diagnostics. A noteworthy advancement occurred in April when the FDA granted authorisation for IDx-DR, an innovative medical device developed by IDx (FDA, 2018). This system utilises sophisticated deep-learning algorithms as an independent diagnostic tool for ocular conditions in diabetic adults, specifically focusing on diabetic retinopathy (Abràmoff, M.D. et al., 2018). In recognition of its potential to improve treatment for critical illnesses, the FDA granted the coveted breakthrough device designation to an AI system engineered explicitly for cancer diagnosis. The regulatory acknowledgement of this phenomenon enables a streamlined evaluation procedure, emphasising the crucial significance of artificial intelligence in enhancing diagnostic capacities for severe and irreversible medical conditions. The recent FDA approvals indicate the advancement and widespread acceptance of diagnostic tools powered by artificial intelligence (AI), marking a significant milestone in integrating technology and healthcare.

2020: AI in the Fight Against COVID-19

AI has played a crucial role in understanding the virus and accelerating medical research. Text and data mining tools, driven by artificial intelligence technologies, provide valuable insights into the virus's historical background, transmission patterns, diagnostic methods, and treatment approaches.

These tools utilise knowledge gained from previous epidemics. Deep learning models have made significant contributions by predicting potential drugs and treatments for COVID-19 (Nguyen, T.T. et al., 2020; Wang, L. et al. 2020). The capabilities of AI extend to the forefront of detection and diagnosis, as evidenced by its predictive abilities in revealing the emergence of an unfamiliar strain of pneumonia in China prior to the global recognition of the COVID-19 threat. In addition to its diagnostic capabilities, AI was pivotal in preventing and mitigating viral transmission by facilitating efficient surveillance and contact tracing (Shahroz, M. et al., 2021). AI plays a crucial role in addressing the health crisis by enabling the dissemination of personalised information and facilitating learning. Machine learning technology, AI, serves as the foundation for the transition towards remote communication, telemedicine, and the preservation of food security. Furthermore, the impact of AI can be observed in its role in monitoring the recovery phase and improving early warning tools. In the face of the significant challenges presented by the global pandemic, AI emerges as a steadfast ally, making substantial contributions to the collective endeavours aimed at comprehending, mitigating, and ultimately surmounting the intricacies associated with COVID-19.

2021: AlphaFold by DeepMind

AlphaFold, an innovative artificial intelligence system developed by DeepMind, made significant advancements in protein structure prediction in 2021, ushering in a paradigm shift in computational biology (AlphaFold, 2020). AlphaFold, a computational system developed by DeepMind, demonstrates remarkable proficiency in accurately predicting the three-dimensional structure of proteins, comparable to experimental results. AlphaFold, trained on an extensive dataset of approximately 100,000 established protein sequences and structures, exhibits an exceptional capacity to rapidly and accurately forecast atomic-level protein conformations (Jumper, J. et al., 2021). In July 2021, DeepMind made a significant contribution to scientific transparency by publishing the comprehensive code and methodology of AlphaFold 2 in the prestigious journal Nature (Jumper, J. et al., 2021). Furthermore, the joint effort between the European Molecular Biology Laboratory (EMBL) and our organisation resulted in the establishment of the AlphaFold Protein Structure Database, a significant repository containing more than 200 million predictions of protein structures (Structure Database, A.P., n.d). The openly accessible database discussed herein enhances scientific inquiry by providing a comprehensive repository that aligns with UniProt, the widely repository for protein sequences and annotations. Significantly, AlphaFold exhibits extensive predictive capabilities, as it has successfully envisioned the structures of nearly all 20,000 proteins expressed in the human body. In a rigorous benchmark evaluation, where predictions were contrasted with established structures, AlphaFold demonstrated a notable accuracy rate of 95%. This achievement signifies a significant advancement in integrating computational predictions and experimental validations, thereby reducing the disparity between the two. The advancements made by AlphaFold have had a significant impact on the AI landscape, leading to accelerated growth and progress in the field. The advancements made by AlphaFold in protein structure prediction contribute to expanding knowledge of biological complexities. Additionally, it serves as evidence of the significant synergies that can be achieved by integrating artificial intelligence and molecular biology.

References

A. Newell and H. Simon, "The logic theory machine—A complex information processing system," in IRE Transactions on Information Theory, vol. 2, no. 3, pp. 61–79, September 1956, doi: 10.1109/TIT.1956.1056797.

A.L. Samuel, "Some Studies in Machine Learning Using the Game of Checkers," in IBM Journal of Research and Development, vol. 3, no. 3, pp. 210–229, July 1959, doi: 10.1147/rd.33.0210.

Abràmoff, M.D., Lavin, P.T., Birch, M. Pivotal trial of an autonomous AI-based diagnostic system for detection of diabetic retinopathy in primary care offices. npj Digital Med 1, 39 (2018). https://doi.org/10.1038/s41746-018-0040-6

Allen Newell, Cliff Shaw, Herbert Simon (1958). Chess Playing Programs and the Problem of Complexity. IBM Journal of Research and Development, Vol. 4, No. 2, pp. 320–335. Reprinted (1963) in Computers and Thought (eds. Edward Feigenbaum and Julian Feldman), pp. 39–70. McGraw-Hill, New York, N.Y. pd.

ALPAC (1966) Languages and machines: computers in translation and linguistics. A report by the Automatic Language Processing Advisory Committee, Division of Behavioral Sciences, National Academy of Sciences, National Research Council. Washington, D.C.: National Academy of Sciences, National Research Council, 1966. (Publication 1416.) 124pp.

AlphaFold: a solution to a 50-year-old grand challenge in biology. (2020, November 30). Google DeepMind. https://deepmind.google/discover/blog/alphafold-a-solution-to-a-50-year-old-grand-challenge-in-biology/

Artificial Intelligence (AI) Coined at Dartmouth | Dartmouth. (n.d.). Artificial Intelligence (AI) Coined at Dartmouth | Dartmouth. https://home.dartmouth.edu/about/artificial-intelligence-ai-coined-dartmouth

Campbell, M., Hoane, A. and Hsu, F.H. (2002, January). Deep Blue. Artificial Intelligence, 134(1–2), 57–83. https://doi.org/10.1016/s0004-3702(01)00129-1.

Clancey, W.J. (1983, May). The epistemology of a rule-based expert system —a framework for explanation. Artificial Intelligence, 20(3), 215–251. https://doi.org/10.1016/0004-3702(83)90008-5.

Computer-Based Medical Consultations: Mycin. (1976). https://doi.org/10.1016/b978-0-444-00179-5.x5001-x.

Dean, J., Corrado, G., Monga, R., Chen, K., Devin, M., Mao, M., ... and Ng, A.Y. (2012). "Large Scale Distributed Deep Networks." Advances in Neural Information Processing Systems, 1223–1231.

Edward Feigenbaum, Joshua Lederberg and Bruce G. Buchanan (1978, August). Dendral and meta-dendral: Their applications dimension. Artificial Intelligence, 11(1–2), 5–24. https://doi.org/10.1016/0004-3702(78)90010-3.

FDA permits marketing of artificial intelligence-based device to detect certain diabetes-related eye problems. (2018, April 11). FDA Permits Marketing of Artificial Intelligence-based Device to Detect Certain Diabetes-related Eye Problems | FDA. https://www.fda.gov/news-events/press-announcements/fda-permits-marketing-artificial-intelligence-based-device-detect-certain-diabetes-related-eye

IBM Watson Health and Pfizer Inc. Collaboration. (2016). Press Release.

James Lighthill, 'Artificial intelligence: a general survey', published as part of Science Research Council, Artificial Intelligence: A Paper Symposium, London: SRC, 1973, pp. 1–21.

Jumper, J., Evans, R., Pritzel, A. Highly accurate protein structure prediction with AlphaFold. Nature 596, 583–589 (2021). https://doi.org/10.1038/s41586-021-03819-2.

Kawamoto, K., Houlihan, C.A., Balas, E.A. and Lobach, D.F. (2005). Improving clinical practice using clinical decision support systems: a systematic review of trials to identify features critical to success. BMJ (Clinical research ed.), 330(7494), 765. https://doi.org/10.1136/bmj.38398.500764.8F.

LeCun, Y., Bengio, Y. and Hinton, G. Deep learning. Nature 521, 436–444 (2015). https://doi.org/10.1038/nature14539

Litjens, G., Kooi, T., Bejnordi, B.E., Setio, A.A.A., Ciompi, et al. (2017). A survey on deep learning in medical image analysis. Medical image analysis, 42, 60–88. https://doi.org/10.1016/j.media.2017.07.005.

Maddison, C.J., Huang, A., Sutskever, I. and Silver, D. (2014). Move Evaluation in Go Using Deep Convolutional Neural Networks. https://doi.org/10.48550/ARXIV.1412.6564.

McCarthy, J. (1958). "Programs with Common Sense." Proceedings of the Symposium on Mechanization of Thought Processes, 77–84.

Metz, C. (2016). "In Two Moves, AlphaGo and Lee Sedol Redefined the Future." Wired.

Minsky, M. and Papert, S. (1969). "Perceptrons: An Introduction to Computational Geometry." MIT Press.

Newell, A., Shaw, J.C. and Simon, H.A. (1959). "Report on a general problem-solving program." Proceedings of the International Conference on Information Processing, 256–264.

Newell, A., Shaw, J.C., and Simon, H.A. (1956). A Proposal for the Dartmouth Summer Research Project on Artificial Intelligence. In J. McCarthy (Ed.), Proceedings of the Dartmouth Conference on Artificial Intelligence (pp. 128–135). Dartmouth College.

Nguyen, T.T., Nguyen, Q.V.H., Nguyen, D.T., Yang, S., Eklund, P.W., Huynh-The, T., Nguyen, T.T., Pham, Q.-V., Razzak, I. and Hsu, E.B. (2020). Artificial Intelligence in the Battle against Coronavirus (COVID-19): A Survey and Future Research Directions. https://doi.org/10.48550/ARXIV.2008.07343.

Rosenblatt, F. (1958). The perceptron: A probabilistic model for information storage and organization in the brain. Psychological Review, 65(6), 386–408. https://doi.org/10.1037/h0042519.

Samuel, A.L. (1960). Programming Computers to Play Games. Advances in Computers Volume 1, 165–192. https://doi.org/10.1016/s0065-2458(08)60608-7.

Shahroz, M., Ahmad, F., Younis, M.S., Ahmad, N., Kamel Boulos, et al. (2021). COVID-19 digital contact tracing applications and techniques: A review post initial deployments. Transportation Engineering, 5, 100072. https://doi.org/10.1016/j.treng.2021.100072.

Silver, D., Huang, A., Maddison, C. Mastering the game of Go with deep neural networks and tree search. Nature 529, 484–489 (2016). https://doi.org/10.1038/nature16961.

Structure Database, A.P. (n.d.). AlphaFold Protein Structure Database. AlphaFold Protein Structure Database. https://alphafold.ebi.ac.uk/

Swartout, W.R. (1985, July). Rule-based expert systems: The mycin experiments of the stanford heuristic programming project. Artificial Intelligence, 26(3), 364–366. https://doi.org/10.1016/0004-3702(85)90067-0.

Topol, E. (2019, March 12). Deep Medicine: How Artificial Intelligence Can Make Healthcare Human Again.

Walker. (1976, October 1). Speech Understanding Research. In Stanford Research Inst Menlo Park CA (No. ADA034606). Defense Technical Information Center. https://apps.dtic.mil/sti/pdfs/ADA034606.pdf.

Wang, L., Lin, Z.Q. and Wong, A. (2020). COVID-Net: A tailored deep convolutional neural network design for detection of COVID-19 cases from chest X-ray images. Scientific Reports, 10(1), 1-12. https://doi.org/10.1038/s41598-020-76550-z.

CHAPTER

3

Futuristic Application of Artificial Intelligence

Romy Roy,[1] *Soniya Shende,*[1] *V.P Thejashwini,*[1] *Kamlesh Yadav*[2*] and *Khader Shameer*[3*]

1. Introduction

Artificial Intelligence (AI) is poised to redefine the way we diagnose, treat, and manage diseases. Beyond the theoretical promises, AI is now being applied in critical areas such as personalized medicine, predictive analytics, and real-time patient monitoring. This evolution is supported by a convergence of technologies like block chain, the Internet of Things (IoT), and virtual reality, which are enhancing the precision and scope of AI applications. While these advancements hold the potential to revolutionize patient care, they also introduce complex ethical, legal, and social challenges. Issues such as data privacy, AI accountability, and the equitable distribution of AI's benefits require careful consideration. As this chapter explores, the future of healthcare will not only depend on the successful integration of AI but also on navigating the responsibilities that come with its power. The journey ahead promises not just improved healthcare outcomes but a transformation of the healthcare landscape itself.

2. Emerging Trends and Future Applications

2.1 Personalized Medicine

2.1.1 What is Personalized Medicine?

Personalized medicine, also known as precision medicine, is a medical approach that tailors disease prevention, diagnosis, and treatment to the individual characteristics of each patient. This approach considers the genetic makeup, lifestyle, and environmental factors unique to each individual to develop more effective and tailored healthcare strategies. Personalized medicine moves away

[1] Type 3 Civilization Technologies Pvt Ltd, Kerala, India.
[2] School of Engineering Medicine, Texas A&M University, Houston, USA.
[3] Faculty of Medicine, School of Public Health, Imperial College of London.
* Corresponding author: romyroyz@gmail.com

from the one-size-fits-all approach, focusing instead on how individual differences impact disease progression and treatment response. (Roy et al., 2022)

2.1.2 Modalities of Personalized Medicine

2.1.2.1 Genomic Analysis

Genomic analysis is a critical component of personalized medicine, offering profound insights into the genetic underpinnings of diseases and individual responses to treatment. This scientific process involves examining a patient's complete set of DNA (genome) to identify genetic variations that contribute to health and disease. The primary goal of genomic analysis is to understand how genetic variations influence an individual's susceptibility to diseases, their progression, and response to specific treatments. By analyzing these genetic factors, healthcare providers can develop more effective, tailored treatment strategies.The field has rapidly evolved due to advancements in genetic sequencing technologies, bioinformatics, and computational biology. These advancements have made genomic analysis more accessible, accurate, and cost-effective. (Lappalainen et al., 2019)

2.1.2.1.1 Types of Genomic Analysis

2.1.2.1.1.1 Whole Genome Sequencing (WGS)

Involves sequencing the entire genome of an individual. WGS provides a comprehensive view of the genome, including all genes and the non-coding regions of DNA. It's particularly useful for identifying rare genetic disorders, understanding complex diseases, and exploring new therapeutic targets. (Yu et al., 2014)

2.1.2.1.1.2 Whole Exome Sequencing (WES)

Focuses on sequencing only the exome, which consists of all the exons in a genome. Exons are the coding regions of DNA that translate into proteins. WES is effective in identifying genetic variants within protein-coding regions, which are often responsible for many genetic diseases. (Jelin & Vora, 2018)

2.1.2.1.1.3 Targeted Gene Sequencing

Involves sequencing specific genes or regions of interest within the genome. This method is more focused and less resource-intensive compared to WGS and WES. Commonly used for diagnosing diseases with known genetic markers, pharmacogenomics (how genes affect a person's response to drugs), and cancer genomics. (Martin et al., 2016)

2.1.2.1.1.4 Genotyping

Refers to the process of determining differences in genetic make-up (genotype) by examining a small percentage of an individual's DNA. Unlike sequencing, genotyping is usually focused on known areas of variation. Widely used in personalized medicine for disease risk assessment, determining drug responses, and identifying carriers of genetic disorders. (Miller et al., 2008)

2.1.2.1.1.5 Comparative Genomic Analysis

Involves comparing the genomes of different species or individuals to identify genetic similarities and differences. This comparative approach can provide insights into evolutionary biology and disease mechanisms. Useful in research for understanding disease evolution, drug resistance patterns, and identifying potential therapeutic targets. (Premzl, 2020)

2.1.2.1.2 Role of AI in Genomic Analysis

The role of Artificial Intelligence (AI) in genomic analysis is transformative, offering unprecedented capabilities in understanding complex genetic data. AI's integration into genomics represents a significant leap forward in personalized medicine, enabling deeper insights into genetic factors that influence health and disease (Figure 1).

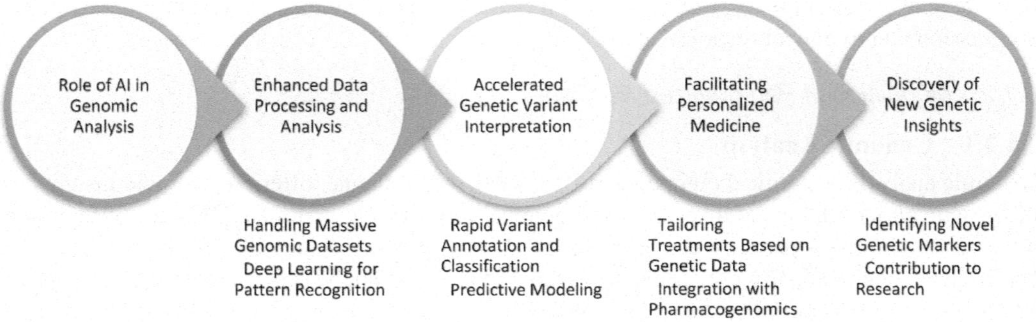

Fig. 1 Role of AI in Genomic Analysis

2.1.2.1.2.1 Enhanced Data Processing and Analysis

Handling Massive Genomic Datasets

Genomic data is incredibly vast and complex. AI, particularly machine learning algorithms, can process and analyze these large datasets more efficiently than traditional methods. AI can quickly sift through millions of genetic sequences to identify patterns and correlations (*Artificial Intelligence, Machine Learning and Genomics*, n.d.)

Deep Learning for Pattern Recognition

Deep learning models, such as Convolutional Neural Networks (CNNs), are adept at recognizing intricate patterns in genomic data. They can detect subtle genetic variations that might be linked to specific diseases or conditions. (*Artificial Intelligence, Machine Learning and Genomics*, n.d.)

2.1.2.1.2.2 Accelerated Genetic Variant Interpretation

Rapid Variant Annotation and Classification: AI streamlines the process of annotating genetic variants, categorizing them based on their potential impact on diseases. This rapid classification is crucial for understanding genetic risks and informing clinical decisions. (O'Brien et al., 2022)

Predictive Modeling: AI models can predict the potential impact of genetic mutations on protein functions, helping in assessing disease risk and guiding treatment strategies. (Jamal et al., 2020)

2.1.2.1.2.3 Facilitating Personalized Medicine

Tailoring Treatments Based on Genetic Data: AI's ability to analyze a patient's genetic makeup enables the development of personalized treatment plans. It can predict how individual patients will respond to various medications, thus informing personalized therapy choices. (Johnson et al., 2021)

Integration with Pharmacogenomics: AI enhances pharmacogenomic approaches by correlating genetic variants with drug efficacy and safety, leading to more effective and safer drug prescriptions. (Johnson et al., 2021)

2.1.2.1.2.4 Discovery of New Genetic Insights

Identifying Novel Genetic Markers: AI algorithms can uncover new genetic markers that are associated with diseases. These discoveries can lead to the development of new diagnostic tests and targeted therapies. (Lacalamita et al., 2023)

Contribution to Research: AI aids researchers in understanding the genetic basis of complex diseases, such as cancer, cardiovascular diseases, and rare genetic disorders.

2.1.2.2 Pharmacogenomics

Pharmacogenomics is a branch of genetics that studies how an individual's genetic makeup affects their response to drugs. This field combines pharmacology (the science of drugs) and genomics (the study of genes and their functions) to develop effective, safe medications and doses tailored to a person's genetic composition. The main objective is to optimize drug therapy, ensuring maximum efficacy with minimal adverse effects. It helps in understanding why some medications work effectively for some individuals but not for others, or why certain individuals experience severe side effects while others do not. Pharmacogenomics is integral to personalized medicine. By analyzing a patient's genetic profile, healthcare providers can predict their response to various drugs, thus personalizing treatment plans. (Roden et al., 2019)

2.1.2.2.1 Types of Pharmacogenomic Analysis

2.1.2.2.1.1 Single Gene Pharmacogenomics

It focuses on the variation in single genes and how these variations affect drug response. Single gene analysis is often used when a specific gene is known to significantly influence the response to a particular medication. In medications where specific gene variants are known to influence drug metabolism, such as the TPMT gene in thiopurine metabolism or the VKORC1 gene in warfarin dosing. (Haidar et al., 2022)

2.1.2.2.1.2 Multigene or Panel Testing

This involves examining multiple genes simultaneously to understand how they collectively influence drug response. This is more comprehensive and can provide a broader understanding of a patient's pharmacogenomic profile. Used for complex drug regimens where multiple drugs are prescribed, or in cases where various genetic factors contribute to drug metabolism and efficacy. (Swen et al., 2023)

2.1.2.2.1.3 Whole Genome Pharmacogenomics

It analyzes the entire genome to assess how numerous genetic variants may impact drug response. This approach is more exploratory and is generally used in research settings. Useful in uncovering novel genetic variants that affect drug response, leading to the discovery of new pharmacogenomic markers. (Katsila & Patrinos, 2015)

2.1.2.2.1.4 Population Pharmacogenomics

This studies pharmacogenomic variations across different populations. It recognizes that genetic variations affecting drug response can vary significantly between ethnic and racial groups.It is important in ensuring that drug therapy is effective and safe across diverse populations, leading to more equitable healthcare. (Jordan et al., 2023).

2.1.2.2.2 Role of AI in Pharmacogenomics

The integration of Artificial Intelligence (AI) in pharmacogenomics is reshaping the landscape of drug development and personalized medicine. Pharmacogenomics, the study of how genes affect a person's response to drugs, is a critical field where AI is making significant contributions by enhancing the understanding of genetic influences on drug efficacy and safety (Figure 2).

2.1.2.2.2.1 Enhanced Drug Response Prediction

Predictive Analytics: AI algorithms, particularly machine learning and deep learning models, are used to predict individual responses to various drugs based on genetic data. These models analyze complex genetic information and correlate specific genetic markers with drug responses. (Vora et al., 2023)

Tailoring Medication Regimens: AI aids in developing personalized medication regimens by analyzing a patient's genetic profile. This approach reduces the trial-and-error process in finding effective medications, leading to more efficient and successful treatments. (Roy et al., 2022)

2.1.2.2.2.2 Improved Understanding of Drug Interactions

Analyzing Complex Genetic Data: AI can process and analyze large datasets to identify how different genetic variations might influence the metabolism and efficacy of multiple drugs. This is

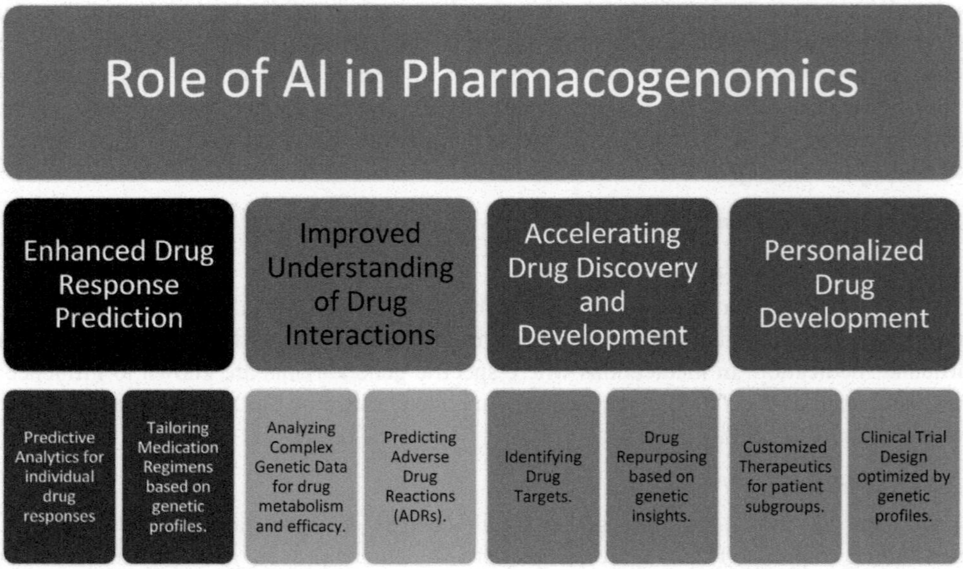

Fig. 2 Role of AI in Pharmacogenomics

crucial in polypharmacy, where patients are prescribed multiple drugs simultaneously. (Dey et al., 2018)

Predicting Adverse Drug Reactions (ADRs): AI models can help predict the likelihood of adverse drug reactions by analyzing genetic predispositions. This predictive ability is vital for enhancing patient safety. (Dey et al., 2018)

2.1.2.2.2.3 Accelerating Drug Discovery and Development

Identifying Drug Targets: AI accelerates the process of identifying new drug targets by analyzing genetic data associated with diseases. This approach leads to more targeted and effective drug development. (Vora et al., 2023)

Drug Repurposing: AI models can suggest existing drugs that might be effective for new therapeutic uses based on genetic insights, speeding up the drug discovery process. (Vora et al., 2023)

2.1.2.2.2.4 Personalized Drug Development

Customized Therapeutics: AI contributes to the development of customized therapeutics by identifying patient subgroups that are likely to benefit from specific drugs based on their genetic makeup. (Roy et al., 2022)

Clinical Trial Design: AI assists in designing more effective clinical trials by identifying suitable candidates based on genetic profiles, increasing the likelihood of successful trial outcomes. (Roy et al., 2022)

2.1.2.3 Lifestyle and Environmental Factors in Personalized Medicine

In personalized medicine, the consideration of lifestyle and environmental factors is crucial in understanding health and treatment outcomes. These factors play a significant role in determining an individual's risk of developing certain diseases, their progression, and response to treatments. Personalized medicine adopts a holistic approach to health, recognizing that genetic factors alone do not determine health outcomes. Lifestyle choices and environmental exposures are equally important in shaping an individual's overall health. Lifestyle and environmental factors can influence the onset and progression of diseases, interact with genetic predispositions, and affect the efficacy and toxicity of medications.

2.1.2.3.1 *Types of Lifestyle and Environmental Factors*

2.1.2.3.1.1 Diet and Nutrition

Dietary habits significantly influence health. Certain diets may increase the risk of diseases like diabetes, cardiovascular diseases, and some cancers. Nutrigenomics, a branch of nutritional genomics, studies how food interacts with genes to affect health. This leads to personalized dietary recommendations based on genetic makeup. (Roy et al., 2022)

2.1.2.3.1.2 Physical Activity and Exercise

Regular physical activity is known to reduce the risk of several chronic diseases and can influence treatment outcomes. Personalized Exercise Plans: Exercise recommendations can be tailored based on an individual's genetic predisposition to certain conditions and their response to different types of physical activity. (Roy et al., 2022)

2.1.2.3.1.3 Smoking and Alcohol Consumption

Smoking and excessive alcohol consumption are significant risk factors for numerous health conditions. Treatment Interaction: These habits can also affect how the body metabolizes medications, influencing treatment effectiveness and side effects. (Mukamal, 2006)

2.1.2.3.1.4 Exposure to Pollutants and Toxins

Exposure to pollutants, chemicals, and toxins in the environment can contribute to various health conditions, including respiratory diseases and cancers. The impact of environmental toxins can be influenced by genetic factors, affecting an individual's susceptibility to harm from these exposures. (Lavezzi & Ramos-Molina, 2023)

2.1.2.3.1.5 Psychosocial Stressors

Chronic stress, anxiety, and depression can have significant impacts on physical health and are associated with several chronic conditions. Mental health and stress levels can influence an individual's response to medical treatments, including medication effectiveness and recovery rates. (Mariotti, 2015)

2.1.2.3.1.6 Sleep Patterns and Quality

Sleep quality and circadian rhythms are increasingly recognized for their roles in overall health and the development of diseases. Genetic predispositions can influence sleep patterns, which are also affected by environmental and lifestyle factors. (Foster, 2020)

2.1.2.3.2 *Role of AI in Lifestyle and Environmental Factors in Personalized Medicine*

The application of Artificial Intelligence (AI) in assessing lifestyle and environmental factors marks a significant advancement in personalized medicine. AI technologies offer sophisticated means to analyze how these non-genetic factors influence individual health and treatment outcomes.

2.1.2.3.2.1 Comprehensive Lifestyle Analysis

Data Integration and Pattern Recognition: AI algorithms integrate and analyze data from various sources, including wearable health trackers, mobile apps, and patient self-reports. These algorithms are adept at identifying patterns and correlations between lifestyle factors like physical activity, diet, sleep patterns, and health outcomes. (Oftedal et al., 2019)

Behavioral Prediction and Modification: Machine learning models can predict the impact of lifestyle behaviors on health and are increasingly used in developing personalized lifestyle modification plans. For instance, AI can suggest tailored dietary or exercise plans based on individual health goals and preferences. (Dogan et al., 2023)

2.1.2.3.2.2 Environmental Health Impact Assessment

Analyzing Environmental Exposures: AI plays a crucial role in assessing the impact of environmental factors, such as air quality, pollution levels, and exposure to toxins, on individual health. By processing large environmental datasets, AI models can identify potential health risks associated with specific environmental exposures. (Subramaniam et al., 2022)

Geospatial Analysis: AI, coupled with geospatial technologies, can map environmental risk factors to specific locations, providing insights into how geographic and environmental variables contribute to health disparities. (Kamel Boulos et al., 2019)

2.1.2.3.2.3 Personalized Health Risk Assessments

Risk Factor Identification: AI models can analyze combined lifestyle and environmental data to identify individual risk factors for various diseases. This includes assessing the risk of chronic diseases like diabetes, cardiovascular diseases, and certain cancers. (Anil et al., n.d.)

Predictive Modeling for Disease Prevention: AI's predictive modeling capabilities are used to forecast future health risks based on current lifestyle and environmental exposures, enabling proactive and preventive healthcare interventions. (Anil et al., n.d.)

2.1.2.3.2.4 Enhancing Disease Management and Prevention

Tailoring Prevention Strategies: By understanding how lifestyle and environmental factors affect health, AI enables the development of personalized prevention strategies that are more likely to be effective and sustainable. (Dogan et al., 2023)

Monitoring and Adjusting Treatment Plans: *AI systems can continually monitor the impact of lifestyle changes on health outcomes and adjust treatment plans accordingly, ensuring that they remain effective over time. (Dogan et al., 2023)*

2.1.2.4 Biomarker Testing in Personalized Medicine

Biomarker testing is a crucial aspect of personalized medicine, offering insights into the biological state of an individual at a molecular level. Biomarkers, short for biological markers, are objective measures or indicators of normal biological processes, pathogenic processes, or pharmacological responses to a therapeutic intervention. They can be DNA, RNA, proteins, metabolites, or other molecules that provide critical information about a person's health status, disease risk, progression, and response to treatment. Biomarker testing allows healthcare providers to diagnose diseases more accurately, predict disease progression, and tailor treatments based on individual responses. It helps in identifying the right drug for the right patient at the right time. (Strimbu & Tavel, 2010)

2.1.2.4.1 Types of Biomarker Testing

2.1.2.4.1.1 Diagnostic Biomarkers

These are used to detect or confirm the presence of a disease or condition. Diagnostic biomarkers can help in early disease detection or confirm a diagnosis when symptoms are present. (Zygulska & Pierzchalski, 2022)

2.1.2.4.1.2 Prognostic Biomarkers

These biomarkers provide information about the likely course of a disease in an individual, regardless of the treatment. They can predict disease progression and overall survival. Example: Gene expression profiles in certain types of cancer like breast cancer, indicating the likelihood of disease recurrence. (Group, 2016b)

2.1.2.4.1.3 Predictive Biomarkers

These biomarers indicate the likelihood of response to a specific treatment. Predictive biomarkers are crucial in the era of targeted therapies and personalized medicine.Example: HER2/neu in breast cancer for determining the potential efficacy of trastuzumab (Herceptin) treatment. (Group, 2016a)

2.1.2.4.1.4 Pharmacodynamic Biomarkers

They measure the biological response to a therapeutic intervention. They are used to determine whether a drug is hitting its intended target and eliciting a biological response. For example, levels of certain enzymes or proteins after administering a drug indicate the drug's effectiveness. (Guo et al., 2019)

2.1.2.4.1.5 Safety Biomarkers

These are used to assess the safety and potential toxic effects of a therapeutic intervention. They are critical in the early detection of adverse responses to a drug. Example: Biomarkers indicating liver or kidney injury following drug administration. (Group, 2016c)

2.1.2.4.1.6 Companion Diagnostics

These are specific types of biomarkers used to determine the appropriateness of a particular therapeutic treatment for a patient. They are often developed alongside specific drugs. Example: Genetic tests to identify specific mutations in cancer cells that can be targeted by certain cancer drugs. (Valla et al., 2021)

2.1.2.4.2 Role of AI in Biomarker Testing

The integration of Artificial Intelligence (AI) in biomarker testing represents a significant leap forward in diagnostics and personalized medicine. Biomarkers, which are biological molecules found in blood, other body fluids, or tissues, serve as indicators of normal or pathological processes or responses to therapies. AI technologies are increasingly used to enhance the identification, analysis, and application of biomarkers in healthcare (Figure 3).

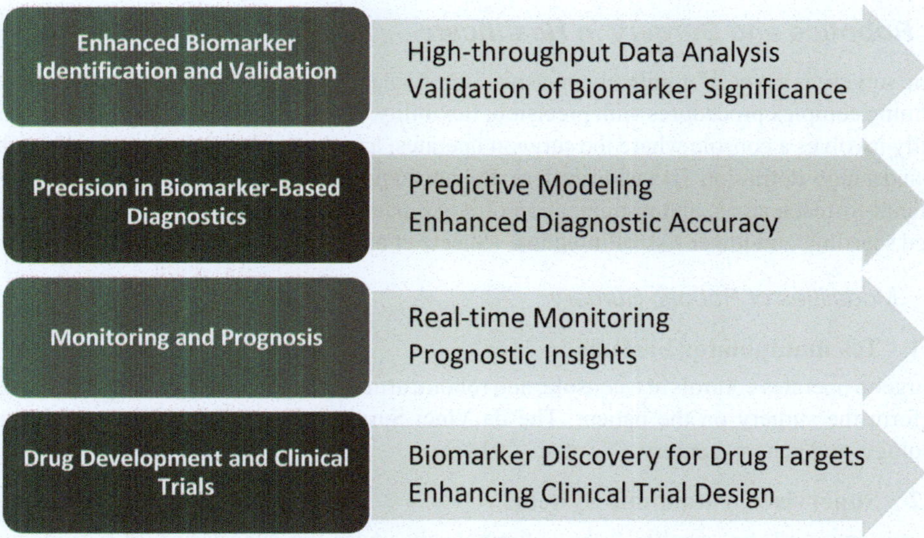

Fig. 3 Role of AI in Biomarker Testing

2.1.2.4.2.1 Enhanced Biomarker Identification and Validation

High-throughput Data Analysis: AI algorithms, particularly machine learning models, can process and analyze high-throughput data (such as genomics, proteomics, and metabolomics) to identify potential biomarkers. This includes detecting patterns and correlations in complex datasets that might be overlooked by traditional methods. (Mann et al., 2021)

Validation of Biomarker Significance: AI helps in validating the clinical significance of identified biomarkers, determining their reliability and effectiveness in diagnosing diseases or monitoring treatment responses. (Mann et al., 2021)

2.1.2.4.2.2 Precision in Biomarker-Based Diagnostics

Predictive Modeling: AI models can predict disease presence or progression based on the levels and patterns of specific biomarkers. This predictive capability is crucial for early disease detection, improving the chances of successful treatment. (Prelaj et al., 2024)

Enhanced Diagnostic Accuracy: By integrating biomarker data with other clinical information, AI improves the accuracy and specificity of diagnostic tools. This precision is especially vital in conditions where early detection can significantly alter the disease outcome. (Prelaj et al., 2024)

2.1.2.4.2.3 Monitoring and Prognosis

Real-time Monitoring: AI systems can monitor changes in biomarker levels over time, providing real-time insights into a patient's response to treatment or disease progression. (Arya et al., 2023)

Prognostic Insights: AI aids in prognostication by analyzing biomarker trends, helping predict the likely course of a disease and informing decisions about ongoing treatment strategies. (Arya et al., 2023)

2.1.2.4.2.4 Drug Development and Clinical Trials

Biomarker Discovery for Drug Targets: AI accelerates the discovery of new biomarkers that can serve as drug targets, facilitating the development of novel therapeutics. (Pun et al., 2023)

Enhancing Clinical Trial Design: In clinical trials, AI models utilize biomarkers to identify suitable candidates for specific drugs, ensuring that trial populations are more likely to show beneficial responses. (Pun et al., 2023)

2.2 Robotics and Surgery in Healthcare

Robotic surgery is a type of minimally invasive surgery that uses robotic systems to assist surgeons in performing complex procedures with precision, flexibility, and control beyond human capabilities. It typically involves a console where the surgeon operates, a patient-side robotic cart with mechanical arms, and a high-definition 3D vision system. Enhanced precision, reduced tremor, improved access to difficult-to-reach areas, and minimization of large incisions, leading to quicker patient recovery, reduced scarring, and lower risk of infection. (Sheetz et al., 2020)

2.2.1 Modalities of Robotic Surgery

2.2.1.1 Telemanipulator Systems

The surgeon operates controls at a console, and robotic arms replicate the surgeon's hand movements to perform the surgery on the patient. The da Vinci Surgical System is widely used in urology, gynecology, and general surgery. (Yu et al., 2014)

2.2.1.2 Supervised Autonomic Systems

These systems perform specific tasks autonomously under the supervision of a surgeon. They are often used for precise, repetitive tasks. The ROBODOC system was initially developed for orthopedic surgery. (Liow et al., 2017)

2.2.2 How is Robotic Surgery Conducted?

2.2.2.1 Preoperative Planning

Imaging and Mapping: Advanced imaging techniques are used to create detailed 3D maps of the patient's anatomy, which guide the robotic system during surgery. (Du et al., 2017; Lei et al., 2022; Rodler et al., 2023; Su et al., 2022)

2.2.2.2 Setting Up the Robotic System

Installation: The robotic system is set up in the operating room, with robotic arms positioned to access the surgical site effectively. (Du et al., 2017; Lei et al., 2022; Rodler et al., 2023; Su et al., 2022)

Calibration: The system is calibrated, and instruments are attached to the robotic arms. (Du et al., 2017; Lei et al., 2022; Rodler et al., 2023; Su et al., 2022)

2.2.2.3 Surgical Procedure

Console Operation: The surgeon operates from a console, viewing a high-definition, magnified 3D image of the surgical site. The surgeon's hand movements are translated into precise movements of the robotic instruments. (Du et al., 2017; Lei et al., 2022; Rodler et al., 2023; Su et al., 2022)

Robotic Arm Manipulation: The robotic arms carry out the surgery with high precision, stability, and control. The system filters out hand tremors and allows for scaled movements. (Du et al., 2017; Lei et al., 2022; Rodler et al., 2023; Su et al., 2022)

2.2.2.4 Monitoring and Assistance

Surgical Team: A trained surgical team assists in the operating room, managing the robotic system and providing assistance as needed. (Du et al., 2017; Lei et al., 2022; Rodler et al., 2023; Su et al., 2022)

Real-time Adjustments: The surgeon can make real-time adjustments based on the surgical progress. (Du et al., 2017; Lei et al., 2022; Rodler et al., 2023; Su et al., 2022)

2.2.2.5 Postoperative Management

Recovery Monitoring: Patients typically benefit from faster recovery times due to smaller incisions and reduced trauma compared to traditional surgery. (Du et al., 2017; Lei et al., 2022; Rodler et al., 2023; Su et al., 2022)

Follow-up: Postoperative care and follow-ups are conducted to monitor the patient's recovery. (Du et al., 2017; Lei et al., 2022; Rodler et al., 2023; Su et al., 2022)

2.2.3 Role of AI in Robotics and Surgery

The integration of Artificial Intelligence (AI) in robotics and surgery marks a pivotal advancement in healthcare. AI enhances robotic surgical systems, making them more intelligent, responsive, and capable of supporting surgeons in complex procedures.

2.2.3.1 Enhancing Surgical Precision and Decision-Making

AI-Powered Surgical Planning: AI algorithms can analyze preoperative medical images to create detailed 3D models of the patient's anatomy. This helps in planning the surgical procedure with higher precision, determining the best approach for incisions, and identifying potential risks. (Hashimoto et al., 2018; Mithany et al., 2023; Pakkasjärvi et al., 2023)

Machine Learning for Predictive Analytics: Machine learning models can predict various aspects of surgery outcomes, including potential complications and the estimated time of surgery. This helps surgeons prepare more effectively for the procedure. (Hashimoto et al., 2018; Mithany et al., 2023; Pakkasjärvi et al., 2023)

Real-Time Data Analysis: During surgery, AI systems can process real-time data, such as visual and sensory information, to provide critical insights. This includes monitoring vital signs, blood loss, and other physiological parameters, aiding in immediate decision-making. (Hashimoto et al., 2018; Mithany et al., 2023; Pakkasjärvi et al., 2023)

2.2.3.2 Improving Surgical Techniques and Training

Robotic Technique Optimization: AI can optimize the movements of robotic arms, ensuring smoother, more precise, and tremor-free maneuvers. This is crucial for delicate surgeries where precision is paramount. (Hashimoto et al., 2018; Mithany et al., 2023; Pakkasjärvi et al., 2023)

Surgical Training and Simulation: AI-driven surgical simulations offer an advanced training platform for surgeons. These simulations use real patient data to create realistic scenarios, allowing surgeons to practice and refine their skills in a virtual environment. (Hashimoto et al., 2018; Mithany et al., 2023; Pakkasjärvi et al., 2023)

Feedback and Learning: AI systems can provide feedback to surgeons by analyzing their performance during surgeries. This continuous learning loop can lead to improved surgical techniques and outcomes over time. (Hashimoto et al., 2018; Mithany et al., 2023; Pakkasjärvi et al., 2023)

2.2.3.3 Autonomous and Semi-Autonomous Surgical Robots

Autonomous Surgical Tasks: In some cases, AI enables robotic systems to perform certain surgical tasks autonomously, such as suturing, cutting, or tying knots, under the supervision of a surgeon. (Hashimoto et al., 2018; Mithany et al., 2023; Pakkasjärvi et al., 2023)

Collaborative Robotics: AI enhances the collaboration between robotic systems and human surgeons. The system can assist in ways that complement the surgeon's skills, leading to a more efficient surgical process. (Hashimoto et al., 2018; Mithany et al., 2023; Pakkasjärvi et al., 2023)

2.3 IoT in Healthcare

The Internet of Things (IoT) in healthcare refers to the interconnected network of medical devices and applications that collect, exchange, and analyze data. This ecosystem leverages internet-connected devices to improve healthcare outcomes, enhance patient experiences, and optimize operational efficiency.

2.3.1 Modalities of IoT in Healthcare

2.3.1.1 Wearable Health Monitors

Devices like fitness trackers, smartwatches, and heart rate monitors collect health data like heart rate, sleep patterns, and physical activity. They enable continuous monitoring of patient health, facilitating early detection of potential health issues. (Al-Kahtani et al., 2022)

2.3.1.2 Remote Patient Monitoring (RPM) Systems

RPM systems allow healthcare providers to monitor patients' health data remotely, using devices that collect data on vital signs, blood sugar levels, and other health metricsIt. improves the management of chronic diseases, reduces the need for hospital visits, and provides real-time data for better patient care. (Salehi et al., 2020)

2.3.1.3 Smart Medical Devices

These are advanced medical devices, such as smart inhalers, insulin pumps, and connected pacemakers, that provide precise treatment and real-time monitoring.They enhance the efficacy of treatments and provide critical data to healthcare providers for informed decision-making. (Paik & Kim, 2019)

2.3.1.4 IoT in Hospital Management

IoT devices are used for tracking hospital assets, managing inventory, and monitoring environmental conditions in healthcare facilities. They improve hospital efficiency, optimize resource usage, and ensure the safety and comfort of patients. (Al-Kahtani et al., 2022)

2.3.2 How is IoT Implemented in Healthcare?

2.3.2.1 Installation of IoT Devices

Setup: This involves deploying IoT devices like wearables, sensors, and medical equipment within the healthcare ecosystem. (Al-Kahtani et al., 2022; Kelly et al., 2020)

Integration: This involves e nsuring that these devices are integrated with existing healthcare IT systems for seamless data exchange. (Al-Kahtani et al., 2022; Kelly et al., 2020)

2.3.2.2 Data Collection and Transmission:

Continuous Monitoring: IoT devices continuously collect health data from patients, whether in a clinical setting or remotely. (Al-Kahtani et al., 2022; Kelly et al., 2020)

Secure Data Transmission: The collected data is transmitted securely to a central system for analysis, often using cloud-based solutions. (Al-Kahtani et al., 2022; Kelly et al., 2020)

2.3.2.3 Data Analysis and Application

Advanced Analytics: The data is analyzed using advanced algorithms to derive actionable insights. (Al-Kahtani et al., 2022; Kelly et al., 2020)

Decision Support: Healthcare providers use these insights for patient monitoring, diagnosis, treatment adjustments, and preventive care measures. (Al-Kahtani et al., 2022; Kelly et al., 2020)

2.3.2.4 Patient Engagement and Interaction

Patient Apps and Platforms: Patients use these to interact with their data and manage their health using dedicated apps and platforms. (Al-Kahtani et al., 2022; Kelly et al., 2020)

Feedback and Communication: IoT systems enable two-way communication between patients and healthcare providers, enhancing patient engagement and adherence to treatment plans. (Al-Kahtani et al., 2022; Kelly et al., 2020)

2.3.3 Role of AI in IoT in Healthcare

The integration of Artificial Intelligence (AI) with the Internet of Things (IoT) in healthcare creates a powerful synergy that enhances patient care, improves healthcare operations, and offers advanced analytics capabilities. AI leverages the vast amount of data generated by IoT devices to bring intelligent insights and automation into healthcare processes.

2.3.3.1.1 Advanced Data Processing

AI algorithms analyze the large and diverse datasets collected by IoT devices, such as wearables and monitoring sensors. This data includes vital signs, patient activity, and environmental conditions. This enables the extraction of meaningful insights from vast amounts of data, facilitating informed decision-making and proactive healthcare interventions. (Wang et al., 2023)

2.3.3.1.2 Predictive Analytics for Patient Monitoring

AI utilizes historical and real-time data to predict health events or risks, such as potential falls, heart failure, or diabetic episodes. It helps in early intervention and prevention, significantly improving patient outcomes and reducing hospital readmissions. (Al-Kahtani et al., 2022; Kelly et al., 2020)

2.3.3.2 Personalized Patient Care

2.3.3.2.1 Customized Treatment Plans

AI algorithms tailor treatment plans based on data gathered from patients' daily activities, physiological parameters, and medical history. They provide personalized healthcare experiences, enhancing the effectiveness of treatments and patient satisfaction. (Johnson et al., 2021)

2.3.3.2.2 Adaptive Support Systems

AI dynamically adapts support and recommendations based on continuous feedback from IoT devices, adjusting treatment plans as patient conditions change. These systems ensure that patient care is responsive and adaptive to changing health needs. (Bohr & Memarzadeh, 2020)

2.3.3.3 Operational Efficiency in Healthcare Facilities

2.3.3.3.1 Resource Management and Optimization

AI optimizes hospital operations by analyzing data from IoT devices used in asset tracking, inventory management, and environmental monitoring. ItI improves resource allocation, reduces costs, and enhances the overall efficiency of healthcare services. (Kelly et al., 2020)

2.3.3.3.2 Automated Alerts and Responses:

AI systems provide automated alerts based on data from IoT devices, facilitating immediate responses to critical patient needs or operational issues. It enhances patient safety and ensures timely intervention in emergencies. (Abdulmalek et al., 2022; Pradhan et al., 2021)

2.3.3.3.3 Enhancing Remote Healthcare

2.3.3.3.3.1 Telemedicine and Remote Consultations

AI enhances telemedicine by analyzing data from patients' IoT devices, providing healthcare professionals with detailed insights during remote consultations. It makes remote healthcare more effective and accessible, especially for patients in remote or underserved areas. (Abdulmalek et al., 2022; Pradhan et al., 2021)

2.3.3.3.3.2 Continuous Remote Monitoring

AI enables continuous and real-time monitoring of patients' health status remotely, alerting healthcare providers to any concerning changes. It allows for constant vigilance of high-risk patients and those with chronic conditions, reducing the need for frequent hospital visits. (Abdulmalek et al., 2022; Pradhan et al., 2021)

2.4 Blockchain, Cryptography, and Web3 in Healthcare

2.4.1 What is Blockchain, Cryptography, and Web3 in Healthcare?

2.4.1.1 Block chain in Healthcare

Blockchain is a decentralized and distributed ledger technology (DLT) that ensures the integrity, transparency, and security of data transactions. In healthcare, it offers a way to securely and efficiently manage health records, supply chains, and other sensitive data. Its main features are immutability, transparency, decentralization, and security. (Hasselgren et al., 2020)

2.4.1.2 Cryptography in Healthcare

Cryptography involves secure communication techniques to protect data privacy and integrity. In healthcare, it is used to encrypt patient data, ensuring that sensitive information is securely transmitted and stored. Its main features include encryption, secure hashing, and digital signatures. (Jayanthi et al., 2020)

2.4.1.3 Web3 in Healthcare

Web3, or the decentralized web, utilizes blockchain, cryptocurrencies, and token-based economics. In healthcare, Web3 proposes a patient-centric model of health data management and accessibility. Its features include decentralization, tokenization, and smart contracts. (Li et al., 2023)

2.4.2 Modalities of Blockchain, Cryptography, and Web3 in Healthcare

2.4.2.1 Secure Health Records Management

Blockchain: Used for creating tamper-proof, decentralized electronic health records (EHRs). (Han et al., 2022)

Cryptography: It ensures that patient data in EHRs is encrypted and only accessible to authorized individuals. (Kruse et al., 2017)

Web3: It facilitates patient-controlled EHRs where patients can control who accesses their data using blockchain-based applications. (Giustini, 2007)

2.4.2.2 Supply Chain Transparency and Integrity

Blockchain: It tracks medical supplies and pharmaceuticals from manufacturers to patients, ensuring transparency and reducing counterfeit products. (Hasselgren et al., 2020)

Web3: It implements smart contracts for automatic verification and tracking in supply chains. (Giustini, 2007)

2.4.2.3 Clinical Trials and Research

Blockchain: This provides a secure platform for storing and sharing clinical trial data, maintaining data integrity and participant privacy. (Benchoufi et al., 2019)

Web3: It facilitates decentralized and transparent data sharing among research institutions, potentially enhancing collaboration and innovation. (Benchoufi et al., 2019)

2.4.3 How is it Implemented in Healthcare?

2.4.3.1 Development and Integration of Blockchain Systems

Building Infrastructure: This involves developing blockchain-based platforms customized for healthcare applications, like secure EHR systems. (Hasselgren et al., 2020; Jayanthi et al., 2020; Li et al., 2023)

Integration: This involves integrating these systems with existing healthcare IT infrastructures while ensuring compliance with healthcare regulations. (Hasselgren et al., 2020; Jayanthi et al., 2020; Li et al., 2023)

2.4.3.2 Implementing Cryptography Solutions

Data Encryption: This involves encrypting patient data both at rest and in transit using advanced cryptographic techniques. (Hasselgren et al., 2020; Jayanthi et al., 2020; Li et al., 2023)

Access Control: Access control includes implementing cryptographic keys and digital signatures to control access to sensitive health data. (Hasselgren et al., 2020; Jayanthi et al., 2020; Li et al., 2023)

2.4.3.3 Leveraging Web3 Technologies

Patient Empowerment: Developing patient-centric applications that allow patients to own and control their health data using blockchain and tokenization. (Hasselgren et al., 2020; Jayanthi et al., 2020; Li et al., 2023)

Smart Contracts for Healthcare Services: Utilizing smart contracts for automated and transparent healthcare services, such as insurance claims processing. (Hasselgren et al., 2020; Jayanthi et al., 2020; Li et al., 2023)

2.4.4 Role of AI in Blockchain, Cryptography, and Web3 in Healthcare

Artificial Intelligence (AI) plays a transformative role when integrated with Blockchain, Cryptography, and Web3 technologies in healthcare. This synergy enhances data security, optimizes workflows, and facilitates advanced data analysis, leading to improved healthcare delivery and patient outcomes.

2.4.4.1 Enhancing Data Security and Integrity

2.4.4.1.1 AI in Blockchain for Healthcare

Data Verification and Management: AI algorithms can be used to efficiently manage and verify large volumes of healthcare data on blockchain networks, ensuring data integrity and reducing errors. (Tagde et al., 2021)

Smart Contract Automation: AI can automate and optimize smart contract protocols on blockchain, ensuring they respond dynamically to complex healthcare scenarios and data inputs. (Vargas & Mira da Silva, 2023)

2.4.4.1.2 AI in Cryptographic Applications

Advanced Encryption Techniques: AI can aid in developing advanced encryption techniques, enhancing the security of sensitive healthcare data. (Hasselgren et al., 2020)

Decryption and Threat Detection: AI algorithms are effective in identifying and responding to cybersecurity threats, safeguarding encrypted data against unauthorized access. (Hasselgren et al., 2020; Jayanthi et al., 2020; Li et al., 2023)

2.4.4.2 Optimizing Healthcare Operations

2.4.4.2.1 Workflow Automation and Efficiency

AI-Driven Process Optimization: Integrating AI with blockchain-based systems can streamline various healthcare processes, such as patient data management, billing, and supply chain logistics. (Hasselgren et al., 2020; Jayanthi et al., 2020; Li et al., 2023)

Predictive Analytics: AI enhances the capability of blockchain and Web3 platforms to perform predictive analytics, enabling proactive healthcare management and resource allocation. (Hasselgren et al., 2020; Jayanthi et al., 2020; Li et al., 2023)

2.4.4.2.2 Personalized Patient Care and Engagement

Customized Health Solutions: AI can analyze patient data stored on blockchain to provide personalized healthcare recommendations and treatment plans. (Hasselgren et al., 2020; Jayanthi et al., 2020; Li et al., 2023)

Enhanced Patient Interaction: In Web3 platforms, AI can facilitate more interactive and personalized patient engagement tools, such as AI chatbots or personalized health trackers. (Hasselgren et al., 2020; Jayanthi et al., 2020; Li et al., 2023)

2.4.4.2.3 Facilitating Advanced Healthcare Research

2.4.4.2.3.1 Data Analysis and Research

Analyzing Blockchain Data: AI can analyze data from blockchain systems for research purposes, uncovering insights into patient outcomes, treatment effectiveness, and epidemiological trends. (Hasselgren et al., 2020; Jayanthi et al., 2020; Li et al., 2023)

Clinical Trial Optimization: AI algorithms can assist in analyzing and selecting suitable candidates for clinical trials based on blockchain-stored patient data, enhancing the efficiency and efficacy of trials. (Hasselgren et al., 2020; Jayanthi et al., 2020; Li et al., 2023)

2.4.4.2.4 Innovative Healthcare Models

Decentralized Healthcare Models: AI and Web3 can work together to create decentralized healthcare models that promote data sharing and collaboration while ensuring data privacy and security. (Puri et al., 2021)

AI-Enhanced Telemedicine: In Web3 environments, AI can enhance telemedicine services, providing more accurate and efficient remote patient care.

2.5 Virtual Reality in Healthcare

Virtual Reality (VR) in healthcare involves the use of VR technology to create simulated environments for various medical purposes. This immersive technology allows users - whether they are healthcare professionals or patients - to interact with three-dimensional environments in a way that simulates real-life scenarios or creates therapeutic settings. (Pensieri & Pennacchini, 2016)

2.5.1 Modalities of Virtual Reality in Healthcare

2.5.1.1 Medical Training and Education

Simulation of Surgical Procedures: VR is used to simulate surgeries, allowing medical students and surgeons to practice complex procedures in a risk-free, controlled environment. (Jiang et al., 2022; Pensieri & Pennacchini, 2016)

Anatomical Education: Interactive VR modules enable medical students to explore human anatomy in 3D, enhancing their understanding in a more engaging and detailed manner than traditional methods. (Jiang et al., 2022; Pensieri & Pennacchini, 2016)

2.5.1.2 Patient Treatment and Rehabilitation

Pain Management: VR experiences can distract patients from pain, especially in burn treatment and during certain medical procedures, reducing the perception of pain and the need for medications. (Novo et al., 2021)

Physical Rehabilitation: VR is used in physiotherapy to create engaging and interactive environments for patients undergoing rehabilitation for injuries or neurological conditions, like stroke. (Novo et al., 2021)

2.5.1.3 Mental Health Treatment

Exposure Therapy: In mental health, VR is utilized for exposure therapy, particularly for treating phobias and PTSD, by gradually exposing patients to their fear triggers in a controlled and safe virtual environment. (Bell et al., 2020)

Cognitive Behavioral Therapy: VR supports cognitive-behavioral therapy by simulating situations where patients can practice coping strategies. (Bell et al., 2020)

2.5.1.4 Surgical Planning

Preoperative Planning: Surgeons use VR to plan complex surgeries. By creating a 3D model of the patient's anatomy, surgeons can explore various approaches to optimize surgical outcomes. (Reinschluessel et al., 2022)

2.5.2 How is Virtual Reality Implemented in Healthcare?

2.5.2.1 Development of VR Content and Applications

Content Creation: Developing VR content specific to healthcare needs, such as surgical simulations, educational modules, or therapeutic environments. (Jiang et al., 2022; Pensieri & Pennacchini, 2016)

Application Development: Building VR applications tailored for specific medical or therapeutic purposes. (Jiang et al., 2022; Pensieri & Pennacchini, 2016)

2.5.2.2 Integration with Medical Data

Data Visualization: Incorporating medical imaging data (like MRI, CT scans) into VR to create accurate and interactive 3D models. (Jiang et al., 2022; Pensieri & Pennacchini, 2016)
Customization: Customizing VR experiences based on individual patient data for treatments like physical rehabilitation or mental health therapy. (Jiang et al., 2022; Pensieri & Pennacchini, 2016)

2.5.2.3 Hardware and Software Setup

VR Headsets and Accessories: Setting up VR hardware, including headsets, gloves, and motion trackers, in hospitals, clinics, or patient homes. (Bell et al., 2020; Jiang et al., 2022; Pensieri & Pennacchini, 2016; Reinschluessel et al., 2022)
Software Installation: Installing and configuring VR software compatible with healthcare IT systems.

2.5.2.4 Training and User Adoption

Healthcare Professional Training: Training doctors, nurses, and therapists to use VR technology effectively in their practice. (Bell et al., 2020; Jiang et al., 2022; Pensieri & Pennacchini, 2016; Reinschluessel et al., 2022)

Patient Orientation Educating patients on how to use VR tools, especially for at-home treatments or therapy sessions. (Bell et al., 2020; Jiang et al., 2022; Pensieri & Pennacchini, 2016; Reinschluessel et al., 2022)

2.5.2.5 Monitoring and Evaluation

Effectiveness Assessment: Monitoring the effectiveness of VR applications in treatment and training, and collecting feedback for continuous improvement. (Bell et al., 2020; Jiang et al., 2022; Pensieri & Pennacchini, 2016; Reinschluessel et al., 2022)

Clinical Research: Conducting research and clinical trials to validate the efficacy of VR applications in healthcare. (Bell et al., 2020; Jiang et al., 2022; Pensieri & Pennacchini, 2016; Reinschluessel et al., 2022)

2.5.3 Role of AI in Virtual Reality in Healthcare

Artificial Intelligence (AI) significantly enhances the capabilities and applications of Virtual Reality (VR) in healthcare. By integrating AI with VR, healthcare professionals can offer more personalized, efficient, and effective treatments and training experiences (Figure 4).

2.5.3.1 Personalized Patient Experiences and Treatments

2.5.3.1.1 Customized Therapy Sessions

AI algorithms analyze patient data to tailor VR therapy sessions to individual needs, particularly in mental health and physical rehabilitation. Ensures that VR therapy is more effective by aligning with the specific therapeutic requirements and progress of each patient. (Liaw et al., 2023; Pensieri & Pennacchini, 2016; Usmani et al., 2022)

2.5.3.1.2 Adaptive Learning Environments

In medical training, AI adapts VR simulations to the learning pace and style of individual students, enhancing their educational experience. Improves the learning process by providing personalized feedback and challenges, leading to more effective training outcomes. (Liaw et al., 2023; Pensieri & Pennacchini, 2016; Usmani et al., 2022)

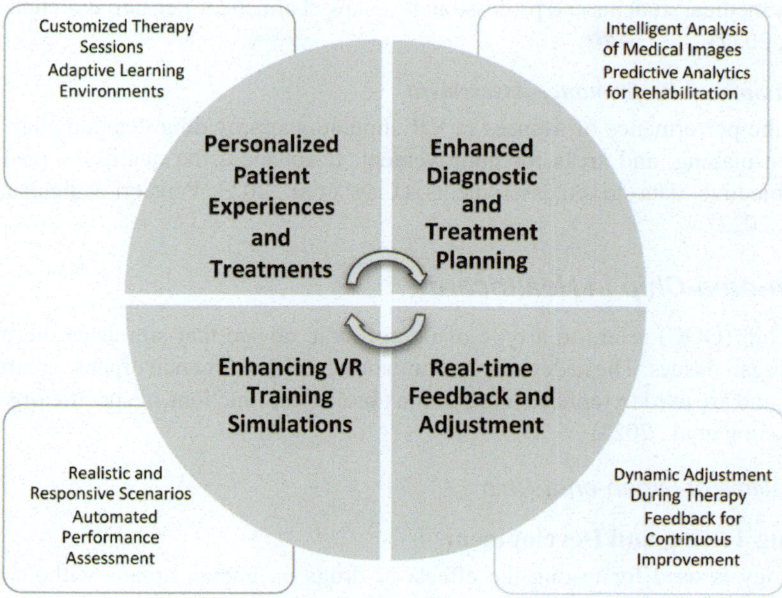

Fig. 4 Role of AI in Virtual Reality Healthcare

2.5.3.2 Enhanced Diagnostic and Treatment Planning

2.5.3.2.1 Intelligent Analysis of Medical Images

AI processes and interprets medical images, integrating them into VR environments for detailed and interactive exploration, beneficial in surgical planning and diagnostics. Surgeons can plan surgeries more effectively with a comprehensive 3D understanding of the patient's anatomy. (Liaw et al., 2023; Pensieri & Pennacchini, 2016; Usmani et al., 2022)

2.5.3.2.2 Predictive Analytics for Rehabilitation:

AI uses predictive analytics in VR-based rehabilitation to forecast patient progress and optimize therapy regimens. Enables therapists to anticipate and address potential challenges, enhancing the effectiveness of rehabilitation programs. (Liaw et al., 2023; Pensieri & Pennacchini, 2016; Usmani et al., 2022)

2.5.3.3 Real-time Feedback and Adjustment

2.5.3.3.1 Dynamic Adjustment During Therapy

AI systems provide real-time adjustments to VR therapy based on patient responses and progress, especially in physical therapy and mental health treatment. It ensures that therapy remains challenging yet achievable, maximizing therapeutic benefits. (Liaw et al., 2023; Pensieri & Pennacchini, 2016; Usmani et al., 2022)

2.5.3.3.2 Feedback for Continuous Improvement

AI gathers and analyzes feedback from VR sessions to continuously improve therapy or training protocols. Leads to an evolving and improving VR experience, ensuring it remains up-to-date with the latest medical standards and patient needs. (Liaw et al., 2023; Pensieri & Pennacchini, 2016; Usmani et al., 2022)

2.5.3.4 Enhancing VR Training Simulations

2.5.3.4.1 Realistic and Responsive Scenarios

AI generates realistic and responsive VR simulations for medical training, creating lifelike scenarios that react in real-time to the trainee's actions. It provides a more immersive and practical learning experience for medical students and professionals. (Liaw et al., 2023; Pensieri & Pennacchini, 2016; Usmani et al., 2022)

2.5.3.4.2 Automated Performance Assessment

AI evaluates the performance of trainees in VR simulations, providing detailed analytics on their skills, decision-making, and areas for improvement. It enhances the quality of medical training by offering objective, data-driven assessments. (Liaw et al., 2023; Pensieri & Pennacchini, 2016; Usmani et al., 2022)

2.6 Organ-on-a-Chip in Healthcare

Organ-on-a-Chip (OOC) refers to a type of microfluidic device that simulates the physiological responses of organ tissues. These devices are miniature models of human organs, created on a chip-like platform, and are used to replicate the complex biological functions of specific organs. (Koyilot et al., 2022; Leung et al., 2022)

2.6.1 Modalities of Organ-on-a-Chip

2.6.1.1 Drug Testing and Development

OOC technology is used for testing the effects of drugs on human organs without the need for animal testing. This includes assessing the efficacy and toxicity of new pharmaceutical compounds. Provides more accurate and human-relevant data compared to traditional in vitro studies or animal models. (Koyilot et al., 2022; Leung et al., 2022)

2.6.1.2 Disease Modeling and Research

Creating models of human organs on chips to study the progression and treatment of diseases. This is particularly useful for understanding complex diseases that involve multiple organ systems. Enables detailed study of disease mechanisms and the testing of potential treatments in a controlled environment. (Koyilot et al., 2022; Leung et al., 2022)

2.6.1.3 Personalized Medicine

Developing patient-specific organ models using cells from the patient to test how they would respond to different treatments. It alows for the customization of medical treatments, ensuring higher efficacy and lower risk of adverse reactions. (Koyilot et al., 2022; Leung et al., 2022)

2.6.1.4 Toxicology Testing

Assessing the toxicity of chemicals and environmental factors on human tissues and organs. Provides a more ethical and reliable method for toxicity testing than traditional animal testing. (Koyilot et al., 2022; Leung et al., 2022)

2.6.2 How is Organ-on-a-Chip Done?

2.6.2.1 Design and Fabrication of the Chip

Microengineering: Creating a microfluidic chip that incorporates channels and chambers designed to simulate the physical and biological environment of human organs. (Cho et al., 2023)

Materials: Using biocompatible materials that can support living human cells and mimic tissue and organ structures. (Cho et al., 2023)

2.6.2.2 Cell Culture and Integration

Cell Sourcing: Obtaining human cells, either from cell lines or patient biopsies. (Cho et al., 2023)

Seeding the Chip: Placing cells into the chip where they grow and differentiate to mimic the function of organ tissues. (Cho et al., 2023)

2.6.2.3 Creating the Organ Environment

Microfluidics: Utilizing microfluidic technology to circulate air and nutrient-rich fluids through the chip, replicating blood flow and other bodily processes. (Cho et al., 2023)

Mechanical and Electrical Stimulation: Applying mechanical forces or electrical stimuli to mimic physiological conditions that organs experience in the body. (Cho et al., 2023)

2.6.2.4 Monitoring and Analysis

Data Collection: Using sensors and imaging technology to monitor how the cells on the chip respond to various stimuli, such as drug compounds. (Cho et al., 2023)

Data Analysis: Analyzing the collected data to draw conclusions about the efficacy and safety of treatments, or to gain insights into disease mechanisms. (Cho et al., 2023)

2.6.3 Role of AI in Organ-on-a-Chip Technology in Healthcare

Artificial Intelligence (AI) significantly enhances the capabilities and applications of Organ-on-a-Chip (OOC) technology in healthcare. By integrating AI, researchers and clinicians can derive deeper insights from OOC systems, leading to more accurate disease models, personalized treatments, and efficient drug development processes.

2.6.3.1 Enhancing Data Analysis and Interpretation

2.6.3.1.1 Complex Data Analysis

AI algorithms analyze the vast and complex data generated by OOC systems, identifying patterns and correlations that might be invisible to the human eye. This includes cellular responses, biomarker expressions, and tissue interactions under various conditions.Enables a deeper understanding of organ functions, disease mechanisms, and the effects of pharmaceutical compounds on human tissues. (Dai et al., 2023; De Chiara et al., 2021; Deng et al., 2023)

2.6.3.1.2 Predictive Modelling

AI models use data from OOC experiments to predict biological responses to new drugs, chemicals, or disease conditions. This can include predicting drug efficacy, toxicity, and potential side effects. Accelerates the drug development process by predicting outcomes without the need for extensive in vivo testing. (Dai et al., 2023; De Chiara et al., 2021; Deng et al., 2023)

2.6.3.2 Personalized Medicine and Disease Modelling

2.6.3.2.1 Patient-specific Disease Models

AI can tailor OOC systems to model specific patient conditions by analyzing patient data and simulating how individual cellular and tissue characteristics might influence disease progression and treatment responses. They facilitate the development of personalized treatment plans and targeted therapies, improving patient outcomes. (Dai et al., 2023; De Chiara et al., 2021; Deng et al., 2023)

2.6.3.2.2 Disease Progression and Treatment Outcome Simulations

AI simulates disease progression and treatment responses in OOC models, allowing researchers to explore various therapeutic strategies and their long-term effects. Provides valuable insights into the most effective treatment options and helps in understanding complex disease dynamics. (Dai et al., 2023; De Chiara et al., 2021; Deng et al., 2023)

2.6.3.3 Operational Efficiency and Innovation

2.6.3.3.1 Automation and High-throughput Screening

AI algorithms automate the operation of OOC systems, enabling high-throughput screening of drugs and compounds. This includes automated imaging, data capture, and analysis. It increases the efficiency and throughput of drug testing, reducing time and costs associated with drug development. (Dai et al., 2023; De Chiara et al., 2021; Deng et al., 2023)

2.6.3.3.2 Design and Optimization of OOC Systems

AI aids in the design and optimization of OOC devices, using computational models to predict how changes in design parameters affect organ behavior and experiment outcomes. Leads to more effective and reliable OOC systems that accurately replicate human organ functions. (Dai et al., 2023; De Chiara et al., 2021; Deng et al., 2023)

3. Ethical, Legal, and Social Implications of AI in Healthcare

The integration of Artificial Intelligence (AI) in healthcare brings a host of ethical, legal, and social implications that are crucial to consider. These aspects are fundamental to ensuring that the use of AI in healthcare is safe, fair, and beneficial for all stakeholders involved.

3.1 Ethical Considerations

In the contemporary landscape of healthcare technology, the integration of Artificial Intelligence (AI) introduces complex ethical considerations, particularly concerning privacy and data security. The expansive collection and processing of sensitive patient data by AI systems necessitate a rigorous examination of privacy and security measures to safeguard against potential breaches and misuse.

The potential for data breaches and misuse in AI systems, which handle vast repositories of healthcare data, magnifies the risk of cybersecurity threats. This vulnerability is accentuated by the concerns surrounding the secondary use of healthcare data for purposes not initially consented to by patients, such as targeted marketing or unauthorized research. The cornerstone of data security in AI-driven healthcare is the implementation of advanced encryption techniques. Encryption of healthcare data, both at rest and in transit, is vital to stave off unauthorized access and cyber threats. End-to-end encryption ensures that data is accessible only to authorized entities. However, the challenge of balancing data utility and anonymity persists, as excessive anonymization can diminish the effectiveness of data for analytical purposes. (Farhud & Zokaei, 2021). Access control mechanisms are crucial in mitigating unauthorized data access. Implementing stringent user authentication protocols and role-based access systems, governed by the principle of least privilege, restricts access to the minimum necessary information for any given role. These controls are pivotal in preserving the integrity of patient data.The anonymization of patient data, including data masking and pseudonymization, is especially crucial in research contexts where patient identification is not required. Anonymization techniques, while safeguarding patient privacy, must be carefully balanced with the utility of data for research and analysis. (Farhud & Zokaei, 2021; Varkey, 2021)

The ethical framework for patient consent in AI-driven healthcare encompasses an informed and dynamic consent model. Patients must be transparently informed about how their data will be used, ensuring full awareness and comprehension. The consent process should be voluntary, free from coercion, and adaptable to the evolving nature of AI technologies. Dynamic consent models, facilitated by digital platforms, offer patients continuous control over their data preferences.In ensuring patient comprehension in the consent process, the complexities inherent in AI systems must be effectively communicated. This process should be culturally sensitive and tailored to diverse patient demographics, respecting linguistic and educational differences. (Farhud & Zokaei, 2021; Varkey, 2021)

3.2 Legal Challenges and Regulatory Frameworks

The integration of Artificial Intelligence (AI) into healthcare not only ushers in a new era of medical innovation but also brings to the forefront a range of legal challenges and regulatory complexities. Central to these challenges is the evolving nature of AI technologies, which often outpaces the existing legal and regulatory frameworks, creating a landscape marked by legal uncertainties and the need for adaptive governance. (Naik et al., 2022). One of the primary legal challenges in implementing AI in healthcare revolves around the issues of liability and accountability. In scenarios where AI systems contribute to clinical decisions, questions arise as to who is liable for the outcomes of those decisions, especially in cases of misdiagnosis or treatment errors. This ambiguity in liability necessitates a clear legal framework that distinctly addresses the accountability of AI systems, healthcare providers, and the manufacturers of these technologies. As AI systems in healthcare can make complex decisions, often without human intervention, establishing a clear legal pathway for liability becomes imperative for patient safety and trust. (Farhud & Zokaei, 2021; Naik et al., 2022)

Another critical aspect is the compliance of AI in healthcare with existing health data protection regulations. In various jurisdictions, there are stringent laws governing the use of personal health information, such as the Health Insurance Portability and Accountability Act (HIPAA) in the United States and the General Data Protection Regulation (GDPR) in the European Union. AI systems in healthcare must be designed and operated in ways that comply with these regulatory standards, ensuring the protection of patient data privacy and security. This compliance is challenging, given the vast amounts of data processed by AI systems and their capacity for deep learning and pattern recognition. (Reddy, 2023). The approval processes for AI applications in healthcare are complex and multifaceted. Regulatory bodies like the U.S. Food and Drug Administration (FDA) are tasked with ensuring that AI applications are not only effective but also safe for patient use. However, the traditional pathways for medical device approval may not be entirely suited to AI technologies, which are characterized by their ability to learn and adapt over time. This dynamic nature of AI requires a regulatory approach that can accommodate continuous learning and adaptation, ensuring that AI systems remain safe and effective throughout their life cycle. (Davenport & Kalakota, 2019; Reddy, 2023)

3.3 Impact on Healthcare Professionals and Patient Relationships

The advent of Artificial Intelligence (AI) in healthcare is reshaping the dynamics between healthcare professionals and patients, impacting the nature of medical practice and the patient care experience. While AI presents opportunities for enhanced care and efficiency, it also poses challenges and considerations in the healthcare professional-patient relationship.

3.3.1 Augmentation of Healthcare Professionals' Roles

The introduction of AI in healthcare is not about replacing healthcare professionals but rather augmenting their capabilities. AI systems provide support through advanced diagnostics, predictive analytics, and personalized treatment suggestions, which can enhance the decision-making process for doctors and other healthcare providers. However, this also introduces a requirement for healthcare professionals to acquire new skills and understanding of AI technologies. The need for continuous education and training in AI is paramount to ensure that healthcare providers can effectively integrate these technologies into their practice. (Hazarika, 2020)

3.3.2 Changes in Clinical Workflow

AI applications streamline numerous processes in clinical settings, from diagnostics to treatment plans and patient monitoring. This shift can significantly alter clinical workflows, requiring adjustments from healthcare professionals in terms of how they interact with technology and patients. While AI can reduce the burden of administrative tasks and allow more time for direct patient care, it also necessitates a change in how healthcare providers approach their day-to-day activities, balancing AI insights with their clinical judgement. (Blezek et al., 2021)

3.3.3 Building Trust in AI Among Patients

For patients, trust in AI technologies is crucial. Healthcare providers play a key role in educating and communicating with patients about how AI is used for care. This includes discussing the benefits and limitations of AI, addressing any concerns about data privacy, and ensuring that patients feel their personal and medical information is handled securely and ethically. The transparency in how AI is utilized and its role in decision-making processes is essential for maintaining trust and a strong patient-provider relationship. (Asan et al., 2020)

3.3.4 Personalization vs. Human Touch

AI offers the potential for highly personalized medical care, tailoring treatments and interventions to the individual characteristics of each patient. However, it's vital to maintain the human element in healthcare. The empathy, compassion, and understanding that healthcare professionals provide cannot be replicated by AI. Balancing the precision and personalization offered by AI with the human touch that is central to patient care is a crucial consideration. (Asan et al., 2020)

4. Preparing for the AI-Driven Healthcare Future

4.1 Education and Training for Healthcare Professionals

As the healthcare sector progressively embraces Artificial Intelligence (AI), significant shifts in education and training for healthcare professionals are imperative to navigate this AI-driven future effectively. The integration of AI into healthcare demands a workforce that is not only proficient in clinical expertise but also adept in understanding and working alongside advanced technologies. (Paranjape et al., 2019)

4.1.1 Evolving Educational Curricula

The foundation for preparing healthcare professionals for an AI-driven future lies in the evolution of medical and healthcare-related educational curricula. Traditional medical education needs to be augmented with courses and modules that provide a fundamental understanding of AI, its applications in healthcare, and the ethical considerations involved. This education should not only be limited to the technology itself but also extend to data analytics, interpretation of AI-driven insights, and understanding the limitations and potential biases of AI systems. Such an interdisciplinary approach, blending medical knowledge with AI literacy, is crucial in developing a workforce capable of leveraging AI effectively in clinical settings. (Paranjape et al., 2019)

4.1.2 Continuous Professional Development

In addition to formal education, continuous professional development plays a critical role. Healthcare is a rapidly evolving field, and AI technologies continue to advance at a remarkable pace. As such, ongoing training and professional development opportunities are essential to ensure that healthcare professionals remain up-to-date with the latest advancements in AI. This could take the form of workshops, seminars, online courses, and hands-on training, focusing on the practical aspects of implementing and working with AI in healthcare. (Hazarika, 2020)

4.1.3 Simulation-Based and Experiential Learning

Emphasizing simulation-based and experiential learning can greatly aid in preparing healthcare professionals for an AI-driven future. Simulation labs equipped with AI technologies, virtual reality, and other digital tools can provide hands-on experience in a controlled environment. This experiential learning approach helps in understanding the practical implications of AI in clinical decision-making, patient interactions, and healthcare delivery. (Komasawa & Yokohira, n.d.)

4.1.4 Fostering a Culture of Lifelong Learning

Lastly, fostering a culture of lifelong learning is vital in an era where technological advancements continually reshape the healthcare landscape. Healthcare professionals must be encouraged to view learning as a continuous journey, where adapting to new technologies and methodologies is an integral part of their professional growth and commitment to patient care. (Naik et al., 2022; Reddy, 2023)

4.2 Policy Recommendations and Strategic Planning

As healthcare systems globally adapt to the rapid integration of Artificial Intelligence (AI), comprehensive policy recommendations and strategic planning become indispensable. These policies and plans should aim to maximize the benefits of AI while addressing potential challenges and ensuring equitable, ethical, and efficient healthcare delivery. (Naik et al., 2022; Reddy, 2023)

4.2.1 Developing Comprehensive AI Integration Policies

Establishing Regulatory Standards: There's a need for clear and robust regulatory frameworks that govern the development, deployment, and use of AI in healthcare. These standards should ensure patient safety, data security, and compliance with ethical norms. (Naik et al., 2022; Reddy, 2023)

Data Governance and Privacy: Policies must be developed to govern the collection, storage, use, and sharing of healthcare data. These should prioritize patient privacy, consent mechanisms, and data security, adhering to regulations like GDPR and HIPAA. (Naik et al., 2022; Reddy, 2023)

Ensuring Ethical AI Use: Establishing ethical guidelines for the use of AI in healthcare, addressing issues like transparency, accountability, and the avoidance of bias in AI algorithms. (Naik et al., 2022; Reddy, 2023)

4.2.2 Promoting Research and Development

Fostering AI Innovation: Encourage research and development in AI through funding, partnerships between academia and industry, and by providing platforms for innovation. (Mahardhani, 2023)

Clinical Trials and Research: Implement policies that support the integration of AI into clinical trials and research, ensuring that these technologies are rigorously tested for safety and efficacy. (Mahardhani, 2023)

4.2.3 Education and Workforce Development

Training Healthcare Professionals: Develop education and training programs for healthcare workers focusing on AI literacy, including understanding AI tools, interpreting AI-driven insights, and integrating AI into clinical practice. (Hazarika, 2020)

Interdisciplinary Collaboration: Promote collaboration between healthcare professionals, AI technologists, data scientists, and ethicists to foster a comprehensive understanding of AI applications in healthcare. (Hazarika, 2020)

4.2.4 Patient-Centered AI Deployment

Patient Engagement and Education: Implement strategies to educate patients about AI in healthcare, addressing concerns and expectations, and ensuring transparency about how AI is used in their care. (Sauerbrei et al., 2023)

Patient-Centric AI Tools: Develop AI tools and applications that are patient-centric, focusing on improving patient outcomes, enhancing the patient experience, and ensuring accessibility for all patient populations. (Sauerbrei et al., 2023)

4.2.5 Strategic Planning for AI Implementation

Long-term Vision and Roadmap: Create a long-term strategic vision for AI in healthcare, outlining goals, key areas for AI application, and steps for implementation. (Mahardhani, 2023)

Evaluation and Adaptation: Regularly evaluate the impact of AI technologies in healthcare settings and adapt strategies as needed to address emerging challenges and integrate new advancements. (Mahardhani, 2023)

Public-Private Partnerships: Encourage partnerships between public healthcare entities and private AI technology firms to facilitate the sharing of knowledge, resources, and best practices. (Mahardhani, 2023)

4.3 Fostering Collaborations between Tech Companies, Researchers, and Healthcare Providers

Fostering collaborations between tech companies, researchers, and healthcare providers is a critical step towards the successful integration and optimization of Artificial Intelligence (AI) in healthcare. Such collaborations are essential in bridging the gap between technological advancements and practical, clinical applications. They ensure that AI tools and systems are not only technologically advanced but also clinically relevant and focused on improving patient outcomes.

4.3.1 Building Collaborative Frameworks

The establishment of collaborative frameworks involves creating formal partnerships and alliances among tech companies, academic institutions, healthcare providers, and research organizations. These collaborations are designed to facilitate the sharing of knowledge, expertise, and resources. For instance, tech companies bring innovative AI solutions and computational resources, researchers contribute with cutting-edge scientific knowledge and methodologies, and healthcare providers offer practical insights and clinical expertise. (Bhatt & Bae, 2023; Javed et al., 2020)

4.3.2 Advancing Research and Development

Collaborative efforts significantly boost research and development in AI for healthcare. They enable the practical application of AI research in clinical settings, ensuring that technological developments are aligned with real-world healthcare needs. Joint research initiatives can focus on developing AI solutions for diagnostic accuracy, treatment personalization, disease prediction, and healthcare management, ensuring that these solutions are rigorously tested and validated in clinical environments. (Bhatt & Bae, 2023; Javed et al., 2020)

4.3.3 Bridging the Gap Between Technology and Clinical Practice

One of the key benefits of fostering collaborations is bridging the gap between AI technology and clinical practice. Through these partnerships, AI tools can be designed and refined based on direct feedback from healthcare practitioners, ensuring they meet the practical needs of clinicians and patients. These collaborations also facilitate the translation of academic research into commercially viable AI products that can be deployed in healthcare settings. (Bhatt & Bae, 2023; Javed et al., 2020)

4.3.4 Enhancing Education and Training

Collaborations can also play a pivotal role in education and training. Joint educational programs and workshops that bring together technology developers, researchers, and healthcare providers can be established. These programs aim to enhance the understanding of AI applications in healthcare, addressing both the technological aspects and the ethical, legal, and social implications of AI in healthcare. (Bhatt & Bae, 2023; Javed et al., 2020)

Conclusion

The exploration of AI in healthcare reveals a future replete with innovation, efficiency, and enhanced patient care. The chapter's in-depth analysis underscores AI's transformative impact on healthcare delivery, from diagnostic precision to personalized treatments and efficient healthcare operations. It highlights AI's role in responding to global health challenges, improving patient outcomes, and paving the way for more advanced medical technologies. As AI continues to evolve, it promises to further revolutionize the healthcare landscape, offering unprecedented opportunities for advancements in patient care and medical research.

References

Abdulmalek, S., Nasir, A., Jabbar, W.A., Almuhaya, M.A.M., Bairagi, A.K. et al. (2022). IoT-Based Healthcare-Monitoring System towards Improving Quality of Life: A Review. *Healthcare*, *10*(10), 1993. https://doi.org/10.3390/healthcare10101993

Al-Kahtani, M.S., Khan, F. and Taekeun, W. (2022). Application of Internet of Things and Sensors in Healthcare. *Sensors (Basel, Switzerland)*, *22*(15), 5738. https://doi.org/10.3390/s22155738

Anil, S., Porwal, P. and Porwal, A. (n.d.). Transforming Dental Caries Diagnosis Through Artificial Intelligence-Based Techniques. *Cureus*, *15*(7), e41694. https://doi.org/10.7759/cureus.41694

Artificial Intelligence, Machine Learning and Genomics. (n.d.). Retrieved January 31, 2024, from https://www.genome.gov/about-genomics/educational-resources/fact-sheets/artificial-intelligence-machine-learning-and-genomics

Arya, S.S., Dias, S.B., Jelinek, H.F., Hadjileontiadis, L.J. and Pappa, A.-M et al. (2023). The convergence of traditional and digital biomarkers through AI-assisted biosensing: A new era in translational diagnostics? *Biosensors and Bioelectronics*, *235*, 115387. https://doi.org/10.1016/j.bios.2023.115387

Asan, O., Bayrak, A.E. and Choudhury, A. (2020). Artificial Intelligence and Human Trust in Healthcare: Focus on Clinicians. *Journal of Medical Internet Research*, *22*(6), e15154. https://doi.org/10.2196/15154

Bell, I.H., Nicholas, J., Alvarez-Jimenez, M., Thompson, A. and Valmaggia, L. et al. (2020). Virtual reality as a clinical tool in mental health research and practice. *Dialogues in Clinical Neuroscience*, *22*(2), 169–177. https://doi.org/10.31887/DCNS.2020.22.2/lvalmaggia

Benchoufi, M., Altman, D. and Ravaud, P. (2019). From Clinical Trials to Highly Trustable Clinical Trials: Blockchain in Clinical Trials, a Game Changer for Improving Transparency? *Frontiers in Blockchain*, *2*. https://www.frontiersin.org/articles/10.3389/fbloc.2019.00023

Bhatt, A.B. and Bae, J. (2023). Collaborative Intelligence to catalyze the digital transformation of healthcare. *Npj Digital Medicine*, *6*(1), Article 1. https://doi.org/10.1038/s41746-023-00920-w

Blezek, D.J., Olson-Williams, L., Missert, A. and Korfiatis, P. (2021). AI Integration in the Clinical Workflow. *Journal of Digital Imaging*, *34*(6), 1435–1446. https://doi.org/10.1007/s10278-021-00525-3

Bohr, A. and Memarzadeh, K. (2020). The rise of artificial intelligence in healthcare applications. *Artificial Intelligence in Healthcare*, 25–60. https://doi.org/10.1016/B978-0-12-818438-7.00002-2

Cho, S., Lee, S. and Ahn, S.I. (2023). Design and engineering of organ-on-a-chip. *Biomedical Engineering Letters*, *13*(2), 97–109. https://doi.org/10.1007/s13534-022-00258-4

Dai, M., Xiao, G., Shao, M. and Zhang, Y.S. (2023). The Synergy between Deep Learning and Organs-on-Chips for High-Throughput Drug Screening: A Review. *Biosensors*, *13*(3), Article 3. https://doi.org/10.3390/bios13030389

Davenport, T. and Kalakota, R. (2019). The potential for artificial intelligence in healthcare. *Future Healthcare Journal*, *6*(2), 94–98. https://doi.org/10.7861/futurehosp.6-2-94

De Chiara, F., Ferret-Miñana, A. and Ramón-Azcón, J. (2021). The Synergy between Organ-on-a-Chip and Artificial Intelligence for the Study of NAFLD: From Basic Science to Clinical Research. *Biomedicines*, *9*(3), Article 3. https://doi.org/10.3390/biomedicines9030248

Deng, S., Li, C., Cao, J., Cui, Z., Du, J. et al. (2023). Organ-on-a-chip meets artificial intelligence in drug evaluation. *Theranostics*, *13*(13), 4526–4558. https://doi.org/10.7150/thno.87266

Dey, S., Luo, H., Fokoue, A., Hu, J. and Zhang, P. et al. (2018). Predicting adverse drug reactions through interpretable deep learning framework. *BMC Bioinformatics*, *19*(21), 476. https://doi.org/10.1186/s12859-018-2544-0

Dogan, A., Li, Y., Peter Odo, C., Sonawane, K., Lin, Y. et al. (2023). A utility-based machine learning-driven personalized lifestyle recommendation for cardiovascular disease prevention. *Journal of Biomedical Informatics*, *141*, 104342. https://doi.org/10.1016/j.jbi.2023.104342

Du, Z., Wang, W., Wang, W. and Dong, W. (2017). Preoperative planning for a multi-arm robot-assisted minimally invasive surgery system. *SIMULATION*, *93*(10), 853–867. https://doi.org/10.1177/0037549717719336

Farhud, D.D. and Zokaei, S. (2021). Ethical Issues of Artificial Intelligence in Medicine and Healthcare. *Iranian Journal of Public Health*, *50*(11), i–v. https://doi.org/10.18502/ijph.v50i11.7600

Foster, R.G. (2020). Sleep, circadian rhythms and health. *Interface Focus*, *10*(3), 20190098. https://doi.org/10.1098/rsfs.2019.0098

Giustini, D. (2007). Web 3.0 and medicine. *BMJ (Clinical Research Ed.)*, *335*(7633), 1273–1274. https://doi.org/10.1136/bmj.39428.494236.BE

Group, F.-N. B.W. (2016a). Predictive Biomarker. In *BEST (Biomarkers, EndpointS, and other Tools) Resource [Internet]*. Food and Drug Administration (US). https://www.ncbi.nlm.nih.gov/books/NBK402283/

Group, F.-N. B.W. (2016b). Prognostic Biomarker. In *BEST (Biomarkers, EndpointS, and other Tools) Resource [Internet]*. Food and Drug Administration (US). https://www.ncbi.nlm.nih.gov/books/NBK402289/

Group, F.-N. B.W. (2016c). Safety Biomarker. In *BEST (Biomarkers, EndpointS, and other Tools) Resource [Internet]*. Food and Drug Administration (US). https://www.ncbi.nlm.nih.gov/books/NBK402287/

Guo, X., Wang, S., Godwood, A., Close, D., Ryan, P.C. et al. (2019). Pharmacodynamic biomarkers and differential effects of TNF- and GM-CSF-targeting biologics in rheumatoid arthritis. *International Journal of Rheumatic Diseases*, *22*(4), 646–653. https://doi.org/10.1111/1756-185X.13395

Haidar, C.E., Crews, K.R., Hoffman, J.M., Relling, M.V. and Caudle, K.E. et al. (2022). Advancing Pharmacogenomics from Single-Gene to Preemptive Testing. *Annual Review of Genomics and Human Genetics*, *23*, 449–473. https://doi.org/10.1146/annurev-genom-111621-102737

Han, Y., Zhang, Y. and Vermund, S.H. (2022). Blockchain Technology for Electronic Health Records. *International Journal of Environmental Research and Public Health*, *19*(23), 15577. https://doi.org/10.3390/ijerph192315577

Hashimoto, D.A., Rosman, G., Rus, D. and Meireles, O.R. (2018). Artificial Intelligence in Surgery: Promises and Perils. *Annals of Surgery*, *268*(1), 70–76. https://doi.org/10.1097/SLA.0000000000002693

Hasselgren, A., Kralevska, K., Gligoroski, D., Pedersen, S.A. and Faxvaag, A. et al. (2020). Blockchain in healthcare and health sciences—A scoping review. *International Journal of Medical Informatics*, *134*, 104040. https://doi.org/10.1016/j.ijmedinf.2019.104040

Hazarika, I. (2020). Artificial intelligence: Opportunities and implications for the health workforce. *International Health*, *12*(4), 241–245. https://doi.org/10.1093/inthealth/ihaa007

Jamal, S., Khubaib, Mohd., Gangwar, R., Grover, S., Grover, A. et al. (2020). Artificial Intelligence and Machine learning based prediction of resistant and susceptible mutations in Mycobacterium tuberculosis. *Scientific Reports*, *10*, 5487. https://doi.org/10.1038/s41598-020-62368-2

Javed, A.R., Sarwar, M.U., Beg, M.O., Asim, M., Baker, T. et al. (2020). A collaborative healthcare framework for shared healthcare plan with ambient intelligence. *Human-Centric Computing and Information Sciences*, *10*(1), 40. https://doi.org/10.1186/s13673-020-00245-7

Jayanthi, P., Iyyanki, M., Jayanthi, P. and Iyyanki, M. (2020). *Cryptography in the Healthcare Sector With Modernized Cyber Security* (cryptography-in-the-healthcare-sector-with-modernized-cyber-security) [Chapter]. Https://Services.Igi-Global.Com/Resolvedoi/Resolve.Aspx?Doi=10.4018/978-1-7998-2253-0.Ch008; IGI Global. https://www.igi-global.com/gateway/chapter/www.igi-global.com/gateway/chapter/248157

Jelin, A.C. and Vora, N. (2018). Whole Exome Sequencing: Applications in Prenatal Genetics. *Obstetrics and Gynecology Clinics of North America*, *45*(1), 69–81. https://doi.org/10.1016/j.ogc.2017.10.003

Jiang, H., Vimalesvaran, S., Wang, J.K., Lim, K.B., Mogali, S.R. et al. (2022). Virtual Reality in Medical Students' Education: Scoping Review. *JMIR Medical Education*, *8*(1), e34860. https://doi.org/10.2196/34860

Johnson, K.B., Wei, W., Weeraratne, D., Frisse, M.E., Misulis, K. et al. (2021). Precision Medicine, AI, and the Future of Personalized Health Care. *Clinical and Translational Science*, *14*(1), 86–93. https://doi.org/10.1111/cts.12884

Jordan, I.K., Sharma, S. and Mariño-Ramírez, L. (2023). Population Pharmacogenomics for Health Equity. *Genes*, *14*(10), 1840. https://doi.org/10.3390/genes14101840

Kamel Boulos, M.N., Peng, G. and VoPham, T. (2019). An overview of GeoAI applications in health and healthcare. *International Journal of Health Geographics*, *18*(1), 7. https://doi.org/10.1186/s12942-019-0171-2

Katsila, T. and Patrinos, G.P. (2015). Whole genome sequencing in pharmacogenomics. *Frontiers in Pharmacology*, *6*, 61. https://doi.org/10.3389/fphar.2015.00061

Kelly, J.T., Campbell, K.L., Gong, E. and Scuffham, P. (2020). The Internet of Things: Impact and Implications for Health Care Delivery. *Journal of Medical Internet Research*, *22*(11), e20135. https://doi.org/10.2196/20135

Komasawa, N. and Yokohira, M. (n.d.). Simulation-Based Education in the Artificial Intelligence Era. *Cureus*, *15*(6), e40940. https://doi.org/10.7759/cureus.40940

Koyilot, M.C., Natarajan, P., Hunt, C.R., Sivarajkumar, S., Roy, R. et al. (2022). Breakthroughs and Applications of Organ-on-a-Chip Technology. *Cells*, *11*(11), Article 11. https://doi.org/10.3390/cells11111828

Kruse, C.S., Smith, B., Vanderlinden, H. and Nealand, A. (2017). Security Techniques for the Electronic Health Records. *Journal of Medical Systems*, *41*(8), 127. https://doi.org/10.1007/s10916-017-0778-4

Lacalamita, A., Serino, G., Pantaleo, E., Monaco, A., Amoroso, N. et al. (2023). Artificial Intelligence and Complex Network Approaches Reveal Potential Gene Biomarkers for Hepatocellular Carcinoma. *International Journal of Molecular Sciences*, *24*(20), 15286. https://doi.org/10.3390/ijms242015286

Lappalainen, T., Scott, A.J., Brandt, M. and Hall, I.M. (2019). Genomic analysis in the age of human genome sequencing. *Cell*, *177*(1), 70–84. https://doi.org/10.1016/j.cell.2019.02.032

Lavezzi, A.M. and Ramos-Molina, B. (2023). Environmental Exposure Science and Human Health. *International Journal of Environmental Research and Public Health*, *20*(10), 5764. https://doi.org/10.3390/ijerph20105764

Lei, K., Liu, L., Yang, P., Xiong, R., Yang, L. et al. (2022). Robotics versus personalized 3D preoperative planning in total knee arthroplasty: A propensity score-matched analysis. *Journal of Orthopaedic Surgery and Research*, *17*, 227. https://doi.org/10.1186/s13018-022-03115-3

Leung, C.M., de Haan, P., Ronaldson-Bouchard, K., Kim, G.-A., Ko, J. et al. (2022). A guide to the organ-on-a-chip. *Nature Reviews Methods Primers*, *2*(1), Article 1. https://doi.org/10.1038/s43586-022-00118-6

Li, Y., Gunasekeran, D.V., RaviChandran, N., Tan, T.F., Ong, J.C.L. et al. (2023). The next generation of healthcare ecosystem in the metaverse. *Biomedical Journal*, 100679. https://doi.org/10.1016/j.bj.2023.100679

Liaw, S.Y., Tan, J.Z., Lim, S., Zhou, W., Yap, J. et al. (2023). Artificial intelligence in virtual reality simulation for interprofessional communication training: Mixed method study. *Nurse Education Today*, *122*, 105718. https://doi.org/10.1016/j.nedt.2023.105718

Liow, M.H.L., Chin, P.L., Pang, H.N., Tay, D.K.-J. and Yeo, S.-J et al. (2017). THINK surgical TSolution-One® (Robodoc) total knee arthroplasty. *SICOT-J*, *3*, 63. https://doi.org/10.1051/sicotj/2017052

Mahardhani, A.J. (2023). The Role of Public Policy in Fostering Technological Innovation and Sustainability. *Journal of Contemporary Administration and Management (ADMAN)*, *1*(2), Article 2. https://doi.org/10.61100/adman.v1i2.22

Mann, M., Kumar, C., Zeng, W.-F. and Strauss, M.T. (2021). Artificial intelligence for proteomics and biomarker discovery. *Cell Systems*, *12*(8), 759–770. https://doi.org/10.1016/j.cels.2021.06.006

Mariotti, A. (2015). The effects of chronic stress on health: New insights into the molecular mechanisms of brain–body communication. *Future Science OA*, *1*(3), FSO23. https://doi.org/10.4155/fso.15.21

Martin, D.P., Miya, J., Reeser, J.W. and Roychowdhury, S. (2016). Targeted RNA Sequencing Assay to Characterize Gene Expression and Genomic Alterations. *Journal of Visualized Experiments: JoVE*, *114*, 54090. https://doi.org/10.3791/54090

Miller, M.B., Schwander, K. and Rao, D.C. (2008). Genotyping errors and their impact on genetic analysis. *Advances in Genetics*, *60*, 141–152. https://doi.org/10.1016/S0065-2660(07)00406-3

Mithany, R.H., Aslam, S., Abdallah, S., Abdelmaseeh, M., Gerges, F. et al. (2023). Advancements and Challenges in the Application of Artificial Intelligence in Surgical Arena: A Literature Review. *Cureus*, *15*(10), e47924. https://doi.org/10.7759/cureus.47924

Mukamal, K.J. (2006). The Effects of Smoking and Drinking on Cardiovascular Disease and Risk Factors. *Alcohol Research & Health*, *29*(3), 199–202.

Naik, N., Hameed, B.M.Z., Shetty, D. K., Swain, D., Shah, M. et al. (2022). Legal and Ethical Consideration in Artificial Intelligence in Healthcare: Who Takes Responsibility? *Frontiers in Surgery*, *9*. https://www.frontiersin.org/articles/10.3389/fsurg.2022.862322

Novo, A., Fonsêca, J., Barroso, B., Guimarães, M., Louro, A., Fernandes, H. et al. (2021). Virtual Reality Rehabilitation's Impact on Negative Symptoms and Psychosocial Rehabilitation in Schizophrenia Spectrum Disorder: A Systematic Review. *Healthcare*, *9*(11), 1429. https://doi.org/10.3390/healthcare9111429

O'Brien, T.D., Campbell, N.E., Potter, A.B., Letaw, J.H., Kulkarni, A. et al. (2022). Artificial intelligence (AI)-assisted exome reanalysis greatly aids in the identification of new positive cases and reduces analysis time in a clinical diagnostic laboratory. *Genetics in Medicine: Official Journal of the American College of Medical Genetics*, *24*(1), 192–200. https://doi.org/10.1016/j.gim.2021.09.007

Oftedal, S., Vandelanotte, C. and Duncan, M.J. (2019). Patterns of Diet, Physical Activity, Sitting and Sleep Are Associated with Socio-Demographic, Behavioural, and Health-Risk Indicators in Adults. *International Journal of Environmental Research and Public Health*, *16*(13), 2375. https://doi.org/10.3390/ijerph16132375

Paik, S.-H. and Kim, D.-J. (2019). Smart Healthcare Systems and Precision Medicine. *Advances in Experimental Medicine and Biology*, *1192*, 263–279. https://doi.org/10.1007/978-981-32-9721-0_13

Pakkasjärvi, N., Luthra, T. and Anand, S. (2023). Artificial Intelligence in Surgical Learning. *Surgeries*, *4*(1), Article 1. https://doi.org/10.3390/surgeries4010010

Paranjape, K., Schinkel, M., Nannan Panday, R., Car, J. and Nanayakkara, P. et al. (2019). Introducing Artificial Intelligence Training in Medical Education. *JMIR Medical Education*, *5*(2), e16048. https://doi.org/10.2196/16048

Pensieri, C. and Pennacchini, M. (2016). Virtual Reality in Medicine. In Y. Sivan (Ed.), *Handbook on 3D3C Platforms: Applications and Tools for Three Dimensional Systems for Community, Creation and Commerce* (pp. 353–401). Springer International Publishing. https://doi.org/10.1007/978-3-319-22041-3_14

Pradhan, B., Bhattacharyya, S. and Pal, K. (2021). IoT-Based Applications in Healthcare Devices. *Journal of Healthcare Engineering*, *2021*, 6632599. https://doi.org/10.1155/2021/6632599

Prelaj, A., Miskovic, V., Zanitti, M., Trovo, F., Genova, C. et al. (2024). Artificial intelligence for predictive biomarker discovery in immuno-oncology: A systematic review. *Annals of Oncology: Official Journal of the European Society for Medical Oncology*, *35*(1), 29–65. https://doi.org/10.1016/j.annonc.2023.10.125

Premzl, M. (2020). Comparative genomic analysis of eutherian interferon genes. *Genomics*, *112*(6), 4749–4759. https://doi.org/10.1016/j.ygeno.2020.08.029

Pun, F.W., Ozerov, I.V. and Zhavoronkov, A. (2023). AI-powered therapeutic target discovery. *Trends in Pharmacological Sciences*, *44*(9), 561–572. https://doi.org/10.1016/j.tips.2023.06.010

Puri, V., Kataria, A. and Sharma, V. (2021). Artificial intelligence-powered decentralized framework for Internet of Things in Healthcare 4.0. *Transactions on Emerging Telecommunications Technologies*, *n/a*(n/a), e4245. https://doi.org/10.1002/ett.4245

Reddy, S. (2023). Navigating the AI Revolution: The Case for Precise Regulation in Health Care. *Journal of Medical Internet Research*, *25*, e49989. https://doi.org/10.2196/49989

Reinschluessel, A.V., Muender, T., Salzmann, D., Döring, T., Malaka, R. et al. (2022). Virtual Reality for Surgical Planning – Evaluation Based on Two Liver Tumor Resections. *Frontiers in Surgery*, *9*. https://www.frontiersin.org/articles/10.3389/fsurg.2022.821060

Roden, D.M., McLeod, H.L., Relling, M.V., Williams, M.S., Mensah, G.A. et al. (2019). Pharmacogenomics. *Lancet (London, England)*, *394*(10197), 521–532. https://doi.org/10.1016/S0140-6736(19)31276-0

Rodler, S., Kidess, M.A., Westhofen, T., Kowalewski, K.-F., Belenchon, I.R. et al. (2023). A Systematic Review of New Imaging Technologies for Robotic Prostatectomy: From Molecular Imaging to Augmented Reality. *Journal of Clinical Medicine*, *12*(16), 5425. https://doi.org/10.3390/jcm12165425

Roy, R., Marakkar, S., Vayalil, M.P., Shahanaz, A., Anil, A.P. et al. (2022). Drug-food Interactions in the Era of Molecular Big Data, Machine Intelligence, and Personalized Health. *Recent Advances in Food, Nutrition & Agriculture*, *13*(1), 27–50. https://doi.org/10.2174/2212798412666220620104809

Salehi, S., Olyaeemanesh, A., Mobinizadeh, M., Nasli-Esfahani, E. and Riazi, H. et al. (2020). Assessment of remote patient monitoring (RPM) systems for patients with type 2 diabetes: A systematic review and meta-analysis. *Journal of Diabetes and Metabolic Disorders*, *19*(1), 115–127. https://doi.org/10.1007/s40200-019-00482-3

Sauerbrei, A., Kerasidou, A., Lucivero, F. and Hallowell, N. (2023). The impact of artificial intelligence on the person-centred, doctor-patient relationship: Some problems and solutions. *BMC Medical Informatics and Decision Making*, *23*(1), 73. https://doi.org/10.1186/s12911-023-02162-y

Sheetz, K.H., Claflin, J. and Dimick, J.B. (2020). Trends in the Adoption of Robotic Surgery for Common Surgical Procedures. *JAMA Network Open*, *3*(1), e1918911. https://doi.org/10.1001/jamanetworkopen.2019.18911

Strimbu, K. and Tavel, J.A. (2010). What are Biomarkers? *Current Opinion in HIV and AIDS*, *5*(6), 463–466. https://doi.org/10.1097/COH.0b013e32833ed177

Su, P., Li, J., Yue, C., Liu, T., Liu, B. et al. (2022). Preoperative positioning planning for a robot-assisted minimally invasive surgical system based on accuracy and safety. *The International Journal of Medical Robotics and Computer Assisted Surgery*, *18*(4), e2405. https://doi.org/10.1002/rcs.2405

Subramaniam, S., Raju, N., Ganesan, A., Rajavel, N., Chenniappan, M., Prakash, C., Pramanik, A., Basak, A.K. and Dixit, S. (2022). Artificial Intelligence Technologies for Forecasting Air Pollution and Human Health: A Narrative Review. *Sustainability*, *14*(16), Article 16. https://doi.org/10.3390/su14169951

Swen, J.J., van der Wouden, C.H., Manson, L.E., Abdullah-Koolmees, H., Blagec, K., Blagus, T., Böhringer, S., Cambon-Thomsen, A., Cecchin, E., Cheung, K.-C., Deneer, V.H., Dupui, M., Ingelman-Sundberg, M., Jonsson, S., Joefield-Roka, C., Just, K.S., Karlsson, M.O., Konta, L., Koopmann, R., … Ubiquitous Pharmacogenomics Consortium. (2023).

Swen, J.J., van der Wouden, C.H., Manson, L.E., Abdullah-Koolmees, H., Blagec, K. et al. (2023). A 12-gene pharmacogenetic panel to prevent adverse drug reactions: An open-label, multicentre, controlled, cluster-randomised crossover implementation study. *Lancet (London, England)*, *401*(10374), 347–356. https://doi.org/10.1016/S0140-6736(22)01841-4

Tagde, P., Tagde, S., Bhattacharya, T., Tagde, P., Chopra, H. et al. (2021). Blockchain and artificial intelligence technology in e-Health. *Environmental Science and Pollution Research International*, *28*(38), 52810–52831. https://doi.org/10.1007/s11356-021-16223-0

Usmani, A., Imran, M. and Javaid, Q. (2022). Usage of artificial intelligence and virtual reality in medical studies. *Pakistan Journal of Medical Sciences*, *38*(4Part-II), 777–779. https://doi.org/10.12669/pjms.38.4.5910

Valla, V., Alzabin, S., Koukoura, A., Lewis, A., Nielsen, A.A. et al. (2021). Companion Diagnostics: State of the Art and New Regulations. *Biomarker Insights*, *16*, 11772719211047763. https://doi.org/10.1177/11772719211047763

Vargas, C. and Mira da Silva, M. (2023). Case studies about smart contracts in healthcare. *Digital Health*, *9*, 20552076231203571. https://doi.org/10.1177/20552076231203571

Varkey, B. (2021). Principles of Clinical Ethics and Their Application to Practice. *Medical Principles and Practice: International Journal of the Kuwait University, Health Science Centre*, *30*(1), 17–28. https://doi.org/10.1159/000509119

Vora, L.K., Gholap, A.D., Jetha, K., Thakur, R.R.S., Solanki, H.K. et al. (2023). Artificial Intelligence in Pharmaceutical Technology and Drug Delivery Design. *Pharmaceutics*, *15*(7), 1916. https://doi.org/10.3390/pharmaceutics15071916

Wang, C., He, T., Zhou, H., Zhang, Z. and Lee, C. et al. (2023). Artificial intelligence enhanced sensors—Enabling technologies to next-generation healthcare and biomedical platform. *Bioelectronic Medicine*, *9*(1), 17. https://doi.org/10.1186/s42234-023-00118-1

Yu, J., Wang, Y., Li, Y., Li, X., Li, C. et al. (2014). The safety and effectiveness of Da Vinci surgical system compared with open surgery and laparoscopic surgery: A rapid assessment. *Journal of Evidence-Based Medicine*, *7*(2), 121–134. https://doi.org/10.1111/jebm.12099

Zygulska, A.L. and Pierzchalski, P. (2022). Novel Diagnostic Biomarkers in Colorectal Cancer. *International Journal of Molecular Sciences*, *23*(2), 852. https://doi.org/10.3390/ijms23020852

CHAPTER

4

Comparison of Human Intelligence with Artificial Intelligence

Shajil UK,[1] *Sujith A*[2] and *Baiju G Nair*[1*]

1. Introduction

1.1 How to Define Intelligence

The ability to study, comprehend, and form reasonable thoughts or conclusions is intelligence makes tremendous sense. It's a complicated characteristic that affects social interactions, emotional depth, and cognitive functions in humans. Conversely, artificial intelligence, which originated in computer science, is commonly characterized as a machine's or program's capacity for independent thought, learning, and decision-making. The framework for a complex comparison is established by this similar yet different concept. Intelligence is the overall mental capacity of an individual to adapt to new situations, learn from experience, solve problems, and reason effectively (Wechsler, D. [1939]). Anyhow intelligence is a multifaceted complex phenomenon that has been observed, experienced, and discussed by philosophers and scientists for centuries. As far as the common man is concerned about his challenges and experiences he feels it. Obviously it is hard to define intelligence as it is a magnificent biological phenomenon. However, it is generally accepted that it encompasses the ability to learn, understand, reason, question, solve problems, adapt, and survive in all situations.

Intelligence is a trait, and it is more advanced in humans when compared to other living species. Even though they may exhibit cognitive abilities, humans have high levels of abstract thought, creativity, and problem-solving abilities which make them unique. "Intelligence is the global capacity of a person to deal effectively with the environment." This definition highlights the broad scope of intelligence, encompassing various cognitive skills used for adaptation. (Wechsler, D. [1939]) Adaptation is considered as a specific trait that has been processed through natural selection that results in functional significance for existence and survival on earth.

[1] Department of Bioscience and Engineering, National Institute of Technology, Calicut, Kerala India
[2] Department of Chemistry, National Institute of Technology, Calicut, India.
* Corresponding author: bgnair@nitc.ac.in

2. AI's Historical Context

The idea of automating intelligence has ancient roots, with references to non-human thinking machines existing in ancient texts like Homer's Iliad and Thomas Hobbes' Leviathan (Oliveira, A.L. and Figueiredo, M.A.T. [2023]). The history of artificial intelligence is a fascinating tapestry woven with innovative theories and bright minds. In the middle of the 20th century, it was just an idea, best summed up by Alan Turing's groundbreaking research that raised the controversial topic of whether machines are capable of thinking. Artificial Intelligence (AI) has progressed from the invention of basic algorithms to sophisticated neural networks with deep learning capabilities. It now mimics human intelligence and, in certain situations, even questions its superiority (Turing, A.M. [1950]).

3. Knowledge of Human Intelligence

Human intelligence is a complex textile woven over centuries. Philosophy, psychology, neuroscience, and cognitive science have contributed to our understanding of human intelligence (Gardner, H. [1983]). This ongoing pursuit creates a nuanced depiction of this trait. The comprehension of human intelligence, that mysterious yet defining quality of our species, is a complex tapestry crafted over countless centuries. Its strands extend back to the emergence of philosophy, where the likes of Plato and Aristotle grappled with the essence of thought, reason, and comprehension. Throughout the course of history, these fundamental strands have been enriched by the findings of psychology, unraveling the intricate mechanisms of the human mind. Neuroscience, delving into the very circuits of thought, has added further depth and intricacy to our understanding (Sternberg, R.J. [1998]), and in more recent times, cognitive science has joined the pursuit, shedding light on the processes through which we acquire knowledge, retain information, and solve problems. This interwoven narrative, shaped by diverse fields of study and ongoing discoveries, continues to develop, creating an increasingly nuanced and comprehensive depiction of human intelligence (Sternberg, R.J. [1998]).

4. The Goal of the Comparative

In a time when artificial intelligence is permeating every aspect of our life, a comparison between human and artificial intelligence is not only a necessary academic exercise but also a necessary investigation." Through this comparison, we can better comprehend AI's potential and limitations, ethical ramifications, and ways in which it can enhance rather than completely replace human talents. It also forces us to reconsider the fundamentals of human creativity and cognition, which leads to a deeper knowledge of our intelligence.

5. Overview and Scope

It is a journey to analyze and contrast the many aspects of AI and human intelligence. It will look at the foundations of each, contrast how well they can learn, solve problems, and be creative, and assess the ethical issues surrounding AI. It envisages a thorough examination that both challenges and educates our views of intelligence by utilizing theoretical ideas, practical applications, and current research" (Bostrom, N. [2014]). Before delving into this comparison, we first examine the basic traits of human intelligence, such as the subtle differences between emotional and social intelligence and the complex functions of the brain. This lays the groundwork for a more thorough comprehension of AI and enables a useful comparison between these two types of intelligence."

5.1 Human Intelligence Evolution

The progression of human intelligence is a protracted process that is synchronized with the development of humanity. Firth (2012) put forth the proposition that the genesis of human culture stems from our cognitive capabilities, which enable us to exchange information (Firth, R. [2012]).

This is also the reason why we assert that human cognition is the outcome of the evolution of human communities. Tomasello, in 1999, mentioned that what sets human cognition apart is the presence of culture and social context (Tomasello, M. [1999]). Furthermore, even within a specific social context, various social attributes such as race, language, class, and education can also contribute to a markedly distinct form of cognition. In addition to this, the abilities possessed by human intelligence also play a role in the progression of society. The capacity for communication and learning that humans possess lays the groundwork for the development and modification of diverse forms of social culture, including the early stages of language acquisition (Carpenter, M. et al., 1998). It indicates a reciprocal cycle where the human community is also an outcome of human intelligence, and the combination of the human capacity for learning and the acceptance of the environment has brought about the evolution of human intelligence.

6. The Foundations of Human Intelligence

6.1 Cognitive Psychology: Understanding Perception, Memory, and Learning

The process of gathering, deciphering, and organizing sensory data is known as perception in humans. Due to the influence of personal experiences, convictions, and expectations, it is quite subjective. Perception surpasses mere sensory reception, coordinating a symphony of stimuli from the bottom-up and expectations from the top-down to construct a complex feeling of reality. Sensory threads intertwine-vision, hearing, and smell-as previous knowledge guides the interpretation of raw information, converting light and darkness into recognizable faces, and a buzzing sound into either nature's lullaby or an alarm in a bedroom. This dynamic interaction continuously fine-tunes our perception, from anticipating potential danger to finding comfort in a familiar fragrance (Neisser, U. [2014]).

6.2 Memoria

Let's consider the three stages of the memory model: sensory, short-term, and long-term memory (Young, J.E. [2016]). The sensations that are perceived by the senses are filtered by sensory memory. Information is manipulated for cognitive tasks by short-term memory, also known as working memory. The long-term retention of knowledge is handled by long-term memory (Hong Zhibin et al., 2015). Figure 1 denotes Atkinson-Shiffrin memory model consisting of three distinct memory stores: Sensory Register, Short-Term Memory (STM), and Long-Term Memory (LTM). Sensory Register holds fleeting information, STM holds information for 15-30 seconds, and LTM holds indefinite information. A neural network with three layers represents this model, with input, hidden, and output layers. This model emphasizes the flow of information and processing, but it acknowledges that memory is likely more complex and interconnected than this model suggests (Hong Zhibin et al., 2015).

6.3 Acquiring Knowledge

In humans, learning entails gaining information or abilities through study, instruction, or experience. It's a multifaceted procedure that involves several cognitive functions. The acquisition of knowledge occurs through the process of learning, and it is well-established that this learning process is shaped by various theories (Watson, J.B. [1913]). The following are some of the key theories that contribute to our understanding of how knowledge is gained:

6.3.1 Behaviorism

Imagine Pavlov's dogs salivating at the sound of a bell. Behaviorism focuses on this type of observable learning, where behaviors are shaped through reinforcement-either adding pleasant consequences

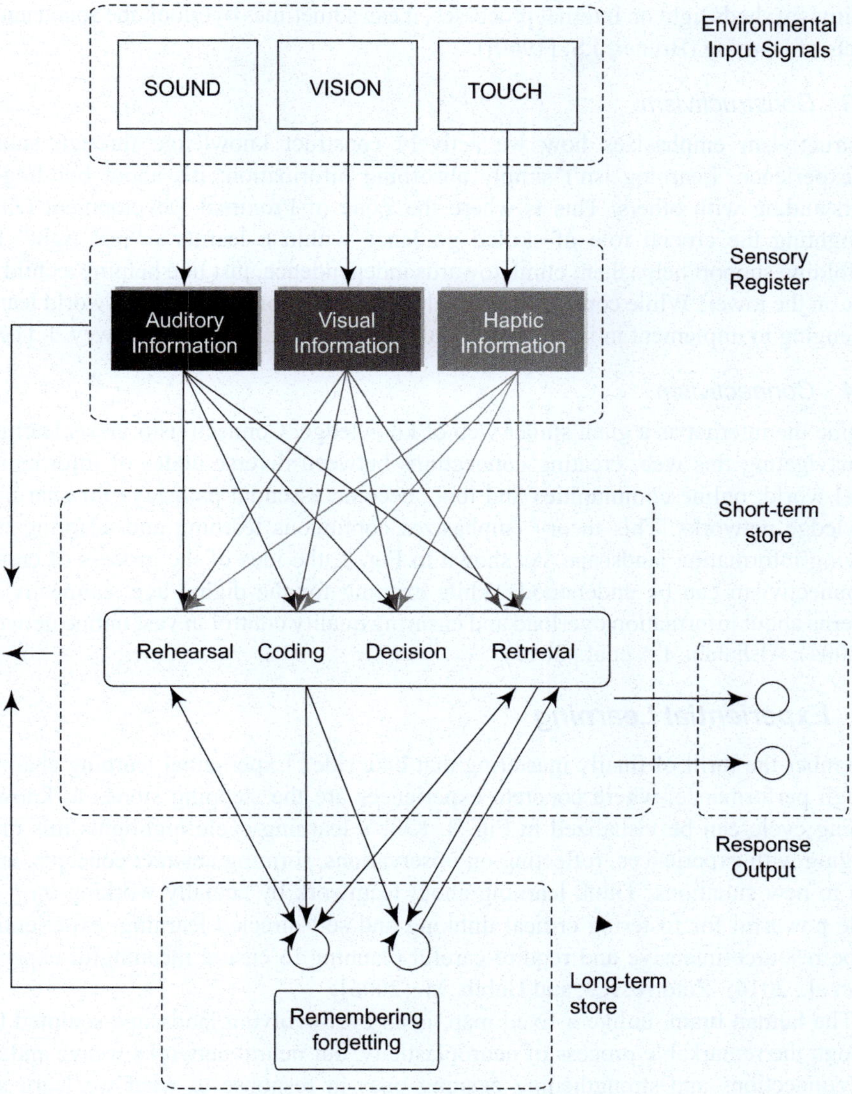

Fig. 1 The Atkinson-Shiffrin memory model, represented by a neural network, illustrates the possible structure of the neural network in the human brain.

(positive) or removing unpleasant ones (negative). Think of praise for good behavior or avoiding detention for completing homework. While its emphasis on measurable outcomes makes it valuable for training animals and shaping specific behaviors, behaviorism often gets criticized for neglecting internal mental processes and potentially encouraging rote learning (Watson, J.B. (1913); Skinner, B.F. [1938]).

6.3.2 Cognitivism

This theory delves into the fascinating world within our minds. A cognitivist views learning as an active process of information processing, which stores, retrieves, and manipulates information using mental structures called schemata (Piaget, J. [1970]). Like building with Lego blocks, new information fits into existing frameworks, shaping how we understand and learn new things. Remember struggling with algebra as a kid? Your existing math schemata likely played a role! While cognitivism sheds light on internal processes, it can sometimes overlook the social and collaborative aspects of learning (Bruner, J.S. [1960]).

6.3.3 Constructivism

Constructivism emphasizes how we actively construct knowledge through social interaction and experience. Learning isn't simply absorbing information; it's about building meaning and understanding with others. This is where the Zone of Proximal Development (ZPD) comes in, highlighting the crucial role of skilled guidance within a learner's "just right" learning zone. Scaffolding support helps them climb towards independence, just like helping a child reach a higher block on the tower! While constructivism celebrates collaboration and real-world learning, it can be challenging to implement in large, traditional classrooms (L.S. [1978]; Dewey, J. [1938]).

6.3.4 Connectivism

Imagine the internet as a giant spider web of knowledge. Connectivism views learning as building and navigating this web, creating connections between diverse nodes of information. In today's digital world, online communities and tools become essential pathways for sharing and forming knowledge networks. This theory emphasizes continuous learning and adapting to a constantly evolving information landscape. As shown in Fig. 2, the idea of the process of creating networks in connectivism can be understood. While exciting for the digital age, connectivism also raises concerns about information overload and ensuring quality control in vast online networks (Siemens, G. (2004); Alshalabi, I.A et al., 2013).

6.4 Experiential Learning

Remember the thrill of finally mastering that bike ride? Experiential learning champions learning through performance, where concrete experiences are the stepping stones to knowledge. Kolb's learning cycle can be visualized in Fig. 3. Kolb's learning cycle highlights this process actively engaging with experiences, reflecting on observations, forming abstract concepts, and then testing them in new situations. Think learning about teamwork by actually working on a group project. While powerful for fostering critical thinking and self-directed learning, experiential approaches can be resource-intensive and require careful planning to ensure meaningful experiences (Poore, J.A. et al., 2014; Zatorre, R.J. and Habib, M. [2006]).

The human brain, unlike a fixed map, is an ever-evolving landscape sculpted by experience. Through the remarkable process of neuroplasticity, our neural networks weave and dance, forging new connections and strengthening existing ones in response to what we learn and encounter. As we navigate a new skill, master a foreign language, or etch an emotional memory onto our minds, neurons fire together, sparking a symphony of chemical signals that solidify these pathways (Ericsson, K.A. et al., 1996). The intricate web of synapses, the brain's communication junctions, grows denser, allowing information to flow more effortlessly along established routes. Conversely, unused pathways gradually fade, pruned back by the brain's relentless optimization process. This dynamic dance of creation and refinement is the very essence of learning, shaping our thoughts, behaviors, and our very selves. Every experience, from the joy of a breakthrough to the sting of a loss, leaves its mark on this intricate neuronal tapestry, continuously sculpting who we are and who we might become (Firth, R. [2012]).

7. Brain Structure and Neuronal Networks in Neurobiology

The brain, a master of multitasking, isn't one uniform lump but a mosaic of specialized regions. Deep within, the limbic system orchestrates our emotional tapestry, from the amygdala's primal urges to the hippocampus' memory whispers. Above, the wrinkled cerebral

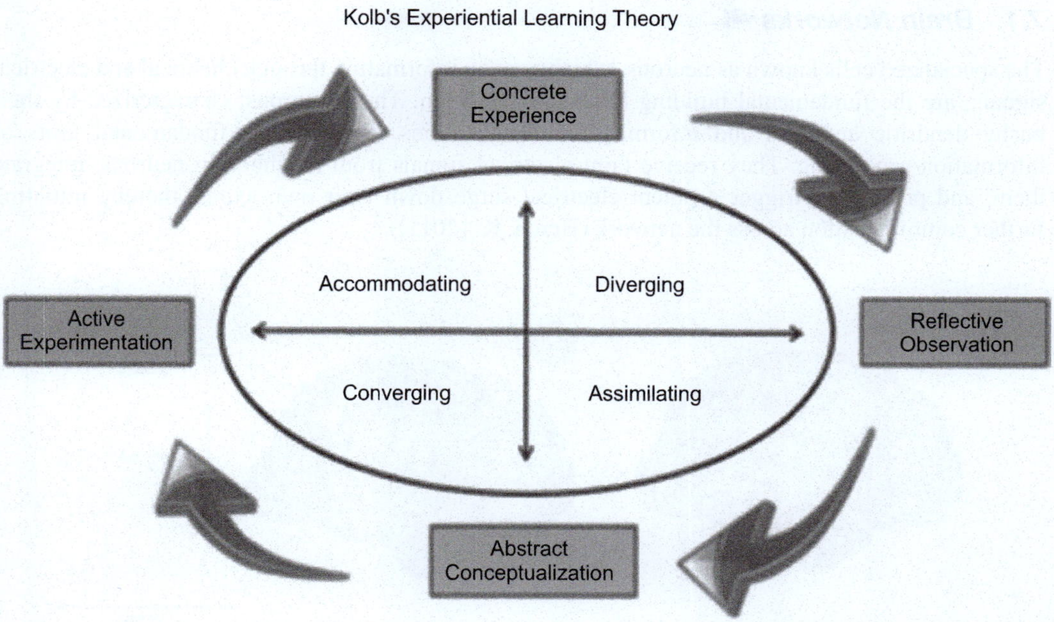

Fig. 2 Kolb's Experiential Learning Theory. Adapted with permission from Ref. Poore, J.A. et al., 2014

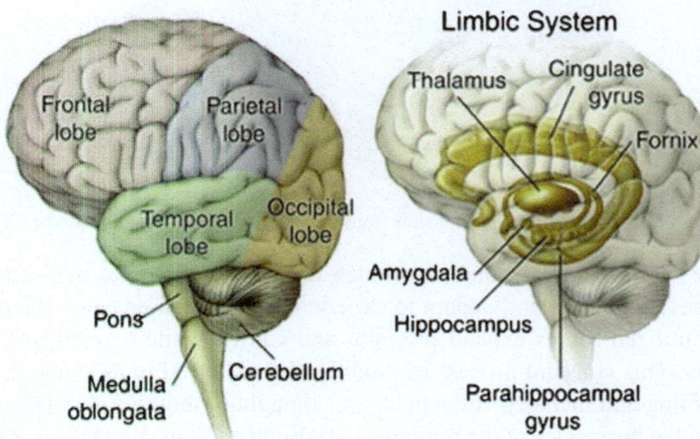

Fig. 3 Limbic system adapted with permission from (Braun, K. [2011])

Cortex handles the intellectual heavy lifting, with specialized lobes like the frontal cortex planning strategies and the temporal and occipital lobes deciphering the world's sights and sounds. Finally, the cerebellum, our "little brain," ensures smooth movements, whether it's balancing on a tightrope or mastering a backflip (LeDoux, J.E. [2012]). This functional specialization – like using the right tool for the job – maximizes efficiency, allows for complex cognitive feats, and even offers backup when one region faces setbacks. From the limbic system's emotional echoes to the cerebellum's finely tuned movements, each brain region plays a unique role in the symphony of our being, enabling us to experience the world in all its vibrant chaos (Figure 4) (Braun, K. [2011]).

7.1 Brain Networks

The specialized cells known as neurons, which convey information through chemical and electrical signals, are the fundamental building blocks of the brain. These neurons, characterized by their bushy dendritic antennas and information-carrying axons, serve as the fundamental units of information processing. They receive faint electrical signals from neighboring neurons, integrate them, and potentially trigger a potent electrical surge down their own axons, thereby initiating further communication across the network (Braun, K. [2011]).

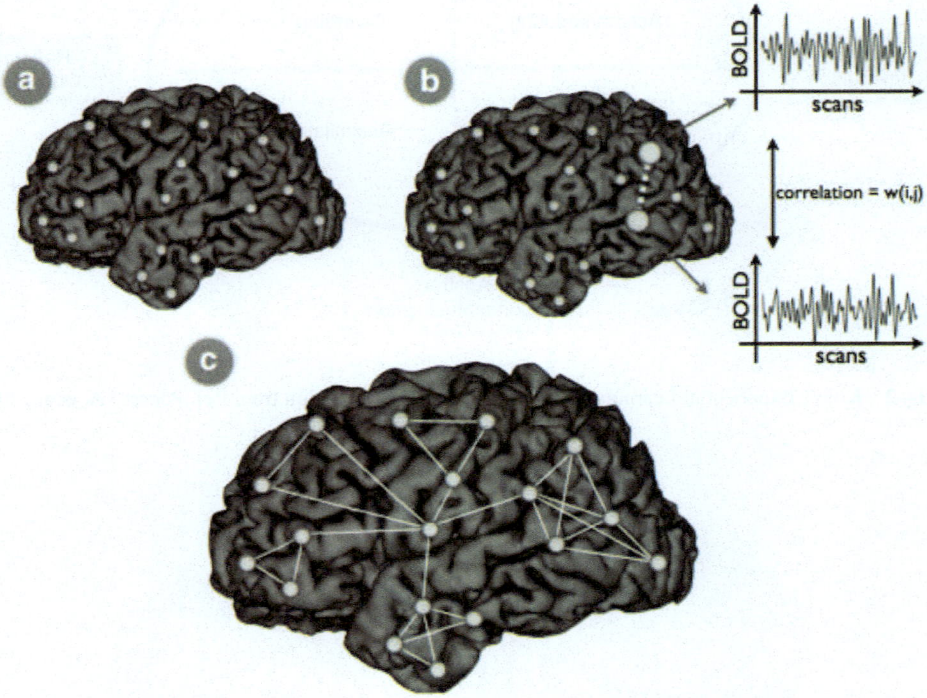

Fig. 4 Brain networks. Adapted with permission from van den Heuvel, M.P. and Hulshoff Pol, H.E. (2010)

This intricate web of connections constitutes the neural networks of the brain, which are dynamic landscapes that consistently adapt to experiences. One can envision them as living maps, where frequently utilized routes expand in width and clarity, while less utilized paths gradually fade into obscurity. This constant process of modification, referred to as synaptic plasticity, is the foundation of learning and memory, influencing our thoughts, emotions, and behaviors. These can be visualized from the intricacies of the brain network Figure 4 (van den Heuvel, M.P. and Hulshoff Pol, H.E. [2010]). Every new experience induces ripples within these networks, fortifying existing connections and forging novel ones. Analogous to a city reconstructing itself after a storm, the brain redirects traffic, constructing neural shortcuts and detours to efficiently handle familiar inputs. This remarkable ability to adapt is the key element of human intelligence, enabling us to acquire knowledge, develop, and flourish within a complex and ever-evolving world (van den Heuvel, M.P. and Hulshoff Pol, H.E. [2010]).

7.2 The Creative and Emotional Intelligence

Beneath the surface of logic and reasoning lies a hidden superpower: emotional intelligence, the ability to navigate the tumultuous seas of human feeling.The ability to identify, comprehend, and control our own emotions as well as those of others is a necessary component of emotional intelligence (Yang, L. and Duan, M. [2023]).

Navigating the social landscape with grace and aplomb isn't just about charm; it's about mastering the art of emotional intelligence (EI). This potent quartet of skills – relationship management, social awareness (empathy), self-regulation, and self-awareness – empowers us to connect authentically, build fulfilling relationships, and achieve success in all aspects of life (Goleman, D. [2006]). Social awareness, like a finely tuned antenna, allows us to pick up subtle cues, decipher emotions expressed both verbally and nonverbally, and tune into the unspoken needs of others. This fuels empathic communication, fostering collaboration and authentic connections. Think of a manager attuned to their team's anxieties, offering support before tensions escalate. Self-awareness then acts as an internal compass, guiding us to recognize our own emotions, biases, and triggers. This introspection allows for self-regulation, enabling us to navigate challenging situations with poise and avoid impulsive reactions. Picture a doctor, calmly de-escalating a distraught patient by acknowledging their frustrations and offering a compassionate response.

Finally, relationship management brings these elements together, enabling us to build and maintain positive connections. From navigating conflicts with compromise to celebrating successes with warmth, EI equips us to navigate the complex terrain of human interactions. Imagine a couple resolving a disagreement by calmly expressing their needs and finding a mutually agreeable solution (Salovey, P. & Mayer, J.D. [1990]). This potent cocktail of skills isn't just for personal harmony; it's a gold standard for professional success. Studies have shown that individuals with high EI are more likely to land job offers, excel in leadership roles, and foster collaborative workplaces. They're adept at conflict resolution, able to bridge communication gaps and navigate even the most challenging interpersonal dynamics (Salovey, P. and Mayer, J.D. [1990]).

7.3 Inventivity

Human creativity is the capacity to generate original, worthwhile ideas. It entails both convergent (combining those ideas into the best outcome) and divergent thinking (creating many original ideas). The notion of creativity is intricate and varied, and it has been thoroughly examined in the field of philosophical inquiry into creativity. It involves the generation of novel and valuable ideas, products, or solutions. Creativity is not limited to the realm of art; it encompasses various domains, including science. The definition of creativity is a subject of debate, but it generally involves the ability to think outside the box, make connections between seemingly unrelated concepts, and produce original and innovative outcomes. Imagination plays a crucial role in creativity, as it allows individuals to envision new possibilities and explore different perspectives. The creative process is not always rational, and it can involve intuitive insights and leaps of imagination. Additionally, creativity is influenced by factors such as knowledge, tradition, and the aesthetic value attributed to the creative output. Computational and Darwinian theories have also been proposed to explain creativity. Overall, creativity is a complex phenomenon that encompasses various dimensions and is influenced by psychological, philosophical, and cultural factors (Gaut, B. [2010]).

Our exploration has revealed the breathtaking versatility and intricate complexity of human intelligence, a kaleidoscope of capabilities far exceeding the current state of AI. We've examined its analytical prowess, meticulously dissecting data like a master surgeon. We've marveled at its emotional intelligence, delving into the nuanced world of feelings with an empathy no machine can replicate. And finally, we've encountered its boundless creativity, soaring into the ethereal realms of the new and imaginative, where AI often stumbles. These diverse facets, interwoven into a singular tapestry, stand as a testament to the uniqueness of human intelligence. Unlike AI, which often excels in isolated domains, we seamlessly switch between analytical rigor and emotional understanding, seamlessly transitioning from dissecting equations to deciphering the whispers of a tear. It is this inherent adaptability, this symphony of logic and intuition, that truly sets us apart. While AI may one day master specific tasks, replicating the multifaceted dance of human intelligence, with its intricate blend of analysis, empathy, and creativity, remains a distant dream. Our minds, in their glorious complexity, offer a viewpoint as rich and varied as the world itself, a perspective no single algorithm can currently hope to capture.

8. Principles of Artificial Intelligence

A Brief Overview of Artificial Intelligence: Definition and Range

In computer science, artificial intelligence (AI) is the field that focuses on developing systems that can carry out activities that normally require human intelligence. Learning, solving problems, identifying patterns, comprehending language, and making decisions are some of these tasks. However, in order to comprehend the realistic capabilities of AI, it is crucial to differentiate between three categories of AI: (i) Artificial narrow intelligence (ANI) pertains to systems that possess the ability to execute a specific task within a confined context, such as a self-driving car; (ii) Artificial general intelligence (AGI) pertains to systems that exhibit intelligence akin to human intelligence; and (iii) Artificial superintelligence (ASI) pertains to systems that have the capacity to surpass human intelligence. AGI and ASI, at present, do not exist, and acknowledging this fact presents a more realistic portrayal of what AI, specifically ANI, can achieve (Russell, S.J. and Norvig, P. [2020]).

8.1. The Past

Although the idea of artificial intelligence (AI) has its roots in fables from antiquity, the field of AI as we know it today was founded in the middle of the 20th century by visionaries like Alan Turing and John McCarthy. The advent of the earliest artificial intelligence (AI) systems, such as ELIZA, a computer program for natural language processing, and the most recent developments in machine learning and neural networks are noteworthy turning points. The Turing Test is a tool used to evaluate a machine's capacity for intelligent behaviour (Siau, K.L. and Yang, Y. [2017]; Panetta, K. [2018, July 9]).

9. Foundations of Machine Learning and Deep Learning

9.1 Machine Instruction

AI's machine learning subfield involves teaching robots to learn from data. The process entails supplying substantial datasets to algorithms, which employ statistical methods to identify trends and arrive at conclusions or forecasts.

Some important forms of machine learning are:

- *Supervised Learning:* It utilizes labeled data as training material; the algorithm learns to predict results from input data.
- *Unsupervised Learning:* In this mode of learning an algorithm receives unlabelled data i.e data without clear instructions. The algorithm learns to identify patterns and structures within the data.
- *Reinforcement Learning:* Using feedback from its own actions and experiences, the algorithm learns by making mistakes (Russell, S.J. and Norvig, P. [2016]; McCarthy, J. [2005]).

9.2 In-depth Education

Artificial neural networks with numerous layers, or "deep networks," are a subtype of machine learning that have the ability to learn and make deft decisions on their own. These networks are organized into layers of nodes, or "neurons," with each layer taking in information from the one above, processing it, and then sending it further. They can recognize patterns and extract features on a complex level. Major advances in AI, such as complex speech and picture recognition systems and advanced natural language processing, have been made possible thanks in a large part to deep learning (Goodfellow, I et al., 2016).

10. Computer Models and Algorithms

10.1 AI Algorithmic Work

The AI system's processing and learning mechanisms are determined by algorithms, which are at its core. Simple decision-tree algorithms and sophisticated neural networks are among them. In machine learning, some popular AI algorithms are back propagation in neural networks, k-nearest neighbors, and clustering algorithms, along with linear regression (Massachusetts Institute of Technology. [n.d.]).

10.2 Computational Frameworks

AI computational models are structures or programs that simulate how the human brain processes, evaluates and makes inferences from data. Computational complexity, or the resources needed for an algorithm to solve a problem, is one of the difficulties these models address. This is especially true as datasets get bigger. A computational framework is a software environment that provides a structured way to build, train, and deploy computational models, often for machine learning, deep learning, or other computationally intensive tasks. It acts as a foundation or scaffold, offering pre-defined components, tools, and guidelines to streamline the development process (Heaton, J. [2018]).

11. AI Uses in Diverse Industries

11.1 Medications

Applications of artificial intelligence (AI) in healthcare include robotic surgery systems, algorithms for individualized therapy suggestions, and diagnostic systems that evaluate medical images. AI is being employed within the pharmaceutical sector for diverse purposes, encompassing the analysis and interpretation of data, the exploration and progression of medications, and efficiency enhancement. The automatization of tasks, cost reduction, and elevation of data analysis precision are among the capabilities that AI technology brings. Moreover, it adeptly manages the surging volume of scientific information and studies, which would pose a formidable challenge for humans to process promptly. AI-based apparatuses, exemplified by the GI Genius-AI based endoscope, are currently utilized to identify indications of severe ailments. The utilization of AI within the pharmaceutical industry is anticipated to reconfigure the sector and engender progressions in both healthcare and marketing. Insider Intelligence has forecasted a 48% annualized growth rate for artificial intelligence in healthcare during the period spanning, 2017 to 2023 (A.K. Sharma., 2022; Hamet, P. and Tremblay, J. [2017, April]).

11.2 Finances

Algorithmic trading, which utilizes algorithms to conduct trades at high speeds and volumes, fraud detection systems that examine transaction patterns, and risk management tools-all contribute to the transformation of finance through artificial intelligence. Artificial intelligence (AI) is rapidly transforming the financial landscape, injecting automation, precision, and personalization into a traditionally human-centric domain. By leveraging machine learning, natural language processing, and other cutting-edge techniques, AI tackles a diverse range of financial challenges, from guarding against fraudulent transactions (Figure 1) to crafting personalized investment portfolios through robo-advisors (Akter, S. et al., 2023). Efficiency gains are a major draw of AI in finance. Mundane tasks like data entry and report generation are readily automated, freeing human workers for higher-level analysis and strategic decision-making. Beyond saving time and resources, AI's analytical prowess shines in its ability to sift through vast datasets, uncovering hidden patterns and trends that might elude even the most seasoned finance professionals. This enhanced precision translates to improved risk assessments, more accurate credit scoring, and ultimately, smarter investment strategies.

But AI's impact extends beyond just crunching numbers. The rise of robo-advisors powered by AI algorithms is democratizing wealth management, offering personalized investment advice previously reserved for high-net-worth individuals. These AI-driven platforms tailor investment portfolios to each user's unique risk tolerance and financial goals, making sophisticated financial planning accessible to a wider audience (Miau, L.D. and Ravi, S. [2023]). While the benefits of AI in finance are undeniable, navigating its ethical implications is crucial. Concerns around algorithmic bias, data privacy, and potential job displacement necessitate responsible development and deployment. Ensuring transparency and explainability in AI decision-making processes is critical to mitigating bias and fostering trust. Additionally, protecting sensitive financial data and upskilling workforces to adapt to changing landscapes are essential considerations for smooth and equitable AI integration into the financial sector (Akter, S. et al., 2023; Miau, L.D. and Ravi, S. [2023]). The ongoing AI revolution in finance promises a future of streamlined processes, data-driven insights, and personalized financial solutions. By addressing ethical concerns and fostering responsible development, AI has the potential to unlock a new era of financial inclusivity and prosperity for all (Miau, L.D. and Ravi, S. [2023]; Swan, M. [2023]).

11.3 Automobiles and Mobility

Artificial Intelligence is transforming the automobile sector, especially with the introduction of driverless cars that utilize AI to sense their surroundings and make judgment calls. Autonomous vehicles: Driven by artificial intelligence (AI) and a multitude of sensors, these automobiles possess superhuman agility to maneuver through urban and rural areas, significantly lowering the risk of collisions and anxiety. (Imagine an elegant, future vehicle doing a graceful dash through traffic, a la "I, Robot.") (Sadaf, M. et al., 2023); more intelligent assistants With features like lane departure alerts, adaptive cruise control that maintains a safe distance, and blind-spot monitoring that functions like a sixth sense, advanced driver-assistance systems (ADAS) effectively become your driving angels (Swan, M. (2023); Sadaf, M. et al., 2023). Road conditions and traffic information are shared in real-time by cars communicating with one another, forming a kind of friendly informant network known as "traffic whisperers." As a result, there is less traffic, commutes go more smoothly, and maybe drivers are happy. Customized rides: Your vehicle may create a unique driving experience that resembles an extension of your living room by remembering your favorite music, preferred routes, and perfect temperature. The device's mechanic: Analyzing vehicle data, AI may anticipate any issues before they happen, much like a doctor identifying a sickness. You'll drive with confidence because there won't be as many problems or unforeseen repair costs. It's obvious that there are advantages: everyone can move about independently, commutes run more smoothly, and driving is just more fun.

There are, however, detours to take into account, just like on any vacation. Obstacles we must overcome include the need to update infrastructure for connected vehicles, ethical questions around AI decision-making, and technical limitations preventing self-driving cars from being perfect. Self-driving cars powered by AI and sensors can reduce accidents and stress on the road. Advanced driver-assistance systems (ADAS) provide smarter assistants, while traffic whisperers allow vehicles to share real-time information. Personalized rides and AI analysis can improve the driving experience. However, ethical concerns, technological hurdles, and infrastructure upgrades are challenges. Despite these, AI has the potential to transform our relationship with cars, making them intelligent companions towards a smarter, safer, and more enjoyable future of mobility (Pilot, C.V.).

11.4 Client Support

AI has had a significant impact on customer service with the introduction of chatbots and automated systems that can answer questions from customers, make reservations, and fix issues-often in several languages and around-the-clock. AI enhances efficiency by automating routine tasks, providing 24/7 availability, delivering personalized services, enhancing customer satisfaction, and providing proactive support through predictive solutions, while also providing valuable data-driven insights (Calderon, T.G. and Ofobike, E. [2007]). The manner in which Artificial Intelligence (AI) is revolutionizing a wide range of sectors and the rapid advancements being achieved in technologies like deep learning and machine learning are of great significance. It is imperative to consider the potential advantages and challenges that lie ahead, including ethical dilemmas associated with the implementation of AI and the necessity for robust and morally sound AI development in order to ensure favorable outcomes for the society.

12. Evaluation via Comparison

12.1 Adaptability and Learning

Human learning is a dynamic process rooted in experience, social interactions, and cognitive flexibility. It is a dynamic tapestry that adapts, evolves, and navigates the ever-changing landscapes of life. Each interaction shapes our understanding of the world, and social learning, the sharing of knowledge through observation, imitation, and collaboration, enriches our knowledge. Cognitive flexibility, our ability to adapt and think outside the box, keeps the tapestry dynamic and ever-evolving with changing times. AI researchers are exploring the possibility of capturing this intricate process of human learning, but the challenge lies in mimicking the nuanced structure of human learning. Understanding these elements allows us to appreciate the wonder of human adaptability and its potential to shape a brighter future (Bruner, J.S. [1990]). Machine learning (ML) is a technology that enables computers to learn without explicit programming. It excels in tasks like image recognition, recommendation systems, fraud detection, and natural language processing. However, ML has limitations such as dependence on data, limited understanding, the black box effect, and job displacement. ML algorithms are only as good as the data they are trained on, and they may struggle to explain why patterns exist or to apply them in novel situations. Additionally, ML algorithms can operate as "black boxes," making it difficult to understand their decision-making processes. Job displacement concerns arise as ML automates tasks previously done by humans. Despite these limitations, ML remains a powerful tool with immense potential, and responsible development is crucial for harnessing its power for a better future (Calderon, T.G. and Ofobike, E. [2007]).

Machine learning models have limitations in their application to society, including quantification commitments, unmodeled dependencies, and optimistic cross-validation. They may not account for constructs, affecting predictive and external validity. False predictions can support regressive political agendas. It is crucial to be aware of these limitations and conduct real-world testing before making claims (Calderon, T.G. and Ofobike, E. [2007]). AI's promise lies in its flexibility and adaptability, particularly through reinforcement learning systems. These systems thrive on feedback loops, constantly iterating and improving, enabling AI to master complex environments, personalize interactions, and discover creative solutions. They can adjust responses based on context and real-time feedback, becoming trusted confidantes or even designing materials with unprecedented properties. However, AI's ability to adapt within its training data and reward systems can be limited by unexpected and novel situations, such as encountering a dragon or a talking dragon (Bruner, J.S. [1990]). Human learning is unparalleled in its depth and breadth, encompassing not just facts and patterns but also emotions, intuitions, and complex social nuances. This allows us to apply knowledge in diverse situations, surpassing the capabilities of AI. Human curiosity knows no bounds, as we learn not just for practical utility but for the joy of discovery. This boundless human ability to weave knowledge into creative ways is a testament to its adaptitude and adaptability (Whiten, A. [2017]).

12.2 Solving Problems and Making Decisions: A Human Perspective

Human problem-solving is a complex process that involves creativity, intuition, and rationality. Unlike machines, humans approach problems with a spark of imagination, generating unconventional solutions and connecting seemingly disparate ideas. Intuition, a blend of experience, subconscious processing, and emotional cues, guides us even when logic falters. Our minds are not solely governed by cold logic, but also by feelings and subjective experiences. Empathy can lead us to prioritize others' well-being, while biases can cloud our judgment. The true magic lies in finding the harmony between logic, intuition, and emotion, resulting in solutions that resonate with our values and humanity. This delicate dance between logic and heart is what makes us unique, allowing us to navigate life's complexities with both reason and heart (Amabile, T.M. [1996]).

12.3 AI Decision-Making and Problem-Solving

AI and human intelligence are powerful problem-solving tools. AI is an algorithmic powerhouse that can analyze vast amounts of data at high speeds, making consistent decisions based on established patterns. It also uses data-driven insights to uncover hidden patterns. However, AI's strengths include data dependence, the "black box effect," and an emotional blind spot. Human problem-solving is characterized by creativity, contextual savviness, and learning from errors. However, human strengths also have downsides, such as inconsistent outcomes, cognitive limitations, and time constraints (Russell, S.J. and Norvig, P. [2021]). The most effective approach to problem-solving often involves a synergy between AI and human intelligence. AI's speed, consistency, and data-driven insights can be complemented by human creativity, contextual awareness, and emotional intelligence. By leveraging the strengths of each approach, we can create a powerful problem-solving symphony that tackles challenges with efficiency, nuance, logic, and empathy. The future of problem-solving lies in harnessing the unique strengths of AI and human intelligence to create a dynamic and adaptable problem-solving ecosystem (Russell, S.J. and Norvig, P. [2021]).

Integrating AI and human intelligence into problem-solving can lead to a symphony of solutions that can tackle challenges in ways neither could alone. AI can process vast amounts of data, spotting hidden patterns, and generate countless options for humans to consider. Human intuition guides the machine, ensuring human values and empathy remain at the heart of decision-making. AI's ability to explore vast solution spaces can spark human creativity, leading to groundbreaking solutions. However, merging human and AI intelligence comes with ethical considerations, trust gaps, and the need for skills and education. Despite these challenges, the potential rewards are vast: enhanced solutions, augmented human capabilities, and a collective intelligence that surpasses the limitations of either alone. By working together, we can build a future where technology acts as a tool to amplify human potential, leading to a world where innovation and progress become collaborative endeavors (Simon, H.A. [1996]).

12.4 Originality and Sensitivity to Emotions

Human Originality and Emotional Intelligence

Human intelligence is an intricate embroidery crafted from creativity and emotional intelligence. Creativity is a kaleidoscope that shapes our artistic endeavors, daily problem-solving, mental landscapes, and existence. It defies singular definitions, allowing us to break free from the mundane and imagine what could be. Emotional intelligence, on the other hand, is the unseen guide, providing context and nudging us towards responsible destinations. It plays a fundamental role in our decisions and interactions, allowing us to navigate social interactions with empathy and grace (Russell, S.J. and Norvig, P. [2021]). (Emotional drivers shape our decisions, motivations, and worldviews. While AI has made strides in recent years, replicating the intricacies of human creativity and emotional intelligence remains a distant dream. Current AI models exhibit a degree of originality within predefined parameters, but lack the spark of true imagination and the ability to break free from established patterns. AI's emotional intelligence, while processing and analyzing emotional data, lacks the true understanding of subjective experiences, empathy, and the nuanced social awareness that defines human emotional intelligence. The future unfolds, and embracing both strengths and limitations will be key to unlocking the full human potential (Floridi, L. [2020]; Pink, D.H. [2012]).

AI is making significant strides in music and art, analyzing vast amounts of data to generate original compositions, perform, create music loops, and personalize recommendations. In art, AI can learn visual styles and create original paintings, sculptures, or photographs. It can also respond to real-time input, creating dynamic art experiences. AI can also analyze art collections and identify patterns, helping museums personalize exhibits. However, AI cannot innovate in the same way humans do, and its true potential lies in collaboration. By leveraging AI's efficiency and pattern recognition with human originality and emotional intelligence, AI can usher in a new era of artistic expression (Floridi, L. [2020]). Affective computing, a field combining AI and psychology, aims to create intelligent systems that can recognize and mirror human emotions, paving the way for a future where machines understand our emotions as well as we do. Current advancements include facial recognition, text analysis, and physiological sensing. However, replicating genuine emotional understanding remains a challenge due to subjectivity, context, and the embodied experience of emotions. The future of affective computing lies in collaboration, not competition, and incorporating human expertise in psychology and ethics into AI development. Transparency and trust are crucial, and AI should not replace human emotional connection but complement it, enhancing communication and understanding. The ultimate goal is to create AI systems that augment human understanding, not replace it, fostering emotional well-being and deepening connections with each other (Pink, D.H. [2012]).

12.5 In Contrast

In the comparison between AI's utilization of data-driven creativity and simulated emotional responses and the inherent human capacity for abstract cognition and profound emotional comprehension, notable distinctions become apparent. The creative abilities of humans, which are bolstered by intuition and personal experiences, surpass the pattern-based approach employed by AI, particularly within disciplines such as art. Within therapeutic contexts, human therapists capitalize on genuine emotional comprehension and adaptability, attributes that AI lacks. The comprehension of the human spirit, which is influenced by a multitude of factors, represents a unique strength of humans that AI struggles to replicate. Furthermore, human abstract cognition, which plays a pivotal role in intellectual endeavors, outshines AI's algorithmic reasoning. When integrating AI into these domains, it is imperative to acknowledge these differences and recognize the complementary nature of human and artificial intelligence (Goleman, D. [1995]).

AI and human intelligence are not a binary battleground, but a dance of strengths and limitations. AI excels in computational speed, predictive power, and automation of routine tasks, while human intelligence dominates in abstract thinking, creativity, emotional intelligence, and ethical judgment. The future lies in complementarity, where AI's computational power amplifies human creativity, AI's analytical precision guides intuition, and AI's tireless efficiency allows humans to focus on deeper pursuits. This collaborative future requires mutual respect and ongoing dialogue, ensuring AI technology is developed and deployed ethically, with human values at its core. By harnessing the distinct strengths of both AI and human intelligence, we can create a future of unparalleled progress and shared prosperity, where AI serves as a partner rather than a rival (Collins, K. [2019]).

12.6 Strengths and Limitations

Mental Flexibility and Original Approaches to Problem-Solving

Human creativity is a unique ability that transcends logic and data, allowing us to tackle complex issues, create new worlds, and craft tools. This human spark has contributed to significant advancements in science, art, and technology. For example, Rosalind Franklin's Double Helix revolutionized genetics through a leap of imagination, while Dmitri Mendeleev arranged the periodic table based on predicted properties of undiscovered elements. Pablo Picasso's Guernica showcased the power of human emotion and imagination, while Beethoven's Fifth Symphony evoked human emotions through a sonic question mark. The printing press, invented by Johannes Gutenberg, democratized knowledge and paved the way for the information age. The Post-it Note, invented by Arthur Fry, revolutionized communication and note-taking by connecting adhesive to paper. These examples demonstrate the strength of human creativity, allowing us to transcend the known, bend ideas, and transform the world. While AI can be a powerful tool, it is the human mind that remains the driving force behind groundbreaking scientific discoveries, captivating art, and groundbreaking technology (Gary Marcus and Ernest Davis, 2019).

13. Emotional Intelligence and Interpersonal Communication

Emotional intelligence (EQ) is a crucial aspect of human intelligence, encompassing empathy, social skills, self-awareness, and self-regulation. It forms the basis of effective interpersonal interactions, teamwork, and leadership. Emotional intelligence allows individuals to emphatize witn others, fostering deeper connections and trust. Social skills enable smooth communication, facilitating smoother interactions in teamwork. Self-awareness guides our emotional landscape, allowing us to identify our emotions and anticipate their impact on ourselves and others. Self-regulation helps manage our emotions, ensuring they don't hijack our actions or words. High EQ individuals build stronger relationships, foster trust, and facilitate smooth communication. They create a collaborative environment that is morale boosting and productive. Leaders who connect emotionally with their team become role models, inspiring them and guiding them towards shared goals. Research shows that organizations with high EQ leaders experience lower employee turnover, higher customer satisfaction, and greater success. EQ is a multifaceted concept that can vary across cultures, situations, and personalities (Goleman, D. [2006]).

Human generalization is a unique trait that sets humans apart from AI. Unlike AI, which requires extensive training and vast datasets, humans excel in adaptability, making sense of incomplete information, and applying knowledge across diverse contexts. This is due to their inherent flexibility, which allows them to navigate novel situations and learn from limited information. They also excel at making sense of incomplete information, allowing them to infer, hypothesize, and fill the gaps. Humans also have the ability to apply knowledge across diverse domains, allowing them to make unexpected connections and solve problems creatively. This is a hallmark of human intelligence, unlike AI, which often struggles with transferring knowledge between domains. Human learning is a dynamic dance of trial and error, allowing for flexible reasoning skills and adaptability. This generalization has profound implications, fueling creativity, social intelligence, and scientific breakthroughs. While AI offers immense potential in specific domains, the human brain's ability to generalize remains unmatched (James Manyika et al., 2017).

14. The Advantages of Machine Learning

Processing Data Quickly and Efficiently

Machine learning (ML) has the potential to process and analyze massive datasets with speed and efficiency, making it a game-changer in fields like genetics, astronomy, and climate science. AI algorithms can decode the genome, peer into the universe, and predict climate change. For example, DeepMind's AlphaFold can predict protein structures with remarkable accuracy, paving the way for new drug discoveries and disease prevention strategies. AI can also peer through vast datasets from telescopes and satellites, revealing the secrets of the universe. AI can also predict climate change by analyzing satellite data to track deforestation and forest loss, aiding conservation efforts. Additionally, AI can provide consistent and unbiased results, unlike humans, which can be biased

(Mitchell, M. [2019]). AI systems offer unwavering consistency, making them invaluable in areas like medical diagnosis, financial trading, and scientific research. However, this consistency can also lead to biases. AI systems are designed and trained by humans, which can influence their decision-making. Biased training data can lead to biased outputs. To combat AI bias, transparency and accountability are crucial. This involves identifying and addressing potential biases through clear documentation of training data, algorithms, and decision-making processes. Including diverse data sets and involving people from different backgrounds in development can minimize unintentional biases. Regular monitoring and evaluation can help assess AI systems for bias and implement corrective measures. By promoting diversity, ensuring transparency, and promoting data scrutiny, we can harness AI's power while ensuring fair, equitable, and beneficial outcomes (Gary Marcus and Ernest Davis, 2019).

15. The Capability to Scale and Repeat

AI solutions are revolutionizing industries like manufacturing and digital services due to their scalability. These models can be deployed across multiple platforms and locations without additional training, allowing for efficient and adaptable solutions. In manufacturing, AI models can predict equipment failures, maintain quality standards, and optimize processes by analyzing data from multiple factories. In digital services, AI can provide personalized customer experiences, detect cyber threats, and prevent fraud by analyzing financial transactions from across the globe. This scalability transcends geographical boundaries and domain specificity, unlocking unprecedented potential for efficiency, adaptation, and innovation in these sectors. As AI scalability continues to evolve, we stand at the precipice of a transformative era where intelligent solutions can adapt and thrive in our ever-evolving global landscape (Burrell, J. [2016]).

The human brain, despite its remarkable intelligence, has inherent limitations that impact our thinking, decision-making, and interactions with the world. These limitations include processing boundaries, such as information overload and time constraints, and cognitive biases, which can lead to suboptimal choices. These biases include availability heuristics and confirmation bias, which can lead to clinging to inaccurate beliefs despite contradictory data. Decision fatigue and anchoring bias can also contribute to these limitations. Recognizing these limitations is a crucial step towards self-awareness and improvement. By understanding our processing boundaries and our susceptibility to cognitive biases, we can develop strategies to mitigate their impact, making more informed decisions, learning effectively, and navigating the world with greater clarity and awareness. This understanding can also lead to better collaboration with AI, where our strengths complement the efficiency and objectivity of machines (Kusiak, A. [2018]).

16. Decision-Making Affected by Emotions

Stress and emotional distress can significantly impair our ability to make sound decisions and form accurate judgments. Stress hormones like cortisol flood brain regions responsible for logical reasoning, leading to immediate reactions and cognitive biases. Emotions also trigger mental shortcuts, such as negativity bias, which can lead to pessimistic assessments. Distress narrows our attention, making it difficult to consider the full picture. Studies show that people in stressful situations make impulsive investment choices, leading to poor outcomes. AI, while excelling at numbers and identifying patterns, lacks common sense and generalization, leading to overfitting, biased outputs, and lack of context. Acknowledging our emotional biases and limitations can help us make more informed choices and develop more robust and ethical algorithms (Nickerson, R.S. [1998]). AI has revolutionized technology, but it often struggles with generalization and common sense in real-world scenarios. AI models excel at recognizing patterns and making predictions within their training data, but in the dynamic and unpredictable world, situations often deviate from the rigid data sets. This generalization gap leaves AI struggling to adapt and apply its knowledge to novel situations. Beyond generalization, AI lacks the human capacity for "common sense", the intuitive understanding of everyday situations and implicit knowledge shared amongst a community. This includes physical reasoning, social awareness, intuition, and inference. AI's dependence on data can exacerbate its limitations, leading to biases, data scarcity, and limitations in encoding human experiences and knowledge. The paper concludes that AI's limitations in generalizing beyond data and lacking "common sense" highlight the need for continued development in algorithms and training methods. The focus should shift from crunching data to fostering AI that learns and adapts like humans, incorporating elements of common sense and intuitive reasoning into its decision-making processes (Loewenstein, G. and Lerner, J.S. [2003]).

AI's capabilities are heavily reliant on data, which is crucial for its performance and limitations. The quality of AI models is directly influenced by the volume and quality of training data. Data dependence can lead to biased, incomplete, or inaccurate models, perpetuating bias and causing societal inequalities. Gathering reliable, impartial, and representative data for AI training presents challenges such as bias infiltration, data scarcity, and privacy concerns. Strategies to ensure quality and reliability include data diversity, data augmentation, and human-in-the-loop processes (LeCun, Y. et al., 2015). To address these challenges, AI development should prioritize high-quality, representative data, ongoing research into bias, privacy, and human-AI collaboration. By building a solid foundation of data integrity and ethical awareness, AI can unlock its true potential to benefit humanity in a just and equitable manner (Zittrain, J. et al., 2019).

16.1 Absence of Creativity and Emotional Intelligence

AI has made significant strides in various aspects of our lives, but it still struggles in areas that require emotional intelligence and creativity. AI's limited perception of emotions, lack of intuition, social awkwardness, and ability to build relationships are some of the areas where it falls short. AI can recognize basic emotions but lacks the deeper understanding of emotional contexts and subtleness. It also struggles with social awkwardness and understanding non-verbal cues. AI can create unique art but lacks the originality and emotional depth that characterize human-created art. It also lacks the subjective experience and emotional context that inform human creativity. AI's comprehension of text is shallow, and it lacks critical thinking skills. Future directions include integrating emotional data, developing hybrid systems, and fostering ethical AI (Bostrom, N. [2014]). AI and human intelligence are not a duel, but rather a complementary force. AI excels in data-driven processing, algorithmic prowess, and adaptive learning. However, humans possess unique advantages such as common sense, creativity, and social intelligence. By combining AI and human intelligence, we can unlock enhanced decision-making, augmented creativity, and humanized AI. This harmonious collaboration can create a symphony of intelligence, tackling challenges and creating a better world. By embracing the strengths of both, we can unleash a symphony of intelligence capable of tackling challenges and creating a future that benefits all. This conclusion emphasizes the key differences between human and AI intelligence and offers a hopeful outlook for the future (Bickmore, L.W. and Picard, R.W. [2005]).

17. Case Studies and Real-World Examples

17.1 AI Is More Advanced Than Humans

Artificial Intelligence (AI) is revolutionizing language translation by mimicking the human brain's ability to learn and understand language. Google's Neural Machine Translation (NMT) system, for example, can translate languages at lightning speed, processing millions of words in seconds. This results in real-time communication, instant access to information in foreign languages, and potential for faster translation services on a global scale. However, NMT systems still struggle with complex sentence structures, rare languages, and nuanced cultural references. The true potential lies in combining the strengths of AI and human intelligence, allowing NMT to handle the bulk of translation tasks with incredible speed and accuracy, freeing up human translators to focus on complex cases, nuanced revisions, and tasks requiring cultural understanding. The implications of these advancements extend beyond translation, bridging language barriers and unlocking access to information and collaboration across cultures (Bickmore, L.W. and Picard, R.W. [2005]).

17.2 Financial Trading with AI

High-frequency trading (HFT) is being transformed by AI algorithms, enabling real-time analysis of vast datasets, predicting market movements, and adapting to changing market conditions. AI algorithms can process vast datasets, identify hidden patterns, minimize slippage, and scale up operations. However, human oversight and expertise are crucial for understanding market psychology, interpreting news events, and adapting to unforeseen circumstances. The future of HFT lies in collaboration, where AI acts as a powerful tool augmenting human judgment and decision-making. While AI has increased market efficiency, liquidity, and speed, it has also raised concerns about market manipulation, systemic risk, and potential inequities. To ensure a fairer, more efficient, and sustainable financial system for the future, it is essential to strike a balance between embracing AI's benefits and mitigating its potential downsides. By adopting responsible AI practices and fostering human-AI collaboration, we can ensure AI's influence in HFT leads to a more equitable and sustainable financial system (Bentivoglio, P. et al., 2019).

17.3 Astronomical AI

Artificial Intelligence (AI) is revolutionizing astronomical research by aiding in groundbreaking discoveries and revealing hidden patterns. AI algorithms like NASA's Deep Learning Transit Detection (DLT) can identify exoplanet candidates with superhuman accuracy, accelerating the rate of exoplanet discoveries. AI also helps uncover hidden patterns in massive datasets from telescopes like the Hubble and Vera C. Rubin Observatory, revealing hidden populations and suggesting new galaxy types. AI's capabilities extend beyond visible light, processing radio telescope data to identify faint radio sources, revealing hidden pulsars, black holes, and the Big Bang. Collaboration, not competition, is the true power of AI, as AI acts as a tireless assistant, sifting through data and highlighting potential discoveries. The future of astronomical discoveries is bright, with AI expected to help detect fainter exoplanets, delve deeper into dark matter and dark energy, and potentially catch glimpses of the first stars ever born (Srinivas, S. [2023, July 25]).

There exist situations where human Intelligence is essential for human judgment in ethical conundrums. AI algorithms struggle to understand the nuance of ethical dilemmas, as they lack empathy and understanding for nuanced moral judgment. They also struggle to navigate complex situations, as human minds can approach problems from different angles and create innovative solutions (Hutson, R.A. et al., 2023). Accountability is crucial in moral decisions, as AI systems operate as black boxes, making it difficult to understand their decision-making process. Maintaining human values is essential, and AI should be used as a tool to empower human decision-making. AI can aid in legal systems by analyzing case histories and social policy by identifying potential biases. By fostering a collaborative approach, we can harness AI's power while safeguarding human judgment in moral decision-making (Johnson, R.E. et al., 2020).

18. Education through Human Interaction

The rise of AI and automation has led to a future where machines handle mundane tasks, leaving humans free to focus on the irreplaceable. In education and proficiency in variable environments, the human touch remains essential. AI can offer tools like adaptive practice exercises and personalized learning programs, but it cannot replicate the context-sensitive, individualized, and compassionate instruction provided by human educators. A skilled teacher can adapt their approach based on a student's emotional state, cultural background, and individual learning style, while AI algorithms struggle with unexpected learning roadblocks or hidden potential. Compassion and motivation are essential for learning, and AI can provide data-driven insights into student progress. In the unpredictable real world, human competence and agility shine in emergency response and battlefield scenarios. The future lies in harnessing the complementary strengths of humans and AI, focusing on creative problem-solving, emotional intelligence, and adaptability in unpredictable situations (Floridi, L. [2020]).

19. The Coexistence of AI and Human Intelligence

Research on Climate Change Using AI

Artificial Intelligence (AI) is a powerful tool in combating climate change. It enables human scientists to analyze complex data and build precise models for forecasting future scenarios. AI algorithms can analyze weather patterns, ocean currents, and atmospheric composition, revealing hidden patterns and relationships. Deep learning models can process vast datasets of satellite imagery and aerial photographs, providing insights into deforestation, glacial melt, and land-use changes. AI also assists in building and refining climate models, reducing uncertainty, and simulating local impacts. It can also inform mitigation and adaptation strategies, optimize renewable energy production and grid management, and identify vulnerable ecosystems. However, AI is not a replacement for human expertise. The true power lies in harnessing the unique strengths of both: human scientists providing context, interpreting data, and setting the research agenda, and AI algorithms processing massive datasets, identifying hidden patterns, and refining climate models (Bakhshi, H. and Schneider, E.W. [2023]).

20. AI for Urban Design

Artificial Intelligence (AI) is revolutionizing urban planning and development by providing data-driven insights and guiding developmental decisions. AI can analyze population density, traffic patterns, and sustainable resource management, enabling planners to understand spatial dynamics and allocate resources efficiently. It can also simulate scenarios and predict outcomes, allowing for informed decisions in urban design and development. AI can also predict disaster preparedness and resilience, ensuring citizens' safety and ensuring equitable resource distribution. Human-AI collaboration is crucial, with AI serving as a data processor and pattern identifier, while human planners provide context and interpretation. The AI revolution in urban planning is still in its infancy, but it holds the potential to personalize service delivery and provide adaptive infrastructure. By harnessing AI's data-driven insights, predictive capabilities, and optimization potential, cities can design more sustainable and livable environments. AI is not a replacement for human ingenuity, but a powerful collaborator in crafting smarter, more sustainable, and livable cities (National Academies of Sciences, Engineering, and Medicine [2019]; Jackson, M.B., 2017).

21. Personalized Health Care

Applications of AI in addressing biological challenges include drug discovery, disease diagnosis, and understanding complex biological processes. AI systems analyze vast datasets, identify patterns, and make predictions, facilitating researchers in unraveling the mysteries of the human body and other living organisms. AI and human expertise are revolutionizing medicine by enabling personalized treatment. AI can predict cancer risk, optimize drug dosage, and predict treatment response. It can analyze complex genetic data to identify individuals at an elevated risk of specific cancers, allowing for proactive screening and prevention measures. AI can also optimize drug dosage by analyzing a patient's genetic, physiological, and medical history, preventing under- or over-dosing and maximizing therapeutic benefits. It can also predict treatment response by analyzing vast datasets of clinical trials and patient records. The synergy between AI and human expertise is crucial, as AI's data processing prowess and human understanding of medical science and patient circumstances inform physicians' judgment. As AI capabilities evolve, there will be more remarkable advancements in early disease detection and prevention, targeted therapies, improved treatment efficacy, reduced side effects, and enhanced patient-physician communication and engagement. By embracing the collaborative potential of AI and human expertise, we can create a healthier and happier world for healthcare (Shafiq, M. et al., 2022).

22. AI in Typical Life

22.1 AI in Online and Retail

AI is revolutionizing retail by enhancing efficiency, allowing for automatic inventory management, personalized product recommendations, dynamic pricing, and reimagining customer experiences. AI algorithms analyze past sales data, real-time trends, and weather patterns to predict demand with accuracy, reducing costs and ensuring the appropriate products are on shelves. Personalized recommendations boost conversion rates, customer satisfaction, and brand loyalty. Dynamic pricing allows retailers to adjust prices in real time, maximizing profits and offering competitive deals. Virtual assistants and chatbots provide 24/7 customer service, while AI-powered search engines provide relevant results and intuitive filtering options. Augmented Reality and Virtual Reality (VR) offer immersive product demonstrations and virtual store tours, engaging customers and boosting sales. However, the human touch remains crucial, as the retail landscape is evolving rapidly. As AI capabilities advance, retailers can expect more innovations in predictive maintenance, hyper-personalized marketing campaigns, and delivery drones and autonomous robots (Beam, A.L. and Kohane, I.S. [2017]).

22.2 Using AI to Create Content

Artificial intelligence (AI) is transforming various fields, including video games, music composition, journalism, and farming. AI algorithms can generate branching narratives, procedural content, and dynamic NPCs, creating immersive and personalized experiences for gamers. In music composition, AI can generate new melodies, harmonies, and styles, allowing composers and musicians to spark creative inspiration. In journalism, AI can generate automated news content, allowing journalists to focus on in-depth investigations and human-interest stories. It can also help journalists verify claims and combat falsehoods by analyzing vast amounts of information. AI can also deliver personalized news delivery by analyzing user preferences and interests (McKinsey and Company [2019]). However, AI is not a replacement for human creativity. Instead, it serves as a powerful tool, providing artists, musicians, and journalists with new ways to explore their craft, push boundaries, and reach wider audiences. As AI technologies continue to evolve, more exciting possibilities will emerge across various artistic domains. AI is also making significant strides in the agricultural sector, optimizing crop yields, predicting weather patterns (Boden, M.A. [2020]).

In order to give farmers insights for increasing crop yields and decreasing waste, AI systems examine data on crop health, soil conditions, and weather patterns. This is an example of how AI is being used in precision agriculture. Summarize how AI interacts with humans and improves their capabilities across a range of disciplines to close. As you consider the partnership's future course, highlight the possible advantages of a well-rounded strategy that makes use of both artificial and human intelligence's advantages.

22.3 Wrap-Up and Prospective Routes

The interaction between AI and humans is a complex dance, enhancing each other's strengths across various disciplines. AI offers computational prowess, enabling faster and more accurate predictions, while humans can focus on higher-order thinking, creativity, and strategic decision-making. AI also offers enhanced understanding, personalization, and customization, leading to more effective outcomes. A well-rounded strategy combining AI and human intelligence is crucial for a harmonious problem-solving approach, unlocking human potential, and building trust and transparency. The future lies in embracing the transformative potential of their partnership, navigating complexities, building a more equitable world, and unlocking the full potential of human and artificial intelligence (AI Today, 2018).

22.4 Emphasizing the Equilibrium

The future of AI and human intelligence is a delicate equilibration act, where both partners enhance each other, creating a symphony of augmented capabilities. Human intelligence fosters AI's growth by providing the data compass, moral anchor, creative spark, and cognitive boost. AI amplifies human potential by analyzing vast datasets, automating repetitive tasks, and providing personalized mentoring. The equilibrium between HI and AI is not just about individual benefits but also about collective evolution. By combining human intuition and emotional intelligence with AI's computational prowess, we can tackle complex challenges like climate change or global pandemics with unprecedented effectiveness. The synergy between HI's creativity and AI's rapid processing speeds us towards breakthroughs in diverse fields. By leveraging AI's potential for personalization and adaptation, we can create a more inclusive future where everyone has the opportunity to thrive, guided by human values. The future belongs not to AI or HI alone, but to their harmonious equilibrium (Lee, I., and Seo, M.H. [2018]).

23. Possible Future AI Developments and Technological Progress

Artificial intelligence (AI) is a vast and exciting field with numerous potential advancements. These include enhanced neural networks, modular neural networks, explainable AI, evolving machine learning, and federal learning. The next generation of reinforcement learning algorithms could improve AI's ability to adapt to changing environments, while transfer learning on steroids will make learning more efficient. Federated learning will enable collaborative learning across decentralized devices, protecting individual privacy. The marriage of AI and quantum computing holds immense potential for solving problems in fields like materials science, drug discovery, and financial modeling. Natural language processing (NLP) breakthroughs will enable seamless human-computer interaction and true machine translation. Explainable robotics will ensure trust and safety in robotic systems. These advancements will have profound consequences across various domains, including automation, ethical considerations, and human-AI collaboration. As AI becomes more prevalent, it is crucial to embrace a collaborative spirit and stay informed about the evolving landscape of AI (Moravec, H. [2019]).

24. Complementarity with Society

Artificial Intelligence (AI) is rapidly integrating into various sectors, including employment, education, healthcare, entertainment, and societal shifts. AI will augment jobs, enhance productivity and efficiency, and provide new job opportunities. It will also enable personalized learning, global accessibility, advanced diagnostics and treatment, robotic-assisted surgery, drug discovery and development, and immersive entertainment experiences. In healthcare, AI algorithms can analyze medical images and data to identify diseases, personalize treatment plans, and predict potential health risks. Robotic systems guided by AI can perform minimally invasive surgeries, while AI-powered virtual assistants can monitor patients remotely and offer personalized health advice. AI can also accelerate the development of new medications and personalized therapies. In entertainment, AI algorithms can suggest movies, music, and games, thereby, creating immersive experiences. AI-powered virtual reality and augmented reality will blur the lines between the virtual and real worlds (Russell, S.J. and Tegmark, M. [2023]).

AI, a powerful technology, presents numerous ethical challenges. Privacy and surveillance concerns arise due to AI's vast data collection and analysis capabilities. Autonomy and bias issues arise from AI's potential to perpetuate biases and unfair outcomes. Moral dilemmas arise from AI judgments, particularly in autonomous vehicles and criminal justice. To address these, robust ethical standards must be developed, prioritizing transparency, fairness, and human oversight. Comprehensive legal frameworks and regulations are needed to address data privacy, bias mitigation, and liability for AI-driven harms. Public education and dialogue about AI's ethical implications are crucial. International cooperation in AI governance is necessary to address global challenges. Fair access to AI technology and benefits is also essential. Accountability and transparency mechanisms are necessary to hold developers accountable for ethical violations. Collective action is needed to shape AI into a force for good and a brighter future for all (Brynjolfsson, E., and McAfee, A. [2016]).

25. The Synergistic Potential of Human-AI Collaboration

The future of AI and humans is a dance floor where human intelligence and artificial intelligence work in synergy. AI can augment human capabilities by analyzing vast data, automating repetitive tasks, and providing real-time insights. However, AI lacks inherent ethics, and its development should align with humanity's best interests. Human creativity fuels innovation, unlocking groundbreaking solutions like sustainable energy and personalized medicine. AI can help in sustainable development, healthcare, and climate change action, but human leadership, policy changes, and collective action are crucial. Ensuring human oversight is essential, with transparency and explainability allowing humans to identify and address potential biases. AI systems should be designed with built-in human oversight mechanisms to maintain control and prevent unintended consequences. AI needs continuous learning and adaptation to changing circumstances. The future belongs to the alliance between humans and AI, leveraging human intuition and creativity to tackle complex challenges and build a brighter future (Floridi, L. [2023]).

26. Getting Ready for Workforce Development and Education in the Future

The rise of artificial intelligence (AI) is driving a shift in workforce development and education, necessitating a reskilling of skills beyond technical proficiency. Soft skills like creativity, critical thinking, communication, and emotional intelligence will be crucial for collaborating with AI and adapting to changing contexts. Lifelong learning is becoming a vital aspect of education, with individuals needing to develop adaptability and self-directed learning skills. Human-AI harmony is also essential, with AI enhancing human capabilities. The labor market will need reskilling and redistribution, and social safety nets and support systems will be crucial. The debate around Universal Basic Income (UBI) is also gaining momentum. The journey towards an AI-integrated future requires collaborative policymaking, investing in lifelong learning, and empowering individuals. By preparing our workforce and societies, we can create a more equitable, prosperous, and fulfilling world where humans and AI work in harmony (Bostrom, N. [2014]).

27. Considering How Artificial Intelligence and Human Intelligence Relate

The interaction between human and artificial intelligence (AI) presents both opportunities and challenges. AI can enhance human capabilities by boosting productivity, creativity, and problem-solving skills, while AI can tackle global challenges like climate change, poverty, and disease. It can also bridge geographical and socioeconomic divides by providing access to educational resources and personalized learning tools. However, ethical considerations such as bias, privacy, and accountability are crucial. The job landscape is also a concern, with AI automating tasks and causing fear of job displacement. To ensure a smooth transition, reskilling and upskilling initiatives, lifelong learning, and robust social safety nets are needed. Trust and understanding between humans and AI are essential, and clear communication protocols, transparency, and public education are necessary (World Economic Forum. (2020); Malone, T.W. [2018]).

Collaboration across various disciplines is essential for responsible AI development. Interdisciplinary research, open dialogue, and continuous learning are key to developing safe, reliable, and ethical AI applications. Open dialogue involving academics, decision-makers, and the public is essential for building trust and shaping AI's future. As AI evolves, so must our understanding and approach. As the curtain falls on this exploration of the intricate dance between Minds of Machine and Man, we stand at the crossroads. The path ahead shimmers with the promise of unprecedented collaboration, where AI amplifies human potential and tackles global challenges. Yet, shadows of ethical quandaries and societal shifts linger, urging us to tread carefully. This dance demands more than technological feats; it requires a delicate choreography of collective wisdom, responsible development, and open dialogue. Only then can the waltz of AI and HI resonate with the harmony of a future where technology elevates our humanity, not replaces it. Let us step forward, hand in hand, and embrace the thrilling, yet demanding, challenge of shaping a world where the brilliance of both machine and man illuminates the path to a brighter, shared tomorrow.

References

AI Today: The Rise of Machine Intelligence" by James Manyika, Michael Chui, and Michael Osborne (2018).

A.K. Sharma, "A study on the applicability of AI in Pharmaceutical Industry," 2022 1st International Conference on Computational Science and Technology (ICCST), CHENNAI, India, 2022, pp. 501–505, doi: 10.1109/ICCST55948.2022.10040339.

Akter, S., Ahamed, S. and Uysal, I. (2023). Artificial intelligence in finance: A systematic review and bibliometric analysis. arXiv preprint arXiv:2307.03565.

Alshalabi, I.A., Hamada, S.E. and Elleithy, K.M. (2013). Research Learning Theories that Entail M-Learning Education Related to Computer Science and Engineering Courses - TI Journals. *International Journal of Engineering Science*. Kolb, D.A. (1984). Experiential learning: Experience as the source of learning and development. Prentice Hall.

Amabile, T.M. (1996). Creativity in context: Update and extensions of the theory of creativity. New York: Springer..

Bakhshi, H. and Schneider, E.W. (2023). Artificial intelligence in education: An ethical and critical look. Journal of Educational Technology, 44(2), 33–46.

Beam, A.L. and Kohane, I.S. (2017). Big data in health care: The path to personalized medicine. Health affairs, 36(7), 1106–1113.

Bentivoglio, P., Bisazza, A., Ciotti, M., Gatti, M., Murgia, A. et al. (2019). On the limitations of neural machine translation. arXiv preprint arXiv:1703.03906

Bickmore, L.W. and Picard, R.W. (2005). Establishing and maintaining long-term human-computer relationships. ACM Transactions on Computer-Human Interaction, 12(2), 293–327

Boden, M.A. (2020). Artificial intelligence and creativity. Oxford University Press.

Bostrom, N. (2014). Superintelligence: Paths, dangers, strategies. Oxford University Press

Braun, K. (2011). The Prefrontal-Limbic System: Development, Neuroanatomy, Function, and Implications for Socioemotional Development. *Clinics in Perinatology*

Bruner, J.S. (1960). The process of education. Harvard University Press.

Bruner, J.S. (1990). Acts of meaning: Perspectives on the human mind. Harvard University Press.

Brynjolfsson, E. and McAfee, A. (2016). The second machine age: Work, progress, and prosperity in a time of brilliant technologies. W.W. Norton & Company.

Burrell, J. (2016). How the algorithm sees you: A critical introduction to the social psychology of the algorithm. Oxford University Press.

Calderon, T.G. and Ofobike, E. (2007). Determinants of client-initiated and auditor-initiated auditor changes. *Managerial Auditing Journal*, *23*(1), 4–25.

Carpenter, M., Nagell, K. and Tomasello, M. (1998). Social cognition, joint attention, and communicative competence from 9 to 15 months of age. Monographs of the Society for Research in Child Development, 63(4), 176. https://doi.org/10.2307/1166214

Collins, K. (2019). Creative machines: An assembly of artists and inventors in the age of AI. MIT Press.

Dewey, J. (1938). Experience and education. Macmillan.

Ericsson, K.A., Krampe, R.T. and Tesch-Römer, C. (1996). The role of deliberate practice in the acquisition of expert performance. Psychological Review, 103(3), 453–474.

Firth, R. (2012). Evolution and religion: Culture, cognition, and information flow. Cambridge University Press

Floridi, L. (2020). Artificial intelligence: A very short introduction. Oxford University Press.

Floridi, L. (2023). Artificial intelligence: A very short introduction (3rd ed.). Oxford University Press.

Gardner, H. (1983). Frames of mind: The theory of multiple intelligences. Basic Books.

Gary Marcus and Ernest Davis, “Why Machines Can’t Be Creative,” The Atlantic (2019).

Gaut, B. (n.d.). (2010): The Philosophy of Creativity. Philosophy Compass 5/12 1034–1046, 10.1111/j.1747-9991.2010.00351.

Goleman, D. (1995). Emotional intelligence: Why it can matter more than IQ. Bantam Books.

Goleman, D. (2006). Emotional intelligence: Why it can matter more than IQ. Bantam Books.

Goodfellow, I., Bengio, Y. and Courville, A. (2016). Deep learning. MIT Press.

Hamet, P. and Tremblay, J. (2017, April). Artificial intelligence in medicine. In *ScienceDirect* (Vol. 69, Supplement). https://www.sciencedirect.com/science/article/pii/S0378378220301481

Heaton, J. (2018). Deep learning with Python. Manning Publications.

Hong Zhibin, Zhe Chen, Chaohui Wang, Xue Mei, Danil Prokhorov, et al. MUlti-Store Tracker (MUSTer): a Cognitive Psychology Inspired Approach to Object Tracking. IEEE Conference on Computer Vision and Pattern Recognition (CVPR), Jun 2015, Boston, United States.

Hutson, R.A., Masters, D., Brown, M.E. and Shirley, M. (2023). AI and machine learning in astronomy. Physics Today, 76(3), 20–27.

Jackson, M.B. (2017). Big data and the future of climate change impact and adaptation. Annual Review of Environment and Resources, 42(1), 169–202.

James Manyika, Michael Chui, and Michael Osborne, “Artificial intelligence: The next frontier?”, McKinsey Global Institute (2017).

Johnson, R.E., Chen, K. and Wright, E.L. (2020). Galaxy morphology classification via machine learning: Galaxy Zoo 2. Monthly Notices of the Royal Astronomical Society, 495(3), 3255-3271.

Kusiak, A. (2018). Smart manufacturing. Springer.

LeCun, Y., Bengio, Y. and Hinton, G. (2015). Deep learning. Nature, 521(7553), 436-444.

LeDoux, J.E. (2012). A fearful brain: The neurosciences of fear and anxiety. Guilford Publications.

Lee, I. and Seo, M.H. (2018). Human-AI interaction: Development and applications. Springer.

Loewenstein, G., & Lerner, J. S. (2003). The role of affect in decision making. In R.J. Davidson, K.R. Scherer and H.H. Goldsmith (Eds.), *Handbook of affective sciences* (pp. 619–642). Oxford University Press.

Malone, T.W. (2018). Superminds: How Humans, Algorithms, and Collaborative Intelligence Are Creating the Future. Norton.

Massachusetts Institute of Technology. (n.d.). 6.S191: Introduction to Deep Learning. https://deeplearning.mit.edu/

McCarthy, J. (2005). Reflections on the history of artificial intelligence. Cognitive Science.

McKinsey and Company (2019). Artificial intelligence in retail: Opportunity and challenges.

Miau, L.D. and Ravi, S. (2023). Artificial intelligence in finance: Current applications and future potential. Journal of Risk and Financial Management, 16(4), 265.

Mitchell, M. (2019). Artificial intelligence: A guide for thinking humans. Viking.

Moravec, H. (2019). AI and the future of humanity. Springer.

National Academies of Sciences, Engineering, and Medicine (2019). Artificial intelligence for the future of weather and climate forecasting. The National Academies Press.

Neisser, U. (2014). Cognitive psychology. Psychology press.

Nickerson, R.S. (1998). Confirmation bias: A ubiquitous phenomenon in many guises. Review of General Psychology, 2(2), 175–201.

Oliveira, A.L. and Figueiredo, M.A.T. (2023). Artificial intelligence: Historical context and state of the art. History and Theory, 61(4), 107–118.

Panetta, K. (2018, July 9). Widespread artificial intelligence, biohacking, new platforms and immersive experiences dominate this year’s Gartner Hype Cycle. Gartner. https://www.gartner.com/smarterwithgartner/5-trends-emerge-in-gartner-hype-cycle-for-emerging-technologies-2018/

Piaget, J. (1970). Structuralism.

Pilot, C.V. Connected Vehicle Pilot Deployment Program: United States Department of Transportation (ITS).

Pink, D.H. (2012). Drive: The surprising truth about what motivates us. Viking.

Poore, J.A., Cullen, D.L. and Schaar, G.L. (2014). Simulation-based interprofessional education guided by Kolb’s experiential learning theory. Clinical Simulation in Nursing, 10(5), 280–287.

Roine, U., Roine, T., Salmi, J., Könönen, M., Kankaanpää, A. et al. (2015). Abnormal wiring of the connectome in adults with high-functioning autism spectrum disorder. Molecular Autism, 6(1), 65.
Russell, S.J. and Norvig, P. (2016). Artificial intelligence: A modern approach (3rd ed.). Pearson Education.
Russell, S.J. and Norvig, P. (2020). Artificial intelligence: A modern approach (4th ed.). Pearson.Shanahan, M. (2015). The technological singularity. MIT Press.
Russell, S.J. and Tegmark, M. (2023). Superintelligence: Paths, dangers, strategies (3rd ed.). Oxford University.
Russell, S.J. and Norvig, P. (2021). Artificial intelligence: A modern approach (4th ed.). Pearson.
Sadaf, M., Iqbal, Z., Javed, A.R., Saba, I., Krichen, M. et al. Connected and Automated Vehicles: Infrastructure, Applications, Security, Critical Challenges, and Future Aspects. *Technologies* 2023, *11*, 117. https://doi.org/10.3390/technologies11050117
Salovey, P. and Mayer, J.D. (1990). Emotional intelligence, imagination, and creativity. Educational Psychologist, 25(1), 115–134.
Shafiq, M., Alam, F. and Salahuddin, S. (2022). Artificial intelligence for smart city development: A systematic literature review. Journal of Cleaner Production, 290, 125817.
Siau, K.L. and Yang, Y. (2017). Impact of artificial intelligence, robotics, and machine learning on sales and marketing. In Midwest United States for Information Systems Conference Proceedings (MWAIS 2017), 48. https://aisel.aisnet.org/mwais2018/
Siemens, G. (2004). Connectivism: A learning theory for the digital age. International Journal of Learning, 10(5), 5–12.
Simon, H.A. (1996). The sciences of the artificial (3rd ed.). MIT Press.
Skinner, B.F. (1938). The behavior of organisms: An experimental analysis. Appleton-Century.
Srinivas, S. (2023, July 25). High Frequency Trading & AI.
Sternberg, R.J. (1998). Cognitive psychology (4th ed.). Harcourt Brace College Publishers.
Swan, M. (2023). AI and finance: How artificial intelligence is transforming the financial industry. Princeton University Press.
Tomasello, M. (1999). The cultural origins of human cognition. Harvard University Press.
Turing, A.M. (1950). Computing machinery and intelligence. Mind, 59(236), 433–460.
van den Heuvel, M.P. and Hulshoff Pol, H.E. (2010). Exploring the brain network: A review on resting-state fMRI functional connectivity. *European Neuropsychopharmacology*, 20(8), 519–534.
Vygotsky, L.S. (1978). Mind in society. Harvard University Press.
Watson, J.B. (1913). Psychological Review, 20(2), 151–177.
Wechsler, D. (1939). *The measurement of adult intelligence.* Williams & Wilkins Co.
Whiten, A. (2017). The evolution of human life history: A coevolutionary perspective. Proceedings of the National Academy of Sciences, 114(30), 7892–7899.
World Economic Forum. (2020). The Future of Jobs Report 2020.
Yang, L. and Duan, M. (2023). The role of emotional intelligence in EFL learners' academic literacy development. Heliyon, 9(1), e13110.
Young, J.E. (2016). The stages of memory: Reflections on memorial art, loss, and the spaces between. University of Massachusetts Press.
Zatorre, R.J. and Habib, M. (2006). Plasticity of the auditory cortex: Implications for music learning and performance. Annals of the New York Academy of Sciences, 1098(1), 157–164.
Zittrain, J. Leman, D. and Bafna, V. (2019). Data and discrimination: Three ways algorithmic systems can harm and what we can do about it. Harvard Law Review, 133(6), 1109–1240.

CHAPTER

5

Advantages and Disadvantages of Artificial Intelligence

Alex Thomas[1] and *Pratap Reddy Kuppasani*[2*]

1. Introduction

In the dynamic realm of biological sciences, the integration of Artificial Intelligence (AI) stands as a transformative force, reshaping how we approach research and analysis. This chapter explores the dual nature of this technological revolution, examining both the profound advantages and stimulating disadvantages associated with the strategic use of AI in the pursuit of biological knowledge. In the space where intelligent machines and living things come together, we have an opportunity to speed up discoveries. However, it's crucial to be cautious and carefully consider ethical concerns. This chapter explores the many aspects of AI in biological sciences and guides us through the delicate equilibrium between advancing scientific knowledge and using powerful technologies responsibly.

2. Advantages of Artificial Intelligence

2.1 Accelerating Data Analysis in Biological Research

2.1.1 Big Data Challenges

The advent of new technologies has facilitated the cost-effective generation of comprehensive profiles for biological systems, resulting in a substantial influx of biological data. This transition is propelling us into the era of "big data," underscoring the pressing need to bridge the gap between the rapid development of high-throughput technologies and our capacity to efficiently manage, analyze, and integrate this wealth of biological big data. To fully harness the potential of this information, it is crucial to foster expertise in the management and analysis of big data. As the genomic era unfolds, scientists worldwide are gaining access to these extensive datasets for their research, while clinicians and patient advocates utilize the data to tailor medical treatments. Additionally, with the gradual reduction in the cost of next-generation sequencing, the scientific community is

1 Toxway Research Solutions, Mumbai, Maharshtra, India.
2 Health Life Sciences Business Unit, Medical Electronics, Tata Elxsi Limited, Pune, maharashtra, India.
* Corresponding author: prathapkuppasani@gmail.com; alexthomas.a@gmail.com

grappling with significant challenges related to storing and analyzing high-throughput, large-scale omics data, thereby complicating the interpretation of results (Pal et al., 2020). Several models based on artificial intelligence (AI) have been developed for the processing of textual sequence data and chemical information. The dna2vec model, for instance, transforms DNA sequences into a continuous vector space of 100 dimensions. Similarly, the Mol2vec model represents chemical formulas using 300-dimensional vector embeddings, while Mut2Vec captures cancerous mutations in 300-dimensional vectors. These models are built upon established AI-based word2vec models, which encode English words into 300-dimensional vectors (Pal et al., 2020). In a comprehensive review by Eraslan et al. (2019), various deep learning techniques applied to genomics data have been highlighted. However, it's noteworthy that all the examples discussed thus far operate within a centralized environment. The current landscape lacks dependable tools in deep learning capable of functioning on distributed frameworks designed for big data. This deficiency arises from the growing complexities associated with executing computational tasks in a decentralized environment (Pal et al., 2020).

AI/ML and data analytics are gaining growing significance in biotechnology due to their capacity to analyze extensive datasets and forecast outcomes in intricate biological systems. This involves employing AI techniques to scrutinize various biological data, such as genomic and proteomic data, encompassing different omics, to enhance comprehension of the fundamental mechanisms of diseases and pinpoint potential therapeutic targets (Holzinger et al., 2023).

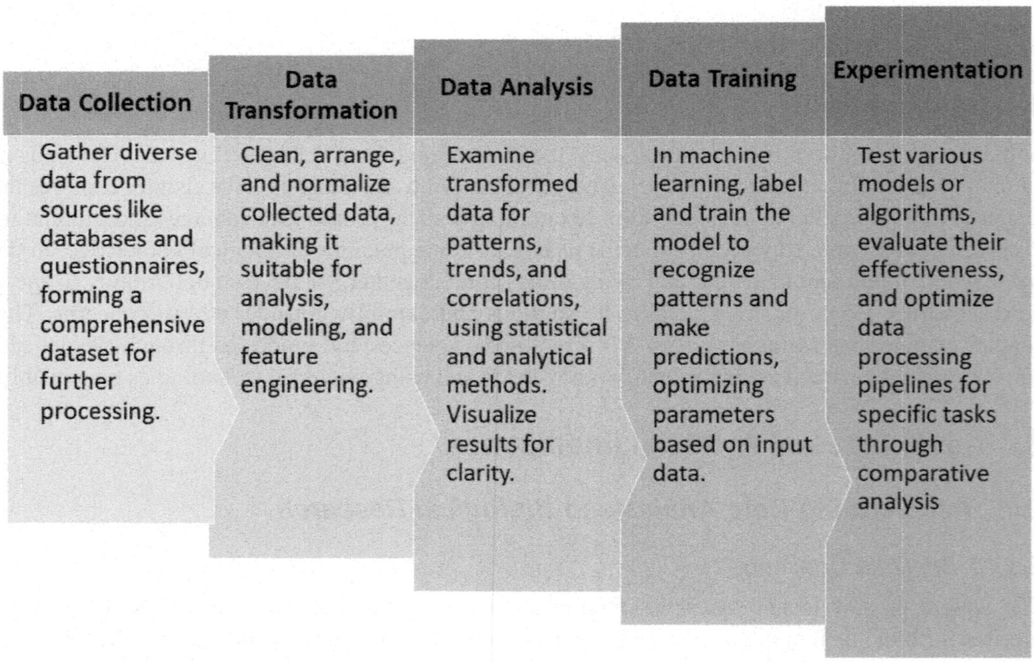

Fig. 1 Steps involved in data processing using AI *(Tetrick, 2023)*

In contemporary settings, Artificial Intelligence (AI) and Machine Learning (ML) have emerged as the predominant choices for data mining and big data analysis. Equipped with adaptive learning models, they offer solutions across various engineering applications, employing techniques like Artificial Neural Network modeling, reasoning-based decision algorithms, simulation models, DNA computing, and Quantum computing. The integration of AI in biomedical research has notably reduced the challenges associated with handling fuzzy and random data. Ongoing technological progress has further enhanced the efficacy of AI techniques in managing such data, presenting a comprehensive overview of machine learning, AI computing models, advanced data analytics, and optimization approaches in Bioengineering domains such as Drug Designing and Analysis, Medical Imaging, biologically inspired learning, and adaptive analytics (Chakraborty et al., 2017).

3. AI-Powered Data Processing

3.1 Role of AI in Extracting Meaningful Insights from Complex Datasets

At the core of this metamorphosis lies the pivotal role of data. Data serves as the lifeblood for training and evaluating AI models, forming an indispensable element in their development. The efficacy of AI models hinges on extensive datasets, enabling them to discern intricate patterns and uncover trends that elude conventional data analysis techniques. Consequently, AI models acquire the ability to learn and generate predictions founded on the wealth of data upon which they have undergone training (Aldoseri et al, 2023).

4. Data Learning Approaches

4.1 Real-time Analysis and Decision-making in Biological Research

The progress of this field in the coming years hinges significantly on the adept utilization of advanced AI solutions by biotechnology researchers. Presently, the biotechnology sector heavily depends on proficient management of data, encompassing storage, filtering, analysis, and sharing. Globally, biotechnology companies and diverse healthcare organizations manage extensive databases. Various processes such as drug manufacturing, chemical analysis of compounds, RNA and DNA sequencing, enzyme studies, and analogous biological procedures demand robust support from AI software solutions to expedite processes and minimize manual errors (Holzinger et al., 2023). The European In Vitro Diagnostics Regulation (IVDR) explicitly incorporates software and AI algorithms within its stipulations, presenting substantial challenges for in vitro diagnostics (IVD) companies utilizing AI for data analysis and decision support. Nevertheless, if ethical and legal considerations are thoroughly addressed, the potential for AI to revolutionize medical biotechnology is substantial, facilitating quicker, more accurate, and cost-effective identification and development of new drugs (Holzinger et al., 2023).

5. Revolutionizing Drug Discovery

5.1 AI-Driven Target Identification

5.1.1 Identifying Potential Drug Targets with Precision and Efficiency

Utilizing AI methods like machine learning and deep learning, large-scale genomic datasets can be employed to pinpoint potential drug targets. Through the examination of genetic data from individuals with diseases, AI algorithms can prioritize genes or proteins crucial to disease pathways, facilitating the exploration and identification of innovative therapeutic targets (Anonymous, 2023).

5.1.2 Reducing the Time and Cost of Early Drug Discovery Stages

AI can assess data on the activity of potential drugs against different targets, identifying those with the highest likelihood of effectiveness. Machine learning algorithms can predict a drug's efficacy based on its characteristics and the characteristics of the target. AI can analyze data from diverse sources, such as genomic and protein-protein interaction data, to pinpoint potential therapeutic targets for disease treatment. Machine learning algorithms can unveil patterns and correlations not immediately evident to humans (Holzinger et al., 2023).

6. Predictive Modeling for Drug Design

6.1 Utilizing AI Algorithms for Drug Design and Optimization

In the pharmaceutical industry, AI finds diverse applications, and one notable initiative is "Open Targets." This strategic effort aims to explore the intricate relationships between drug targets, diseases, and the genetic links associated with certain diseases. Another AI technique, SPIDER, is under development to elucidate the role of natural products in the realm of drug discovery. Moreover, quantitative structure–activity relationship (QSAR) studies prove instrumental in rapidly creating novel and effective medications through computer simulations. A recent study employed a QSAR model based on a radial basis function (RBF) artificial neural network (ANN), trained using the particle swarm optimization (PSO) technique, to predict the pKa values of 74 different types of drugs. In the field of medicine, three key areas of AI advancement are evident: natural language processing (NLP), machine learning (ML), and robotic process automation, each contributing significantly to the evolution of medical practices (Bhardwaj et al., 2022). The amalgamation of omics data with bioinformatics and ML holds the promise of transitioning from explanatory data to practical applications, spanning fields like medicine, agriculture, and forestry. In the realm of drug discovery and development, AI proves instrumental in analyzing extensive datasets to discern patterns and relationships that may elude human observation. Its application extends to the identification of novel drugs and drug targets, along with the optimization of existing therapeutic approaches (Holzinger et al., 2023).

6.2 Accelerating the Development of Novel Therapeutics

Utilizing AI-driven computer models trained on biological datasets and laboratory procedures, it becomes possible to design experiments aimed at optimizing particular conditions, such as enhancing the growth rate of organisms. In various engineering and scientific applications, the emphasis is often on conducting experiments and developing empirical models to enhance specific responses of interest. The use of AI to automate certain procedures that used to require hands-on, tacit knowledge could expand the pool of individuals who can participate in the biological sciences. AI finds application throughout the scientific research and development (R&D) process, encompassing tasks like designing laboratory experiments, establishing protocols for operating specific equipment, and addressing various "de-skilling" aspects of scientific research. The integration of AI with other biological technologies has the potential to reduce technical and knowledge barriers, broadening access to capabilities for a greater number of participants in the field (Kuiken, 2023).

In the era of Industry 4.0, a remarkable achievement is the capability of machines to replicate the functions of living systems, particularly human intelligence. The crucial attributes of biological systems, such as object recognition and decision-making, are now within the grasp of AI, which leverages cognitive and perceptual abilities akin to those found in living systems. This potential of AI extends to various sectors, including medical research, agriculture, and bio-based industries, contributing to a sustainable way of life. Noteworthy successes include the efficient production of bio-enzymes, highlighting how AI applications can transform industries like biotechnology by mitigating production costs, a significant challenge faced by the industry (Bhardwaj et al., 2022).

6.3 Streamlining Clinical Trials with AI

Predicting patient outcomes in clinical trials is integral to precision medicine and trial design. AI can simulate data for efficient statistical outcome measures, potentially leading to shorter trials. It aids in predicting participant outcomes, identifying fast progressors, and assessing the likelihood of trial dropouts. Machine Learning (ML) models reduce cancer mortality and offer clinical outcome predictions based on environmental and genetic attributes. In drug development, ML predicts molecular features, target sensitivity, and toxicity, influencing trial success and regulatory approval.

ML algorithms play a crucial role in reshaping clinical trial design. They assist in predicting disease progression, enabling the creation of virtual control arms, reducing budgets, and addressing ethical concerns. Recruitment challenges are addressed by AI tools matching patients with complex inclusion criteria, improving trial access and awareness. AI aids in investigator ranking, expediting site initiation, and positively impacting recruitment. In trial conduct, digital health technologies enhance safety oversight, while AI streamlines the analysis of medical images. AI identifies effect heterogeneity, imputes missing data, and supports data extraction for analysis tools. Regulatory bodies in the EU and the US are developing frameworks to assess and validate AI in clinical trials. Challenges include ethical use, governance, data inclusiveness, and access control (Askin et al., 2023).

7. Genomics and Precision Medicine

7.1 AI Applications in Genomic Sequencing

7.1.1 Enhancing the Speed and Accuracy of Genome Analysis

AI methodologies, including machine learning and deep learning, have the capacity to be employed on extensive genomic datasets for the identification of potential drug targets. Through the analysis of genetic data from individuals with diseases, AI algorithms can prioritize genes or proteins integral to disease pathways, facilitating the exploration and discovery of innovative therapeutic targets. AI can play a role in advancing the development of gene-editing technologies such as CRISPR. Through the analysis of genomic data, AI algorithms have the capability to foresee the outcomes of gene-editing experiments and enhance the design of guide RNAs, ensuring precise and efficient gene editing (Anonymous, 2023). AI is deemed crucial for analyzing the rapidly expanding genetic sequence data, as suggested by experts. The National Security Commission on Artificial Intelligence emphasizes the indispensability of AI in comprehending the intricate interactions between genetic code and biological processes. With the capability to process extensive biological data, particularly genetic sequences from diverse sources, AI plays a vital role in deciphering complex biological systems. This role has become imperative in various biological research fields, including engineering biology. For instance, researchers leverage AI to analyze genomic datasets, identifying the genetic basis of specific traits and uncovering potential genetic markers associated with those traits. AI is versatile in handling various biological data types, including sequence data, molecular structures, images, time-series, and omics data, making it an essential tool for advancing biological research and design (Kuiken, 2023).

Artificial intelligence (AI) has become an integral tool in the field of biotechnology, addressing an array of challenges. Its applications span diverse areas such as drug discovery, ensuring drug safety, advancing functional and structural genomics, facilitating proteomics, delving into metabolomics, enhancing pharmacology, exploring pharmacogenetics, and delving into pharmacogenomics. The scope of AI's impact extends to numerous other domains (Holzinger et al., 2023). The progress of this field in the coming years hinges significantly on the adept utilization of advanced AI solutions by biotechnology researchers. Presently, the biotechnology sector depends heavily on proficient management of data, encompassing storage, filtering, analysis, and sharing. Globally, biotechnology companies and diverse healthcare organizations manage extensive databases. Various processes such as drug manufacturing, chemical analysis of compounds, RNA and DNA sequencing, enzyme studies, and analogous biological procedures demand robust support from AI software solutions to expedite processes and minimize manual errors (Holzinger et al., 2023).

7.2 Personalized Medicine and Tailored Treatment Approaches

Emerging research, exemplified by a study featured in Nature Medicine in 2023 titled "Enhancing Metabolic Health through Personalized Exercise Prescription Using Muscle Response Biomarkers," reveals a groundbreaking shift toward personalized exercise plans tailored to genetic data. The findings highlight a transformative trajectory, envisioning a future where our genetic makeup and gut ecosystem serve as guiding principles for achieving customized well-being, signaling a paradigm shift in healthcare approaches (Rakshita, 2024).

Personalized medicine can benefit from AI by harnessing DNA sequencing data. Through the integration of genomic information with patient health records, AI algorithms have the capability to pinpoint genetic predispositions to diseases, anticipate responses to treatments, and offer customized healthcare recommendations (Anonymous, 2023). In the domain of personalized medicine, AI plays a pivotal role in scrutinizing an individual's genomic and other health-related data. This analysis facilitates the creation of personalized treatment plans customized to meet specific needs. Machine learning algorithms are employed to predict an individual's response to a particular treatment and identify potential adverse reactions (Holzinger et al., 2023).

The implementation of an AI-driven personalized medicine system not only lowers treatment expenses but also mitigates drug side effects for patients. Beyond time savings and enhanced patient care, AI streamlines processes such as gene editing, radiography, and drug management planning. Additionally, evidence-based clinical decision support systems can enhance electronic health records (EHRs) (Bhardwaj et al., 2022).

AI finds application in predicting post-surgery outcomes, particularly in facial and cleft palate procedures. It aids in evaluating and visualizing the expected results of these surgeries (NIH, 2023).

7.3 Clinical Care

Within the healthcare sector, AI plays a significant role in clinical decision support systems. These systems leverage health observations and case knowledge to assist healthcare professionals in making informed treatment decisions (NIH, 2023).

7.4 Telehealth

The surge in telehealth persists, providing convenient and easily accessible healthcare via platforms such as Practo and Zyla Health. A study published in The BMJ in 2023 revealed that telehealth consultations for chronic conditions were equally effective compared to in-person visits. As telehealth becomes more integrated into healthcare systems, it democratically extends access to quality healthcare, reaching remote communities and individuals facing mobility constraints (Rakshita, 2024). Wearable devices enable continuous patient monitoring, allowing the detection of physiological changes. This constant surveillance serves as an early warning system, particularly in anticipating events like asthma attacks (NIH, 2023). In an effort to improve the management of blood glucose levels, researchers are developing machine learning models utilizing data collected from wearable sensors. Advanced portable sensing technologies offer continuous monitoring of various parameters, such as heart rate, skin conductance, temperature, and body movements. This data is then utilized to train an artificial intelligence network, enabling the prediction of impending changes in blood glucose levels. The proactive identification and prevention of blood glucose control issues are expected to enhance patient safety and mitigate the risk of costly complications (NIH, 2023).

In a bid to alleviate the widespread issue of low back pain, researchers are actively working on the development of smart, cyber-physically assistive clothing (CPAC). Recognizing that forces exerted on back muscles and discs during routine activities pose significant risks for back pain and injury, the research team is compiling a comprehensive public dataset comprising over 500 movements from each participant. This dataset will be utilized to train a machine learning algorithm, informing the creation of assistive clothing capable of identifying unsafe conditions and intervening to safeguard the well-being of the lower back. The overarching goal is to establish smart clothing that can monitor lumbar loading, instill safe movement patterns, actively assist wearers in minimizing the occurrence of low back pain, and consequently, reduce healthcare expenses and productivity losses associated with missed work (NIH, 2023).

8. Enhanced Imaging and Diagnostics

8.1 AI in Advanced Imaging Techniques

8.1.1 Improving Resolution and Interpretation of Biological Imaging

In Radiology, AI offers a valuable capability in interpreting imaging results, enhancing the detection of subtle changes in images that may go unnoticed by clinicians (NIH, 2023).

AI can scrutinize medical images like CT scans and MRI images to detect abnormalities and diagnose diseases. Deep learning algorithms can automatically segment and classify structures in medical images (Holzinger et al., 2023).

8.1.2 Early Detection of Diseases Through AI-enhanced Diagnostics

Advancements in medical science and biotechnology have paved the way for the development of medications and antibiotics, and the pharmaceutical industry stands to gain significantly from the vast potential of AI. AI facilitates the discovery of novel therapeutic molecules based on known target structures. Particularly, machine learning (ML), a subset of AI, is widely employed in disease diagnosis, leveraging diagnostic testing outcomes to enhance result accuracy. AI addresses complex challenges in quantitative and predictive epidemiology, precision-based medicines, and host–pathogen interactions. It contributes to disease detection, diagnosis, and makes computer code more accessible to non-technical individuals. In the realm of medical science, AI holds promise for early disease prediction and identification, enabling precise treatment even in asymptomatic conditions. This not only has the potential to save lives but also to reduce medical costs significantly (Bhardwaj et al., 2022).

The incorporation of artificial intelligence (AI) into the realm of healthcare is transforming futuristic concepts into tangible realities. Noteworthy examples include Qure.ai, an AI-driven platform specializing in cardiovascular disease diagnosis. This platform leverages AI to analyze ECGs and echocardiograms, predicting heart failure risks and tailoring treatment plans to individual patients. A Lancet Digital Health study from 2023, titled "Prospective Multicenter Study on Deep Learning–Based Breast Cancer Detection on Mammograms," showcased an AI-based algorithm surpassing human doctors in the accurate diagnosis of breast cancer from mammograms. The continuous evolution of AI in healthcare is giving rise to commonplace features such as personalized treatment strategies, proactive disease detection, and AI-driven virtual therapists, blurring the traditional boundaries between humans and machines in the healthcare domain (Rakshita, 2024).

Predictive epidemiology, individual-based precision medicine, and the analysis of host–pathogen interactions stand out as research areas poised to benefit from breakthroughs in machine and deep learning. These approaches improve disease diagnosis, individual case identification, result in more accurate forecasts, fewer errors, faster decision-making, and enhanced risk analysis. The increasing complexity of tissue biomarkers and their evaluations is effectively managed with AI-based techniques. These AI-driven biomarkers assist physicians in predicting and analyzing diagnoses, patient responses to treatment, and patient survival. Knowledge representation and reasoning modeling enable the creation of more realistic models for complex socio-biological systems. ML-based methods further enhance the efficiency and reliability of epidemiological models. Notably, ML advancements have led to the development of algorithmic program-based models incorporating ten cellular parameters, accurately distinguishing between benign and malignant tumors (Bhardwaj et al., 2022).

AI can analyze data from diverse sources, including electronic health records and wearable devices, to predict individual health outcomes. Machine learning algorithms can forecast the likelihood of developing a specific disease or the effectiveness of a particular treatment (Holzinger et al., 2023). Machine learning algorithms can undergo training on extensive genomic datasets to aid in the diagnosis of genetic diseases. Through the comparison of an individual's DNA sequence with reference genomes, AI systems can discern disease-causing mutations or genetic variations associated with specific conditions (Anonymous, 2023). In a recent study, the application of an AI-based approach demonstrated notable benefits in the early identification, diagnosis, prognosis, and treatment of myopia. Deep learning algorithms, particularly in cardiology, dermatology, and oncology, have exhibited superior diagnostic capabilities compared to physicians. For instance, computer algorithms achieved over 91 percent accuracy in detecting metastatic breast cancer in sentinel lymph node biopsies from full slide images. This accuracy rate increased to 99.5 percent with the incorporation of physician inputs. An established application of AI in risk analysis involves diagnosing heart malfunctions through cardiovascular imaging. This encompasses automated monitoring of deviations from normal conditions based on image processing, myocardial function assessment, and detection and analysis of coronary atherosclerotic plaques. Notably, the YOLOV3 algorithm was employed for AI-based medical image segmentation, specifically for 3D printing and naked-eye 3D visualization to detect the prostate in T2-weighted MRI images (AIMIS3D) (Bhardwaj et al., 2022).

The process of neurological degeneration associated with AD initiates long before the manifestation of clinical symptoms. Functional MRI (fMRI) neuroimaging data provides valuable information capable of identifying changes in brain tissue during the early phases of AD, presenting an opportunity for early detection and intervention. Researchers are integrating the sensitivity of fMRI in capturing subtle brain changes with the analytical power of machine learning to examine multiple brain changes over time. This innovative approach aims to enhance the early detection of AD and other neurological disorders such as schizophrenia, autism, and multiple sclerosis (NIH, 2023). Scientists are working on an innovative image scanning system designed to significantly improve the sensitivity and specificity of detecting colon cancers. The research involves the development of deep neural networks capable of analyzing a broader field within radiographic images obtained during surgery. These extended scans encompass not only suspected lesion areas but also a more comprehensive view of surrounding tissue. The neural networks are programmed to compare patient images with a database of previously diagnosed cases. It is anticipated that this system will surpass the capabilities of existing computer-aided diagnostic systems in identifying colorectal lesions. Widespread adoption could revolutionize cancer prevention and early diagnosis (NIH, 2023).

It is reported that an AI-based predictive model was able to predict the progression of AMD (Age-related macular degeneration) with high accuracy. The AI-model was able to accurately detect and predict the progression of Alzheimer's disease with an accuracy of 93.95% in the first layer and 87.08% in the second layer. The AI based model achieved predicting severe COVID-19 cases with an accuracy of 70–80%. An accuracy of 93% was achieved in predicting the survival of ovarian cancer patients, and 77% accuracy was achieved in predicting the surgical outcome. LCP-CNN (Lung Cancer Prediction Convolutional Neural Network) was able to predict the malignancy of pulmonary nodules with higher accuracy and lower false negative results than the Brock model. IAT-BPNN was able to predict an influenza-like illness in a large population size with a high accuracy (Bhardwaj et al., 2022). AI offers efficient analysis of various variables, such as determining antibiotic resistance in specific conditions. This analysis can support physicians, leading to a significant reduction in unnecessary testing and overall costs in medical care. In the realm of diagnostics and disease prediction, AI proves invaluable in analyzing data from diverse sources, including electronic health records and wearable devices. By identifying patterns and correlations, AI can signal the potential presence of a specific disease. This capability enhances diagnostic accuracy and enables timely interventions to impede the progression of diseases (Holzinger et al., 2023).

Wearable devices are evolving into health predictors, surpassing basic step counting to actively monitor real-time sleep patterns, stress levels, and heart health. Leading companies like Fitbit and Garmin offer sophisticated devices that provide comprehensive insights into overall well-being. A study published in JAMA Internal Medicine in 2023, titled "Wearable Device–Based Heart Rate Variability for the Early Detection of Heart Failure," demonstrated the effectiveness of wearable devices in early heart failure detection through heart rate variability monitoring. Transforming into comprehensive health dashboards, wearables now offer continuous feedback, empowering individuals to make well-informed decisions about their overall health (Rakshita, 2024).

8.2 Integrating AI for Improved Accuracy

8.2.1 Reducing Diagnostic Errors in Medical Imaging

In the field of biomedical image analysis, AI is employed to scrutinize medical images like CT scans and MRI images. Its applications encompass the identification of abnormalities and disease diagnosis. This involves leveraging deep learning algorithms to autonomously segment and categorize structures within medical images (Holzinger et al., 2023).

8.2.2 AI Applications in Pathology and Radiology

With the recent achievements of artificial intelligence (AI) in computer vision applications, pathologists are optimistic about AI's potential to aid them in various digital pathology tasks. The significant progress in deep learning has facilitated a harmonious integration with AI, enabling image-based diagnosis within the realm of digital pathology. Ongoing initiatives are focused on creating AI-based tools aimed at saving time for pathologists and minimizing errors in the diagnostic process (Kim et al., 2022). Digital pathology has emerged as a crucial field for enhancing diagnostic accuracy, particularly in addressing challenges associated with intra- and interobserver disagreements among pathologists during traditional visual examinations of pathology slides. Automated computational methods have become essential for pathological diagnosis, with whole-slide imaging (WSI) standing out as an advanced technology in pathology. WSI demonstrates excellent correlation with traditional light microscopy diagnosis, offering advantages such as portability, easy image sharing, and efficient task management. This digital pathology environment has given rise to computational pathology (CPATH), enabling direct disease investigation through computational analysis of pathology slide images. Artificial intelligence (AI) approaches, often utilize deep learning, play a pivotal role in analyzing entire slide images computationally, surpassing the capabilities of human observation and setting the current standard in digital pathology (Kim et al., 2022).

Conventional pathological digital image machine learning methods require trained pathologists to manually categorize abnormal picture attributes before incorporating them into algorithms, leading to time-consuming and labor-intensive processes with significant disagreements among pathologists. The translation of human-extracted visual features into numerical forms for computer algorithms proved challenging, especially in complex diseases. Previous studies relied on handcrafted features for algorithm development, resulting in limited generalization and high false-positive rates. Deep learning has revolutionized this landscape, allowing computers to autonomously extract feature vectors from pathology image data and construct optimal algorithms. It has demonstrated superior performance, even outperforming physicians in some cases, marking it a cutting-edge machine learning method in medical clinical practice. Diverse deep architectures, trained with extensive image datasets, contribute to biological informatics discoveries and exceptional object recognition (Kim et al., 2022).

AI holds significant potential in transforming pathology practice, offering the promise of error reduction, improved reproducibility, and enhanced expert communication, addressing challenges inherent in traditional microscopic glass slides. The current focus in AI application trends emphasizes affordability, practicality, interoperability, explainability, generalizability, manageability, and reimbursement feasibility. Despite the optimism surrounding AI, currently there are limited AI-driven software tools in pathology, emphasizing the need for pathologist involvement in algorithm development to ensure efficacy. The quantitative nature of computational pathology (CPATH) opens avenues for transforming pathology laboratory and clinical practices, enabling case stratification, expedited review, and the development of meaningful models for treatment guidance and predictive insights. As the field evolves, collaboration among researchers, doctors, industry, regulatory agencies, and patient advocacy groups is crucial for refining AI applications in pathology and establishing robust and standardized practices across multiple sites (Kim et al., 2022).

9. Automation in Biological Laboratories

9.1 Robotics and AI Integration

9.1.1 Streamlining Laboratory Workflows for Efficiency

Robotics and Robotic Process Automation (RPA) have transformed laboratory workflows and sample handling, ensuring precise and consistent task execution while minimizing human error and variability. Robotic Process Automation takes automation to the next level by efficiently automating repetitive tasks across various software systems. The integration of RPA in laboratories facilitates automation of data entry, report generation, and seamless data transfer between different software applications, elevating overall efficiency and accuracy. Artificial Intelligence (AI) and Machine Learning (ML) algorithms play a vital role in data analysis and interpretation within laboratory settings. Utilizing these technologies enables laboratories to analyze complex datasets more effectively, identify patterns, and extract valuable insights. The implementation of intelligent automation, driven by AI and ML, enables real-time monitoring and control of laboratory processes. AI systems, continuously learning from data, can optimize experimental parameters, detect anomalies, and provide timely alerts, contributing to enhanced efficiency and accuracy in laboratory automation (Kewaunee International Group, 2024).

The future of laboratory automation will be instrumental in enhancing efficiency and accuracy in research and development processes. Current technologies, such as robotic systems and workflow automation tools, offer significant benefits but come with limitations. Process optimization, data integration, and standardization are crucial for further improvement. Emerging technologies like robotics, Artificial Intelligence (AI), Machine Learning (ML), and the Internet of Things (IoT) have transformative potential. Advancements in laboratory robotics, high-throughput screening, and Laboratory Information Management Systems (LIMS) demonstrate positive impacts on efficiency and accuracy. Future trends envision autonomous laboratories combining robotics with AI capabilities and the integration of cloud computing and big data analytics. Personalized and precision medicine is advancing through automation, yet challenges like data privacy and regulatory compliance must be addressed. Staying informed and embracing new technologies is essential for laboratory professionals to drive the future of automation and enhance research outcomes (Kewaunee International Group, 2024).

9.2 Enhancing Reproducibility and Reliability in Experiments

9.2.1 Minimization of Human Errors

One notable advantage of Artificial Intelligence lies in its capacity to markedly diminish errors, thereby enhancing overall accuracy and precision. AI-driven decision-making relies on a foundation of pre-existing information and specific algorithms. With meticulous programming, the occurrence of errors can be virtually eliminated. Illustratively, the application of AI in robotic surgery systems exemplifies this reduction in human errors. These systems adeptly execute intricate procedures with exceptional precision, mitigating the potential for human errors and thereby elevating the level of patient safety within the realm of healthcare (Duggal, 2023).

In summary, the field of biological research is experiencing a profound transformation driven by the integration of artificial intelligence (AI) and machine learning (ML) technologies. The emergence of big data in biological research necessitates efficient strategies for managing, analyzing, and integrating vast datasets, and AI proves to be a key solution. AI models, such as dna2vec and Mol2vec, demonstrate the ability to process textual sequence data and chemical information, offering new insights into biological systems. Despite challenges in developing dependable deep learning tools for distributed frameworks tailored for big data, AI plays a growing role in various domains of biotechnology, from drug designing to medical imaging. The application of AI in drug discovery, personalized medicine, clinical care, and laboratory automation showcases its potential to revolutionize scientific exploration, drug development, and healthcare. As we enter the era of Industry 4.0, the combination of AI with biological technologies holds promise for overcoming barriers and advancing our understanding of complex biological phenomena. Despite existing challenges, the accelerated data analysis facilitated by AI is reshaping the landscape of biological sciences, offering new avenues for innovation and progress.

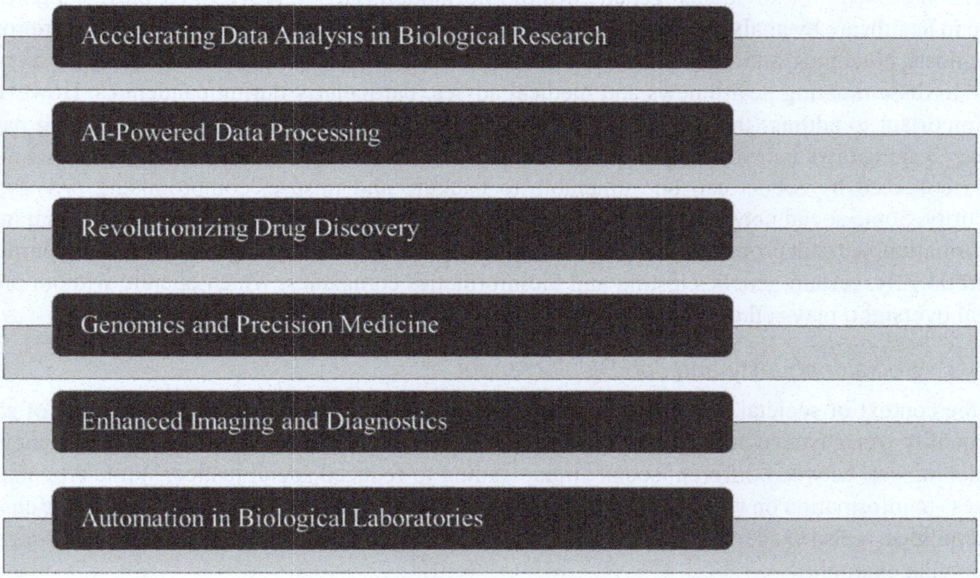

Fig. 2 Advantages of AI

9.3 Disadvantages of Artificial Intelligence

9.3.1 Ethical Concerns in Artificial Intelligence

Artificial Intelligence (AI) applications have revolutionized the healthcare industry, transforming various aspects such as medical imaging, electronic medical records (EMR), laboratory diagnosis, treatment, and even augmenting the intelligence of physicians. Additionally, AI has played a significant role in new drug discovery, preventive and precision medicine, extensive data analysis, and streamlining processes for health organizations. However, the integration of AI into healthcare also presents ethical and legal challenges. One of the key challenges is the accessibility of AI technologies. While AI has made remarkable progress in certain communities, many low-income, developing countries still lack access to these advanced technologies. This creates a disparity in healthcare capabilities and resources. Furthermore, ethical dilemmas arise in the use of AI in healthcare. Issues such as privacy and data protection, informed consent, social gaps, medical consultation, empathy, and sympathy need to be carefully addressed. These challenges require practitioners and specialists to consider the four principles of medical ethics - autonomy, beneficence, nonmaleficence, and justice-in all aspects of healthcare when integrating artificial intelligence into the system. In conclusion, AI applications have brought significant advancements to the healthcare field, but they also bring forth ethical and legal considerations. It is crucial for healthcare professionals to navigate these challenges and ensure that AI is integrated in a manner that upholds the principles of medical ethics and promotes equitable access to healthcare technologies (Scalzo, 2020; Pasquale, 2020; Farhud et al., 1970; Farhud, 2019; Farhud and Zokaei, 2019).

9.3.2 Privacy and Data Security Risks

The General Data Protection Regulation (GDPR) was initially implemented by the European Union (EU) to amend privacy legislation in various countries, including the US and Canada. These regulations aim to protect personal data and activities of foreign communities and companies by ensuring that data processors or controllers based in the EU provide sufficient protection for the information of individuals (EU Regulation 2016/679, 2016)). In the United States, the Genetic Information Non-discrimination Acts (GINA) prohibits employers from making discriminatory decisions based on individuals' genetic health information (GINA: ASHG). AI plays a significant role in healthcare by analyzing consumer health data and medical device images, thereby improving diagnosis, outcomes, and accelerating health research. Social media, as part of AI, also play a crucial role in disseminating health news and medical advice, particularly during pandemics. However, it is important to address the potential risks associated with AI, particularly in safeguarding patient data. Existing laws in healthcare may not be adequate to protect individuals' health data, as clinical data collected by robots can be vulnerable to hacking and misuse, compromising privacy and security. Some social networks gather and store large amounts of user data, including mental health information, without proper consent, which can be exploited for marketing and advertising purposes. Additionally, certain genetics testing and bioinformatics companies, which operate without proper legal oversight, may sell customer data to pharmaceutical and biotechnology companies.

9.3.3 Social Inequality and Job Displacement

In the context of societal development alongside AI, one significant concern is the issue of social inequality. As advancements in AI continue to unfold, societies worldwide are experiencing a widening gap between different social strata, leading to reduced social justice. While AI enhances access to information on various subjects such as science, technology, global events, climate change, and politics, it also exacerbates social disparities (Nordling, 2019). This can be observed through the following examples:

1. Automation and growth of advanced economies have contributed to an increased divide between developing and advanced countries. As robots become more prevalent and sophisticated, many individuals find themselves displaced from their jobs. This phenomenon affects various professions, including bookkeepers and managers within different communities. The rise of automated systems often leads to job losses and a significant decline in salaries.
2. The emergence of surgical robots and robotic nurses within the healthcare sector poses a threat to the future employment prospects of surgeons and nurses. These automated systems have the potential to perform surgical procedures and provide patient care, potentially replacing human professionals in these roles.
3. It is crucial to address these social gap issues associated with AI development to ensure a more equitable and inclusive society. By acknowledging and actively working towards mitigating these challenges, we can strive for a future where the benefits of AI are accessible to all members of society, without exacerbating social inequalities.

10. Human Emotions

In the realm of healthcare, the integration of artificial intelligence (AI) across all areas presents challenges that may appear difficult or even impossible to overcome. The unique nature of human emotions suggests that the evolution of human and medical robots together may not occur within a short timeframe. In this context, healthcare professionals, including physicians, must rely on consultation with their colleagues, a process that may not be feasible within autonomous or robotic systems. Furthermore, it is unlikely that patients will readily accept "machine-human" medical relationships as a substitute for "human-human" interactions. The expectation for doctors and nurses to provide treatment within an empathetic and compassionate environment plays a significant role in the healing process of patients. Robotic physicians and nurses are unable to replicate human attributes such as compassion, resulting in a loss of empathy, kindness, and appropriate behavior when patients interact with them. This absence of human qualities represents one of the most significant drawbacks of artificial intelligence in the field of medical science.

Addressing these challenges requires careful consideration and balancing of the benefits and limitations of AI in healthcare. While AI can offer valuable support and advancements, it is crucial to ensure that the human element, including empathy and compassion, remains integral to the patient experience. By striking a balance between technological advancements and human-centered care, we can strive for a healthcare system that maximizes the benefits of AI while preserving the essential human qualities that contribute to effective medical treatment. For example, in the case of Obstetrics and Gynecology, the essence of any clinical examination lies in the cultivation of compassion and empathy, qualities that cannot be replicated by robotic doctors. The interactions between healthcare professionals and children in medical settings are often fraught with fear and anxiety. The typical behavioral responses exhibited by children, such as lack of cooperation, withdrawal, and even uncontrollable aggression, pose significant challenges when confronted with the prospect of interaction with the emerging field of robotic medicine. The integration of medical robots, particularly in psychiatric hospitals, raises concerns about the potential adverse effects on patients grappling with severe psychiatric disorders (The Guardian, 2020).

10.1 Potential Risks and Dangers

10.1.1 Threats associated with Misuse of Artificial Intelligence

In this segment, we outline a couple of categories of threats associated with the misuse of AI, encompassing deliberate actions, negligence, accidental occurrences, and the failure to anticipate and adapt to AI's transformative impacts on society. The initial set of threats arises from AI's remarkable capacity to swiftly clean, organize, and analyze vast datasets containing personal information, including images captured by omnipresent cameras. This capability allows for the creation of highly personalized and targeted marketing and information campaigns, as well as the expansion of extensive surveillance systems. While this potential can be harnessed for positive purposes, such as enhancing access to information or countering terrorism, it also poses significant risks when misused (Federspiel et al., 2023). The utilization of AI-driven capabilities to generate commercial revenue for platforms like social media has contributed to the escalation of polarization and extremist views globally. Moreover, this power has been exploited by commercial entities to establish a formidable personalized marketing infrastructure capable of manipulating consumer behavior. Experimental evidence reveals the substantial impact of AI on social media platforms, providing political candidates with a potent tool for manipulation, as witnessed in various elections worldwide. Instances of AI-driven subversion include the 2013 and 2017 Kenyan elections, the 2016 US presidential election, and the 2017 French presidential election (Lorenz-Spreen et al., 2023; Agudo and Matute, 2021; Bond et al., 2012).

When coupled with the advancing ability to distort reality through deep fakes, AI-driven information systems may further jeopardize democracy by eroding trust, fostering social division, and causing public health ramifications. Governments and powerful entities can also exploit AI-driven surveillance to directly control and oppress individuals, exemplified by China's Social Credit System. This system combines facial recognition technology with the analysis of extensive personal data to evaluate individual behavior and trustworthiness, resulting in automatic sanctions for perceived misconduct. Such AI applications may exacerbate social and health inequalities, perpetuating existing socioeconomic disparities. The proliferation of AI surveillance is not limited to China, with at least 75 countries, spanning various political systems, expanding similar systems. The formidable capabilities of AI make it easier for authoritarian regimes to be established or strengthened, enabling targeted persecution and oppression (Creemers, 2018; Federspiel et al., 2023). Another set of threats revolves around the development of Lethal Autonomous Weapon Systems (LAWS). While AI applications in military and defense systems can have security and peace-promoting functions, the risks associated with LAWS outweigh potential benefits. LAWS are autonomous in their ability to locate, select, and engage human targets without human supervision, constituting the third revolution in warfare after gunpowder and nuclear arms. These weapons include explosives attached to small, mobile devices with self-piloting capabilities, perception, and navigation skills. The potential mass production and deployment of such weapons raise concerns about their potential use in industrial-scale killings (Federspiel et al., 2023; Javorsky et al., 2019). Similar to chemical, biological, and nuclear weapons, LAWS present a new form of mass destruction, one that is relatively inexpensive and capable of selective targeting. This poses profound implications for the conduct of armed conflict and overall international, national, and personal security. Ongoing debates revolve around strategies to prevent the proliferation of LAWS and whether these systems can be safeguarded against cyber-infiltration or inadvertent and intentional misuse.

10.1.2 Biosafety and Biosecurity Concerns

In the context of utilizing GenAI for biological design, there are concerns about its potential repurposing or misuse, leading to heightened biosafety and biosecurity apprehensions (Sean Ekins et al., 2023). As part of an intriguing "thought experiment" with potential implications for international agreements such as the Chemical and Biological Weapons Conventions (Spiez Convergence Conference, 2021), Collaborations Pharmaceuticals, Inc. decided to repurpose its AI model for drug development. Instead of designing molecules to counteract toxicity, the AI model was retrained to specifically design molecules for toxicity (Fabio Urbina et al., 2021). Ordinarily, the AI model penalizes predicted toxicity while rewarding predicted target activity (Fabio Urbina et al., 2022). However, by inverting this logic, the AI model was directed to simultaneously favor toxicity and bioactivity (Congressional Research Service, 2023). The retrained AI model was then exposed to a public database of drug-like molecules and their corresponding bioactivities. Remarkably, within a span of less than six hours, the AI model generated a staggering 40,000 molecules that aligned with the predefined thresholds for both toxicity and bioactivity. Notably, this set of generated molecules included known chemical warfare agents like the nerve agent VX, as well as novel compounds predicted to exhibit even higher toxicities. It is crucial to emphasize that these new molecules were not part of the datasets used to train the AI model, and despite their generation, there was no attempt by the company to synthesize or produce them.

The capability of AI to create molecules with similar properties but distinct chemical structures or DNA sequences raises legitimate concerns. There is a worry that existing detection methods and screening protocols, relying on specific DNA sequences or other identifiable features, might struggle to identify these newly generated toxins. For instance, researchers postulate that using AI, they could computationally alter the active site of a known toxin like ricin, removing its surrounding stabilizing protein structure. By then employing a generative algorithm to envision and imprint a new proteinaceous structure around the active site, they could potentially create a molecule that evades current screening methods reliant on DNA sequences (HHS or USDA: potential threat to public health and safety; (Ekins et al., 2024); International Gene Synthesis Consortium). It's important to note, however, that the mere use of AI to identify and design novel molecules or DNA sequences doesn't guarantee their practical synthesis or the creation of a functional living organism based on AI-generated components. Overcoming additional scientific, technical, and ethical challenges, such as synthesizing the DNA and ensuring its proper functioning in a living organism, is essential to actualize the biosecurity risks associated with this technology (John et al., 2020).

11. Security Issues

In a perspectives report by RAND authored by Osoba and Welser (2017), a comprehensive overview of security concerns associated with AI is presented. The report underscores the risks linked to fully automated decision-making, emphasizing the potential for costly errors and even fatalities. It raises alarms about the utilization of AI weapons without human intervention, highlighting the need for ethical considerations in the deployment of such technology. Addressing issues related to AI vulnerabilities in cybersecurity, the report sheds light on the potential risks posed by the application of AI to surveillance and cybersecurity for national security. One notable concern is the emergence of a new attack vector based on 'data diet vulnerability,' indicating the susceptibility of AI systems to malicious exploitation.

The report further delves into the use of network intervention methods by foreign-deployed AI, introducing a larger scale and more strategic version of the current advanced targeting of political messages on social media. These sophisticated methods underscore the evolving landscape of cyber threats and the potential misuse of AI in influencing public opinion and political processes (Rodrigues, 2020).

Domestic security-related issues are also identified in the report, particularly the increasing deployment of artificial agents for civilian surveillance by governments, exemplified by predictive policing algorithms. Critically, these applications are acknowledged for their potential infringement on fundamental citizens' rights, as highlighted by Couchman (2019). The significance of these issues is underscored by their capacity to expose critical infrastructures to harm, leading to severe societal and individual impacts. The report emphasizes the potential threats to life, human security, and access to resources that may arise from the misuse or inadequately regulated deployment of AI in security contexts. Moreover, the report draws attention to the concealed nature of cybersecurity vulnerabilities, which often become apparent only after damage has already occurred. This delayed awareness further compounds the challenges associated with addressing and mitigating the impacts of AI-related security breaches (Rodrigues, 2020).

11.1 Concerns Associated with Clinical Implementation of AI

The primary challenge hindering the successful deployment of AI-based medications lies in the absence of empirical data validated through planned clinical trials (Khan et al., 2023). The majority of AI research has been conducted within business settings, leaving a significant gap in our understanding of its impact on patient outcomes. Notably, healthcare AI studies have predominantly taken place in non-clinical environments, making it difficult to generalize research findings. Traditional gold-standard medical practices, such as randomized controlled studies, have struggled to demonstrate the benefits of AI in healthcare due to the lack of practical data and the uneven quality of research. This has led to hesitancy among businesses to implement AI-based solutions (Alami et al., 2021). The potential integration of artificial intelligence into medical processes for increased efficiency has been hindered by various factors. One key consideration is the need for AI-based treatments to seamlessly integrate into existing workflows without impeding clinicians during examinations or when exploring electronic medical data. The investment of time and resources required to train medical professionals to effectively use AI technology further adds to the reluctance in adopting these solutions. While there have been a few instances of successful integration of AI into clinical therapy, they mostly remain in the experimental phase, emphasizing the challenges faced in practical implementation (Denti and Hemlin, 2012).

Stakeholder participation during the development phase emerges as a critical barrier to the successful integration of AI in healthcare. In numerous examples of innovation adoption, the lack of input from a diverse range of individuals has impeded the creation of solutions that can be seamlessly incorporated into clinical practice. Despite significant advancements in AI spurred by events like the SARS and Ebola pandemics, their practical utility in healthcare is limited by the necessity for seamless integration into existing procedures. This is especially crucial to avoid confusion or slowdowns for clinicians lacking training in AI. Additionally, clinical research faces challenges related to the algorithms supporting these advancements (Damschroder et al., 2009; Davenport and Kalakota, 2019).

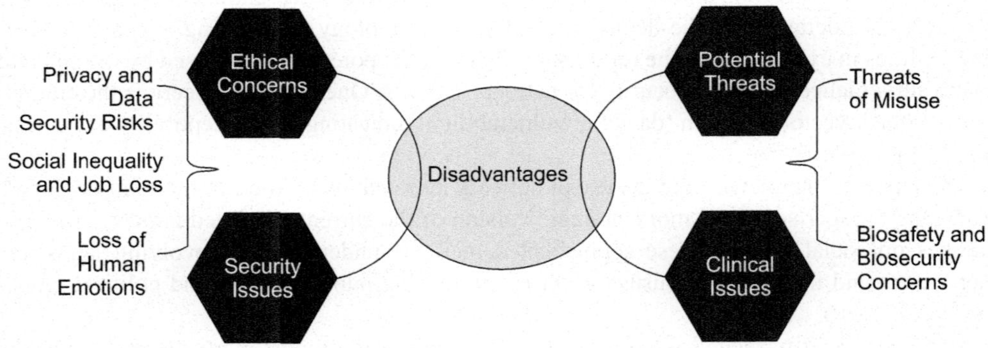

Fig. 3 Disadvantages of AI

Conclusion

In conclusion, the integration of AI in biological sciences heralds a new era of innovation and discovery. The advantages explored in this chapter underscore the transformative potential of AI technologies, offering a glimpse into the future of biotechnological advancements. As we navigate the ethical considerations and challenges, the promise of AI in biological sciences remains a beacon guiding us toward unprecedented scientific achievements.

References

Agudo, U. and Matute, H. (2021). The influence of algorithms on political and dating decisions. PLoS One, 16, e0249454. https://doi.org/10.1371/journal.pone.0249454

Alami, H., Lehoux, P., Denis, J.-L., Motulsky, A., Petitgand, C et al. (2021). Organizational readiness for artificial intelligence in health care: insights for decision-making and practice. Journal of Health Organization and Management, 35(1), 106–114. https://doi.org/10.1108/JHOM-03-2020-0074

Aldoseri, A., Al-Khalifa, K.N. and Hamouda, A.M. (2023). Re-Thinking Data Strategy and Integration for Artificial Intelligence: Concepts, Opportunities, and Challenges. Applied Sciences, 13(12), 7082. https://doi.org/10.3390/app1312708

Anonymous. (2023). 2023 Sandip University. Artificial Intelligence meets Biology: Current Applications and Job Opportunities. Sandip University. Retrieved September 12, 2023, from https://www.sandipuniversity.edu.in/blog/artificial-intelligence-meets-biology/

Askin, S., Burkhalter, D., Calado, G. and El Dakrouni, S. (2023). Artificial Intelligence Applied to clinical trials: opportunities and challenges. Health Technol (Berl), 13(2), 203-213. https://doi.org/10.1007/s12553-023-00738-2

Bhardwaj, A., Kishore, S. and Pandey, D.K. (2022). Artificial Intelligence in Biological Sciences. Life, 12(9), 1430. https://doi.org/10.3390/life12091430

Bond, R.M., Fariss, C.J., Jones, J.J. (2012). A 61-million-person experiment in social influence and political mobilization. Nature, 489, 295–8. https://doi.org/10.1038/nature11421

Chakraborty, I., Choudhury, A. and Banerjee, T. (2017). Artificial Intelligence in Biological Data. International Journal of Information Technology, 7. doi:10.4172/2165-7866.1000207

Congressional Research Service. (2023, November 22). Artificial Intelligence in the Biological Sciences: Uses, Safety, Security, and Oversight. https://crsreports.congress.gov/product/pdf/R/R47849

Couchman, H. (2019, July 8). Policing by machine: predictive policing and the threat to our rights. London: Liberty.

Creemers, R. (2018). China's social credit system: an evolving practice of control. SSRN Journal. 10.2139/ssrn.3175792.

Damschroder, L.J., Aron, D.C., Keith, R.E., Kirsh, S.R., Alexander, J.A. et al. (2009). Fostering implementation of health services research findings into practice: a consolidated framework for advancing implementation science. Implementation Science, 4(50). doi: 10.1186/1748-5908-4-50.

Davenport, T. and Kalakota, R. (2019). The potential for artificial intelligence in healthcare. Future Healthc. J., 6(2), 94–98. https://doi.org/10.7861/futurehosp.6-2-94

Denti, L. and Hemlin, S. (2012). Leadership and innovation in organizations: a systematic review of factors that mediate or moderate the relationship. International Journal of Innovation Management, 16(03), 1240007. doi: 10.1142/S1363919612400075.

Duggal, N. (2023, November 24). Advantages and Disadvantages of Artificial Intelligence [AI]. Simplilearn. Available from: https://www.simplilearn.com/advantages-and-disadvantages-of-artificial-intelligence-article

Ekins, S., Lane, T.R., Urbina, F. and Puhl, A.C. (2024). In silico ADME/tox comes of age: twenty years later. Xenobiotica, 54(7), 352–358.

Eraslan, G., Avsec, Z., Gagneur, J. and Theis, F.J. (2019). Deep learning: new computational modelling techniques for genomics. Nat Rev Genet, 20(7), 389-403. https://doi.org/10.1038/s41576-019-0122-6

Farhud, D.D. (2019). Epigenetic and Ethics: How are Ethical Traits Inherited? International Journal of Ethics & Society (IJES), 1, 1–4.

Farhud, D.D. and Zokaei, S. (2021, November). Ethical Issues of Artificial Intelligence in Medicine and Healthcare. Iran Journal of Public Health, 50(11), i-v. https://doi.org/10.18502/ijph.v50i11.7600

Federspiel, F., Mitchell, R., Asokan, A., Umana, C. and McCoy, D et al. (2023, May). Threats by artificial intelligence to human health and human existence. BMJ Global Health, 8(5), e010435.

Future of Life Institute. (2015). Autonomous weapons: an open letter from AI & robotics researchers. Available at: https://futureoflife.org/open-letter-autonomous-weapons/ [Accessed 21 May 2021].

Holzinger, A., Keiblinger, K., Holub, P., Zatloukal, K. and Müller, H et al. (2023). AI for life: Trends in artificial intelligence for biotechnology. New Biotechnology, 74, 16-24. https://doi.org/10.1016/j.nbt.2023.02.001

Javorsky, E., Tegmark, M. and Helfand, I. (2019). Lethal autonomous weapons. BMJ, 364, l1171. https://doi.org/10.1136/bmj.l1171

Kewaunee International Group. (2024). Future of Laboratory Automation: Enhancing Efficiency and Accuracy. Retrieved January 4, 2024, from https://www.kewaunee.in/blog/future-of-laboratory-automation-enhancing-efficiency-and-accuracy/#:~:text=Intelligent%20automation%20powered%20by%20AI,and%20accuracy%20in%20laboratory%20automation

Khan, B., Fatima, H., Qureshi, A., Kumar, S., Hanan, A. et al. (2023, February 8). Drawbacks of Artificial Intelligence and Their Potential Solutions in the Healthcare Sector. Biomed Mater Devices. https://doi.org/10.1007/s44174-023-00063-2

Kim, I., Kang, K., Song, Y. and Kim, T.J. (2022, November 15). Application of Artificial Intelligence in Pathology: Trends and Challenges. Diagnostics (Basel), 12(11), 2794. https://doi.org/10.3390/diagnostics12112794

Kuiken, T. (2023). Artificial Intelligence in the Biological Sciences: Uses, Safety, Security, and Oversight. Congressional Research Service. Retrieved from https://crsreports.congress.gov/product/pdf/R/R47849#:~:text=This%20has%20enabled%20research%20and,markers%20linked%20with%20those%20traits

Lorenz-Spreen, P., Oswald, L., Lewandowsky (2023). A systematic review of worldwide causal and correlational evidence on digital media and democracy. Nat Hum Behav, 7, 74–101. https://doi.org/10.1038/s41562-022-01460-1

National Institutes of Health (NIH). (2023, December 21). Artificial intelligence. Science education. Available at: https://www.nibib.nih.gov/science-education/science-topics/artificial-intelligence-ai

Nordling, L. (2019). A fairer way forward for AI in health care. Nature, 573, S103–s105.

Osoba, O.A. and Welser, W. (2017, December). The risks of artificial intelligence to security and the future of work. Santa Monica, CA: RAND.

Pal, S., Mondal, S., Das, G., Khatua, S. and Ghosh, Z. (2020). Big data in biology: The hope and present-day challenges in it. Gene Reports, 21, 100869. ISSN 2452-0144.

Pasquale, F. (2020, November 9). When medical robots fail: Malpractice principles for an era of automation. The Brookings Institution. https://www.brookings.edu/techstream/when-medical-robots-fail-malpractice-principles-for-an-era-of-automation

Rakshita, A. (2024, January 5). Personalised Care: Redefining Healthcare Trends In 2024. Business World Healthcare. https://bwhealthcareworld.businessworld.in/article/-Personalised-Care-Redefining-Healthcare-Trends-In-2024/05-01-2024-504708/

Regulation (EU) 2016/679 of the European Parliament and of the Council of 27 April 2016 on the protection of natural persons with regard to the processing of personal data and on the free movement of such data, and repealing Directive 95/46/EC. (2016). Available at: https://eur-lex.europa.eu/eli/reg/2016/679/oj

Rodrigues, R. (2020, December 1). Legal and human rights issues of AI: Gaps, challenges and vulnerabilities. Journal of Responsible Technology, 4, 100005.

Russel, S. (2018). The new weapons of mass destruction? Available at: https://www.the-security-times.com/wp-content/uploads/2018/02/ST_Feb2018_Doppel-2.pdf

Tetrick, M. (2023). The Importance Of Data Processing In Machine Learning & AI. Rely Services. Available at: https://www.relyservices.com/blog/importance-data-processing-in-machine-learning. Retrieved December 6, 2023.

The Guardian Gd (2020). RoboDoc: how India's robots are taking on Covid patient care. https://www.theguardian.com/global-development/2020/ dec/02/robodoc-how-india- robots-are-taking-on-covid-patient-care-mitra

CHAPTER

6

Comparison of Machine Learning and Deep Learning

Priyanka Narad,[1] *Alakto Choudhury,*[2] *Kritika Shukla,*[2] *Mannat Singh,*[2] *Deepak Modi*[3] and *Abhishek Sengupta*[2*]

1. Introduction

The interaction between Artificial Intelligence (AI) and human cognition in the digital age has made it possible for machines to carry out tasks that were previously limited to human intellect. Unlike conventional algorithms which are deterministic, rule-based operations, it is a highly disruptive technological breakthrough that comprises two main subsets namely, Machine Learning (ML) and Deep Learning (DL). Both have had a significant impact on how current technology is developed, empowering everything ranging from self-driving cars to recommendation systems, and transforming everything from healthcare to banking. Artificial intelligence encompasses any method that enables computers to emulate human behavior and decision-making capabilities with as little human intervention as possible (Janiesch, Zschech and Heinrich 2021). Often, such high-complexity tasks need extensive knowledge, planning, communication, and learning by the intelligent system (Artificial Intelligence: A Modern Approach, 4th US ed.).

Machine Learning is a subfield of Artificial Intelligence that, as quoted by Arthur Samuel, gives "computers the ability to learn without being explicitly programmed." (Samuel 1967). It primarily deals with the exploration of data and the construction of algorithms that can learn from the data and make predictions based on it and is also closely associated and overlaps with the domain of computational statistics (Jordan and Mitchell 2015a; Louridas and Ebert 2016). ML algorithms do not follow strict static program instructions and can make data-driven predictions or decisions, by creating models from sample inputs (Ongsulee 2018). ML models are usually unsupervised (Hastie, Tibshirani and Friedman 2009) or supervised (Cunningham, Cord and Delany 2008); the former is essentially a system that learns by itself by identification of patterns in data, it establishes baseline behavior of various entities in data and finds meaningful anomalies within it (Hastie, Tibshirani and

[1] Division of Biomedical Informatics (BMI), Indian Council of Medical Research, Ansari Nagar, New Delhi, India.
[2] Systems Biology and Data Analytics Lab, Amity Institute of Biotechnology, Amity University, Noida, India.
[3] Department of Molecular and Cellular Biology, National Institute for Research in Reproductive Health and Child Health, Mumbai, India.
* Corresponding author: asengupta@amity.edu

Friedman 2009) whereas the latter is trained with labeled subsets of data and learns by comparing new unlabeled input with the known labeled data that it is trained on using methods like classification, regression, gradient boosting and prediction (Cunningham, Cord and Delany 2008).

Deep Learning, also referred to as hierarchical learning, is an advanced machine learning concept based on artificial neural networks (ANNs) (Janiesch, Zschech and Heinrich 2021). It is an approach within ML that focuses on learning complex representations directly from data (Xin *et al.* 2018). It primarily deals with the study of ANNs and related algorithms which contain more than one hidden layer. These methods use a cascade of layers of nonlinear processing units for feature extraction and transformation. Each layer takes the output from its preceding layer as input. The algorithms may be supervised or unsupervised. DL algorithms learn multiple levels of representations corresponding to different levels of abstraction, thus forming a hierarchical representation of concepts (Deng and Yu 2013).

Computer science has long been built on traditional algorithms, which include techniques like sorting, searching, and optimization. These algorithms yield predictable outcomes and are very efficient and well-understood. Their deterministic, rule-based operation makes them perfect for activities requiring accuracy and a well-defined flow of actions. However, the landscape of problem-solving has undergone a significant transformation with the emergence of ML and DL algorithm-based approaches, coinciding with the artificial intelligence movement. This book chapter aims to give a thorough understanding of the concepts of ML and DL algorithms and compares their differences. Our goal is to clarify their advantages, disadvantages, and situations in which one strategy might be more practical to employ as compared to another. The readers shall be able to comprehend the AI and ML domains by delving into the underlying distinctions and uses of these two paradigms. This will ultimately assist them in selecting the appropriate techniques for a problem statement keeping all nuances and peculiarities of those issues into consideration.

1. Machine Learning

The machine learning approach originated in the early 1950s as a part of the artificial intelligence movement, it enables systems to solve problems and learn without being explicitly programmed (Samuel 1959). It also enables the program to progressively improve over time with experience and hence is a good means to automate tasks and to view relationships and hidden insights between the data that is used as the input (Janiesch et al., 2021). Generally, the machine learning structure involves the application of algorithms on some historical or series data which is used for training to form a predictive model which is then tested on an unused portion of the historical data or new data; the algorithms used for this purpose can be classified into four broad categories based on the input data and given problem. These are supervised learning, semi-supervised learning, unsupervised learning and reinforcement learning (Sarker 2021).

1.1 Machine Learning Algorithms

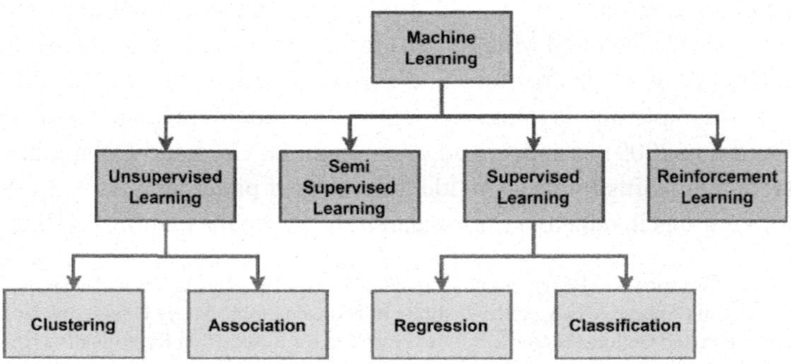

Fig. 1 Categories of Machine Learning

1.2 Unsupervised Learning

When an algorithm is trained on unlabeled data, that is, data without particular target labels or categories, it is referred to as unsupervised learning in the context of machine learning. Unsupervised learning's main objective is to find hidden structures, correlations, or patterns in the data. Unsupervised learning algorithms shift through the data on their own to uncover significant patterns (Jordan and Mitchell 2015b; Louridas and Ebert 2016).

1.3 Semi-Supervised Learning

A machine learning paradigm known as semi-supervised learning blends aspects of unsupervised and supervised learning. The technique is trained on a dataset including both labeled and unlabeled data in semi-supervised learning. In a semi-supervised learning scenario, you have a dataset with a smaller percentage of unlabeled examples (where the target labels are absent) and a smaller number of examples with labeled data (where the inputs are associated with specific target labels). Semi-supervised learning blends unsupervised learning with supervised learning. By utilizing both kinds of data, the model's performance is to be enhanced. One popular method in semi-supervised learning is to train a model on a limited quantity of labeled data, and then use the model to infer labels from the unlabeled data. We call these anticipated labels "pseudo-labels." Next, utilizing both the labeled and pseudo-labeled data, the model is adjusted (Louridas and Ebert 2016).

1.4 Supervised Learning

In supervised learning, an algorithm learns to make predictions or choices by analyzing a labeled dataset that includes examples or input-output pairs. In the context of supervised learning, "supervision" refers to the presence of a person or "oracle" who gives the algorithm the right responses during the training process. To enable the algorithm to generate precise predictions on fresh, unseen data, supervised learning aims to create a mapping from the input to the output data (Louridas and Ebert 2016).

1.5 Reinforcement Learning

Within the discipline of machine learning, reinforcement learning (RL) involves an agent interacting with its surroundings and learning how to make decisions in a manner that maximizes a cumulative reward. It is frequently applied in scenarios where an agent must make a series of decisions over time to accomplish a certain objective, and it draws inspiration from behavioral psychology. Reinforcement learning is extensively utilized throughout diverse fields such as robots, gaming, recommendation systems, and self-governing systems (Barto 1997). Several essential ideas and elements are included in reinforcement learning (RL), which is applied to interactions between agents and their surroundings. In reinforcement learning, an agent, or decision-making entity, interacts with its surroundings and makes choices that optimize cumulative rewards. States, which depict the current state of affairs; actions, which outline options; policies, which direct the agent's conduct; rewards, which provide comments on actions; and value functions, which gauge the desirability of states or state-action pairings, are important constituents. Tradeoffs between exploration and exploitation are crucial because agents have to strike a balance between pursuing fresh insights and utilizing tried-and-true tactics. This procedure is designed to simulate the dynamics of transitions using Markov Decision Processes (MDPs). Different reinforcement learning (RL) methods, such as Q-learning and Deep Q-Networks, are used to train agents for a variety of applications, including gaming and robotics (Barto 1997).

2. Deep Learning

2.1 Origin of Deep Learning

The inspiration to create an artificial nerve network inspired by brain and biological receptors was implemented in 1957 when Frank Rosenblatt proposed the idea of Perceptron (Perceptron | Encyclopedia of Computer Science). The first generalized, functional learning algorithm was created for a multilayer Perceptron (MLP), and employed a "feedforward design" (Ivachnenko and Lapa 1967). The idea of backpropagation for ANNs was introduced in 1974 by Paul Werbos as a part of his research dissertation, this concept described a method by which errors could be processed at the output (instead of the input) and then distributed backward, traversing the system's layers for learning which was later rediscovered in the year 1986 by Rumelhart et al. and introduced in their book "Learning Internal Representations by Error Propagation" (Rumelhart, Hinton and Williams 1986). In 1979, Kunihiko Fukushima started developing the Neocognitron which was the first design of a DL model using a convolutional neural network which later evolved into computer vision (Fukushima 1980). The concept of DL gained more interest during the 1980s with the introduction of the concept of ANNs by Kohonen (Kohonen 1982).

A neural network comprises three main layers. The first and the last layers in a neural network are the input layer and output layer respectively. The various layers that exist between the two are generally referred to as the hidden layers. Each layer is normally made up of a simple algorithm supporting a single activation function (Bonifazi and Burrascano 1984). The addition of hidden layers could significantly enhance the abilities of their neural networks to evolve into ANNs (Uzair and Jamil 2020).

2.2 Neural Network Architecture

DL is mostly based on neural networks, which emulate the architecture and operation of the human brain (Lecun, Bengio and Hinton 2015). These networks consist of the following interconnected layers.

2.3 Input Layer

This is the input layer where the data is first received and then processed. In this layer, every neuron is associated with a feature or input variable (Perceptron | Encyclopedia of Computer Science).

2.4 Hidden Layers

Neural networks may contain one or more hidden layers in between the input and output layers. Each neuron in the layer above or below is connected to every other neuron in the hidden layer. The network picks up knowledge and derives patterns from the data within these hidden layers (Uzair and Jamil 2020).

2.5 Output Layer

The predictions or outcomes of the network are generated by the output layer. Depending on the task, the output layer's neuron count may vary. A neuron (0 or 1) may exist in binary classification, however, the number of neurons in multi-class classification will be equal to the number of classes (Perceptron | Encyclopedia of Computer Science).

3. Types of Neural Networks

3.1 Perceptron

A perceptron is a simple model of a neuron that can make a binary decision based on weighted inputs and a threshold (Perceptron | Encyclopedia of Computer Science). To simplify the concept,

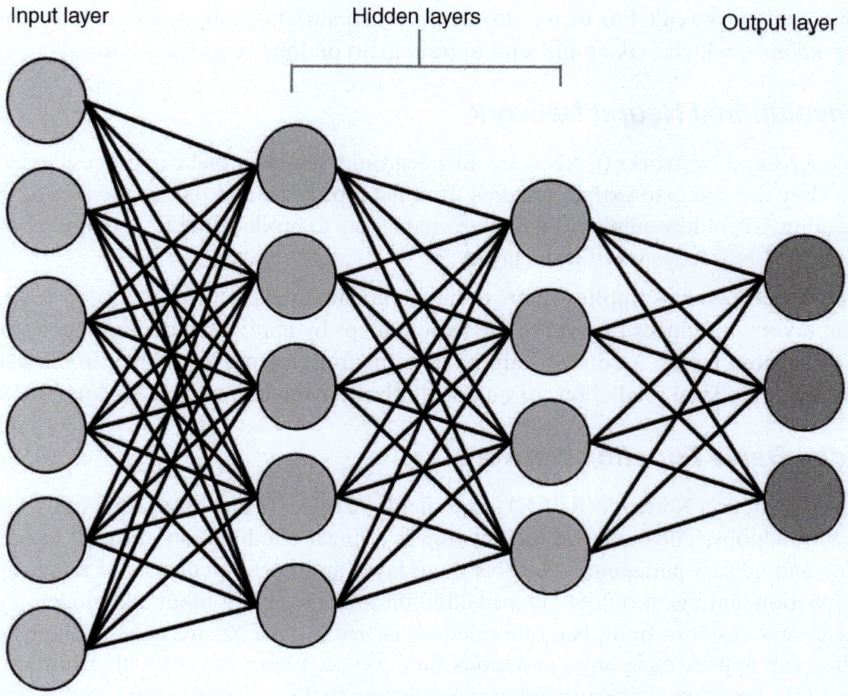

Fig. 2 Neural Network Architecture

imagine that a perceptron is like a system where a decision is passed based on many votes. In a democracy, a vote count of over 50% for a party is the majority and that party is elected (decision is taken). So, by this logic, the perceptron is the election, the input layers are the votes and 50% is the threshold, where anything above it is the majority (output).

3.2 Multilayer Perceptron

A multilayer perceptron (MLP) is a type of ANN that has multiple layers of neurons. Each neuron takes inputs from the previous layer and passes the output onto the next successive layer after applying a nonlinear function to it. MLPs can learn complex patterns and functions from data that are not linearly separable. They use the backpropagation method to adjust the weights of the connections between neurons based on the error between the predicted and actual outputs. MLPs are widely used for classification, regression, and other tasks that involve supervised learning (Ivachnenko and Lapa 1967; Ramchoun *et al.* 2017).

3.3 Feed Forward Neural Network

A feedforward neural network (FFNN) is a type of ANN in which the information flows unidirectionally i.e. from input to output. Therefore, the connections between the neurons in a FFNN do not form cycles, also, the network has no feedback loops either (Feed-forward neural networks | IEEE Journals & Magazine | IEEE Xplore; Svozil, Kvasnička and Pospíchal 1997).

3.4 Recurrent Neural Network

A recurrent neural network (RNN) is a type of ANN that can process sequential data or time series data, such as text, speech, or video. Unlike FFNNs, RNNs have feedback connections between the neurons, which enables them to store and use information from previous inputs in the current computation. This makes RNNs capable of learning long-term dependencies and generating outputs based on the context provided to them. However, it also suffers from issues such as vanishing or exploding gradients, which make it difficult to train them on long sequences (Grossberg 2013).

3.5 Convolutional Neural Network

Convolutional Neural Networks (CNNs) are deep learning networks that can process image, text, or audio data. They use filters to extract features from the input data and reduce the size of the feature maps by pooling. Then they employ fully connected layers to produce the final outputs based on the learned features. CNNs consists of three layers:

- **Convolutional layer:** Applies filters to the input data to extract features.
- **Pooling layer:** Reduces the size of the feature maps by applying a pooling operation.
- **Fully connected layer:** Connects all the neurons from the previous layer to the output layer and produces the final predictions or outputs (Albawi, Mohammed and Al-Zawi 2018).

3.6 Radial Basis Function Network

A Radial Basis Function Network (RBFN) is another type of ANN that uses radial basis functions as its activation functions. The output of the network is a linear combination of radial basis functions of the inputs and neuron parameters. RBFNs three-layer architecture consists of a non-linear RBF activation function and linear output layer in addition to the standard input and hidden layers; they are also faster and easier to train than other neural networks. RBFNs are universal approximators, meaning they can approximate any continuous function on a bounded set with arbitrary precision and are useful for function approximation, time series prediction, classification, and system control (Schwenker, Kestler and Palm 2001).

3.7 Long Short-Term Memory Network

A Long Short-Term Memory Network (LSTM) is a type of RNN that can learn, process, and classify sequential data. LSTM's special structure allows them to remember long-term dependencies between time steps of data and avoid the vanishing gradient problem faced by other RNNs. These are widely used for natural language processing, speech recognition, time series forecasting, and other applications that involve sequential data (Noguchi, Iizuka and Yamamoto 2016).

3.8 Applications of Deep Learning

Since DL models can handle more complex data and perform more complex tasks, it can handle biological data and is widely applicable in the field of healthcare for tasks such as disease prediction (Mukherjee *et al.* 2020; Zeleznik *et al.* 2021), biomarker discovery (Wang *et al.* 2023), drug-discovery (Askr *et al.* 2023) and diagnostics (Computational Intelligence for Cybersecurity Management and Applications - Google Books; Jiao *et al.* 2019; Bharati, Podder and Mondal 2020). It also has a variety of applications in fields requiring a higher level of decision-making. Some of these applications include object detection (Jiao *et al.* 2019), autonomous vehicles (Kuutti *et al.* 2021), speech and audio processing (Wang *et al.* 2022; Weng *et al.* 2023), image recognition (Yolcu *et al.* 2020), fraud detection (Pumsirirat and Yan 2018; Lebichot *et al.* 2020), Natural Language Processing (NLP) (Li *et al.* 2023), Cybersecurity (Computational Intelligence for Cybersecurity Management and Applications - Google Books; Torre, Mesadieu and Chennamaneni 2023), and many more.

4. Machine learning Vs Deep Learning

4.1 How are They Different and which Model to use?

Conventionally, ML methods are employed in computational processes in which designing and programming explicit algorithms for an application with sufficient performance is either difficult or unfeasible. Due to the large volume of structured, unstructured, and semi-structured data generated today, it is easier to build a ML model for a variety of applications, also it is faster, cost-effective and easier to implement and interpret (Attaran and Deb 2018). It is therefore a highly effective method for applications such as email filtering, commercial predictive analysis, data analytics, and studying historical relationship trends in data (Gbenga Dada *et al.* 2017). Since it can work with smaller datasets as well, it can also be highly employable in grouping unknown data to identify patterns, identify data outliers, and item recommendations.

However, creating a ML model involves feature extraction by manual means, on the other hand, DL models enable automated extraction and analysis of useful information from raw data. The results obtained from DL methods are also better than conventional ML approaches (Setiowati *et al.* 2017). This can be attributed to the use of a non-linear model employing multiple hidden layer architecture that makes the system learn about the complexities in the relationship between the input and output (Chauhan and Singh 2019). Due to the multiple levels of abstraction in DL, it can transform original data into higher abstract levels for making decisions; this facilitates easier solution searching for non-linear complex functions (Ramachandran *et al.* 2015). However, such tasks require a huge volume of data to train, which is significantly more than a conventional ML model along with higher processing power often requiring GPU processing as well (Coelho *et al.* 2017; Chauhan and Singh 2019), therefore it is more suitable to use when dealing with larger data and complexity (LeCun, Bengio and Hinton 2015). The differences between ML models and DL models are summarized below in **Table 1.**

Table 1 Differences between ML models and DL models

ML models	*DL models*
Use algorithms to perform a task without being explicitly programmed	Use ANNs to mimic the learning process of the human brain
Need manual extraction of features to be studied.	Automatically extract features from raw data and process them to make decisions .
The main problem is often divided into smaller chunks of sub-problems, solving each successively to obtain the result.	Works on direct end-to-end problem-solving.
Results are easily interpretable due to manual feature selection during the model-building process.	It is difficult to explain and interpret some results since the reasoning and decisions are based on the system's logic.
Require a smaller volume of data to be trained.	Require a higher data volume than ML models for training (Model improves with increase in data volume).
More suitable for structured, semi structured or unstructured data of volume smaller as compared to that used in DL models.	More suitable for visualizing hierarchical data, to establish relationships between complex data and to observe solutions at different levels of abstraction.
It can operate with low processing power.	Generally, need higher processing power to operate.
Less training time is needed.	Longer training time is required.
Longer testing time is required generally.	Lesser time is required for testing. Higher execution power (takes less time to generate results).
Typically used for tasks with lower level of complexity, clustering data, pattern identification, and prediction.	Typically used for more tasks with higher complexity such as speech recognition, image identification, and language translation.

5. Case Studies

The following case studies illustrate how ML and DL methods have been implemented in the past while showcasing their efficacy. We specifically selected examples within a similar field, in this instance, healthcare, to accentuate the strengths of these methods and provide a meaningful contrast between their respective use cases.

5.1 ML for Classification of New Pathogens (Case Study on COVID-19)

The methodology used by Randhawa et al. to classify COVID-19 virus genomes utilizing machine learning-based alignment quickly and accurately-free methodology and a case study centered on the SARS-CoV-2 genomic signature. A dataset of 5000 distinct viral genome sequences was gathered by the researchers, of which 29 SARS-CoV-2 genomic sequences were accessible at the time of the study. Accurate taxonomic categorization of novel diseases, like the COVID-19 virus, was made possible using machine learning for feature extraction and classification of viral genomic sequences. The study examined intrinsic dinucleotide genomic markers for eight pathogenic species, including SARS-CoV-2, using whole genome sequence data. Extreme gradient boosting (XGBoost) was used to convert the genome sequences into dinucleotide relative frequencies for categorization. For genomic analysis, the suggested method integrated supervised machine learning and digital signal processing (MLDSP), with the machine learning component being enhanced via a decision tree approach. The taxonomic procedure was made simpler and the real-time classification of novel viral diseases was made possible by the machine learning-based technique, which made it easier to classify viral genomes quickly and accurately. Various machine-learning models were built to categorize viral genomes at varying taxonomic levels. These models were then employed to forecast the accurate classification of COVID-19 samples across the various taxonomic levels. For result validation, Spearman's rank correlation coefficient analysis was employed in the study. The whole-genome machine-learning approach without alignment was put out as a dependable real-time solution for the taxonomic classification of novel pathogens and viral genome sequences. Novel viral and pathogen genome sequences could now be taxonomically classified in real-time with a dependable option thanks to the alignment-free whole-genome machine-learning approach (Randhawa *et al.* 2020).

5.2 DL-based Prediction and Analysis of COVID-19 Positive Cases

The methodology used by Arora et al. in their work, which involved employing RNN-based LSTM variants for short-term prediction and a DL-based model to estimate the number of COVID-19-positive patients across 32 states and union territories of India: To obtain the required COVID-19 positive case data for 32 Indian states and union territories, the researchers gathered and preprocessed Indian information. A variety of Long Short-Term Memory (LSTM) model variations were utilized to compute one-week forecasts for various Indian states and union territories, including Deep LSTM, Bi-directional LSTM (Bi-LSTM), and Convolutional LSTM (Conv-LSTM). A short-term prediction accuracy evaluation of the suggested strategy produced an error of less than 3% for daily predictions and less than 8% for weekly predictions. To make it easier to identify COVID-19 hotspots, Indian states were divided into zones according to the daily growth rate and the dispersion of positive cases. Based on the predictions, the study recommended preventive actions to lessen the spread in the relevant zones. For short-term COVID-19 case prediction in India, deep learning more specifically, RNN-based LSTM variants were employed to produce precise forecasts for both daily and weekly instances. Based on the spread of positive cases and the daily growth rate, the model made it easier to categorize Indian states into different zones, making it possible to estimate the risk of spread and put preventive measures in place (Arora, Kumar and Panigrahi 2020).

Conclusion

Significant leaps in technological advancements have led to the switch from conventional algorithms to more intelligent systems and notably, AI. As technology evolves, algorithms within AI, such as ML and DL, continually find novel applications and are improved further. These models have become indispensable tools for researchers, developers, and practitioners, acting as disruptive catalysts for progress across various fields of study. In light of the intricacies and growing demand for a nuanced understanding of these transformative technologies, it has become imperative to understand the concepts and ideology behind the development of these tools and to understand the use case and applicability to determine the ideal implementation of each of the methods. Therefore, this chapter discusses the concepts and distinctions between the two subsets of AI, while keeping a keen emphasis on the usability and applicability of the two with the aid of case studies to illustrate better.

References

Albawi S., Mohammed T.A. and Al-Zawi S. Understanding of a convolutional neural network. *Proceedings of 2017 International Conference on Engineering and Technology, ICET 2017* 2018; 2018-January: 1–6.

Arora P., Kumar H. and Panigrahi B.K. Prediction and analysis of COVID-19 positive cases using deep learning models: A descriptive case study of India. *Chaos Solitons Fractals* 2020; 139: 110017.

Artificial Intelligence: A Modern Approach, 4th US ed.

Askr H., Elgeldawi E. and Aboul Ella H. Deep learning in drug discovery: an integrative review and future challenges. *Artif Intell Rev* 2023; 56: 5975–6037.

Attaran M. and Deb P. Machine learning: the new "big thing" for competitive advantage. *International Journal of Knowledge Engineering and Data Mining* 2018; 5: 277.

Barto A.G. Reinforcement Learning. *Neural Systems for Control* 1997: 7–30.

Bharati S., Podder P. and Mondal M.R.H. Hybrid deep learning for detecting lung diseases from X-ray images. *Inform Med Unlocked* 2020; 20: 100391.

Bonifazi G. and Burrascano P. Ceramic powder characterization by multilayer perceptron (MLP) data compression and classification. *Amphibia-Reptilia* 1984; 5: 225–39.

Chauhan N.K. and Singh K. A review on conventional machine learning vs deep learning. *2018 International Conference on Computing, Power and Communication Technologies, GUCON 2018* 2019:347–52.

Coelho I.M., Coelho V.N., Luz E.J. da S.A. GPU deep learning metaheuristic based model for time series forecasting. *Appl Energy* 2017; 201: 412–8.

Computational Intelligence for Cybersecurity Management and Applications - Google Books.

Cunningham P., Cord M. and Delany S.J. Supervised learning. *Cognitive Technologies* 2008:21–49.

Deng L. and Yu D. Deep learning: Methods and applications. *Foundations and Trends in Signal Processing* 2013; 7: 197–387.

Feed-forward neural networks | IEEE Journals & Magazine | IEEE Xplore.

Fukushima K. Neocognitron: a self organizing neural network model for a mechanism of pattern recognition unaffected by shift in position. *Biol Cybern* 1980; 36: 193–202.

Gbenga Dada E., Stephen Bassi J. and Chiroma H. *et al.* Machine learning for email spam filtering: review, approaches and open research problems. 2017, DOI: 10.1016/j.heliyon.2019.e01802.

Grossberg S. Recurrent neural networks. *Scholarpedia* 2013; 8: 1888.

Hastie T., Tibshirani R. and Friedman J. Unsupervised Learning. 2009: 485–585.

Ivachnenko A.G. and Lapa V.G. Cybernetics and forecasting techniques. 1967.

Janiesch C., Zschech P. and Heinrich K. Machine learning and deep learning. *Electronic Markets* 2021; 31: 685–95.

Jiao L., Zhang F. and Liu F. A survey of deep learning-based object detection. *IEEE Access* 2019; 7: 128837–68.

Jordan M.I., Mitchell T.M. Machine learning: Trends, perspectives, and prospects. *Science (1979)* 2015a; 349: 255–60.

Jordan M.I., Mitchell T.M. Machine learning: Trends, perspectives, and prospects. *Science (1979)* 2015b; 349: 255–60.

Kohonen T. Self-organized formation of topologically correct feature maps. *Biol Cybern* 1982; 43: 59–69.

Kuutti S., Bowden R. and Jin Y. A Survey of Deep Learning Applications to Autonomous Vehicle Control. *IEEE Transactions on Intelligent Transportation Systems* 2021; 22: 712–33.

Lebichot B., Le Borgne Y-A, He-Guelton L *et al.* Deep-Learning Domain Adaptation Techniques for Credit Cards Fraud Detection. 2020: 78–88.

Lecun Y., Bengio Y. and Hinton G. Deep learning. *Nature* 2015; 521: 436–44.

Li C., Zhang Y. and Weng Y. Natural Language Processing Applications for Computer-Aided Diagnosis in Oncology. *Diagnostics 2023, Vol 13, Page 286* 2023; 13: 286.

Louridas P. and Ebert C. Machine Learning. *IEEE Softw* 2016; 33: 110–5.

Mukherjee D., Mondal R. and Singh P.K. EnsemConvNet: a deep learning approach for human activity recognition using smartphone sensors for healthcare applications. *Multimed Tools Appl* 2020; 79: 31663–90.

Noguchi W., Iizuka H. and Yamamoto M. Proposing Multimodal Integration Model Using LSTM and Autoencoder. 2016, DOI: 10.4108/eai.3-12-2015.2262505.

Ongsulee P. Artificial intelligence, machine learning and deep learning. *International Conference on ICT and Knowledge Engineering* 2018:1–6.

Perceptron | Encyclopedia of Computer Science.

Pumsirirat A. and Yan L. Credit Card Fraud Detection using Deep Learning based on Auto-Encoder and Restricted Boltzmann Machine. *International Journal of Advanced Computer Science and Applications* 2018; 9: 18–25.

Ramachandran R., Rajeev D. and Krishnan S. Deep learning – An overview. *International journal of applied engineering research* 2015.

Ramchoun H., Idrissi M.A.J. and Ghanou Y. Multilayer perceptron: Architecture optimization and training with mixed activation functions. *ACM International Conference Proceeding Series* 2017; **Part F129474**, DOI: 10.1145/3090354.3090427.

Randhawa G.S., Soltysiak M.P.M. and El Roz H. *et al.* Machine learning using intrinsic genomic signatures for rapid classification of novel pathogens: COVID-19 case study. *PLoS One* 2020; 15: e0232391.

Rumelhart D.E., Hinton G.E. and Williams R.J. Learning representations by back-propagating errors. *Nature 1986 323:6088* 1986; 323: 533–6.

Samuel A.L. Some Studies in Machine Learning Using the Game of Checkers. *IBM J Res Dev* 1959; 3: 210–29.

Samuel A.L. Some Studies in Machine Learning Using the Game of Checkers. *IBM J Res Dev* 1967; 3: 210–29.

Sarker I.H. Machine Learning: Algorithms, Real-World Applications and Research Directions. *SN Comput Sci* 2021; 2: 1–21.

Schwenker F., Kestler, H.A. and Palm, G. Three learning phases for radial-basis-function networks. *Neural Networks* 2001; 14: 439–58.

Setiowati S., Zulfanahri, Franita E.L. A review of optimization method in face recognition: Comparison deep learning and non-deep learning methods. *2017 9th International Conference on Information Technology and Electrical Engineering, ICITEE 2017* 2017; **2018-January**: 1–6.

Svozil D., Kvasnička V. and Pospíchal, J. Introduction to multi-layer feed-forward neural networks. *Chemometrics and Intelligent Laboratory Systems* 1997; 39: 43–62.

Torre D., Mesadieu F. and Chennamaneni A. Deep learning techniques to detect cybersecurity attacks: a systematic mapping study. *Empirical Software Engineering 2023 28:3* 2023; 28: 1–71.

Uzair M. and Jamil N. Effects of Hidden Layers on the Efficiency of Neural networks. *Proceedings - 2020 23rd IEEE International Multi-Topic Conference, INMIC 2020* 2020, DOI: 10.1109/INMIC50486.2020.9318195.

Wang C-W, Muzakky H, Lee Y-C; Annotation-Free Deep Learning-Based Prediction of Thyroid Molecular Cancer Biomarker BRAF (V600E) from Cytological Slides. *International Journal of Molecular Sciences 2023, Vol 24, Page 2521* 2023; 24: 2521.

Wang Y., Wei-Kocsis J. and Springer J.A. Deep Learning in Audio Classification. *Communications in Computer and Information Science* 2022; **1665 CCIS**: 64–77.

Weng Z., Qin Z. and Tao X. Deep Learning Enabled Semantic Communications with Speech Recognition and Synthesis. *IEEE Trans Wirel Commun* 2023, DOI: 10.1109/TWC.2023.3240969.

Xin Y., Kong L. and Liu Z. Machine Learning and Deep Learning Methods for Cybersecurity. *IEEE Access* 2018; 6: 35365–81.

Yolcu G., Oztel I. and Kazan S. Deep learning-based face analysis system for monitoring customer interest. *J Ambient Intell Humaniz Comput* 2020; 11: 237–48.

Zeleznik R., Foldyna B. and Eslami P. Deep convolutional neural networks to predict cardiovascular risk from computed tomography. *Nature Communications 2021 12:1* 2021; 12: 1–9.

CHAPTER

7

Supervised Learning

Arasu Narayan[1*]

1. Introduction

Traditionally, computational systems in biostatistics were designed with static algorithms dictated by inflexible rules to perform tasks with precision. Consider the development of an e-commerce platform managing patient data. Initially, one might conceptualize a framework where:

- Users interface via a web or mobile application.
- The system uses a robust database to track user activity and transaction history.
- The core logic of the system—its 'brain'—is a series of rules correlating every possible event with an explicit response.

For instance, adding an item to a virtual cart would entail an update to the database linking the user ID with the product ID. Developers would iterate over potential anomalies, refining rules to ensure robustness. While effective for specific applications, such a deterministic approach often fails in the dynamic field of biostatistics, where data and outcomes are more easily predicted and controlled. Contrast this with the complexity of tasks like predicting patient outcomes based on historical health data or identifying genetic markers from biomedical imagery. Traditional programming falls short when dealing with the intricate patterns inherent in biological data, which are subject to change and often follow principles not immediately discernible to human logic.

Machine learning, particularly deep learning, emerges as a pivotal field of study in these scenarios. It focuses on algorithms that evolve through exposure to data, enhancing performance, and adapting to new patterns. Unlike the static nature of conventional software, machine learning algorithms in biostatistics continually improve, processing vast amounts of data to unveil insights in fields such as genomics and personalized medicine. This chapter provides a foundational understanding of these techniques, vital for advancing biostatistics and the larger medical community.

In the dynamic world of biostatistics, the approach to problem-solving often mimics the process of requesting directions using a voice assistant like Siri. This seemingly simple task results from complex machine learning models in action. A traditional program would have defined rules, like those used for mapping a user's journey in a health informatics system. However, the complexity of tasks such as voice recognition—where a phrase like "Hey Siri" triggers a sequence

[1] Northeastern University, Seattle, USA.

* Corresponding author: a.narayan@northeastern.edu

of intelligent operations—requires more than predefined rules due to the variability and subtlety of human language and the vast array of possible inputs. Hence, a similar sophistication is needed in biostatistical research to handle diverse datasets and unpredictable variables in health outcomes or genetic research. The essence of machine learning, particularly in biostatistics, is captured by not explicitly programming a computer to recognize patterns but rather by creating a system that learns from data, much like a clinician knows to diagnose from clinical experiences. In biostatistics, this translates to developing models that can, for instance, predict disease outbreaks from health data or identify biomarkers from complex bio-sequences. To operationalize this, a 'learning algorithm' is employed to tune the parameters of a model—a set of data-driven 'knobs'—to the correct settings that yield the most accurate predictions or classifications. The process iterates, using data to adjust these settings, enhancing the model's ability to discern between different health conditions. This iterative process is akin to refining a diagnostic tool to improve its accuracy in patient care. Deep learning, a subset of machine learning, extends these capabilities further by enabling the analysis of even more intricate and layered data structures commonly found in biostatistical research.

In the realm of biostatistics and data science, understanding the critical components of machine learning is crucial. One core concept is supervised learning, which aims to predict unknown labels based on known inputs within a dataset of labeled examples. This is one of several machine learning problem types. Essential elements in tackling these problems include:

- The data itself.
- A model to transform this data.
- An objective function to measure the model's performance.
- An algorithm to optimize this function.

Data, the cornerstone of data science, is a collection of examples needing numerical representation. Features represent the individual attributes of each example, which are used to predict a 'label' or target. The nature of the data varies widely, from fixed-length vectors in image data, represented as grids of pixel brightness values, to varying-length text data in customer reviews. The quantity and quality of data are paramount. More data facilitates the training of more powerful models with fewer assumptions, a key factor in the success of deep learning. However, data must be accurate and representative; otherwise, it risks reinforcing biases or perpetuating errors.

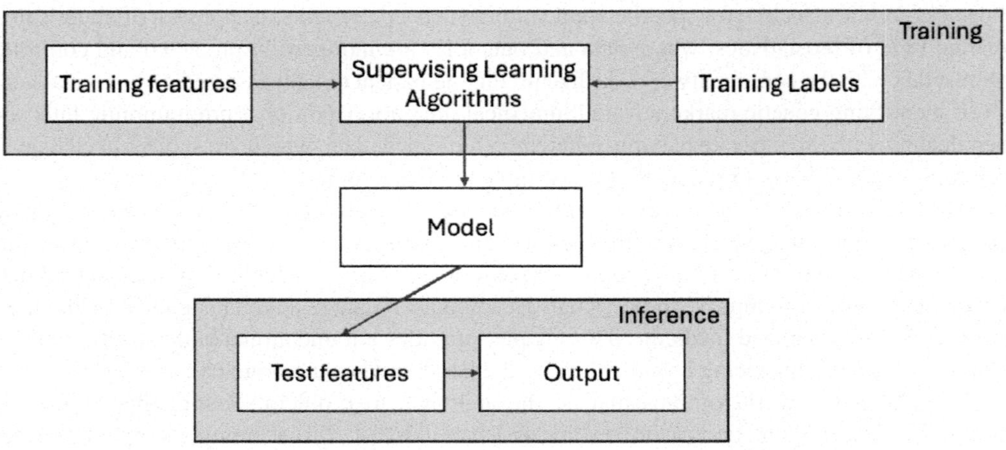

Fig. 1 Supervised Learning

Machine learning fundamentally involves data transformation. Models are the tools for ingesting one data type and outputting predictions of another. Deep learning involves many layered data transformations, hence the term deep. Objective functions are critical for assessing the performance of these models. Typically, these functions are structured so that a lower value indicates better performance. Popular loss functions include squared error for numerical predictions and error rate for classification tasks. Optimization, essential for effective machine learning, involves minimizing the loss function using algorithms like gradient descent. However, success in training data only ensures effectiveness on unseen data, highlighting the importance of model generalization and the avoidance of overfitting. Understanding the various types of machine learning problems, such as the wake word problem, enriches our grasp of the field's diverse applications and challenges.

Supervised learning, a key concept in machine learning and biostatistics, involves predicting an unknown label from known inputs within a labeled dataset. This data-driven approach underpins modern biostatistical methods where predictive models are developed to classify health outcomes or genetic markers accurately. For instance, in a study by Cheung et al. (2017), machine learning algorithms were used to predict patient survival rates from electronic health records, considering a range of features from vital signs to medical history (Cheung, 2017). These models, trained on large and diverse datasets, can handle the complexity of biostatistical data, much like voice recognition software, which interprets a range of human vocal inputs. In the field of biostatistics, the transformation of data is crucial. Models ingest data and output predictions, and these predictions are evaluated against an objective function to determine the model's performance. For example, Weissler et al. (2021) used deep learning models consisting of multiple layers of data transformation to interpret complex medical imaging data. The iterative learning process involves optimizing model parameters through algorithms like gradient descent to minimize loss functions. This methodology not only improves the predictive power of the models but also ensures that they generalize well to new, unseen data, avoiding overfitting.

2. Supervised Learning

Supervised learning, a pivotal concept in biostatistics and machine learning, involves training algorithms on a labeled dataset, where the correct output is known, to make predictions or classifications on new data. It's a method commonly employed in various fields, including medical diagnostics and biological research, to analyze patterns and make informed decisions. This learning paradigm hinges on the algorithm's ability to learn the mapping from inputs to outputs, using examples in the training data, to predict outcomes on unseen data. For instance, in oncology, supervised learning can identify patterns in patient data for early disease detection or treatment prediction. Such applications often use a mix of traditional statistical methods and machine learning techniques to enhance the accuracy and reliability of predictions, integrating conventional statistics with advanced computational models. This approach has been explored extensively, highlighting the interplay between machine learning algorithms and statistical methods in understanding complex biomedical data.

Supervised Learning Regression in the context of biostatistics involves building a model that predicts a continuous outcome variable based on one or more predictor variables. This type of regression analysis is essential in studying the relationships between independent (predictor) and dependent (outcome) variables. For example, in medical research, linear regression might be used to understand how various clinical measurements (like blood pressure and cholesterol levels) can predict a patient's risk of developing a certain disease. This approach helps identify key factors contributing to health outcomes, thus facilitating better disease management and prevention strategies. The effectiveness of regression models in supervised machine learning, particularly in the biostatistics field, is well-documented, offering a robust framework for predictive analysis. On the other hand, classification in supervised learning deals with predicting categorical outcomes from given input data. In biostatistics, this method is often used for diagnostic purposes. For instance, a classification model can be trained on patient data to categorize individuals into risk groups for a particular disease based on their symptoms and test results. This can significantly aid in early diagnosis and tailored treatment plans. Classification models, like logistic regression, are widely utilized in medical research for categorizing patient outcomes, making them indispensable tools in healthcare data analysis. In this section, we delve into the intricacies of the linear regression model, a cornerstone of statistical analysis and a fundamental tool in biostatistics.

3. Linear Regression

Linear regression originated in the 19th century, with the least squares method formally developed by Carl Friedrich Gauss and Adrien-Marie Legendre. It became a cornerstone of statistical analysis and was later adopted in machine learning. Linear regression is a fundamental statistical and machine learning technique used to model the relationship between a dependent variable and one or more independent variables. The dependent variable must be continuous, and the independent variable may be continuous, binary, or categorical. The simplest form is simple linear regression, explained in equation 1.0, where one independent variable predicts a dependent variable. This technique laid the groundwork for understanding and quantifying the strength and form of the relationship between variables, which is essential in numerous fields, including biostatistics, economics, and engineering.

$$Y = \beta 0 + \beta 1 * x1 + \varepsilon \qquad ...(1.0)$$

where:

- Y is the dependent variable representing the outcome,
- $x1$ is the independent or predictor variable,
- $\beta 0$ is the y-intercept,
- $\beta 1$ is the slope of the line, and
- ε is the error term.

For example, drawing on biological insights, it is often posited that certain variables exhibit a predictable relationship-take, for instance, the linear increase of blood pressure with age. Such hypotheses streamline analytical processes and facilitate the extraction of significant insights from the data. Equation 1.1 delineates this linear correlation between blood pressure and age, while Figure 2 graphically represents the linear association between these two continuous variables.

$$\text{blood pressure} = \beta 0 + \beta 1 * \text{age} + \varepsilon \qquad ..(1.1)$$

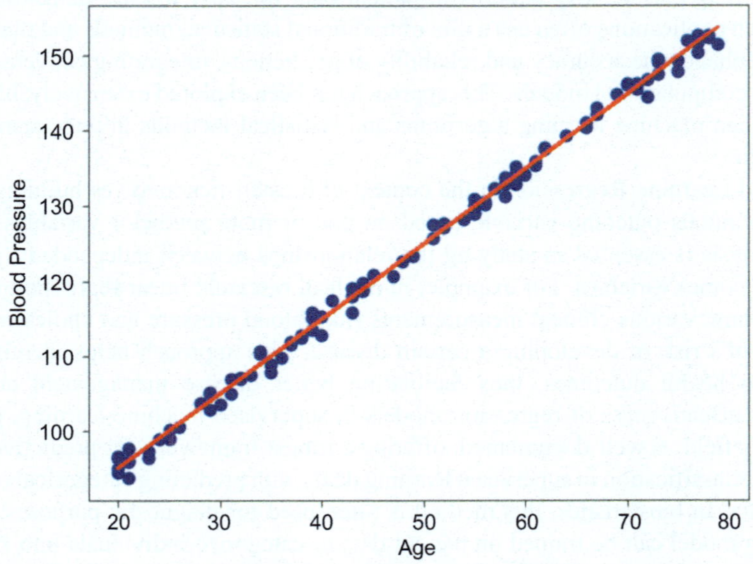

Fig. 2 Scatter plot showing a linear relationship.

Assumptions for Linear Regression: The simple linear regression model relies on a set of fundamental assumptions that are crucial for its validity and reliability. These assumptions are as follows:

Linearity: It is presumed that a linear relationship exists between the independent variable, X, and the dependent variable, Y. This linear correlation is pivotal for the model to quantify the influence of X on Y accurately.

Normal Distribution of Residuals: The model presumes that the residuals, or prediction errors, are normally distributed and independent. This means that the errors are expected to fluctuate randomly around the linear relationship without showing any systematic pattern or dependency on X.

Homoscedasticity: The assumption of homoscedasticity stipulates that the variances of the errors are uniform across all values of X. This is to avoid the complications associated with heteroscedasticity, where error variances might vary.

Representativeness of Sample Data: The model assumes that the sample data are representative of the population from which they were drawn, which is essential for the generalizability of the model's predictions. Here are the sample data from the dataset:

Age	*Blood Pressure*
64	139.4053978
67	142.1186066
73	149.2544141
20	96.41910204
23	97.25614392
79	151.4825629
23	96.49290398
59	135.1490761
29	102.8064217
39	115.1410424
41	117.5094451
70	146.0677751

Equation 1.1 has been redefined using a sample dataset comprising 100 observations. The intercept and slope values indicate the expected blood pressure changes corresponding to age. Specifically, Equation (1.2) delineates the intercept at 75.1397, which theoretically represents the anticipated blood pressure value for an age of zero. Although this may not be clinically applicable, especially for newborns, it is a statistical baseline. The positive slope of 0.9962 implies an incremental increase in blood pressure concomitant with age; thus, for each year that age advances, the blood pressure is forecasted to ascend by 0.9962 units. Expressed in terms of millimeters of mercury (mmHg), this signifies that each additional year of age is associated with an increment of 0.9962 mmHg in blood pressure.

$$\text{blood pressure} = 75.1397 + 0.9962 * \text{age} \quad ...(1.2)$$

Figure 3 showcases the residual plot, which elucidates the discrepancies between the blood pressure values predicted by the linear regression model and the actual observed values. The horizontal axis at zero demarcates the absence of residuals, where points above this axis represent underestimations, and those below signify overestimations by the model.

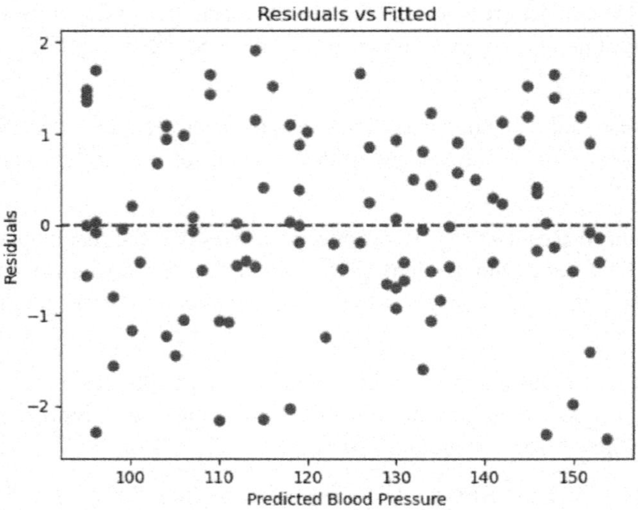

Fig. 3 Residual Plot

The performance of the model is evaluated using the metrics R-squared (R^2), Mean Squared Error (MSE), and Mean Absolute Error (MAE). For the provided sample dataset, the R^2 is 0.9967, indicating that the model accounts for approximately 99.67% of the variance in the dependent variable, signifying an excellent fit. The MSE of 1.0751, which quantifies the average of the squared discrepancies between the predicted and actual values, is considerably low, implying that the model's predictions are generally close to the observed values. Similarly, an MAE of 0.8294 reflects that, on average, the model's predictions are within 0.8294 units of the actual values, denoting high accuracy in prediction. Collectively, these metrics suggest that the model performs well regarding prediction accuracy.

4. Logistic Regression

Logistic regression, developed by statistician David Cox in 1958, revolutionized biostatistical analysis by enabling researchers to handle binary classification problems effectively. This method extends the concept of linear regression to scenarios where the outcome is categorical, particularly binary. Logistic regression is a cornerstone of biostatistics machine learning, providing a framework for analyzing datasets with categorical outcomes, often binary. Unlike linear regression, which predicts a continuous outcome based on the linear relationship between dependent and independent variables, logistic regression estimates the probability of a binary outcome, making it ideal for classification problems. It operates on the logit function, the logarithm of the odds ratio, to model the relationship, which maps any real-valued number into a value between 0 and 1, thus predicting the probability of occurrence of an event. Its historical success in epidemiology, which has been used to link risk factors to health outcomes, underlines its enduring relevance in biostatistics.

The key difference from linear regression is like the dependent variable: logistic regression does not assume a linear relationship between the independent variables and the log-odds of the dependent variable. This non-linear approach allows for a more flexible model structure that can handle changes in directionality and the effects of boundary constraints inherent in probability outcomes. The output of logistic regression is a probability score that reflects the likelihood of the dependent event, such as the presence or absence of a disease, which can be crucial for decision-making in medical research and public health. Logistic regression's strength in biostatistics lies in its ability to provide a probabilistic foundation for binary outcomes, enabling researchers to predict outcomes and understand the influence of several risk factors controlled for one another.

Consequently, logistic regression models are extensively used in medicine and epidemiology, where understanding and predicting dichotomous outcomes is often necessary. Unlike linear regression, logistic regression does not assume a linear relationship between the independent and dependent variables. Instead, it uses the logistic function to squeeze the output of a linear equation between 0 and 1. The logistic function is defined as:

$$p - \frac{1}{1+e^{-(\alpha+\beta X)}} \qquad \text{...(1.3)}$$

where:

- p is the probability of the presence of the characteristic of interest,
- e is the base of the natural logarithm,
- α is the intercept,
- β is the coefficient for the covariate X.

4.1 Assumptions for Logistic Regression

Statology (2020) states the following six assumptions underlying logistic regression.

Binary/Dichotomous Outcome: Logistic regression requires the response variable to be binary, meaning it can only take on two possible outcomes. For example, in a biostatistics study, this could be the presence or absence of a particular disease (e.g., diabetes: yes or no). This dichotomy allows the logistic model to calculate probabilities of occurrence versus non-occurrence based on predictors.

Linear Relationship in Logit: The model assumes a linear relationship between the predictors and the logit of the outcome, which is the log of odds of the dependent variable. For instance, in analyzing patient recovery (recovered or not), the age and dosage of a drug might linearly predict the log-odds of recovery, albeit this relationship is not in the original measurement scale but in a transformed logit scale.

No Multicollinearity: The independent variables should be somewhat correlated. In a study examining risk factors for heart disease, if cholesterol and fat intake are both predictors, they should be more correlated than they provide redundant information, as this would distort the influence of each variable on the outcome.

No Extreme Outliers: Logistic regression is sensitive to outliers, which can disproportionately influence the model's performance. In biostatistical research, outliers could be patients with unusual health characteristics or measurement errors, which, if included, could lead to misleading results in predicting disease risk.

Large Sample Size: A sufficiently large sample size is crucial for the reliability of the maximum likelihood estimation, the method used to estimate the model parameters. In biostatistics, this would mean enough data points to ensure stable probability estimates for the condition being studied, like the sufficient number of cases and controls in a study on the genetic predisposition to a rare disease.

In biostatistics, logistic regression is a powerful tool to elucidate risk factors associated with diseases or ascertain the presence of a specific health-related characteristic, such as disease status. It empowers practitioners to estimate the probability of events, like a cardiac event, factoring in a patient's age, gender, body mass index, and a spectrum of blood test results. Within this text, we reference a dataset comprising age, body weight, height, medication use, time before medical attention, and bone mineral density (BMD), all instrumental in prognosticating bone fracture risks. The table below showcases a subset of the data, exemplifying how logistic regression can harness such variables to predict potential health outcomes.

age	*sex*	*fracture*	*weight_kg*	*height_cm*	*medication*	*waiting_time*	*bmd*
57.0527678	F	0	64	155.5	Anticonvulsant	18	0.8793
75.7412248	F	0	78	162	No medication	56	0.79460001
70.7789	M	0	73	170.5	No medication	10	0.90670002
78.2471753	F	0	60	148	No medication	14	0.7112
54.1918766	M	0	55	161	No medication	20	0.79089999
77.1777515	M	0	65	168	No medication	7	0.73009998
56.1806177	M	0	77	159	No medication	26	1.0096
49.9161399	F	0	59	150	No medication	9	0.73100001
68.4084033	M	0	64	167	Glucocorticoids	6	0.6893

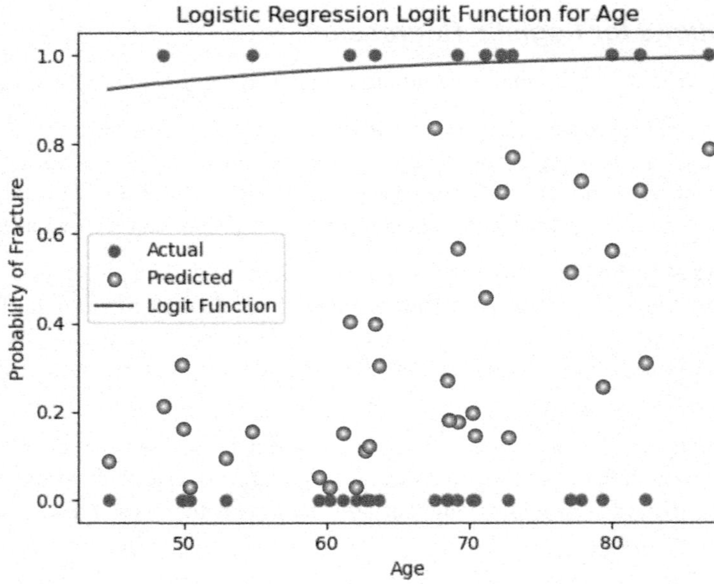

Fig. 4 Logit Function for the feature Age

The logistic regression model has been applied to the simulated dataset. The scatter plot (Figure 4) visualizes the data points with different colors representing the binary outcome of fractures (predicted) and the actual fracture binary feature (actual). The green line represents the logistic regression model's estimated probabilities of fractures as a function of age. As age increases, the model suggests that the likelihood of having a fracture also increases, following the logistic curve. This model can predict the likelihood of fractures based on age, particularly useful in medical studies and healthcare analytics. Equation 1.3 has been recalibrated with the dataset according to the specifications detailed in equation 1.4.

$$p(\text{fracture}) = \frac{1}{1 + e^{-\left(\begin{array}{c}-0.2026 + 0.05696\,x\,\text{age} - 0.0609\,x\,\text{weight}_{\text{kg}} + 0.0052\,x\,\text{height}_{\text{cm}} \\ +0.0069\,x\,bmd - 2.7609\,x\,\text{sex} + 0.4363\,x\,\text{medication}\end{array}\right)}} \quad ...(1.4)$$

Here is the interpretation of the equation 1.4 coefficients:

- $\beta 0$ (Intercept, –0.2026): This is the log odds of a fracture when all other predictor variables are zero.

- $\beta1$ (age, 0.05696): For each one-year increase in age, the log odds of fracture increase by 0.05696, holding all other variables constant.
- $\beta2$ (weight_kg, −0.0609): For each one-kilogram increase in weight, the log odds of having a fracture decrease by 0.0609, indicating that higher weight might be associated with a lower risk of fractures.
- $\beta3$ (height_cm, 0.0052): For each one-centimeter increase in height, the log odds of fracture increase by 0.0052.
- $\beta4$ (bmd, 0.0069): For each unit increase in bone mineral density, the log odds of fracture increase by 0.0069. Typically, you'd expect a negative coefficient here, as higher BMD is usually associated with a lower risk of fractures; this unexpected sign might warrant further investigation.
- $\beta5$ (sex, −2.7609): If sex is coded as 0 for females and 1 for males, this coefficient suggests that being male decreases the log odds of having a fracture by 2.7609 compared to females, all else being equal.
- $\beta6$ (medication, 0.4363): If medication use is coded as 0 for no and 1 for yes, this suggests that taking medication increases the log odds of fracture by 0.4363.

In logistic regression, evaluation metrics are critical for assessing the model's performance in classification tasks. The most common metrics include:

- **Accuracy:** This measures the proportion of true results among the total number of cases examined. It answers the question, "How many predictions were correct?"
- **Precision:** This indicates the proportion of positive identifications that were actually correct and is especially important when the cost of false positives is high.
- **Recall** (Sensitivity): This measures the proportion of actual positives that were correctly identified and is crucial when the cost of false negatives is significant.
- **F1 Score:** This is the harmonic mean of precision and recall, balancing the two with an uneven class distribution.

The model fitted using the sample dataset provides the following performance metrics.

	precision	recall	f1-score	support
0	0.80	0.87	0.83	23
1	0.67	0.55	0.60	11
accuracy			0.76	34
macro avg	0.73	0.71	0.72	34
weighted avg	0.76	0.76	0.76	34

The binary variable 'fracture' classification report reveals a model with moderate predictive performance. For the 'no fracture' class (label 0), the model exhibits a high degree of precision at 80%, indicating a strong ability to identify negative cases correctly. It also shows a recall of 87%, meaning it captures most true negative instances. The F1-score, which balances precision and recall, is robust at 0.83. For the 'fracture' class (label 1), the precision drops to 67%, with a recall of 55%. This signifies that while two-thirds of the model's positive predictions are accurate, it only identifies over half of the positive cases. The corresponding F1-score is 0.60, reflecting the trade-off between precision and recall for this class. The model's overall accuracy is 76%, indicating that it correctly predicts three-quarters of the outcomes across both classes. The macro average scores provide an unweighted mean of the model's performance across the two classes, resulting in 0.73 for precision, 0.71 for recall, and 0.72 for the F1-score. The weighted average, which considers the class imbalance, matches the model's accuracy at 0.76 for all three metrics, suggesting consistent performance relative to the prevalence of each class within the dataset.

5. Decision Tree

Decision trees (Nengsih., 2016) are a fundamental concept in machine learning and biostatistics, used for regression and classification tasks. Their history dates back to the 1960s, with the development of algorithms like ID3 and CART (Classification and Regression Trees). These trees work by repeatedly splitting the data into subsets based on specific criteria, creating a tree-like model of decisions. In biostatistics, decision trees are particularly valuable for their simplicity and interpretability. For instance, in a clinical trial, a decision tree might help categorize patients into risk groups based on their medical history and test results. The tree's branches represent the decision paths, and the leaves represent outcomes or classifications. Decision trees predict continuous values for regression tasks, like estimating a patient's future cholesterol level based on their diet and physical activity. The model does this by segmenting patients into groups with similar outcomes and calculating the average outcome within each group. However, decision trees have their pros and cons. One of their most significant advantages is their ease of interpretation. Unlike many complex models, decision trees can be visualized and understood even by those with limited statistical knowledge, making them particularly useful in biostatistics, where explainability is crucial. For example, a decision tree can clearly show how different treatment options might affect patient outcomes, aiding in patient understanding and clinical decision-making. Additionally, they are non-parametric, meaning they don't require specific assumptions about the data structure. Conversely, decision trees are prone to overfitting, especially with complex data and without proper pruning. They can also be unstable, as small changes in the data might lead to a significantly different tree. Moreover, decision trees are often less accurate than other, more sophisticated algorithms.

The interpretability of decision trees in biostatistics is particularly noteworthy. They can effectively illustrate the decision-making process in medical diagnoses or treatment protocols. For example, a decision tree might be used to decide on the course of treatment for cancer patients, considering factors like age, tumor size, and genetic markers. The tree structure allows for clear visualizations of the paths leading to different treatment recommendations, making it easier for healthcare professionals to explain the rationale behind their decisions to patients. Furthermore, decision trees can be combined with other techniques, such as random forests or boosting, to improve accuracy while maintaining interpretability. These ensemble methods use multiple decision trees to make more robust predictions. However, while they increase accuracy, they can also make the model more complex and less interpretable. This is a crucial consideration in biostatistics, where understanding the model's decision path can be as important as the prediction itself. The ID3 algorithm for decision trees is a classification method that aims to predict categorical outcomes, such as whether a fracture has occurred based on various features. Applying ID3 to a dataset (previous section sample dataset) with attributes like age, sex, weight_kg, height_cm, medication, waiting_time, and bmd to predict 'fracture' status involves several steps. Initially, the algorithm calculates the entropy for the 'fracture' attribute to measure the dataset's disorder. It then computes the information gain for each feature—how much each feature reduces uncertainty about the fracture outcome. The feature with the highest information gain is the tree's root node. For instance, if 'age' offers the highest information gain, the dataset is split based on age groups. This process recursively continues for each subset, with the algorithm selecting the best feature at each node based on the highest information gain, which reduces entropy. The splits stop when the algorithm reaches a leaf node with a homogenous class (all 'fracture' or 'no fracture') or when it meets other stopping criteria, such as a maximum tree depth. It's important to note that while ID3 can handle discrete data directly, continuous variables like age, weight_kg, height_cm, and waiting_time must be discretized before they can be used in ID3. Also, ID3 is not inherently suited for regression tasks (predicting a continuous variable); extensions or variations like the C4.5 or CART algorithms are more appropriate for such tasks, as they can handle continuous target variables.

The classification report showcases the performance of a predictive model on two classes, denoted as '0' and '1'. Class '0' demonstrates high effectiveness with precision, recall, and F1-

score, all at 0.87, indicating strong predictive accuracy and consistency for the 23 instances in this class. Class '1', with 11 instances, shows a moderate performance with precision, recall, and F1-score, all at 0.73, reflecting decent accuracy but lower than that of class '0'. The overall model accuracy stands at 0.82, meaning it correctly predicts the outcome 82% of the time. The macro average, reflecting unweighted mean performance across both classes, is 0.80 for precision, recall, and F1-score, suggesting balanced class performance. The weighted average for all three metrics is 0.82, taking into account the class imbalance, which indicates that the model performs very well overall, especially when recognizing the more frequent class '0'.

Tree pruning is a crucial technique in decision tree algorithms aimed at simplifying the model to prevent overfitting, where a tree might predict training data with high accuracy but fail to generalize to unseen data. In decision trees, pruning can be executed through two main strategies: pre- and post-pruning. Pre-pruning, also known as early stopping, halts the tree's growth prematurely by setting constraints such as maximum depth or minimum gain, thus avoiding fully complex trees. Post-pruning, or cost complexity pruning, involves creating an entire tree and systematically removing the branches that contribute least to the tree's predictive ability, often guided by a complexity parameter penalizing the model for having too many leaves. Pruning aims to strike an optimal balance between the decision tree's depth and accuracy, thereby enhancing the model's generalizability and interpretability. Pre-pruning can save computational resources by stopping the tree's growth early but may risk underfitting if stopped too soon. On the other hand, post-pruning allows the tree to capture the nuances in the data before strategically simplifying it. This step is essential in the model-building process, as it creates a more robust and streamlined decision tree that is easier to understand and performs better on new, unseen datasets, making it highly valuable in practical machine learning applications.

The performance metrics have improved after pruning the decision tree, indicating a more generalized model. Class '0' now has an increased precision of 0.88 and a recall of 0.91, with an F1-score of 0.89 for its 23 instances, signaling enhanced predictive accuracy and consistency. Class '1' also shows improved performance with a precision of 0.80 and a recall of 0.73, leading to an F1-score of 0.76 for its 11 instances, which, while still lower than class '0', reflects better accuracy post-pruning. The overall accuracy of the model has risen to 0.85, meaning it now makes correct predictions 85% of the time. The macro average, which gives an equal weight to each class performance, is now 0.84 for precision and 0.82 for recall, with a F1-score of 0.83, suggesting a more balanced performance across classes. The weighted average has also increased to 0.85 for precision, recall, and F1-score, acknowledging the class imbalance and highlighting that the pruned model has improved its predictive power, particularly in correctly identifying the more prevalent class '0'.

6. K-Nearest Neighbors (KNN)

The *k*-nearest neighbors (KNN) algorithm (Peterson, 2009) is a non-parametric method used for classification and regression, fundamental in machine learning, including biostatistics. It was first mentioned in 1951 by Fix and Hodges and later developed into a more defined algorithm by Cover and Hart in 1967. The premise of KNN is straightforward: it predicts the outcome for a new sample based on the majority vote or average of its '*k*' closest neighbors in the feature space. In classification, KNN assigns a class to the new sample based on the most common class among its nearest neighbors, while in regression, it predicts a continuous value based on the average or median of the values of its neighbors. For example, in biostatistics, KNN can classify patients into risk categories for a particular disease by analyzing the closest neighbors in a multidimensional space of risk factors or predict a patient's future health metrics by averaging the measures of the nearest neighbors in the dataset.

The KNN model is known for its simplicity and effectiveness, especially in scenarios where the decision boundary is irregular. In biostatistics, this could translate to quickly classifying patient outcomes based on similar case profiles or predicting a disease's progression by averaging similar patients' trajectories. It does not make any assumptions about the underlying data distribution, which is a significant advantage when dealing with real-world biostatistical data that often violates the typical assumptions of many parametric methods. Moreover, KNN is a lazy learning algorithm. It does not build a model using the training data but stores the dataset and performs an action on the dataset at the time of classification. However, this also leads to one of the cons of KNN: as the size of the dataset grows, the prediction phase can become computationally expensive. Additionally, the performance of KNN is heavily dependent on the choice of 'k' and the distance metric used to identify the nearest neighbors. Choosing the right parameters and metrics without overfitting requires careful validation, which can be time-consuming. Interpretability is another factor where KNN provides mixed outcomes. On the one hand, the rationale behind the KNN predictions can be pretty transparent-based on resemblance to the closest known examples. This can be particularly illuminating in biostatistics, where understanding why a specific classification or prediction has been made can be as critical as the outcome itself. For instance, a doctor might find it helpful to explain that a patient is classified as high risk for a condition because they share key health parameters with other high-risk patients. On the other hand, when dealing with high-dimensional data, the concept of "nearest" can become abstract and challenging to visualize, making the interpretability less straightforward. Moreover, KNN's sensitivity to irrelevant or redundant features means that feature selection can significantly impact the model's performance and interpretability. Therefore, while KNN can be very powerful and intuitive in many biostatistical applications, it also demands a rigorous process of parameter tuning and understanding of the dataset to ensure that the results are accurate and meaningful.

The KNN algorithm identifies the '*k*' closest data points to a query instance and makes predictions based on these neighboring data points. The steps include: selecting 'k,' the number of neighbors; choosing a distance metric (commonly Euclidean distance: $d(x, y) = \sqrt{\sum_{i=1}^{n} (x_i - y_i)^2}$; finding the '*k*' nearest neighbors to the query point in the training dataset; and then, for classification, determining the majority class among those neighbors or, for regression, calculating the mean or median of the continuous outcome of those neighbors.

The KNN classification model uses the BMD dataset, with k set to 5, based on the features 'age' and 'weight'. The KNN plot visualizes the decision boundaries of a k-Nearest Neighbors classifier with $k = 5$. Here, the background color represents the classification decision areas based on the majority vote of the five closest data points, with the data points themselves plotted as dots. Purple regions indicate one class, while yellow regions indicate the other class, showcasing where a new data point would fall into either category based on its 'Age' and 'Weight_kg' features.

In classification, KNN assigns the class most common among the nearest neighbors to the query point. For instance, in a biostatistics application, if you're classifying patient disease status and choose $k = 3$, the algorithm will find the three patients in the dataset closest to the new patient in terms of similarity metrics like age, weight, or blood markers, and assign the disease status that appears most among these three patients. In regression, KNN predicts a value based on the average (or median) of the '*k*' nearest neighbors. So, if estimating a patient's future blood sugar levels, KNN would average the levels of the '*k*' closest patients in the dataset, providing a prediction for the new patient.

7. Support Vector Machine (SVM)

Support Vector Machines (SVMs) are related to supervised learning methods for classification and regression. As a classification tool, a SVM model represents the examples as points in space, mapped so that the examples from separate categories are divided by a clear gap that is as wide as possible. In regression, SVMs are used to predict the value of a given data point by fitting a line (or hyperplane in higher dimensions) that deviates from the actual observed values by a value no more significant than a specified error margin. The fundamental SVM was originally designed for binary classification, introduced in COLT-92 by Boser, Guyon, and Vapnik. The method was later extended to regression and multi-class classification problems, which are highly relevant in biostatistics for predicting patient outcomes or classifying medical images.

The advantages of SVMs include their effectiveness in high-dimensional spaces and their use of a subset of training points in the decision function, known as support vectors, which makes the method memory efficient. SVMs (Cervantes et al., 2020) are also versatile, as different kernel functions can be specified for the decision function. However, SVMs could improve with large datasets due to their high algorithmic complexity and extensive memory requirements during training. They also require careful parameter tuning and data preprocessing, like feature scaling, which can be a significant disadvantage in biostatistical applications where the interpretability of the model is crucial. In biostatistics, the interpretability of SVMs can be challenging, especially when using non-linear kernels. For example, understanding how a SVM model with a radial basis function kernel has classified different disease genotypes may take much work. However, the models can be powerful for organizing patients into risk groups for heart disease based on various biometric measurements. SVMs can handle the complex relationships in medical data, but the results may require translating into more clinically meaningful terms for practitioners and stakeholders.

SVM for regression and classification operates on fundamental assumptions. Firstly, SVM assumes the data is linearly separable in a high-dimensional space. It relies on kernel functions to map the data into a suitable space for non-linear separability. The core idea is based on finding the optimal hyperplane that maximizes the margin between different classes in classification tasks, following the formula $w \cdot x - b = 0$, where w is the normal to the hyperplane, x represents the feature vectors, and b is the bias. In regression tasks, known as Support Vector Regression (SVR), an attempt is made to fit the best hyperplane within a threshold ε that captures as many data points as possible, which can be expressed as $|y - (w \cdot x + b)| \leq \varepsilon$. SVM assumes that features are on a similar scale, thus requiring feature scaling, and it takes an apparent gap or margin between classes without too much noise. Biostatistics examples include using SVMs to classify patient outcomes based on clinical trial data or to model biological responses to drug doses, where clear margins between responses are assumed. Plots demonstrating SVM classification might depict the separation margin and support vectors, whereas SVR plots could illustrate the ε-insensitive tube and fitted line within.

The SVM is a versatile machine learning algorithm that can be used for regression and classification tasks, though it is commonly used for classification. For both tasks, the SVM constructs a hyperplane in high-dimensional space to separate different classes or to fit the data. The goal of the SVM algorithm is to find a hyperplane with the maximum margin, meaning it is as far away as possible from the nearest data points of any class, known as support vectors. Mathematically, this involves solving an optimization problem to maximize the margin around the hyperplane. The hyperplane can be described by the equation $w \cdot x - b = 0$, where w is the weight vector, x is the feature vector, and b is the bias. SVMs can perform linear classification by finding the straight line (in two dimensions) or hyperplane (in higher dimensions) that best separates two classes. In cases where the data is not linearly separable, SVMs can use kernel functions to project the data into higher dimensions where a hyperplane can separate the classes. The most common kernels are polynomial, radial basis function (RBF), and sigmoid.

In the case of regression (SVR), the SVM algorithm attempts to fit the best hyperplane that has at most ϵ deviation from the actual values of the training dataset for a given part of the data points. Unlike classification, where the motivation is to maximize the margin between two classes, in regression, the objective is to fit as many data points as possible within the ε-margin while limiting the penalties for data points outside the margin. Biostatistical examples of SVM applications include using SVM classification to differentiate between benign and malignant tumor cells based on cellular characteristics or SVR to predict patient recovery times based on treatment regimens and patient demographics.

8. Random Forest

Random Forest is a flexible, easy-to-use machine learning algorithm that usually produces excellent results, even without hyperparameter tuning. It is an ensemble learning method where a group of weak models combine to form a robust model. Introduced by Leo Breiman in 2001, Random Forest can be used for both classification and regression tasks. It builds multiple decision trees and merges them together to get a more accurate and stable prediction. The algorithm starts with a dataset and randomly samples data points with replacement, creating what are known as bootstrap samples. Each tree is grown on these bootstrap samples, but at each split in the learning process, only a subset of all the features is considered for the best split. This "bagging" with "feature randomness" helps create an ensemble of diverse trees and leads to a model with lower variance than a single decision tree, thereby reducing overfitting.

The advantages of Random Forest (Belgiu, et al., 2016) are numerous. It can handle large datasets with higher dimensionality and evaluate each feature's importance in the regression or classification task. It works well with categorical and numerical data and doesn't require feature scaling. Random Forests also handle missing values and maintain accuracy for missing data. One of the downsides, however, is model interpretability: Random Forest models are not easily interpretable compared to Decision Trees, as they are more complex due to the ensemble of trees. They also require more computational resources and are less intuitive. When the data includes groups of correlated features of similar relevance for the output, smaller trees can favor those groups over others. In biostatistics, Random Forest could be applied to classify patients into different health risk categories based on their medical records or predict how other genetic profiles respond to specific treatments. The interpretability can be partially addressed through techniques like feature importance, which gives insights into which variables contribute most to the model's decisions. Despite the complexity, the interpretability of Random Forest can be enhanced with such techniques, allowing biostatisticians to extract meaningful conclusions from the model predictions.

Random Forest, an ensemble learning method that operates by constructing multiple decision trees, relies on a few assumptions about the data. Unlike linear models that require a linear relationship between independent and dependent variables, Random Forest can capture nonlinear relationships. It does not presuppose that the data follows any distribution. The central assumption in Random Forest is that the individual trees in the ensemble should be as independent from one another as possible to reduce the overall variance. It assumes that there are interactions between features, which allows the Random Forest to capture complex structures in the data. For regression, Random Forest believes that the relationships between predictors and the outcome are additive, which allows individual decision trees to estimate partial dependence plots effectively. In biostatistics, Random Forest can be used to identify essential predictors for disease risk factors or treatment efficacy by examining feature importance scores derived from the forest. While individual decision trees can be visualized to understand the decision paths and splits, the collective interpretation of an entire Random Forest requires statistical summaries or aggregations, such as variable importance plots, to interpret the model's predictions. Random Forest is an ensemble learning technique that creates a 'forest' of decision trees and aggregates their predictions. Here is how you would apply it to your dataset for classification:

Bootstrap Dataset Creation: From the previous section's bmd dataset, Random Forest, starts by creating multiple bootstrap datasets. This means it samples from your dataset with replacement, creating new datasets of the same size as the original but with some instances repeated and some omitted.

Building Decision Trees: Random Forest grows a decision tree for each bootstrap dataset. At each decision point within the tree, only a random subset of all possible features is considered to determine the best split. For instance, if you have the features 'age', 'sex', 'weight_kg', 'height_cm', 'medication', 'waiting_time', and 'bmd', a tree might only consider 'age', 'weight_kg', and 'medication' for the first split. This process is repeated at each split point in each tree, resulting in a diverse forest of trees.

Aggregating Predictions: Once the forest of trees is built, predictions are made by having each tree vote for its predicted class (in classification tasks) or by averaging the predicted values (in regression tasks). The class that gets the majority of votes becomes the model's prediction.

To visualize a Random Forest model, you could plot feature importance scores, which quantify the contribution of each feature to the prediction. However, plotting the decision boundaries of a Random Forest is more complex due to the complexity of the model. However, the importance of features can provide insights into the predictive power of each variable in your dataset.

Now we look at the relative importance of each feature when predicting the target variable using a Random Forest model. In this context, 'bmd' (bone mineral density) is the most significant predictor, which is intuitive given its direct clinical relevance to fracture risk. It has the highest score, indicating that changes in 'bmd' values are highly influential in the model's decision-making process. 'Age' is the second most important feature, which aligns with the medical understanding that fracture risk increases with age. The 'weight_kg,' 'waiting_time,' and 'height_cm' have moderate importance, suggesting that while they play a role in fracture prediction, their impact is less than 'bmd' and 'age.' 'Medication' and 'sex' hold the most minor importance in this model. This could be because they vary less in the population studied, or other features overshadow their predictive power. It is critical to note that while 'sex' has low importance in this model, it does not imply that sex is not an important factor in fracture risk; rather, it may be that the information it provides is redundant in the presence of other, more dominant features.

Random Forest is a robust ensemble machine learning algorithm renowned for its versatility, as it can be used for both classification and regression tasks. One of its main advantages is that it can handle large datasets with high dimensionality and is relatively immune to overfitting due to its mechanism of averaging multiple decision trees. It also performs well with missing values and can maintain accuracy when significant data is missing. Random Forest provides a good indicator of the importance of the feature, which can be valuable in biostatistical analysis. However, Random Forest comes with its disadvantages. It can be computationally intensive and require more resources, making it less ideal for real-time predictions. Despite its robustness, it can still be challenging to interpret compared to simpler models such as decision trees, as understanding the collective decision-making process of hundreds of trees is not straightforward. This could be a significant drawback in biostatistics, where interpretability can be as crucial as predictive power. Additionally, Random Forest models can take a long time to create, mainly when dealing with large datasets, which may need to be more practical in time-sensitive environments.

The classification report for the Random Forest model reveals its superior performance over other models like Logistic Regression, Decision Trees, SVM, and KNN, particularly in the biostatistical context. Class '0' shows a high precision of 0.91, indicating 91% accuracy in predictions, and a recall of 0.87, reflecting the model's ability to identify 87% of actual instances correctly. The F1-score of 0.89 suggests a strong balance between precision and recall. For the more challenging class '1', the model achieves a precision of 0.75 and a recall of 0.82, with a F1-score of 0.78, indicating effective identification of class '1' instances despite its lower representation in the dataset. The overall accuracy stands at 0.85, meaning the model correctly predicts outcomes 85% of the time. Both the macro and weighted average hover around 0.83-0.85, illustrating the model's consistent and balanced performance across classes. These results underscore the Random Forest model's enhanced capability to handle complex, non-linear relationships in data, making it a more effective tool for predictive analytics in biostatistics compared to other standard machine learning models.

9. Neural Network

Artificial Neural Networks (ANN) are inspired by the biological neural networks that constitute animal brains. Both artificial and biological neural networks are composed of basic units called neurons. A biological neuron has three main parts: the soma, cell body, and dendrites that branch out like trees and become finer with each division to receive signals from other neurons, and a single axon that transmits the neuron's output to the synapses of other neurons; often covering more distance than the dendrites.

Neural Networks, or ANNs, are computational models that have evolved from the biological concept of neural networks in living organisms. Their history dates back to the 1940s with the introduction of the McCulloch-Pitts neuron model, which laid the groundwork for developing ANNs. The first practical application came with the perceptron in the 1950s, developed by Frank Rosenblatt. However, it was not until the 1980s, with the advent of the backpropagation algorithm, that neural networks truly gained momentum, allowing for the training of multi-layer networks. These networks were found to be particularly good at pattern recognition tasks, leading to widespread adoption in various fields, including biostatistics. Today, deep learning, a subset of machine learning based on ANNs with representation learning, has become a significant area of research and application, contributing to advances in image and speech recognition, natural language processing, and medical diagnosis. Neural Networks offer several advantages, making them popular for complex tasks. Given enough neurons and layers, according to the universal approximation theorem, they can approximate any continuous function. ANNs are particularly powerful for modeling non-linear relationships in data, which are common in biostatistics, such as gene expression patterns or the progression of diseases. They can handle large volumes of data and find patterns that may be too complex for a human researcher to identify. Moreover, ANNs have been pivotal in advancing the field of predictive analytics in medicine, where they are used to predict patient outcomes, drug responses, and disease spread. However, their computation-intensive nature requires significant processing power and time, particularly for training deep learning models.

On the downside, neural networks can be seen as black boxes, offering little explanatory insight into how they reach their conclusions. This lack of interpretability can be a significant issue in biostatistics, where understanding the decision-making process is crucial for clinical acceptance. ANNs can also overfit the training data if not correctly regularized or if the network is too complex relative to the amount of training data. This can lead to poor generalization of unseen data. Despite these cons, ANNs have been utilized effectively in biostatistics, for example, in developing personalized medicine, where they process complex patient data to tailor treatment plans. Nevertheless, the field continues to evolve with efforts to make neural networks more interpretable and to balance model complexity with predictive performance.

In a neural network designed for biostatistical analysis, the input layer receives data points such as 'age,' 'sex,' 'weight_kg,' 'height_cm,' 'medication,' 'waiting_time,' and 'bmd' as individual nodes. These inputs are then transmitted to the hidden layers, where each neuron processes the inputs using weighted connections and biases through activation functions that introduce non-linearity, allowing the model to learn complex patterns. The output layer consists of a single neuron for the binary classification of 'fracture,' predicting the likelihood of a fracture occurrence.

For instance, in a clinical study dataset, the neural network would take patient demographics ('age,' 'sex'), physiological measurements ('weight,' height,' 'BMD'), treatment data ('medication,' 'waiting_time'), and learn to predict fracture risk. It would adjust its internal weights through training to minimize prediction errors, using techniques like backpropagation and gradient descent. The 'fracture' node in the output layer would provide the probability of a fracture, aiding in patient risk stratification and management. This neural network model is a powerful tool for uncovering intricate biostatistical relationships and enhancing decision-making in healthcare.

The Neural Network model's classification report reflects its substantial capability, particularly compared to traditional models like Logistic Regression, Decision Trees, SVM, and KNN. Class '0' shows impressive precision (0.89) and recall (0.96), resulting in a high F1-score of 0.93, indicating the model's effectiveness in correctly predicting and identifying class '0' instances. In contrast, for class '1', the precision is slightly lower at 0.83, and the recall drops to 0.62, leading to a F1-score of 0.71, which suggests a modest decline in accuracy for this class. Overall, the model achieves an accuracy of 0.88, with a macro average of 0.82 and a weighted average of 0.88, reflecting a balanced and robust performance across both classes. This level of accuracy and precision in the model's predictions, particularly for class '0', showcases its superiority in handling complex datasets over the conventional machine learning models, affirming its suitability for more intricate predictive analytics tasks.

This chapter delved into various fundamental supervised learning techniques, including Linear Regression, Logistic Regression, Decision Trees, Support Vector Machines, K-nearest neighbors, Random Forests, and Neural Networks, thoroughly examining their advantages and limitations. The advent of advanced supervised learning methods has significantly transformed the fields of biostatistics and medical science, enabling more precise and in-depth analysis of multifaceted biological and medical data. Techniques such as Deep Learning and various Ensemble Models, notably Random Forests and Gradient Boosting Machines, have been pivotal in extracting complex patterns and insights from large datasets, surpassing the capabilities of traditional statistical approaches. Deep Learning, particularly with the application of Convolutional Neural Networks (CNNs) and Recurrent Neural Networks (RNNs), has been instrumental in advancing medical imaging and genomic sequence analysis. Ensemble models, renowned for their robust accuracy and proficiency in managing imbalanced datasets, are vital in forecasting disease outcomes and evaluating treatment effectiveness. Their proficiency in dealing with the high dimensionality of medical data renders these methods indispensable for advancing precision medicine and tailoring individualized healthcare solutions.

References

Belgiu, M. and Drăguţ, L. (2016). Random forest in remote sensing: A review of applications and future directions. *ISPRS journal of photogrammetry and remote sensing*, *114*, 24–31.

Cervantes, J., Garcia-Lamont, F., Rodríguez-Mazahua, L. and Lopez, A. (2020). A comprehensive survey on support vector machine classification: Applications, challenges and trends. *Neurocomputing*, *408*, 189–215.

Cheung, J.H., Kulasegaram, K.M. and Woods, N.N. (2017) Columbia Precision Medicine Initiative. (n.d.). Biostatistics. Retrieved from https://www.precisionmedicine.columbia.edu/content/biostatistics

Iniesta, R., Stahl, D. and McGuffin, P. (2016). Machine learning, statistical learning and the future of biostatistics. Biostatistics, 17(3), 589–613. https://doi.org/10.1093/biostatistics/kxw014

Nengsih, W. (2016). Fundamental Concept of Decision Tree. Advanced Science Letters, 22(10), 3173–3175.

Peterson, L.E. (2009). K-nearest neighbor. *Scholarpedia*, *4*(2), 1883.

Statology. (2020, October 13). The 6 Assumptions of Logistic Regression (With Examples).

Weissler, E.H., Naumann, T., Andersson, T., Fergus, P. and Dunger-Baldauf, C et al. (2021). The role of machine learning in clinical research. Trials, 22(1), 1–10. https://doi.org/10.1186/s13063-021-05489-x

CHAPTER

8

Unsupervised Learning

Arasu Narayan[1*]

1. Introduction

Unsupervised learning is a type of machine learning that deals with finding patterns and relationships in data that are not labeled or classified. Unlike supervised learning, where the model is trained on a predefined set of data with known outcomes, unsupervised learning algorithms explore the intrinsic structure of the data to uncover hidden patterns without prior knowledge of the outcomes. In the biostatistics space, unsupervised learning is beneficial for discovering novel patterns or subgroups within complex biological data, such as genomics, proteomics, or metabolomics data, where the sheer volume and complexity of the data make manual analysis impractical. Historically, the application of unsupervised learning in bioinformatics has evolved with the increasing availability of large-scale biological data sets, as demonstrated in studies such as "Unsupervised Machine Learning in Bioinformatics" and "Machine Learning: Algorithms, Real-World Applications." These methods have facilitated advancements in understanding biological processes and disease mechanisms, such as identifying new biomarkers or therapeutic targets in the absence of biases introduced by supervised methods, thereby accelerating discovery in biostatistics.

2. Unsupervised Learning

Unsupervised learning is a branch of machine learning that involves finding patterns and structures in data without labeled responses. It is instrumental in biostatistics for its ability to discern underlying structures in complex biological data. It can lead to insights into genomics, proteomics, or disease patterns without the need for predefined categories or labels. Cluster analysis is a method of unsupervised learning that groups a set of objects so that objects in the same group are more similar to each other than those in other groups. It's valuable in biostatistics for categorizing patient subgroups based on genetic or phenotypic data without prior knowledge of the grouping. Dimensionality reduction simplifies large sets of data into fewer dimensions while retaining significant information, aiding in the visualization and analysis of high-dimensional biostatistical data. Recommendation engines use unsupervised learning to predict user preferences, which could be adapted in biostatistics for personalized medicine approaches. Anomaly detection

[1] Northeastern University, Seattle, USA.
* Corresponding author: a.narayan@northeastern.edu

identifies unusual patterns that do not conform to expected behavior, which helps identify outliers or anomalies in biomedical datasets. Lastly, self-organizing maps are an artificial neural network that uses unsupervised learning to produce a low-dimensional representation of the input space, helping to map complex biological data into simpler geometries for analysis. In this section, we explore the complexities of clustering models, which are essential to the framework of unsupervised learning.

3. Clustering Algorithm

Clustering in biostatistics is an essential unsupervised machine learning technique. It involves analyzing biostatistical data without predefined labels, allowing the algorithm to identify natural groupings or clusters. These clusters represent data points that share similarities, providing insights into patterns or features inherent in the biological or medical data. This method is instrumental in exploratory data analysis, pattern recognition, and feature extraction in biostatistics.

Types of Clustering Algorithms in Biostatistics

1. **Density-Based Clustering:** In this type of clustering, data points in datasets are clustered into areas of high density. These clusters can assume various shapes and help identify dense regions in genetic data or patient populations. Outliers, which are points in low-density regions, are typically not included in clusters, making this method robust for anomaly detection in biostatistical data.
2. **Distribution-Based Clustering:** In biostatistics, distribution-based clustering assigns data points to clusters based on the likelihood of belonging to a particular distribution. This method suits datasets where the biological or medical data points form distinct distributions, as in the analysis of disease prevalence across different demographics.
3. **Centroid-Based Clustering:** Widely used in biostatistics, centroid-based clustering assigns data points to clusters based on their proximity to a central point or centroid. It's efficient for large biostatistical datasets but sensitive to initial conditions. This method effectively groups patients based on similar clinical characteristics or genetic markers.
4. **Hierarchical-Based Clustering:** This method suits biostatistics data with inherent hierarchical structures, such as phylogenetic trees or patient hierarchies in a healthcare database. It organizes data into a tree-like structure, facilitating the analysis of nested biostatistical data.

In biostatistics, clustering algorithms are invaluable for tasks like anomaly detection, where they help identify unusual data points, such as rare diseases in patient datasets. Clustering assists in feature selection by uncovering patterns in biostatistical data, which is crucial for developing predictive models in medical research. It is beneficial for exploratory analysis of unknown datasets, providing insights into potential relationships or previously unknown groupings. Real-world applications include genetic clustering for personalized medicine, patient segmentation in healthcare services, and epidemiological studies for public health planning.

4. K Mean Clustering

The K-means algorithm was first proposed by Stuart Lloyd at Bell Labs in 1957 as a technique for pulse-code modulation, but it was not published in a journal until 1982. Since then, K-means clustering has been applied in various fields due to its simplicity and efficiency. The algorithm's popularity stems from its ease of implementation, speed in processing large datasets, and adaptability to different data types. K-means clustering is a widely used unsupervised machine learning algorithm for partitioning a dataset into K distinct, non-overlapping subgroups (clusters) where each data point belongs to only one group (Sinaga & Yang, 2020). It aims to minimize the variance within each cluster. The process begins by randomly selecting K points as initial centroids. Each data point is then assigned to the nearest centroid, and the centroid of each cluster is recalculated. This assignment and updating process continues iteratively until the centroids stabilize, meaning their positions no longer change significantly, indicating that the clusters are as compact and distinct as possible.

In biostatistics, K-means clustering is employed for various analyses , such as gene expression profiling or patient stratification. For instance, in a dataset containing patient medical records, K-means can be used to group patients with similar symptoms or diagnoses, thereby aiding in identifying common characteristics or treatments. In genetic studies, it can cluster genes with similar expression patterns, which may indicate similar functions or shared regulatory mechanisms. This section performed K-mean clustering using the 'RNA-Seq (HiSeq) PANCAN' dataset. It is a well-recognized collection in the field of genomics, consisting of gene expression profiles from patients diagnosed with various types of tumors, including Breast Invasive Carcinoma (BRCA), Kidney Renal Clear Cell Carcinoma (KIRC), Colon Adenocarcinoma (COAD), Lung Adenocarcinoma (LUAD), and Prostate Adenocarcinoma (PRAD). Each entry in this dataset represents the expression levels of thousands of genes, providing a comprehensive snapshot of the active biological processes within the cancerous cells of each tumor type.

The K-means clustering model applied to this dataset aims to partition the high-dimensional gene expression data into distinct groups or clusters, each corresponding to different tumor profiles. By doing so, the model provides insights into the underlying patterns and similarities between various cancer types. The clustering algorithm iteratively assigns each gene expression profile to the nearest cluster centroid, which is recalculated until convergence, leading to the formation of clusters that minimize within-cluster variance. Figure 1 illustrates the distribution of gene expression profiles grouped into five distinct clusters, each marked by a unique color and a red '*' denoting the centroid. The clustering likely reflects intrinsic biological similarities between the gene expressions of the tumors. For example, closer clusters may indicate tumor types with similar gene expression patterns, suggesting shared pathways or origins. The separation and density of clusters can also

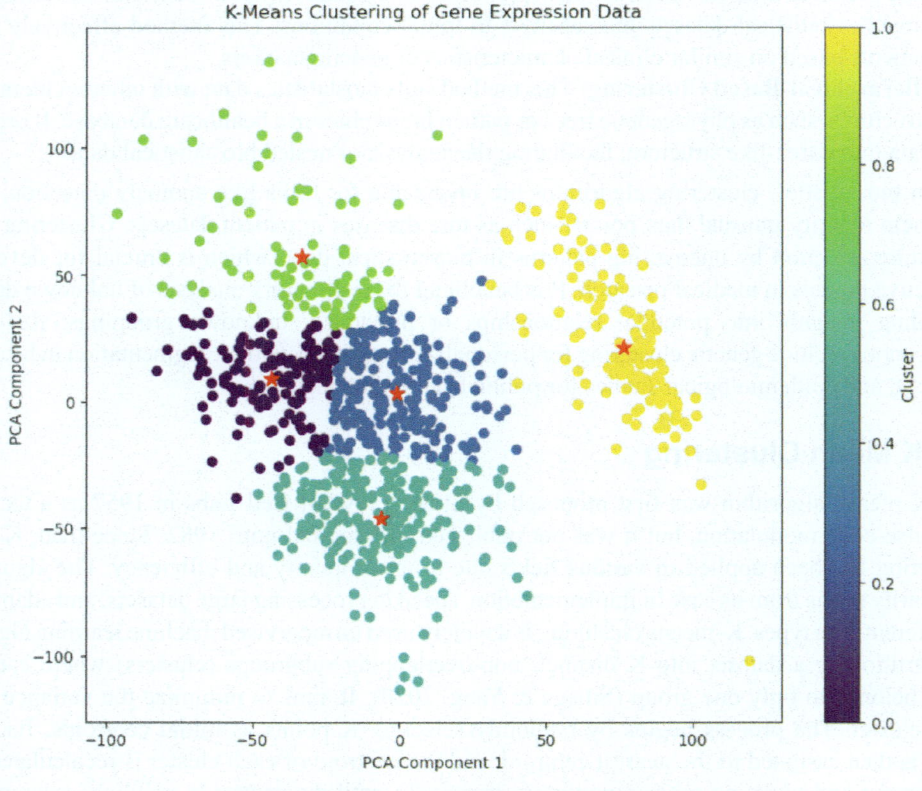

Fig. 1 K means Clustering for Gene *Dataset*

provide an indirect measure of the variability within and between different cancer types, offering potential pathways for further biomedical research and personalized medicine approaches.

Pros and Cons of K-Means Clustering

Pros:

1. **Efficiency:** K-means is computationally less intensive, making it suitable for large datasets.
2. **Simplicity:** Its ease of implementation and understanding makes it accessible for various applications.
3. **Adaptability:** K-means can be used with various data types and effectively identifies distinct clusters.

Cons:

1. **Selection of K:** The need to pre-specify the number of clusters (K) can be a drawback as it may only sometimes be evident.
2. **Sensitivity to Initial Centroids:** The final clusters can vary significantly based on the initial choice of centroids.
3. **Difficulty with Non-Convex Shapes:** K means clustering struggles with clusters with non-spherical shapes or varying sizes.
4. **Outliers Influence:** Outliers can skew the clustering significantly, leading to misleading results.

With its straightforward approach, K-means clustering has become a fundamental tool in data analysis, including biostatistics, despite some limitations that necessitate careful application and interpretation of results.

5. DBSCAN Clustering

The DBSCAN algorithm was proposed by Martin Ester, Hans-Peter Kriegel, Jörg Sander, and Xiaowei Xu in 1996. It was introduced as a solution to the limitations of earlier clustering algorithms, like K-means, that had difficulty with clusters of arbitrary shapes and varying densities. DBSCAN's ability to find clusters based on the density of data points made it suitable for applications such as identifying regions of high traffic in transportation systems, grouping stars in astronomy, and segregating types of crops in agriculture. DBSCAN (Density-Based Spatial Clustering of Applications with Noise) is an unsupervised machine learning algorithm used for clustering tasks (Shen et al., 2016). It is particularly adept at identifying clusters of varying shapes and sizes. The key concept behind DBSCAN is that a cluster is defined as a high-density area surrounded by an area of low density. It starts with an arbitrary point, and a cluster is formed if this point has a sufficient number of neighbors within a given radius (ε). The process continues by adding all density-reachable points to the cluster. This expands the cluster until no new points can be added. Points that do not belong to any cluster are labeled as noise. Applying DBSCAN to a biostatistics dataset, such as one containing gene expression data, could enable the identification of complex patterns. For instance, it could be used to detect outlier genes with unique expression profiles compared to the rest of the dataset or to find clusters of genes that behave similarly across different conditions or treatments.

The advantages of DBSCAN include its ability to form clusters of arbitrary shapes and sizes, its robustness to outliers, and its independence from prior knowledge of the number of clusters. It also requires minimal input parameters and can identify outliers in the clustering process. However, DBSCAN has some limitations as well. It can need help with datasets where the density varies significantly across clusters, as it may merge distinct clusters or split a single cluster into multiple parts. The choice of ε and the minimum number of points can also be challenging to set without domain knowledge and may require trial and error. Moreover, DBSCAN's performance can degrade with high-dimensional data due to the curse of dimensionality. DBSCAN, a density-based clustering algorithm, is distinguished by its use of two parameters: 'eps' (epsilon) and 'MinPts' (Minimum

Points). The eps parameter specifies the neighborhood's radius around a data point, with points falling within this radius deemed neighbors. This setting is crucial as it determines the granularity of clustering; a small 'eps' can classify numerous points as noise, while a large eps might amalgamate distinct clusters. The optimal eps is often deduced from the k-distance graph, which illustrates the inflection point in the distance to the k-th nearest neighbor across the dataset. Meanwhile, 'MinPts' dictates the least number of neighbors required within the eps radius for a point to be considered a core point, which serves as the nucleus for cluster formation. The dataset's density dictates the 'MinPts' value, with denser datasets demanding a higher threshold, typically set at one more than the dataset's dimensionality (MinPts $\geq$ D + 1). However, a minimum of three is standard to begin identifying substantial clusters.

The DBSCAN algorithm first employs these parameters to pinpoint core points, which must have fewer than 'MinPts' within their 'eps' vicinity, then forms clusters by recruiting reachable points. Border points fall short of the 'MinPts' criterion but lie within a core point's neighborhood, and noise points, which fail to meet either condition, are also identified. By iteratively linking points that are density-connected via a chain of core points within the eps parameter, DBSCAN can expand clusters. This process allows for detecting clusters with diverse shapes and sizes and skillfully manages outliers and areas of varying density within the dataset. Points not assimilated into a cluster are categorized as noise, an essential step in DBSCAN's robust clustering methodology.

The clustering plot in Figure 2 illustrates the output of the DBSCAN algorithm applied to a dataset with two features: 'Annual Income' and 'Healthcare Spending.' The points are distributed in a two-dimensional space where each point represents a data instance characterized by these two metrics.

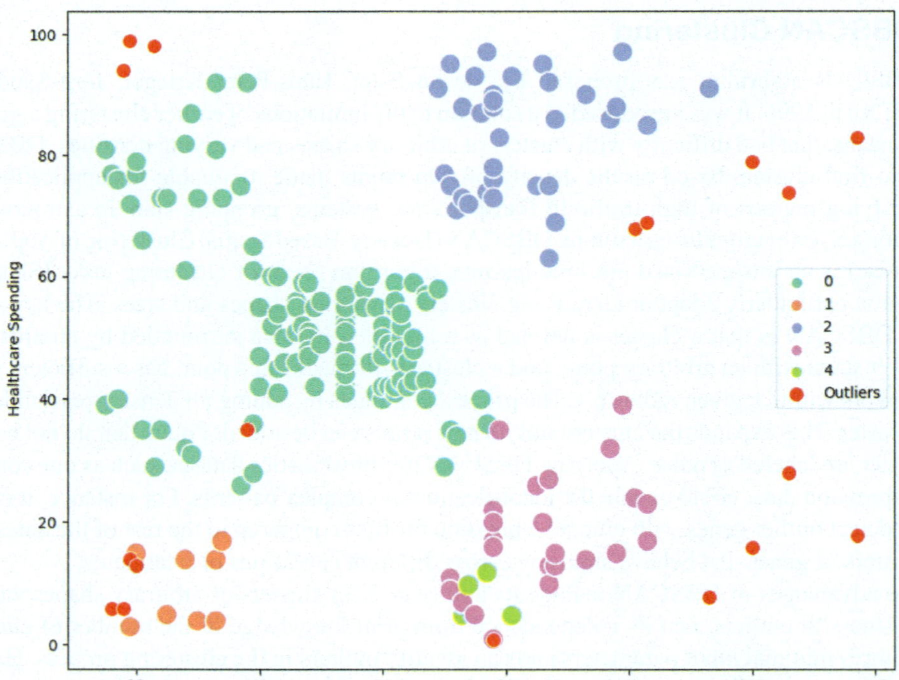

Fig. 2 DBSCAN Clustering for Income vs Healthcare Spending

In the plot, different colors represent distinct clusters identified by DBSCAN, ranging from cluster 0 to cluster 4. Each cluster groups data points densely packed together within the specified 'eps' radius, with a minimum number of points (MinPts) in their neighborhood. The DBSCAN algorithm has determined that these clusters are distinct enough from each other based on density criteria. The clusters may represent groupings of individuals with similar income and spending patterns, indicative of similar financial behaviors or socio economics status.

The red points are labeled as outliers, data instances that did not fit into the density criteria of any of the clusters. These outliers are significant as they represent unusual cases that stand apart from the general patterns observed in the dataset. In a real-world context, these could be individuals with exceptionally high or low healthcare spending relative to their income, which might warrant further investigation for anomalies or unique circumstances. This plot is valuable for identifying natural groupings within the data and detecting data points that deviate from the norm, providing an opportunity for deeper analysis into why those deviations might occur.

6. Agglomerative Hierarchy Clustering

Agglomerative Hierarchical clustering (Sasirekha and Baby, 2013) is a method of cluster analysis that seeks to build a hierarchy of clusters. The history of hierarchical clustering dates back to the early 20th century. However, it became more defined as a strategy in the 1950s and 1960s through the work of sociologists, psychologists, and statisticians, who sought to classify information systematically. This clustering approach has been widely used in various fields, including biology for genetic and species data, business for customer segmentation, and information science for document clustering. The main idea is to start by treating each data point as a single cluster and then successively merge pairs of clusters until all points have been merged into a single remaining cluster. The algorithm works by first calculating the proximity matrix for all data points. At each step, the two clusters that are closest to each other are merged together, which updates the proximity matrix. This process can be visualized using a dendrogram - a tree-like diagram that records the sequences of merges and shows the arrangement of the clusters formed at each stage. There are different metrics to measure the distance between clusters, such as single linkage (minimum distance), complete linkage (maximum distance), and average linkage (average distance). The choice of linkage criteria can significantly affect the structure of the formed clusters.

The advantages of Agglomerative Hierarchical clustering include the intuitive dendrogram representation that provides a detailed view of the data's clustering structure and the flexibility of not needing to specify the number of clusters in advance. Additionally, it can reveal exciting data patterns and is particularly useful for smaller datasets. However, this algorithm has limitations, such as its quadratic time complexity, which makes it less scalable to large datasets. It is also sensitive to noise and outliers, which can lead to clusters being merged incorrectly. Moreover, once clusters are merged, they cannot be separated in subsequent steps, potentially leading to suboptimal solutions.

Figure 3 represents a dendrogram from hierarchical clustering on a gene expression cancer dataset. It visualizes the outcomes of an agglomerative clustering process applied to RNA-Seq data from patients with different types of tumors. The construction of this dendrogram begins with each sample starting as its cluster. As we move up the tree, the algorithm iteratively combines the two nearest clusters into one, based on a distance metric and linkage criterion—typically, the distance represents dissimilarity between clusters, and the linkage criterion could be the average, maximum, or minimum distance between points in each cluster. In this case, the 'ward' method was likely used to minimize the total within-cluster variance. At each level, the vertical lines represent the merged clusters, with the line height proportional to the distance or dissimilarity between those clusters. The process continues until all samples are merged into a single cluster, capturing the hierarchy of clusters within the data.

The colors in the dendrogram do not inherently represent different types of tumors but are often used to visually distinguish different levels of the hierarchy for ease of interpretation. Each color change on the dendrogram marks a level where clusters have merged, with the color transition indicating a significant increase in the distance at that point of merging—this can suggest a natural division within the data. If we cut the dendrogram at a particular height, the colors below that cut

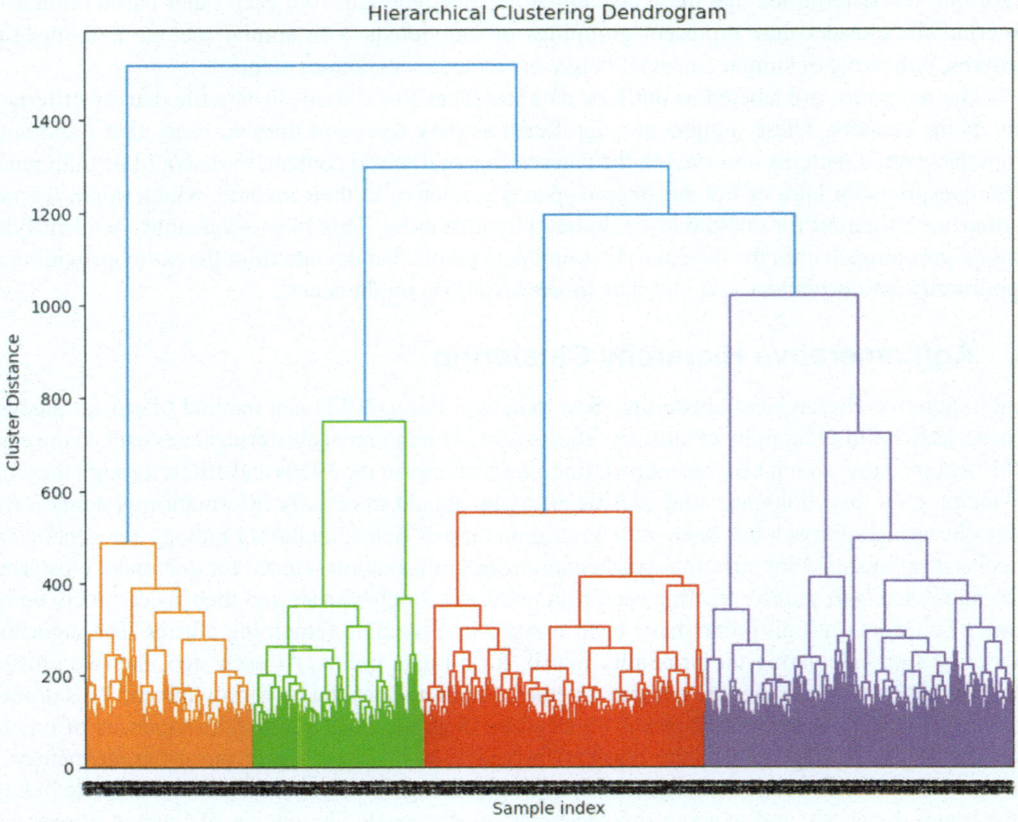

Fig. 3 Hierarchical Clustering

could indicate potential cluster groupings, which might correspond to different tumor types if the gene expression patterns are distinct enough. For instance, blue and red clusters merging at the highest level of the tree indicate that these are the most dissimilar groupings in the dataset, suggesting that the underlying gene expressions are significantly different, potentially corresponding to different tumor types. However, without further domain-specific analysis and labels provided, it is not possible to directly correlate the colors with particular tumor types. The dendrogram guides researchers to investigate the cluster compositions further and determine their biological significance.

7. Dimensionality Reduction

Dimensionality in data science refers to the number of variables or features in a dataset. High dimensionality means a dataset has many features, which can complicate the analysis, often leading to the phenomenon known as the "curse of dimensionality." This term was coined by Richard Bellman when considering problems in dynamic optimization. However, it has since become a central issue in machine learning and statistics, particularly in biostatistics, where large datasets are standard due to the comprehensive measurements taken at the molecular level. The history of dimensionality reduction techniques dates back to the early 20th century. The first formalized method was Principal Component Analysis (PCA), introduced by Karl Pearson in 1901. PCA aims to reduce the feature space by transforming the original variables into a set of new, uncorrelated variables (principal components) that capture the most variance in the data. Over time, various other methods have been developed and applied across different domains, including biostatistics, where they are used to analyze and visualize complex data such as genetic sequences or protein structures.

Dimensionality reduction works by finding a lower-dimensional space that captures the essence of the data. This can be achieved through feature selection, where irrelevant or redundant features are removed, or through feature extraction, where new features are created from combinations of the original ones. In biostatistics, dimensionality reduction is essential for overcoming computational challenges, avoiding overfitting, and improving the interpretability of models. Techniques like PCA transform high-dimensional biological data into fewer dimensions while retaining the variation in the dataset that differentiates observations, such as distinguishing between different types of cell expression profiles. The main advantage of reducing dimensionality is the simplification of models, which often leads to better performance and less computational cost. It enables visualization of complex data, facilitates noise reduction, and helps to avoid the curse of dimensionality. However, it may also result in loss of information, potential misinterpretation of compressed data, and the challenge of choosing the appropriate method and parameters. For example, mean dimensionality reduction, which could refer to methods averaging out less significant features, may only sometimes capture non-linear relationships in biological data.

Some of the top dimensionality reduction methods include:

- **Principal Component Analysis (PCA):** Transforms data to a new coordinate system, capturing the most significant variance by any data projection.
- **t-Distributed Stochastic Neighbor Embedding (t-SNE):** Non-linear technique ideal for embedding high-dimensional data for visualization in a low-dimensional space.
- **Linear Discriminant Analysis (LDA):** Finds the linear combinations of features that best separate two or more classes of objects or events.
- **Autoencoders:** Neural networks used for unsupervised learning of efficient codings by minimizing the reconstruction errors.
- **Uniform Manifold Approximation and Projection (UMAP):** A non-linear dimensionality reduction technique that is particularly good at preserving the data's local and global structure.

Principal Component Analysis (PCA)

This section discusses Principal Component Analysis (PCA). It is a statistical procedure that has become a foundational tool for dimensionality reduction in data analysis. The origins of PCA trace back to the early 20th century with the work of Karl Pearson in 1901, followed by contributions from Harold Hotelling in the 1930s. They sought an efficient approach to transform correlated variables into a set of uncorrelated variables called principal components, thereby simplifying the complexity of multivariate data. PCA has since been applied across diverse fields, from finance to social sciences, and notably in biostatistics, where it is used to interpret genetic data, patient responses to treatments, and other multivariate biological phenomena.

PCA identifies the directions, or 'principal components,' in which the data varies the most (Tharwat, 2016). It begins by calculating the covariance matrix of the data to understand the interrelationships between variables. This matrix derives the eigenvalues and eigenvectors, which inform the components and their explained variance. The first principal component accounts for the most significant possible variance, and each succeeding component has the highest variance possible under the constraint that it is orthogonal to the preceding components. By projecting the original data along these new orthogonal axes, PCA effectively reduces the number of dimensions while preserving as much variability as possible. The benefits of PCA are numerous, particularly in its capacity to overcome the curse of dimensionality and to uncover hidden patterns in data. PCA can simplify complex datasets by reducing the number of variables, making them easier to explore and visualize. It also aids in the removal of noise and improves the efficiency of other algorithms by minimizing computational costs. However, PCA does have its challenges. The interpretation of principal components can be non-intuitive, as they are linear combinations of the original variables. PCA is also sensitive to the scaling of the variables, and it assumes that the principal components with the highest variance are the most important, which may only sometimes be the case. Furthermore, PCA is a linear algorithm and may not effectively capture the underlying structure in data with nonlinear relationships.

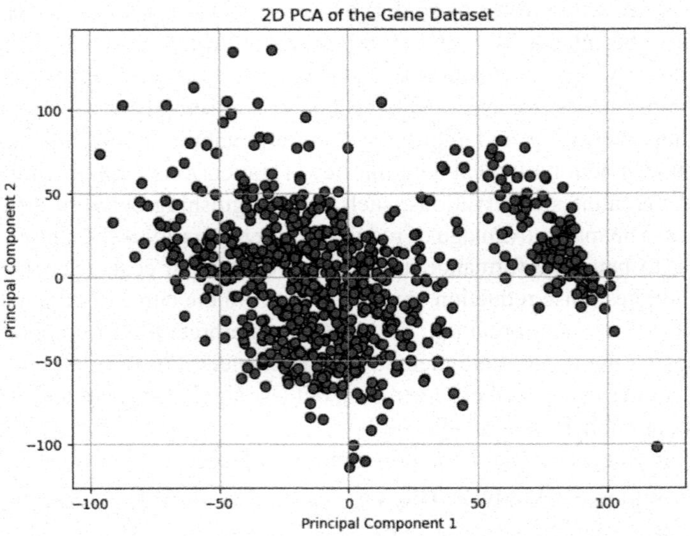

Fig. 4 PCA for the Gene Dataset

The Figure 4 PCA plot provided is a visual representation derived from the application of Principal Component Analysis to the RNA-Seq (HiSeq) PANCAN dataset, which comprises a vast array of gene expressions from patients with different types of tumors, including BRCA, KIRC, COAD, LUAD, and PRAD. This dataset initially contains 16,383 features, signifying a high-dimensional space that poses challenges for analysis, interpretation, and visualization. PCA addresses these challenges by transforming the data into a new coordinate system where the first axis (Principal Component 1) accounts for the maximum variance in the data, and each subsequent axis (like Principal Component 2) captures the remaining variance under the constraint of being orthogonal to the previous axes. The result is a 2D scatter plot where each point corresponds to a patient's gene expression profile projected onto the first two principal components. This enables us to observe patterns and potential groupings previously obscured in the high-dimensional space.

The plot shows that the first principal component (x-axis) and the second principal component (y-axis) display a scatter of points that do not form distinct, well-separated clusters. This suggests that while PCA has reduced the dimensions significantly, the variance captured by the first two principal components is not sufficient to separate the different types of tumors in the dataset distinctly. There appear to be some regions where data points are more densely packed, indicating potential subgroups within the data. However, with additional information, such as the variance ratio represented by each principal component or cluster label, it is easier to draw definitive conclusions about the types of tumors or to identify specific patterns related to the underlying biology. The overlap and spread of the points imply either a high degree of similarity in gene expression profiles across different tumors or that more than two principal components are required to capture the complex structures and relationships in the data. Other non-linear dimensionality reduction techniques or feature selection methods might more effectively reveal the distinctions between the different tumor types in this dataset.

t-Distributed Stochastic Neighbor Embedding (t-SNE)

t-Distributed Stochastic Neighbor Embedding (t-SNE) is a sophisticated machine learning algorithm for dimensionality reduction that is highly effective for visualizing high-dimensional data (Cieslak et al., 2020). It was developed by Laurens van der Maaten and Geoffrey Hinton in 2008, building on earlier techniques such as Stochastic Neighbor Embedding (SNE). The algorithm has gained significant traction in various fields that handle complex datasets, including biostatistics, where it is utilized to visualize gene expression profiles, protein interactions, or the genetic makeup of populations. The main goal of t-SNE is to translate high-dimensional data into a lower-dimensional space (typically two or three dimensions) for visualization, preserving the local structure of the data and revealing the inherent clustering patterns.

t-SNE works by converting the high-dimensional Euclidean distances between points into conditional probabilities that represent similarities. The similarity of datapoint x_j to datapoint x_i is the probability, $P_{j|i}$, that x_i would pick x_j as its neighbor if neighbors were picked in proportion to their probability density under a Gaussian centered at x_i. For the low-dimensional counterpart y_i and y_j, a similar probability $Q_{j|i}$ is calculated using a Student-*t* distribution rather than a Gaussian. The t-SNE algorithm then iteratively minimizes the Kullback-Leibler divergence between the two probability distributions P (high-dimensional) and Q (low-dimensional), with respect to the positions of the points in the map. This optimization is performed using gradient descent. The result is a map that reflects the original data's clusters with high fidelity in a reduced dimension, where similar items are modeled by nearby points and dissimilar items are modeled by distant points. The benefits of t-SNE are particularly evident in its ability to retain the local structure and reveal the data's inherent clusters at the expense of the global structure. It is powerful for exploratory data analysis and has been a significant advancement in the visualization of complex datasets, such as those encountered in biostatistics. The algorithm is adept at unfolding data manifolds, making it an invaluable tool for detecting patterns and structures that other dimensionality reduction techniques might overlook. However, t-SNE is computationally intensive, particularly for large datasets, and its results can be sensitive to the choice of parameters. The algorithm also tends to form clusters even when they might not be present, which can mislead interpretations. Furthermore, t-SNE maps are only sometimes interpretable straightforwardly, as they do not preserve distances or the relative positions of clusters, and the output can vary significantly depending on the random seed used during initialization.

RNA-Seq dataset after applying t-SNE

This technique is particularly adept at handling the complexity of high-dimensional data, such as the 16,383 features present in this dataset. The t-SNE algorithm has reduced the data into a two-dimensional space where similar data points are placed closer together while dissimilar points are placed further apart. The plot shows distinct clusters that have emerged, likely corresponding to the various types of tumors represented in the dataset (BRCA, KIRC, COAD, LUAD, and PRAD). The density and separation of the clusters suggest that t-SNE has effectively captured the intrinsic groupings within the gene expression data. Each cluster could represent unique gene expression profiles associated with each tumor type, revealing differences and similarities in their genetic characteristics. Outlying points could indicate data points that do not conform neatly to the identified clusters, potentially representing atypical expressions or subtypes within the tumors.

The t-SNE visualization provides a more segregated and defined clustering structure than the previous PCA plot. While PCA aimed to preserve the global structure and maximize variance, it resulted in a plot where the clusters were less distinct and somewhat overlapping. In contrast, t-SNE focuses on preserving local structures, often leading to more clearly separated clusters, as seen in the attached plot. This characteristic of t-SNE is beneficial in identifying subtle structures within complex datasets, such as distinguishing between different types of cancer based on gene expression. However, it is worth noting that t-SNE plots can sometimes exaggerate clusters, and distances may not correspond to actual dissimilarities as linearly as they do in PCA. Therefore, while t-SNE offers a more nuanced view of local relationships and can separate clusters more effectively, PCA provides a more interpretable and globally consistent data representation. The specific analytical goals and the nature of the dataset should guide the choice between PCA and t-SNE.

8. Recommendation Engine

A *recommendation engine* is an information filtering system that seeks to predict and show users' preferences for a wide range of items. They are primarily used in commercial applications, most notably in *e*-commerce and content delivery platforms like Amazon and Netflix. The goal is to personalize the user experience by suggesting products, services, or content that is relevant to the user's preferences and past behavior. The history of recommendation engines dates back to the early days of the Internet, where systems were relatively simple and based on explicit user feedback. The 1990s saw the development of collaborative filtering, a technique that automatically predicts user interests by collecting preferences from many users. The rise of machine learning and big data in the 2000s significantly advanced recommendation systems, allowing more sophisticated algorithms to handle vast amounts of implicit data, such as clicks, views, and purchase histories. In biostatistics, while not a traditional application area for recommendation engines, analogous systems could recommend similar genetic profiles, relevant clinical trials, or compatible treatment plans based on patient data.

Recommendation engines typically operate by collecting information on a user's behavior and additional information on the items themselves. There are two main types of recommendation systems: collaborative filtering and content-based filtering. Collaborative filtering makes recommendations based on the collective preferences of the user community, operating under the assumption that users who agreed in the past will agree in the future. Conversely, content-based filtering makes recommendations by comparing the content of the items with a user profile, which contains the items the user has liked in the past. These systems often use machine learning models to predict user preferences. Despite being data-driven, recommendation engines are only sometimes considered unsupervised learning models. They can employ supervised learning to predict user ratings, unsupervised learning to discover patterns in user-item interactions, or even reinforcement learning to adapt to a user's changing preferences continuously.

The advantages of recommendation engines include increased user engagement, personalized user experience, and enhanced opportunities for product discovery. They can drive sales and content consumption by connecting users with items they are likely to enjoy but might not have found on their own. However, there are also challenges, such as the cold start problem, where new users or items with no history need something on which to base recommendations. There is also the risk of creating a "filter bubble," limiting user exposure to a narrow range of items. Privacy concerns arise with the collection and use of personal data. Finally, the accuracy of recommendations depends heavily on the quality and quantity of the data, which can be a significant limitation. In biostatistics, recommendation techniques are leveraged to predict outcomes, suggest patient treatments, identify potential drug interactions, or even recommend research pathways. Here are several recommendation techniques that could be applied within biostatistics, each with an example:

Collaborative Filtering

Collaborative filtering in biostatistics could be used to recommend personalized treatment plans. For instance, if multiple patients with similar genetic backgrounds and health histories have responded well to a particular treatment, collaborative filtering could recommend this treatment to a new patient with similar attributes. By analyzing the treatment outcomes and patient characteristics across large datasets, the system identifies correlations that may not be immediately apparent to healthcare providers, thus aiding decision-making.

9. Content-Based Filtering

This method could be used to suggest relevant research articles or clinical trials to biostatisticians and medical researchers. For example, if a researcher is studying a specific gene mutation's effect on breast cancer, content-based filtering could analyze the content of numerous research papers and recommend those that focus on similar mutations or cancers. This process would involve natural language processing to understand the content of research papers and match them with the researcher's areas of interest.

Hybrid Systems

Hybrid systems combine collaborative and content-based filtering to provide robust recommendations. Biostatistics could employ a hybrid system in a clinical decision support system to recommend patient referrals. It would consider the success rates of specialists (from collaborative data) and the specific medical requirements of the patient (content data) to recommend the most suitable specialist for a given condition.

Association Rule Mining

In pharmacogenomics, association rule mining could be used to recommend drug combinations based on patient genetic data. It would analyze databases of patient responses to drug regimens to find associations between genetic markers and positive drug responses. This can be especially useful in precision medicine to recommend a cocktail of drugs tailored to a patient's unique genetic makeup.

Matrix Factorization

Matrix factorization techniques could be used in genomic data analysis. By decomposing large genetic interaction matrices, matrix factorization could reveal latent structures that suggest how different genes contribute to traits or diseases. An example would be predicting how genetic variations contribute to complex diseases like type 2 diabetes, which could inform personalized prevention strategies.

Deep Learning

Deep learning methods, particularly neural networks, could be utilized to predict disease progression by learning complex patterns in longitudinal health data. For instance, a neural network could be trained on patient data, including genetic information, lifestyle factors, and disease markers, to predict the likelihood of disease onset or progression. This could help in early diagnosis and the identification of at-risk individuals. Each of these recommendation techniques could significantly enhance decision-making in biostatistics and personalized medicine. They leverage the power of statistical analysis and machine learning to provide actionable insights from complex biological and medical data. However, the success of these methods hinges on the availability of high-quality, comprehensive datasets and careful consideration of the ethical implications of algorithm-based recommendations in healthcare.

References

Cieslak, M.C., Castelfranco, A.M., Roncalli, V., Lenz, P.H. and Hartline, D.K. et al. (2020). t-Distributed Stochastic Neighbor Embedding (t-SNE): A tool for eco-physiological transcriptomic analysis. *Marine genomics*, *51*, 100723.

Sasirekha, K. and Baby, P. (2013). Agglomerative hierarchical clustering algorithm-a. *International Journal of Scientific and Research Publications*, *83*(3), 83.

Shen, J., Hao, X., Liang, Z., Liu, Y., Wang, W. et al. (2016). Real-time superpixel segmentation by DBSCAN clustering algorithm. *IEEE transactions on image processing*, *25*(12), 5933–5942.

Sinaga, K.P. and Yang, M.S. (2020). Unsupervised K-means clustering algorithm. IEEE access, 8, 80716–80727.

Tharwat, A. (2016). Principal component analysis-a tutorial. *International Journal of Applied Pattern Recognition*, *3*(3), 197–240.

CHAPTER

9

Ethics in Artificial Intelligence

Athira SS,[1] *Megha KB*,[1] *Greeshma N*,[2] *Aswathy V*,[2] *Manjoosha RY*,[2] *Renu John*[2] and *Mohanan PV*[1*]

1. Introduction

The far-reaching advancements in AI and machine learning hold immense scope in various sectors. However, there has been a parallel concern over the ethical challenges imparted by AI, which need to be overviewed. The possible applications of AI range from modest therapeutic procedures to extremely multifaceted robotic science chores (Alemzadeh, H. et al., 2016). Most of the capabilities are more proficient than humans can ever envisage. In the meantime, enhancing ethical concerns in AI are significant, ranging from algorithmic bias to severe healthcare issues (Vinuesa, R. et al., 2020). As per researchers, AI had its flag off around 1953, and the ethical alarms started coming up by the middle of the 20th century (McCarthy, J. et al., 2006). Active debates on these AI ethical issues have been occurring globally, nourishing the development of improved AI algorithms. Ethical issues in AI further the potential of increased computing resource availability and mark furtherance in data transparency worldwide (Gambelin, O. 2021).

Conceptually, an ethic can be defined as a moral principle that regulates an individual's behavior, quality of thoughts, or their decision to act out. AI ethics thus connect the safety and efficiency of manufacturers, developers, or distributors to common users during its propagation. Unfortunately, a perfectly scrutinizing technology is yet to be unveiled, resulting in poor design, execution, or inappropriate application of AI technologies (Khakurel, J. et al., 2018). Over the past five years, AI has substantially shifted from academia to various sectors including healthcare, finance, politics, industrial and judicial. This underlines how AI generates a massive impact in various zones of society. Moreover, a rapid involvement of AI-based products and technologies in our day-to-day lives, in turn, accelerates the misuse of AI in the mentioned sectors. For example, smartphones and laptops. (Arnold, T. and Scheutz, M. 2018). The expansion of technology and related products has led to the emergence of specific AI, such as machine learning (ML). ML is considered as one of the most rapidly growing areas; it works based on principles acquired from computer science, statistics,

[1] Division of Toxicology, Sree Chitra Tirunal Institute of Medical Science and Technology, Thiruvananthapuram, Thiruvananthapuram, India.

[2] Department of Biomedical Engineering, Indian Institute of Technology, Hyderabad, India.

* Corresponding author: mohananpv10@gmail.com

AI and data science (Decatur, S et al., 1997). Additional development in ML technology has led to the advent of a deep learning approach. The basic principle behind this approach is artificial neural networks, which resemble interconnected neuronal networks in the human brain. This has further created ethically problematic disputes, including unfair and illegal discrimination (Bengio, Y et al., 2021).

According to AI developers, most challenging ethical concern in this era is the opacity of AI-driven technologies. Despite a particular well-designed and organized AI algorithm, there is a fog of unawareness about how AI works in reality and associated ethical hitches. This further necessitates the discovery of not only a powerful and ascendable AI, but also one that is 'transparent' in all senses (Good, I.J. 1966). AI has taken charge of many humanoid assignments which emphasizes the thriving of modern technologies. For instance, the Human Resource Management (HRM) division has transformed into an utmost advanced form in which AI regulates its key functionalities. One set of experts around the world view this as a revolutionary discovery, since it remarkably lowers fraud and bribery in hiring employees. Besides, it ensures pure talent based recruitment and hence showcases eminence of the employer or company (Chen, Z. 2023). Certain contradictory opinions emanated in view of this, as AI works on the basis of pre-designed algorithms and initial inputs provided. If this initial signal is improper, it can generate serious ethical discrepancies, including socio-economic discrimination. According to professionals, such unethical logic can lead to widespread inequality and expose inconsistencies. Also, such social injustice is capable of causing moral as well as economic injuries to those who are victims of AI ethical discrimination. The ultimate result of such events can be global economic instability. This further puts forward a point of caution that an employer or any other active AI operators should have comprehensive knowledge of the particular AI algorithm they are developing, and the future ethical consequences. In the context of AI in HRM, the programmed system should circumvent even a minuscule level of racial or gender-based bias during recruitment (Santos Botelho, C. 2023).

Considering the inevitable requirement of the topic, the chapter collectively presents various facts, underlying theories and associated dilemmas behind AI ethics in detail. The structure of the chapter is as follows: The first section narrates five major principles of AI ethics followed by current issues arising in different AI platforms. The third and fourth sections deal with some key insights on illegal discrimination and certain ethical case studies respectively. Major AI ethical guidelines and standards, AI governance and bias management are covered in fifth and sixth sections. The requirement of responsible AI in future has been recounted in the final part of the chapter.

2. Five Key Principles of AI Ethics

AI ethics account for a group of principles and guidelines which developers and users should pay attention to constantly. Various global organizations have formulated these principles as 'key principles of AI ethics' together. They comprise; transparency, impartiality, accountability, reliability, security and privacy. Each of these principles is detailed below (Figure 1) (Cortese, J.F. N.B. et al., 2023).

2.1 Transparency

Indeed, transparency is the most vital tool for prohibiting the intrinsic opaqueness of AI technologies. Researchers consider transparency as a major factor which is closely associated with AI sustainability and thus the most studied AI ethical feature over the past decade. Papadouli et al., (2022) have proposed that the most challenging mission that AI driven systems are experiencing nowadays is termed as, 'black box effect' or 'black box problem'. The explanation for the effect can be drawn in many ways; the collective meaning can be the lack of clarity in providing the principle of operation by an autonomous AI system (Papadouli, V. 2022). Also there is a gap in the proper justification for each result the AI system provides. Such a gap is more likely to create a trust issue in humans and

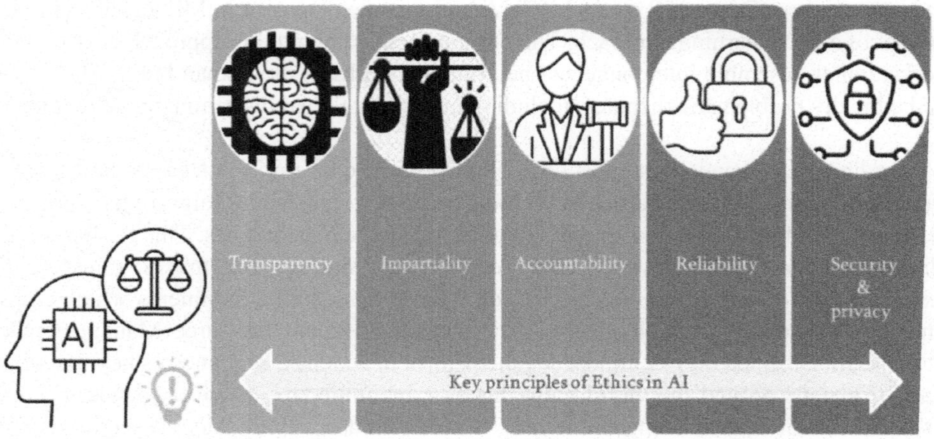

Fig. 1 Key ethical principles linked with AI.

therefore results in a problem regarding AI sustainability. As per scientific experts, AI transparency is the most appropriate remedy for counteracting this dilemma. Being a multidisciplinary concept, transparency varies in accordance with the area of application. For example, transparency in public governance refers to the concept of decreasing unpredictability, dishonesty as well as inducement in the political sector. This is entirely different from AI transparency in the scientific community. In the latter, transparency displays the actual infrastructure of the procedure and hence gaining public acceptance and confidence. From all these concepts, AI transparency accounts for the ability to arrive at a decision or outcome in such a way that it is interpretable and comprehensible by humans who hold the right to question and appeal to them (Lepri, B. et al., 2018).

2.2 Impartiality

The broader meaning of AI impartiality can be suitably recited as; every element of a particular AI system such as inputs, algorithms, features, intentions, outputs etc. should be provided with utmost equality, integrity and justice. As it is obvious from the underlying concept of AI that it is purely dependent or entirely programmed by man power. Hence, it is our responsibility to consider that, at what stage we make known to the black mark of inherent biases or unfairness to AI systems (Lin, P. et al., 2011). There are three basic features of AI impartiality: independence, fairness and context. The primary feature; AI independence can be well-explained from the concept by the renowned Swedish philosopher Nick Bostrom in his book; 'Superintelligence: Paths, Dangers, Strategies'. According to the book, AI independence is nothing but a dangerous; yet life threatening concept which would find all the possible routes to bring out the ultimate result regardless of the hidden dangers. For instance, if it is provided with a data instruction to make paper clips, superintelligent AI would utilize every single resource; even atoms from human bodies for making the clips. In due course, it results in the end of the world (Richmond, S. 2016). Fairness refers to the perception that AI technologies should be treated with equality without any bias in the context of personal information.

2.3 Accountability

Accurate governance of AI technologies is mainly guided by the accountability of a particular system. It is referred to as the foundation of AI supremacy and it deals with the matter of liability between two parties (Han, Y. 2023). Modestly it can be exemplified like, consider two parties A and B in which A is responsible for performing a task and the report needs to be submitted to B for verification and approval. Only if B verifies and approves the task, then A would be able to publish or stream the result in the public sector. Therefore, accountability strictly scrutinizes whether AI systems perform in alliance with ethical values and legal standards. Bovens et al., in 2010 (Bovens, M. 2010) proposed that AI accountability lies in close association with responsiveness and a 'sense of responsibility'. Besides, it clearly showcases the willingness to act in a more transparent, liable, fair and reasonable way.

2.4 Reliability

The core of AI techniques relies on reliability of the wholesome including the design of the machine, working principle, accuracy of the underlying algorithm and moreover, the extent of ethical acceptability of the system's output. The key factors driving interpersonal relationships and trust among humans are; aptitude, integrity and generosity. When it comes to the human-technology relationship, these terms differ slightly since technology is incapable of delivering outcomes that meet human expectation levels. Therefore the aforementioned terms in the context of AI reliability can be replaced by functionality, trustworthiness and effectiveness. A reliability concern often can be summed up by transparency, explainability as well as accuracy of the system (Creel, K.A. 2020). McKnight et al., (2011) (Mcknight, D.H. et al., 2011) reviewed AI reliability or trust, as if, it is qualitatively different from a human's level of trust since it does not possess moral or preference values. Reliability aspect offers consistency of the AI principles and performance as long as it is technically error free.

2.5 Security and Privacy

While AI is growing quickly and is about to replicate humans in decision making, researchers make a point that there must be an approach which ensures the safety and privacy of humankind. The Japanese Society of AI (JSAI) in 2018 put forward the notion on improvising security and privacy of AI users' techniques effectively. They proposed the 'effective and transparent communication' strategy among AI developers, scientists and engineers to seek protection of personal information and data of common technology users (Kozuka, S. 2018). Compared to other innovations, the risks accompanying AI techniques are far more thoughtful since it can assimilate several issues associated with basic human rights. Certain examples of AI-involved risk scenarios include; infrastructures like transportation, AI in the educational sector, therapeutic procedures like robotics-assisted surgeries, legal services, management and recruitment of employees by organizations, to name a few. All these platforms represent potential risk targets that AI focuses on to generate safety to lives (Alemzadeh, H. et al., 2016).

3. Current Issues Arising in Various AI Platforms

The expansion and organization of AI systems requires access to data in a profound sense. Such a detailed admittance even into privacy information of various sectors of human life poses a noteworthy level of risk factors. A detailed note on such risk factors in various sectors is narrated below:

3.1 Socio-economic Impacts

An explosive growth in AI technologies put man in a groove of concern over replacement of workers by AI. Studies have predicted such a scenario occurs due to a rapid boom in automation, computing, AI robotics etc., in almost every sector of society. The communal effect of such innovations is more likely to create a lack of job opportunities and hence an imbalanced labor market as a whole (Autor, D.H. 2015). Leontief in 1983 (Leontief, Wassily, 1983) closely observed the magical processing power of computer chips and raised the concern that humans can be replaced by machine power in the near future. Also the observation did compare the picture of such an engrossment with the auxiliary entry of internal combustion engines over horses in the history of modernization. Predictive mass unemployment across widespread sectors of occupations has already raised concerns among people which would harness sharing of employment in the next two decades. Some of the major areas wherein AI imprints its intervention include; advanced manufacturing, funding, space surveys, transportation (unmanned vehicles and autonomous drones), business etc. (Frey, C.B. and Osborne, M.A. 2017).

Economic progression by the involvement of AI and associated technologies is another imperative area that economists actively focus on. As per a monetary survey in 2015, there occurred a 0.4 percentage (%) addition in the annual GDP and labor productivity across 17 countries within 14 years (between 1993 and 2007) by the engrossment of robotics. In the meantime, another population of economists reviewed that there would not be as much revolution by AI related techniques as we anticipate. Instead, the human brain remains the most powerful source of new revolutionary rationale that can introduce more occupational possibilities. In a study, the team surveyed observations of 1,986 economic experts about the influence of modern techniques in society. Around 48% of the members opinionated that contemporary information technologies in this digital era can considerably displace many of the 'blue' and 'white-collar' jobs, creating a mass financial inequality. Nevertheless, the remaining 52% of the participating experts suggested that there would not be any additional unemployment scenario beyond 2025 as human intelligence can productively bring out more job opportunities by beating robots or any digital agent (Graetz, G. and Michaels, G. 2015).

3.2 Effects in Human life Privacy, Rights and Stateliness

Being a system that does not possess any moral restraints, AI techniques nowadays can be misused for personal harassment. More precisely, AI can efficiently be used for creating fake videos of people in which one person's face or another's body parts can be switched with another person's. This imposes a very dangerous consequence as innumerable pornographic contents are being created illegally without any personal interests. As AI is incapable of understanding humane values or emotions, it cannot resist itself until the given data or algorithm instructs it (McAllister, A. 2016). Another AI compelled criminal activity includes; manipulation of people's psychological state to commit a crime or even persuading suicide. For example; AI based games, sexbots etc. Regardless of the purpose, the AI user gradually turns himself/herself into an addiction phase and unintentionally reveals all the personal information into the system. Robots even can regularize a criminal offense and continue to perpetuate doing personal harm to people in this way. Researchers suggest that strict liability rules can significantly lower such offenses. However in reality, the question that still remains is; 'Who should be punished for such an AI-allied offense'? (De Angeli, A. 2009).

3.3 Consequences on Democracy

As per the conventional approach route, politicians do campaigning for electoral tasks using barely available anecdotal evidence from voters. This can to some extent turn into beneficial results by collecting interests from a specific group of voters. However, by the growth of AI in the political sector, the campaigning was made easier by accessing a tremendous amount of voter's information and their background of political beliefs and views. Such innovations eventually created a revolutionary breakthrough in the elections to gain votes irrespective of the political groups who follow AI tools (Hess, A. 2016). This represents one of the dangerous faces of AI powered procedures, that are likely to support a particular political group without giving any relevance to citizen's interest. Two popular incidents which gained widespread attention in this context include;the presidential election of the United States in 2016 and the Brexit vote in the United Kingdom in 2018. Both events faced fraudulent involvement of 'Bots' or autonomous accounts (most of them were proven to have originated from Russia later) which circulated false impressions and news about the political parties involved through Twitter. These bots are marked as highly pulsating in spreading biases, fake or spam updates about political candidates selectively via amplifying artificial likes, shares and retweets in social media. Altogether, such AI-assisted political dealings execute serious consequences in support of democratic discrepancies (Gorodnichenko, Y. et al., 2021).

3.4 Effects on the Natural Ecosystem

Driverless vehicles or unmanned cars and multipurpose drones comprise two major contributions of the AI revolt. Pure automation in these entities can offer several benefits over current transportation routines. Conversely, huge amounts of fuel consumption and exhaust spitting from these automated systems can be a potential threat to living creatures including animals and humans. Numerous numbers of illegal activities through such automated vehicles have already been reported globally; such as, AI-equipped drug trafficking and smuggling (Sharkey, N., 2010). Natural resources are highly vulnerable to draining of fuel and liquid waste exhausts which all originate due to industrialization. The AI era further adds such waste tremendously and hence raises widespread concern. Remote-controlled cocaine-trafficking through submarine routes were reported in 2010 which in turn extends the health concern of marine life as well. Autonomous vehicles used for such illegal submarine trafficking are termed as; unmanned underwater vehicles (UUVs). UUVs have gained significant attention among scientists as they are considered as a dreadful source destroying the entire ecosystem (Gogarty, B. and Hagger, M. 2008).

3.5 Trust Issues

The development of AI techniques primarily aims to improve the quality of human livelihood and welfare in every aspect. Some such major aspects can be drawn e.g. transportation, industries, healthcare, academics, amusements etc. Meanwhile, the parent AI algorithm should be designed or programmed in such a way that it protects basic human rights, principles and construct liability. A consistent monitoring of AI programs during each usage is perhaps an unmanageable task. Also, a continuous supervision of digital technologies can consume a remarkable amount of time and may result in boycotting of the entire system. But on the contrary, if the AI possesses utmost clarity in its processing and outcomes, then trust would definitely not be a concern among people (Floridi, L. and Taddeo, M. 2016). Taddeo in 2017 suggested a short term and a long term remedy for building trust in AI techniques. The short term remedy is to set up a pop-up messaging system that is automatically generated every time the particular AI program is about to cause trust conflicts. This message must be accessible to every user, so that they can be more cautious and predictive about the futuristic threat. The long term remedy consists of the development of an infrastructure which can offer 100% transparency and liability by including humans into the AI programming loop (Taddeo, M. 2017).

4. Unfair and Illegal Discrimination and Bias

A concern about discrimination denotes the most discussed social issue all over the world, even with non-AI platforms. Considering the discrimination existing within AI systems, the origin of cause can be evident as real life data is used for system training. If the input data contains any discriminative content related with a current societal scenario, it would be clearly reflected in the AI result received. For example, if the training data encompasses any gender, skin color, or race-based groupings which discriminate against each other, AI would recognize the data as real and proceed accordingly. The consequence of such discrimination is certainly problematic and has the potential to light an ethical fire in society (Pagallo, U. 2018). A common nature of discrimination events mentioned above is the violation of fundamental human rights. Since AI indicates an entirely man-driven system, any human made biases introduced in the form of training data can persist unless the developers recognize and work on it to clarify. The extent of AI bias consequences can range from errors occurring in bank credit checks to dishonest imprisonment of people by unfair bias generated by AI involved in a country's federal law (O'Neil, C. 2016). Certain real life incidents of AI bias which have received immense global attention are discussed in the following section.

5. AI Ethical Concerns: An Emphasis on Medical Perspective

5.1 Major Ethical Principles Guiding AI in Healthcare Research

Regardless of whether it is AI-based or not, healthcare research policies should be in strict compliance with basic principles that ensure respect for humans, beneficence, non-malfeasance as well as uniformity in justice to all. According to ICMR (Indian Council of Medical Research) National Ethical guidelines 2017, there is a categorization of AI ethical principles, particularly for healthcare research; this includes ten basic principles: Non-discrimination and Fairness, Optimization and data quality, Accessibility and equity, Risk minimization and safety, Collaboration, Autonomy, Data privacy, Accountability and liability, Data security and trustworthiness and Validity (Mathur, R. and Swaminathan, S. 2018). Unlike other sectors wherein AI technology finds its role, healthcare requires foremost attention since it directly connects with people's health and well-being. Therefore it becomes a necessity to discuss the ethical principles at their core as capsulized in Table 1.

Table 1 Major AI ethical principles applicable to healthcare research. *HMSC – Health Ministry Screening Committee, *HITL – 'Human in the Loop'.

No.	*Ethical principle*	*Key points*	*Reference*
1	Non-discrimination and fairness	• The researcher must ensure accuracy of AI training data • End user's feedback should be considered for minimizing possible biases • Acknowledgement of possible discrepancies by AI developers • Superior care on under-represented, vulnerable or minor populations like children, women, disabled ones etc. • AI decision should be reversed in case of any patient harm	(Wachter, S. et al., 2021) (Giovanola, B. and Tiribelli, S. 2023)
2	Optimization and data quality	• Adequate representation of data inclining to a patient population • Multiple replicate studies should be performed for higher output quality • Laborious clinical validation is needed • A 'pre-deployment system' is required to monitor data collection methodologies	(Halle, A. et al., 2021) (Khadem, M. et al., 2023)
3	Accessibility and equity	• Fair distribution of AI technology • Appropriate governance of AI accessibility to minor communities • AI design must consider a varied range of social, cultural, economic and racial characteristics of end users	(Thomasian, N.M. et al., 2021)
4	Risk minimization and Safety	• Extreme scrutinization of deliberate or non-deliberate exploitation of patient data • Implementation of AI secure softwares in hospitals • De-linking of personal data from global technology network • AI benefit should outweigh possible patient harm • Ethical committee/ regulatory body assessment on patient security measures	(Choudhury, A. and Asan, O. 2020) (Darwiesh, A. et al.,2022)
5	Collaboration	• Encouragement of interdisciplinary collaboration in patient care • International collaborations related to medical research or sharing of data/bio-samples require HMSC's* (Govt. of India) approval	(Nalluri, M. et al., 2024)

Contd.

Table 1 *Contd.*

No.	*Ethical principle*	*Key points*	*Reference*
6	Autonomy	• Careful control over AI-based decision making in healthcare without interrupting patient autonomy • HITL* model in AI technologies nourishes its transparency in performance • Patient consent is mandatory before usage of any AI-based techniques during diagnosis/treatment • Complete freedom for patients whether to agree or disagree in using AI techniques	(Tiribelli, S. 2023)
7	Data privacy	• AI-users should have complete access to the data • Users must be aware about the safety level offered by AI technique • Precautionary measures are required to prevent leakage of human biometric details	(Bak, M et al., 2022)
8	Accountability and liability	• Active participation of a health care professional during the entire AI application process • Open source softwares are needed to ensure ethical guidelines are not compromised • Consistent supervision is mandatory in order to prevent data misinterpretations • Accurate legal credentials are required for all the AI entities involved • **Responsible entities for harm by AI-tools** – Flaws in machine functionality: designer/developer/manufacturer – Technology implementation: End user/organization	(Bottomley, D. and Thaldar, D. 2023)
9	Data security and trustworthiness	• A trustworthy AI algorithm should be lawful • Deployment of AI must create a complete trust among people by upholding ethical principles • AI-driven tools must sync with the working of conventional methods • AI results on patient diagnosis must be explainable • Opacity in AI performance may doubt its applicability in long term clinical assessments	(Cutillo, C.M. et al., 2020)
10	Validity	• Clinical validation of AI techniques is required before application in patients • Continuous feedbacks should be send to developers for preventing huge variations in medical outputs with regard to trained algorithm • AI-tools are recommended to have suitable clinical validation updates	(Azzam, T. 2023)

5.2 AI Governance in Healthcare and Research in India

Development and execution of a governing framework for AI-based techniques in the healthcare sector is still in its budding stage; even in developed countries like the United States and United Kingdom. The members of the 'Legal, Ethical and Accountable Digital Society (LEADS) at the University of Birmingham put forward a response to the European Union's (EU) proposal on regulations of AI ethics (Smuha, N.A. et al., 2021). In the document, LEADS members reviewed the EU's proposal as a sensational approach to prevent ethical harms linked to AI techniques through a legal perspective. This legal approach guarantees protection of fundamental rights of humans and ensures upholding of ethical values for strengthening the acceptability of AI tools all over the world.

As a mission to reinforce involvement of AI techniques along with other sectors, the Government of India also had set up a streamline across the country. For the establishment of the National Digital Health Authority, Government of India started an effort through the National Health Policy in 2017. This assignment envisions the integration of digital healthcare platforms with AI policies for the upgradation of the national healthcare system (Ministry of Health and Family Welfare, Government of India. National Health Policy-2017). Similarly, another legal service was established in 2019 named; National Digital Health Blueprint (NDHB 2019) aiming digital recording of patient health data providing international ethical standards. In addition, NDHB focuses on establishment of a data ownership conduit with an emphasis on data anonymization (Ranganathan, S. 2020). DISHA (Digital Information Security in Healthcare Act) was established in 2018; for the launching of national and state e-Health authorities and exchanges for the assortment of public health information (Kapoor, S.K. 2020). The medical device rule was established in 2017 which accurately defines the possibility of any medical software or device in the medical field (Sharma, D. and Chandra, A. 2020).

6. Addressing AI Bias with Real World Examples

6.1 Gender Bias in Recruitment Sector

Recruitment of employees by an organization or institute denotes a tedious procedure which requires a considerable amount of time and effort. The Human Resource (HR) team takes up the role during most of the cases; this helps recruitment to be more organized. However, if the number of applicants is unexpectedly higher and extremely difficult for short listing or sorting, the importance of AI arises. As discussed earlier in this chapter, engrossment of AI in this recruitment aspect also can create bias (Bacciarelli, A. 2023). For instance, in 2014, Amazon began to actively apply AI tools by replacing the role of the HR team for more convenient screening of applicants. The AI tool applied for this purpose has been trained for reviewing Curriculum Vitae (CV) of applicants that are almost 10 years old. The AI training data had clear predominance of male applicants; which occurred intentionally or unintentionally in the company. The AI followed the conventional selection procedure and resulted in significant reduction in the number of females in the final list. A huge controversy began to rise in view of this, which turned into a clear example of gender discrimination in the recruitment sector. In the case of Amazon, the company had to consciously abandon the use of AI after the controversy (Heilweil, R. 2019).

6.2 Discrimination on the Basis of Skin Color and Personality

In 2016, the New Zealand Department of Internal Affairs rejected the passport renewal request of an Engineering student from the Asian continent. The underlying reason causing such a disappointing rejection was that; the passport photo which the student used in the request did not contain his eyes. The student later raised his voice in a media conference and expressed his feeling of dissatisfaction and commented that he has very small eyes since he is from the Asian race. The automated AI tool could not recognize his eyes properly and caused the rejection (Stahl, B.C. et al., 2022). Another example of color-based discrimination was noticed in 2020 in the UK wherein; dark skinned women got mercilessly rejected during passport related matters. The unacceptable justification for the incident was that, photo clarity of a dark-skinned person can be four-times poorer than that of a light-skinned one (Ahmed, M. 2020). Although a conspicuous voice has never been heard as louder as that of the Amazon case, another incident occurred during the time of COVID-19 lockdown in 2019. A huge hike in the usage of online meetings and interviews became a mainstay during that time and so did the recruitment procedures. In accordance with the fed instructions, automated AI tools did measure the facial expressions and mannerisms of applicants and erroneously interpreted as ineligible for a particular job of interest. Likewise, AI can effectively; yet unethically sense the psychological states of people for making a decision (Stahl, B.C. et al., 2023).

6.3 Inadequate Data in Global Health Care Sector

Despite the astounding application possibilities of AI in the healthcare sector, ethical concerns and disparities exist in this regard. These disparities can occur from basic clinical measurements to tedious surgical procedures which are likely to affect a large population of patients. For example, an oxygen saturation level in a patient's body during its administration is conventionally performed by an oximeter; which works by passing infrared light through the skin. By the aid of AI, the oximeter efficiency increases multifold than manual or ordinary instrumental support. However, the reading often is highly dependent on the skin color; wherein dark skin can make erroneously higher measurements. This can lead to occult hypoxia in dark colored people; which is absolutely a risk setup causing lethal consequences (Sjoding, M.W., Dickson, 2020). Through a consistent innovative thought, FDA put forward the concept of Artificial Intelligence/Machine learning (AI/ML)-based software as a Medical Device (SaMD) Action Plan in 2019. As per the SaMD plan, any medical device released into the market should have proper information about the possible bias it may cause or a new ethical disparity if any.

One of the major drawbacks of AI in the healthcare context is termed as, 'distribution shift'. This indicates a substantial shift in the trained data of the AI system and the operational data that the system relies on. This can be due to either some missing element in the trained data, or change in patient's health status over time. For instance, prediction of acute kidney damage can be erroneously diagnosed or predicted by AI-driven tools due to shifts in the disease pattern over an extensive time period. Some of the key terms on safety issues associated with AI driven medical practices are illustrated in Figure 2 (Davis, S.E., Lasko, T.A., 2017). The evident medical AI bias further necessitates discovery of safety measures or it emphasizes the need of addressing safety hazards during AI-based clinical procedures.

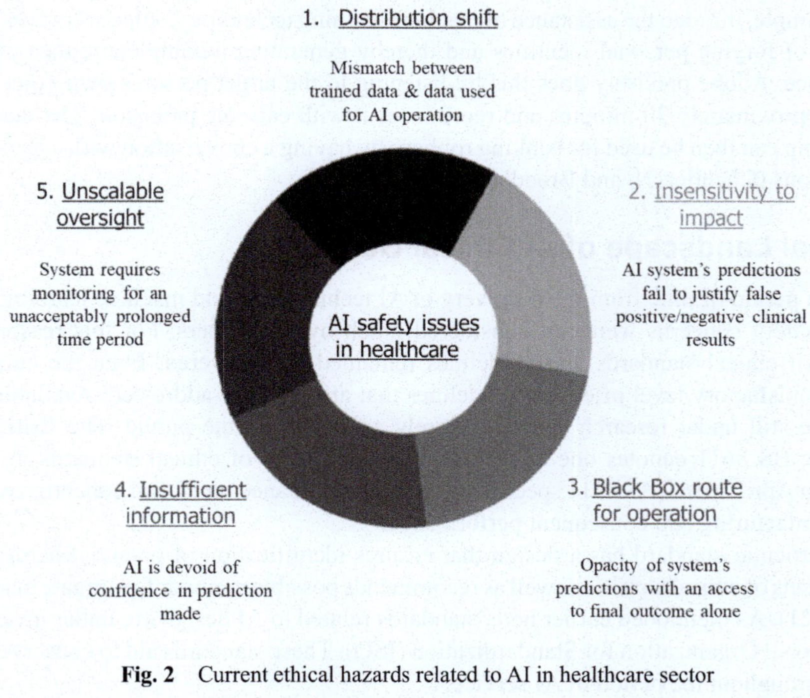

Fig. 2 Current ethical hazards related to AI in healthcare sector

6.4 Online Advertising and Image Representation (Deep Fake AI)

Deep fake AI is relatively a new term which has got popularized within recent years. It did exist for a long period of time, nonetheless, high sophistication has started merging into deep fake concepts making it a concealed regular practice nowadays. Since the core feature of deep fake AI is creating videos or images using fake/subject profiles instead of a 'target profile' by placing it accurately without any flaws. Such synthetic videos or images carrying realistic footage of incidents are considered a serious offense against human rights as it has never occurred. The foulest part of such a technique is that, it requires only limited involvement of humans to execute the intention using popular applications like Face2Face, Neural Textures, FaceSwap etc. (Hussain, S. et al., 2021). Art museums for instance exhibit a symbiotic relationship with deep fake AI applications with an aim of advertising, gaining public interest, educational outreach purposes etc. In most such cases, museums have pre-set collections of images used for exhibition, as well as interactive session images with fake personalities. One of such art-based deep fake includes; animated Mona Lisa's deep fake image which gathered immense attention globally in 2019. The AI algorithm behind such faking has used facial landmarks for creating animations or movements in the facial features in order to turn the famous portraits more interesting and realistic (Mihailova, M. 2021). Deep fake' is a combinatorial word that comes from deep learning and fake. As the notion can generate hyper-realistic fake videos in which people can be displayed doing things which they actually do not commit. In 2017, a particular case of deep fake came up in which; a fake video of a celebrity's sexual content started appearing in social media. The video was posted publicly by a Reddit user by faking someone's face in the video with the target person's. Media experts in the meantime, proposed some elusive methods to identify deep fake videos effectively and are depicted in Figure 3 (Westerlund, M. 2019). Such methods aid social media users to discriminate between original content and fake videos.

6.5 Fraudulence in Banking

AI as mentioned in previous sections is exceptionally proficient in snatching people's very personal information including their voice, identity, relationships, commitments, transactions etc. In the banking sector, this fraudulence can build a perfidious situation which may result in huge financial loss. For example, through the assistance of machine learning technology, some softwares like Adobe are capable of forging personal identities and thereby generating a complete replica of a specific person's voice. Adobe precisely does this by listening to the target person's own voice for a time period of approximately 20 minutes and reproduces it with extreme precision. The newly created fake voice clip can then be used for banking robbery by having a conversation with a banking officer for transactions (Chantler, N. and Broadhurst, R. 2008).

7. Global Landscape of AI Ethical Guidelines

Over a great period of time from the discovery of AI technologies and machine learning, its ethical as well as safety concerns were not considered much by researchers. For this reason, a proper framework of ethical standards and guidelines remained undiscovered. Even the current digital era lacks a satisfactory level of ethical guidelines that are properly addressed. Available standards therefore are still under research and not entirely accessible to the public. The British standard on Robotics; BS 8611 denotes one of the primitive categories of ethical standards in this regard (published in April 2016). BS 8611 specifically provides guidance on ethical concerns over robotics design, manufacturing and consequent performance.

This particular standard has a design that ensures identification of potential harm in robotics based on 20 sets of ethical hazards as well as recommends possible methods to mitigate bias (Kapeller, A. et al., 2021). As mentioned earlier here, standards related to AI design are under investigation at the International Organization for Standardization (ISO). These standards aid to conserve the ethical principles throughout the period of AI service.

As a response to societal anxiety on the AI ethical arena, several national and international organizations laid their foundation with a set of policies on account of safety levels of AI. Certain committees were set up; i.e. the High-level Expert Group on Artificial Intelligence (by the European

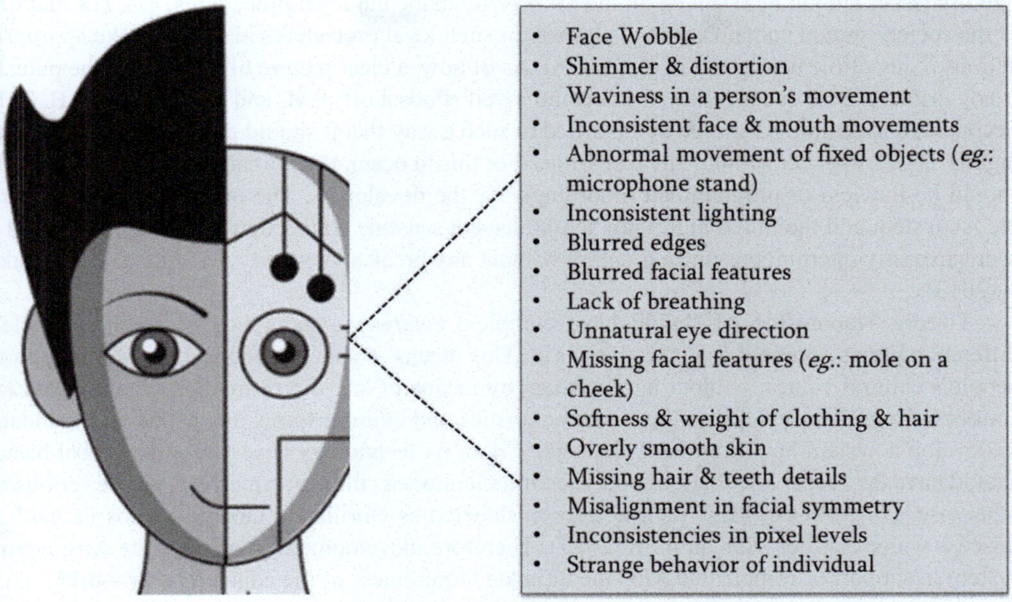

Fig. 3 Subtle methods used for the identification of deep fake videos and images.

commission), Society of the Organization for Economic Cooperation and Development (OECD), The advisory Council on the Ethical Use of Artificial Intelligence (Singapore) and Committee on Artificial Intelligence in United Kingdom etc. (Cath, C. et al., 2018). In recent years, a number of private companies have also started providing AI ethical instructions such as those released by Google and SAP, Association of Computing Machinery (ACM), Access Now, and Amnesty International. There are an enormous number of AI ethical guidelines and policy documents available now; though, a complete narration of all of those guidelines is beyond the scope of this chapter.

8. Responsible AI: The Future of AI Ethics

The rapidly propagating waves of AI and related technologies have significantly generated hot discussions among people in the society since AI is capable of eliminating current job opportunities in the near future. Numerous organizations and researchers are raising a warning on the negative impacts and ethical concerns that AI can impart unlike other technologies discovered till date. Currently available study reports in this context indicate that, the relative percentage of job replacements with AI would be as high as 47% and 35% in the United States and United Kingdom respectively. As a remark on the topic of this chapter, we can make a point that the development of responsible AI may be a cure to the possible negatives. We can introduce AI design with more intelligence only if it has humane moral values. Moreover, responsibility is the prime moral feature it should embrace. In other words, responsibility is the key to intelligence and a system can never be intelligent unless it understands responsibility (Frey, C.B. and Osborne, M.A. 2017).

Responsible AI resides on three basic pillars of fact; each one of them is relevant in the same way. Firstly, society as a whole or the users of AI should have a comprehensive idea about the responsibilities of a citizen or human being. This is exceptionally imperative since building responsibility in people is the first pillar of the well-being of the social order. The education curriculum of every nation should contain information on human rights and values, emphasizing the responsibilities of a citizen. This would undeniably blow up humanity from childhood itself. The creators of AI like the designers, executors, programmers, organizers and of course the distributors should be aware of the actual meaning of responsibility. In addition, they should be aware of how liability issues should be resolved on the basis of fundamental legal procedures. The law and order of the society should contain separate clauses for such legal procedures in order to take appropriate actions if any crime occurs in the chain of AI. As of now, a clear picture of who should be punished if any crime or bias occurs with AI is behind a veil (Podsakoff, P.M. and MacKenzie, S.B. 2014). Secondly, responsible AI should be modified in such a way that it should protect human values and dignity in every aspect without any disparities. For this to occur, the primary algorithm or input data should be flawless or programmed accordingly by the developers. The output data received from the AI system and the initial algorithm should have a sensible connection with each other; and the user can easily interpret the entire program without any break and opacity (Modgil, S. and Prakken, H. 2013).

Thirdly, responsible AI should have complete awareness about how it reaches people of different cultures and beliefs across the world. This means, it should understand, favor and protect people's cultural values; without hurting them by means of any discrimination. People from each continent have their own phenotypic, characteristics and cultural forms and hence it is mandatory to develop a system honoring them. Especially if an AI technology is serving on a global basis, it should have far deeper knowledge about the aforementioned cultures; which can induce serious bias otherwise. For instance, black people were mislabeled as Gorillas in Google photos in 2015 and raised a wider conflict (Benjamin, R. 2023). Therefore, development of an apposite AI-governing system in support of responsible AI is the ultimate requirement of the contemporary world.

Conclusion

As AI plays a crucial role in various domains of our day-to-day life, it can certainly bring imports in smaller or greater ways. The magnitude of such imports is highly dependent on the ethical conflicts that AI generates. The present chapter discussed various aspects of AI ethics which highlights the requirement of knowledge on ethical concerns. In every sector of application, AI must be trained with proper data that must consider moral values, fundamental human rights and legal actions associated with all these facts. An AI-driven system must be conscious of adding the key ethical principles to its priority list before commencing the programmed action. Foremost categories of mentioned ethical principles are; transparency, impartiality, accountability, reliability, security and privacy. From the period of discovery itself, AI developers are working on improvising its intelligence level or an attempt to mimic the human brain. Deep learning and neural networks in AI started blooming as a result of such research. However, it started nourishing the 'black box' mode of data processing which presents an opaque nature of AI in every way. This is the underlying cause of AI-related ethical disputes all round the world. The ongoing AI bias in various sectors of life is said to be netting in different levels such as socio-economic, human life privacy, democratic, environmental and liability. As per expert views, by putting the rule of ethical principles in the core of AI, we can remarkably lower the ethical disparities. Robots and intelligent artifacts in the market nowadays are incredibly dependent on or controlled by their own developers. Therefore, the ethical dilemmas are forcefully; yet intentionally regulated or manipulated by those developers on their own. This further necessitates the emergence of 'Responsible AI'; which rigidly provides importance on the concept in AI culture. The foremost fact we all should be aware of is that, AI holds an array of beneficial services for human welfare; provided only if it is devoid of any data black boxing.

References

Ahmed, M. 2020. UK Passport Photo Checker Shows Bias against Dark-Skinned Women. BBC News, 7.

Alemzadeh, H., Raman, J., Leveson, N., Kalbarczyk, Z. and Iyer, R.K et al. 2016. Adverse events in robotic surgery: a retrospective study of 14 years of FDA data. PloS one, *11*(4), e0151470.

Arnold, T. and Scheutz, M. 2018. The "big red button" is too late: an alternative model for the ethical evaluation of AI systems. Ethics Inf. Technol. *20*: 59–69.

Autor, D.H. 2015. Why are there still so many jobs? The history and future of workplace automation. J. Econ. Perspect. *29*(3): 3–30.
Azzam, T. 2023. Artificial intelligence and validity. New Directions for Evaluation, 2023(178–179): 85–95.
Bacciarelli, A. 2023. The Toronto Declaration: Protecting the right to equality and non-discrimination in machine learning systems.
Bak, M., Madai, V.I., Fritzsche, M.C., Mayrhofer, M.T. and McLennan, S et al. 2022. You can't have AI both ways: balancing health data privacy and access fairly. Front. Genetics. 13: 1490.
Bengio, Y., Lecun, Y. and Hinton, G. 2021. Deep learning for AI. Commun. ACM, *64*(7): 58–65.
Benjamin, R. 2023. Race after technology. In Social Theory Re-Wired. 405–415. Routledge.
Bottomley, D., and Thaldar, D. 2023. Liability for harm caused by AI in healthcare: an overview of the core legal concepts. Front. Pharmacol. 14.
Bovens, M. (2010). Two concepts of accountability: Accountability as a Virtue and as a mechanism. West Euro. Pol. *33*(5): 946–967.
Cath, C., Wachter, S., Mittelstadt, B., Taddeo, M. and Floridi, L et al. 2018. Artificial intelligence and the 'good society': the US, EU, and UK approach. Sci.Eng.Ethics, *24*: 505–528.
Chantler, N. and Broadhurst, R. 2008. Social engineering and crime prevention in cyberspace. Proceed. Korean Inst. Crim. 65–92.
Chen, Z. 2023. Ethics and discrimination in artificial intelligence-enabled recruitment practices. Humanit. Soc. Sci. Commun., *10*(1): 1–12.
Choudhury, A. and Asan, O. 2020. Role of artificial intelligence in patient safety outcomes: systematic literature review. JMIR Med. Info. *8*(7): e18599.
Cortese, J.F.N.B., Cozman, F.G., Lucca-Silveira, M.P. and Bechara, A.F. 2023. Should explainability be a fifth ethical principle in AI ethics?. AI Ethics. *3*(1): 123–134.
Creel, K.A. 2020. Transparency in complex computational systems. Philos. Sci. *87*(4), 568–589.
Cutillo, C.M., Sharma, K.R., Foschini, L., Kundu, S., Mackintosh, M. et al. and MI in Healthcare Workshop Working Group Beck Tyler 1 Collier Elaine 1 Colvis Christine 1 Gersing Kenneth 1 Gordon Valery 1 Jensen Roxanne 8 Shabestari Behrouz 9 Southall Noel 1. 2020. Machine intelligence in healthcare—perspectives on trustworthiness, explainability, usability, and transparency. NPJ Digital Med. *3*(1), 47.
Darwiesh, A., El-Baz, A.H., Abualkishik, A.Z. and Elhoseny, M. 2022. Artificial Intelligence Model for Risk Management in Healthcare Institutions: Towards Sustainable Development. Sustainability, *15*(1): 420.
Davis, S.E., Lasko, T.A., 2017. Calibration drift in regression and machine learning models for acute kidney injury. J. American. Med. Info. Assoc., *24*(6): 1052–1061.
De Angeli, A. 2009. Ethical implications of verbal disinhibition with conversational agents. PsychNology Journal, *7*(1).
Decatur, S., Goldreich, O. and Ron, D. 1997. Computational sample complexity. In Proceedings of the tenth annual conference on Computational learning theory. 130–142.
Floridi, L. and Taddeo, M. 2016. What is data ethics?. Philosophical Transactions of the Royal Society A: Mathematical, Physical and Engineering Sciences, *374*(2083), 20160360.
Frey, C.B. and Osborne, M.A. 2017. The future of employment: How susceptible are jobs to computerization? Technological forecasting and social change, *114*: 254–280.
Gambelin, O. 2021. Brave: what it means to be an AI Ethicist. AI Ethics, *1*(1): 87–91.
Giovanola, B. and Tiribelli, S. 2023. Beyond bias and discrimination: redefining the AI ethics principle of fairness in healthcare machine-learning algorithms. AI Soc. *38*(2): 549–563.
Gogarty, B. and Hagger, M. 2008. The laws of man over vehicles unmanned: The legal response to robotic revolution on sea, land and air. JL Inf. & Sci. *19*, 73.
Good, I.J. 1966. Speculations concerning the first ultraintelligent machine. Adv. Comput. *6*: 31–88.
Gorodnichenko, Y., Pham, T. and Talavera, O. (2021). Social media, sentiment and public opinions: Evidence from# Brexit and# USElection. European Economic Review, 136, 103772.
Graetz, G. and Michaels, G. 2015. Robots at work. Centre for Economic Performance. CEP Discussion Paper, 1335.
Habli, I., Lawton, T. and Porter, Z. 2020. Artificial intelligence in health care: accountability and safety. Bulletin of the World Health Organization, *98*(4), 251.
Halle, A., Campanile, L.F. and Hasse, A. 2021. An Artificial Intelligence–Assisted Design Method for Topology Optimization without Pre-Optimized Training Data. App. Sci. *11*(19), 9041.
Han, Y. 2023. The contribution of evaluation to accountability mechanisms. In *Handbook of Public Policy Evaluation*. 104–116
Heilweil, R. 2019. Artificial intelligence will help determine if you get your next job. *vox*.
Hess, A. 2016. On Twitter, a Battle Among Political Bots. *International New York Times*, NA-NA.
Hussain, S., Neekhara, P., Jere, M., Koushanfar, F. and McAuley, J. et al. 2021. Adversarial deepfakes: Evaluating vulnerability of deepfake detectors to adversarial examples. In Proceedings of the IEEE/CVF winter conference on applications of computer vision. 3348–3357.

Kapeller, A., Felzmann, H., Fosch-Villaronga, E., Nizamis, K. and Hughes, A.M. et al. 2021. Implementing ethical, legal, and societal considerations in wearable robot design. Appl. Sci. 11(15): 6705.

Kapoor, S.K. 2020. New face of indian governance sector with artificial intelligence. IME J., *14*(1): 97–110.

Khadem, M., Khadem, A., and Khadem, S. 2023. Application of Artificial Intelligence in Supply Chain Revolutionizing Efficiency and Optimization. Inter. J. Ind. Eng. Oper. Res. *5*(1), 29–38.

Khakurel, J., Penzenstadler, B., Porras, J., Knutas, A. and Zhang, W. et al. 2018. The rise of artificial intelligence under the lens of sustainability. Technologies. *6*(4): 100.

Kozuka, S. 2018. Japan's Response to New Technologies: Draft Artificial Intelligence Research and Development Guidelines for International Discussions. Zeitschrift für Japanisches Recht, *23*(46): 3–18.

Leontief, Wassily, 1983. National Perspective: The Definition of Problems and Opportunities.. The Long-Term Impact of Technology on Employment and Unemployment. Washington, DC: The National Academies Press.

Lepri, B., Oliver, N., Letouzé, E., Pentland, A. and Vinck, P. et al. 2018. Fair, transparent, and accountable algorithmic decision-making processes: The premise, the proposed solutions, and the open challenges. Philos. Technol. *31*: 611–627.

Lin, P., Abney, K. and Bekey, G. 2011. Robot ethics: Mapping the issues for a mechanized world. Artif. Intell. *175*(5–6): 942–949.

Mathur, R. and Swaminathan, S. 2018. National ethical guidelines for biomedical & health research involving human participants, 2017: A commentary. Indian J. Med. Res. *148*(3): 279.

McAllister, A. 2016. Stranger than science fiction: The rise of AI interrogation in the dawn of autonomous robots and the need for an additional protocol to the UN convention against torture. Minn. L. Rev., *101*, 2527.

McCarthy, J., Minsky, M.L., Rochester, N. and Shannon, C.E. 2006. A proposal for the dartmouth summer research project on artificial intelligence. 1955. AI magazine, *27*(4): 12–12.

Mcknight, D.H., Carter, M., Thatcher, J.B. and Clay, P.F. 2011. Trust in a specific technology: An investigation of its components and measures. ACM Transactions on management information systems (TMIS), *2*(2): 1–25.

Mihailova, M. 2021. To dally with Dalí: Deepfake (Inter) faces in the art museum. Convergence, *27*(4): 882–898.

"Ministry of Health and Family Welfare, Government of India. National Health Policy – 2017. New Delhi: MoHFW;," 2017

Modgil, S., and Prakken, H. 2013. A general account of argumentation with preferences. Artif. Intell. *195*, 361–397.

Nalluri, M., Babu Mupparaju, C., Rongali, A.S. & Polireddi, N.S.A. 2024. HUMAN-AI COLLABORATION IN HEALTHCARE STUDYING THE IMPACT OF AI ON HEALTHCARE PROFESSIONALS DECISION-MAKING PROCESSES. Pak. Heart J. *57*(1): 69–77.

O'Neil, C. 2016. Weapons of math destruction: How big data increases inequality and threatens democracy. SCIENTIFIC AMERICAN, *315*(2): 74–74.

Pagallo, U. 2018. Apples, oranges, robots: four misunderstandings in today's debate on the legal status of AI systems. Philosophical Transactions of the Royal Society A: Mathematical, Physical and Engineering Sciences, *376*(2133): 20180168.

Papadouli, V. 2022. Transparency in Artificial Intelligence: A Legal Perspective. J.Ethics Legal Technol. *4*(1): 25-40.

Podsakoff, P.M., and MacKenzie, S.B. 2014. Impact of organizational citizenship behavior on organizational performance: A review and suggestions for future research. Organ. Citizen. Behavior Context. Perform. 133–151.

Ranganathan, S. 2020. Towards a Holistic Digital Health Ecosystem in India. Observer Research Foundation, 2.

Richmond, S. 2016. Superintelligence: Paths, Dangers, Strategies. By Nick Bostrom. Oxford University Press, Oxford, 2014, pp. xvi+ 328. ISBN: 9780199678112. Philos. *91*(1): 125–130.

Santos Botelho, C. 2023. The End of the Deception?–Counteracting Algorithmic Discrimination in the Digital Age. Forthcoming in Giovanni De Gregorio, Oreste Pollicino & Peggy Valcke (eds.), The Oxford Handbook on Digital Constitutionalism Oxford University Press, 2024.

Sharkey, N., Goodman, M. and Ross, N. 2010. The coming robot crime wave. Computer, *43*(8): 115–116.

Sharma, D., and Chandra, A. 2020. Medical Device Rules-2017, India: An Insight. Appl. Clin. Res. Clin. Trials Regul. Aff. *7*(2): 126–134.

Sjoding, M.W., Dickson, 2020. Racial bias in pulse oximetry measurement. New Eng. J. Med. *383*(25): 2477–2478.

Smuha, N. A., Ahmed-Rengers, E., Harkens, A., Li, W., MacLaren, J et al. 2021. How the EU can achieve legally trustworthy AI: a response to the European Commission's proposal for an artificial intelligence act. *Available at SSRN 3899991.*

Stahl, B.C., Schroeder, D. and Rodrigues, R. 2023. Ethics of Artificial Intelligence: Case Studies and Options for Addressing Ethical Challenges.116. Springer Nature.

Stahl, B.C. Schroeder, D. and Rodrigues, R. 2022. Unfair and illegal discrimination. In Ethics of artificial intelligence: Case studies and options for addressing ethical challenges. 9-23. Cham: Springer International Publishing.

Taddeo, M. 2017. Trusting digital technologies correctly. Minds. Mach. 27, 565–568.

Thomasian, N.M., Eickhoff, C. and Adashi, E.Y. 2021. Advancing health equity with artificial intelligence. J.Pub. Health Pol. 42, 602–611.

Tiribelli, S. 2023. The AI ethics principle of autonomy in health recommender systems. Argumenta, 16, 1–18.

Vinuesa, R., Azizpour, H., Leite, I., Balaam, M., Dignum, V. et al. 2020. The role of artificial intelligence in achieving the Sustainable Development Goals. Nat. Commun. *11*(1): 1–10.

Wachter, S., Mittelstadt, B. and Russell, C. 2021. Why fairness cannot be automated: Bridging the gap between EU non-discrimination law and AI. Comput. Law. Secur. Rev. *41*: 105567.

Westerlund, M. 2019. The emergence of deep fake technology: A review. Technology innovation management review, *9*(11).

CHAPTER

10

Learning Anatomy using Virtual Dissections: Previous Experience and Future Directions Integrating Artificial Intelligence and Multiple Approaches

Miguel Pais-Vieira[1*]

1. Introduction

In the broader context of biological education, the topic of the present book, the instruction of human anatomy holds particular significance for the integration of artificial intelligence (AI). This is due to a distinct set of challenges that are frequently absent in other disciplines. On one hand, the syllabus is inherently extensive, while on the other, there exists an intrinsic tridimensional complexity within each studied system, as well as in the spatial relations among different systems (Davis et al., 2014; Pereira et al., 2007; Santos et al., 2022), that must be mastered by students. Thus, identifying which parts of the syllabus may benefit from a teaching approach based on the spatial relation within the system or between different systems requires the analysis of large amounts of data and, therefore, can benefit from the use of AI.

Human anatomy is a fundamental building block in undergraduate education in the field of health; however, the large number of structures to be memorized makes it a difficult and often underappreciated subject (Davis et al., 2014; Pereira et al., 2007; Santos et al., 2022). The main variables to be considered in the study of anatomy are the curriculum organization, the educators,

[1] iBiMED – Institute of Biomedicine, Department of Medical Sciences, University of Aveiro Portugal.
* Corresponding author: miguelpaisvieira@ua.pt

the students, and the equipment/installations. Traditionally, the anatomy curriculum involves either a systematic regional approach, where specific systems are studied in the whole body, or a regional approach where all systems in each region are studied (Wang et al., 2010; Pais and Moxham, 2013; Attardi and Rogers, 2015). Both approaches seem to have benefits that depend on multiple variables such as time and resources (Estai and Bunt, 2016; Costa et al., 2023). These variables are relevant because the access to human cadavers is difficult in many universities (Habicht et al., 2018) and this necessarily conditions the regional approach.

As an alternative, there have been great efforts in an attempt to improve human anatomy learning through the use of supporting or replacing technologies. More precisely, the appearance of technologies that allow for novel forms of interaction between students and the educational content (that is, the anatomical structures or their functions) have always been the main contenders for the replacement of human cadavers. This has occurred with the generalized use of information systems (Pereira et al., 2007), 3D printing (Ye et al., 2020; Brumpt et al., 2023), virtual dissection software (Triepels et al., 2020; Bartoletti-Stella et al., 2021; Raja et al., 2022); virtual and augmented reality (Ma et al., 2016; Moro et al., 2017), and more recently, with Artificial Intelligence (Chan and Zary, 2019; Abdellatif et al., 2022; Castellano et al., 2023).

Altogether, these advances have brought a large number of new approaches for the study of anatomy which have the potential to be used alone or in combination in an attempt to adjust the content presentation to the specific needs of the user. The present chapter has two main goals: (1) To describe findings from the continuous experience of four years using virtual dissection software in different anatomy courses (Biomedical sciences, Biomedical engineering, Post-graduate Nursing); and (2) To propose a future approach combining multiple methods and AI to improve both group and self-learning of anatomy.

The use of virtual dissection for the study of anatomy was significantly potentiated during the COVID-19 pandemic, which enforced distant learning in many institutions (Papa et al, 2022). The present chapter summarizes the experiences of anatomy teachers during the period of 2020-2023 and proposes potential pathways for future studies. A description of the main findings resulting from the use of virtual dissections, rodent dissections, and the use of textbook figures combined with plastic models, will be made. Most importantly, instead of advocating a regional or a systemic approach to anatomy teaching, or in favor of the use of plastic models or cadaver dissection, we will describe which techniques have been reported by students and educators as being most useful for each specific system/region.

2. Report of a Previous Experience with Virtual Dissections and other Approaches

The qualitative results presented here are from an experience of four years (2020-2023) with 1st year Biomedical Engineering students engaged in two courses of Anatomy and Physiology (Anatomy and Physiology 1 and 2, one semester each). The findings presented here will be complemented with additional findings from pre-graduate students in Biomedical Sciences and from post-graduate Health students to discuss and generalize the findings from the larger sample of Biomedical Engineering students.

2.1 Curriculum and Approach

The curriculum included the main topics in anatomy and physiology. In semester 1, the topics included, basic anatomy nomenclature and orientation, cell physiology, histology, integumentary system, muscular-skeletal system, and nervous system (including the senses). In semester 2, topics involved were the endocrine system, reproductive system, cardiovascular system, respiratory system, excretory system, and immune system. The present description will focus solely on the anatomy component. The curriculum involved a mixed systemic and regional anatomy approach. The theoretical seminars were mostly about physiology and were taught with a systematic approach (i.e., one system was presented and, only then, the next system would be presented) through an oral seminar. Meanwhile, in the practical classes, a regional approach was (Fig. 1). Here, students were given a worksheet with questions that needed to be answered by groups of 2-3 students by the end of the class. These questions required students to learn new concepts, identify anatomical structures, relate structures from different systems, and relate the anatomical component with the physiology previously learned. Students were required to complete the worksheet, which involved a combination of manipulating the software Human Biodigital[1] and multiple searches in textbooks, websites, among others. All the work was supervised by the instructor with solutions being provided at the end of the session. Throughout the session, if a group had already found the solution for a problem, they were encouraged to explain it to other groups that were unable to find the solution. At the end of the session, a brief presentation summarizing the most relevant aspects was made by the instructor.

Each of the different approaches is detailed below.

To answer the worksheet, students were required to use the virtual dissection software and perform multiple searches in websites, scientific manuscripts, and books. When in doubt, students were asked to confirm information using three different sources.

Regarding the virtual dissections, the Human Biodigital software was used. Students should use the software under the following recommendations.

(I) Students should not go directly to the structure that they want to study. They should always start from the most superficial structure and remove the necessary structures that impede the observation of the structure of interest.
(II) Before removing a structure, students were required to indicate, in advance, which structures they would encounter after removing that structure.
(III) If students failed to indicate the structures to be encountered before removing a structure, they were required to put the structure back and, once again, predict which structures they would encounter after removing the structure (see algorithm below).

In specific cases (also see details below) recommendations to remove or include specific systems were made.

Throughout the year, two different rabbit dissections were performed (one at the end of each semester). These dissections had two main goals. First to identify general anatomical structures, and second, to allow students to explore a topic of their interest (e.g., following the abdominal aorta, dissecting the sciatic nerve). In addition to the remaining resources, multiple plastic models (SOMSO®, Germany) were available for some of the classes (2-3 times per semester), with students being encouraged to interact with them. These models allowed the removal of organs, muscles, as well as some bones.

Lastly, a group project was assigned. In the first semester, students were required to deliver a brief presentation on a biomedical engineering problem with a significant anatomy component. The themes included: vertebral artery stent, 3D-printed scapula, wheelchair with upright position, and dorsal column stimulation for chronic pain relief. Students were required to know the technology used, describe how it interacted with the human body, and lastly, discuss the implications regarding surgery/treatment for the different anatomical structures involved. In the second semester, students were required to make a small presentation analyzing the same topic that they had analyzed in the first semester, but now considering which changes should be implemented if the biomedical engineering solution were applied to an elderly patient, a pregnant woman, or a child.

---

1 https://www.biodigital.com

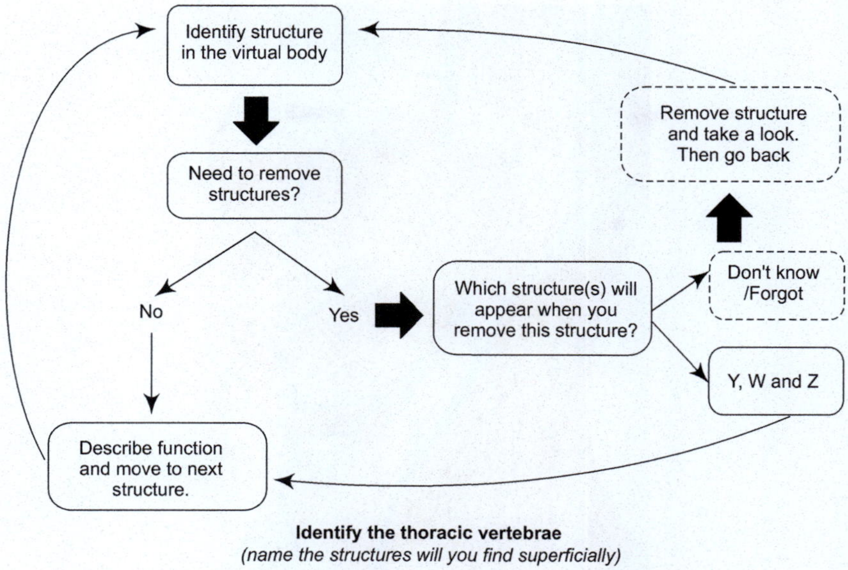

Fig. 1 Algorithm for use of Visible human Body software

2.2 Qualitative Description of the use of Virtual Dissections

Throughout the two anatomy-physiology courses, all students were able to learn the minimum contents required to be approved in the theoretical and practical components, suggesting that the method used was sufficient for them to acquire the basic concepts of anatomy and physiology.

2.3 General Reactions

The students' reaction to the virtual dissection software was generally positive, indicating that it allowed for 3D visualization and manipulation of the different structures. Due to their previous experience in high school with plastic models (where they were often not allowed to touch them or otherwise were available only for a limited period), students also reported that being allowed to manipulate as they pleased, without having time limits and without being worried about "breaking something", allowed them to become very confident with the software, facilitating the study.

2.4 Positive Experiences with Virtual Dissections

Students considered the virtual model to be far superior for the study of the bones, muscles, most of the blood and lymphatic vessels (see exceptions below), and most structures of the digestive system, as well as to establish relations between structures of different systems (Figure 2).

For example, students indicated unawareness of the depth of muscle tissue that was present between the skin and the body of the vertebrae. This became clear on removing one muscle at a time in the virtual model. They also reported that the segmental muscles of the back became more obvious in the software model, where they could be observed simultaneously and/or separately.

When studying the brain, students reported that the relations between superior temporal gyrus, insula, putamen, Globus pallidus, internal capsule, lateral ventricle, caudate nucleus, thalamus, and third ventricle could easily be understood through the process of sequentially removing structures from the most superficial towards the deeper structures. Similarly, the comprehension

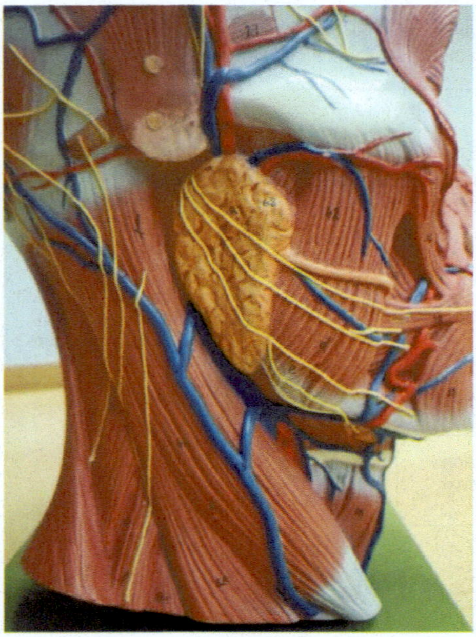

Fig. 2 (A) The use of the virtual model facilitated learning of neck structures, especially those presenting relations with structures from other systems (e.g., ansa cervicalis and internal jugular vein)

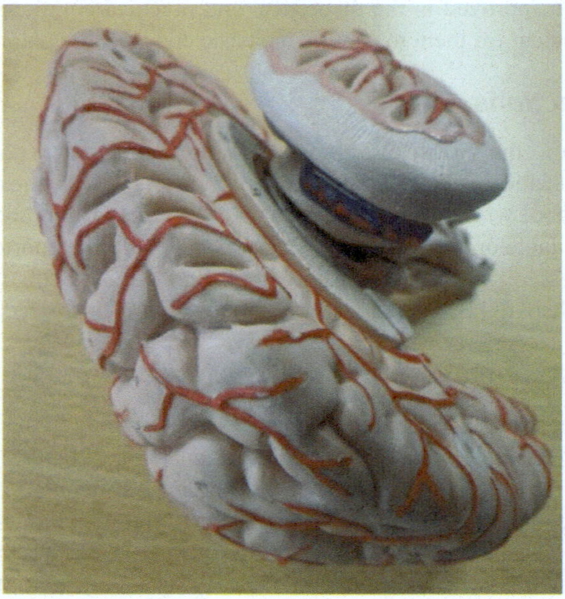

Fig. 3 The use of the virtual model facilitated learning the relations between basal ganglia and other structures.

of the C-shaped neural structures (e.g., caudate nucleus, hippocampus) was also considered for improvement through the utilization of software (Figure 3).

One particularly interesting case was the larynx cartilages, where students considered the software to be the best resource to understand how they affected each other and moved. However, a full comprehension of the position of the vocal folds and the vestibular folds was best achieved through the use of the plastic model. It is not absolutely clear why this occurred, since images and the virtual model allowed performing the exact same operations as the plastic model. However, the manipulation of the model seems to have had an effect on students' perception of the relative position of these two structures.

2.5 Negative Experiences with Virtual Dissections

The use of the virtual model was not as useful as some of the other approaches in a small number of topics, which are highlighted below so that improvements in teaching these topics can be implemented in the future.

The study of the celiac trunk was actually impaired by the use of the virtual dissection model. This trunk has such a large number of branches that their representation in the virtual model tends to confuse students. Meanwhile, in textbook images and in the plastic models, this representation tends to be simplified, allowing the immediate identification of the three main arteries. This was the only case where it was directly recommended for students not to use the model to study anatomical structures.

The continuous use of the virtual model was useful, but it may have led to biases in learning of specific items. For example, it was found that multiple students were not decipher whether the trachea is anterior or posterior to the esophagus. Even though students interacted with these structures in the virtual model, they typically did so in different classes, and therefore, when questioned, they expressed doubts. To prevent this, in the following year, a question asking students to specifically relate the position of the trachea to the position of the esophagus was included in the exercises performed in classes. Despite some improvement, it was found that manipulating the plastic model (i.e., inserting and removing the respiratory and digestive tracts) contributed to students reporting increased confidence in this specific topic.

Students also presented difficulty with the distinction between naso-/oro-/laringo-pharynx and related structures. It was found that the use of figures was helpful to identify some of the major structures (e.g., the constrictor muscles), but the plastic model significantly improved/complemented the student's perception. This was probably because the plastic model allowed touching the opening of the auditory tube (Eustachian tube), as well as the salpingopharyngeal and the salpingopalatine folds, therefore complementing the knowledge regarding the other structures.

The use of the plastic model was also relevant in the acquisition of knowledge regarding the relative position of the liver and the spleen. Specifically, students could not tell if the liver expanded beyond the right hypochondrium and were confused about the relative positions of the spleen and the left kidney. This was corrected through the use of the plastic model, which allowed manipulation of these structures, and also through the inclusion of additional images showing this relationship. One exercise that increased the overall confidence of students in the relative position of multiple anatomical structures, and to partially correct for some of the biases reported here, was to approach the transpyloric plane by asking separately where each of the structures of the plane was located, and only in the end showing that all of them were at the same level. This allowed students to explore the position of structures from multiple systems - and which are typically taught separately - in a single exercise. According to students' reports, this exercise was greatly facilitated by the presentation of the plastic model (even though it could be performed with the dissection software).

Students demonstrated difficulty in relating the position of the anal canal and prostate, and they could not establish a relation between the pubic symphysis, the bladder, and the uterus. The use of the plastic models, which often have hemi-sections of the structures greatly improved the students' reported confidence.

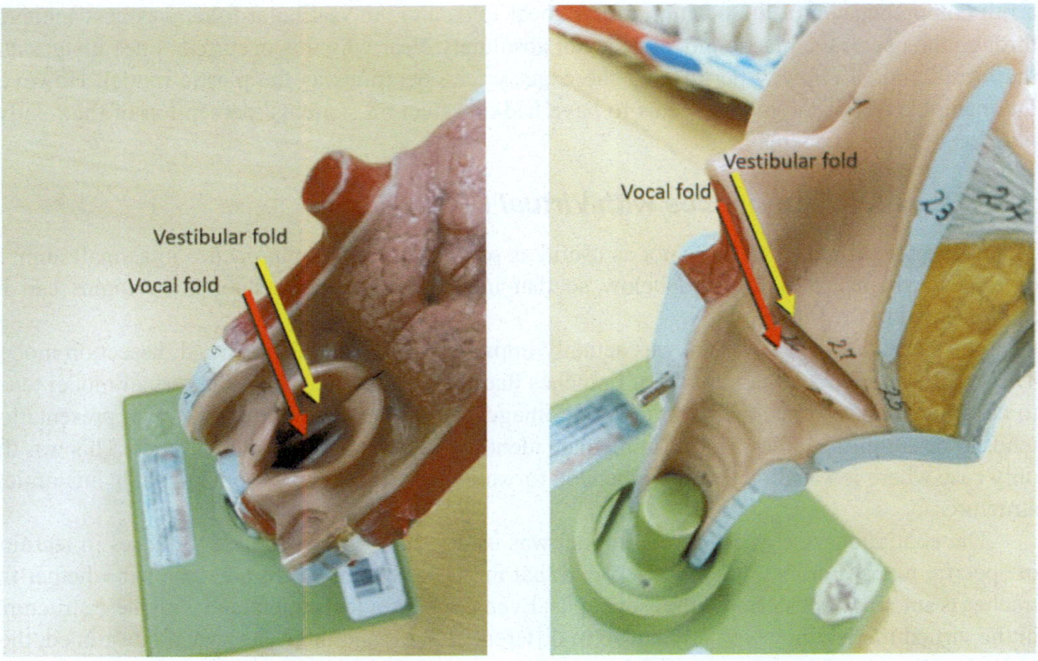

Fig. 4 Students encountered difficulty in learning the difference between the vocal and vestibular folds. The use of the plastic model was considered as a better learning tool for these structures.

Besides the dissection software, students also used the more traditional resources. According to students, the use of books/images continued to be of extreme utility because: (i) the software did not present all the structures that appeared in book images (especially the images from books by Frank Netter and colleagues) (e.g., Netter, 2014), and (ii) being exposed to multiple representations of the same structure contributed to becoming more confident with the knowledge acquired.

2.6 The use of the Plastic Models

The use of the plastic models proved to be highly relevant, especially for the small number of topics reported above. These models were also relevant for cases where the remaining approaches did not allow for a clear observation of the relation between structures or otherwise did not permit the use of physical contact.

2.7 The use of Rodent Dissections

The use of rodent dissections generated an overall division among students, with approximately half of them choosing not to perform this activity (recall that these were biomedical engineering students and therefore some of the students do not have a natural interest in cadaver anatomy). For the students who performed the activity, it was generally considered very motivating and "highly rewarding". This activity allowed students to recognize that anatomical variations are very common and that the images from the textbooks and the virtual model are fairly accurate but cannot account for the complexity of the real models. This aspect is of extreme relevance for biomedical engineers because they will have to: (i) develop devices that can accommodate for anatomical variations, and (ii) work with people (e.g., surgeons) that deal with these variations on a daily basis, and therefore, can make significant contributions to the devices being developed.

2.8 Results from other Courses

The use of virtual dissections in practical classes has also been tested in two other courses with different results. In a short post-graduate course for Health students, virtual dissections were used to review brain structures. Students reported that the software was extremely useful because it allowed them to relate structures that they had previously studied but weren't sure of their location or function. Interestingly, the use of the virtual dissections has also been implemented in a second year undergraduate course in anatomy in the field of biomedical sciences (respiratory, cardiovascular, excretory, digestive systems), but students' responses to the use of virtual dissections were divided. Approximately half of the students indicated that they preferred to memorize the textbook figures and focus solely on learning a single system, instead of being able to relate the different structures present in each region. The other half considered that the virtual dissection allowed them to relate the different topics taught throughout the years, giving them more confidence in the overall subject of anatomy. It is noteworthy that these students study systematic anatomy and physiology (2-3 systems per semester) and may not be motivated to learn/recall beyond what is required for the specific semester. The results from this sample of students suggest that, depending on the course and the student's profile, the use of the virtual model may or may not be motivating.

3. Summary of Results from the use of the Virtual Model

Altogether, the use of the virtual model proved to be successful for multiple anatomy topics, even though it has the potential to introduce some biases (Table 1). Some of these biases were highlighted here for improvement in the near future. Below, a program for an anatomy and physiology course is proposed that includes the use of AI to improve the student's learning process.

Table 1 Summary of topics where the virtual model was insufficient for learning the position and relations of a structure

Difficulty /Bias	*Solution*
Vocal vs vestibular folds	Plastic model
Esophagus vs trachea	Plastic model, videos, images
Peritoneum vs mesentery	Images, videos
Uterus vs bladder	Plastic model, images
Anal canal vs prostate	Plastic model, images, videos
Larynx cartilages	Plastic model, instructor voice modulation
External vs Extreme capsule	Images
Celiac trunk	Images

3.1 Proposal for an AI Supported Multi-approach Anatomy Program

The use of multiple approaches (virtual dissection, plastic models, textbook images, rodent dissection) was associated with positive learning experiences as reported by the Biomedical Engineering students. Here, it is proposed that the use of AI can significantly improve the existing program through an intervention in three different phases, namely: initial evaluation, study materials, and final evaluation.

4. Four-step Approach for Tailored Anatomy Study

The approach proposed here combines the ability of Large Language Models (LLMS) and other AI implementations to adapt the contents of an anatomy course to suit the user's profile and preferences, ensuring that critical topics are appropriately addressed (i.e., topics that students typically struggle with or that tend to be biased when a specific method is used), and promoting advanced knowledge. The algorithm requires some information from the user. This will allow tailoring the contents to ensure that: (i) the contents are at the level required by the course, (ii) the contents are presented (within certain limits) according to the user's preferences, and (iii) to stimulate the user's interest in additional topics (if the user demonstrates such interest and availability). After this initial evaluation, the algorithm will gather relevant sources of information and present definitions, summaries, and examples as appropriate.

4.1 Step 1: General Input

In step 1, the proposed algorithm starts with an initial evaluation of the course objectives, the user profile, and the user preferences. The course objectives vary significantly according to the type of curriculum where the course is included, and therefore, it is of critical importance to ensure that the contents are neither too easy nor too difficult for the user. Introducing the course objectives, reference textbook, and expected difficulty level (e.g., first year anatomy for biomedical engineering students versus advanced surgical hand techniques for medical students) can significantly contribute to this approach. Also, from our previous experience with students from multiple undergraduate and graduate courses, there are students that prefer to read a textbook and then slowly start memorizing anatomical structures, while others prefer to carefully analyze anatomical plates with legends and then start reading a textbook description of the plates. Also, a small number of users preferred to learn new concepts through a practical application or a clinical case, instead of learning a specific set of structures or a system.

4.2 Step 2: Initial Exposure to Contents and Interaction with Different Stimuli

Step 2 is composed of two parts. In the first part, the algorithm will generate learning materials according to the user's preferences. Here, the input gathered in step 1 will be critical to determine the type of materials presented, their sequence, and depth. Even though the user's preference will be respected, the algorithm will still generate[2] additional material in other formats, to prevent biases that may appear due to the use of a particular type of presentation. For specific contents (e.g., the heart), and depending on the initial input given in step 1, the algorithm will suggest the performance of additional activities such as the actual dissection of specific structures. In the second part, small quizzes and exercises will help the user identify the knowledge level acquired from the first part of this step. Namely, the user will be able to identify if the knowledge gathered is sufficient for *remembering*, *understanding*, *applying*, *analyzing*, or *evaluating* (Bloom, 1956; Krathwohl, 2002; Adams, 2015).[3] As it is possible that the student and/or the algorithm identifies themes that need additional study, steps 2 and 3 may require the student to go back and review specific contents. At the end of step 2, the student is expected to have been familiar with all contents and, in addition, is expected to have been separately tested in all themes.

From the point of view of algorithm improvement, step 2 will include information that has been previously gathered (and that will continue to grow as more information is gathered in upcoming years) regarding which topics are often misunderstood by students. For example, as reported in the

[2] The term "generate" material is used here to indicate both the generation of texts as well as the indication of reference books, websites, or scientific manuscripts.

[3] Only in the alternative step 4 the higher level of creation will be proposed.

first part of this chapter, the sole use of virtual dissections as a study method, was associated with biases of students in the identification of components of the celiac trunk, position of the esophagus, identification of the different parts of the larynx, identifying the relative position of the liver and the spleen, anal canal and surrounding structures, as well as the identification of the different structures in the urogenital system. The inclusion of this information in the algorithm, along with the strategies to minimize the biases that may be associated with specific student profiles, will be critical for continuous improvement of the algorithm.

4.3 *Step 3: Testing All Contents Simultaneously*

Step 3 constitutes the final evaluation and will require the student to perform an exam with the duration and complexity expected to be encountered in the actual exam to be performed in the context of the class. Both the duration and complexity of questions will be adjusted to match the student's needs and the course requirements. Also, this final evaluation will include different types of questions (e.g. multiple choice, legends, short answers) to ensure that the student is exposed to multiple stimuli related to the different topics of the course.

4.4 *Step 4: Advanced Goals*

If step 3 is concluded with success, the algorithm will suggest an activity in a relevant topic at the intersection of anatomy and one of the state-of-the-art problems in the area of interest (e.g., biomedical engineering, medicine, biomedical sciences). For this, the information gathered in step 1 (where the user indicates the general purpose of the course), will be combined with the user's performance and preferences to suggest a small research project (theoretical or practical) where the student can expand his/her knowledge towards more advanced questions. This step, although not critical for the user to pass the course, will be included to stimulate users to explore areas of knowledge that they find interesting and may otherwise not be aware of them.

5. Discussion and Final Notes

We have described the results from our experience with the use of virtual dissections in teaching anatomy and, based on that experience, proposed a program to combine multiple approaches with artificial intelligence to improve students' learning. We have not included a discussion of the benefits of human dissection, because it was not included in our teaching program. It is noteworthy to highlight that this is undoubtedly relevant for the study of human anatomy and that we plan to address its contribution to the proposed program in the near future. Altogether, the results from our previous experience suggest that multiple techniques are beneficial for anatomy courses since they seem to have different results for different subjects, as demonstrated by others (Diaz et al., 2015; Estai and Bunt, 2016). Here, we have specifically identified that the study of subjects related to bones, muscles, articulations, and nerves can greatly benefit from regional virtual anatomy dissection and that, if possible, the occasional support of animal dissection can be of great relevance[4]. Students that focused their study solely on virtual anatomical dissection expressed difficulties in a small number of topics. Although previous studies have compared the benefits of using specific techniques for anatomy teaching (Pereira et al., 2007; McMenamin, 2008; Ma et al., 2016: Bartoletti-Stella et al., 2021; Santos et al., 2022), we have focused on identifying parts which can significantly benefit from each approach[5]. Therefore, it is assumed here that all approaches can be beneficial and it is critically important to identify topics that can better suit a specific student profile. The use of the AI-based algorithm is expected to contribute to this tailored approach.

---

[4] As indicated above, human dissection would also be relevant.

[5] It should be highlighted though, that we have reported here our experience, but we have not performed a controlled study to evaluate the effect of each technique in each specific topic.

It was observed that the combination of methods seemed critical for the renal/urinary, and reproductive systems, namely, students benefited the most when plastic models were combined with the virtual dissection (which allowed making the structures partially transparent). This allowed relating the anterograde position of the uterus with the position of the bladder, the urethra, and the associated glands. In addition, the plastic models emphasized the use of touch, which, as reported for other strategies, is likely to have a relevant role in anatomy learning (McMenamin, 2008). Lastly, in all cases, the use of videos and images from textbooks and atlases was considered extremely useful. For the study of the larynx cartilages, a combination of: (i) virtual model dissection, (ii) exercises involving the instructor's voice modulation (e.g., by applying pressure anteriorly to the thyroid cartilage) and plastic models proved to be useful. Notably, most students expressed an understanding of the difference between the vocal and vestibular folds after observing the plastic model from above (which seems to allow them to relate the medial and lower position of the vocal cords). The study of the human brain seems to constitute a special case in the learning process. This is because the structures tend to look similar (i.e., white, and grey matter), and therefore, learning structures was often not related to the visual characteristics of the area but to its function and relative position. Virtual dissection was particularly useful to establish spatial relations between the temporal gyri, insula, external capsule, putamen, Globus pallidus, internal capsule, lateral ventricle, caudate nucleus, thalamus, and third ventricle. In addition, it also allowed identifying the corpus callosum, anterior commissure, and internal capsule with white tracts.

Notably, the virtual model did not represent the external capsule, claustrum, and extreme capsule, and this had to be pointed out to students in textbook images. This situation can be considered from two different points of view. First, one can consider that eventually the software will be able to represent these structures, and this will no longer be a problem. Second, it could be argued that a properly developed AI-based algorithm should be able to detect this type of situation automatically and adjust the learning materials to prevent learning biases. We have also proposed a program for an algorithm to facilitate anatomy learning through the use of AI. The applications and challenges of using AI in medical education have been previously discussed (Chand and Zary, 2019), and include adaptation from both students and instructors. It is expected however that the need for such adaptation will be temporary, and the use of the AI will probably become the norm (Farrow, 2021; Hauptman et al., 2023).

Here, we propose that AI can enhance various aspects of the anatomy learning process. First, the utilization of AI, with its capacity to process large volumes of data, enables the development of approaches that can identify and adapt to student's preferences (Gao et al., 2021). Second, an AI algorithm can access multiple resources and promptly process the relevant data, which is helpful for generating summaries, exercises, and quizzes. Third, the algorithmic nature of the proposed program also allows students to revisit concepts that were unclear initially, utilizing additional resources for both presentation and testing. Finally, the program presented here is specifically designed to foster the further development of students showing interest in a particular area. In other words, the algorithm is expected to be flexible enough to assist students aiming to grasp the basics while also maintaining the interest and curiosity of those seeking to delve beyond the course requirements. In sum, whereas previous algorithms have been developed to identify students' preferences (Lu et al., 2018; Gao et al., 2021), generate summaries (Dash et al., 2019), and create exams (Yang et al., 2023), our proposal integrates these advancements with our previous experience to apply them to the study of anatomy.

Acknowledgements

The authors would like to thank: Human Biodigital for authorizing the use of images from their software as well as Support from the Pedagogical Innovation Project from University of Aveiro, Portugal "Art as a teaching strategy in biomedical sciences: Unveiling new horizons with artificial intelligence", 2023.

References

Abdellatif, H., Al Mushaiqri, M., Albalushi, H., Al-Zaabi, A.A., Roychoudhury, S et al. (2022). Teaching, learning and assessing anatomy with artificial intelligence: The road to a better future. *International Journal of Environmental Research and Public Health*, *19*(21), 14209.

Adams, N.E. (2015). Bloom's taxonomy of cognitive learning objectives. *Journal of the Medical Library Association: JMLA*, *103*(3), 152.

Attardi, S.M. and Rogers, K.A. (2015). Design and implementation of an online systemic human anatomy course with laboratory. *Anatomical sciences education*, *8*(1), 53–62.

Bartoletti-Stella, A., Gatta, V., Mariani, G.A., Gobbi, P., Falconi, M. et al. (2021). Three-dimensional virtual anatomy as a new approach for medical student's learning. *International Journal of Environmental Research and Public Health*, *18*(24), 13247.

Bloom, B.S. (1956). Engelhart MD, Furst EJ, Hill WH, Krathwohl DR. *Taxonomy of educational objectives: Handbook I: Cognitive domain*.

Brumpt, E., Bertin, E., Tatu, L. and Louvrier, A. (2023). 3D printing as a pedagogical tool for teaching normal human anatomy: a systematic review. *BMC Medical Education*, *23*(1), 783.

Chand, K.S. and Zary, N. (2019). Applications and challenges of implementing artificial intelligence in medical education: integrative review. *JMIR medical education*, *5*(1), e13930.

Castellano, M.S., Contreras-McKay, I., Neyem, A., Farfán, E., Inzunza, O. et al. (2023). Empowering human anatomy education through gamification and artificial intelligence: An innovative approach to knowledge appropriation. *Clinical Anatomy*.

Costa, T.G., Campos, M.H., Barros, A.C., Andrade, M., Vancini, R.L. et al. 2023. Regional-based approach during anatomy coursework results in better academic performance than systemic-based approach.

Dash, Abhisek. "Summarizing user-generated textual content: Motivation and methods for fairness in algorithmic summaries." *Proceedings of the ACM on Human-Computer Interaction* 3.CSCW (2019): 1–28.

Davis, C.R., Bates, A.S., Ellis, H. and Roberts, A.M. (2014). Human anatomy: let the students tell us how to teach. *Anatomical sciences education*, *7*(4), 262–272.

Diaz, Claudia M. and Torres Woolley. (2015). "Engaging multidisciplinary first year students to learn anatomy via stimulating teaching and active, experiential learning approaches." *Medical Science Educator 25*: 367–376.

Estai, M. and Bunt, S. (2016). Best teaching practices in anatomy education: A critical review. *Annals of Anatomy-Anatomischer Anzeiger*, *208*, 151–157.

Farrow, E. (2021). Mindset matters: how mindset affects the ability of staff to anticipate and adapt to Artificial Intelligence (AI) future scenarios in organisational settings. *AI & society*, *36*(3), 895–909.

Gao, Zhaolong, Mengvi P. Gatpandan, and Paulino H. Gatpandan. "Classification Decision Tree Algorithm in Predicting Students' Course Preference." *2021 2nd International Symposium on Computer Engineering and Intelligent Communications (ISCEIC)*. IEEE, 2021.

Habicht, J.L., Kiessling, C. and Winkelmann, A. (2018). Bodies for anatomy education in medical schools: an overview of the sources of cadavers worldwide. *Academic Medicine*, *93*(9), 1293.

Hauptman, A.I., Schelble, B.G., McNeese, N.J. and Madathil, K.C. (2023). Adapt and overcome: Perceptions of adaptive autonomous agents for human-AI teaming. *Computers in Human Behavior*, *138*, 107451.

Krathwohl, D.R. (2002). A revision of Bloom's taxonomy: An overview. *Theory into practice*, *41*(4), 212–218.

Lu, Owen HT. "Applying learning analytics for improving students engagement and learning outcomes in an MOOCs enabled collaborative programming course." *Learning analytics*. Routledge, 2018. 78–92.

Ma, M., Fallavollita, P., Seelbach, I., Von Der Heide, A.M., Euler, E. et al. (2016). Personalized augmented reality for anatomy education. *Clinical Anatomy*, *29*(4), 446–453.

McMenamin, P.G. (2008). Body painting as a tool in clinical anatomy teaching. *Anatomical sciences education*, *1*(4), 139–144.

Moro, C., Štromberga, Z., Raikos, A. and Stirling, A. (2017). The effectiveness of virtual and augmented reality in health sciences and medical anatomy. *Anatomical sciences education*, *10*(6), 549–559.

Netter, F.H. (2014). *Atlas of human anatomy, Professional Edition E-Book: including NetterReference. com Access with full downloadable image Bank*. Elsevier health sciences.

Pais, D. and Moxham, B.J. (2013). Should gross anatomy be taught systemically or regionally. *Eur J Anat*, *17*(1), 43–47.

Papa, V., Varotto, E., Galli, M., Vaccarezza, M. and Galassi, F.M et al., 2022. One year of anatomy teaching and learning in the outbreak: Has the COVID-19 pandemic marked the end of a century-old practice? A systematic review. *Anatomical sciences education*, *15*(2), pp. 261–280.

Pereira, J.A., Pleguezuelos, E., Merí, A., Molina-Ros, A., Molina-Tomás, M.C. et al. (2007). Effectiveness of using blended learning strategies for teaching and learning human anatomy. *Medical education*, *41*(2), 189–195.

Raja, B.S., Chandra, A., Azam, M.Q., Das, S. and Agarwal, A. et al. (2022). Anatomage-the virtual dissection tool and its uses: A narrative review. *Journal of Postgraduate Medicine*, *68*(3), 156.

Santos, V.A., Barreira, M.P. and Saad, K.R. (2022). Technological resources for teaching and learning about human anatomy in the medical course: systematic review of literature. *Anatomical Sciences Education*, *15*(2), 403–419.

Triepels, C.P., Smeets, C.F., Notten, K.J., Kruitwagen, R.F., Futterer, J.J et al. 2020. Does three-dimensional anatomy improve student understanding?. *Clinical Anatomy*, *33*(1), pp. 25–33.

Wang, J., Zhang, W., Qin, L., Zhao, J., Zhang, S. et al. (2010). Problem-based learning in regional anatomy education at Peking University. *Anatomical Sciences Education*, *3*(3), 121–126.

Yang, H., 2023. A Study on an Intelligent Algorithm for Automatic Test Paper Generation and Scoring in University English Exams. *Journal of ICT Standardization*, *11*(4), pp. 391–401.

CHAPTER

11

Artificial Intelligence in Medical Sciences

Ravinder Sharma,[1] *Vikas Gupta*,[2] *Subh Naman*,[1] *Gunpreet Kaur*[2] and *Parveen Bansal*[2*]

1. Introduction

Artificial intelligence (AI) has become a transformative force in various industries, but perhaps nowhere is its impact more promising and profound than in the field of medical sciences. With its ability to analyze vast amounts of data, recognize patterns, and make predictions, AI is revolutionizing healthcare by enhancing diagnostics, personalizing treatment plans, and streamlining administrative tasks. This synergy between technology and medicine holds the promise of improving patient outcomes and transforming the way healthcare is delivered (Buchanan & Shortliffe, 1984; Shortliffe & Buchanan, 1990).

2. The Genesis of AI in Medicine

The evolution of Artificial Intelligence (AI) in medicine has been a journey marked by significant advancements, starting from its conceptualization to its integration as a crucial component of modern healthcare systems. In the 1950s and 1960s, the inception of AI sparked interest in applying computational methods to simulate human intelligence. Early pioneers in the field envisioned using computers to solve complex problems, including those in the medical domain. One of the earliest instances was the Dendral project in the 1960s, which aimed to apply AI techniques, particularly expert systems, to interpret chemical mass spectrometry data, leading to the identification of organic compounds (Mitchell T.M., 1997; Shortliffe E., 2012). The 1970s and 1980s saw the development of rule-based expert systems. These systems encoded medical knowledge in the form of rules and logical reasoning to emulate human expertise. MYCIN, developed in the 1970s, was among the

1 Faculty of Pharmaceutical Sciences, ICFAI University, Baddi, India.
2 University Centre of Excellence in Research, Baba Farid University of Health Sciences, Faridkot, India.
* Corresponding author: ucer_bfuhs@rediffmail.com

first expert systems used for diagnosing bacterial infections and recommending antibiotics based on patient symptoms and laboratory data. Although these systems were groundbreaking, they had limitations in handling uncertainty and lacked the ability to learn and adapt from new data.The late 20th century witnessed a shift toward machine learning approaches within AI. Machine learning techniques, such as neural networks, gained prominence for their ability to automatically learn patterns from data. However, due to limitations in computational power and available data, their applications in medicine were initially constrained (Topol E.J., 2019; Litjens G. et al., 2017).

3. Evolution of Artificial Intelligence in Medical Sciences

The 21st century brought about a paradigm shift with the proliferation of digital health records, imaging technologies, and vast repositories of healthcare data. This explosion of data, coupled with advancements in computational power and algorithmic sophistication, enabled AI to make significant inroads into various medical disciplines. The breakthroughs in deep learning, a subset of machine learning involving neural networks with multiple layers, revolutionized AI's role in medicine. Deep learning algorithms demonstrated exceptional performance in image recognition and analysis. In medical imaging, convolutional neural networks (CNNs) became adept at interpreting radiological images, assisting in the detection of diseases like cancer, fractures, and abnormalities with high accuracy. Presently, AI in medicine is not limited to diagnostic imaging; it encompasses predictive analytics, natural language processing for mining clinical texts, drug discovery, personalized medicine, virtual health assistants, and more. AI-driven predictive models aid in identifying potential health risks and customizing treatments based on individual patient characteristics. The future trajectory involves further refinement of AI algorithms, enhanced interpretability of AI-generated insights to gain clinicians' trust, and addressing challenges related to ethical guidelines, privacy, and regulatory frameworks.

4. Integration of Machine Learning and Deep Learning

The integration of Machine Learning (ML) and Deep Learning (DL) in medical sciences has ushered in a new era of advanced data analytics, diagnostic accuracy, and personalized healthcare. These technologies, collectively falling under the umbrella of Artificial Intelligence (AI), have shown tremendous promise in revolutionizing various aspects of medical research, diagnosis, treatment, and patient care.

5. Machine Learning in Medical Sciences

5.1 Predictive Analytics and Disease Identification

Machine learning techniques, particularly supervised learning, have been extensively employed in predictive analytics for disease identification and risk stratification. By analyzing large datasets comprising patient records, genetic information, and clinical parameters, ML models can predict the likelihood of developing specific conditions. This aids in early intervention and personalized preventive strategies (Singh P. et al., 2019).

5.2 Decision Support Systems

ML algorithms are integrated into decision support systems to assist healthcare professionals in making informed decisions. These systems analyze complex medical data, provide recommendations for treatment plans, and assist in optimizing healthcare delivery. They prove especially beneficial in handling vast amounts of patient information and staying up-to-date with the latest medical literature (Peiffer-Smadja N. et al., 2020).

5.3 Personalized Medicine

ML enables the development of personalized treatment plans by considering individual patient characteristics, responses to therapies, and genetic variations. This approach is particularly significant in oncology, where ML models can predict the effectiveness of different cancer treatments based on the unique genetic makeup of a patient's tumor (Khosravi P., Schweitzer M., 2023).

5.4 Natural Language Processing (NLP) for Data Mining

In medical research, NLP, a subfield of ML, is employed for mining valuable information from unstructured clinical notes, research papers, and electronic health records. This assists in extracting patterns, trends, and insights that might be challenging to discern manually (Choi E. et al., 2016).

6. Deep Learning in Medical Sciences

6.1 Advanced Image Analysis in Diagnostic Imaging

Deep learning, with its ability to automatically learn hierarchical representations from data, has revolutionized medical imaging. Convolutional Neural Networks (CNNs) excel in tasks such as image segmentation, object detection, and classification. This technology significantly enhances diagnostic accuracy in radiology, pathology, and other imaging modalities (Chan H-P et al., 2020).

6.2 Drug Discovery and Development

Deep learning models contribute to drug discovery by predicting potential drug-target interactions, analyzing molecular structures, and simulating biological processes. This accelerates the identification of promising drug candidates and reduces the time and costs associated with the drug development pipeline (Zhang N. et al., 2015).

6.3 Remote Patient Monitoring and Wearables

Deep learning algorithms are integrated into remote patient monitoring systems, analyzing data from wearables and IoT devices. This facilitates continuous tracking of vital signs, early detection of anomalies, and proactive management of chronic conditions. It empowers patients to actively participate in their healthcare while providing real-time data to healthcare providers (Attia Z.I. et al., 2019).

6.4 Neural Language Models for Clinical Documentation

DL-based models, like transformer-based architectures, are applied to generate clinical documentation from spoken or typed inputs. These models enhance the efficiency of medical transcription, streamline electronic health record (EHR) documentation, and improve overall healthcare workflow.

While ML and DL hold immense promise, challenges such as interpretability, transparency, ethical considerations, and data privacy need careful attention. The continuous integration of AI technologies in medical sciences requires collaborative efforts from healthcare professionals, data scientists, and policymakers. The integration of Machine Learning and Deep Learning in medical sciences has propelled the healthcare industry into a new era of data-driven decision-making, precision medicine, and advanced diagnostics. As these technologies continue to mature, their impact on patient outcomes, disease management, and the overall efficiency of healthcare systems is set to grow, promising a future where AI plays a central role in shaping the landscape of medical sciences, Figure 1 (Wu S. et al., 2020).

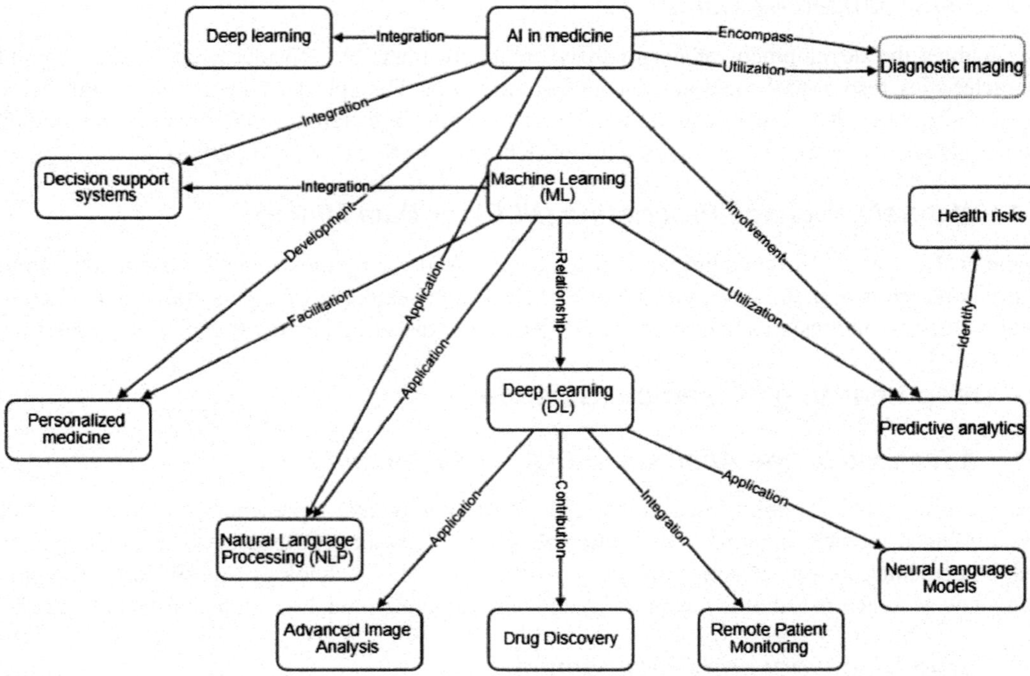

Fig. 1 Interrelationship between artificial intelligence, machine learning and deep learning

7. Advancements in Diagnostic Imaging

As computational capabilities advanced, the 1980s and 1990s witnessed an increased focus on applying AI to diagnostic imaging. Computer-aided diagnosis (CAD) systems, incorporating AI algorithms, began to assist radiologists in interpreting medical images. These early systems aimed to improve the accuracy of image analysis and facilitate early detection of abnormalities. For example, in the late 1980s, the development of CAD systems for mammography gained attention in breast cancer detection. AI algorithms were employed to analyze mammographic images, aiding radiologists in identifying potential malignancies. While these systems were not as sophisticated as contemporary AI applications, they marked a crucial step in demonstrating the potential of AI to enhance diagnostic capabilities (Kononenko I., 2001; Esteva A. et al., 2019).

8. Personalized Medicine and Treatment Optimization

The role of Artificial Intelligence (AI) in personalized medicine and treatment optimization represents a groundbreaking shift in healthcare. By harnessing the power of AI, healthcare professionals can tailor treatments to individual patients, considering their unique characteristics, genetic makeup, and responses to therapies. This approach holds immense promise for improving treatment efficacy, reducing adverse effects, and enhancing overall patient outcomes (Obermeyer Z., Emanuel E.J., 2016; Luz C.F. et al., 2020).

8.1 Genetic and Molecular Profiling

AI plays a pivotal role in analyzing extensive genetic and molecular data to identify patterns and variations associated with diseases. Machine learning algorithms can process large datasets and recognize subtle genetic factors that contribute to individual susceptibility or response to treatments. This enables the identification of specific biomarkers and the development of targeted therapies (Shah N.H., Tenenbaum J.D., 2012).

8.2 Predictive Analytics for Disease Risk

Utilizing predictive analytics, AI models assess an individual's risk of developing certain diseases based on a combination of genetic, lifestyle, and clinical data. This proactive approach allows for early interventions, personalized screening programs, and lifestyle recommendations tailored to an individual's unique risk profile (Saria S., Goldenberg A., 2015; Naylor C.D., 2018; Green E.D., Guyer M.S., 2011).

8.3 Treatment Response Prediction

AI models, particularly machine learning and deep learning algorithms, are employed to predict individual patient responses to specific treatments. By analyzing historical patient data, these models can identify patterns that correlate with treatment outcomes. This information assists healthcare providers in selecting the most effective and least harmful interventions for each patient (Bates D.W., Gawande A.A., 2003).

8.4 Drug Discovery and Development

In the realm of drug discovery, AI expedites the identification of potential drug candidates. Machine learning models analyze vast datasets of chemical structures, pharmacological properties, and historical drug interactions to predict novel compounds with therapeutic potential. This accelerates the drug development pipeline and facilitates the creation of more targeted and effective medications (Collins F.S., Varmus H., 2015).

8.5 Personalized Treatment Plans

AI systems assess diverse patient data, including genetic information, medical history, and lifestyle factors, to generate personalized treatment plans. These plans consider individual responses to medications, potential side effects, and the likelihood of treatment success. Consequently, healthcare providers can optimize treatment strategies, minimizing trial-and-error approaches and improving patient adherence (Denny J.C. et al., 2010).

8.6 Adaptive Clinical Decision Support Systems

AI-driven clinical decision support systems assist healthcare professionals in making real-time decisions based on the latest medical evidence, patient data, and treatment guidelines. These systems analyze patient records, interpret diagnostic results, and recommend tailored interventions, fostering a more adaptive and responsive healthcare environment. Despite the remarkable progress, challenges such as data privacy concerns, ethical considerations, and the need for interpretability in AI models persist. Ensuring the seamless integration of AI into clinical workflows and addressing disparities in access to personalized medicine remain important focal points for the future (Huang R.S. et al., 2007; Liu G., Stokes J.M., 2022; Hossain E. et al., 2019; Zhang Y., Chen X., 2020; Liu M. et al., 2020; Kohane I.S., 1979).

The role of AI in personalized medicine and treatment optimization represents a paradigm shift towards more individualized and precise healthcare. By leveraging AI's analytical capabilities, healthcare providers can unravel the complexities of individual patient profiles, leading to tailored treatment plans that improve therapeutic outcomes while minimizing potential risks. As technology continues to evolve, the synergy between AI and personalized medicine holds the potential to transform healthcare delivery, ushering in an era where treatments are not only more effective but also uniquely suited to each patient's specific needs (Vokinger K.N. et al., 2021; Yu H. et al., 2021; Raghunath S. et al., 2021; Dudley J.T. et al., 2011; Cheng F. et al., 2012; López Martínez F., 2020).

9. Current Applications of AI in Medical Sciences

9.1 Disease Diagnosis and Prediction

AI, particularly deep learning algorithms, have revolutionized diagnostic imaging, enhancing the accuracy and efficiency of disease detection. These algorithms analyze medical images such as X-rays, MRIs, and CT scans, detecting anomalies like tumors, fractures, or lesions with high precision. For instance, in radiology, convolutional neural networks (CNNs) can segment and classify images, aiding radiologists in faster and more accurate diagnoses.

9.2 Predictive Analytics and Disease Risk Assessment

AI models analyze vast amounts of patient data to predict disease risks and aid in early diagnosis. By leveraging machine learning techniques, these models consider diverse data sources, including genetic information, lifestyle factors, and clinical records, to predict an individual's susceptibility to certain diseases. This facilitates proactive measures, personalized screening, and timely interventions for at-risk individuals (Liu X. et al., 2021).

9.3 Accelerated Drug Identification

AI expedites drug discovery by analyzing molecular structures, predicting interactions, and identifying potential drug candidates. Machine learning models process vast datasets to predict compounds with therapeutic potential. These models also assist in repurposing existing drugs for new applications, potentially reducing the time and cost of bringing new treatments to market (Peterson E.D., 2019).

9.4 Personalized Medicine and Treatment Optimization

AI helps tailor treatments by considering individual patient characteristics, genetic makeup, and responses to therapies. By analyzing extensive datasets, AI models predict individual responses to treatments, enabling healthcare providers to select the most effective and least harmful interventions for each patient. This personalized approach enhances treatment efficacy and minimizes adverse effects.

9.5 Health Monitoring and Wearable Technology

Wearable devices and IoT sensors equipped with AI algorithms enable continuous monitoring of vital signs and health parameters. These devices collect real-time data, allowing for remote patient monitoring and early detection of anomalies. Healthcare providers can intervene promptly in case of deviations from normal health parameters, particularly beneficial for managing chronic conditions and elderly care (Das P. et al., 2021).

9.6 AI-Driven Health Apps

AI-powered health apps analyze user-generated data, offering personalized health recommendations and monitoring tools. These apps track fitness activities, diet patterns, sleep cycles, and other health metrics, providing users with insights to improve their overall well-being (Das P, 2017).

10. Clinical Decision Support Systems

10.1 Enhanced Decision-Making for Healthcare Providers

AI-driven clinical decision support systems analyze patient data, medical records, and the latest medical literature to assist healthcare professionals in making informed decisions. These systems provide recommendations for diagnoses, treatment plans, and medication choices, augmenting clinical expertise and reducing diagnostic errors (Noseworthy P.A. et al., 2022).

11. Robotic Surgery and Assistance

11.1 Precision and Minimally Invasive Procedures

AI-enabled robotic systems assist surgeons in performing intricate and minimally invasive procedures with high precision. These systems utilize machine learning to enhance surgical accuracy, allowing for smaller incisions, reduced complications, and faster recovery times. Surgeons control robotic arms to perform surgeries with greater precision than conventional methods (Forni A. et al., 2010).

11.2 Surgical Assistance and Training

AI-driven simulations and virtual reality platforms aid in surgical training and pre-operative planning. Surgeons can practice procedures in simulated environments, improving their skills and allowing for better preparedness for complex surgeries.

The integration of AI in medical sciences across these areas illustrates its transformative potential in revolutionizing disease diagnosis, treatment, monitoring, and surgical interventions, ultimately leading to more efficient, precise, and patient-centered healthcare practices (Peleg M. et al., 2004; Marescaux J. et al., 2001; Patel V.L. et al., 1998; Friedrich G., Zanker M., 2011).

12. Challenges and Limitations in AI integration

12.1 Data Privacy and Security Concerns

One of the primary concerns with AI in healthcare is ensuring the privacy and security of patient data. As AI systems rely heavily on vast amounts of sensitive patient information, maintaining confidentiality and preventing unauthorized access or breaches becomes crucial. Adhering to robust data encryption, access controls, and compliance with data protection regulations (like HIPAA in the US) is essential (Mittelstadt B.D., Floridi L., 2016).

12.2 Vulnerability to Cyber Threats

AI systems are susceptible to cyberattacks, including hacking attempts, data manipulation, or system malfunctions. The interconnectedness of healthcare networks and the reliance on AI for critical decision-making pose significant risks. Addressing vulnerabilities through stringent cybersecurity measures, regular system audits, and continuous updates is imperative to safeguard patient information and maintain system integrity (Ben-Israel D. et al., 2020).

12.3 Ethical Implications and Bias in AI Algorithms

AI algorithms might exhibit biases based on the data they are trained on, leading to disparities in healthcare delivery. Biased algorithms could perpetuate inequalities in diagnoses, treatments, or access to care, particularly for underrepresented or marginalized populations. Striving for algorithmic fairness and mitigating biases through diverse and representative datasets, transparency in algorithm design and continuous monitoring is essential.

12.4 Ethical Use of AI in Decision-Making

The use of AI in clinical decision-making raises ethical dilemmas. While AI systems can provide valuable insights, the ultimate responsibility for patient care rests with healthcare professionals. Balancing the autonomy of AI-driven recommendations with the expertise and ethical judgment of healthcare providers becomes pivotal to ensure patient safety and well-being.

13. Integration with Traditional Healthcare Systems

13.1 Compatibility and Interoperability

Integrating AI systems with existing healthcare infrastructure poses challenges due to interoperability issues among disparate systems. Ensuring seamless data exchange between AI platforms and Electronic Health Records (EHRs) while maintaining system compatibility and compliance with healthcare standards requires substantial effort and investment (Tucker K. et al., 2016).

13.2 Cultural Adoption and Training

Embracing AI in healthcare necessitates a cultural shift and adequate training for healthcare professionals. Resistance to change, skepticism, and the need for up-skilling or retraining healthcare staff to effectively utilize AI tools and interpret their outputs pose challenges in the adoption phase (Beesley L. et al., 2018).

13.3 Regulatory Hurdles and Validation

Regulatory frameworks lag behind the rapid advancements in AI technologies. Establishing robust regulatory frameworks specific to AI in healthcare, including guidelines for validation, quality assurance, and standardization, is crucial. Ensuring that AI-driven medical devices and algorithms meet rigorous validation standards before deployment is imperative to guarantee their safety, effectiveness, and reliability, Figure 2 (El Emam K., 2013; Beam A.L., Kohane I.S., 2018).

13.4 Liability and Accountability

Determining liability and accountability in cases of AI-related errors or adverse outcomes remains a complex issue. Defining clear responsibilities between AI systems and healthcare providers, as well

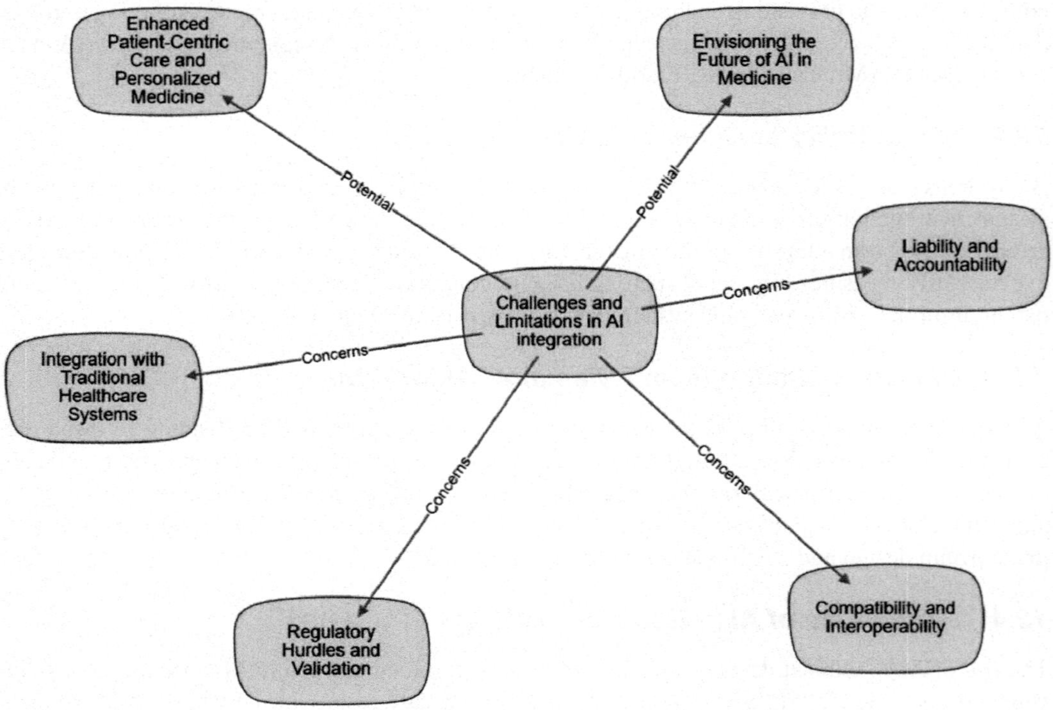

Fig. 2 Artificial integration's obstacles and constraints

as establishing frameworks for liability in case of errors, is essential to address legal and ethical concerns. Navigating these challenges and limitations in AI integration into healthcare requires a concerted effort from policymakers, healthcare providers, technology developers, and regulatory bodies to ensure responsible and ethical implementation while maximizing the benefits of patient care (Mittelstadt B.D. et al., 2016; Kruse C.S. et al., 2016; Kruse C.S. et al., 2018; Murdoch T.B., Detsky A.S., 2013; Kayaalp M., 2018; Bajwa J. et al., 2021).

14. Envisioning the Future of AI in Medicine

14.1 Enhanced Patient-Centric Care and Personalized Medicine

AI advancements pave the way for highly personalized healthcare. By analyzing extensive patient data, including genetics, lifestyle, and medical history, AI can tailor treatments to individual needs. This approach enhances treatment efficacy, reduces adverse effects, and improves patient outcomes (Ghassemi M. et al., 2020).

14.2 Predictive Analytics for Preventive Care

AI models predict potential health risks and diseases based on patient data, enabling proactive interventions and preventive measures. By analyzing patterns and trends, AI assists in early disease detection and recommends lifestyle modifications or interventions to mitigate risks, fostering a preventive healthcare approach (Ting D.S.W. et al., 2019; Price W.N., 2018).

14.3 Patient Engagement and Empowerment

AI-driven tools, such as health apps and virtual assistants, empower patients to actively manage their health. These tools provide personalized health insights, encourage adherence to treatment plans, and facilitate continuous monitoring, fostering a more engaged and informed patient population (Jiang F. et al., 2017).

15. Augmented Research and Development

15.1 Accelerated Drug Discovery

AI expedites drug discovery by analyzing vast datasets and predicting molecular interactions. AI algorithms efficiently identify potential drug candidates and repurpose existing drugs, significantly reducing the time and cost involved in bringing new medications to market (Stokes J.M. et al., 2020).

15.2 Precision Medicine Advancements

AI enables the identification of specific biomarkers and treatment targets for various diseases. This facilitates the development of targeted therapies, enhancing precision medicine approaches for complex conditions such as cancer, cardiovascular diseases, and neurological disorders (Schwamm L.H. Telehealth, 2014; Kourou K. et al., 2015).

16. AI-Powered Drug Customization

16.1 Individualized Treatment Regimens

AI's capacity to process large datasets enables the creation of individualized treatment regimens. By considering a patient's genetic makeup, disease characteristics, and responses to medications, AI assists in optimizing drug dosages and combinations, increasing treatment efficacy and reducing adverse reactions (Kourou K. et al., 2016).

16.2 Pharmacogenomics and Personalized Therapy

AI models analyze genetic data to predict how individuals will respond to certain drugs. This allows healthcare providers to prescribe medications tailored to a patient's genetic profile, improving treatment outcomes while minimizing adverse effects (Pater J. et al., 2022; Huang C. et al., 2017).

17. AI in Global Healthcare Accessibility

17.1 Telemedicine and Remote Care

AI-driven telemedicine solutions facilitate remote consultations, enabling access to healthcare services in underserved or remote areas. These platforms enhance healthcare accessibility, particularly for individuals with limited mobility or in regions with healthcare shortages (Arafah A. et al., 2023). Public Health Management: AI's predictive capabilities aid in disease surveillance, early outbreak detection, and resource allocation in public health systems. By analyzing diverse data sources, AI supports policymakers in making informed decisions to tackle global health challenges efficiently (Nebeker C. et al., 2017). AI's integration into these areas presents vast opportunities to revolutionize healthcare by improving patient-centered care, accelerating research and development, personalizing treatments, and enhancing healthcare accessibility globally. These advancements hold the potential to transform healthcare delivery and improve health outcomes on a global scale (Scheibner J. et al., 2021).

Conclusion

In conclusion, the synergy between AI and medical sciences holds the potential to transform healthcare delivery on a global scale. As technology continues to evolve, the integration of AI into clinical workflows promises to enhance patient outcomes, disease management, and overall healthcare system efficiency. The transformative impact of AI in medical sciences is not only imminent but also crucial for advancing towards a future where treatments are not only more effective but also uniquely tailored to each patient's specific needs. The collaborative efforts of policymakers, healthcare professionals, technology developers, and regulatory bodies are essential in navigating challenges and realizing the full potential of AI in reshaping the landscape of medical sciences.

References

Arafah A., Khatoon S., Rasool I., Khan A., Rather M.A., Abujabal K.A., et al. The future of precision medicine in the cure of Alzheimer's disease. Biomedicines. 2023; *11*(2): 335.

Attia Z.I., Noseworthy P.A., Lopez-Jimenez F., Asirvatham S.J., Deshmukh A.J., Gersh B.J., et al. An artificial intelligence-enabled ECG algorithm for the identification of patients with atrial fibrillation during sinus rhythm: a retrospective analysis of outcome prediction. The Lancet. 2019;394(10201): 861–7.

Bajwa J., Munir U., Nori A., Williams B. Artificial intelligence in healthcare: transforming the practice of medicine. Future Healthc J. 2021; *8*(2): e188.

Bates D.W. and Gawande A.A. Improving safety with information technology. New England journal of medicine. 2003; *348*(25): 2526–34.

Beam A.L. and Kohane I.S. Big data and machine learning in health care. Jama. 2018 Apr 3;319(13):1317-8.

Beesley L., Salvatore M., Fritsche L., Pandit A., Rao A., Brummett C., et al. The Emerging Landscape of Epidemiological Research Based on Biobanks Linked to Electronic Health Records: Existing Resources, Analytic Challenges and Potential Opportunities. 2018.

Ben-Israel D., Jacobs W.B., Casha S., Lang S., Ryu W.H., de Lotbiniere-Bassett M., Cadotte D.W. et al. The impact of machine learning on patient care: a systematic review. Artificial intelligence in medicine. 2020 Mar 1;103:101785.

Buchanan B.G. and Shortliffe E.H. Rule based expert systems: the mycin experiments of the stanford heuristic programming project (the Addison-Wesley series in artificial intelligence). Addison-Wesley Longman Publishing Co., Inc.; 1984.

Chan H-P, Samala R.K., Hadjiiski L.M. and Zhou C. Deep learning in medical image analysis. Deep Learning in Medical Image Analysis: Challenges and Applications. 2020; 3–21.

Cheng F., Liu C., Jiang J., Lu W., Li W., Liu G., et al. Prediction of drug-target interactions and drug repositioning via network-based inference. PLoSComput Biol. 2012; *8*(5):e1002503.

Choi E., Bahadori M.T., Schuetz A., Stewart W.F., Sun J. et al. Doctor ai: Predicting clinical events via recurrent neural networks. In: Machine learning for healthcare conference. PMLR; 2016. p. 301–18.

Collins F.S. and Varmus H. A new initiative on precision medicine. New England journal of medicine. 2015; *372*(9): 793–95.

Das P., Sercu T., Wadhawan K., Padhi I., Gehrmann S., Cipcigan F., Chenthamarakshan V., Strobelt H., Dos Santos C., Chen P.Y., Yang Y.Y. et al. Accelerated antimicrobial discovery via deep generative models and molecular dynamics simulations. Nature Biomedical Engineering. 2021 Jun;5(6): 613–23.

Denny J.C., Ritchie M.D., Basford M.A., Pulley J.M., Bastarache L., Brown-Gentry K., et al. PheWAS: demonstrating the feasibility of a phenome-wide scan to discover gene–disease associations. Bioinformatics. 2010; *26*(9): 1205–10.

Dudley J.T., Sirota M., Shenoy M., Pai R.K., Roedder S., Chiang A.P., et al. Computational repositioning of the anticonvulsant topiramate for inflammatory bowel disease. Sci Transl Med. 2011; *3*(96): 96ra76-96ra76.

El Emam K. Guide to the de-identification of personal health information. CRC Press; 2013.

Esteva A., Robicquet A., Ramsundar B., Kuleshov V., DePristo M., Chou K., et al. A guide to deep learning in healthcare. Nat Med. 2019; *25*(1): 24–29.

Forni A., Chu H.T., Fanikos J. Technology utilization to prevent medication errors. Current Drug Safety. 2010 Jan 1; *5*(1): 13–18.

Friedrich G. and Zanker M. A taxonomy for generating explanations in recommender systems. AI Magazine. 2011 Jun 9; *32*(3): 90–98.

Ghassemi M., Naumann T., Schulam P., Beam A.L., Chen I.Y., Ranganath R. et al. A review of challenges and opportunities in machine learning for health. AMIA Summits on Translational Science Proceedings. 2020; 2020:191.

Green E.D., Guyer M.S., S NHGRIO leadership GEDGM, LC of writing contributions (see A for list of other contributors) MTAPJ. Charting a course for genomic medicine from base pairs to bedside. Nature. 2011; *470*(7333): 204–13.

Hampel H.O., O'Bryant S.E., Castrillo J.I., Ritchie C., Rojkova K., Broich K., Benda N., Nisticò R., Frank R.A., Dubois B., Escott-Price V. et al. Precision medicine-the golden gate for detection, treatment and prevention of Alzheimer's disease. The journal of prevention of Alzheimer's disease. 2016 Dec; *3*(4): 243.

Hossain E., Khan I., Un-Noor F., Sikander S.S. and Sunny M.S.H. et al. Application of big data and machine learning in smart grid, and associated security concerns: A review. Ieee Access. 2019; *7*: 13960–13988.

Huang C., Mezencev R., McDonald J.F. and Vannberg F. Open source machine-learning algorithms for the prediction of optimal cancer drug therapies. PLoS One. 2017; *12*(10):e0186906.

Huang R.S., Duan S., Bleibel W.K., Kistner E.O., Zhang W., Clark T.A., et al. A genome-wide approach to identify genetic variants that contribute to etoposide-induced cytotoxicity. Proceedings of the National Academy of Sciences. 2007; *104*(23): 9758–63.

Jiang F., Jiang Y., Zhi H., Dong Y., Li H., Ma S., et al. Artificial intelligence in healthcare: past, present and future. Stroke Vasc Neurol. 2017; *2*(4).

Kayaalp M. Patient privacy in the era of big data. Balkan Med J. 2018; *35*(1): 8–17.

Khosravi P. and Schweitzer M. Artificial intelligence in neuroradiology: a scoping review of some ethical challenges. Frontiers in Radiology. 2023; *3*:1149461.

Kohane I.S. Ten things we have to do to achieve precision medicine. Science (1979). 2015; 349(6243): 37–38.

Kononenko I. Machine learning for medical diagnosis: history, state of the art and perspective. ArtifIntell Med. 2001; *23*(1): 89–109.

Kourou K., Exarchos T.P., Exarchos K.P., Karamouzis M.V., Fotiadis D.I. et al. Machine learning applications in cancer prognosis and prediction. Comput Struct Biotechnol J. 2015; *13*:8–17.

Kruse C.S., Kristof C., Jones B., Mitchell E. and Martinez A. Barriers to electronic health record adoption: a systematic literature review. J Med Syst. 2016; *40*:1–7.

Kruse C.S., Stein A., Thomas H. and Kaur H. The use of electronic health records to support population health: a systematic review of the literature. J Med Syst. 2018; *42*:1–16.

Litjens G., Kooi T., Bejnordi B.E., Setio A.A.A., Ciompi F., Ghafoorian M., et al. A survey on deep learning in medical image analysis. Med Image Anal. 2017; *42*:60–88.

Liu G. and Stokes J.M. A brief guide to machine learning for antibiotic discovery. Current Opinion in Microbiology. 2022 Oct 1; *69*:102190.

Liu M., Wu Y., Chen Y., Sun J., Zhao Z., Chen X., et al. Large-scale prediction of adverse drug reactions using chemical, biological, and phenotypic properties of drugs. Journal of the American Medical Informatics Association. 2012; *19*(e1):e28–35.

Liu X., Gao K., Liu B., Pan C., Liang K., Yan L., Ma J., He F., Zhang S., Pan S., Yu Y. et al. Advances in deep learning-based medical image analysis. Health Data Science. 2021 May 19.

López Martínez F.E. A Big Data and Machine Learning Model to Improve Medical Decision Support in Population Health Management. 2020;

Luz C.F., Vollmer M., Decruyenaere J., Nijsten M.W., Glasner C., Sinha B. et al. Machine learning in infection management using routine electronic health records: tools, techniques, and reporting of future technologies. Clinical Microbiology and Infection. 2020 Oct 1; *26*(10): 1291–99.

Marescaux J., Leroy J., Gagner M., Rubino F., Mutter D., Vix M., et al. Transatlantic robot-assisted telesurgery. Nature. 2001; *413*(6854): 379–80.

Mitchell T.M. Machine learning. McGraw-hill; 1997.

Mittelstadt B.D., Allo P., Taddeo M., Wachter S., Floridi L. et al. The ethics of algorithms: Mapping the debate. Big Data Soc. 2016; *3*(2): 2053951716679679.

Mittelstadt B.D. and Floridi L. The ethics of big data: current and foreseeable issues in biomedical contexts. The ethics of biomedical big data. 2016; 445–80.

Murdoch T.B. and Detsky A.S. The inevitable application of big data to health care. JAMA. 2013; *309*(13):1351–52.

Naylor C.D. On the prospects for a (deep) learning health care system. Jama. 2018 Sep 18; *320*(11): 1099–100.

Nebeker C., Harlow J., Espinoza Giacinto R., Orozco-Linares R., Bloss C.S., Weibel N. et al. Ethical and regulatory challenges of research using pervasive sensing and other emerging technologies: IRB perspectives. AJOB EmpirBioeth. 2017; *8*(4): 266–76.

Noseworthy P.A., Attia Z.I., Behnken E.M., Giblon R.E., Bews K.A., Liu S., Gosse T.A., Linn Z.D., Deng Y., Yin J., Gersh B.J. et al. Artificial intelligence-guided screening for atrial fibrillation using electrocardiogram during sinus rhythm: a prospective non-randomised interventional trial. The Lancet. 2022 Oct 8; *400*(10359): 1206–12.

Obermeyer Z. and Emanuel E.J. Predicting the future—big data, machine learning, and clinical medicine. N Engl J Med. 2016; *375*(13): 1216.

Patel V.L., Allen V.G., Arocha J.F. and Shortliffe E.H. Representing clinical guidelines in GLIF: individual and collaborative expertise. Journal of the American Medical Informatics Association. 1998 Sep 1; *5*(5): 467–83.

Pater J., Fiesler C. and Zimmer M. No humans here: Ethical speculation on public data, unintended consequences, and the limits of institutional review. Proceedings of the ACM on Human-Computer Interaction. 2022 Jan 14; 6(GROUP): 1–3.

Peiffer-Smadja N., Rawson T.M., Ahmad R., Buchard A., Georgiou P., Lescure F-X, et al. Machine learning for clinical decision support in infectious diseases: a narrative review of current applications. Clinical Microbiology and Infection. 2020; *26*(5): 584–95.

Peleg M., Boxwala A.A., Tu S., Zeng Q., Ogunyemi O., Wang D., et al. The InterMed approach to sharable computer-interpretable guidelines: a review. Journal of the American Medical Informatics Association. 2004; *11*(1): 1–10.

Peterson E.D. Machine learning, predictive analytics, and clinical practice: can the past inform the present?. Jama. 2019 Dec 17; *322*(23): 2283–84.

Price W.N. Big data and black-box medical algorithms. Science translational medicine. 2018 Dec 12; *10*(471): eaao5333.

Raghunath S., Pfeifer J.M., Ulloa-Cerna A.E., Nemani A., Carbonati T., Jing L., vanMaanen D.P., Hartzel D.N., Ruhl J.A., Lagerman B.F., Rocha D.B. et al. Deep neural networks can predict new-onset atrial fibrillation from the 12-lead ECG and help identify those at risk of atrial fibrillation–related stroke. Circulation. 2021 Mar 30; *143*(13): 1287–98.

Rajkomar A., Dean J. and Kohane I. Machine Learning in Medicine. Reply. N Engl J Med. 2019; 380(26).

Saria S. and Goldenberg A. Subtyping: What it is and its role in precision medicine. IEEE Intell Syst. 2015; *30*(4): 70–75.

Scheibner J., Raisaro J.L., Troncoso-Pastoriza J.R., Ienca M., Fellay J., Vayena E., et al. Revolutionizing medical data sharing using advanced privacy-enhancing technologies: technical, legal, and ethical synthesis. J Med Internet Res. 2021; *23*(2): e25120.

Schwamm L.H. Telehealth: seven strategies to successfully implement disruptive technology and transform health care. Health Aff. 2014; *33*(2): 200–206.

Shah N.H. and Tenenbaum J.D. The coming age of data-driven medicine: translational bioinformatics' next frontier. Vol. 19, Journal of the American Medical Informatics Association. BMJ Group; 2012. p. e2–4.

Sharma S., Naman S., Dwivedi J. and Baldi A. Artificial Intelligence-Based Smart Identification System Using Herbal Images: Decision Making Using Various Machine Learning Models. In: Applications of Optimization and Machine Learning in Image Processing and IoT. Chapman and Hall/CRC; p. 123–55.

Shortliffe E. Computer-based medical consultations: MYCIN. Vol. 2. Elsevier; 2012.

Shortliffe E.H. and Buchanan B.G. A model of inexact reasoning in medicine. In: Readings in uncertain reasoning. 1990. p. 259–75.

Singh P., Singh S.P. and Singh D.S. An introduction and review on machine learning applications in medicine and healthcare. In: 2019 IEEE conference on information and communication technology. IEEE; 2019. p. 1–6.

Singh P.A., Bajwa N., Naman S. and Baldi A. A review on robust computational approaches based identification and authentication of herbal raw drugs. Lett Drug Des Discov. 2020; *17*(9): 1066–83.

Stokes J.M., Yang K., Swanson K., Jin W., Cubillos-Ruiz A., Donghia N.M., et al. A deep learning approach to antibiotic discovery. Cell. 2020; *180*(4): 688–702.

Ting D.S.W., Peng L., Varadarajan A.V., Keane P.A., Burlina P.M., Chiang M.F., et al. Deep learning in ophthalmology: the technical and clinical considerations. Prog Retin Eye Res. 2019; 72:100759.

Topol E.J. High-performance medicine: the convergence of human and artificial intelligence. Nat Med. 2019; *25*(1): 44–56.

Tucker K., Branson J., Dilleen M., Hollis S., Loughlin P., Nixon M.J., et al. Protecting patient privacy when sharing patient-level data from clinical trials. BMC Med Res Methodol. 2016; *16*(1): 5–14.

Vamathevan J. and Birney E. A review of recent advances in translational bioinformatics: bridges from biology to medicine. Yearbook of medical informatics. 2017 Aug; 26(01): 178–87.

Vokinger K.N., Feuerriegel S. and Kesselheim A.S. Mitigating bias in machine learning for medicine. Communications medicine. 2021 Aug 23; *1*(1): 25.

Wu S., Roberts K., Datta S., Du J., Ji Z., Si Y., et al. Deep learning in clinical natural language processing: a methodical review. Journal of the American Medical Informatics Association. 2020; *27*(3): 457–70.

Yu H., Yang L.T., Zhang Q., Armstrong D. and Deen M.J. Convolutional neural networks for medical image analysis: state-of-the-art, comparisons, improvement and perspectives. Neurocomputing. 2021 Jul 15; 444: 92–110.

Zhang N., Wang H., Fang Y., Wang J., Zheng X. and Liu X.S. et al. Predicting anticancer drug responses using a dual-layer integrated cell line-drug network model. PLoSComput Biol. 2015; *11*(9): e1004498.

Zhang Y. and Chen X. Explainable recommendation: A survey and new perspectives. Foundations and Trends® in Information Retrieval. 2020; *14*(1): 1–101.

CHAPTER

12

Artificial Intelligence in Biotechnology

Kunal Kumar,[1] *Rajani Sharma,*[1] *Kunal Ranjan*[1] and *Dhananjay K. Pandey*[1*]

1. Introduction

The biological world, once shrouded in mystery, is now being illuminated by the powerful searchlights of artificial intelligence (AI). This confluence not only marks a paradigm shift in our understanding of life, but also reveals fertile ground for transformative advancements in diverse fields, including healthcare, agricultural practices, and sustainability (Bhardwaj et al., 2022; Biswas et al., 2023). This chapter examines the seismic potential of AI in biotechnology, exploring its applications across critical sectors and navigating the ethical considerations inherent in this potent synergy. One of the most significant contributions of AI to biotechnology lies in its ability to transform vast, often messy, biological data into fertile fields of insight. Deep learning algorithms, trained on massive datasets of genes, proteins, and cellular interactions, can unearth hidden patterns and relationships invisible to the human eye (Bhardwaj et al., 2022; Biswas et al., 2023). This "machine-learning magic" allows researchers to decipher the intricate language of life, paving the way for breakthroughs in diverse areas (Holzinger et al., 2023). Drug discovery, once a labyrinthine trek, is now being charted by AI's potent algorithms. By analyzing chemical compounds and biological targets with unparalleled precision, AI can predict potential drug candidates with significantly higher accuracy. This "precision design" approach opens the door for personalized medicine, where treatments are tailored to individual genetic profiles and disease variations (Han et al., 2023; Paul et al., 2021; Sadybekov and Katritch, 2023; Yang et al., 2019). Imagine a future where AI guides the creation of bespoke medications, each a unique melody harmonizing with a patient's specific biological symphony. Similarly, the frontiers of synthetic biology, the engineering of biological systems, are also being reshaped by AI.

[1] Amity Institute of Biotechnology, Amity University Jharkhand, Ranchi, India.
* Corresponding author: dhananjaypandey38@gmail.com

Machine learning algorithms can optimize cell-based factories, designing organisms for the efficient production of biofuels, pharmaceuticals, and even food. This "biomanufacturing revolution" holds immense potential for a more sustainable future, reducing dependence on fossil fuels and paving the way for customized nutrition and biocompatible materials (Kim et al., 2020; Lawson et al., 2021; Parthiban et al., 2023). AI's prowess extends beyond drug discovery and biomanufacturing to the emerald fields of agriculture. By analyzing satellite imagery, soil data, and weather patterns, AI can guide precision agriculture practices, optimizing irrigation, resource use, and crop yields. This symphonic collaboration between AI and farmers can not only boost food production but also minimize environmental impact. Imagine vast farmlands dotted with autonomous robots, guided by AI's watchful eye, nurturing crops with laser-like precision, ensuring bountiful harvests while cherishing the earth's resources (Benos et al., 2021; Dhanaraju et al., 2022; Javaid et al., 2023; Li et al., 2018; Subeesh and Mehta, 2021; Wakchaure et al., 2023). AI's influence extends beyond drug discovery, biomanufacturing, and agriculture. Its keen vision also empowers disease prediction and prevention. By analyzing medical images, genomic data, and patient profiles, AI can identify disease risks and predict clinical outcomes with remarkable accuracy (Bhardwaj et al., 2022). The confluence of AI and biotechnology is not merely a technological leap; it is a transformative dance between the digital and biological worlds. This synergistic waltz promises to reshape the life sciences landscape, offering solutions to global challenges and ushering in a future teeming with possibilities. By harnessing the power of AI with responsible development and ethical principles, we can ensure this groundbreaking convergence leads to a healthier, more sustainable, secure, and equitable future for all, from the bustling cityscapes to the verdant farmlands.

2. Revolutionizing Drug Discovery with AI

AI has equally revolutionized the physical and biological world. As discussed in the above sections, the role of AI as per the current demand is indispensable. Even in the health sector, it has robust demand. Its role is well established in the field of disease identification, prediction and prevention of therapeutics-related issues. This is considered a beneficial tool to rationalize specific drugs and their therapeutic action (Singh et al., 2024). The traditional method involved in drug designing is time consuming and expensive. But the integration of AI with biomedical research under the name 'Big Data' has narrowed down these issues along with the introduction of new opportunities such as discovery of drug targets, physicochemical analysis of drugs and their toxicity analysis. Monitoring the absorption, distribution, metabolism and excretion (ADME) properties and drug repositioning has also been included. The whole world has witnessed the role of drug repositioning during pandemic 2019 as the COVID-19 outbreak. Implementation of computational techniques in the field of pharmaceutical science is promising and increasing their demand. This includes virtual screening (VS), molecular docking and quantitative structure-activity relationship (QSAR) (Jawarkar et al., 2022; Sharma et al., 2022).

3. Virtual Screening and AI-Enabled Compound Prioritization

Virtual Screening of drugs is an *in silico* computer aided drug designing method which screens lead compounds for their potential to be a drug. The screening is based on the structure of ligands (LBVS), receptors (RBVS), and fragments (FBVS) (Oliveira et al., 2023). VS centrally screens the molecules which can present target based biological activity. The process is executed by different algorithms focused on machine learning (ML) and deep learning (DL). SBVS is the best suited method for the molecules whose 3D structure for the target is known. It predicts the most stable ligand receptor orientation, applicable to proteins, carbohydrates as well as DNA. The major advantage of SBVB is that it is not dependent on the physical presence of the molecule. But sometimes it generates false results too (Vázquez et al., 2020).

LBVS provides a platform to design new compounds based on known compounds. There are various ways to compare a compound with the reference compound. Jaccard/Tanimoto coefficient is one of the ways to analyze the similarity between two compounds. There are three dimensions to compare the similarity coefficient of the compounds (Shin et al., 2015):

1. One dimension – This focuses on the molecular properties like weight, number of rotatable bonds, H-bond acceptor and donor and physicochemical properties such as log P, degree of ionization and water solubility (log D).
2. Two dimensions – These compare molecular topology which indicates topological indices (shape and size) and structural descriptors (carbon index, connectivity index, degree of branching and aromatic rings).
3. Three dimensions – These compare compounds based on spatial distribution of atoms, chemical groups, and their molecular properties.

LBVS is considered a better choice than SBVS as this can be used even when little is known about the target molecule. But when SBVS and VBVS are used simultaneously the accuracy of hits is maintained. However, a major disadvantage is that even a modest change in a molecule causes activity cliffs. The FBVS is a screening method to distinguish low molecular weight molecules from micromolar targets. These molecules are the early hits, as low molecular weights do not fix the binding affinity but are well suited for synthesizing the lead compound. Such approaches have a high success rate and hence reduces time and cost (Ilic et al., 2016). For virtual screening there are various algorithms which regulate the accuracy of screening (**Table 1**).

Table 1 Various virtual screening processes and the algorithms used for them.

Virtual screening process	*Algorithm*
Similarity-Based Algorithms	Cosine similarity-measures the angle between two molecules. It ranges from –1 to 1. Here, A_i and B_i denote the components of vectors A and B. $$S^{cos} = \frac{\sum_{i=1}^{n} A_i B_i}{\sqrt{\sum_{i=1}^{n} A_i^2}\sqrt{\sum_{i=1}^{n} B_i^2}}$$
	Correlation – Related pairs of drugs A and B with equal sample sizes, n, and it is given by: $$S^{cor} = \frac{\sum_{i=1}^{n} (A_i - \bar{A})(B_i - \bar{A})}{\sqrt{\sum_{i=1}^{n} (A_i - \bar{A})^2}\sqrt{\sum_{i=1}^{n} (B_i - \bar{B})^2}}$$ $$A = \frac{1}{n}\sum_{i=1}^{n} A_i \text{ and } \bar{B} = \frac{1}{n}\sum_{i=1}^{n} B_i$$
	Hamming similarity – This calculates the similarity between two strings of equal length. It is based on the percentage of common interaction positions for a pair of drugs/targets. Calculated as follows: $$S^{ham} = 1 - \frac{\#(A_i \neq B_i)}{n}$$
	Jaccard similarity – This is defined as the percentage of common non-zero interaction positions for the two given sample sets of drugs/target vectors. $$S^{jac} = \frac{\#[(A_i = B_i) \cap ((A_i \neq 0) \cup (B_i \neq 0))]}{\#[(A_i = 0) \cup (B_i \neq 0)]}$$

Contd.

Table 1 *Contd.*

Virtual screening process	*Algorithm*
QSAR	Simple regression methods
Machine Learning-Based Algorithms	Artificial Neural Networks
	Support Vector Machine (SVM)-Based Techniques
	Bayesian Techniques
	K-Nearest Neighbors
	Kohonen's SOMs
Deep learning	Meta-Heuristic Algorithms
	Tabu Search
	Simulated Annealing (SA)
	Genetic Algorithms

4. Predictive Modeling of Drug-Target Interactions

Drug-Target interaction (DTI) is a major factor to determine the success rate of drug discovery or repurposing. For the ease prediction of DTI several gold standard databases are available where most are available at no cost (**Table 2**). The dataset primarily includes enzymes, ion channels, GPCRs and nuclear receptors **(Table 3)**.

Table 2 Database available to determine Drug-Target interaction.

Database	*Link*
ChEMBL	https://www.ebi.ac.uk/chembl/
Drug Bank	https://go.drugbank.com/
Super Target	https://ngdc.cncb.ac.cn/databasecommons/database/id/564
BioSNAP	https://snap.stanford.edu/biodata/datasets/10002/10002-ChG-Miner.html
Yamanishi_08	http://web.kuicr.kyoto-u.ac.jp/supp/yoshi/drugtarget/
Kyoto Encyclopedia of Genes and Genomes (KEGG)	https://www.genome.jp/kegg/
PubChem	https://pubchem.ncbi.nlm.nih.gov/
UniProt	https://www.uniprot.org/
Pfam	http://pfam.xfam.org/
Manually Annotated Targets and Drugs Online Resource *(MATADOR)*	https://www.hsls.pitt.edu/obrc/index.php?page=URL1209757429
GLIDA	https://newsite.schrodinger.com/platform/products/glide/
Therapeutic target database (TTD)	https://idrblab.net/ttd/
STITCH	http://stitch.embl.de/
Side Effect Resource (SIDER)	http://sideeffects.embl.de/
National Institute for Standards and Technology (NIST) Mass spectral Library	https://chemdata.nist.gov/

Table 3 Tools used for structural and functional analysis of ligands and receptors

Tools	*Language used and feature*	*Link*
ChemCPP	C++ based tool to calculate kernel functions between the chemical compounds like Tanimoto kernelstree.	https://chemcpp.sourceforge.net/html/index.html
EDragon	Evaluating molecular structure activity and similarity coefficient based on topological and geometrical descriptors.	https://vcclab.org/lab/edragon/
CDK Descriptor	Open platform to calculate structure descriptors and various fingerprints like hashed, MACCS etc.	https://egonw.github.io/cdkbook/descriptor.html#:~:text=The%20CDK%20descriptor%20API%20differs,one%20or%20more%20descriptor%20values.
Open Babel	Tools to convert chemical files from one format to another.	https://openbabel.org/index.html
RDKit	Applicable for fingerprint generation, similarity analysis and molecular serialization.	https://www.rdkit.org/
PaDEL	Computes fingerprints.	http://yapcwsoft.com/dd/padeldescriptor/
BlueDesc	Based on JAVA tool calculates geometrical and several topological descriptors. Applicable for various functional groups analysis.	http://www.scbdd.com/blue_desc/index/
ChemDes	Fingerprint generation and similarity calculations etc.	http://www.scbdd.com/chemdes/
Rcpi	R (Rcpi) package is used for complex representations of drugs, proteins, and their interactions. It computes various chemical, physiochemical, and structural descriptors.	https://www.bioconductor.org/packages/release/bioc/vignettes/Rcpi/inst/doc/Rcpi.html
PyDPI	It is based on a python package and computes molecular descriptors for drugs and provides the structural and physicochemical properties of proteins.	https://pypi.org/project/pydpi/
protr	Based on R package computes amino acid composition, Position Specific Scoring Matrix (PSSM), C-Terminal Domain (CTD) etc.	https://cran.r-project.org/web/packages/protr/vignettes/protr.html
ProtDCal	Based on Java generates 3D structure and sequence. Also applicable for similarity computation and function prediction.	https://protdcal-suite.cbe.bci.tu-dortmund.de/
ProtParam	Computes amino acid composition and molecular weight	https://web.expasy.org/protparam/

They involve the Deep Long Short-Term Memory (DeepLSTM) model to represent drug-target pairs and encode fingerprint features for drug and protein sequences based on Legendre Moments (LM) and Position Specific Scoring Matrix (PSSM) respectively. Deep learning is also featured to compress the features of drugs and protein into a uniform vector space using Sparse Principal Component Analysis (SPCA) (Wang et al., 2020).

In assistance with the AI for DTI there are three categories of computational approaches: Ligand based, docking based and chemogenomic based. All three have their own advantages and disadvantages as presented in **Table 4**.

Table 4 Different types of computational approaches and their pros and cons.

Different types of computational approaches for DTI	*Advantages*	*Disadvantages*
Ligand-based approaches	Can be used even when 3D structure is not known.	Highly complex and require a large amount of data.
Docking-based approaches	Model reality more accurately	High computational cost and low scalability
Chemogenomic-based approaches	Detect drug analogs assuring simpler structure–activity relationships.	Prediction of drugs without prior interaction may give false results.

5. Shaping the Future of Drug Development through AI-Driven Approaches

Undoubtedly, advancement of science and technology has brought comfort to our lives. However, on the flip side it has various challenges. Though these challenges have been raised to cover fulfil the alarming population, crowding has evolved various diseases too. Technology is making enough efforts to eradicate daily upcoming hurdles. But the rate of demand is more than the fulfillment rate. In this respect interference of AI has narrowed down the effort, cost as well as time. AI driven approaches have cut down the time to half the duration. In the upcoming days AI would be a major factor to prevent the wastage of resources with more positive results. It would be a major factor in developing drugs even for such diseases whose target structure is still not established.

5.1 Genomics, Personalized Medicine, and AI

The above section has well elaborated the applicability of AI in the field of drug designing and repurposing. But the major requirement for such an approach is the availability of protein structures and genomic sequences. The AI role in this approach is also indispensable.

5.2 AI's Radiance in Medical Imaging and Diagnostics

The role of AI is promising in diagnosis as well. AI is well versed in computing complex images which cannot be done by a human eye.

5.3 Precision Diagnosis through AI-Enhanced Medical Imaging Analysis

The diagnosis of a disease regulates its treatment protocol. Severity occurs when there is delay in diagnosis or wrong intervention. In this respect AI is a cutting-edge technology that analyzes complex data using computer algorithms. Computer aided diagnostics is highly specialized, sensitive, specific, and superior precision. AI is not limited to image processing rather it also reflects the demographic frequency of occurrence of disease. Advancement in the field of diagnosis has been seen with the implementation of machines and deep learning. These are commonly applied to diagnose heart, skin and genetic diseases. The system uses logistic regression, fuzzy logic, Boltzmann machine, K nearest neighbor (KNN), Convolutional neural networks (CNN), Ada Boost, Decision tree, Long short-term memory, among others to present accuracy in the diagnosis (Kumar et al., 2023). **Table 5** represents a few of the success rates of disease diagnosed by AI intervention.

5.4 Optimizing Bioprocesses through AI

Recent events such as the pandemic and climate change have impacted global policies. As a result, governments are prioritizing healthcare, self-sufficiency, sustainable economy practices, and clean industrial development to address these challenges and encouraging bioproducts development

Table 5 Improvement in the diagnosis of diseases due to intervention of AI.

Diseases	*AI intervention in diagnosis*	*Reference*
Hepatitis virus	Recurrent neural network (RNN), achieved 97.59% accuracy	(Ansari et al., 2011)
	Feed-forward neural network, achieved 100% accuracy	(Owais et al., 2019)
Skin disease	Backpropagation neural network	(Dabowsa et al., 2017)
	Combined application of neural networks with MobileNet V2 and Long Short-Term Memory (LSTM), achieved 85% accuracy	(P.N. Srinivasu et al., 2021)
Gastrointestinal disease.	Residual neural network and long short-term memory, achieved 97.057% accuracy	(Owais et al., 2019)
Diabetic	Based on online and offline queries developed Pima Indian Diabetes database.	(Kaur and Kumari, 2022)
	The Naïve Bayes method and RKM achieved 80.55% accuracy	(Aldhyani et al., 2020)
Blood pressure, heart rate, weight, and blood glucose	Analyzed using sensor nodes to a smartphone.	(Rajalakshmi et al., 2018)
Coronary heart disease risk	Supervised using techniques such as Naïve Bayes. Support vector machines, and Decision trees.	(Gonsalves et al., 2019)
Abnormalities in the human liver	Using the AWHARIS algorithm to perform automated segmentation of CT scans, with 78% accuracy.	(P. Srinivasu et al., 2021)
Alzheimer's disease	Electronic medical record (EMR) data. The results acquired showed 90% accuracy on using the SCRL dataset.	(Ljubic et al., 2020)
	deep algorithm has been considered the better solution provider to recognize the ailment at its primary stage with reasonable accuracy based on shrink tissue	(Soundarya et al., 2021)
	Support vector machine (SVM), gradient boosting, K-nearest neighbor, Random Forest, are neural networks that show accuracy rate 97.56%, 97.25%, 95.00%, 97.86% and 98.36%, respectively.	(Tigga and Garg, 2020)
	Deep neural network uses neuroimaging data without pre-processing for feature collection that yields accuracies up to 96.0%.	(Panicker et al., 2018)
Parkinson's disease	Decision tree, support vector machine, ensemble classifier, and Bayes' classifier, are used for statistical and kinematic analysis that predict the severity of Parkinson's disease	(E. et al., 2020)
Cancer	Watershed Gaussian-based profound learning method is used for detection of malignant growth.	(Das et al., 2019)
	Wisconsin breast cancer database using artificial neural networks, support vector machines, decision trees, and k-nearest neighbor	(Memon et al., 2019)
	The DenseNet201 extraction model with the Knearest neighbor classifier, accomplished an accuracy of 96.805%.	(Rodrigues et al., 2020)
	Naïve Bayes method and RKM achieved an accuracy of 97.53%	(Aldhyani et al., 2020)

Contd.

Table 5 *Contd.*

Diseases	*AI intervention in diagnosis*	*Reference*
Tuberculosis	The American Public Health Association used a machine-learning algorithm, performed data preparation, data reduction, and data analysis to relate demographic factors for tuberculosis.	(Romero et al., 2020)
	developed a convolution neural network specifically for the chest X-ray dataset taken from the national institute of health care and centers. In the obtained distribution the area under the curve between 0.9845 and 0.8502 was specified for tuberculosis with a specificity of 82% and sensitivity of 72%.	(Krittanawong et al., 2017)
	Developed automation technique to abnormal chest X-ray images for infiltration, fibrosis, pleural effusion, etc., due to tuberculosis.	(Sathitratanacheewin et al., 2020)

(Mondal et al., 2023). The bioprocessing industry has seen a significant increase in the production of new bioproducts and bioprocesses. This development has led to advancements in various fields such as healthcare, agriculture, and technology. Bioprocessing involves the use of living organisms, such as bacteria, fungi, and plants, to produce valuable products (Barragán-Ocaña et al., 2020). The bioprocessing industry is constantly evolving, and researchers are constantly developing new methods and techniques to improve production processes. Some of the key bioproducts include vaccines, medications, enzymes, and biofuels. The growth of the bioprocessing industry has significant implications for the economy and society, as it offers new opportunities for job creation and opens avenues to address concerns of sustainability. Bioproduct development processes must be planned and executed such that they ensure better commercial profits and focus sustainability (Al-Asheh et al., 2021; Joshi and Mishra, 2022).

In the last few decades, bioprocesses have extensively applied orthogonal experimental design (OED) and response surface methodology (RSM) to determine the best parameters. This integration enhances efficiency and productivity in bioprocessing operations. The integration of artificial neural networks has proven valuable in modeling and optimizing bioprocesses (Cheng et al., 2023). Studies have demonstrated their effectiveness in elucidating complex nonlinear relationships between process variables (Zafar et al., 2012). as well as in optimizing network architecture and learning rules (Abdolrasol et al., 2021). These approaches have significantly impacted the optimization of fermentation processes for the production of various bioproducts, such as poly(3-hydroxybutyrate-co-3-hydroxyvalerate) (PHBV) copolymers (Bolmanis et al., 2023). Likewise, optimized hyaluronic acid production involves radial basis function neural work (RBF-ANN) and particle swarm optimization (PSO). Moreover, real-time monitoring technologies have been facilitating inline biomass concentration detection. Such advancements illustrate the evolving landscape of modeling in biotechnology (Liu et al., 2009). Despite extensive and prominent improvements made in bioprocess optimization, few potential problems will restrict the tapping of potential applications.

For example, it is complicated to correctly forecast difficult nonlinear systems using simple machine learning tools or standard modeling techniques. Furthermore, the added value of some inline detection technologies and equipment, such as viable cell biosensors and in situ microscopy is restricted to lab-based research and not to industries handling large volumes of samples and pilot scale equipment. Here artificial intelligence programs with predetermined rules and data pattern recognition capabilities make conscious decisions to solve problems imitating human intelligence (Khanijou et al., 2022). Recently, a growing number of AI technologies have been employed in bioprocess optimization and control to develop the performance of bioprocesses.

AI-based ML techniques monitor, regulate and optimize bioprocessing systems. AI-ML has shown significant potential in predicting and influencing critical process parameters (CPP) and product critical quality attributes (CQA) in bioprocess development (Nikita et al., 2022). These techniques are efficient at learning complex relationships between these parameters and performance. Additionally, machine learning can play a crucial role in identifying critical quality attributes (CQA) and critical process parameters (CPP) that impact them in process design, monitoring, and control (Puranik et al., 2022). Furthermore, scale-up modeling in bioreactor design involves interconnected relationships between process parameters, hydrodynamic characteristics, and bioreactor design parameters, indicating the broad applications of AI based machine learning in biopharmaceutical processes and other parallel domains of bioprocess industries (Facco et al., 2020). The integration of innovative technologies, including machine learning and artificial intelligence, is driving a digital transformation in industries such as biotech, drug discovery, and healthcare. These technologies enable advancements in bioprocess optimization, drug development, and healthcare applications (Paul et al., 2021).

6. AI-guided Modeling and Optimization of Bioprocess

Artificial intelligence models are increasingly being used in the bioprocess industry. These models can help to optimize the conditions required for different types of bioprocessing, such as the fermentation of microorganisms, to improve efficiency and yield (**Table 6**). AI models can also be used to monitor the conditions and control process variables in real-time, which can lead to better quality output. This technology has the potential to revolutionize the bioprocess industry by reducing costs, improving quality control, and enabling the production of new bioproducts. Few of the notable models in use in bioprocess industries are:

6.1 Artificial Neural Networks

An Artificial Neural Network (ANN) is a computational model inspired by the structure and functioning of biological neural networks in the brain. In bioprocessing, ANNs are used to develop predictive models for various parameters, such as pH, temperature, and viscosity, in complex processes like fermentation or enzymatic reactions. These models help optimize process conditions for maximum efficiency and product yield. ANNs are particularly advantageous in non-linear and complex data-driven processes where traditional statistical methods may struggle. ANNs can be divided into three main types of networks: Feedforward Neural Networks (FFNNs), Convolutional Neural Networks (CNNs), and Recurrent Neural Networks (RNNs). FFNNs are the simplest type of ANN where information flows in one direction only, from input to output, through multiple layers of interconnected neurons. CNNs are designed for image or video recognition tasks and have a specialized architecture that incorporates convolutional layers. Recurrent Neural Networks (RNNs) are a subtype of neural networks commonly used in natural language processing and time series analysis (Cheng et al., 2023).

6.2 Support Vector Machine (SVM)

A support vector machine (SVM) is a type of supervised learning algorithm used for classification and regression tasks. It works by finding the optimal boundary between different classes in a high-dimensional space. SVMs are effective in handling large datasets, non-linear relationships, and high-dimensional data. They are also known for their ability to handle noise and outliers in the data. SVMs are often used in image classification, text classification, and bioinformatics (Cortes and Vapnik, 1995).

6.3 Multi-objective Model

A multi-objective model is a type of mathematical model that seeks to optimize multiple conflicting objectives simultaneously. It is commonly used in decision-making processes where there are multiple factors to consider, and the goal is to find the best possible outcome that satisfies all involved stakeholders. Multi-objective models use various algorithms and techniques, such as linear programming, integer programming, and genetic algorithms, to identify the optimal solution. They are widely used in bioprocess optimization problems, especially determining biomass and bioproducts concentrations and yields. For Example: Cyanobacterial phycobiliproteins (PBPs) are high-value-added products with therapeutic properties, and Nostoc sp. CCC-403 is a source of biofuels. A recent study aimed to enhance the production of PBPs in Nostoc sp. CCC-403 by optimizing biomass production using multi-objective models (Saini et al., 2021). This research is significant as it aims to achieve economic viability for industrial production of both biomass and PBPs, leveraging the pharmaceutical and biofuels industries.

6.4 Fuzzy Intelligence

Fuzzy intelligence (FL) is a form of artificial intelligence that deals with the processing of incomplete, uncertain, and imprecise data, also known as fuzzy data. FL systems employ techniques such as fuzzy logic, takagi-sugeno fuzzy modeling, and neural networks to approximate and model the behavior of complex systems with a high degree of accuracy (Mullai et al., 2022). Unlike traditional AI systems, FL systems are able to handle real-world problems that are not well-suited to crisp, precise mathematical modeling, such as agricultural decision support systems or weather prediction. The FL (artificial intelligence) is capable of logical thinking and has been successfully applied to various bioprocesses. These processes may involve the use of biocatalysts, bioreactors, and other biotechnological techniques to carry out chemical or biological transformations. The application of artificial intelligence in bioprocesses can lead to improved efficiency, reduced costs, and enhanced productivity. For example: the projection of metabolic flow in Chlorella sp. was effectively forecasted based on a FL model (Bhola et al., 2017). Likewise, FL based algorithms were used to determine the optimal nutrient composition for producing phycobiliproteins from cyanobacteria (Saini et al., 2021).

6.5 Genetic algorithm (GA)

GA, or Genetic Algorithm, is a powerful optimization technique that is used to find the best values for a set of hyperparameters and process parameters. It works by simulating the process of natural selection, where solutions that are most fit are selected and used to create new solutions in the population. Over time, the population adapts and evolves towards the best possible solution for the given problem, leading to an efficient and effective optimization of the parameters (Renata et al., 2015). GA has applications in various fields, including machine learning, artificial intelligence, and mathematical optimization. For example, lipid production was optimized from cellulosic ethanol wastewater employing SVM and GA (Zhang et al., 2020). In the context of poly (3HB-co3HV) production, (Zafar et al., 2012), GA used a combination of ANN (Artificial Neural Network) and RSM (Response Surface Methodology) simulations to optimize the production process.

7. AI in Agricultural and Plant Biotechnology

7.1 Advancing Crop Breeding and Genomic Selection with AI

The increasing global population confronts an ongoing and persistent obstacle in providing sustenance to all individuals. The problem is further complicated by climate change, resource scarcity, and pests and diseases (Chakraborty and Newton, 2011). Conventional crop breeding

Table 6 AI tools employed in various Bioprocess and Biochemical Methods (Cheng et al., 2023; Khanal et al., 2023; Mondal et al., 2023).

Bioprocess & Biochemical Methods	*AI Model*	*Advantages*
Anaerobic digestion	Multivariate statistical analysis, Tree based models, Support Vector Machine, Artificial neural network	Anaerobic co-digestion process optimization, prediction and control, initial identification of system instability, prediction of unknown input/ output variables, and real-time monitoring.
Biohydrogen production via dark fermentation	SVM, random forest (RF), ANN, and genetic algorithm (GA), as well as hybrid modeling	Forecasting and optimizing the generation of biohydrogen while making appropriate decisions
Thermochemical conversion	Multi-layer perceptron (MLP), adaptive-network-based fuzzy inference systems (ANFIS), SVM and RF	Assist in many process outputs. Estimate the yield of gaseous products, biochar, and bio-oil from the conversion of waste from various streams.
Composting	ANN, RF, ANFIS, SVM, and deep neural networks (DNNs)	In composting operations, parameter optimization, missing data prediction, non-conformity detection, and complex variable management are all important, AI improving the composting process's sustainability, quality, and efficiency in the process.
Microalgae identification	Image-based cytometric techniques such as the CytoSense, Flowcam, and Flowcytobot integrated with Keras, PyTorch, Caffe, Deeplearning4j, and TensorFlow	Identification of microalgae species, segmentation, classification, and object detection
Fungi bioprocessing	AI-based tools and systems, including automated sensors and bioprocess controls, automated microscopic image analysis, robotic systems, and smart bioreactor concepts. Besides, automated fungal cell counting using ilastik and ImageJ, estimation of fungal biomass using multiphase ANN	Boost data dependability, expand process flexibility, and boost output of fungal-based bioprocesses that include solid-state fermentation (SSF) and submerged fermentation.
Wastewater treatment	ANN, fuzzy logic (FL), RF, and long short-term memory (LSTM)	Forecast important variables, including energy usage, gas emission, microbial contamination, organic solids, nutrients, metal ions, and hydraulic properties
Bioprocess productions of enzyme consortium	ANN and multi-objective GA, SVM, and fuzzy intelligence; Particle swarm optimization (PSO) and control through ANN- or FL-based controllers. Integrating genome-scale metabolic models and computational fluid dynamics in AI-based system	Boost the efficiency of bioprocesses, greater comprehension of the effects of environmental gradients and turbulence on cell efficiency in bioprocesses.

techniques, although efficient, are frequently time-consuming and demanding. This is where artificial intelligence (AI) comes in, delivering a powerful set of tools to transform crop breeding and genetic/genomic selection. Genomic selection is the central component of AI-powered crop breeding. This technique examines the genetic composition (genome) of a plant to anticipate its advantageous characteristics, such as productivity, resistance to diseases, and ability to withstand drought. Historically, breeders would assess these characteristics by means of field experiments, which were laborious and demanded substantial resources (Budhlakoti et al., 2022). AI algorithms have the ability to analyze large quantities of genomic data at a rapid pace and with more efficiency (Bhardwaj et al., 2022). This enables them to quickly discover potential candidates for breeding in the early stages of the selection process. The AI toolbox for crop breeding is diverse and ever-expanding. Machine learning algorithms, such as random forests and support vector machines, can utilize historical breeding data to make predictions about the future performance of subsequent generations (Yoosefzadeh Najafabadi et al., 2023).

Deep learning with highly sophisticated neural networks has the ability to analyze intricate connections between genes and phenotypes, while also considering environmental influences (Pérez-Enciso and Zingaretti, 2019). Moreover, computer vision techniques enable the automated evaluation of plant phenotypes, such as physical traits, from field photos or drone footage, thereby enhancing the efficiency of the selection process (Li et al., 2020) (**Table 6**). Additionally, by reducing the duration of breeding cycles, AI can facilitate the creation of crops that are more suited to evolving climates and consumer preferences (Biswas et al., 2023). AI can boost stress tolerance in crops by using machine learning to discover genes related to stress tolerance, enabling crops to better tolerate drought, pests, and diseases (Chowdhury et al., 2023). It can also enhance resource efficiency by optimizing the utilization of fertilizers and water through customizing these inputs based on the individual requirements of each plant (Talaviya et al., 2020). Furthermore, It can facilitate the development of crops with augmented amounts of vitamins, minerals, and other necessary nutrients, hence enhancing their nutritional worth (Bhardwaj et al., 2022). Although AI-powered crop breeding is now in its nascent phase, there exist multiple instances of success that illustrate its capacity to transform agriculture.

In recent years, the machine learning approach has been employed to examine extensive databases of genetic and weather information to discover potential genes that show promise in enhancing resistance to drought and creating hybrid maize plants that might produce significantly better yields in challenging environments (Babaie Sarijaloo et al., 2021; Baio et al., 2022; Kumar et al., 2022). Similarly, deep learning algorithms are in practice to examine photographs of wheat leaves afflicted by stripe rust, a highly destructive fungal disease with the objective to develop wheat types with rapid resistance to stripe rust (Genaev et al., 2021; Li et al., 2022). Such technological interference has capabilities to develop a predictive model with more than 90% accuracy for disease resistance, allowing breeders to allocate resources towards promising lines and introduce new resistant kinds considerably faster than conventional approaches (Mi et al., 2020). Employing the fusion of computer vision and machine learning techniques to scrutinize tomato pictures and genomes to develop tomatoes with enhanced flavor, texture, and resistance to cracking, highly desired highly desired by both consumers and retailers resulted in the discovery of crucial genes associated with desirable characteristics. These may be utilized to develop novel tomato cultivars with increase sugar content, enhanced firmness, and prolonged shelf life, resulting in larger profit margins for farmers (Bautista et al., 2020; J. Lee et al., 2020; Zhao et al., 2023; Zuo, 2022).

Similarly, fusion of machine learning and genome editing technology as a solution to address the issue of rice blast, a prominent fungal disease that causes substantial reductions in crop productivity is an important area of application. Utilizing this approach key genes linked to blast resistance have been identified and CRISPR-Cas9 technology can be employed to insert these genes into different types of rice, leading to the development of strains with significantly reduced disease incidence (Bhardwaj et al., 2022; Chakraborty et al., 2023; Chithambarathanu and Jeyakumar, 2023; Shanthi et al., 2023). Moreover, with the objective of enhancing cassava productivity and resilience against cassava mosaic disease in African countries, the International Institute of Tropical Agriculture (IITA) in collaboration with other organizations created a mobile application that utilizes machine learning techniques. The app is utilized by farmers to snap photographs of cassava plants, which are then examined for signs of illness and assessed for their prospective production. The application offers immediate suggestions for enhancing crop management and breeding choices, enabling farmers to make well-informed decisions and enhance cassava production (Mrisho et al., 2020). The International Crops Research Institute for the Semi-Arid Tropics (ICRISAT) utilized genomic selection, employing machine learning algorithms to analyze extensive datasets of genomic and phenotypic data of chickpeas with the objective to create chickpea cultivars that exhibit resilience to both drought and heat stress, which is crucial for farmers residing in arid regions grappling with the effects of climate change. Under this initiative, genetic indicators linked to resistance to drought have been discovered and enhanced chickpea varieties developed with a significant increase in productivity under stress. This research facilitated the introduction of climate-resilient chickpea varieties that have been embraced by numerous farmers throughout India (Li et al., 2018; Rani et al., 2020; Roorkiwal et al., 2020; Varshney et al., 2021).

Similarly, CIMMYT utilizes machine learning to expedite the process of wheat breeding with the objective to streamline the process of evaluating extensive wheat populations for favorable characteristics such as resistance to diseases and high grain quality. Under this initiative they have constructed a prognostic algorithm for detecting potential wheat varieties with a high precision rate, thereby minimizing the requirement for manual field evaluation and expediting the breeding procedure (Montesinos-López et al., 2023; Togninalli et al., 2023). This method has optimized wheat breeding programs, resulting in expediting the release of enhanced cultivars. An Indian startup 'AgNext' has developed an AI-based platform called 'TragNext' to identify the quality of tea leaves and other agricultural products. IBM has also developed AI-driven advisory services for farmers, designed to enhance crop yield and the income of Indian agricultural practitioners (Thakur et al., 2021). Moreover, to overcome the issue of stagnant yield and improving the quality of rice, AI algorithms and cloud computing in the field of rice farming have been applied with great success. The researchers use genetic and field data to ascertain the factors associated with high crop output and improved grain quality (Sethy et al., 2021; Teng et al., 2023). Additionally, AI-powered Digital Agriculture Platform, 'CropIn' empowers farmers by reducing the knowledge disparity through equipping farmers with data-based insights to enhance crop management (https://www.cropin.com/blogs/accelerating-farm-productivity-using-ai, n.d.). They offer a digital agriculture platform that utilizes artificial intelligence and remote sensing technology. Agriculturists utilize mobile applications to obtain tailored suggestions on meteorological conditions, watering techniques, pest management, and crop choice, all depending on their own circumstances and up-to-date information. The Indian agricultural landscape is highly suitable for increased integration of artificial intelligence (AI), presenting significant opportunities to revolutionize the sector and guarantee food security for its expanding population.

7.2 Utilizing Artificial Intelligence to Monitor and Improve Plant Health

The escalating food requirements of the world's population necessitate a crucial transition in agricultural methods. Conventional approaches for monitoring and maintaining plant health frequently demonstrate unsustainability, require significant labor, and have restricted scope. Artificial intelligence (AI) offers a groundbreaking solution that revolutionizes the way we monitor, diagnose, and enhance plant health in real-time, leading to a new era of precise agriculture (Bhardwaj et al., 2022). Artificial Intelligence has now a significantly valuable toolkit for monitoring and improving the health of plants. The toolkit includes tools such as deep learning algorithms, which have been trained on extensive datasets of high-resolution images, enabling rapid and precise identification of initial indications of diseases and pest infestations, including those that are imperceptible to human observation (**Table 6**). Convolutional neural networks (CNNs) are highly proficient in detecting tiny patterns in the color, texture, and shape of leaves (Shoaib et al., 2023). This ability allows for early intervention and reduces the amount of crop loss. This exceeds the speed of human observation, possesses objectivity, and has the capability to detect minute symptoms. Similarly, AI models utilize environmental data collected from sensors and weather stations, in addition to plant responses obtained through hyperspectral imaging or chlorophyll fluorescence measurements. (Lu et al., 2020; Nguyen et al., 2021). This enables them to anticipate abiotic stressors such as drought, nutrient shortages, or heat stress prior to the manifestation of observable symptoms. Machine learning algorithms such as linear regression or random forests can assess intricate correlations between environmental factors and plant physiology (H. Lee et al., 2020). This enables farmers to preemptively modify irrigation, fertilization, and other management practices, thereby reducing the adverse stress effects on plant growth and yield. Furthermore, AI-driven systems, using geospatial data and real-time monitoring through drones or aerial imagery, can accurately map the occurrence of pests and diseases with a high level of detail. This allows for precise interventions using drones equipped with specialized application technologies, delivering insecticides or fungicides exclusively to the required areas (Abbas et al., 2023; Hafeez et al., 2023). This highly localized strategy minimizes the environmental footprint, decreases the use of chemicals, and maximizes economic efficiency.

Additionally, AI can enhance resource utilization by analyzing real-time sensor data to determine the specific requirements of each plant and the prevailing environmental circumstances. This enables AI to provide tailored recommendations for watering, fertilization, and pest control tactics for individual plants or small sections within a field (Balaska et al., 2023; Javaid et al., 2023; Subeesh and Mehta, 2021). By utilizing algorithms such as fuzzy logic or Bayesian networks, this highly localized method reduces resource wastage, maximizes efficiency, and guarantees that each plant receives the most suitable amount of care. AI algorithms that integrate data from several sources, including weather, soil, and plant information, may effectively forecast and estimate agricultural yields with enhanced accuracy (Benos et al., 2021; Kale and Patil, 2019) (**Table 6**). This enables farmers to optimize their planting plans, harvest timings, and market decisions by relying on accurate projections. Long Short-Term Memory (LSTM) networks, which are advanced statistical models, can effectively analyze temporal data patterns. This allows for accurate projections of crop yields and helps to reduce the risks associated with unpredictable weather events (Bhimavarapu et al., 2023; Zhang et al., 2018).

Based on application and utilization of AI based techniques several real-life problems related to plant health have been identified and addressed. A team of researchers from the University of Florida has created an artificial intelligence program that examines hyperspectral photographs of citrus leaves (Abdulridha et al., 2020). The model exhibits very high accuracy in identifying initial indications of the disease, even prior to the manifestation of visible symptoms. This allows for timely intervention by applying specific pesticides or removing infected trees, so reducing the spread of diseases and preserving citrus groves. Similarly, a team of researchers from Brazil has developed AI algorithms based on an android application that employs image recognition technology to diagnose coffee rust with more than 97% accuracy based on images supplied by users (Esgario et al., 2022). This enables farmers to promptly identify the disease and swiftly apply methods to manage it, safeguarding their crops and money. Recently, artificial intelligence models integrate weather data, satellite imaging, and historical outbreak information to accurately forecast the likelihood of late blight with a significantly higher success rate (Anim-Ayeko et al., 2023; Fenu and Malloci, 2020). This enables tomato and potato cultivators to implement proactive actions such as applying fungicides and protecting crops in regions with a high risk of damage, hence reducing the amount of crop loss. Similarly, different researchers worldwide have focused on artificial intelligence algorithms that use past weather data and current weather forecasts to accurately anticipate the risk of plant diseases with a comparatively higher accuracy rate (Garrett et al., 2022).

In India, farmers in distant places face limited access to professional guidance and early identification of plant diseases. To tackle this challenge, 'Plantix' provides a platform to the farmers with easily accessible knowledge, allowing them to intervene early and minimize crop losses (Plantix, 2024). This study exemplifies the efficacy of AI in delivering geographically specific agricultural information and assistance. For this, 'Plantix' uses image recognition technology to analyze plant diseases based on images supplied by farmers. The platform offers treatment and preventative recommendations, as well as access to agricultural specialists (Siddiqua et al., 2022). Therefore, artificial intelligence is transforming the way plant health is monitored and improved, providing a promising approach to achieving sustainable and resilient agriculture. By leveraging artificial intelligence to enhance human expertise, we can guarantee the cultivation of more robust crops, optimize resource utilization, and ultimately, protect food security for future generations (Taneja et al., 2023). AI will undeniably have a pivotal role in determining the future of precision agriculture, revolutionizing methods for methods for food crop cultivation and overseeing them.

8. Sustainable Agriculture: The Role of AI in Precision Farming and Optimizing Resources

Sustainable agriculture is a vital solution that seeks to enhance food production while minimizing harm to the environment and conserving resources for future generations. Artificial intelligence (AI) offers a groundbreaking tool in this endeavor, allowing for precise farming and optimal resource utilization with unparalleled accuracy and efficiency. Big data is the essential component of AI-powered agriculture (Wolfert et al., 2017). Sensors, drones, satellites, and agricultural equipment amass huge quantities of data regarding soil quality, meteorological patterns, crop development, and market trends. Artificial intelligence systems, such as machine learning and deep learning, examine these databases to reveal concealed patterns and connections (Janiesch et al., 2021; Mahmud et al., 2021). Selecting the most suitable planting schedules, densities, and varieties based on anticipated weather patterns, soil conditions, and market requirements is important in agricultural practices. Supervised learning models, such as those used in agriculture, have the ability to forecast precise yield results for various crop kinds and planting densities (Kumar et al., 2020). This empowers farmers to make well-informed decisions that optimize profitability. Furthermore, accurate water distribution real-time sensor data is used to monitor soil moisture levels and dynamically change irrigation schedules. Adaptive algorithms have the capability to consider real-time weather updates and the specific requirements of plants, thereby reducing water wastage and maximizing water utilization for each individual plant (Dhanaraju et al., 2022; Talaviya et al., 2020) (**Table 7**). Moreover, precise application of fertilizers according to soil nutrient conditions and crop growth requirements might be precisely regulated by applying such technologies. Artificial intelligence has the capability to examine the composition of soil and plant tissue samples for accurate identification of specific fertilizer requirements (Thorat et al., 2023). This helps to minimize the use of unneeded nutrients and reduces environmental contamination caused by runoff. On the other hand, utilizing AI-powered picture identification and remote sensing assisted agricultural practitioners in early disease detection and taking precise actions. Computer vision-enabled drones possess the capability to accurately detect and identify particular pests and illnesses, allowing for targeted application of insecticides and fungicides exclusively in the affected areas (Abbas et al., 2023). This approach reduces the utilization of chemicals, safeguards advantageous insects, and enhances the well-being of the environment.

AI models are utilized to estimate the most favorable periods for harvesting by analyzing factors such as crop maturity, weather forecasts, and market considerations. This enhances the quality of the harvest, minimizes post-harvest losses, and streamlines the process of storing and transporting the crops. In addition to the above, AI extends beyond the realm of data processing, exerting its influence on the tangible aspects of agriculture by means of autonomous robotics and machine vision (Wakchaure et al., 2023). Robots outfitted with artificial intelligence and visual sensors carry out many duties, such as, employing computer vision for accurate differentiation of weeds and crops, allowing them to effectively eliminate undesired vegetation while minimizing any harm to desired plants (Fountas et al., 2022). This obviates the necessity for chemical herbicides and diminishes labor expenses. Automated robots, directed by GPS and AI, can efficiently plant seeds with precise spacing and depth. This ensures that plants are evenly established and maximizes the potential for high crop yield. Autonomous drones with multispectral sensors can be used for crop monitoring. These drones are capable of mapping fields, detecting early symptoms of stress or illness, and offering real-time information to farmers (Botta et al., 2022). This allows farmers to take proactive measures to address any issues that arise. AI enhances sustainable agriculture by optimizing resource utilization and minimizing environmental harm across various crucial domains. By implementing precise methods for controlling pests and diseases and applying fertilizers only where needed, the reliance on chemical inputs is greatly reduced.

Table 7 Distinct components of algorithms that are employed in AI based smart agriculture.

S.N.	*Types*	*Algorithm*	*Purpose/Application*	*Reference*
1	Supervised Learning	Linear regression, decision trees, random forests, support vector machines	• Predicting crop yields based on weather, soil, and management data. • Identifying optimal planting times and densities • Classifying plant diseases and pests from images • Recommending fertilizer application rates • Estimating harvest times and yields	(Manish Lad et al., 2022)
2	Unsupervised Learning	K-means clustering, principal component analysis, anomaly detection	• Grouping plants with similar characteristics for targeted management • Detecting anomalies in crop health or growth patterns • Identifying patterns in large datasets to uncover hidden relationships • Uncovering soil fertility patterns for variable-rate fertilization	(Chaudhry et al., 2023)
3	Deep Learning	Convolutional neural networks (CNNs), Recurrent neural networks (RNNs)	• Image recognition for plant disease and pest detection • Object detection for weed identification in fields. • Yield prediction from satellite imagery • Analysis of complex sensor data for crop health monitoring	(Boulent et al., 2019)
4	Reinforcement Learning	Q-learning, deep Q-networks	• Robotic control for autonomous weeding and harvesting • Dynamic irrigation scheduling based on real-time sensor data • Optimizing agricultural supply chains and logistics	(de Sousa Bezerra et al., 2023)

Contd.

Table 7 *Contd.*

S.N.	*Types*	*Algorithm*	*Purpose/Application*	*Reference*
5	Natural Language Processing (NLP)	Text classification, sentiment analysis, topic modeling	• Analyzing agricultural research papers and reports to extract insights. • Mining social media data to understand farmer sentiment and trends. • Providing personalized recommendations and decision support to farmers	(Cao et al., 2022)
6	Predictive Analytics	Time-series forecasting, regression analysis, ensemble methods	• Predicting crop yields, weather patterns, and market prices • Identifying potential risks and opportunities for farmers • Optimizing resource allocation and financial planning	(Hasan et al., 2023; Patrick et al., 2023; Paul et al., 2022)
7	Computer Vision	Object detection, image segmentation, feature extraction	• Analyzing drone and satellite imagery for crop health assessment • Identifying pests and diseases in plants • Monitoring livestock health and behavior • Evaluating soil quality and nutrient levels	(Dhanya et al., 2022)
8	Genetic Algorithms	Combines multiple AI models for improved accuracy and robustness	• Optimizing crop breeding programs for drought resistance • Developing pest-resistant varieties • Customizing fertilizer recommendations for specific soil types	(Roy and De, 2022)
9	Bayesian Networks	Represent probabilistic relationships between variables for uncertainty modeling	• Assessing risk factors for crop failure or disease outbreaks • Modeling the spread of pests and diseases to predict their trajectories. • Optimizing decision-making under uncertain conditions, such as weather variability	(Abbaszadeh et al., 2022; Bazrafshan et al., 2022)
10	Transfer Learning	Applies knowledge learned from one task to a different but related task	• Adapting models trained on large datasets to specific farms or crops with limited data. • Improving image recognition accuracy for rare diseases or pests by leveraging models trained on common ones. • Accelerating development of new AI models for agricultural applications	(Al Sahili and Awad, 2022; Fraiwan et al., 2022; Gautam et al., 2022; Tamilvizhi et al., 2022)

Contd.

Table 7 *Contd.*

S.N.	*Types*	*Algorithm*	*Purpose/Application*	*Reference*
11	Explainable AI (XAI)	Provides insights into how AI models make decisions, enhancing transparency and trust.	• Understanding why a model recommends a particular fertilizer application. • Identifying the factors that contribute to a disease prediction. • Building trust among farmers in AI-powered recommendations	(Cartolano et al., 2024)
12	Generative Adversarial Networks (GANs)	Generate realistic synthetic data for training AI models when real data is scarce.	• Creating images of diseased plants to augment datasets for disease identification models. • Generating virtual fields for testing precision farming strategies • Simulating different climate scenarios to evaluate crop resilience.	(Lu et al., 2022)

This approach helps to minimize pollution caused by runoff and preserves the health of the soil. Similarly, by utilizing real-time data, precise irrigation techniques are employed to achieve optimal water usage, hence reducing water wastage, and preserving this valuable resource. Precision farming techniques decrease reliance on wide-ranging insecticides, safeguarding advantageous insects and fostering a wider range of ecosystems. Utilizing data-driven nutrient management and minimizing chemical usage result in improved soil health, fertility, and sustainable agricultural productivity in the long run (Bhardwaj et al., 2022; Biswas et al., 2023; Dhanaraju et al., 2022; Javaid et al., 2023). Therefore, AI is leading the way in revolutionizing agriculture for a more sustainable future. By utilizing its analytical prowess, automation functionalities, and data-derived insights, we can enhance the efficiency of resource utilization, mitigate environmental repercussions, and guarantee long-term food security. By persistently conducting research, engaging in responsible development, and fostering collaboration, we can fully harness the potential of artificial intelligence to revolutionize the way we cultivate, oversee, and assess our food supplies.

9. Ethics and Regulations

AI plays a transformative role in the field of biotechnology, offering innovative solutions and accelerating research and development processes. AI has the potential to influence diverse fields including biotechnology. So, ethical considerations are not merely a choice but an essential requirement of the framework that ensures the responsible use of AI in sensitive and impactful areas. The ethics of AI not only concerns the legality but also questions of discrimination, fairness, and humanity. Data privacy, bias, transparency and security are core elements of the ethical issues framework (Gerke et al., 2020; Holzinger et al., 2023). With the use of advanced technology, biotechnology studies generate vast amounts of data which can be sensitive personal data, particularly in areas like genetic testing and personalized medicine. It becomes imperative to implement robust security measures to safeguard against unauthorized access, data breaches, or misuse (Maserat, 2022; Schork, 2019). AI algorithms designed to analyze genetic information for medical purposes may inadvertently expose individuals to privacy breaches if not adequately secured. Striking a balance between the benefits of AI in biotechnology, such as personalized medicine and disease diagnosis, and safeguarding individuals' privacy rights is a complex challenge. The European Union was the first to implement the General Data Protection Regulation, followed by the US and Canada. Despite this, these laws are not sufficient to safeguard individuals' data (GDPR, 2016).

AI models can carry human biases on a large scale, often linked to the data they are trained on rather than the algorithms themselves. The data might include human decisions or reflect social and historical inequities. The collection and use of data can also introduce bias. Currently, there are no established guidelines to report or compare these models. Future efforts should focus on developing such standards to guide researchers and clinicians (Nelson, 2019; Parikh et al., 2019).

As AI becomes an integral part of digital systems, it is crucial to ensure ethical decision-making and eliminate unfair biases. Responsible AI systems should be transparent, explainable, and accountable. The increasing use of AI in healthcare for improving patient pathways and surgical outcomes underscores its importance (Parikh et al., 2019; Shah et al., 2020). Biotechnology companies allocate money and other resources heavily to research and development. Intellectual properties are the core element of their business framework. A single data compromise can lead to the theft of confidential and sensitive information, potentially causing detrimental repercussions for the company. According to the report, the average expenditure with a single data breach in the US is approximately $9.45 million. This cost includes direct expenses such as legal expenses, regulatory penalties and investigation costs, and indirect costs such as customer loss and reputation. On average, it takes 327 days to detect and mitigate data breaches (IBM, 2023). A data breach in healthcare costs more than any industry, estimated at approximately 10.1 million dollars (Davis, 2022). With the lack of robust AI and cybersecurity, there is a risk of manipulating and misusing the algorithm which leads to the breach of the integrity of genomic datasets. This misuse could give rise to hybrid cybersecurity threats, posing potential dangers such as widespread data harm, research, and industrial damage, and compromised governance systems. These threats have significant implications for health, food security, and civilian safety, emphasizing the importance of a robust AI and cybersecurity infrastructure (Pauwels, 2023).

Article 12 of the UN Convention on Electronic Communications emphasizes that AI tools do not have a will of their own. Their decision is based on the collected data and program. It cannot be legally responsible for the harm it causes. So, an individual or company, on behalf of whom it was created should be responsible for the actions of AI tools. In the US, HIPAA's (Health Insurance Portability and Accountability Act) Privacy Rule governs patient information privacy, outlining permitted and restricted uses (Magrani, 2019). Despite having their own AI policy, EU countries signed a "Declaration of Cooperation on AI" in 2018. The high-level expert group of the European Commission first first-ever legal framework on AI in 2021 (EuropeanCommission, 2021). As per Article 1(2) of Directive 93/42/EEC, two regulations, Medical Devices Regulation (MDR) and the *In Vitro* Diagnostic Medical Device Regulation (IVDR) reformed the regulatory framework w.e.f. 26 May 2020 and 26 May 2022, respectively (Pesapane et al., 2018). MIT media Lab and FDA collaborated under "Health 0.0" to advance AI and ML research in healthcare. The primary objective is to enhance health outcomes by establishing a regulatory framework. Key contributors to this effort include stakeholders from various sectors such as life sciences, biotechnology, foundations, universities, and patient advocacy groups (Shah et al., 2019).

In India, Ministry of Electronics & Information Technology (MeitY) identified four committees namely, (1) platforms and data for AI, (2) Leveraging AI for identifying National Missions in Key Sectors, (3) Mapping Technological capabilities, Key policy enablers, Skilling, Re-skilling, R&D and (4) Cybersecurity, Safety, Legal and Ethical Issues. These committees were created to form a policy framework for AI. Various ministries and agencies, such as the Department of Agriculture Cooperation, the Department of Revenue, and Niti Aayog, are actively utilizing several areas including healthcare. The state of Kerala's 'e-Health Project' has stored electronic health records for improved healthcare accessibility, while Wadhwani AI, established in 2018, focuses on AI solutions for social good in healthcare, education, agriculture, and infrastructure (Aayog, 2018; Mabiyan, 2020; Roy and Malhotra, 2019). AI regulatory frameworks are in the nascent stage but are developing rapidly. Global regulatory bodies recognize the potential of AI in biotechnology and healthcare. The establishment of comprehensive frameworks and guidelines is essential to ensure ethical practices, data security, and the responsible use of AI technologies (Murgia, 2020; Sharma and Manchikanti, 2020).

10. Horizons and Challenges: AI in Biotech's Future

Artificial intelligence (AI) has the potential to revolutionize the biotechnology industry. AI can be used to improve drug discovery and development, as well as to improve the accuracy and efficiency of genetic testing and diagnostics. It can also be used to optimize laboratory automation and process control, and to develop personalized medicine for individual patients. In addition, AI can be used to analyze large amounts of data from various sources, such as genetic databases, to identify patterns and predict outcomes. This information can be used to inform research and development efforts and improve patient outcomes (Holzinger et al., 2023; Paul et al., 2021). While there are still challenges to overcome, the potential benefits of AI in biotechnology are vast, and the field is likely to see continued growth and innovation in the coming years. In recent years, there has been significant progress in the development of advanced technologies such as CRISPR, and nanotech. These cutting-edge technologies have the potential to revolutionize various industries, from healthcare and medicine to energy. AI, on the other hand, is also an amalgamation of innovative technologies, and it has already begun to have a significant impact on various sectors (Chehelgerdi et al., 2023; Dixit et al., 2024). Combining these technologies can provide a powerful synergy, allowing for more efficient and effective solutions to complex problems. For instance, AI can be used to analyze large amounts of data generated by these technologies, helping to refine their performance and optimize their use. As these technologies continue to evolve, their potential to transform industries in future is eagerly awaited. AI can be revolutionary in terms of improving certain aspects of our lives, but it also presents numerous technical, ethical, and societal challenges that need to be overcome. Some examples of these challenges include bias, privacy, security, job displacement, and the potential of AI to be used for evil purposes. To address these challenges, researchers, policymakers, and industry leaders need to work together and develop AI systems that prioritize fairness, transparency, and accountability. Additionally, robust legal frameworks and ethical guidelines can be established to ensure that the development and deployment of AI benefits society as a whole.

Conclusion

In this exploration into the intersection of artificial intelligence (AI) with biological sciences and biotechnology, we find ourselves in a groundbreaking era that is reshaping scientific inquiry. The integration of AI has brought about a revolution in drug discovery, alleviating the burdens of high costs and lengthy timelines through the implementation of virtual screening and predictive modeling. This breakthrough not only propels personalized medicine forward but also tailors treatments based on individual genetic profiles. The profound impact of AI extends into genomics, where it unravels the complexities of the human genome and addresses ethical considerations associated with personalized medicine. In the realm of bioprocess optimization, artificial neural networks are streamlining intricate processes, thereby contributing to the sustainability and economic viability of biomanufacturing. In the agricultural domain too, AI is a catalyst for advancements, assuming a pivotal role in crop breeding, optimizing yields, and safeguarding plant health through precision farming techniques. As we navigate through this transformative era, the multifaceted impact of AI in biological sciences and biotechnology becomes increasingly apparent. While the undeniable benefits propel scientific innovation to new heights, the responsible deployment of AI calls for the evolution of ethical considerations and regulatory frameworks. Embracing the vast potential of AI, we embark on a journey to unravel the mysteries of life, shaping a future where technology and responsibility converge to redefine the scientific landscape.

References

Aayog, N., 2018. National Strategy for Artificial Intelligence. Niti Aayog.

Abbas, A., Zhang, Z., Zheng, H., Alami, M.M., Alrefaei, A.F., Abbas, Q., Naqvi, S.A.H., Rao, M.J., Mosa, W.F.A., Abbas, Q., Hussain, A., Hassan, M.Z., Zhou, L., 2023. Drones in Plant Disease Assessment, Efficient Monitoring, and Detection: A Way Forward to Smart Agriculture. Agronomy 13, 1524.

Abbaszadeh, P., Gavahi, K., Alipour, A., Deb, P., Moradkhani, H. et al., 2022. Bayesian Multi-modeling of Deep Neural Nets for Probabilistic Crop Yield Prediction. Agric. For. Meteorol. 314, 108773.

Abdolrasol, M.G.M., Hussain, S.M.S., Ustun, T.S., Sarker, M.R., Hannan, M.A., Mohamed, R., Ali, J.A., Mekhilef, S., Milad, A., 2021. Artificial Neural Networks Based Optimization Techniques: A Review. Electronics 10.

Abdulridha, J., Ampatzidis, Y., Qureshi, J., Roberts, P., 2020. Laboratory and UAV-Based Identification and Classification of Tomato Yellow Leaf Curl, Bacterial Spot, and Target Spot Diseases in Tomato Utilizing Hyperspectral Imaging and Machine Learning. Remote Sens. 12, 2732.

Al-Asheh, S., Bagheri, M., Aidan, A., 2021. Membrane bioreactor for wastewater treatment: A review. Case Stud. Chem. Environ. Eng. 4, 100109.

Al Sahili, Z., Awad, M., 2022. The power of transfer learning in agricultural applications: AgriNet. Front. Plant Sci. 13.

Aldhyani, T.H., Alshebami, A.S., Alzahrani, M.Y., 2020. Soft Clustering for Enhancing the Diagnosis of Chronic Diseases over Machine Learning Algorithms. J. Healthc. Eng. 2020, 1–16.

Anim-Ayeko, A.O., Schillaci, C., Lipani, A., 2023. Automatic blight disease detection in potato (Solanum tuberosum L.) and tomato (Solanum lycopersicum, L. 1753) plants using deep learning. Smart Agric. Technol. 4, 100178.

Ansari, S., Shafi, I., Ansari, A., Ahmad, J., Shah, S.I. et al., 2011. Diagnosis of liver disease induced by hepatitis virus using Artificial Neural Networks. In: 2011 IEEE 14th International Multitopic Conference. IEEE, pp. 8–12.

Babaie Sarijaloo, F., Porta, M., Taslimi, B., Pardalos, P.M., 2021. Yield performance estimation of corn hybrids using machine learning algorithms. Artif. Intell. Agric. 5, 82–89.

Baio, F.H.R., Santana, D.C., Teodoro, L.P.R., Oliveira, I.C. de, Gava, R., de Oliveira, J.L.G., Silva Junior, C.A. da, Teodoro, P.E., Shiratsuchi, L.S., 2022. Maize Yield Prediction with Machine Learning, Spectral Variables and Irrigation Management. Remote Sens. 15, 79.

Balaska, V., Adamidou, Z., Vryzas, Z., Gasteratos, A., 2023. Sustainable Crop Protection via Robotics and Artificial Intelligence Solutions. Machines 11, 774.

Barragán-Ocaña, A., Silva-Borjas, P., Olmos-Peña, S., Polanco-Olguín, M., 2020. Biotechnology and Bioprocesses: Their Contribution to Sustainability. Processes 8.

Bautista, J.F., Ocena, C.D., Cabreros, M.J., Alagao, S.P.L., 2020. Automated Sorter and Grading of Tomatoes using Image Analysis and Deep Learning Techniques. In: 2020 IEEE 12th International Conference on Humanoid, Nanotechnology, Information Technology, Communication and Control, Environment, and Management (HNICEM). IEEE, pp. 1–6.

Bazrafshan, O., Ehteram, M., Moshizi, Z.G., Jamshidi, S., 2022. Evaluation and uncertainty assessment of wheat yield prediction by multilayer perceptron model with bayesian and copula bayesian approaches. Agric. Water Manag. 273, 107881.

Benos, L., Tagarakis, A.C., Dolias, G., Berruto, R., Kateris, D., Bochtis, D. et al. 2021. Machine Learning in Agriculture: A Comprehensive Updated Review. Sensors 21, 3758.

Bhardwaj, A., Kishore, S., Pandey, D.K., 2022. Artificial Intelligence in Biological Sciences. Life 12, 1430.

Bhimavarapu, U., Battineni, G., Chintalapudi, N., 2023. Improved Optimization Algorithm in LSTM to Predict Crop Yield. Computers 12, 10.

Bhola, V., Swalaha, F.M., Nasr, M., Bux, F. et al., 2017. Fuzzy intelligence for investigating the correlation between growth performance and metabolic yields of a Chlorella sp. exposed to various flue gas schemes. Bioresour. Technol. 243, 1078–1086.

Biswas, A., Kumari, A., Gaikwad, D.S., Pandey, D.K., 2023. Revolutionizing Biological Science: The Synergy of Genomics in Health, Bioinformatics, Agriculture, and Artificial Intelligence. Omi. A J. Integr. Biol. 27, 550–569.

Bolmanis, E., Dubencovs, K., Suleiko, A., Vanags, J., 2023. Model Predictive Control—A Stand Out among Competitors for Fed-Batch Fermentation Improvement. Fermentation 9.

Botta, A., Cavallone, P., Baglieri, L., Colucci, G., Tagliavini, L., Quaglia, G., 2022. A Review of Robots, Perception, and Tasks in Precision Agriculture. Appl. Mech. 3, 830–854.

Boulent, J., Foucher, S., Théau, J., St-Charles, P.-L., 2019. Convolutional Neural Networks for the Automatic Identification of Plant Diseases. Front. Plant Sci. 10.

Budhlakoti, N., Kushwaha, A.K., Rai, A., Chaturvedi, K.K., Kumar, A., Pradhan, A.K., Kumar, U., Kumar, R.R., Juliana, P., Mishra, D.C., Kumar, S., 2022. Genomic Selection: A Tool for Accelerating the Efficiency of Molecular Breeding for Development of Climate-Resilient Crops. Front. Genet. 13.

Cao, Y., Sun, Z., Li, L., Mo, W., 2022. A Study of Sentiment Analysis Algorithms for Agricultural Product Reviews Based on Improved BERT Model. Symmetry (Basel). 14, 1604.

Cartolano, A., Cuzzocrea, A., Pilato, G., 2024. Analyzing and assessing explainable AI models for smart agriculture environments. Multimed. Tools Appl.

Chakraborty, B., Banerjee, S., Samanta, S., Debangshi, U., Yadav, S.V., Khaire, P.B., Shelar, V.B., Bansode, G.D. and Landage, K.B., 2023. Detection of Rice Blast Disease (Magnaporthe grisea) Using Different Machine Learning Techniques. Int. J. Environ. Clim. Chang. 13, 2256–2264.

Chakraborty, S., Newton, A.C., 2011. Climate change, plant diseases and food security: an overview. Plant Pathol. 60, 2–14.

Chaudhry, M., Shafi, I., Mahnoor, M., Vargas, D.L.R., Thompson, E.B., Ashraf, I., 2023. A Systematic Literature Review on Identifying Patterns Using Unsupervised Clustering Algorithms: A Data Mining Perspective. Symmetry (Basel). 15, 1679.

Chehelgerdi, Mohammad, Chehelgerdi, Matin, Allela, O.Q.B., Pecho, R.D.C., Jayasankar, N., Rao, D.P., Thamaraikani, T., Vasanthan, M., Viktor, P., Lakshmaiya, N., Saadh, M.J., Amajd, A., Abo-Zaid, M.A., Castillo-Acobo, R.Y., Ismail, A.H., Amin, A.H., Akhavan-Sigari, R., 2023. Progressing nanotechnology to improve targeted cancer treatment: overcoming hurdles in its clinical implementation. Mol. Cancer 22, 169.

Cheng, Y., Bi, X., Xu, Y., Liu, Y., Li, J., Du, G., Lv, X., Liu, L., 2023. Artificial intelligence technologies in bioprocess: Opportunities and challenges. Bioresour. Technol. 369, 128451.

Chithambarathanu, M., Jeyakumar, M.K., 2023. Survey on crop pest detection using deep learning and machine learning approaches. Multimed. Tools Appl. 82, 42277–42310.

Chowdhury, R.H., Eti, F.S., Ahmed, R., Gupta, S. Das, Jhan, P.K., Islam, T., Bhuiyan, M.A.R., Rubel, M.H., Khayer, A., 2023. Drought-responsive genes in tomato: meta-analysis of gene expression using machine learning. Sci. Rep. 13, 19374.

Cortes, C., Vapnik, V., 1995. Support-vector networks. Mach. Learn. 20, 273–297.

Dabowsa, N.I.A., Amaitik, N.M., Maatuk, A.M., Aljawarneh, S.A. et al., 2017. A hybrid intelligent system for skin disease diagnosis. In: 2017 International Conference on Engineering and Technology (ICET). IEEE, pp. 1–6.

Das, A., Acharya, U.R., Panda, S.S., Sabut, S., 2019. Deep learning based liver cancer detection using watershed transform and Gaussian mixture model techniques. Cogn. Syst. Res. 54, 165–175.

Davis, J., 2022. Healthcare data breaches cost an average of $10.1M, more than any other industry. SC Media.

de Sousa Bezerra, C.D., Teles Vieira, F.H., Queiroz Carneiro, D.P., 2023. Autonomous Robotic Navigation Approach Using Deep Q-Network Late Fusion and People Detection-Based Collision Avoidance. Appl. Sci. 13, 12350.

Dhanaraju, M., Chenniappan, P., Ramalingam, K., Pazhanivelan, S., Kaliaperumal, R. et al., 2022. Smart Farming: Internet of Things (IoT)-Based Sustainable Agriculture. Agriculture 12, 1745.

Dhanya, V.G., Subeesh, A., Kushwaha, N.L., Vishwakarma, D.K., Nagesh Kumar, T., Ritika, G., Singh, A.N., 2022. Deep learning based computer vision approaches for smart agricultural applications. Artif. Intell. Agric. 6, 211–229.

Dixit, S., Kumar, A., Srinivasan, K., Vincent, P.M.D.R., Krishnan, N.R. et al., 2024. Advancing genome editing with artificial intelligence: opportunities, challenges, and future directions. Front. Bioeng. Biotechnol. 11.

E., B., D., B., R., B., 2020. Supervised machine learning based gait classification system for early detection and stage classification of Parkinson's disease. Appl. Soft Comput. 94, 106494.

Esgario, J.G.M., de Castro, P.B.C., Tassis, L.M., Krohling, R.A., 2022. An app to assist farmers in the identification of diseases and pests of coffee leaves using deep learning. Inf. Process. Agric. 9, 38–47.

EuropeanCommission, 2021. Proposal for a regulation on a European approach for artificial intelligence.

Facco, P., Zomer, S., Rowland-Jones, R.C., Marsh, D., Diaz-Fernandez, P., Finka, G., Bezzo, F., Barolo, M., 2020. Using data analytics to accelerate biopharmaceutical process scale-up. Biochem. Eng. J. 164, 107791.

Fenu, G., Malloci, F.M., 2020. Artificial Intelligence Technique in Crop Disease Forecasting: A Case Study on Potato Late Blight Prediction. pp. 79–89.

Fountas, S., Malounas, I., Athanasakos, L., Avgoustakis, I., Espejo-Garcia, B. et al., 2022. AI-Assisted Vision for Agricultural Robots. AgriEngineering 4, 674–694.

Fraiwan, M., Faouri, E., Khasawneh, N., 2022. Classification of Corn Diseases from Leaf Images Using Deep Transfer Learning. Plants 11, 2668.

Garrett, K.A., Bebber, D.P., Etherton, B.A., Gold, K.M., Plex Sulá, A.I., Selvaraj, M.G., 2022. Climate Change Effects on Pathogen Emergence: Artificial Intelligence to Translate Big Data for Mitigation. Annu. Rev. Phytopathol. 60, 357–378.

Gautam, V., Trivedi, N.K., Singh, A., Mohamed, H.G., Noya, I.D., Kaur, P., Goyal, N., 2022. A Transfer Learning-Based Artificial Intelligence Model for Leaf Disease Assessment. Sustainability 14, 13610.

GDPR, 2016. REGULATION (EU) 2016/679 OF THE EUROPEAN PARLIAMENT AND OF THE COUNCIL; Protection of natural persons with regard to the processing of personal data and on the free movement of such data, and repealing Directive 95/46/EC (General Data Protection Regulatio.

Genaev, M.A., Skolotneva, E.S., Gultyaeva, E.I., Orlova, E.A., Bechtold, N.P., Afonnikov, D.A., 2021. Image-Based Wheat Fungi Diseases Identification by Deep Learning. Plants 10, 1500.

Gerke, S., Minssen, T., Cohen, G., 2020. Ethical and legal challenges of artificial intelligence-driven healthcare. In: Artificial Intelligence in Healthcare. Elsevier, pp. 295–336.

Gonsalves, A.H., Thabtah, F., Mohammad, R.M.A., Singh, G., 2019. Prediction of Coronary Heart Disease using Machine Learning. In: Proceedings of the 2019 3rd International Conference on Deep Learning Technologies. ACM, New York, NY, USA, pp. 51–56.

Hafeez, A., Husain, M.A., Singh, S.P., Chauhan, Anurag, Khan, M.T., Kumar, N., Chauhan, Abhishek, Soni, S.K., 2023. Implementation of drone technology for farm monitoring & pesticide spraying: A review. Inf. Process. Agric. 10, 192–203.

Han, R., Yoon, H., Kim, G., Lee, H., Lee, Y. et al., 2023. Revolutionizing Medicinal Chemistry: The Application of Artificial Intelligence (AI) in Early Drug Discovery. Pharmaceuticals 16, 1259.

Hasan, M., Marjan, M.A., Uddin, M.P., Afjal, M.I., Kardy, S., Ma, S., Nam, Y., 2023. Ensemble machine learning-based recommendation system for effective prediction of suitable agricultural crop cultivation. Front. Plant Sci. 14.

Holzinger, A., Keiblinger, K., Holub, P., Zatloukal, K., Müller, H. et al., 2023. AI for life: Trends in artificial intelligence for biotechnology. N. Biotechnol. 74, 16–24.

https://www.cropin.com/blogs/accelerating-farm-productivity-using-ai, n.d. Accelerating Farm Productivity Using AI, https://www.cropin.com/blogs/accelerating-farm-productivity-using-ai [WWW Document]. URL https://www.cropin.com/blogs/accelerating-farm-productivity-using-ai (accessed 1.15.24).

IBM, 2023. Cost of a Data Breach Report 2023. IBM.

Ilic, S., Akabayov, S.R., Arthanari, H., Wagner, G., Richardson, C.C., Akabayov, B., 2016. Identification of DNA primase inhibitors via a combined fragment-based and virtual screening. Sci. Rep. 6, 36322.

Janiesch, C., Zschech, P., Heinrich, K., 2021. Machine learning and deep learning. Electron. Mark. 31, 685–695.

Javaid, M., Haleem, A., Khan, I.H., Suman, R., 2023. Understanding the potential applications of Artificial Intelligence in Agriculture Sector. Adv. Agrochem 2, 15–30.

Jawarkar, R.D., Bakal, R.L., Zaki, M.E.A., Al-Hussain, S., Ghosh, A., Gandhi, A., Mukerjee, N., Samad, A., Masand, V.H., Lewaa, I., 2022. QSAR based virtual screening derived identification of a novel hit as a SARS CoV-229E 3CLpro Inhibitor: GA-MLR QSAR modeling supported by molecular Docking, molecular dynamics simulation and MMGBSA calculation approaches. Arab. J. Chem. 15, 103499.

Joshi, S., Mishra, S., 2022. Recent advances in biofuel production through metabolic engineering. Bioresour. Technol. 352, 127037.

Kale, S.S., Patil, P.S., 2019. Data Mining Technology with Fuzzy Logic, Neural Networks and Machine Learning for Agriculture. pp. 79–87.

Kaur, H., Kumari, V., 2022. Predictive modelling and analytics for diabetes using a machine learning approach. Appl. Comput. Informatics 18, 90–100.

Khanal, S.K., Tarafdar, A., You, S., 2023. Artificial intelligence and machine learning for smart bioprocesses. Bioresour. Technol. 375, 128826.

Khanijou, J.K., Kulyk, H., Bergès, C., Khoo, L.W., Ng, P., Yeo, H.C., Helmy, M., Bellvert, F., Chew, W., Selvarajoo, K., 2022. Metabolomics and modelling approaches for systems metabolic engineering. Metab. Eng. Commun. 15, e00209.

Kim, G.B., Kim, W.J., Kim, H.U., Lee, S.Y., 2020. Machine learning applications in systems metabolic engineering. Curr. Opin. Biotechnol. 64, 1–9.

Krittanawong, C., Zhang, H., Wang, Z., Aydar, M., Kitai, T. et al., 2017. Artificial Intelligence in Precision Cardiovascular Medicine. J. Am. Coll. Cardiol. 69, 2657–2664.

Kumar, A., Singh, V.K., Saran, B., Al-Ansari, N., Singh, V.P., Adhikari, S., Joshi, A., Singh, N.K., Vishwakarma, D.K., 2022. Development of Novel Hybrid Models for Prediction of Drought- and Stress-Tolerance Indices in Teosinte Introgressed Maize Lines Using Artificial Intelligence Techniques. Sustainability 14, 2287.

Kumar, Y., Koul, A., Singla, R., Ijaz, M.F., 2023. Artificial intelligence in disease diagnosis: a systematic literature review, synthesizing framework and future research agenda. J. Ambient Intell. Humaniz. Comput. 14, 8459.

Kumar, Y.J.N., Spandana, V., Vaishnavi, V.S., Neha, K., Devi, V.G.R.R., 2020. Supervised Machine learning Approach for Crop Yield Prediction in Agriculture Sector. In: 2020 5th International Conference on Communication and Electronics Systems (ICCES). IEEE, pp. 736–741.

Lawson, C.E., Martí, J.M., Radivojevic, T., Jonnalagadda, S.V.R., Gentz, R., Hillson, N.J., Peisert, S., Kim, J., Simmons, B.A., Petzold, C.J., Singer, S.W., Mukhopadhyay, A., Tanjore, D., Dunn, J.G., Garcia Martin, H., 2021. Machine learning for metabolic engineering: A review. Metab. Eng. 63, 34–60.

Lee, H., Wang, J., Leblon, B., 2020. Using Linear Regression, Random Forests, and Support Vector Machine with Unmanned Aerial Vehicle Multispectral Images to Predict Canopy Nitrogen Weight in Corn. Remote Sens. 12, 2071.

Lee, J., Nazki, H., Baek, J., Hong, Y., Lee, M. et al., 2020. Artificial Intelligence Approach for Tomato Detection and Mass Estimation in Precision Agriculture. Sustainability 12, 9138.

Li, B., Zhang, N., Wang, Y.-G., George, A.W., Reverter, A., Li, Y., 2018. Genomic Prediction of Breeding Values Using a Subset of SNPs Identified by Three Machine Learning Methods. Front. Genet. 9.

Li, Y., Qiao, T., Leng, W., Jiao, W., Luo, J., Lv, Y., Tong, Y., Mei, X., Li, H., Hu, Q., Yao, Q., 2022. Semantic Segmentation of Wheat Stripe Rust Images Using Deep Learning. Agronomy 12, 2933.

Li, Z., Guo, R., Li, M., Chen, Y., Li, G. et al., 2020. A review of computer vision technologies for plant phenotyping. Comput. Electron. Agric. 176, 105672.

Liu, L., Sun, J., Xu, W., Du, G., Chen, J. et al., 2009. Modeling and optimization of microbial hyaluronic acid production by Streptococcus zooepidemicus using radial basis function neural network coupling quantum-behaved particle swarm optimization algorithm. Biotechnol. Prog. 25, 1819–1825.

Ljubic, B., Roychoudhury, S., Cao, X.H., Pavlovski, M., Obradovic, S., Nair, R., Glass, L., Obradovic, Z., 2020. Influence of medical domain knowledge on deep learning for Alzheimer's disease prediction. Comput. Methods Programs Biomed. 197, 105765.

Lu, B., Dao, P., Liu, J., He, Y., Shang, J., 2020. Recent Advances of Hyperspectral Imaging Technology and Applications in Agriculture. Remote Sens. 12, 2659.

Lu, Y., Chen, D., Olaniyi, E., Huang, Y., 2022. Generative adversarial networks (GANs) for image augmentation in agriculture: A systematic review. Comput. Electron. Agric. 200, 107208.

Mabiyan, R., 2020. India bullish on AI in healthcare without electronic health records. Econ. Times.

Magrani, E., 2019. New perspectives on ethics and the laws of artificial intelligence. Internet Policy Rev. 8.

Mahmud, M., Kaiser, M.S., McGinnity, T.M., Hussain, A., 2021. Deep Learning in Mining Biological Data. Cognit. Comput. 13, 1–33.

Manish Lad, A., Mani Bharathi, K., Akash Saravanan, B., Karthik, R., 2022. Factors affecting agriculture and estimation of crop yield using supervised learning algorithms. Mater. Today Proc. 62, 4629–4634.

Maserat, E., 2022. Integration of Artificial Intelligence and CRISPR/Cas9 System for Vaccine Design. Cancer Inform. 21, 11769351221140102.

Memon, M., Li, J., Haq, A., Memon, M., 2019. Breast cancer detection in the Iot health environment using modified recursi ve feature selection. Wirel Commun Mob 19.

Mi, Z., Zhang, X., Su, J., Han, D., Su, B. et al., 2020. Wheat Stripe Rust Grading by Deep Learning With Attention Mechanism and Images From Mobile Devices. Front. Plant Sci. 11.

Mondal, P.P., Galodha, A., Verma, V.K., Singh, V., Show, P.L., Awasthi, M.K., Lall, B., Anees, S., Pollmann, K., Jain, R., 2023. Review on machine learning-based bioprocess optimization, monitoring, and control systems. Bioresour. Technol. 370, 128523.

Montesinos-López, A., Rivera, C., Pinto, F., Piñera, F., Gonzalez, D., Reynolds, M., Pérez-Rodríguez, P., Li, H., Montesinos-López, O.A., Crossa, J., 2023. Multimodal deep learning methods enhance genomic prediction of wheat breeding. G3 Genes|Genomes|Genetics.

Mrisho, L.M., Mbilinyi, N.A., Ndalahwa, M., Ramcharan, A.M., Kehs, A.K., McCloskey, P.C., Murithi, H., Hughes, D.P., Legg, J.P., 2020. Accuracy of a Smartphone-Based Object Detection Model, PlantVillage Nuru, in Identifying the Foliar Symptoms of the Viral Diseases of Cassava–CMD and CBSD. Front. Plant Sci. 11.

Mullai, P., Vishali, S., Sobiya, E., 2022. Experiments and adaptive-network-based fuzzy inference system modelling in a hybrid up-flow anaerobic sludge blanket reactor to assess industrial azadirachtin effluent quality. Bioresour. Technol. 358, 127395.

Murgia, M., 2020. AI-designed drug to enter human clinical trial for first time. Financ. Times.

Nelson, G.S., 2019. Bias in Artificial Intelligence. N. C. Med. J. 80, 220–222.

Nguyen, C., Sagan, V., Maimaitiyiming, M., Maimaitijiang, M., Bhadra, S., Kwasniewski, M.T., 2021. Early Detection of Plant Viral Disease Using Hyperspectral Imaging and Deep Learning. Sensors 21, 742.

Nikita, S., Thakur, G., Jesubalan, N.G., Kulkarni, A., Yezhuvath, V.B., Rathore, A.S., 2022. AI-ML applications in bioprocessing: ML as an enabler of real time quality prediction in continuous manufacturing of mAbs. Comput. Chem. Eng. 164, 107896.

Oliveira, T.A. de, Silva, M.P. da, Maia, E.H.B., Silva, A.M. da, Taranto, A.G. et al., 2023. Virtual Screening Algorithms in Drug Discovery: A Review Focused on Machine and Deep Learning Methods. Drugs Drug Candidates 2, 311–334.

Owais, Arsalan, Choi, Mahmood, Park, 2019. Artificial Intelligence-Based Classification of Multiple Gastrointestinal Diseases Using Endoscopy Videos for Clinical Diagnosis. J. Clin. Med. 8, 986.

Panicker, R.O., Kalmady, K.S., Rajan, J., Sabu, M.K., 2018. Automatic detection of tuberculosis bacilli from microscopic sputum smear images using deep learning methods. Biocybern. Biomed. Eng. 38, 691–699.

Parikh, R.B., Teeple, S., Navathe, A.S., 2019. Addressing Bias in Artificial Intelligence in Health Care. JAMA 322, 2377.

Parthiban, S., Vijeesh, T., Gayathri, T., Shanmugaraj, B., Sharma, A., Sathishkumar, R., 2023. Artificial intelligence-driven systems engineering for next-generation plant-derived biopharmaceuticals. Front. Plant Sci. 14.

Patrick, S., Mirau, S., Mbalawata, I., Leo, J., 2023. Time series and ensemble models to forecast banana crop yield in Tanzania, considering the effects of climate change. Resour. Environ. Sustain. 14, 100138.

Paul, D., Sanap, G., Shenoy, S., Kalyane, D., Kalia, K., Tekade, R.K., 2021. Artificial intelligence in drug discovery and development. Drug Discov. Today 26, 80–93.

Paul, R.K., Yeasin, M., Kumar, Pramod, Kumar, Prabhakar, Balasubramanian, M., Roy, H.S., Paul, A.K., Gupta, A., 2022. Machine learning techniques for forecasting agricultural prices: A case of brinjal in Odisha, India. PLoS One 17, e0270553.

Pauwels, E., 2023. How to Protect Biotechnology and Biosecurity from Adversarial AI Attacks? A Global Governance Perspective. In: Cyberbiosecurity. Springer International Publishing, pp. 173–184.

Pérez-Enciso, Zingaretti, 2019. A Guide for Using Deep Learning for Complex Trait Genomic Prediction. Genes (Basel). 10, 553.

Pesapane, F., Volonté, C., Codari, M., Sardanelli, F., 2018. Artificial intelligence as a medical device in radiology: ethical and regulatory issues in Europe and the United States. Insights Imaging 9, 745–753.

Plantix, 2024. Plantix | #1 FREE app for crop diagnosis and treatments [WWW Document]. URL https://plantix.net/en/ (accessed 1.17.24).

Puranik, A., Dandekar, P., Jain, R., 2022. Exploring the potential of machine learning for more efficient development and production of biopharmaceuticals. Biotechnol. Prog. 38, e3291.

Rajalakshmi, R., Subashini, R., Anjana, R.M., Mohan, V., 2018. Automated diabetic retinopathy detection in smartphone-based fundus photography using artificial intelligence. Eye 32, 1138–1144.

Rani, A., Devi, P., Jha, U.C., Sharma, K.D., Siddique, K.H.M., Nayyar, H., 2020. Developing Climate-Resilient Chickpea Involving Physiological and Molecular Approaches With a Focus on Temperature and Drought Stresses. Front. Plant Sci. 10.

Renata, H., Wang, Z.J., Arnold, F.H., 2015. Expanding the Enzyme Universe: Accessing Non-Natural Reactions by Mechanism-Guided Directed Evolution. Angew. Chemie Int. Ed. 54, 3351–3367.

Rodrigues, D. de A., Ivo, R.F., Satapathy, S.C., Wang, S., Hemanth, J., Filho, P.P.R., 2020. A new approach for classification skin lesion based on transfer learning, deep learning, and IoT system. Pattern Recognit. Lett. 136, 8–15.

Romero, M.P., Chang, Y.-M., Brunton, L.A., Parry, J., Prosser, A., Upton, P., Rees, E., Tearne, O., Arnold, M., Stevens, K., Drewe, J.A., 2020. Decision tree machine learning applied to bovine tuberculosis risk factors to aid disease control decision making. Prev. Vet. Med. 175, 104860.

Roorkiwal, M., Bharadwaj, C., Barmukh, R., Dixit, G.P., Thudi, M., Gaur, P.M., Chaturvedi, S.K., Fikre, A., Hamwieh, A., Kumar, S., Sachdeva, S., Ojiewo, C.O., Tar'an, B., Wordofa, N.G., Singh, N.P., Siddique, K.H.M., Varshney, R.K., 2020. Integrating genomics for chickpea improvement: achievements and opportunities. Theor. Appl. Genet. 133, 1703–1720.

Roy, S., Malhotra, N., 2019. National Artificial Intelligence Mission. Invest India.

Roy, S.K., De, D., 2022. Genetic Algorithm based Internet of Precision Agricultural Things (IopaT) for Agriculture 4.0. Internet of Things 18, 100201.

Sadybekov, A. V., Katritch, V., 2023. Computational approaches streamlining drug discovery. Nature 616, 673–685.

Saini, D.K., Rai, A., Devi, A., Pabbi, S., Chhabra, D., Chang, J.-S., Shukla, P., 2021. A multi-objective hybrid machine learning approach-based optimization for enhanced biomass and bioactive phycobiliproteins production in Nostoc sp. CCC-403. Bioresour. Technol. 329, 124908.

Sathitratanacheewin, S., Sunanta, P., Pongpirul, K., 2020. Deep learning for automated classification of tuberculosis-related chest X-Ray: dataset distribution shift limits diagnostic performance generalizability. Heliyon 6, e04614.

Schork, N.J., 2019. Artificial Intelligence and Personalized Medicine. In: Precision Medicine in Cancer Therapy. pp. 265–283.

Sethy, P.K., Behera, S.K., Kannan, N., Narayanan, S., Pandey, C. et al., 2021. Smart paddy field monitoring system using deep learning and IoT. Concurr. Eng. 29, 16–24.

Shah, M., Naik, N., Somani, B.K., Hameed, B.M.Z., 2020. Artificial intelligence (AI) in urology-Current use and future directions: An iTRUE study. Türk Üroloji Dergisi/Turkish J. Urol. 46, S27–S39.

Shah, P., Kendall, F., Khozin, S., Goosen, R., Hu, J., Laramie, J., Ringel, M., Schork, N., 2019. Artificial intelligence and machine learning in clinical development: a translational perspective. npj Digit. Med. 2, 69.

Shanthi, D., Madhuravani, B., Kumar, A. (Eds.), 2023. Handbook of Artificial Intelligence. BENTHAM SCIENCE PUBLISHERS.

Sharma, Ashwani, Virmani, T., Pathak, V., Sharma, Anjali, Pathak, K., Kumar, G., Pathak, D., 2022. Artificial Intelligence-Based Data-Driven Strategy to Accelerate Research, Development, and Clinical Trials of COVID Vaccine. Biomed Res. Int. 2022, 1–16.

Sharma, K., Manchikanti, P., 2020. Regulation of Artificial Intelligence in Drug Discovery and Health Care. Biotechnol. Law Rep. 39, 371–380.

Shin, W.-H., Zhu, X., Bures, M., Kihara, D., 2015. Three-Dimensional Compound Comparison Methods and Their Application in Drug Discovery. Molecules 20, 12841–12862.

Shoaib, M., Shah, B., EI-Sappagh, S., Ali, A., Ullah, A., Alenezi, F., Gechev, T., Hussain, T., Ali, F., 2023. An advanced deep learning models-based plant disease detection: A review of recent research. Front. Plant Sci. 14.

Siddiqua, A., Kabir, M.A., Ferdous, T., Ali, I.B., Weston, L.A. et al., 2022. Evaluating Plant Disease Detection Mobile Applications: Quality and Limitations. Agronomy 12, 1869.

Singh, S., Gupta, H., Sharma, P., Sahi, S., 2024. Advances in Artificial Intelligence (AI)-assisted approaches in drug screening. Artif. Intell. Chem. 2, 100039.

Soundarya, S., Sruthi, M.S., Sathya Bama, S., Kiruthika, S., Dhiyaneswaran, J. et al., 2021. Early detection of Alzheimer disease using Gadolinium material. Mater. Today Proc. 45, 1094–1101.

Srinivasu, P., Ahmed, S., Alhumam, A., Kumar, A., Ijaz, M. et al., 2021. An AW-HARIS Based Automated Segmentation of Human Liver Using CT Images. Comput Mater Contin 69, 3303–3319.

Srinivasu, P.N., SivaSai, J.G., Ijaz, M.F., Bhoi, A.K., Kim, W., Kang, J.J., 2021. Classification of Skin Disease Using Deep Learning Neural Networks with MobileNet V2 and LSTM. Sensors 21, 2852.

Subeesh, A., Mehta, C.R., 2021. Automation and digitization of agriculture using artificial intelligence and internet of things. Artif. Intell. Agric. 5, 278–291.

Talaviya, T., Shah, D., Patel, N., Yagnik, H., Shah, M. et al., 2020. Implementation of artificial intelligence in agriculture for optimisation of irrigation and application of pesticides and herbicides. Artif. Intell. Agric. 4, 58–73.

Tamilvizhi, T., Surendran, R., Anbazhagan, K., Rajkumar, K., 2022. Quantum Behaved Particle Swarm Optimization-Based Deep Transfer Learning Model for Sugarcane Leaf Disease Detection and Classification. Math. Probl. Eng. 2022, 1–12.

Taneja, A., Nair, G., Joshi, M., Sharma, Somesh, Sharma, Surabhi, Jambrak, A.R., Roselló-Soto, E., Barba, F.J., Castagnini, J.M., Leksawasdi, N., Phimolsiripol, Y., 2023. Artificial Intelligence: Implications for the Agri-Food Sector. Agronomy 13, 1397.

Teng, Z., Chen, J., Wang, J., Wu, S., Chen, R., Lin, Y., Shen, L., Jackson, R., Zhou, J., Yang, C., 2023. Panicle-Cloud: An Open and AI-Powered Cloud Computing Platform for Quantifying Rice Panicles from Drone-Collected Imagery to Enable the Classification of Yield Production in Rice. Plant Phenomics 5.

Thakur, V., Bhatia, K., Jha, B., Gupta, S. et.al., 2021. 75@75 INDIA'S AI JOURNEY.

Thorat, T., Patle, B.K., Kashyap, S.K., 2023. Intelligent insecticide and fertilizer recommendation system based on TPF-CNN for smart farming. Smart Agric. Technol. 3, 100114.

Tigga, N.P., Garg, S., 2020. Prediction of Type 2 Diabetes using Machine Learning Classification Methods. Procedia Comput. Sci. 167, 706–716.

Togninalli, M., Wang, X., Kucera, T., Shrestha, S., Juliana, P., Mondal, S., Pinto, F., Govindan, V., Crespo-Herrera, L., Huerta-Espino, J., Singh, R.P., Borgwardt, K., Poland, J., 2023. Multi-modal deep learning improves grain yield prediction in wheat breeding by fusing genomics and phenomics. Bioinformatics 39.

Varshney, R.K., Roorkiwal, M., Sun, S., Bajaj, P., Chitikineni, A. et al., 2021. A chickpea genetic variation map based on the sequencing of 3,366 genomes. Nature 599, 622–627.

Vázquez, J., López, M., Gibert, E., Herrero, E., Luque, F.J. et al., 2020. Merging Ligand-Based and Structure-Based Methods in Drug Discovery: An Overview of Combined Virtual Screening Approaches. Molecules 25, 4723.

Wakchaure, M., Patle, B.K., Mahindrakar, A.K., 2023. Application of AI techniques and robotics in agriculture: A review. Artif. Intell. Life Sci. 3, 100057.

Wang, Y.-B., You, Z.-H., Yang, S., Yi, H.-C., Chen, Z.-H., Zheng, K., 2020. A deep learning-based method for drug-target interaction prediction based on long short-term memory neural network. BMC Med. Inform. Decis. Mak. 20, 49.

Wolfert, S., Ge, L., Verdouw, C., Bogaardt, M.-J., 2017. Big Data in Smart Farming – A review. Agric. Syst. 153, 69–80.

Yang, X., Wang, Y., Byrne, R., Schneider, G., Yang, S. et al., 2019. Concepts of Artificial Intelligence for Computer-Assisted Drug Discovery. Chem. Rev. 119, 10520–10594.

Yoosefzadeh Najafabadi, M., Hesami, M., Eskandari, M., 2023. Machine Learning-Assisted Approaches in Modernized Plant Breeding Programs. Genes (Basel). 14, 777.

Zafar, M., Kumar, S.S., Kumar, S.S., Dhiman, A.K., 2012. Artificial intelligence based modeling and optimization of poly(3-hydroxybutyrate-co-3-hydroxyvalerate) production process by using Azohydromonas lata MTCC 2311 from cane molasses supplemented with volatile fatty acids: A genetic algorithm paradigm. Bioresour. Technol. 104, 631–641.

Zhang, J., Zhu, Y., Zhang, X., Ye, M., Yang, J. et al., 2018. Developing a Long Short-Term Memory (LSTM) based model for predicting water table depth in agricultural areas. J. Hydrol. 561, 918–929.

Zhang, L., Chao, B., Zhang, X., 2020. Modeling and optimization of microbial lipid fermentation from cellulosic ethanol wastewater by Rhodotorula glutinis based on the support vector machine. Bioresour. Technol. 301, 122781.

Zhao, M., Cang, H., Chen, H., Zhang, C., Yan, T., Zhang, Y., Gao, P., Xu, W., 2023. Determination of quality and maturity of processing tomatoes using near-infrared hyperspectral imaging with interpretable machine learning methods. LWT 183, 114861.

Zuo, H., 2022. Analysis and Detection of Tomatoes Quality using Machine Learning Algorithm and Image Processing. Int. J. Adv. Comput. Sci. Appl. 13.

CHAPTER

13

Artificial Intelligence in Drug Discovery

Grace R. Raji,[1#] ***Arya Vinod***[1#] **and** ***V.B. Sameer Kumar***[1*]

1. Introduction

Though discovering drugs from natural and synthetic precursors started way back in history, modern drug discovery research dates back to the early 1900s. The drug discovery process, from initial drug design to market launch, incurs an approximate cost of $1 billion and spans a duration of 14 to 15 years. The journey from designing a drug to its ultimate execution in the clinical milieu is a long, tedious, and expensive process involving labor intensive trial and error techniques with multiple potential phases of failure. The development of a new drug usually involves multiple stages like (1) The pre-discovery stage to identify the possible therapeutic target associated with disease (2) The discovery stage involves identification of small molecules or therapeutic strategies that can alter the function of the proposed target and cure disease (3) The preclinical stage during which potential drugs are checked in various *in vitro* and *in vivo* models, to validate their efficiency (4) The clinical stage involves human trials (5) The final stage involves conducting pharmacovigilance studies. The discovery paradigm of drugs took a leap in the early 1980s, with the development of computers and crystallography assisted structure-based drug discovery (Batool et al., 2019), which was followed by high-throughput screening of drug design which involves scanning of datasets as large as 10^6–10^7 molecules for analyzing their binding efficacy with a given protein (Gong Z. et al., 2017). The entire drug discovery and development generates huge amounts of data collected from high-throughput technologies. Drug design can be automated via computer-aided approaches to speed up the process and prevent failures. Currently, due to the emergence of high-performance computing and advanced graphical processing units for collateral processing, and the highly sophisticated technology in bioinformatics, this process is now getting revolutionized by artificial intelligence (AI) (Marr B., 2019).

[1] School of Biological Sciences, Central University of Kerala, India.
# Contributed equally
* Corresponding author: sameerkumarvb@gmail.com; skumarvb@cukerala.ac.in

AI is a technology-based system that employs various advanced tools and networks that can imitate human intelligence. AI utilizes several inbuilt softwares that can interpret the input data to make individualistic decisions for executing specific tasks. In the recent past, there has been a sudden expansion in the applications of AI methods in multiple fields of the pharmaceutical industry, including drug discovery and development, drug repurposing, pre-clinical and clinical trials (Lamberti M.J., 2019). AI guided automation involves a fundamental paradigm of machine learning (ML). ML utilizes algorithms that can identify patterns within a set of data and can be processed through artificial neural networks (ANNs) (Beneke F., Mackenrodt M.). The basic unit of ANNs is known as 'perceptron', which perceives the data signals from external sources and from other perceptrons and processes (Steels L., Brooks R., 2018). ANNs are mathematical models comprising a set of highly complex interconnected networks of computing elements analogous to neural circuits within the human body. ANNs comprise a set of nodes, which receive an input, and convert them to output, either singly or multi-linked using algorithms to resolve problems. ANNs are of various forms, which include multilayer perceptron (MLP) networks, recurrent neural networks (RNNs), and convolutional neural networks (CNNs), which employ either supervised or unsupervised training techniques (Bielecki A., 2019).

The applications of MLP include pattern recognition, optimization aids and process identification. These are controlled by supervised training procedures operating unidirectionally. MLPs can be used as universal pattern classifiers (Da Silva I.N., 2017). RNNs are closed-loop networks, involving time sequences of events and ordered data such as characters in words, having the potential to memorize and store information, such as Boltzmann constants and Hopfield networks (Medsker L., Jain L.C., 1999). CNNs constitute a class of recurring nonlinear and locally integrated networks of identical dynamical cells that function in parallel. ANALOG chips are distinguished by their topology, and find application in the modeling of biological systems, processing compound brain functions, image and video processing, pattern recognition, and comlpex signal processing (Hänggi M., Moschytz G.S., 2000). One important tool developed based on the principal architecture of AI technology is the International Business Machine (IBM) Watson supercomputer (IBM, New York, USA). It was designed to aid in the analysis of a patient's medical records and its link with multiple databases, and help in deciding treatment strategies for cancer. This tool can also be used for the rapid diagnosis of diseases.

Implementation of AI based technologies in the field of pharmaceutical chemistry has created a metamorphic impact on the pharmaceutical landscape right from the bench to the bedside which involves: aiding decision making; making decisions on the right treatment strategies including personalized medicines; and utilization of the generated clinical data for future drug development. The revolutionary impact of AI in the pharma field is also evidenced by the groundbreaking patents in this field which include automated cell positioning, prime editing efficiency prediction, predictive modeling and control of cell culture (Lindevall, 2021). This chapter provides an overview of various AI-based technologies in drug discovery and development right from the design of a drug to the post marketing stage, highlighting their benefits and drawbacks.

2. Applications of AI in Drug Discovery and Development

The advancement of AI based technology has invaded almost all the dimensions of drug development. AI is being used extensively to upgrade design techniques and the time required for drug development. The AI technology is used in each step of the drug designing procedure, right from de novo drug design, activity scoring, virtual screening, and in silico analysis of the ADME properties of a drug molecule (Sliwoski G. et al., 2014).

2.1 AI in Target Prediction and Validation

Target prediction during drug discovery involves the identification of molecules, usually proteins, whose activity is modulated during the disease state. Altered protein targets can be identified from different types of data, including gene expression profiles, protein-protein interaction networks, and genomic and proteomic data using machine learning algorithms (Costa P.R. et al., 2010). GNNs or tree-based methods are used to identify causal relationships between diseases and genes and to predict morbidity-associated druggable genes. Decision trees also provide data on gene regulation by multiple transcription factors, their role in metabolic pathways, and extracellular location (Khan J.Y., 2020). Text mining and Natural Language Processing (NLP) approaches can also be employed to identify relevant target-disease pairs from literature. Databases like BeFree (Bravo À. et al., 2015), PKDE4J and other deep learning-based tools can be used to extract articles to identify drug–disease, gene-disease, and target–drug associations (Song M. et al., 2018). Predicting the 3D structures of potential target proteins is essential for drug discovery, and AI can assist in structure-based drug discovery with AlphaFold2. The AI tool, AlphaFold, is based on DNNs, used to identify the distance between the adjacent amino acids and the corresponding bond angles (Jumper J. et al., 2021).

2.2 AI in de Novo Drug Design and Optimization

AI can assist in virtual screening through the development of predictive models that can recognize compounds with a high probability of binding to protein targets (Álvarez-Machancoses Ó., Fernández-Martínez J.L., 2019). AI can also be used to predict most suitable routes for chemical synthesis and develop intuitions into the reaction process of drugs to identify potentially unwanted associations with other molecules. AI based deep neural networks (DNN) focus on rules of organic chemistry and retrosynthesis. Coley *et al.* constructed a framework in which ML was used to identify the dominant product based on a score given by the NNs (Chan H.S., 2019). A DNN architecture called the reinforced adversarial neural computer (RANC) based on RL for *de novo* design of small organic molecules was developed by Putin *et al.* (Putin E., 2018). This platform was trained with molecules identified as SMILES strings, which then generated molecules with predefined chemical descriptors. Recursive neural networks (RNN) used for de novo design take sequential information as input and generate chemical structures (Bengio Y., 2009). The same approach was used for the generation of novel peptide structures (Muller A.T. et al., 2018). Reinforcement learning was employed to bias the generated compounds towards desired properties (Olivecrona M. et al., 2017). Candidate structures of drugs will be purified and modified to enhance target specificity and selectivity. Multi-objective optimization methods are utilized to tune molecules in a desired direction. MD simulation and docking methods are employed to model the orientation, stability, and dynamics of the compounds (Nicolaou C.A., Brown N., 2013).

2.3 AI in Predicting Drug-protein Interaction

The prediction of drug-protein interaction is necessary to understand its efficiency, permit the repurposing of drugs, and eliminate polypharmacology (Wan F., Zeng J., 2016). Several AI methods have been useful in the prediction of drug-protein interactions. Wang *et al.* reported a model using the SVM approach, which was developed based on primary protein sequences and structural characteristics of small molecules to identify nine new compounds and their interaction with four important targets (Wang F., 2011). While repurposing a drug, logistic regression platforms, such as PREDICT, SPACE, and other ML approaches, are employed to assess drug-drug, disease-disease similarity, the similarity between target molecules, chemical structure, and gene expression profiles (Park K., 2019). AI-based tools such as SOM, are used to correlate several compounds to numerous targets and off-targets to avoid off-target adverse effects (Achenbach J., 2011). Li *et al.* reported the use of an AI-based tool KinomeX, using DNNs for the detection of polypharmacology of kinases based on their chemical structures. This platform employs DNNs trained with ~14 000 bioactivity data points developed based on >300 kinases (Li Z., 2019).

2.4 AI in Pre-clinical and Clinical Studies

Generative and prediction-based AI, ML, and reasoning techniques can be employed for preclinical compound discovery, compound-target testing, and defining lead compounds for clinical trials. This may permit lead candidates that have a higher chance of success to be chosen during clinical trials, and the elimination of those with a higher risk of failure before they enter the clinical phase (Harrer S. et al., 2019). AI can aid in choosing only a specific diseased population for recruitment in Phases II and III of clinical trials by using patient-specific genome–exposome profile analysis. AI based tools can also be used to enhance patient cohort selection. AI and ML based deep reinforcement learning algorithms may also be used to predict the risk of dropout for a specific patient (Fogel D.B., 2018).

2.5 AI in FDA Approval and Post-market Analysis

AI based Natural Language Processing (NLP) can be used to retrieve scientific literature to report conflicting effects, such as drug toxicity, drug resistance and prepare automated evaluations for regulatory (FDA) approval or for patent application (Shi Y. et al., 2021). NLP-based sentiment analysis techniques can be used to propose drugs. Machine learning-based systems are capable of predicting likely sales of a product thus helping pharmaceutical companies optimize their business resources (Khalil Zadeh N. et al., 2014).

3. AI Based Drug Discovery in Communicable and Non-communicable Diseases

3.1 Artificial Intelligence in Cancer Drug Discovery

Artificial intelligence has many applications in cancer research, such as imaging and classification of cancer cells, prediction of target protein structures, and prediction of drug-protein interactions (Wang, L. et al., 2023). Drug–target interactions (DTIs) are the initial step in the design of anti-cancer drugs. Peng *et al.* developed a novel end-to-end learning-based framework based on heterogeneous graph convolutional networks for DTI prediction called end-to-end graph (EEG)-DTI (Peng, J. et al., 2021). Yang et al. proposed a drug–target interaction prediction method based on mutual learning mechanisms without 3D structural information along with an explanation (Shao, K. et al., 2022). A Bayesian-based machine learning method (BANDIT), with approximately 90% target prediction accuracy on more than 2000 small molecules has been developed by integrating six types of data (Yang, Z. et al., 2021). Drugnome AI has been developed for predicting the durability of drug targets in cancer (Madhukar, N.S. et al., 2019). Artificial Intelligence also contributes to the screening of Anti-Cancer drug Compounds. High-throughput screening is the method to screen the drug compound, which includes structure-based and ligand-based screening. Fragment-based screening methods are also effective for the discovery of hit compounds. Lu et al. constructed a deep-learning model to predict molecular docking scoring (Olayan, R.S. et al., 2018). Yasuo *et al.* propose a new structure-based virtual screening method for hit compounds, called SIEVE-Score, which provides substantial improvements over virtual screening methods (Raies, A. et al., 2022). AI also helps in the identification of anti-drug repositioning based on drug–disease interaction. The prediction of cancer drug reactions also can be demonstrated using artificial intelligence. The methods and AI tools used for various drug development processes are given below (Table 1).

3.2 AI in Drug Discovery for Neurological Disorders

Neurological disorders or central nervous system (CNS) disorders such as Alzheimer's-type dementia, Parkinson's, Schizophrenia, and Huntington's diseases demand high medical attention, with relatively fewer drugs available. The option of a new drug needs long term clinical experimentation from discovery to regulatory approvals. To overcome this issue, researchers and

Table 1 List of methods and AI tools used for cancer drug development process

Model	*Data Source*	*References*
1. Anti-cancer drug target identification methods		
EEG-DTI	Luo dataset, Yamanishi dataset	(Peng, J. et al, 2021)
DTI-HETA	Yamanishi dataset	(Shao, K. et al., 2022)
ML-DTI	Metz dataset, KIBA dataset, Davis dataset,	(Yang, Z., 2021)
DDR	Yamanishi dataset, KEGG BRITE, BRENDA, SuperTarget, DrugBank	(Raies, A., 2022)
DrugnomeAI	TCRD, StringDB, CTDbase,InterPro, OMIM	(Raies, A. et al., 2022)
KG4SL	SynLethDB	(Wang, S. et al., 2021)
2. Anti-cancer drug hit compounds screening methods		
SECSE	PDB	(Lu, C. et al., 2022)
SIEVE-Score	ChEMBL, ZINC	(Yasuo, N., and Sekijima M. 2019)
DENVIS	PDB, DUD-E, LIT-PCBA	(Krasoulis, A. et al., 2022)
DMMFP	ChEMBL, DUD-E, ZINC	(Hutter, M.C., 2022)
POEM	ChEMBL, ZINC, PDB	(Eguida, M. et al., 2022)
SyntaLinker	ChEMBL	(Yang, Y. et al., 2020)
3. De novo anti-cancer drug design Methods		
PaccMannRL	TCGA, ChEMBL, GDSC, CCLE	(Born, J. et al., 2021)
ACGT	QM9, ZINC	(Hong, S. et al., 2019)
NEVAE	QM9, ZINC	(Samanta, B. et al., 2020)
BIMODAL	ChEMBL	(Grisoni, F. et al., 2020)
Mol-CycleGAN	ChEMBL, ZINC	(Maziarka, Ł. et al., 2020)
GAN-Drug-Generator	ChEMBL, ZINC	(Abbasi, M. et al., 2022)
ReLeaSE	ChEMBL, ZINC	(Popova, M. et al., 2018)
MoleGuLAR	ChEMBL, ZINC	(Goel, M. et al., 2021)
4. Anti-cancer drug repurposing Methods		
GPSnet	DrugBank, TTD, PharmGKB, ChEMBL,BindingDB, UniProt, TCGA	(Cheng, F. et al., 2019)
DeepDRK	CTRP, GDSC, TCGA, DrugBank, KEGG	(Wang, Y. et al., 2021)
MBiRW	Drugbank, OMIM	(Luo, H. et al., 2016)
DR-HGNN	Drugbank, CTD, SIDER	(Sadeghi, S. et al., 2022)
GDRnet	Drugbank, Hetionet, GNBR, STRING, IntAct, DGIdb	(Doshi, S., and Chepuri, S.P., 2022)
5. Cancer drug reactions prediction Methods		
DeepCarc	CPDB, Pubchem, Drugbank	(Li, T. et al., 2021)
Precily	CCLE, MSigDB, GDSC, Pubche	(Chawla, S. et al., 2022)
DRUML	PharmacoDB, DepMap portal, PRIDEdataset, DrugBank, ChEMBL	(Gerdes, H. et al., 2021)
MGPred	SIDER, STITCH, DrugBank, PubChem	(Zhao, H. et al., 2021)
DLADE	cTAKES, EHR and PubMed article	(Wunnava, S. et al., 2019)
MADRL	KEEG, SIDER, CTD, DrugBank	(Zhu, J. et al., 2021)
DDIMDL	DrugBank, KEGG	(Deng Y. et al., 2020)
3DGT-DDI	DrugBank, DDI extraction 2013	(He, H. et al., 2022)

biopharmaceutical companies have developed different methods and databases with the help of artificial intelligence. Different devices are already available in the market to monitor the activity of the patients as well as the neuron signaling pathways with the use of AI. Other than the monitoring and screening devices various methods are available to discover and develop efficient therapeutic options to overcome obstacles (Gautam, V. et al., 2023). AI helps in predicting the structure, component analysis, clustering, and statistical analysis of the available drugs with the support of large databases. A Naïve Bayesian model chemogenomic database WOMBAT (World of Molecular BioAcTivity) was developed in 2005 for the rapid identification of potential targets for compounds based on chemical information of compounds from 964 target classes (Olah, M. et al., 2004). In light of WOMBAT there are different tools developed using AI for drug designing and analysis. AI/ML methods hold a promising characterization of treatment effects from the neuroimaging data sets and thereby improve the understanding of the drug mechanism in the brain.

3.3 AI in Cardiovascular Drug Discovery

One of the earliest applications of AI was in cardiovascular medicine. Deep learning with AI systems has potential innovative applications in drug discovery, personalized drug therapy, and precision medicine in cardiovascular therapy. The dosage system and screening including statistical analysis can be carried out with various algorithms. Zellweger *et al* studied the Role of AI as a non-invasive tool for the diagnosis of coronary artery disease with the use of AI-based mimetic pattern-based algorithm (MPA) (Zellweger, M.J. et al., 2018). The AI-ECG technology facilitated the development of novel wearable devices such as implantable loop recorder (ILR) and ECG-compatible smartphones/smart watches which can be used to predict the high-risk groups for cardiovascular disease in a general population (Nakamura, T. and Sasano, T., 2022). AI Advances the opportunities in *in vitro* human modeling of cardiovascular disease. Recently a team of scientists has developed a bioengineered human cardiac organoid platform that provides functional contractile tissue with biological properties similar to native heart tissue (Joddar, B. et al., 2022).

3.4 AI in Drug Discovery for Infectious Diseases

AI has been implemented in both the screening and drug discovery of various infections. During COVID-19 pandemic, AI/ML was used to develop programs such as MARIA (major histocompatibility complex analysis with recurrent integrated architecture) and MoDec that predicts antigen presentation to understand natural immunity (Arora, G. et al., 2021). Arshadi *et al.* developed an AI tool with a collection of compounds, peptides, and epitopes related to COVID-19 called Corona-DB-AI (Keshavarzi Arshadi, A. et al., 2020). Cunha *et al* identified certain metabolic fingerprints using spectroscopy based on ML technology to analyze and predict the mechanism of action and potency of different antibiotics (Ribeiro da Cunha, B. et al., 2021). ML in combination with genetic algorithms has been developed to assess the intrinsic activity and efficacy of compounds. Feng *et al.* classified the antibacterial peptides using primary sequence information and sequence pattern recognition called Increment of Diversity with Quadratic Discriminant (IDQD) analysis. (Feng, P. et al., 2017).

4. Advanced Applications of AI based Drug Discovery

Nanorobots for drug delivery is an important application, in which AI based nanorobots constituting integrated circuits, sensors and power supply are used for targeted and controlled drug delivery. Nanorobots are programmed to steer to the target site based on physiological conditions, improving the drug efficacy and finally excreting the drug from its body. These nanorobots are controlled by AI tools like NNs, fuzzy logic, and integrators (Feng, P., 2019). AI-based tools like ANNs, logistic regression, and network-based modeling are employed for screening drug combinations and improving overall dose regimens in combination therapy (Feng et al., 2017). Li *et al.* developed a synergistic drug combination model based on gene expression profiles and various networks, using RF for the prediction of synergistic anticancer drug combinations (Calzori and Li, 2017). A combination of nanotechnology and AI could provide solutions to many problems in formulation development. A nanosuspension of methotrexate was computationally formulated by studying the energy generated during drug interaction, and monitoring the conditions that could lead to the aggregation of the nano formulation. Software such as LAMMPS and GROMACS 4 can be used to check the impact of surface chemistry on the internalization of nanoparticles into cells (Mehta C.H., 2019).

5. Challenges and limitations of using AI in Drug Discovery

The overall success of AI depends on the availability of a considerable amount of data as it is for further training provided to the system. Data accession from various database providers can put extra costs on a firm, and it should also be reliable and high in quality to ensure accurate result prediction. Therefore formulating a system, learning the true representation, and labeling the data are major challenges for the success of AI in the drug discovery domain (Hutson M. 2018). Another challenge includes the lack of skilled personnel to operate AI-based platforms, limited budget for small organizations, apprehension of replacing manpower leading to unemployment, suspicion about the data generated by AI among other concerns. AI based technologies in the drug discovery domain also face other challenges, such as data depiction, data labeling, discrepancy among labels, small sample sizes, data privacy, ethical issues, learning paradigms, and model interpretations. The black-box nature of deep learning models is yet another challenge. One cannot entirely rely on mere predictions, without understanding the underlying biological and chemical reasons. Precise and transparent deep learning models with a clear grasp of the accuracy of the model, in order to accurately exploit the available data (Lamberti M.J., 2019).

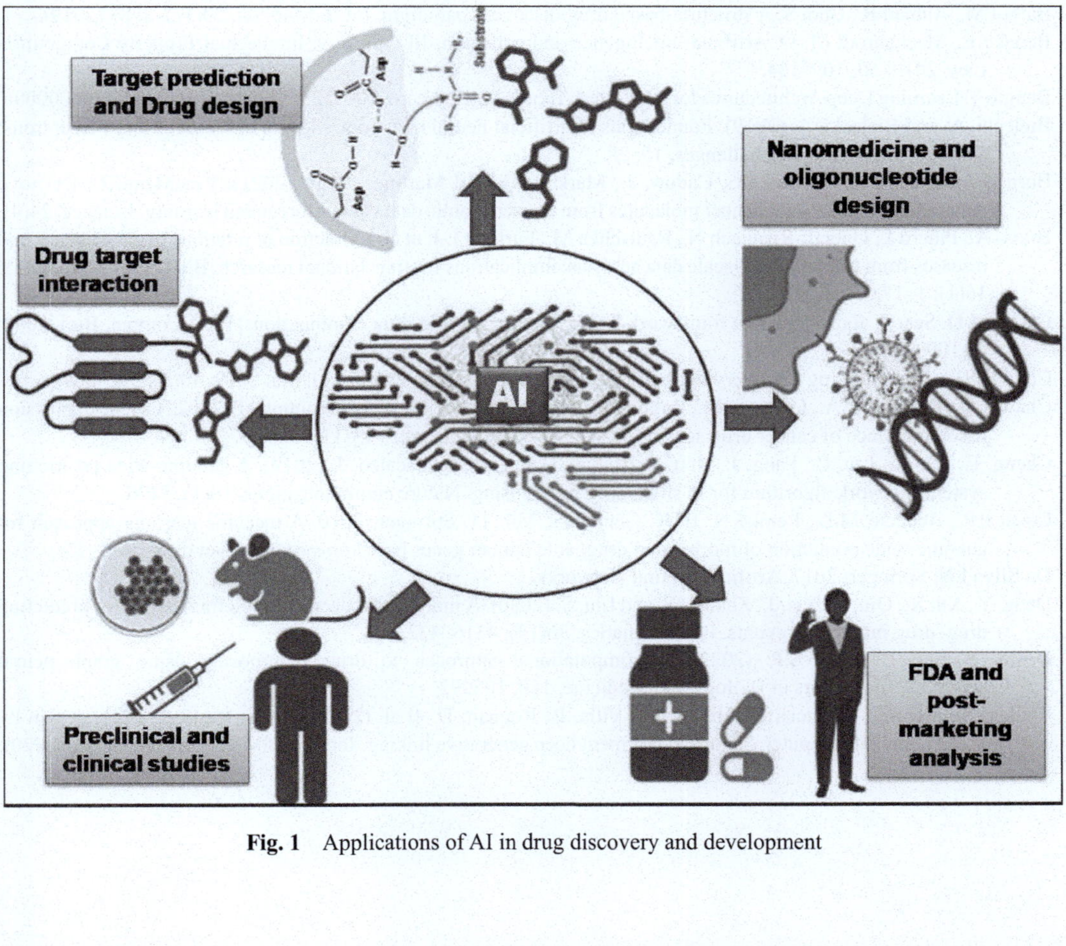

Fig. 1 Applications of AI in drug discovery and development

Conclusion and Future Prospects

The progression of AI, along with its remarkable tools, continuously aims to reduce challenges faced by pharmaceutical companies, influencing the drug discovery process. Currently the pharmaceutical sector is facing several complex challenges, including increased cost of drugs and therapies. With the advent of AI in drug discovery (Figure 1), personalized medications with desired dosages, release parameters, and other required aspects can be manufactured based on the requirements of individual patients. Using the latest AI-based technologies will not only speed up the time required for the products to be launched into the market, but will also improve the quality of products and the overall safety of the production process, and provide better utilization of available resources along with cost effectiveness, thereby increasing the importance of automation. AI can significantly contribute to the developed drug in its correct dosage and optimization, leading to quicker manufacturing of best-quality products along with assurance of batch-to-batch consistency. AI can also help in maintaining the safety and efficacy of the product in clinical trials, and ensure proper positioning and costing in the market. It is likely that AI would become a valuable tool in understanding human biology, and combating human diseases, with collaborative efforts from multiple disciplines in near future.

References

Abbasi, M., Santos, B.P., Pereira, T.C., Sofia, R., Monteiro, N.R., Simões, C.J., ... and Arrais, J.P. (2022). Designing optimized drug candidates with Generative Adversarial Network. Journal of cheminformatics, 14(1), 40.

Achenbach J. Computational tools for polypharmacology and repurposing. Fut. Med. Chem. 2011; 3: 961–968.

Álvarez-Machancoses Ó., Fernández-Martínez J.L. Using artificial intelligence methods to speed up drug discovery. Expert Opin. Drug Discov. 2019; 14(8): 769–77.

Arora, G., Joshi, J., Mandal, R.S., Shrivastava, N., Virmani, R. and Sethi, T. (2021). Artificial intelligence in surveillance, diagnosis, drug discovery and vaccine development against COVID-19. Pathogens, 10(8), 1048.

Batool M., Ahmad B., Choi S. A structure-based drug discovery paradigm. Int. J. Mol. Sci. 20(11), 2783 (2019).

Beneke F., Mackenrodt M.-O. Artificial intelligence and collusion. IIC Int. Rev. Intellectual Property Competition Law. 2019; 50: 109–134.

Bengio Y. Learning Deep Architectures for AI. Found. Trends Mach. Learn. 2009; 2: 1–127. doi: 10.1561/2200000006.

Bielecki, A. and Bielecki, A. (2019). Foundations of artificial neural networks. Models of Neurons and Perceptrons: Selected Problems and Challenges, 15–28.

Born, J., Manica, M., Oskooei, A., Cadow, J., Markert, G. and Martínez, M.R. (2021). PaccMannRL: De novo generation of hit-like anticancer molecules from transcriptomic data via reinforcement learning. Iscience, 24(4).

Bravo À., Piñero J., Queralt-Rosinach N., Rautschka M., Furlong L.I. et al. Extraction of relations between genes and diseases from text and large-scale data analysis: implications for translational research. BMC Bioinform. 2015; 16(1): 1–17.

Calzolari D. Search algorithms as a framework for the optimization of drug combinations. PLoSComput. Biol. 2008; 4:e1000249.

Chan H.S. Advancing drug discovery via artificial intelligence. Trends Pharmacol. Sci. 2019; 40(8): 592–604.

Chawla, S., Rockstroh, A., Lehman, M., Ratther, E., Jain, A., Anand, A., ... and Sengupta, D. (2022). Gene expression based inference of cancer drug sensitivity. Nature communications, 13(1), 5680.

Cheng, F., Lu, W., Liu, C., Fang, J., Hou, Y., Handy, D.E., ... and Loscalzo, J. (2019). A genome-wide positioning systems network algorithm for in silico drug repurposing. Nature communications, 10(1), 3476.

Costa P.R., Acencio M.L., Lemke N. BMC Genomics. Vol. 11. Springer; 2010. A machine learning approach for genome-wide prediction of morbid and druggable human genes based on systems-level data; pp. 1–15.

Da Silva I.N. Springer; 2017. Artificial Neural Networks.

Deng, Y., Xu, X., Qiu, Y., Xia, J., Zhang, W. and Liu, S. (2020). A multimodal deep learning framework for predicting drug–drug interaction events. Bioinformatics, 36(15), 4316–4322.

Doshi, S. and Chepuri, S.P. (2022). A computational approach to drug repurposing using graph neural networks. Computers in Biology and Medicine, 150, 105992.

Eguida, M., Schmitt-Valencia, C., Hibert, M., Villa, P., Rognan, D. et al. (2022). Target-focused library design by pocket-applied computer vision and fragment deep generative linking. Journal of Medicinal Chemistry, 65(20), 13771–13783.

Feng, P., Wang, Z. and Yu, X. (2017). Predicting antimicrobial peptides by using increment of diversity with quadratic discriminant analysis method. IEEE/ACM Transactions on Computational Biology and Bioinformatics, 16(4), 1309–1312.

Fogel D.B. Factors associated with clinical trials that fail and opportunities for improving the likelihood of success: a review. Contemp. Clin. Trials Commun. 2018; 11: 156–164.

Gautam, V., Gaurav, A., Masand, N., Lee, V.S., Patil, V.M. et al. (2023). Artificial intelligence and machine-learning approaches in structure and ligand-based discovery of drugs affecting central nervous system. Molecular Diversity, 27(2), 959–985.

Gerdes, H., Casado, P., Dokal, A., Hijazi, M., Akhtar, N., Osuntola, R., ... and Cutillas, P. R. (2021). Drug ranking using machine learning systematically predicts the efficacy of anti-cancer drugs. Nature communications, 12(1), 1850.

Goel, M., Raghunathan, S., Laghuvarapu, S. and Priyakumar, U.D. (2021). MoleGuLAR: molecule generation using reinforcement learning with alternating rewards. Journal of Chemical Information and Modeling, 61(12), 5815–5826.

Gong Z., Hu G., Li Q. et al. Compound libraries: recent advances and their applications in drug discovery. Curr. Drug Discov. Technol. 14(4), 216–228 (2017).

Grisoni, F., Moret, M., Lingwood, R. and Schneider, G. (2020). Bidirectional molecule generation with recurrent neural networks. Journal of chemical information and modeling, 60(3), 1175–1183.

Hänggi M., Moschytz G.S., editors. Cellular neural networks: analysis, design and optimization. Springer Science & Business Media; 2000 Aug 31.

Harrer S., Shah P., Antony B., Hu J. Artificial intelligence for clinical trial design. Trends in pharmacological sciences. 2019 Aug 1; 40(8): 577–91.

Hassanzadeh P. The significance of artificial intelligence in drug delivery system design. Adv. Drug Delivery Rev. 2019; 151: 169–190.

He, H., Chen, G. and Yu-Chian Chen, C. (2022). 3DGT-DDI: 3D graph and text based neural network for drug–drug interaction prediction. Briefings in Bioinformatics, 23(3), bbac134.

Hong, S.H., Ryu, S., Lim, J. and Kim, W.Y. (2019). Molecular generative model based on an adversarially regularized autoencoder. Journal of chemical information and modeling, 60(1), 29–36.

Hutson M. 2018. Artificial intelligence faces reproducibility crisis.

Hutter, M.C. (2022). Differential Multimolecule Fingerprint for Similarity Search—Making Use of Active and Inactive Compound Sets in Virtual Screening. Journal of Chemical Information and Modeling, 62(11), 2726–2736.

Joddar, B., Natividad-Diaz, S.L., Padilla, A.E., Esparza, A.A., Ramirez, S.P., Chambers, D.R. and Ibaroudene, H. (2022). Engineering approaches for cardiac organoid formation and their characterization. Translational Research.

Jumper J., Evans R., Pritzel A., Green T., Figurnov M., Ronneberger O., Tunyasuvunakool K., Bates R., Žídek A., Potapenko A., et al. Highly accurate protein structure prediction with alphafold. Nature. 2021; 596(7873): 583–589.

Keshavarzi Arshadi, A., Webb, J., Salem, M., Cruz, E., Calad-Thomson, S., Ghadirian, N., ... and Yuan, J.S. (2020). Artificial intelligence for COVID-19 drug discovery and vaccine development. Frontiers in Artificial Intelligence, 65.

Khalil Zadeh N., Sepehri M.M., Farvaresh H. Intelligent sales prediction for pharmaceutical distribution companies: a data mining based approach. Math. Probl. Eng. 2014; 2014.

Khan J.Y., Khondaker M.T.I., Hoque I.T., Al-Absi H.R., Rahman M.S., Guler R., Alam T., Rahman M.S. Toward preparing a knowledge base to explore potential drugs and biomedical entities related to Covid-19: automated computational approach. JMIR Med. Inform. 2020; 8(11).

Krasoulis, A., Antonopoulos, N., Pitsikalis, V. and Theodorakis, S. (2022). DENVIS: scalable and high-throughput virtual screening using graph neural networks with atomic and surface protein pocket features. Journal of Chemical Information and Modeling, 62(19), 4642–4659.

Lamberti M.J. A study on the application and use of artificial intelligence to support drug development. Clin. Ther. 2019; 41: 1414–1426.

Li X. Prediction of synergistic anticancer drug combinations based on drug target network and drug induced gene expression profiles. Artif. Intell. Med. 2017; 83: 35–43.

Li Z. KinomeX: a web application for predicting kinome-wide polypharmacology effect of small molecules. Bioinformatics. 2019; 35: 5354–5356.

Li, T., Tong, W., Roberts, R., Liu, Z., Thakkar, S. et al. (2021). DeepCarc: Deep learning-powered carcinogenicity prediction using model-level representation. Frontiers in artificial intelligence, 4, 757780.

Lindevall, M. (2021). Artificial Intelligence in the Pharmaceutical Industry: Manufacturing, Product Development, and Quality Control-A Systematic Review.

Lu, C., Liu, S., Shi, W., Yu, J., Zhou, Z., Zhang, X., ... and Wang, Y. (2022). Systemic evolutionary chemical space exploration for drug discovery. Journal of Cheminformatics, 14(1), 19.

Luo, H., Wang, J., Li, M., Luo, J., Peng, X., Wu, F.X. and Pan, Y. (2016). Drug repositioning based on comprehensive similarity measures and bi-random walk algorithm. Bioinformatics, 32(17), 2664–2671.

Madhukar, N.S., Khade, P.K., Huang, L., Gayvert, K., Galletti, G., Stogniew, M., ... and Elemento, O. (2019). A Bayesian machine learning approach for drug target identification using diverse data types. Nature communications, 10(1), 5221.

Marr B. The 9 biggest technology trends that will transform medicine and healthcare in 2020 (2019).

Maziarka, Ł., Pocha, A., Kaczmarczyk, J., Rataj, K., Danel, T. and Warchoł, M. (2020). Mol-CycleGAN: a generative model for molecular optimization. Journal of Cheminformatics, 12(1), 1–18.

Medsker L., Jain L.C. CRC Press; 1999. Recurrent Neural Networks: Design and Applications.

Mehta C.H. Computational modeling for formulation design. Drug Discovery Today. 2019; 24: 781–788.

Muller A.T., Hiss J.A., Schneider G. Recurrent Neural Network Model for Constructive Peptide Design. J. Chem. Inf. Model. 2018; 58: 472–479. doi: 10.1021/acs.jcim.7b00414.

Nakamura, T., and Sasano, T. (2022). Artificial intelligence and cardiology: Current status and perspective. Journal of Cardiology, 79(3), 326–333.

Nicolaou C.A., Brown N. Multi-objective optimization methods in drug design. Drug Discovery Today. Technologies. 2013; 10(3): e427–e435.

Olah, M., Mracec, M., Ostopovici, L., Rad, R., Bora, A., Hadaruga, N., ... and Oprea, T.I. (2004). WOMBAT: world of molecular bioactivity. Chemoinformatics in drug discovery, 1.

Olayan, R.S., Ashoor, H. and Bajic, V.B. (2018). DDR: efficient computational method to predict drug–target interactions using graph mining and machine learning approaches. Bioinformatics, 34(7), 1164–1173.

Olivecrona M., Blaschke T., Engkvist O., Chen H. Molecular de-novo design through deep reinforcement learning. J. Cheminform. 2017; 9: 48. doi: 10.1186/s13321-017-0235-x.

Park K. A review of computational drug repurposing. Transl. Clin. Pharmacol. 2019; 27: 59–63.

Peng, J., Wang, Y., Guan, J., Li, J., Han, R., Hao, J., ... and Shang, X. (2021). An end-to-end heterogeneous graph representation learning-based framework for drug–target interaction prediction. Briefings in bioinformatics, 22(5), bbaa430.

Popova, M., Isayev, O. and Tropsha, A. (2018). Deep reinforcement learning for de novo drug design. Science advances, 4(7), eaap7885.

Putin E. Reinforced adversarial neural computer for de novo molecular design. J. Chem. Inform. Modeling. 2018; 58: 1194–1204.

Raies, A., Tulodziecka, E., Stainer, J., Middleton, L., Dhindsa, R.S., Hill, P., ... and Vitsios, D. (2022). DrugnomeAI is an ensemble machine-learning framework for predicting druggability of candidate drug targets. Communications Biology, 5(1), 1291.

Ribeiro da Cunha, B., Fonseca, L.P. and Calado, C.R. (2021). Simultaneous elucidation of antibiotic mechanism of action and potency with high-throughput Fourier-transform infrared (FTIR) spectroscopy and machine learning. Applied Microbiology and Biotechnology, 105, 1269–1286.

Rouse M. 2017. IBM Watson Supercomputer. https://searchenterpriseai.techtarget.com/definition/IBM-Watson-supercomputer. Accessed 13 October 2020.

Sadeghi, S., Lu, J. and Ngom, A. (2022). An Integrative Heterogeneous Graph Neural Network–Based Method for Multi-Labeled Drug Repurposing. Frontiers in Pharmacology, 13, 908549.

Samanta, B., De, A., Jana, G., Gómez, V., Chattaraj, P., Ganguly, N. and Gomez-Rodriguez, M. (2020). Nevae: A deep generative model for molecular graphs. Journal of machine learning research, 21(114), 1–33.

Shao, K., Zhang, Y., Wen, Y., Zhang, Z., He, S. and Bo, X. (2022). DTI-HETA: prediction of drug–target interactions based on GCN and GAT on heterogeneous graph. Briefings in Bioinformatics, 23(3), bbac109.

Shi Y., Ren P., Zhang Y., Gong X., Hu M., Liang H. Information extraction from fda drug labeling to enhance product-specific guidance assessment using natural language processing. Front. Res. Metr. Anal. 2021; 6.

Sliwoski G., Kothiwale S., Meiler J., Lowe E.W. Computational methods in drug discovery. Pharmacol. Rev. 2014; 66(1): 334–395.

Song M., Kim M., Kang K., Kim Y.H., Jeon S. et al. Application of public knowledge discovery tool (pkde4j) to represent biomedical scientific knowledge. Front. Res. Metr. Anal. 2018; 3: 7.

Steels L., Brooks R. Routledge; 2018. The Artificial Life Route to Artificial Intelligence: Building Embodied, Situated Agents.

Wan F., Zeng J. Deep learning with feature embedding for compound–protein interaction prediction. bioRxiv. 2016; 2016.

Wang F. Computational screening for active compounds targeting protein sequences: methodology and experimental validation. J. Chem. Inf. Model. 2011; 51: 2821–2828.

Wang, L., Song, Y., Wang, H., Zhang, X., Wang, M., He, J., ... and Cao, L. (2023). Advances of Artificial Intelligence in Anti-Cancer Drug Design: A Review of the Past Decade. Pharmaceuticals, 16(2), 253.

Wang, S., Xu, F., Li, Y., Wang, J., Zhang, K., Liu, Y., ... and Zheng, J. (2021). KG4SL: knowledge graph neural network for synthetic lethality prediction in human cancers. Bioinformatics, 37(Supplement_1), i418–i425.

Wang, Y., Yang, Y., Chen, S. and Wang, J. (2021). DeepDRK: a deep learning framework for drug repurposing through kernel-based multi-omics integration. Briefings in Bioinformatics, 22(5), bbab048.

Wunnava, S., Qin, X., Kakar, T., Sen, C., Rundensteiner, E.A. and Kong, X. (2019). Adverse drug event detection from electronic health records using hierarchical recurrent neural networks with dual-level embedding. Drug safety, 42, 113–122.

Yang, Y., Zheng, S., Su, S., Zhao, C., Xu, J. and Chen, H. (2020). SyntaLinker: automatic fragment linking with deep conditional transformer neural networks. Chemical science, 11(31), 8312–8322.

Yang, Z., Zhong, W., Zhao, L. and Chen, C.Y.C. (2021). ML-DTI: mutual learning mechanism for interpretable drug–target interaction prediction. The Journal of Physical Chemistry Letters, 12(17), 4247–4261.

Yasuo, N. and Sekijima, M. (2019). Improved method of structure-based virtual screening via interaction-energy-based learning. Journal of Chemical Information and Modeling, 59(3), 1050–1061.

Zellweger, M.J., Tsirkin, A., Vasilchenko, V., Failer, M., Dressel, A., Kleber, M.E., ... and März, W. (2018). A new non-invasive diagnostic tool in coronary artery disease: artificial intelligence as an essential element of predictive, preventive, and personalized medicine. EPMA Journal, 9, 235–247.

Zhao, H., Zheng, K., Li, Y. and Wang, J. (2021). A novel graph attention model for predicting frequencies of drug–side effects from multi-view data. Briefings in bioinformatics, 22(6), bbab239.

Zhu, J., Liu, Y., Zhang, Y., Chen, Z., Wu, X. et al. (2021). Multi-attribute discriminative representation learning for prediction of adverse drug-drug interaction. IEEE Transactions on Pattern Analysis and Machine Intelligence, 44(12), 10129–10144.

CHAPTER

14

Artificial Intelligence in Biosciences

Zitin Wali,[1] *Neha Pinky*[1] and *Suhel Parvez*[1*]

1. Introduction

Artificial intelligence (AI) is a concept without a precise definition. In general, AI corresponds to computer programs created to replicate human intelligence to accomplish any activity that a human can. AI in biological sciences (BS) has progressed from the conceptual approach, which involved encoding intricate rules in a programming language to allow robots to carry out synchronized actions (Bhardwaj et al., 2022; Richards et al., 2022a). The game of chess, which has very basic principles but enormous potential results after a single move, is a common instance of symbolic AI. Since the rules are established in this instance, a computer may be configured to consider every scenario before making a choice and select the one resulting in the most favorable outcome. Furthermore, in terms of how they operate and function, symbolic AI and biological intelligence are very different from one another. AI is limited to making decisions and following predetermined rules. On the other hand, biological intelligence can learn on the spot and make judgments according to knowledge gained through experience and observation of objects (Ching et al., 2018a; Hassoun et al., 2022).

Deep learning (DL), which consists of many levels of processing in artificial neural networks (NN) intended for identifying trends and modeling complicated interactions between inputs and outputs, has resulted in significant progress in AI recently. Furthermore, DL has improved the possibilities of computerized synthesis in drug development by predicting protein structure, designing molecules, and identifying macromolecule targets (Figure 1) (Jiménez-Luna et al., 2020a). In the last fifty years, both biology and information technology (IT) have advanced significantly. The number of transistors on a chip shall increase (2X) roughly every two years, according to Moore's law. It is both a result of and a catalyst for its explosive progress (Oliveira, 2019). The past few years have seen tremendous advancements in BS and biotechnology sectors, thanks to advances in sequencing along with additional high–throughput approaches. Large volumes of unorganized, raw data may be efficiently stored, processed, and made quickly accessible using AI–based algorithms. This is a

[1] Department of Toxicology, Jamia Hamdard, India.
* Corresponding author: sparvez@jamiahamdard.ac.in

prerequisite for developing intelligent computer systems that can make complicated decisions. The establishment of an extensive variety of goods and services across several industries, which include biosciences, is made possible by such advancements in the production of data, preservation, and interpretation (Sellwood et al., 2018; Williams et al., 2018). Automating robotic processes appears to have a wide range of uses and could significantly alter healthcare in the recent years to come when combined with machine learning (ML) and natural language processing (NLP) (Figure 1) (Bhatnagar, 2020). Even with the amazing progress we have seen, much work needs to be conducted until AI–based treatment is widely used. To extract valuable knowledge from data, the application of AI may help in understanding intricate connections between various types of undetectable data. Because of this, the application of AI techniques is now frequently seen in BS and is anticipated to grow as this field of study advances in the coming years (Xu et al., 2021).

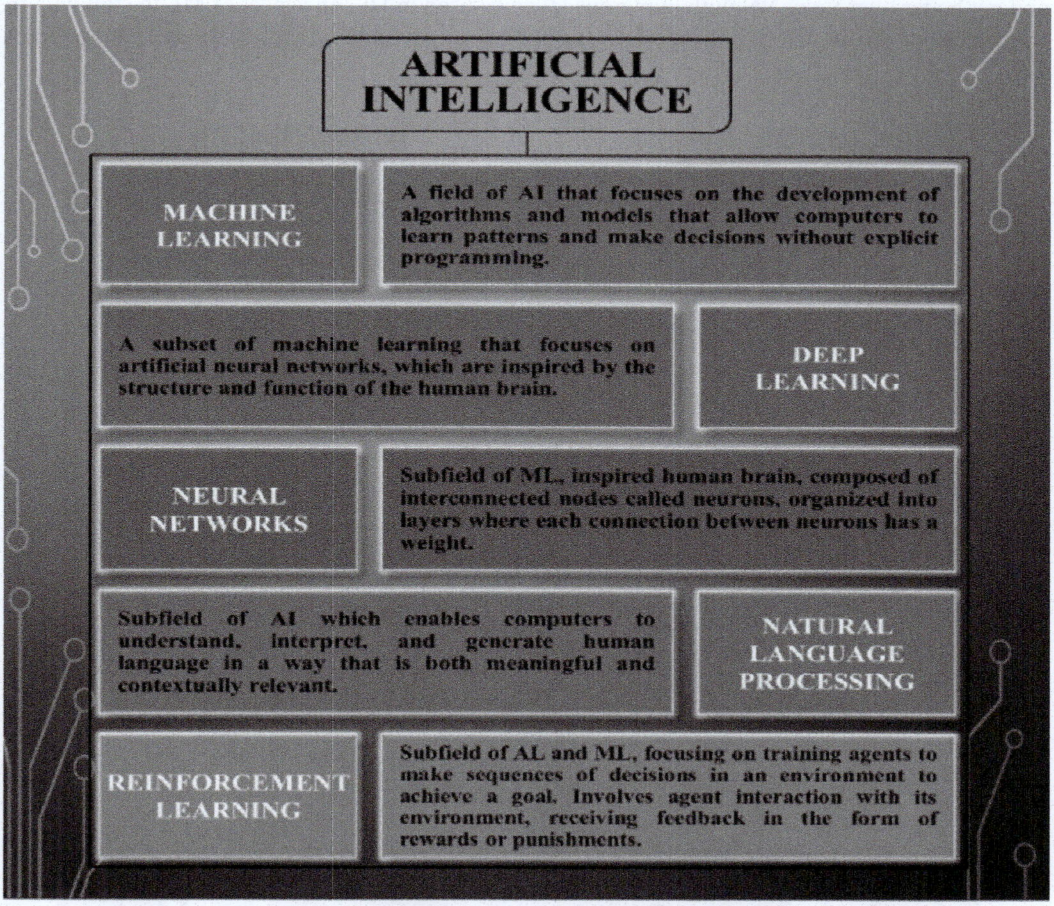

Fig. 1 Representative image is an informative graphic about Artificial Intelligence (AI) and its various subfields. Different AI subfields: Machine Learning, Deep Learning, Neural Networks, Natural Language Processing, and Reinforcement Learning have been described briefly.

2. Artificial Intelligence

The multidisciplinary computational discipline of AI aims to build machines that can carry out operations that are normally done under human expertise. Training, logic, solving issues, vision, language comprehension, and even creative thinking are all included in these tasks. The primary objective of AI is to create robots that can mimic human cognitive processes, enabling them to perform complicated tasks and adjust to varying circumstances. Theoretical debates about the definition of intellect and the viability of building artificial beings may be found in ancient history, where AI first emerged. On the other hand, the mid-1900s saw the start of the current AI age. John McCarthy first used the phrase "artificial intelligence" in 1959 when he organized the Dartmouth Conference, which launched AI as a separate academic discipline (Dai et al., 2020; Zhai, et al., 2021). AI can be divided into two main categories, general or strong AI and narrow or weak AI. Systems developed and honed for a specific purpose, such as image categorization or detection of speech, are referred to as narrow AI systems. Conversely, general AI seeks to create machines that are reminiscent of humans where they can comprehend, acquire, and utilize expertise in a variety of contexts. AI has advanced significantly due to advances in processing capacity, NN, and ML. As a branch of AI, ML concentrates on creating algorithms that let computers gain insights into the data and get better over time. NN with numerous layers is used in DL, an aspect of ML to stimulate the framework of the human brain (Figure 2) (Enholm et al., 2022; Fjelland, 2020; Mikalef & Gupta, 2021).

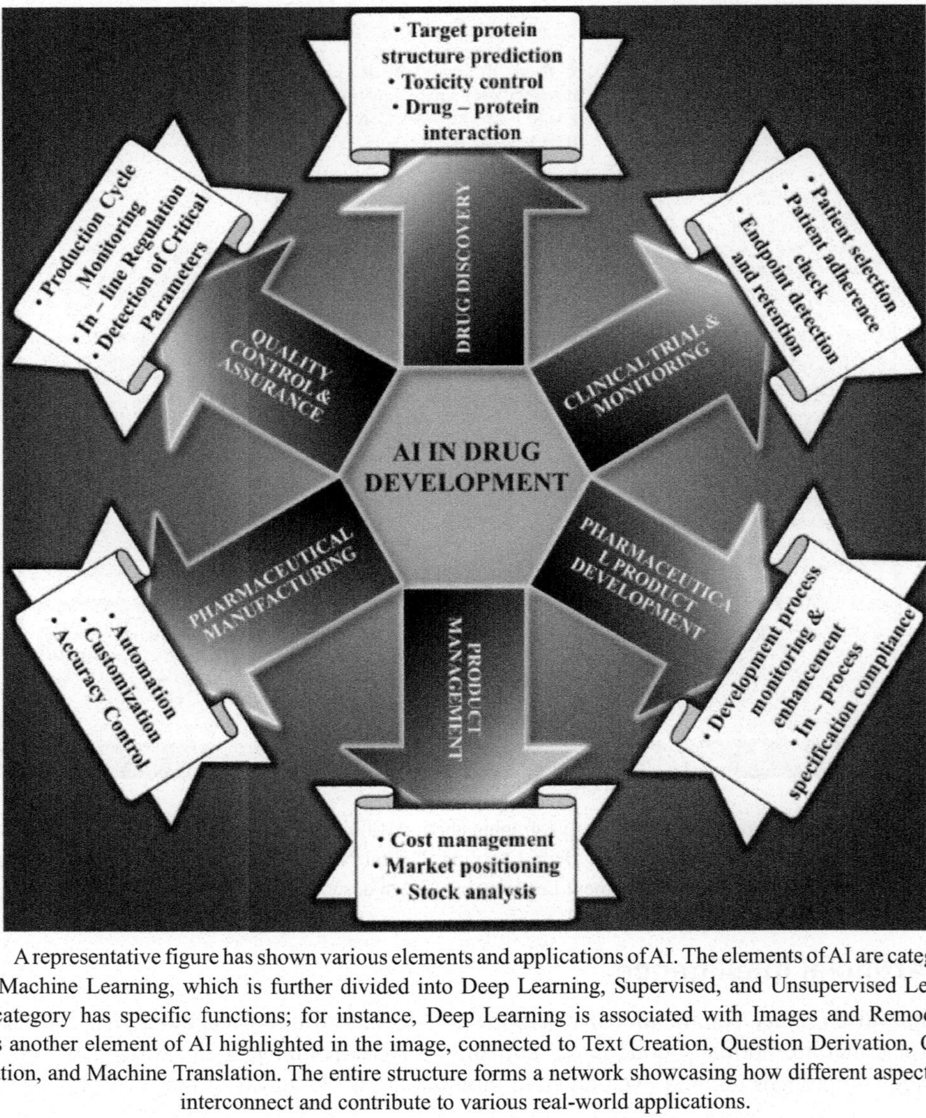

Fig. 2 A representative figure has shown various elements and applications of AI. The elements of AI are categorized under Machine Learning, which is further divided into Deep Learning, Supervised, and Unsupervised Learning. Each category has specific functions; for instance, Deep Learning is associated with Images and Remodelling. NLP is another element of AI highlighted in the image, connected to Text Creation, Question Derivation, Context Derivation, and Machine Translation. The entire structure forms a network showcasing how different aspects of AI interconnect and contribute to various real-world applications.

One important area of AI is ML, which focuses on creating algorithms that let computers acquire knowledge from data. In ML, supervised, unsupervised, and reinforcement learning are popular techniques. The development of DL, a branch of ML, that uses multilayered NN, has made progress in the identification of intricate patterns and making decisions easier (Webb et al., 2021; Zhai, et al., 2021). Another important field is NLP, which gives computers the ability to comprehend, decode, and produce human language. Computer vision has applicability, ranging from image interpretation to self–driven cars, by enabling robots to assess and draw judgments via imagery. Robotics uses AI to give machines the ability to sense their surroundings and make wise judgments, transforming sectors, notably the commercial industry and healthcare. Computer vision, robotics, reinforcement learning, and cognitive computing are other key aspects in the field of AI (Figure 3) (Brown et al., 2020; Gervini and Khanal, 2019; He et al., 2020; Khurana et al., 2023; Nian et al., 2020; Yang et al., 2019).

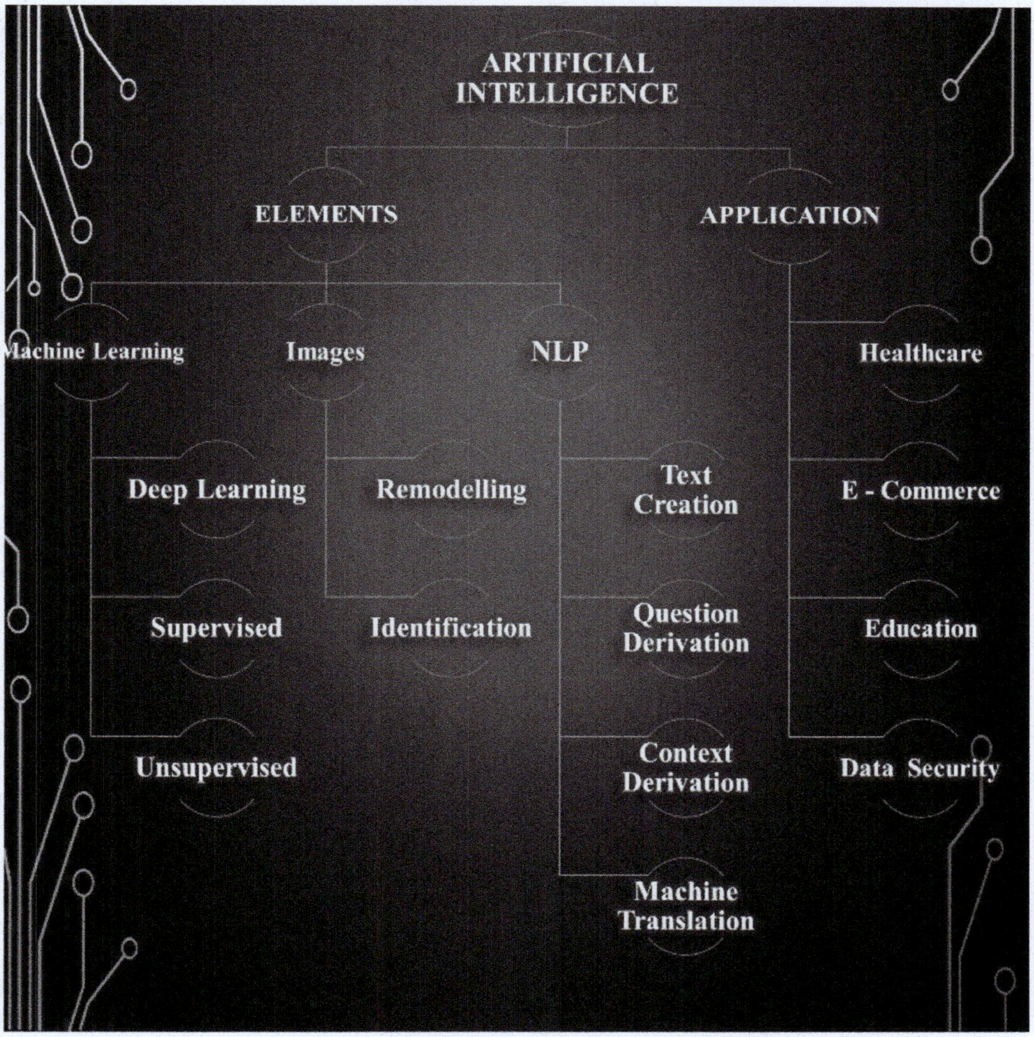

Fig. 3 A schematic image is an infographic that illustrates various components and applications of AI. These include Robotics, Perception, Learning, Natural Language, Reasoning, multi-agent systems, and their connections to fields like Health and Engineering. Each of these categories is detailed with specific examples or elements that fall under them.

3. Artificial Intelligence in Drug Discovery

The use of ML techniques in computer–aided drug development has grown significantly (Jiménez-Luna et al., 2021a). From hit–to-lead, the development of pharmaceuticals remains an arduous undertaking involving the concurrent optimization of multiple parameters, ranging from chemical activity and efficacy maximization to toxicity and side effect minimization. Thanks to recent advances in AI, medications can now be effectively designed in silico before being chemically synthesized and subjected to clinical trials (Lo et al., 2020a). Various fields in the process of drug discovery require AI approaches such as finding materials and consumables, chemical synthesis, data mining and ontology, biological research, lead and biomarker discovery, preclinical/clinical development, pharmacovigilance, and drug repurposing. AI has been incorporated to develop these factors to enhance the pharmaceutical industry. Every phase aims to demonstrate the efficacy of the medication (Dara et al., 2022). The data generated using "omics" and "smart automation tools" is analyzed and evaluated accurately, providing more windows for improving the industry. This data can be in the form of text, images, biometrics, or assay information. The labor and energy needed to get medicine from the laboratory to the patient's bed should also be significantly decreased by employing AI to concurrently solve the pharmacokinetic and pharmacodynamic concerns. In this sense, the application of AI has emerged as a crucial instrument to speed up the procedure of finding new drugs (Figure 2) (Lo et al., 2020b).

The following are a few crucial aspects of AI in drug discovery:

1. **Target Identification** Enormous genome, proteomes, and transcriptome datasets can be analyzed by AI algorithms to find possible therapeutic targets. For example, IBM Watson for therapeutic research analyses patents, clinical trial results, and scientific research to find possible therapeutic targets using ML and NLP. AI technologies help scientists identify potential biological targets that are critical to understanding disease mechanisms by sorting through tremendous quantities of data (Bender and Cortés-Ciriano, 2021).
2. **Quantum Computing** Because quantum computers can replicate intricate chemical reactions, they are becoming more and more popular in drug discovery. Pharmaceutical firms and quantum computing startups, such as D–wave, work together to investigate quantum methods for modeling chemical structures and streamlining drug discovery procedures. Quantum computers can solve computationally complex issues that traditional computers are unable to handle, opening up new possibilities for drug development and optimization (Zinner et al., 2021).
3. **Predictive Toxicology** Here, ML algorithms are used to forecast the toxicity and possible negative impacts of drug candidates, lowering the possibility of unfavorable reactions at a subsequent phase of development. A joint endeavor by multiple federal agencies in the United States, Tox21 uses AI and high-throughput screening to evaluate the toxicity of medicinal compounds, assisting in the discovery of safer therapeutic options (Tice et al., 2013).
4. **Clinical Trial Optimization** AI is essential for the formulation of protocols and choosing recipients in clinical trial optimization. A technology startup called Tempus analyses clinical and genomic data using ML to find people who might be good candidates for clinical trials. By doing this, studies are done with cohorts that are predicted to respond favorably, which enhances trial outcomes. AI also helps with the formulation of protocols by optimizing research parameters based on past trial data, which increases their effectiveness (Askin et al., 2023).
5. **Drug Repurposing** AI has also made advances in medicine repurposing-the procedure for finding novel applications for already approved medications. Benevolent AI uses AI to examine the results of clinical trials, biological literature, and more data to find possibilities for repurposing current medications for alternative applications. This strategy uses molecules that are already on the market and have established safety profiles to accelerate the development process (Chen et al., 2021).

6. **Personalized Medicine** AI programs that evaluate unique patient records, including genetic data, to customize medication regimes, are becoming more and more beneficial for personalized medicine. AI is used by Foundation Medicine, a business that specializes in molecular information approaches, to understand genomic data and help healthcare providers choose individual therapies according to every individual's distinct genetic composition (Rashidi et al., 2019).
7. **Cheminformatics** Cheminformatics uses computer algorithms to handle and analyze molecular data, which is critical to drug discovery. The preservation, access, and evaluation of chemical information are made easier by chemicals and visualization, which are essential components for the same. Chemical formats such as SMILES (Simplified Molecular Input Line Entry System), InChI (International Chemical Identifier), MOL (MDL Molfile), SDFile, and SMARTS (Smiles ARbitrary Target Specification) are used. This data is then generated in chemical representations such as 2D/3D structures, fingerprints, and graph-based – or docking representations (Lovrić et al., 2019)
8. **De novo Drug Design** The creation of new molecules from scratch that exhibit the desired pharmacological attributes, is one of the most difficult computer–aided drug discovery missions because of the large number of chemical space molecules that resemble drugs (Jiménez-Luna et al., 2021b). De novo drug designing includes generative models such as variational autoencoders (VAEs) and generative adversarial networks (GANs). They can generate novel molecules that are chemically feasible and have the desired properties. Deep reinforcement learning (DRL) and reinforcement learning (RL) algorithms can guide the generation and optimization of molecular structures by iteratively improving them through trial and error. Graph NN is well suited for handling molecular graphs, where atoms are nodes and bonds are edges (Blaschke et al., 2020; Button et al., 2019; Humbeck et al., 2018).

4. Artificial Intelligence in Medical Diagnostics

The goal of the fast–developing computational domain i.e., AI is to build systems that are capable of activities that usually necessitate human intelligence. AI is developing at an accelerated rate, which opens possibilities for its use in medical settings and might completely transform healthcare delivery. To offer healthcare professionals the data and tools they need to successfully integrate AI into medical treatment, it is critical to record and share data concerning the technology's application for the same (Alowais et al., 2023; Suleimenov et al., 2020).

Plenty of research has proven that AI algorithms are either as good as or superior to humans at several tasks. An important example in this field includes the analysis of health-related imagery and the correlation of indicators and biomarkers from electronic medical records (EMRs) with the diagnosis and prognosis w.r.t. the ailment (Miller & Brown, 2018). One major factor driving the adoption of AI applications is the potential reduction in expenses for the medical industry that AI can provide. By 2026, AI is predicted to reduce US healthcare expenditure by USD 150 billion, annually (Bohr & Memarzadeh, 2020). The shift in the healthcare system from a reactive to a proactive one, emphasizing managing one's health over illness therapy, is largely responsible for these expenditure reductions. It is anticipated that this will lead to a decrease in hospital stays, physician appointments, and therapy. AI–based technology will play a significant role in assisting individuals and maintaining their health through ongoing guidance and surveillance. It will also guarantee faster diagnosis, and customized therapies and enhance subsequent visits. The hardest phase of all is diagnosing an illness, which is crucial for a healthcare provider to do before coming to a judgment. The diagnosis procedure may be quite laborious and intricate. The most basic diagnosis process involves four steps i.e., (A) Medical history analysis, (B) Physical examination, (C) Performing diagnostic tests and (D) Drawing Conclusions. Healthcare professionals gather empirical information to determine a patient's illness to reduce the complexity associated with medical diagnosis. If the diagnosis procedure is flawed, the patient may have major health problems and have proper treatment delayed or neglected. Sadly, not every medical professional is an expert in every area of the profession. Therefore, an autonomous diagnosis system that benefited from both human experience and machine precision was required (Kaur et al., 2020).

Computer systems now possess new capabilities that were previously unimaginable thanks to ML. It is an area of AI that allows machines to learn from examples, allowing them to assess how various models function in ML without the need for human judgment. The most basic diagnosis process using ML involves five steps i.e., (Marsland, 2014; Suzuki, 2017)

1. Compile test results and patient information.
2. Collect features to select traits that are helpful in diagnosis.
3. Choose and handle that dataset after choosing those attributes.
4. The pre–process dataset can be subjected to a variety of classification techniques to assess the precision of the condition's diagnosis.
5. The evaluation of several classifiers against one another to determine which performs best and has the highest accuracy.

5. Artificial Intelligence in Bioinformatics

AI technologies have significantly advanced bioinformatics, the field that uses computational tools and procedures to study biological data. Among numerous additional uses, this multidisciplinary field is essential for comprehending genetic variance, decoding complicated biological mechanisms, and speeding up the development of new drugs (Choi et al., 2023). Finding significant trends, connections, and biological knowledge from the enormous and varied biological data produced by clinical and scientific studies is the main objective of bioinformatics. Scientists use computational algorithms to help them comprehend the functional relevance of genetic data by performing tasks including gene estimation, motif exploration, and sequence alignment. The 3D arrangements of biomolecules, including proteins and nucleic acids, are the subject of structural biology. Protein designs are predicted, protein–ligand interactions are examined, and the molecular mechanisms that underlie diverse biological operations are comprehended through the applications of bioinformatics tools and simulations (Kilanko, 2023).

AI–driven methods for data analysis, like classification and grouping algorithms, facilitate the identification of gene activity patterns linked to illness, developmental phases, or environmental factors. To obtain a comprehensive understanding of biological systems, systems biology, attempts to combine and simulate large–scale biological data. AI techniques including network evaluation and ML are applied to rebuild intricate biological networks and forecast the impact of perturbations in a single element on other components. Numerous bioinformatics operations, like evaluation of gene expression, protein structure estimation, and drug development, have been effectively completed with the use of ML techniques, including support vector machines (SVMs), random forests, and NN (Gupta et al., 2023; Liang et al., 2023) (Figure 4).

The primary goal of bioinformatics is to identify the distinctive adaptability and variations among various populations, and to gain a deeper understanding of the disorder, resulting from a genetic problem. Usually, a bioinformatics approach involves the following four main steps: (Majhi & Paul, 2020a)

- Gathering statistical information from the various biological subjects.
- Creating an algorithm or computational tool that utilizes the characteristics retrieved from the biological subject.
- Applying the created method to a simulation using the statistical data.
- Testing the method using the simulation output.

The DENDRAL project was the first to use an autonomous data collection system from the mass spectrometric repository and a task–dependent information-based challenge to recognize organic compounds. Two decades later, high-resolution NMR was used to identify the initial 3D structure of proteins as a part of the PROTEAN project. Here, too, several information–based systems that were capable of deciphering crystallographic data were effectively constructed. The ML methodology was first implemented in a practical setting with the Meta–DENDRAL project (Majhi & Paul, 2020b).

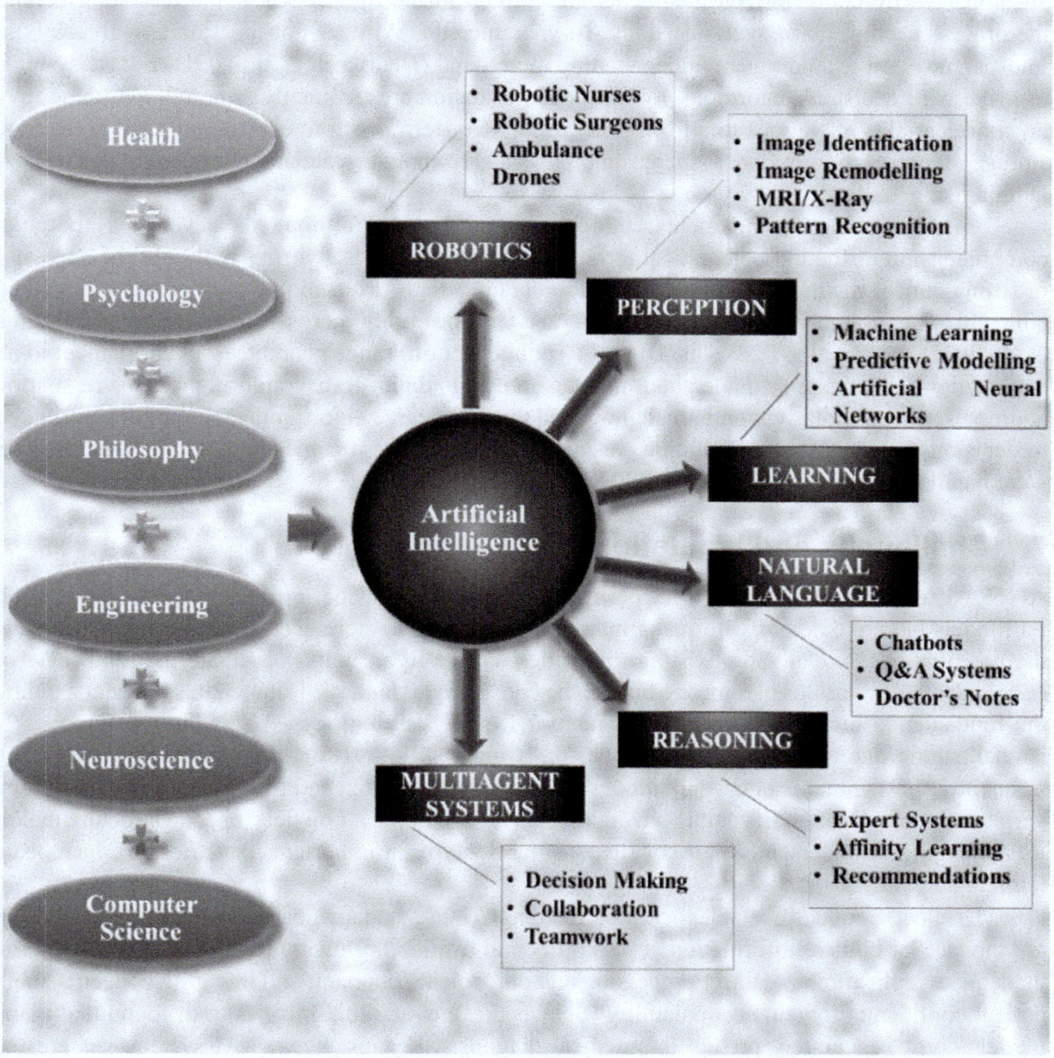

Fig. 4 Representative images explain the role of AI in drug development. It outlines various stages and aspects where AI can be instrumental in drug development, describing the importance of AI and the advancement it provides.

6. Ethical Considerations in Artificial Intelligence

Unprecedented prospects for creativity and industry transformation are being provided by AI, which has truly evolved as a revolutionary force. Yet, there are serious ethical questions raised by its incorporation into many facets of society. Justice, transparency, responsibility, confidentiality, prejudice, and the wider societal impacts of AI–driven judgments are a few of the intricate concerns that AI presents (Deng et al., 2023; Jobson et al., 2022). The problem of prejudice or bias represents one of the main ethical issues with AI and ML. When AI systems use prior information to gain insights, they may reinforce, or perhaps worsen, pre–existing biases in the data. Predictive algorithms that unjustly target neighborhoods or computerized employment systems that discriminate towards demographics are just two examples of how these biases can appear (Fjeld et al., 2020; John, 2022; O'Reilly-Shah et al., 2020). An AI system's internal operations are frequently convoluted and unclear. There may be serious ethical issues with the absence of transparency. Individuals who are impacted by computerized judgments are entitled to the right to know the reasoning behind the process used to make such decisions. To guarantee that AI does not function as a 'black box' transparency and technological understanding are crucial (Henriksen et al., 2021). For AI systems to make wise decisions, enormous volumes of information are frequently required. Yet, there may be confidentiality violations due to this utilization. Concern over the ethical and legal collection, storage and utilization of data is increasing, particularly about private and confidential data (Díaz-Domínguez, 2020).

By tackling these fundamental ethical issues, there is a hope to develop a thorough comprehension of the intricate ethical terrain in AI and ML. These worries highlight the necessity of strong ethical regulations and standards to guarantee that AI techniques are created and applied ethically, encouraging equality, openness, and responsibility across their existence (Bogina et al., 2022; Naik et al., 2022). The Digital Republic Act of France guarantees an entitlement to an explanation for judgments made about the person using administrative algorithms. This regulation addresses several topics which include how and to what degree algorithmic computing contributed to decision-making; what information was analyzed and where it came from; and how parameters were handled and weighed (Lo Piano, 2020).

7. Challenges and Limitations

AI–based technologies have not yet permeated daily life, considering their enormous potential. AI models have the potential to make many biological fields more accessible, but they also have the potential to make already existing disparities worse. To accurately evaluate the robustness of certain deep NN, various parameters need to be considered (Hanna & Hanna, 2022). Understanding the fundamental biological processes of many AI models in biology – particularly DL models – is challenging due to a lack of comprehension (Ching et al., 2018b). since scientific information frequently shows notable inter–individual variations, developing generalized models that work for a variety of populations is difficult (Matzaraki et al., 2017). Some AI models' opaque structure makes it challenging to understand their conclusions, particularly in important industries like healthcare (Holzinger et al., 2019).

The most frequent obstacles to implementing AI in standard healthcare and dental settings are the lack of clinical information accessibility and availability owing to organizational policies, inadequate replication in dataset processing and outcome assessment, and lingering concerns about patient confidentiality and accountability (Schwendicke et al., 2020). It has been found that many models do not accurately predict the medical diagnosis. For instance, an AI system was created that used NLP in radiological data to interpret and categorize chest X–rays (Paeng et al., 2017). Additionally, it has been noted that although ML techniques may outperform alternative research methods aimed at creating models for illness risk estimation, the information's layout may be more intricate. Also, the size of the data may affect how much processing time ML techniques demand (Goldstein et al., 2016). It is imperative to recognize that the applications of AI–based techniques may not invariably result in random classification or more accurate forecasting techniques. AI is a tool that ought to be used in the right situation to answer a relevant query or deal with a big problem (Hashimoto et al., 2020).

8. Future Prospects

AI is affecting all scientific fields by offering new instruments for high–dimensional data analysis. The incorporation of AI with BS can revolutionize research and innovations. AI is establishing

Table 1 The representative table presents a diverse range of medical research studies leveraging AI's applications in neurological, oncological, cardiac, respiratory, and infectious diseases. Investigating novel technologies, for instance, wearable biomarkers and AI–assisted diagnostics, the studies aim to enhance detection, prediction, and therapeutic response in various conditions, contributing to modern healthcare.

S.No.	*Disease*	*Title*	*Condition*	*Location*	*Status*	*Last Verified*
1	Neurological	A Novel Wearable Digital Biomarker for Detecting Changes in Multiple Sclerosis (MS) Condition	Multiple Sclerosis	• The Ottawa Hospital Ottawa, Ontario, Canada, • The Royal London Hospital London, United Kingdom	Recruiting	2023
2	Neurological	Evaluating the Efficacy of Artificial Intelligence –Computer-Aided Diagnosis Software That Assists in Determining Whether or Not to Conduct Amyloid PET for the Diagnosis of Alzheimer's Disease by Predicting Amyloid PET Positivity in Mild Cognitive Impairment Patients	Alzheimer's Disease, Mild Cognitive Impairment	Chonnam National University Hospital Gwangju, Korea	Completed	2022
3	Neurological	MAP THE SMA: a Machine-learning Based Algorithm to Predict THErapeutic Response in Spinal Muscular Atrophy	Spinal Muscular Atrophy	Fondazione Policlinico Universitario Agostino Gemelli IRCCS Roma, Italy	Recruiting	2023
4	Oncological	Artificial Intelligence in Breast Cancer Screening Programs in Córdoba (AITIC)	Breast Cancer	Hospital Universitario Reina Sofia Córdoba, Spain	Recruiting	2023
5	Oncological	Predictors of Ovarian Cancer and Endometrial Cancer for Artificial-Intelligence-Based Screening Tools	Ovarian Cancer, Endometrial Cancer and Hyperplasia	Hasanuddin University Hospital Makassar, South Sulawesi, Indonesia	Recruiting	2023
6	Oncological	Development of Artificial Intelligence Models for Segmentation and Characterization of Prostate Cancer: A Single–Centered Retrospective Observational Study.	Prostate Cancer	Irccs San Raffaele Milano, Italy	Completed	2023
7	Cardiac	Electrocardiogram-based Artificial Intelligence-assisted Detection of Heart Disease (ECG-AID)	Atrial Fibrillation, Structural Heart Disease	• Corewell Health Grand Rapids, Michigan, United States • TriHealth Cincinnati, Ohio, United States • Geisinger Medical Center Danville, Pennsylvania, United States	Recruiting	2023

Contd.

Table 1 *Contd.*

S.No.	*Disease*	*Title*	*Condition*	*Location*	*Status*	*Last Verified*
8	Cardiac	Detection of Heart Conditions with Single Lead ECG Using Artificial Intelligence (DAVID1)	Left Ventricular Dysfunction	• Los Alamitos Cardiovascular Los Alamitos, California, United States • MedStar Cardiovascular Network Washington, District of Columbia, United States • Albert Einstein Medical Center Philadelphia, Pennsylvania, United States	Completed	2022
9	Cardiac	MRI-based Artificial Intelligence Software (NNS-SOT) for Onset Time Estimation in Lesions with Acute Ischemic Stroke	Acute Ischemic Stroke	Asan Medical Centre Seoul, Korea	Completed	2023
10	Respiratory	Artificial Intelligence-assisted Evaluation of Pulmonary Hypertension (AIPHY)	Pulmonary Hypertension, Pulmonary Arterial Hypertension	Fuwai Hospital, Chinese Academy of Medical Sciences, and Peking Union Medical College Beijing, Beijing, China	Recruiting	2023
11	Respiratory	Point of Care Ultrasound Lung Artificial Intelligence (AI) Validation Data Collection Study (AI)	Pleural Effusion, Lung Consolidation	• Arizona Emergency Medical Research Center Tucson, Arizona, United States • University of Colorado Hospital Aurora, Colorado, United States • Yale School of Medicine New Haven, Connecticut, United States • Beth Israel Deaconess Medical Center Boston, Massachusetts, United States • Prisma Health Greenville Memorial Hospital Greenville, South Carolina, United States	Completed	2023
12	Infectious	Application of Artificial Intelligence on the Diagnosis of Helicobacter Pylori Infection and Premalignant Gastric Lesion	Helicobacter Pylori Infection, Premalignant Lesion	Yi-Chia Lee Taipei, Taiwan	Recruiting	2023
13	Infectious	COVID-19 Infection and Machine Learning Using Artificial Intelligence (AI)	COVID – 19, COVID, SARS – CoV 2	East Suffolk and North Essex NHS Foundation Trust Ipswich, United Kingdom	Completed	2023

Contd.

Table 1 *Contd.*

S.No.	*Disease*	*Title*	*Condition*	*Location*	*Status*	*Last Verified*
14	Infectious	Evaluation of the AudibleHealth Dx AI/ML-Based Dx SaMD Using FCV-SDS in the Diagnosis of COVID-19 Illness: Clinical Validation	SARS-CoV-2 Infection	Sunrise Research Institute, Sunrise, Florida, United States	Completed	2022
15	Infectious	Artificial Intelligence for Diagnosing Drug-Resistant Tuberculosis	MDR Tuberculosis, Resistance to Tuberculostatic Drugs	• Kanudjoso Djatiwibowo General Hospital Balikpapan, East Kalimantan, Indonesia • Tarakan General Hospital Tarakan, North Kalimantan, Indonesia • Labuang Baji General Hospital Makasar, South Sulawesi, Indonesia	Completed	2020

itself as a transformative force in several disciplines. AI is turning into a vital tool for evaluating complicated biological data, forecasting molecular reactions, and quickening the rate of scientific discoveries as technology develops (Richards et al., 2022b). Although the full effects of advances in AI are still unknown, it appears probable that applications of AI will rank among the greatest educational technology concerns for the next 20 years. AI–based resources and services offer a great deal of promise to help administrators, teachers, and students at every stage of the student lifecycle (Zawacki-Richter et al., 2019). AI is transforming the medical field by simplifying movement from a universal to a precision medicine model. To find individualized treatment alternatives, ML algorithms can examine large datasets, such as genetic data, medical records, and external variables. By reducing side effects and enhancing patient outcomes, predictive models assist doctors in making well–informed judgments regarding the most effective medicines (Topol, 2019).

Conventional drug discovery is an expensive and laborious procedure. By assessing biological data to forecast possible potential drugs, finding new targets, and optimizing molecular frameworks, AI systems can significantly accelerate this process. By speeding up the drug development process, this strategy may save costs and increase the likelihood of new medications being successfully introduced to the market (Jiménez-Luna et al., 2020b). Since technologies for high–throughput sequencing became available, the amount of information available in the domain of genomics has increased dramatically. AI techniques are essential for organizing and deciphering this vast amount of genetic data. ML models can recognize trends in DNA sequences, forecast the actions of genes, and find genetic variants linked to certain illnesses. The combination of AI with bioinformatics improves our comprehension of the complex interactions found in the genome, leading to advances in molecular biology and genetics (Zhang et al., 2022). The application of AI to BS represents an evolution in how we do research, identify problems, and provide care. AI in biology has a bright future, with possible applications in many different subfields. AI has the potential to revolutionize BS by improving research efficiency and opening new avenues for comprehending and modifying living systems. Examples of this include personalized medicine, drug discovery, genomics, and predictive diagnostics.

Conclusion

This chapter offers a thorough summary of how AI is being used in BS, including bioinformatics, drug discovery, and medical diagnostics. From symbolic techniques to DL's recent supremacy, AI has evolved making a revolutionary journey in simulating human intelligence for complicated tasks. The convergence of biology and IT, spurred by sequencing breakthroughs and Moore's Law has made AI a vital tool for managing massive amounts of material in the biosciences. AI is having an enormous impact on medication development and is completely changing the pharmaceutical sector. AI helps with target recognition, cheminformatics, clinical trial optimization, predictive toxicology, drug repurposing, personalized medicine, and even denovo- drug creation with ML approaches. The efficacy of drug development is increased by the capacity to analyze massive datasets and anticipate molecular interactions, which may result in a reduction in the time and cost needed to introduce novel drugs to the market. AI-powered medical diagnostics have the potential to revolutionize the healthcare industry. AI offers chances for cost savings and a move from reactive to proactive treatment methods, as it can execute jobs like analyzing medical pictures and connecting symptoms with prognosis for diseases better than human capabilities. This study highlights how crucial AI is to enhancing early diagnosis, lowering uncertainty, and enabling individualized treatment plans via ongoing monitoring. AI–enabled bioinformatics is essential for the analysis of biological information. AI-powered algorithms make pattern identification easier, enabling scientists to better grasp variations in genes and decipher intricate biological processes in fields ranging from structural biology to gene prediction and sequence alignment. NN and SVMs are two examples of ML techniques that help in drug discovery, protein structure prediction, and gene expression research.

Some ethical issues are brought up by the incorporation of AI into numerous facets of the community, notably prejudice, confidentiality, accountability, openness, and justice. To solve these issues, the paper addresses the necessity of ethical structures and ideals. To guarantee the responsible growth and implementation of AI technologies, it is necessary to carefully negotiate obstacles such as prejudice in AI systems, inadequate transparency, and confidentiality violations. Although there is much potential, the paper notes that using AI in biology has its share of difficulties. Significant obstacles include DL models' lack of comprehension, the challenge of developing generalized models because of biological information's inter–individual variability, and worries about data accessibility and responsibility. It underlies that the use of AI should be prudent, considering the situation and suitability for dealing with certain problems or queries. AI in BS has bright future possibilities. AI is expected to become a vital instrument for high–dimensional data analysis, impacting scientific fields. AI is predicted to continue revolutionizing BS investigation, diagnostics, and therapy from transforming medicine through precise techniques to accelerating drug development and improving genomics research.

Funding

The authors would like to thank DST FIST (SR/FST/LS-I/2017/05(C)) and DST PURSE grant (SR/PURSE Phase 2/39(C)) for their support. Prof. Suhel Parvez appreciates the Central Council for Research in Unani Medicine (File. No.3-51/2022-CCRUM/Tech) for their support. Ms. Neha received the Senior Research Fellowship from DST through the INSPIRE program (File No. DST/INSPIRE/03/ 2019/001769. IVR No. 201900031405). The authors would like to thank the funding agencies, ICMR Senior Research Fellowship (File No. 45/14/2020/PHA/BMS) awarded to Ms. Pinky.

Declaration of Competing Interest

The authors declare that they have no known competing financial interests or personal relationships that could have appeared to influence the work reported in this paper.

Data Availability Statement

All data generated or analyzed during this study are included in this published article.

Conflict of Interest Statement

There is no conflict of interest for any of the authors.

Abbreviations

AI – Artificial Intelligence
BS – Biological Science
DL – Deep Learning
DRL – Deep Reinforcement Learning
EMR – Electronic Medical Records
GAN – Generative Adversarial Networks
InChI – International Chemical Identifier
IT – Information Technology
ML – Machine Learning
MOL – MDL Molfile
NLP – Natural Language Processing
NMR – Nuclear Magnetic Resonance
NN – Neural Networks
RL – Reinforcement Learning
SMARTS – Smiles ARbitrary Target Specification
SMILES – Simplified Molecular Input Line Entry System
VAE – Variational Autoencoder

References

Alowais, S.A., Alghamdi, S.S., Alsuhebany, N., Alqahtani, T., Alshaya, A.I., Almohareb, S.N., Aldairem, A., Alrashed, M., Bin Saleh, K., Badreldin, H.A., Al Yami, M.S., Al Harbi, S. and Albekairy, A.M. (2023). Revolutionizing healthcare: the role of artificial intelligence in clinical practice. *BMC Medical Education*, *23*(1), 689. https://doi.org/10.1186/s12909-023-04698-z.

Askin, S., Burkhalter, D., Calado, G. and El Dakrouni, S. (2023). Artificial Intelligence Applied to clinical trials: opportunities and challenges. *Health and Technology*, *13*(2), 203–213. https://doi.org/10.1007/s12553-023-00738-2.

Bender, A. and Cortés-Ciriano, I. (2021). Artificial intelligence in drug discovery: what is realistic, what are illusions? Part 1: Ways to make an impact, and why we are not there yet. *Drug Discovery Today*, *26*(2), 511–524. https://doi.org/10.1016/j.drudis.2020.12.009.

Bhardwaj, A., Kishore, S. and Pandey, D.K. (2022). Artificial Intelligence in Biological Sciences. *Life*, *12*(9), 1430. https://doi.org/10.3390/life12091430.

Bhatnagar, N. (2020). *Role of Robotic Process Automation in Pharmaceutical Industries* (pp. 497–504). https://doi.org/10.1007/978-3-030-14118-9_50.

Blaschke, T., Engkvist, O., Bajorath, J. and Chen, H. (2020). Memory-assisted reinforcement learning for diverse molecular de novo design. *Journal of Cheminformatics*, *12*(1), 68. https://doi.org/10.1186/s13321-020-00473-0.

Bogina, V., Hartman, A., Kuflik, T. and Shulner-Tal, A. (2022). Educating Software and AI Stakeholders About Algorithmic Fairness, Accountability, Transparency and Ethics. *International Journal of Artificial Intelligence in Education*, *32*(3), 808–833. https://doi.org/10.1007/s40593-021-00248-0.

Bohr, A. and Memarzadeh, K. (2020). The rise of artificial intelligence in healthcare applications. In *Artificial Intelligence in Healthcare* (pp. 25–60). Elsevier. https://doi.org/10.1016/B978-0-12-818438-7.00002-2.

Brown, T.B., Mann, B., Ryder, N., Subbiah, M., Kaplan, J., Dhariwal, P., Neelakantan, A., Shyam, P., Sastry, G., Askell, A., Agarwal, S., Herbert-Voss, A., Krueger, G., Henighan, T., Child, R., Ramesh, A., Ziegler, D. M., Wu, J., Winter, C., … Amodei, D. (2020). Language models are few-shot learners. *Advances in Neural Information Processing Systems*, *2020-December*.

Button, A., Merk, D., Hiss, J.A. and Schneider, G. (2019). Automated de novo molecular design by hybrid machine intelligence and rule-driven chemical synthesis. *Nature Machine Intelligence*, *1*(7). https://doi.org/10.1038/s42256-019-0067-7.

Chen, Z., Liu, X., Hogan, W., Shenkman, E., Bian, J. et al. (2021). Applications of artificial intelligence in drug development using real-world data. *Drug Discovery Today*, *26*(5), 1256–1264. https://doi.org/10.1016/j.drudis.2020.12.013.

Ching, T., Himmelstein, D.S., Beaulieu-Jones, B.K., Kalinin, A.A., Do, B.T., Way, G.P., Ferrero, E., Agapow, P.-M., Zietz, M., Hoffman, M.M., Xie, W., Rosen, G.L., Lengerich, B.J., Israeli, J., Lanchantin, J., Woloszynek, S., Carpenter, A.E., Shrikumar, A., Xu, J., … Greene, C.S. (2018a). Opportunities and obstacles for deep learning in biology and medicine. *Journal of The Royal Society Interface*, *15*(141), 20170387. https://doi.org/10.1098/rsif.2017.0387.

Choi, B., Lee, Y., Kyung, Y. and Kim, E. (2023). ALBERT with Knowledge Graph Encoder Utilizing Semantic Similarity for Commonsense Question Answering. *Intelligent Automation and Soft Computing*, *36*(1). https://doi.org/10.32604/iasc.2023.032783.

Dai, Y., Chai, C.-S., Lin, P.-Y., Jong, M. S.-Y., Guo, Y. and Qin, J. (2020). Promoting Students' Well-Being by Developing Their Readiness for the Artificial Intelligence Age. *Sustainability*, *12*(16), 6597. https://doi.org/10.3390/su12166597.

Dara, S., Dhamercherla, S., Jadav, S.S., Babu, C.M., Ahsan, M.J. et al. (2022). Machine Learning in Drug Discovery: A Review. *Artificial Intelligence Review*, *55*(3), 1947–1999. https://doi.org/10.1007/s10462-021-10058-4.

Deng, W.H., Yildirim, N., Chang, M., Eslami, M., Holstein, K. and Madaio, M. (2023). Investigating Practices and Opportunities for Cross-functional Collaboration around AI Fairness in Industry Practice. *2023 ACM Conference on Fairness, Accountability, and Transparency*, 705–716. https://doi.org/10.1145/3593013.3594037.

Díaz-Domínguez, A. (2020). How Futures Studies and Foresight Could Address Ethical Dilemmas of Machine Learning and Artificial Intelligence. *World Futures Review*, *12*(2), 169–180. https://doi.org/10.1177/1946756719894602.

Enholm, I.M., Papagiannidis, E., Mikalef, P. and Krogstie, J. (2022). Artificial Intelligence and Business Value: a Literature Review. *Information Systems Frontiers*, *24*(5), 1709–1734. https://doi.org/10.1007/s10796-021-10186-w.

Fjeld, J., Achten, N., Hilligoss, H., Nagy, A., Srikumar, M. et al. (2020). Principled Artificial Intelligence: Mapping Consensus in Ethical and Rights-Based Approaches to Principles for AI. *SSRN Electronic Journal*. https://doi.org/10.2139/ssrn.3518482.

Fjelland, R. (2020). Why general artificial intelligence will not be realized. *Humanities and Social Sciences Communications*, *7*(1), 10. https://doi.org/10.1057/s41599-020-0494-4.

Gervini, D. and Khanal, M. (2019). Exploring patterns of demand in bike sharing systems via replicated point process models. *Journal of the Royal Statistical Society. Series C: Applied Statistics*, *68*(3). https://doi.org/10.1111/rssc.12322.

Goldstein, B.A., Navar, A.M. and Carter, R.E. (2016). Moving beyond regression techniques in cardiovascular risk prediction: applying machine learning to address analytic challenges. *European Heart Journal*, ehw302. https://doi.org/10.1093/eurheartj/ehw302.

Gupta, A., Grattoni, C. and Gupta, A. (2023). Determining Chess Piece Values Using Machine Learning. *Journal of Student Research*, *12*(1). https://doi.org/10.47611/jsrhs.v12i1.4356.

Hanna, M.G. and Hanna, M.H. (2022). Current applications and challenges of artificial intelligence in pathology. *Human Pathology Reports*, *27*, 300596. https://doi.org/10.1016/j.hpr.2022.300596.

Hashimoto, D.A., Witkowski, E., Gao, L., Meireles, O., Rosman, G. et al. (2020). Artificial Intelligence in Anesthesiology. *Anesthesiology*, *132*(2), 379–394. https://doi.org/10.1097/ALN.0000000000002960.

Hassoun, S., Jefferson, F., Shi, X., Stucky, B., Wang, J. and Rosa, E. (2022). Artificial Intelligence for Biology. *Integrative and Comparative Biology*, *61*(6), 2267–2275. https://doi.org/10.1093/icb/icab188.

He, K., Gkioxari, G., Dollár, P. and Girshick, R. (2020). Mask R-CNN. *IEEE Transactions on Pattern Analysis and Machine Intelligence*, *42*(2). https://doi.org/10.1109/TPAMI.2018.2844175.

Henriksen, A., Enni, S. and Bechmann, A. (2021). Situated Accountability: Ethical Principles, Certification Standards, and Explanation Methods in Applied AI. *Proceedings of the 2021 AAAI/ACM Conference on AI, Ethics, and Society*, 574–585. https://doi.org/10.1145/3461702.3462564.

Holzinger, A., Langs, G., Denk, H., Zatloukal, K., Müller, H. et al. (2019). Causability and explainability of artificial intelligence in medicine. *WIREs Data Mining and Knowledge Discovery*, *9*(4). https://doi.org/10.1002/widm.1312.

Humbeck, L., Weigang, S., Schäfer, T., Mutzel, P., Koch, O. et al. (2018). CHIPMUNK: A Virtual Synthesizable Small-Molecule Library for Medicinal Chemistry, Exploitable for Protein–Protein Interaction Modulators. *ChemMedChem*, *13*(6). https://doi.org/10.1002/cmdc.201700689.

Jiménez-Luna, J., Grisoni, F. and Schneider, G. (2020a). Drug discovery with explainable artificial intelligence. *Nature Machine Intelligence*, *2*(10), 573–584. https://doi.org/10.1038/s42256-020-00236-4.

Jiménez-Luna, J., Grisoni, F. and Schneider, G. (2020b). Drug discovery with explainable artificial intelligence. *Nature Machine Intelligence*, *2*(10), 573–584. https://doi.org/10.1038/s42256-020-00236-4.

Jiménez-Luna, J., Grisoni, F., Weskamp, N. and Schneider, G. (2021a). Artificial intelligence in drug discovery: recent advances and future perspectives. *Expert Opinion on Drug Discovery*, *16*(9), 949–959. https://doi.org/10.1080/17460441.2021.1909567.

Jiménez-Luna, J., Grisoni, F., Weskamp, N. and Schneider, G. (2021b). Artificial intelligence in drug discovery: recent advances and future perspectives. *Expert Opinion on Drug Discovery*, *16*(9), 949–959. https://doi.org/10.1080/17460441.2021.1909567.

Jobson, D., Mar, V. and Freckelton, I. (2022). Legal and ethical considerations of artificial intelligence in skin cancer diagnosis. *Australasian Journal of Dermatology*, *63*(1). https://doi.org/10.1111/ajd.13690.

John, T. (2022). The Ethical Considerations of Artificial Intelligence in Clinical Decision Support. *Proceedings of the Wellington Faculty of Engineering Ethics and Sustainability Symposium*. https://doi.org/10.26686/wfeess.vi.7649.

Kaur, S., Singla, J., Nkenyereye, L., Jha, S., Prashar, D., Joshi, G.P., El-Sappagh, S., Islam, Md. S. and Islam, S. M.R. (2020). Medical Diagnostic Systems Using Artificial Intelligence (AI) Algorithms: Principles and Perspectives. *IEEE Access*, *8*, 228049–228069. https://doi.org/10.1109/ACCESS.2020.3042273.

Khurana, D., Koli, A., Khatter, K. and Singh, S. (2023). Natural language processing: state of the art, current trends and challenges. *Multimedia Tools and Applications*, *82*(3), 3713–3744. https://doi.org/10.1007/s11042-022-13428-4.

Kilanko, V. (2023). Leveraging Artificial Intelligence for Enhanced Revenue Cycle Management in the United States. *International Journal Of Scientific Advances*, *4*(4). https://doi.org/10.51542/ijscia.v4i4.3.

Liang, Y., Liang, W. and Jia, J. (2023). Structural Vibration Signal Denoising Using Stacking Ensemble of Hybrid CNN-RNN. *Advances in Artificial Intelligence and Machine Learning*, *3*(2). https://doi.org/10.54364/AAIML.2023.1165.

Lo Piano, S. (2020). Ethical principles in machine learning and artificial intelligence: cases from the field and possible ways forward. *Humanities and Social Sciences Communications*, *7*(1), 9. https://doi.org/10.1057/s41599-020-0501-9.

Lo, Y.-C., Ren, G., Honda, H. and L. Davis, K. (2020a). Artificial Intelligence-Based Drug Design and Discovery. In *Cheminformatics and its Applications*. IntechOpen. https://doi.org/10.5772/intechopen.89012.

Lo, Y.-C., Ren, G., Honda, H. and L. Davis, K. (2020b). Artificial Intelligence-Based Drug Design and Discovery. In *Cheminformatics and its Applications*. IntechOpen. https://doi.org/10.5772/intechopen.89012.

Lovrić, M., Molero, J.M. and Kern, R. (2019). PySpark and RDKit: Moving towards Big Data in Cheminformatics. *Molecular Informatics*, *38*(6). https://doi.org/10.1002/minf.201800082.

Majhi, V. and Paul, S. (2020a). *Artificial Intelligence in Bioinformatics* (pp. 177–190). https://doi.org/10.1007/978-981-15-2620-6_12.

Majhi, V. and Paul, S. (2020b). *Artificial Intelligence in Bioinformatics* (pp. 177–190). https://doi.org/10.1007/978-981-15-2620-6_12.

Marsland, S. (2014). *Machine Learning*. Chapman and Hall/CRC. https://doi.org/10.1201/b17476.

Matzaraki, V., Kumar, V., Wijmenga, C. and Zhernakova, A. (2017). The MHC locus and genetic susceptibility to autoimmune and infectious diseases. *Genome Biology*, *18*(1), 76. https://doi.org/10.1186/s13059-017-1207-1.

Mikalef, P. and Gupta, M. (2021). Artificial intelligence capability: Conceptualization, measurement calibration, and empirical study on its impact on organizational creativity and firm performance. *Information & Management*, *58*(3), 103434. https://doi.org/10.1016/j.im.2021.103434.

Miller, D.D. and Brown, E.W. (2018). Artificial Intelligence in Medical Practice: The Question to the Answer? *The American Journal of Medicine*, *131*(2), 129–133. https://doi.org/10.1016/j.amjmed.2017.10.035.

Naik, N., Hameed, B.M.Z., Shetty, D.K., Swain, D., Shah, M., Paul, R., Aggarwal, K., Ibrahim, S., Patil, V., Smriti, K., Shetty, S., Rai, B.P., Chlosta, P. and Somani, B.K. (2022). Legal and Ethical Consideration in Artificial Intelligence in Healthcare: Who Takes Responsibility? *Frontiers in Surgery*, *9*. https://doi.org/10.3389/fsurg.2022.862322.

Nian, R., Liu, J. and Huang, B. (2020). A review On reinforcement learning: Introduction and applications in industrial process control. *Computers & Chemical Engineering*, *139*, 106886. https://doi.org/10.1016/j.compchemeng.2020.106886.

Oliveira, A.L. (2019). Biotechnology, Big Data and Artificial Intelligence. *Biotechnology Journal*, *14*(8). https://doi.org/10.1002/biot.201800613.

O'Reilly-Shah, V.N., Gentry, K.R., Walters, A.M., Zivot, J., Anderson, C.T. and Tighe, P.J. (2020). Bias and ethical considerations in machine learning and the automation of perioperative risk assessment. *British Journal of Anaesthesia*, *125*(6), 843–846. https://doi.org/10.1016/j.bja.2020.07.040.

Paeng, K., Hwang, S., Park, S. and Kim, M. (2017). *A Unified Framework for Tumor Proliferation Score Prediction in Breast Histopathology* (pp. 231–239). https://doi.org/10.1007/978-3-319-67558-9_27.

Rashidi, H.H., Tran, N.K., Betts, E.V., Howell, L.P., Green, R. et al. (2019). Artificial Intelligence and Machine Learning in Pathology: The Present Landscape of Supervised Methods. *Academic Pathology*, *6*, 2374289519873088. https://doi.org/10.1177/2374289519873088.

Richards, B., Tsao, D. and Zador, A. (2022a). The application of artificial intelligence to biology and neuroscience. *Cell*, *185*(15), 2640–2643. https://doi.org/10.1016/j.cell.2022.06.047.

Richards, B., Tsao, D. and Zador, A. (2022b). The application of artificial intelligence to biology and neuroscience. *Cell*, *185*(15), 2640–2643. https://doi.org/10.1016/j.cell.2022.06.047.

Schwendicke, F., Samek, W. and Krois, J. (2020). Artificial Intelligence in Dentistry: Chances and Challenges. *Journal of Dental Research*, *99*(7), 769–774. https://doi.org/10.1177/0022034520915714.

Sellwood, M.A., Ahmed, M., Segler, M.H. and Brown, N. (2018). Artificial intelligence in drug discovery. *Future Medicinal Chemistry*, *10*(17), 2025–2028. https://doi.org/10.4155/fmc-2018-0212.

Suleimenov, I.E., Vitulyova, Y.S., Bakirov, A.S. and Gabrielyan, O.A. (2020). Artificial Intelligence. *Proceedings of the 2020 6th International Conference on Computer and Technology Applications*, 22–25. https://doi.org/10.1145/3397125.3397141.

Suzuki, K. (2017). Overview of deep learning in medical imaging. *Radiological Physics and Technology*, *10*(3), 257–273. https://doi.org/10.1007/s12194-017-0406-5.

Tice, R.R., Austin, C.P., Kavlock, R.J. and Bucher, J.R. (2013). Improving the Human Hazard Characterization of Chemicals: A Tox21 Update. *Environmental Health Perspectives*, *121*(7), 756–765. https://doi.org/10.1289/ehp.1205784.

Topol, E.J. (2019). High-performance medicine: the convergence of human and artificial intelligence. *Nature Medicine*, *25*(1), 44–56. https://doi.org/10.1038/s41591-018-0300-7.

Webb, M.E., Fluck, A., Magenheim, J., Malyn-Smith, J., Waters, J., Deschênes, M. and Zagami, J. (2021). Machine learning for human learners: opportunities, issues, tensions and threats. *Educational Technology Research and Development*, *69*(4), 2109–2130. https://doi.org/10.1007/s11423-020-09858-2.

Williams, A.M., Liu, Y., Regner, K.R., Jotterand, F., Liu, P., Liang, M. et al. (2018). Artificial intelligence, physiological genomics, and precision medicine. *Physiological Genomics*, *50*(4), 237–243. https://doi.org/10.1152/physiolgenomics.00119.2017.

Xu, Y., Liu, X., Cao, X., Huang, C., Liu, E., Qian, S., Liu, X., Wu, Y., Dong, F., Qiu, C.-W., Qiu, J., Hua, K., Su, W., Wu, J., Xu, H., Han, Y., Fu, C., Yin, Z., Liu, M., … Zhang, J. (2021). Artificial intelligence: A powerful paradigm for scientific research. *The Innovation*, *2*(4), 100179. https://doi.org/10.1016/j.xinn.2021.100179.

Yang, Z., Dai, Z., Yang, Y., Carbonell, J., Salakhutdinov, R. and Le, Q. V. (2019). XLNet: Generalized autoregressive pretraining for language understanding. *Advances in Neural Information Processing Systems*, *32*.

Zawacki-Richter, O., Marín, V.I., Bond, M. and Gouverneur, F. (2019). Systematic review of research on artificial intelligence applications in higher education – where are the educators? *International Journal of Educational Technology in Higher Education*, *16*(1), 39. https://doi.org/10.1186/s41239-019-0171-0.

Zhai, X., Chu, X., Chai, C.S., Jong, M.S.Y., Istenic, A., Spector, M., Liu, J.-B., Yuan, J. and Li, Y. (2021). A Review of Artificial Intelligence (AI) in Education from 2010 to 2020. *Complexity*, *2021*, 1–18. https://doi.org/10.1155/2021/8812542.

Zhai, X., Shi, L. and Nehm, R. H. (2021). A Meta-Analysis of Machine Learning-Based Science Assessments: Factors Impacting Machine-Human Score Agreements. *Journal of Science Education and Technology*, *30*(3), 361–379. https://doi.org/10.1007/s10956-020-09875-z.

Zhang, J., Fei, Y., Sun, L. and Zhang, Q. C. (2022). Advances and opportunities in RNA structure experimental determination and computational modeling. *Nature Methods*, *19*(10), 1193–1207. https://doi.org/10.1038/s41592-022-01623-y.

Zinner, M., Dahlhausen, F., Boehme, P., Ehlers, J., Bieske, L. and Fehring, L. (2021). Quantum computing's potential for drug discovery: Early stage industry dynamics. *Drug Discovery Today*, *26*(7), 1680–1688. https://doi.org/10.1016/j.drudis.2021.06.003.

CHAPTER

15

Artificial Intelligence in System Biology

Pankhi Vatsa[1,2] and *AB Pant*[1,2*]

1. Introduction

Artificial intelligence, or AI, is the imitation of human intelligence in robots designed to possess thought processes and behaviours like *Homo sapiens*. This may include having the cognitive ability to learn, reason, solve problems, perceive, and understanding of languages. The process of computer designing or making a robot, or piece of software thinks logically like a human mind. Therefore, it can be called artificial intelligence. (Górriz, J.M. et al., 2020; Hassani, H. et al., 2020). The conceptualization of the idea behind artificial intelligence goes way back to 700 B.C. Greek mythology literature holds evidence of the birth of the idea of artificial intelligence. As the saying goes, the giant Talus, a figurine created out of bronze rather than born, used to circle the island thrice a day to protect King Minos' mother Europa from kidnappers and invaders (Hassoun, S. et al., 2021).

At a conference in Dartmouth, the term Artificial intelligence or AI was first coined by John McCarthy et al. in the year 1956 (McCarthy, J. 2022). Alan Mathison Turing, a mathematician, made a substantial contribution to AI in its initial years. He presented his theories on the idea of self-learning and self-instructed machines that learn from their own experiences in a public speech in London, just like a human being does (Bhardwaj, A. et al., 2022). Alan Turing, recognized as the founding father of artificial intelligence and contemporary computer science, made early observations and conceptualizations of facts concerning sentient technology. He was among the first to advocate that the human brain operates like a digital computer (Luger, G.F. 2021). He devised the "Turing Test" experiment, a crucial phase in advancing artificial intelligence. His seminal work, "Computing Machinery and Intelligence," explored the notion of a non-living computer capable of human-like thought (Turing, I.B.Y.A.M.). These events were followed by a series of advancements in the development of artificial intelligence. The ground-breaking works of A. Samuel for the

[1] Systems Toxicology Group, CSIR-Indian Institute of Toxicology Research Vishvigyan Bhavan, Mahatma Gandhi Marg, Lucknow, India.

[2] Academy of Scientific and Innovative Research (AcSIR), Ghaziabad, India.

* Corresponding author: abpant@iitr.res.in; abpant@rediffmail.com

International Business Machines Corporation (IBM) in 1952, wherein he designed the IBM 701 prototype and another by J. Holland, who trained a virtual rodent model to pass along a predefined route designed on the basis of a neural network, contributed significantly to the development of AI (Chen, R.H., and Chen, C.C. 2022; Holland, J.H. 1986). In 1945, John von Neuman proposed a computer paradigm in which program instructions and data are stored in the random-access memory (RAM). This design was the predecessor of the modern computer, but AI became an operational possibility with the advent of the fast microprocessor.

AI has grown and encountered failures since its beginning as an area of research and development in 1956 until it eventually flowered with successful applications in academia and industry in the early twenty-first century (Hassoun, S. et al, 2021). Japanese engineers constructed the first humanoid robot in 1973, featuring several unique features like the capacity to carry items, walk upright, and communicate in Japanese (Newton, D.E. 2018). The combination of fresh methodologies, powerful computers, and vast amounts of data resulted in significant investment and widespread interest in AI. Since AI was initially introduced as an area of research and development in 1956, it had advanced and encountered obstacles until the early years of the twenty-first century when it finally bloomed with successful applications in both academia and industry. AI has attracted substantial attention and investment due to the availability of powerful computers, innovative methods, and vast amounts of data. AI has been used practically everywhere because of its nature, whether in industry, academia, or research. Its effect is thus apparent throughout several domains and subdomains. After its discovery, the applications of AI stagnated for around 50 years, but gradually, as digitalization spread worldwide, gigabytes and terabytes of data began to be created. Therefore, AI became essential for using and managing massive amounts of data, and with time, more AI models developed, enabling its extensive application in multiple sectors of the economy.

Today, AI is regarded as one of the most significant innovations of the Fourth Industrial Revolution (Schwab, K. 2017). AI is now known to drive new frontiers every other day in biology. Whether solving challenging issues or deciphering complex neural networks, researchers increasingly rely on artificial intelligence to answer the most complex hypotheses. In the context of biological research, the Human Genome Project (HGP), completed in the early 2000s, produced an enormous amount of biological data that helped establish bioinformatics. Examining these enormous data sets has uncovered important facets of human genes and increased the anticipation regarding the complexity of genomes. Thus, the multidisciplinary discipline of bioinformatics-which combines computer science, biology, physics, chemistry, mathematics, and statistics-is an essential component of contemporary biological research (Biswas, N. and Chakrabarti, S. 2020; Branco, I. and Choupina, A. 2021; Athanasopoulou, K. et al., 2022). AI can assist in managing and analyzing vast amounts of data generated with each passing day along with the advancement of sciences, thus allowing researchers to concentrate on the overall picture instead of delving into hundreds or thousands of images or data. The computational and mathematical evaluation and simulation of intricate biological systems is known as systems biology. It is a multidisciplinary area of study with a foundation that employs a holistic approach to biological research to focus on intricate interconnections within biological systems. Modelling and discovering emergent properties of cells, tissues, and organisms operating as a system theoretically can only be made feasible by methods unique to systems biology, which is one of the goals of the field. Systems biology aims to create a model of the interactions within a system. Hence, system-wide, maximally comprehensive experimental approaches are most appropriate. Therefore, transcriptomics, metabolomics, proteomics, and high-throughput methods are employed to gather quantitative data for model building and validation (Starr, T.K. et al., 2009).

AI encompasses several niches such as knowledge illustration, reasoning, looking for solutions, and machine learning (ML). Machine learning (ML) employs algorithms that identify patterns in data collection once it has been further categorized. Deep learning (DL), a branch of ML, utilizes Artificial Neural Networks (ANNs which are basically a network of interconnected multifaceted computing gears called "perceptons," similar to human neurons, and replicate the transmission mechanism of electrical impulses to the brain. An artificial neural network (ANN) comprises several nodes that take individual inputs and use algorithms to solve issues to turn them into outputs that may be single or multi-linked. ANNs are available in several forms, such as recurrent neural networks (RNNs), convolutional neural networks (CNNs), and multilayer perceptron (MLP) networks. These networks may be trained using supervised or unsupervised methods (Paul, D. et al., 2021). AI has begun to make its way into majority biological system applications, and this chapter will address how AI has accomplished the challenging goal of comprehensively deciphering their intricacies.

2. System Biology and Artificial Intelligence

According to Ronald Germain, NIH, system biology can be described as a "scientific approach that combines the principles of engineering, mathematics, physics, and computer science with extensive experimental data to develop a quantitative as well as a deep conceptual understanding of biological phenomena, permitting prediction and accurate simulation of complex (emergent) biological behaviours." AI has paved its way much deeper into the field of biological sciences and has demonstrated so by showcasing its innovative and cutting-edge approaches to the subject. Moreover, advances in information technology have also made it possible to generate and store huge amounts of data in the past decade. The complex intricacies of computers have allowed researchers to create complex algorithms that can perform well-coordinated tasks with minimum errors. The oldest example is the 1997 IBM Deep Blue Computer that defeated the chess world champion Garry Kasparov. Over the past few years, the applications of AI in biology have grown exponentially.

The algorithms currently used in biology generate massive data that requires storage and methods to process the unstructured data into formats ready for analysis and interpretation (Williams, A.M. et al., 2017; Oliveira, A.L., 2019). Experiments that would have taken years to execute and analyze were often quite expensive are more feasible now with the advances in AI. The generated data can be comprehended using the multi-omics approach involving branches like transcriptomics, proteomics, genomics, or metabolomics. AI-based algorithms are being utilized to study the genetic markers that can be potentially linked to specific traits of a disease and can be used as a diagnosis or prognosis marker (Chakraborty, I. et al., 2017; Dias, R. and Torkamani, A. 2019). Drug designing or medical imaging programs are also constantly evolving with the advances in AI algorithms that have made it possible to develop personalized and precise medicines/treatments (Ahmed, Z. et al., 2020). With the advancements in artificial intelligence and its growing use in system biology, drug development, image processing, and virtual drug testing, it may become a reality in medical sciences.

3. AI-based Diagnostic Tools: Imaging and Computer-aided Diagnosis

The modern-day diagnosis of various diseases or health conditions majorly relies on techniques such as X-ray, Magnetic Resonance Imaging (MRI), computed Tomography (CT scan), positron emission tomography (PET scan), and more. Previously radiologists used to analyze the imaging data generated based on their acquired expertise. However, with the introduction of AI in imaging devices, these are pre-annotated with data and related algorithms that can generate results for any corresponding ailment without any human expertise in a time-efficient manner (Figure 1) (Habuza, T., 2021). One of the first AI-based diagnostic models, the IDx-DR, was created to detect diabetic retinopathy. This integrated system could detect diabetic retinopathy with a sensitivity of 87.2%. The results were also 90.7% specific and had up to 96.1% imageability. This system had minimal chances of error as it was trained with data from human subjects of various races, ages, and sexes (Abràmoff, M.D. et al., 2018).

The three main areas for advancements in AI include machine learning (ML), Natural Language Processing (NLP), and automated robotic processes. NLP has been used to improve the analysis of colonoscopy, thus leading to improved and more accurate detection of polyps and adenomas (Nayor, J. et al., 2018; Leite, M.L., 2021; Zou, K.H., 2020). The ML aspect has been utilized in improving

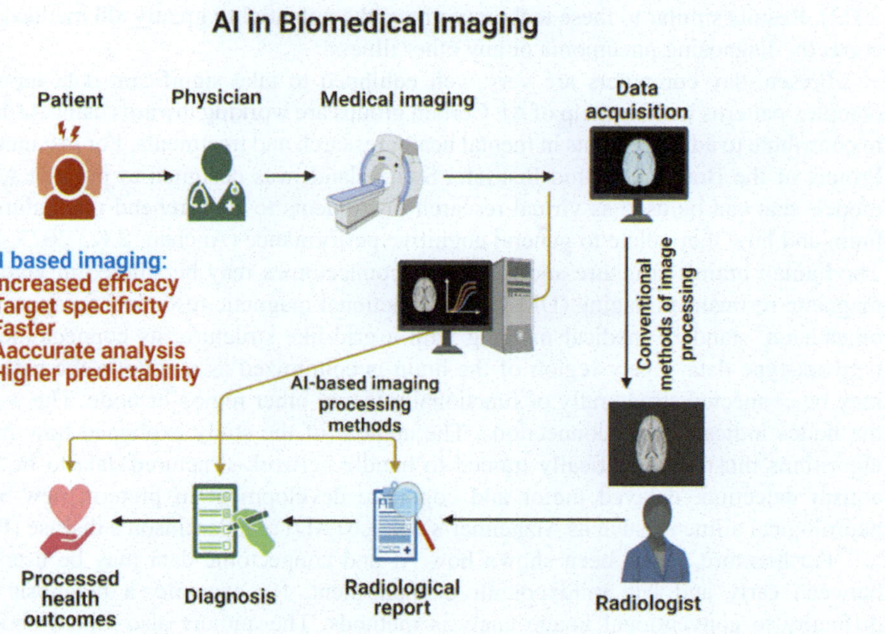

Fig. 1 Illustration depicting the involvement of artificial intelligence-based tools in the different levels of modern biomedical imaging.

the cardiovascular disorder prognosis. The "heart-murmur-detecting technology" or "eMurmur®" received FDA clearance in 2019, and the algorithms can identify whether a murmur in a heart sound recording is problematic or likely benign in a matter of seconds, with expert-level precision. Accurate heart rate and S1/S2 heart sound indicators are also attributes of this technology (Lai, L.S. et al., 2016). The ML approach has also been utilized in detecting urinary tract infections (UTIs). Taylor et al. devised models employing six machine learning algorithms to predict UTIs based on data from the principal complaint, prior medical history, vital signs, test results, and organized historical and physical exam findings. Outperforming numerous proxies for provider judgment and previously created models in the literature, the top-performing machine learning system, XGBoost, correctly predicted positive urine culture findings. However, the study requires further validation (Taylor, R.A. et al., 2018).

Many random clinical trials are also under experimentation that utilize the concept of AI as diagnostic and prognostic tools. Many such results revealed that automated detection outperformed manual detection. A randomized controlled trial assessed the effect of a computerized polyp identification system based on deep learning on polyp detection accuracy and adenoma detection rates (ADRs) (registration number: ChiCTR-DDD-17012221). In this randomized controlled trial, individuals were randomized to get a colonoscopy with or without the automated polyp identification model, which offered a simultaneous visual and auditory alarm upon polyp detection. The automated AI-based detection approach produced better results than the control cohorts for ADR and the average number of polyps and adenomas found during a colonoscopy (Bhardwaj, A. et al., 2022). A study by Gupta et al. showed pneumonia detection from an X-ray or CT scan of a patient's lungs. Using chest X-rays, the Neural Architecture Search (NAS) technique developed by the group locates the optimal convolutional architecture for pneumonia detection. Inspired by the teaching-driven learning approach from humans, they devised the Learning by a Teaching framework and performed experiments on a chest X-ray dataset with more than five thousand pictures related to pneumonia. Their suggested approach was able to detect pneumonia with an area under the ROC curve (AUC) of 97.6%, outperforming earlier NAS approaches by 5.1% (absolute) (Gupta, A. et al.,

2022). Results similar to these techniques have the potential to greatly aid medical professionals in correctly diagnosing pneumonia or any other illness.

Present-day computers are very well equipped to take significant data inputs and generate complex patterns with the help of AI. Certain groups are working towards using AI-based algorithms to contribute to advancements in mental health research and treatments. For instance, the Blue Brain Project of the Brain and Mind Institute, Switzerland, was designed to provide AI-based inspired models that can be used as virtual research instruments to comprehend the features of the human brain and how they relate to general cognitive performance (Anchan, S.K., 2023; Sy, M.F., 2023). The human brain's structure and functional connectomes may be studied in vivo using diffusion magnetic resonance imaging (DMRI) and functional magnetic resonance imaging (FMRI). On the other hand, standard medical imaging with a grid-like structure; its connectome data resembles network-type data. Every region of the brain is epitomized as a node in the network. Each node may be connected structurally or functionally to any other region or node. The weighted edges of the nodes indicate these connections. The authors of the study explained how machine learning algorithms must be specifically trained to handle network-structured data to be applied to early autism detection, delayed motor and cognitive development in preterm new-borns, and other neurological ailments such as Alzheimer's disease (AD) and Parkinson's disease (PD).

Furthermore, it has been shown how AI and connectome data may be used to differentiate between early and late mild cognitive impairment, for example, a diagnosis that presents a difficulty to conventional image analysis methods. The authors also talked about how AI may identify significant links and subnetworks, which would help us understand the genesis of disease (Brown, C.J. and Hamarneh, G. 2016). Depending on the scan resolution, MRI scans might include hundreds of layers. As a result, manual segmentation is difficult, subjective, time-consuming, and inappropriate for large-scale neuroimaging investigations. AI has significantly improved MRI data acquisition, thus reducing the time involved in MRI data acquisition (Liang, N. 2021), and has also helped enhancing the resolution of MRI-generated compressed images (Zhao, C. et al., 2019). AI reduce the signal-to-noise ratio of MRI scans and aided in the conversion of low-resolution images to higher resolutions. Some of the techniques implementing AI include "noise and artifact reduction," "super-resolution", and "image acquisition and reconstruction" (Dong, C. et al, 2015; Higaki, T. et al., 2019; Jiang, D. et al., 2018). Furthermore, AI-based techniques have been used to forecast radiation doses. AI makes it possible to analyze low-dose MRI images, which can be used for those with allergies to contrast materials or renal disorders (Fan, J. et al., 2019). There has been discussion of AI-based dynamic brain imaging techniques that can convert EEG data into neural network circuit activity without human intervention (Ye, S., 2022). Deep learning models operate directly on the multivariate time-series raw data from EEG, extracting features with the necessary pre-processing and transformation that are then used for various analytical tasks. In less than 40 seconds, "Fast-Aid Brain," an effective 2.5D-based deep learning tool, uses a GPU to autonomously segment a human brain MRI picture into 132 cortical and subcortical areas (Ghazi, M.M. and Nielsen, M. 2022). To study brain aging using AI, the models are taught to generate anatomically precise brain maps showing distinct aging trends. Next, the real and biological ages calculated from the picture are contrasted. A higher discrepancy between the two indicates a higher chance of Alzheimer's disease in the individual (Levakov, G. et al., 2020; Irimia, A., 2015).

4. AI based Drug Discovery and Development

Historically, finding new drugs has required laborious and complex screenings to find possible chemicals in a library that have the desired effect; for example, an antibiotic that kills multiple microbes is safer for humans and shows the least signs of developing drug-resistant microbes. However, it is challenging for individual researchers or laboratories to perform this individually due to the breadth and complexity of the libraries utilized to screen for novel drug candidates. Here, machine learning (ML) and AI can be useful. The process of discovering and developing new

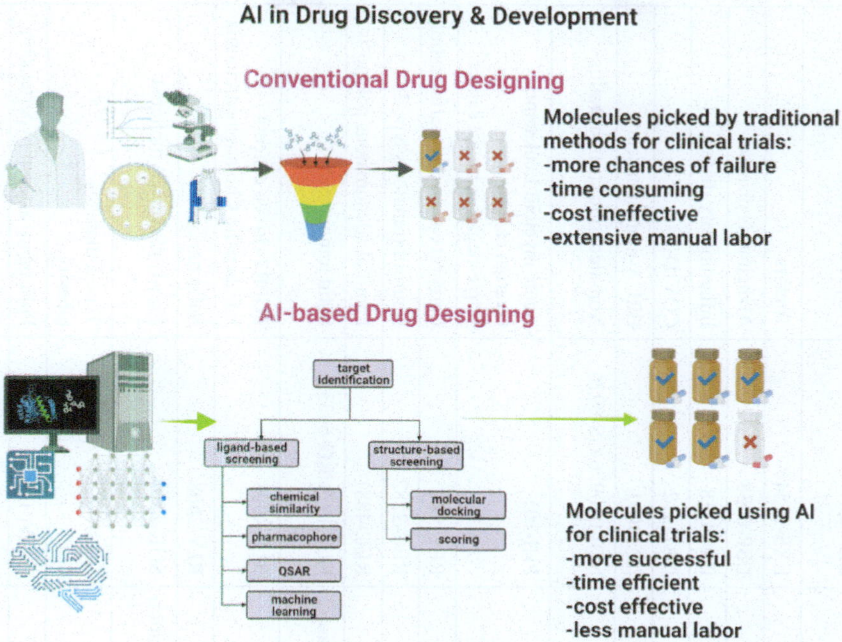

Fig. 2 An overview of artificial intelligence driven applications in drug discovery and development

drugs can be time-consuming and costly. However, AI's computational power and ability to process large datasets have become valuable tools in accelerating this process. AI algorithms can efficiently analyze massive amounts of biological and chemical data, identifying potential drug candidates and predicting their effectiveness (Figure 2).

AI can make precise predictions about a possible compound's characteristics. The advantages are obvious: money and effort are not spent on substances unlikely to produce the intended results. AI can aid in synthesizing new compounds by generating molecules with characteristics that are expected to be successful, significantly enhancing the process of finding and producing novel, potent medications. When determining a medicine's efficacy, AI can eliminate repetitive work in the drug discovery process (Preuer, K. et al., 2019). Recently, scientists have leveraged artificial intelligence (AI) to find new peptides. For example, Yan and his fellow researchers created a DL-based technology called Deep-AmPEP30 to identify short antimicrobial peptides (AMPs) (Yan, J. et al., 2020). A CNN-driven program called Deep-AmPEP30 (https://cbbio.online/AxPEP/) anticipates short AMPs from DNA sequence data. Yan et al. used Deep-AmPEP30 to identify new AMPs from the genomic sequence of the GI tract-associated fungal pathogen C. glabrate. Furthermore, Kavousi et al. created IAMPE (http://cbb1.ut.ac.ir/), a web server for anti-microbial peptide discovery, which combines physicochemical and 13CNMR-based peptide properties as input to machine learning algorithms to find new AMPs. Yi et al. also developed ACP-DL (https://github.com/haichengyi/ACP-DL), a DL-based method for discovering new anticancer peptides, in 2019.

The LSTM technique, an enhanced variant of the recursive neural network (RNN), is employed by ACP-DL to distinguish anticancer peptides from non-anticancer fragments. Additionally, Yu et al. presented DeepACP, a model for finding anticancer peptides based on deep recurrent neural networks. Similarly, Tyagi et al. (2013) created a platform based on SVM for finding novel anti-cancer peptides (Gupta, R. et al., 2021). Similarly, Wu et al. introduced PTPD, a CNN and word2vec-based technique for finding new peptides for medicinal use. An AI-based platform called AI-PRS was created by Shen et al. to help discover the best medication combinations and dosages for HIV treatment using antiretroviral therapy. AI-PRS is a neural network-based method that uses a parabolic response curve (PRS) to link dose and medication combinations to effectiveness.

Table 1 AI-derived drugs in clinical trials

S.No.	*Name of Pharma Company*	*Identified Target*	*Prospective Compounds*	*Target Diseases*
1.	BenevolentAI (https://www.benevolent.com/)	Tropomyosin receptor kinase (TRK) family	BEN-2293	Atopic dermatitis
2.	Exscientia (https://www.exscientia.ai/)	Adenosine 2A receptor (A2AR)	EXS-21546	Tumors
		Serotonin transporter (5-HT1A/2A)	DSP-0038	Alzheimer's disease
		Protein kinase C theta (PKC-θ)	EXS4318	Inflammatory diseases
		3-chymotrypsin like protease (3CLPro)	ISM3312	COVID-19
		Ubiquitin Specific Peptidase 1(USP1)	ISM3091	BRCA-mutant cancer
4.	Nimbus Therapeutics (https://www.nimbustx.com/)	Acetyl-CoA carboxylase (ACC)	NDI-010976/GS-0976	Non-alcoholic steatohepatitis
5.	Pharos iBio (https://www.pharosibio.com/en/)	FMS-like tyrosine kinase 3 (FLT3)	PHI-101	Acute myeloid Leukaemia Ovarian cancer Triple-negative breast cancer
6.	Recursion Pharmaceuticals (https://www.recursion.com/)	Cerebral cavernous malformation protein 2 (CCM2)	REC-994	Cerebral cavernous malformation
		Histone deacetylase (HDAC)	REC-2282	Neurofibromatosis type 2
		Mitogen Activated Protein Kinase Kinase1/2 (MEK1/2)	REC-4881	Familial adenomatous polyposis
7.	Relay Therapeutics (https://relaytx.com/)	Src homology-2 domain containing protein tyrosine phosphatase-2 (SHP2)	RLY-1971/RG-6433	Solid tumors
		Fibroblast growth factor receptor 2 (FGFR2)	RLY-4008	FGFR2-driven cancers Intrahepatic Cholangiocarcinoma
		Phosphoinositide 3-kinase alpha (PI3Kα)	RLY-2608	Solid tumors
8.	Schrödinger (https://newsite.schrodinger.com/)	Mucosa-associated lymphoid tissue lymphoma translocation protein 1 (MALT1)	SGR-1505	Non-Hodgkin's lymphoma
9.	Structure Therapeutics (https://structuretx.com/)	Glucagon-like peptide-1 receptor (GLP1R)	GSBR-1290	Type 2 diabetes Obesity
		Apelin receptor (APLNR)	ANPA-0073	Pulmonary arterial hypertension
10.	Valo Health (https://www.valohealth.com//)	Sphingosine-1-Phosphate (S1P1)	OPL-0301	Post-Myocardial Infarctions Kidney injuries
		Rho-associated kinases 1/2 (ROCK1/2)	OPL-0401	Diabetic Retinopathy Complications in diabetes

Tenofovir, efavirenz, and lamivudine were given to ten HIV patients in Shen et al.'s trial. Eventually, they discovered that tenofovir doses could be lowered by 33% of the initial dose without resulting in viral recurrence, utilizing the PRS approach. Therefore, the ideal medication dose for various disorders may also be determined using AI-PRS (Gupta, R. et al., 2021). Biomarker and target identification, indication prioritization, drug-like molecule design, pharmacokinetics prediction, drug–target interaction, and clinical trial design are only a few of the significant advancements made by AI (Pun, F.W. et al., 2023). AI-derived medications are emerging more frequently in clinical studies despite being in the early stages of clinical trials (Table 1), as demonstrated by INS018_055 for idiopathic pulmonary fibrosis, EXS-21546 for solid tumors, and GS-0976 for the treatment of non-alcoholic steatohepatitis-the first AI-derived medication to show positive topline results in a Phase 1 clinical trial. Also, an A2 receptor antagonist that helps T cells fight malignancies was developed with the aid of AI (Savage, N. 2021).

Billions of compounds are included in screening chemical libraries that are becoming exponentially larger for drug development efforts. The constant expansion of virtual libraries necessitates allocating sufficient resources to carry out essential functions like virtual screening. Deep docking (DD) is a platform for quick and precise docking of billions of molecules. It created by Gentile et al. DD was built on QSAR deep models trained using docking scores of vast chemical library subsets (samples of one million randomly chosen small compounds). Using the acquired model, the outcomes for raw molecules were estimated by repeatedly eliminating the undesirable ones. When DD and the FRED docking engine were combined, 1.36 billion molecules against 12 targets produced precise docking scores. With less processing power, DD was able to decrease an extremely large docking database by up to 100 times and enrich the top-ranked hits by up to 6000 times (Gentile, F. et al., 2020; Hoffmann, T. and Gastreich, M. 2019; Walters, W.P. 2018). Pun et al. used a combination of numerous druggable genes and identified 18 possible targets for amyotrophic lateral sclerosis (ALS) therapy by combining a range of bioinformatic- and deep learning-based algorithms that were trained using disease-specific multi-omics and text-based data. They were able to find viable treatment targets for ALS using this method (Pun, F.W. et al., 2022). In 2020, Zhang and colleagues released a study on the use of deep learning for medicine repurposing. This work employed a deep learning-based drug-target interaction model called "Molecule Transformer-Drug Target Interaction (MT-DTI)" to analyze chemical sequences called a "simplified molecular-input line-entry system (SMILES) strings" and amino acid sequences as input. Several currently available antiviral medications that may interact with the SARS-CoV-2 proteins were discovered due to this research (Zhang, Y. et al., 2021; Beck, B.R. et al., 2020).

Moinul et al. discovered critical fingerprints required for inhibiting the SGLT2 (Sodium-glucose Cotransporter-2) protein, which allowed the drugs to be repurposed for treating diabetes. By employing ligand-based drug design, fragment-based explicitly in silico drug design, they have sped up the anti-SGLT2 drug development process. The results of studies using recursive and Bayesian partitioning are precious in elucidating the mechanical aspects of the fingerprints supporting inhibitory processes. They also created machine-learning algorithms with statistical backing to weed out possible matches from a list of FDA-approved drugs (Moinul, M. et al., 2022). The combination ML/FPL (Machine Learning/False Pulling Ligand) approach was used to screen many natural compounds in search of acetylcholinesterase (AChE) inhibitors for treating Alzheimer's disease. The binding affinities of the most effective lead compounds were predicted using molecular docking. Next, more atomistic simulations were performed to validate the ML and docking computations (Thai, Q.M. et al., 2022). Ren et al. effectively integrated "AlphaFold" into their AI-driven drug discovery engines, which included the generative chemistry platform "Chemistry42" and the bio-computational platform "PandaOmics." The protein of interest for treating hepatocellular carcinoma (HCC) was given by PandaOmics, and Chemistry42 produced the compounds based on AlphaFold's anticipated structure. They then synthesized the chosen molecules and tested them in biological experiments. This resulted in the identification of a small molecule, "ISM042-2-048," that showed inhibitory activity for cyclin-dependent kinase 20 (CDK20) and also possessed selective anti-proliferative activity (Ren, F. et al., 2023). An essential first stage in the current drug development process is target identification. Effective target-finding techniques are desperately needed, as only a small percentage of human targets that may be druggable have been found. The increasing number of AI-identified targets being confirmed through experimentation emphasizes the advantages of using AI algorithms in target identification to improve the effectiveness of finding new targets and creating innovative treatments.

5. AI-driven Genetic Algorithms

With the primary objective of diagnosing and treating illnesses, AI-based algorithms can also concentrate on analyzing genomic and proteomic data. In particular, ML and DL techniques can pinpoint the illness pathways to improve treatment plans (Figure 3). For example, the identification of numerous host diseases, including toxoplasmosis, irritable bowel syndrome, and Clostridium difficile infection (CDI), has been made possible by bioinformatics based examination of the gut microbiota data from metagenomics studies (Kaur, H. et al., 2021; Gou, W. et al., 2021; Cammarota, G. et al., 2021; de Vos, W.M., 2012). Additionally, predictive machine learning models, like SVMs and Random Forests, and artificial neural network (ANN) algorithms, like DNNs and CNNs, have been established for the detection of microbiome–host interactions (Marya, N.B. et al., 2021; Reiman, D. et al., 2017; Reiman, D. et al., 2021). Many AI-based tools, such as "gutMDisorder" and "DISBIOME", have been used, for example, to establish links between the gut microbiome and human diseases, while freely accessible databases, such as "GMrepo v2 and gcMeta", offer an all-inclusive collection of unprocessed along with processed data from sequencing metagenomic studies. Furthermore, an AI-based technique called PopPhy-CNN is employed to predict the host phenotype using metagenomic datasets. Similarly, Read2Pheno is a deep-network construct that can look for nucleotide sections in the sequencing reads and makes accurate read-level phenotypic predictions for microbiome sequence taxonomy classification (Janssens, Y. et al., 2018; Cheng, L. et al., 2020; Dai, D. et al., 2022; Shi, W. et al., 2019; Reiman, D. et al., 2020; Zhao, Z. et al., 2021).

A plethora of computational genomics techniques have been developed to identify a given DNA sequence, predict genes, ascertain their roles, and approximate the location of protein-coding genes. Surprisingly, after three decades of existence, the Basic Local Alignment Search Tool (BLAST) remains a reliable and straightforward heuristic algorithm for the initial examination of each DNA sequence, identifying similar regions, and assigning sequences to positions within a reference genome (SF, A. 1990). Similarly, raw transcriptome data are increasingly being stored in the Expressed Sequence Tag (EST) databases, and raw sequencing data for cDNA, RNA, and DNA are stored in the well-known Sequence Read Archive (SRA) database (Carter, E.L. et al., 2024; Jongeneel, C.V. (2000); Leinonen, R. et al., 2010). Gene expression data for multiple diseases have been produced in vast quantities by microarray and RNA-seq methods. Some of the large repositories that contain genetic expression data are the NCBI Gene Expression Omnibus (GEO) (https://www.ncbi.nlm.nih.gov/geo/), the Cancer Genome Atlas (https://www.cancer.gov/about-nci/organization/ccg/research/structural-genomics/tcga), and Arrayexpress (https://www.ebi.ac.uk/arrayexpress/) (Gupta, R. et al., 2021). IJzendoorn et al. (2019) identified novel biomarkers and their potential drug targets for soft tissue sarcoma using ML and gene expression data patterns (van IJzendoorn, D.G. et al., 2019). TCGA and the National Cancer Institute Genomic Data Commons (NCIGDC) (https://gdc.cancer.gov/) are cancer-related sequencing data repositories. Han et al. (2019) created DriverML (https://github.com/HelloYiHan/DriverML), a supervised machine-learning tool that can identify driver genes linked to cancer by utilizing big data and artificial intelligence (Gupta, R. et al., 2021).

Advances in machine and deep learning might positively impact various research fields, including individual-based precision medicine, host-pathogen interaction studies, and predictive epidemiology (Zhang, W. et al., 2017). These methods support speedier decision-making, improved risk analysis, more precise forecasting with fewer errors, and illness diagnosis and case identification

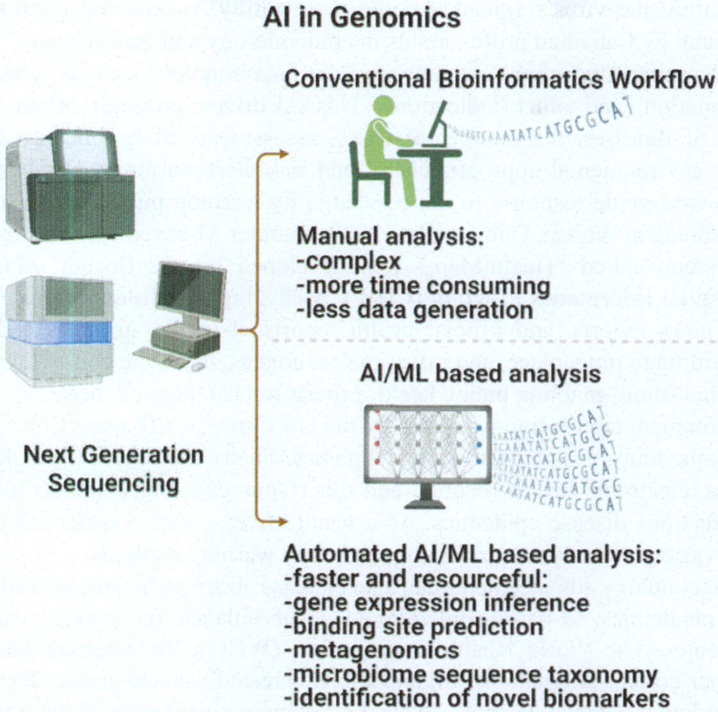

Fig. 3 Depicting the role of artificial intelligence in advanced genomics

for specific individuals. The proliferation of tissue biomarkers and the intricacy of their assessments greatly encourage the application of AI-based methods. Physicians can benefit from these AI-based biomarkers to predict and analyze diagnostic outcomes, patient responses to therapy, and patient survival (Lancellotti, C. et al., 2021). Reasoning modelling and knowledge representation enable more realistic models of complex socio-biological systems (Bedi, G. et al., 2015). Additionally, the effectiveness and dependability of epidemiological models can be increased by applying ML-based techniques (Pinaire, J. et al., 2017).

5.1 AI in Epidemiology

Machine learning-based AI applications have an impactful role in healthcare epidemiology, and they rely on supervised learning. The process of developing tools for applications in epidemiology involves the patient data collected, such as age, gender, disease attributes, etc. This data can be used as a dossier for the algorithms being developed, thus allowing interaction between the computer and the patient. This will assist the computer in assessing the prospective ailments in patients. The resultant algorithm, which has the potential to be the most accurate and precise recollection capacity, is selected and followed by validation. The data obtained from the validated models is compared to the reports from real-world patients. The models with the most successful rate can be used to predict disease patterns alone or in combination with clinical expertise (Hamilton, A.J. et al., 2021). To help people stay informed during crucial times like the COVID-19 pandemic, the Indian government developed "Arogya Setu" in partnership with stakeholders (both public and private), including the "National Informatics Centre (NIC) under the Ministry of Electronics and Information Technology". The application gives users access to self-assessment data and real-time contact tracking. Even though the app has only been downloaded 210 million times in India, its large user base made it possible to gather and analyze a large amount of data, allowing the AI system to accurately advise people and health authorities on tackling the COVID-19 epidemic. Numerous lives were saved throughout India by its AI-driven self-assessment tools and contact tracing features, which also assisted in limiting the virus's spread. Another noteworthy AI-powered platform is "BlueDot," which was created by Canadian professionals in epidemiology and data science.

Machine learning algorithms examine various data sources, such as news articles, airline booking information, and other indications. This AI-driven program offers insights into the characteristics of illnesses, the number of cases, assessments of healthcare capacity, readiness for epidemics, environmental appropriateness, and risk alert warnings. For instance, "BlueDot" influenced the worldwide response to the pandemic by warning public health officials about the COVID-19 outbreak in Wuhan, China, in late 2019. Another AI-based epidemiological information monitoring system, called "HealthMap," was developed by the Boston, Massachusetts-based Children's Hospital Informatics Program (CHIP). Analyzing data from various sources, including social media, news reports, and official health reports, it tracks, arranges and displays illness outbreaks according to time, place, and infectious pathogens. It has the ability to track the spread of COVID-19 in real-time, enabling public health experts to take targeted measures. Furthermore, AI-based epidemiological monitoring was used by the US Center for Disease Control and Prevention (CDC) to track the transmission of infectious illnesses. Their "BioSense" technology analyzes data from ER visits, electronic health records, and other sources using machine learning algorithms to pinpoint infectious disease epidemics. As a result, diverse data sources are used by AI-based software for a variety of analysis, processing, and early warning methods.

AI-based technology advancements can also increase access to healthcare and reduce inequities brought on by pandemics. AI-based epidemiological surveillance, for instance, can guarantee equal access to vaccines. The World Health Organization (WHO), the International Monetary Fund (IMF), and other critical international organizations detected vaccine-access disparities during the COVID-19 pandemic (Galetsi, P. et al., 2022). Following national vaccine distributions, disparities in access to vaccines was predominantly driven by inadequate distribution of vaccines and people's hesitation towards getting vaccinated, despite initially being a supply chain related and production related issue. In turn, issues with uneven vaccination uptake were brought on by a data shortage, accurate information, health archives, education, and digitized monitoring systems (Routledge, I. et al., 2022). To ensure that vaccinations can reasonably reach everyone, AI-powered solutions can expedite vaccine manufacturing, management of supply chains, distribution routes, instructional materials, and data collection depending on target groups' demographic and geographic features. As a result, using AI-based methods can lower vaccination disparities, rate of infections, and vaccine reluctance (Anjaria, P. et al., 2023).

The possibility of computer software producing false positive or false negative test outcomes is one possible drawback. Due to limits in the algorithms or accessible data, AI-based systems may overlook critical signals or lack the ability to predict the trends unrelated to disease outbreaks. This drawback emphasizes the necessity of continuous observation and assessment in order to guarantee the long-term efficacy of AI-based epidemiology surveillance (Bernstein, M.H. et al., 2023). Safeguarding personal data against issues linked to data security or privacy is another hurdle. Sensitive data, like social media activity or personal health information, may be gathered and analyzed by AI systems, for instance. This data needs to be utilized, kept, and safeguarded securely (Murdoch, B. 2021). Should people believe that corporations are not upholding their right to privacy over their data, then public confidence in these systems may be jeopardized. To ensure their appropriate and ethical usage, epidemiological monitoring systems based on AI must be properly regulated and overseen. To preserve public confidence, clear policies pertaining to data gathering, sharing, and analysis are necessary, along with risk reduction techniques (Aldboush, H.H. and Ferdous, M. 2023).

5.2 AI in Bio-robotics

Driven by the staggering volume of biological data, ongoing improvements in processing speed, storage capacity, and revolutionary advances in machine learning techniques, artificial intelligence is shedding its science fiction veneer and, beyond the field of biology, is rapidly permeating medicine. Although the first clinical-decision AI systems were developed in the 1970s, in recent years, deep learning (DL) has emerged as the leading methodology aiming to guarantee precise and timely diagnosis, support decision-making, lower human error, ease medical procedures, and enhance effectiveness of health systems worldwide (Ahmed, Z. et al., 2020; Mukherjee, J. et al., 2023). The widespread applications of AI in applications including image analysis, MRIs and CT scans are discussed in one of the previous sections of the chapter. Further, whole slide imaging (WSI)-digital histopathology, is another rapidly growing application of AI in the field that has helped with the detection, evaluation, and classification of various types of tumors (Kayser, K. et al., 2009; Schüffler, P. et al., 2023; Bera, K. et al., 2019). Moreover, ML and DL algorithms can effectively handle the vast volume of medical data that is accessible to predict outcomes, track patients in real-time, identify risk factors, and reduce medical mistakes in hospitals. For example, AI may determine whether a patient will benefit from a certain medical operation (such as surgery) or therapy (such as chemotherapy) by analyzing information from electronic health records (EHRs), including medical history, drugs, or test results. Additionally, it can predict patient mortality as well as the likelihood of certain infections, septic shock episodes, readmission to the hospital, and post-surgical complications (Gupta, R. et al., 2021).

In the surgical field of medicine, robotics is one of the more quickly emerging uses of AI that has grown in popularity in recent years. Surgical robotic systems have been around since 1985 and are common in operating rooms. Their primary function is to aid with surgical operations and provide stability and precision in surgical motions while still requiring human professionals to manage or oversee them (Peters, B.S. et al., 2018; Beaman, C.B. et al., 2021; Sheth, K.R. and Koh, C.J.). For example, one of the most commonly used robotic systems in minimally invasive surgeries is the "FDA-approved da Vinci® Surgical System (Intuitive Surgical, Sunnyvale, CA, USA)". It makes use of numerous robotic arms that can be remotely controlled by a console to mimic any surgeon's hand actions and provide accuracy in gestures and filters out tremors (Sheth, K.R. and Koh, C.J. (2019); Rassweiler, J.J. et al., 2017; Jiang, B. et al., 2019). Furthermore, significant advancements are being made in creating autonomous robots that can do particular tasks independently, which will benefit surgical practice and save a great deal of time. Amazingly, autonomous robots capable of suturing and cochleostomy have already been created (Athanasopoulou, K. et al., 2022; Wang, T. et al., 2023). A list of applications of artificial intelligence in bio-robotics is given in Table 2.

The current global market for healthcare robots is expected to grow to a value of over £15 billion and is expected to rise at a compound annual growth rate of 35.8% from 2023 to 2030. Surgical support, medicine delivery, medical transportation, telepresence, rehabilitation, and cleanliness have been the main areas of attention for medical robots thus far. Robots are now utilized to increase surgical precision and efficiency, even if they may interfere with anaesthesia by changing the patient position and obstructing communication in the far future (Pauli, H. et al., 2019). Target-controlled anaesthesia with EEG in a feedback loop characterizes pharmaceutical robots. Mechanical robots are more precise and dexterous than humans, and cognitive robots serve as decision-support systems. Pharmaceutical-robots have been utilized for pain management, temperature regulation, hypnosis, and homeostasis. Evidence shows these applications have reduced workloads and improved safety compared to manual methods. The anaesthetists can override systems like McSleepy, which are intended to control hypnosis, analgesia, and neuromuscular blockades on their own (Zaouter, C., 2020). In endoscopy, robotic support is also utilized to deploy rigid or flexible endoscopes and magnetically actuated capsules in gastrointestinal tract imaging, bronchoscopy, and surgery (Li, Z., and Chiu, P.W.Y. 2018). Endoscope manoeuvring is difficult and takes a lot of practice to become proficient at, which is why automated motions are appealing. Methods that rely on learning by example, including behavioural cloning, show promise.

AI approaches are applied in the analysis of the massive picture datasets utilized in diagnostics, as well as in the localization and mapping of endoscopes (Wang, H. et al., 2023; Golland, P. et al, 2014). Some major companies across the globe that are involved in AI-assisted robotics in surgery include "Asensus Senhance (designed by TransEnterix, renamed from the Alf-X in 2016 developed by Italian SOFAR SpA), CMR Versius, Medtronic Hugo, Medicaroid Hinotori, MeerCompany Revo-i, Harbin Sagebot KangDuo, MemicHominis system, and Distalmotion Dexter, Avatera Medical, Medbot Toumai, SSI Mantra, RobSurgical Bitrack, RiverField, Asahi ANSUR, Micro Hand S (Tianjin University), Sina Flex-Sina Robotics, Edge Medical Robotics, Carina-Ronovo Surgical, MIRA surgical system, Luna-Asensus, and Vicarious-Vicarious Surgical". Policies should be created since patient data privacy and confidentiality present severe legal and ethical concerns that need to be considered. This includes everything from electronic health records to intraoperative recordings used in predictive AI models. Additionally, there is a chance of hacking if the physical hard drives are not completely protected or if the platforms used to gather, store, and share data have access to the Internet, such as cloud platforms.

Table 2 The applications AI-based algorithms in the field of bio-robotics

Applications	*AI-based algorithms*	*Detailed use*
Biomedical Imaging and its analysis and interpretation	Arterys Cardio DL	Automatic analysis of MRI scans(cardiac)
	Arterys Oncology DL	MRI and CT scans for the detection of liver lesions and lung nodules
	EchoMD Automated Ejection Fraction Software	Analyzing echocardiograms to determine the left ventricular ejection fraction
	HealthPNX	Analyzing chest X-rays to look for pneumothorax-related symptoms
	Icobrain	Examination of patients' brain injuries and neurological conditions using CT and MRI scans
	IDx-DR	Examination of retinal pictures to identify diabetic retinopathy
	OsteoDetect	Examining wrist X-rays to diagnose fractures
	ProFound™ AI Software V2.1	Using mammography to analyze breast density and find cancerous tumors
	QuantX™	Breast lesion diagnosis using MRI scan analysis
	Viz LVO	CT angiography scans for the identification of stroke symptoms
AI based surgical robotics	Da Vinci® Surgical System	Precision and tremor filtering are provided via a telemanipulated, minimally invasive robotic system with robotic arms that interpret user hand motions.
	Flex® robotic System	Flexible robotic endoscope that enables surgeons to view and access anatomical regions that are often out of reach for minimally invasive procedures
	FreeHand® v1.2	Robotic camera controller that cuts down on operating time, improves control over camera motions, and produces steadier images
	NAVIO™ Surgical System	A portable robotic device that provides real-time bone imaging, implant placement planning, and robotically assisted bone preparation
	NeoGuide™ Endoscopy System	Lower gastrointestinal (GI) tract imaging made possible by a computer-assisted colonoscope that adapts to the form of the colon to reduce looping
	Senhance® Surgical System	Three robotic arms on a remote-controlled robotic device that offers better haptic feedback and visuals
	Sensei® X Robotic Catheter System	Remotely operated cardiac catheter that stabilizes the catheter during placement by translating the surgeon's hand movements

Contd.

Table 2 *Contd.*

Patient monitoring devices/tools	ADAMM Intelligent Asthma Monitoring	Wearable technology worn on the chest that tracks symptoms of asthma and issues warnings when there are noticeable differences
	CardioMEMS™ HF System	Utilizing an implanted PA sensor, a wireless monitoring device measures the pulmonary artery (PA) pressure in heart failure patients. The healthcare provider sends and evaluates the daily data.
	Confirm Rx™ Insertable Cardiac Monitor	An implantable gadget that tracks patients' heart rates, looks for arrhythmia symptoms, and sends information to medical professionals
	ReDS™ System	A tool that examines the absolute fluid content in the lungs of individuals suffering from heart failure quickly and non-invasively.
	Triggerfish®	A contact lens with an integrated microsensor that tracks changes in intraocular pressure in glaucoma patients over the course of 24 hours by measuring changes in ocular dimensions
Personal devices with use remote sensing	Eversense® E3 Continuous Glucose Monitoring (CGM) System	An external transmitter, an implanted glucose sensor that continuously measures blood sugar, and a mobile app that shows data in real time
	Fitbit	A smart wristwatch that tracks exercise, sleep patterns, and atrial fibrillation symptoms
	KardiaMobile® 6L	Cardiac arrhythmias can be detected via a portable cardiac monitor
	Study Watch with Irregular Pulse Monitor	Wearable gadget that tracks biometric data, including heart rate variability and electrical activity

6. AI in Healthcare and Medicine

In precision medicine, including individual variations in genetics, ecology, and lifestyle are critical. Medical professionals know that a person's genetic, physical, physiological, and metabolic characteristics influence how their body reacts to medications. Despite this, we presently use a general strategy wherein the same medication is administered to all patients, irrespective of the differences in their situations. However, a new age of personalized medicine-one in which medications are customized to the body's demands and adaptability-is emerging, primarily because of advancements in artificial intelligence (Figure 4). Despite its seeming simplicity, the change requires substantial data processing, execution, maintenance, and gathering (Mathur, S. and Sutton, J. 2017). In addition, determining which therapeutic candidate molecules are most appropriate for a given situation will include millions of prediction analyses. By employing this method, medical professionals may more accurately anticipate which disease management and prophylactic techniques will work best for specific patient populations. To better visualize the impact of medication dosages on live tissue over time and restructure signalling networks during therapy, researchers might employ AI in DNA, RNA, and protein investigations (Azofeifa, J.G. and Dowell, R.D. (2017); Azofeifa, J.G. et al., 2018).

IBM Watson uses artificial intelligence to help create a patient's most effective treatment strategy based on their genetic composition, medical history, and other personal information (Bhardwaj, A. et al., 2022; Kumar, A. et al., 2021). Precision medicine aims to customize medical care to each patient's particular traits, including genetic composition, way of life, and surroundings. By evaluating enormous datasets and finding patterns that correspond with particular medical problems, artificial intelligence (AI) plays a critical role in allowing customized therapy. This enables medical professionals to give specialized care and interventions, improving patient outcomes while lowering side effects. What the National Research Council first referred to as the creation of "a New

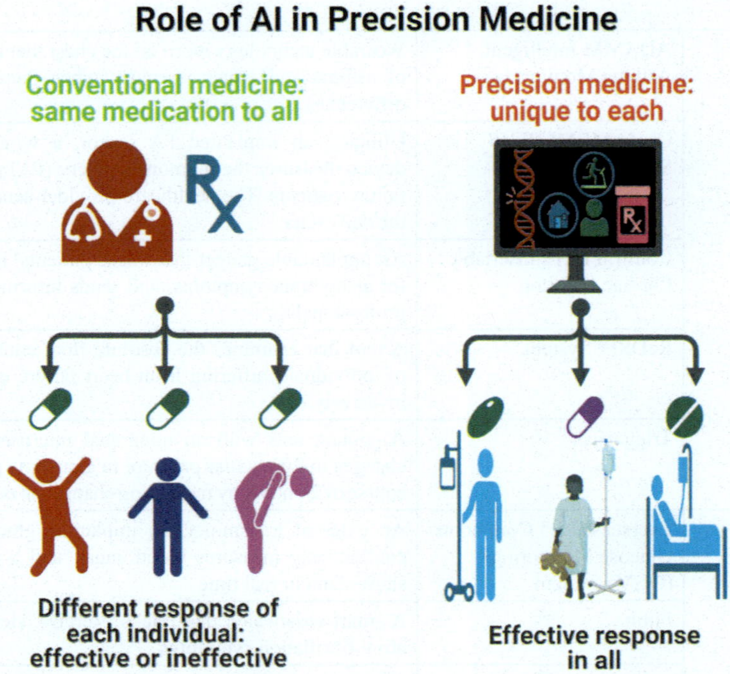

Fig. 4 A representative model of the advantages of artificial intelligence in the field of health and precision medicine

Taxonomy of human disease based on molecular biology" or a revolution in healthcare brought about by insights obtained from sequencing the human genome are probably the best ways to characterize precision medicine. The discipline has subsequently developed to understand how exactly health states, illness states, and treatment choices for afflicted persons are represented by the junction of multi-omics data with medical history, social/behavioural variables, and environmental information (National Research Council. (2011); Ziegelstein, R.C. 2017). Genotype-guided therapy is arguably the most thoroughly researched use of precision medicine in healthcare today. Genotype information has been used by clinicians as a reference to assist them in identifying the appropriate warfarin dosage. Guidelines for genotype-based medication therapy were released by the Clinical Pharmacogenetics Implementation Consortium to assist physicians in maximizing medicinal treatments in conjunction with genetic test results. Targeted treatment regimens for individuals with lung or breast cancer can be informed by the genomic profiling of their malignancies (Hartmaier, R.J. et al., 2017; Jorgensen, A.L. et al., 2019; E. Caudle, K. et al., 2014).

Previously, neural networks were utilized through a learning process to link multi signal combinations and their effects. This led to the discovery that a smooth quadratic surface represents the connection between dosage and treatment efficacy. A phenotypic personalized medicine platform (PPM) comprises data generated using system biology and omics. This combination can determine the drug and its dose based on the patient's medical records or EHRs. The main goal of any PPM is to maximize the desired effects of any drug with minimal dosages. In this aspect, "CURATE.AI" was created by Professor Dean Ho and his team in Singapore; it is an AI-driven approach used for dose selection and even for combinations of drugs-a platform for dose optimization that takes into account individual characteristics to increase the accuracy of medication therapy (Mukhopadhyay, A. et al., 2022). CURATE.AI calibrates each patient's medicine dose based on their unique clinical data by employing a small data method. Different pharmacological doses are administered to each patient, and the patient's reactions to them are recorded. Each patient's digital profile is then created using these datasets with additional pertinent clinical data. CURATE. AI determines the appropriate dosages to maximize treatment results for every digital profile by connecting medication dosage to treatment efficacy and safety. The dosage may change throughout the treatment (Mukhopadhyay, A. et al., 2022; Blasiak, A. et al., 2020). Its potential to modify the combination treatment for individuals has been clinically verified in a prospective pilot trial for mCRPC. This work has extended CURATE and cleared the way for clinical use. AI-enabled dosage optimization for future immunosuppressive therapy (NCT03527238) and multiple myeloma (NCT03759093). Similarly, QPOP has demonstrated validation for a wide range of indications for multiple myeloma, TB, and cancer in several in vivo investigations (Blasiak, A. et al., 2020).

Using the information on patients' gene expression, McDonald et al. trained a support vector machine to forecast the patients' reaction to chemotherapy. Their statistics demonstrate positive results for a variety of medications. Based on the gene-expression profiles (RNA-seq or microarray) of each patient's tumor, their data demonstrated the use of our previously developed open-source support vector machine (SVM)-based algorithm to predict the responses of 175 distinct cancer patients to various standard-of-care chemotherapeutic drugs. The models could predict over 80% of the patients' reactions (Huang, C. et al., 2018).

7. Artificial Intelligence in Toxicology

By combining statistically mined molecular fragments with expert knowledge-based alerts, Chakravarti et al. (2023) created hybrid QSAR models that outperformed models based only on one technique in terms of predictive performance, improving the specificity and exposure of structural alerts *in silico* toxicology (Huang, C. et al., 2018). Inducing hybrid vectors using chemical and bioactivity fingerprints is an approach that Yang et al. presented and has shown to be useful for repeat-dose NOAEL value prediction. They created a read-across method based on machine learning and chemo-informatics, which they applied to predict NOAELs for compounds with little data. They employed a wide range of bisphenols and their metabolites to demonstrate the efficacy of their methodology. Additionally, the scientists provided a framework for evaluation that could effectively rate a variety of chemicals and had room for advancement through the use of artificial intelligence techniques (Yang, C. et al., 2023). The quantitative read-across structure–activity relationship (q-RASAR) approach was recently introduced by Banerjee et al. and improved predictive models across multiple toxicity data sets compared to conventional QSAR models.

Q-RASAR incorporates novel similarity-based functions as additional descriptors (Banerjee, A. and Roy, K. 2023). Lester et al. proposed an approach that quantitatively includes biological and toxicological features using fingerprint keys to compute a similarity score for read-across predictions in toxicological safety assessment. This approach increases regulatory acceptance and improves transparency and consistency among experts (Lester, C. et al., 2023). The research above revealed how useful artificial intelligence techniques like expert systems, structural warnings, and read-across methods are for enhancing prediction performance and developing the area of in silico toxicology. Using multiple instance learning (MIL), Feeney et al. grouped compounds under a single mutagenicity label and their (predicted) xenobiotic metabolites. The MIL models achieved results equivalent to known models in predicting Ames mutagenicity for aromatic amines. According to their results, MIL may be included in current models, particularly for xenobiotic mutagens that are metabolically activated in Ames experiments, including the S9 rodent liver enzyme. Ames mutagenicity has also been studied by Lui et al., who looked at the usefulness of multitask deep learning and toxicological domain knowledge for task grouping in QSAR model development. According to their findings, grouped multitask neural networks based on mechanistic task groupings perform better than single-task models and ungrouped neural networks. This highlights the advantages of combining data-driven and mechanistic modelling in creating toxicological models (Klambauer, G. et al., 2023). In addition to improving prediction skills for a range of toxicity endpoints, applying machine learning and deep learning techniques in QSAR modelling provides insightful information on toxicological processes.

8. Regulations on AI

The first AI-related law in the world was suggested by the European Commission in April 2021 and is known as the Artificial Intelligence Act. The proposed law establishes uniform AI regulations to safeguard ethical standards and further the goal of the European Union being a global leader in developing safe, reliable, and moral AI. Before being put on the market in the European Union, "high-risk" AI systems that pose a serious risk to people's health, safety, or fundamental rights adhere to conformity assessment procedures and mandatory requirements for trustworthiness (European Union (EU), 2021). The UN Educational, Scientific, and Cultural Organization (UNESCO) and its 193 member states adopted an agreement in November 2021 that outlines the common values and principles required to ensure the healthy development of artificial intelligence (European Union (EU), 2021). In June 2021, the World Health Organization (WHO) published a guidance document on "Ethics and Governance of Artificial Intelligence for Health" (European Union (EU), 2021). Additionally, this UNESCO agreement calls on governments to establish a legal framework that outlines a process for conducting ethical impact assessments on AI systems, especially for public authorities. These assessments aim to anticipate outcomes, reduce risks, prevent adverse outcomes, promote citizen participation, and address societal issues.

Artificial intelligence-based systems for diagnosis, epidemiological data generation, etc. all hold certain implications regarding public health. It is first of all very important to define a standard set of regulations and ethical standards that can provide for the assurance of data security and privacy. It is crucial so as to protect huge quantities of sensitive data from numerous data invaders and unauthorized access. Further, the deliberate use of the data produced is still another factor. The implication is that information gathered from many sources for a certain purpose should only be used for that, rather than being misused elsewhere. This calls for generation of stringent guidelines in order to prevent any sort of misuse and data breach. Another aspect in this context is the financial investments in the R&D sector of AI-based technologies. This will encourage fresh innovative concepts resulting in the evolution of existing technologies and enhance effective usage of existing systems. Also, the developers should ensure that while inputting of data for designing their algorithms there should be no bias on the basis of gender, race, age, etc. to ensure equanimity of target populations (Anjaria, P. et al., 2023).

9. Challenges and Future Directions

One significant limitation is the availability and quality of data required for AI algorithms to function optimally. In the realm of biology, acquiring comprehensive and well-annotated datasets can be challenging. Another limitation arises from the interpretability and explainability of AI-driven biological models. Complex deep learning algorithms often act as "black boxes," making it difficult for researchers to understand how decisions are reached. Despite its strength, symbolic AI is confined to systems that follow well-defined rules, which is not always the case in the domain of living systems. Furthermore, there is no resemblance between symbolic AI and biological intelligence in terms of operation and functioning. Symbolic AI can only make decisions based on a predefined set of rules. On the other hand, biological intelligence can learn on the fly and make judgments based on knowledge gained via experience and observation of objects. AI techniques like genetic algorithms and machine learning are revolutionizing biological research. The fusion of AI and biology has led to remarkable advancements in robotics and bioengineering. Bio-robotics explores the creation of robots that mimic living organisms, while prosthetics and exoskeletons enhance human capabilities. Integrating AI and biology can aid environmental scientists and conservationists in monitoring and analyzing ecosystems. The future of AI-biology integration holds promise. Innovations in synthetic biology, neuro-engineering, and brain-computer interfaces can pave the way for groundbreaking discoveries.

Conclusion

The capacity to identify objects and make decisions stands as a pivotal trait within biological systems. Presently, AI has the ability to identify objects and render pronouncements using many of the cognitive and emotive skills inherent in living systems. The potential of AI could find practical applications in the realm of biology, encompassing fields like medical research, farming and agriculture, and bio-based industries, all contributing to our pursuit of a sustainable lifestyle. The merging of AI and biology has restructured scientific research, healthcare, and our understanding of life itself. As AI technologies advance astonishingly, the synergy between AI and biology opens up new avenues for exploration, discovery, and innovation. The convergence of AI and biology carries immense significance and is poised to impact society. Combining AI algorithms' power with biological insights can unlock new possibilities and unravel complex mysteries. This union has the potential to improve medical diagnosis, accelerate drug discovery, enhance conservation efforts, and drive innovation in agriculture. The implications are far-reaching and can transform how we understand and interact with the biological world.

References

Abràmoff, M.D., Lavin, P.T., Birch, M., Shah, N., Folk, J.C. et al. (2018). Pivotal trial of an autonomous AI-based diagnostic system for detection of diabetic retinopathy in primary care offices. NPJ digital medicine, 1(1), 39.

Ahmed, Z., Mohamed, K., Zeeshan, S. and Dong, X. (2020). Artificial intelligence with multi-functional machine learning platform development for better healthcare and precision medicine. Database, 2020, baaa010.

Ahmed, Z., Mohamed, K., Zeeshan, S. and Dong, X. (2020). Artificial intelligence with multi-functional machine learning platform development for better healthcare and precision medicine. Database, 2020.

Aldboush, H.H. and Ferdous, M. (2023). Building Trust in Fintech: An Analysis of Ethical and Privacy Considerations in the Intersection of Big Data, AI, and Customer Trust. International Journal of Financial Studies, 11(3), 90.

Anchan, S.K. A Review on Blue Brain Technology (2023). International Research Journal of Modernization in Engineering Technology and Science, 5(5).

Anjaria, P., Asediya, V., Bhavsar, P., Pathak, A., Desai, D. and Patil, V. (2023). Artificial Intelligence in Public Health: Revolutionizing Epidemiological Surveillance for Pandemic Preparedness and Equitable Vaccine Access. Vaccines, 11(7), 1154.

Anjaria, P., Asediya, V., Bhavsar, P., Pathak, A., Desai, D. and Patil, V. (2023). Artificial Intelligence in Public Health: Revolutionizing Epidemiological Surveillance for Pandemic Preparedness and Equitable Vaccine Access. Vaccines, 11(7), 1154.

Athanasopoulou, K., Daneva, G.N., Adamopoulos, P.G. and Scorilas, A. (2022). Artificial intelligence: the milestone in modern biomedical research. BioMedInformatics, 2(4), 727–744.

Azofeifa, J.G. and Dowell, R.D. (2017). A generative model for the behavior of RNA polymerase. Bioinformatics, 33(2), 227–234.

Azofeifa, J.G., Allen, M.A., Hendrix, J.R., Rubin, J.D., Dowell, R.D. et al. (2018). Enhancer RNA profiling predicts transcription factor activity. Genome research, 28(3), 334–344.

Banerjee, A. and Roy, K. (2023). On some novel similarity-based functions used in the ML-based q-RASAR approach for efficient quantitative predictions of selected toxicity end points. Chemical Research in Toxicology, 36(3), 446–464.

Beaman, C.B., Kaneko, N., Meyers, P.M. and Tateshima, S. (2021). A review of robotic interventional neuroradiology. American Journal of Neuroradiology, 42(5), 808–814.

Beck, B.R., Shin, B., Choi, Y., Park, S., Kang, K. et al. (2020). Predicting commercially available antiviral drugs that may act on the novel coronavirus (SARS-CoV-2) through a drug-target interaction deep learning model. Computational and structural biotechnology journal, 18, 784–790.

Bedi, G., Carrillo, F., Cecchi, G.A., Slezak, D.F., Sigman, M., Mota, N.B. and Corcoran, C.M. (2015). Automated analysis of free speech predicts psychosis onset in high-risk youths. npj Schizophrenia, 1(1), 1–7.

Bera, K., Schalper, K.A., Rimm, D.L., Velcheti, V., Madabhushi, A. et al. (2019). Artificial intelligence in digital pathology-new tools for diagnosis and precision oncology. Nature reviews Clinical oncology, 16(11), 703–715.

Bernstein, M.H., Atalay, M.K., Dibble, E.H., Maxwell, A.W., Karam, A.R., Agarwal, S. and Baird, G.L. (2023). Can incorrect artificial intelligence (AI) results impact radiologists, and if so, what can we do about it? A multi-reader pilot study of lung cancer detection with chest radiography. European Radiology, 1–7.

Bhardwaj, A., Kishore, S. and Pandey, D.K. (2022). Artificial intelligence in biological sciences. Life, 12(9), 1430.

Biswas, N. and Chakrabarti, S. (2020). Artificial intelligence (AI)-based systems biology approaches in multi-omics data analysis of cancer. Frontiers in Oncology, 10, 588221.

Blasiak, A., Khong, J. and Kee, T. (2020). CURATE. AI: optimizing personalized medicine with artificial intelligence. SLAS TECHNOLOGY: Translating Life Sciences Innovation, 25(2), 95–105.

Branco, I. and Choupina, A. (2021). Bioinformatics: new tools and applications in life science and personalized medicine. Applied microbiology and biotechnology, 105, 937–951.

Brown, C.J. and Hamarneh, G. (2016). Machine learning on human connectome data from MRI. arXiv preprint arXiv:1611.08699.

Cammarota, G., Ianiro, G., Ahern, A., Carbone, C., Temko, A., Claesson, M.J. and Tortora, G. (2020). Gut microbiome, big data and machine learning to promote precision medicine for cancer. Nature reviews gastroenterology & hepatology, 17(10), 635–648.

Carter, E.L., Constantinidou, C. and Alam, M.T. (2024). Applications of genome-scale metabolic models to investigate microbial metabolic adaptations in response to genetic or environmental perturbations. Briefings in Bioinformatics, 25(1).

Chakraborty, I., Choudhury, A. and Banerjee, T.S. (2017). Artificial intelligence in biological data. J Inf Technol Softw Eng, 7(4), 207.

Chen, R.H. and Chen, C.C. (2022). Artificial Intelligence: An Introduction for the Inquisitive Reader. CRC Press.

Cheng, L., Qi, C., Zhuang, H., Fu, T., Zhang, X. et al. (2020). gutMDisorder: a comprehensive database for dysbiosis of the gut microbiota in disorders and interventions. Nucleic Acids Research, 48(D1), D554–D560.

Dai, D., Zhu, J., Sun, C., Li, M., Liu, J., Wu, S. and Chen, W.H. (2022). GMrepo v2: a curated human gut microbiome database with special focus on disease markers and cross-dataset comparison. Nucleic acids research, 50(D1), D777–D784.

de Vos, W.M. and de Vos, E.A. (2012). Role of the intestinal microbiome in health and disease: from correlation to causation. Nutrition reviews, 70(suppl_1), S45–S56.

Dias, R. and Torkamani, A. (2019). Artificial intelligence in clinical and genomic diagnostics. Genome medicine, 11(1), 1–12.

Dong, C., Loy, C.C., He, K. and Tang, X. (2015). Image super-resolution using deep convolutional networks. IEEE transactions on pattern analysis and machine intelligence, 38(2), 295–307.

Draft text of the Recommendation on the Ethics of Artificial Intelligence. 2021 [cited 2024 January 12]; https://unesdoc.unesco.org/ark:/48223/pf0000377897.

E Caudle, K., E Klein, T., M Hoffman, J., J Muller, D., Whirl-Carrillo, M., Gong, L. and G Johnson, S. (2014). Incorporation of pharmacogenomics into routine clinical practice: the Clinical Pharmacogenetics Implementation Consortium (CPIC) guideline development process. Current Drug Metabolism, 15(2), 209–217.

European Union (EU). Proposal for a Regulation of The European Parliament and of the Council Laying Down Harmonised Rules on Artificial Intelligence (Artificial Intelligence Act) And Amending Certain Union Legislative Acts. 2021 [Accessed on 2024 January 12]; https://eur-lex.europa.eu/legal-content/EN/TXT/?uri=celex%3A52021PC0206.

Fan, J., Wang, J., Chen, Z., Hu, C., Zhang, Z. and Hu, W. (2019). Automatic treatment planning based on three-dimensional dose distribution predicted from deep learning technique. Medical physics, 46(1), 370–381.

Galetsi, P., Katsaliaki, K. and Kumar, S. (2022). The medical and societal impact of big data analytics and artificial intelligence applications in combating pandemics: A review focused on COVID-19. Social Science & Medicine, 301, 114973.

Gentile, F., Agrawal, V., Hsing, M., Ton, A.T., Ban, F., Norinder, U. and Cherkasov, A. (2020). Deep docking: a deep learning platform for augmentation of structure based drug discovery. ACS central science, 6(6), 939–949

Ghazi, M.M. and Nielsen, M. (2022). FAST-AID Brain: Fast and accurate segmentation tool using artificial intelligence developed for brain. arXiv preprint arXiv:2208.14360.

Golland, P., Hata, N., Barillot, C., Hornegger, J., Howe, R. et al. (2014, January). 17th International Conference on Medical Image Computing and Computer-Assisted Intervention. In Medical Image Computing and Computer-assisted Intervention: MICCAI International Conference on Medical Image Computing and Computer-assisted Intervention,17(2), V–VI

Górriz, J.M., Ramírez, J., Ortíz, A., Martinez-Murcia, F.J., Segovia, F., Suckling, J., Ferrandez, J.M. (2020). Artificial intelligence within the interplay between natural and artificial computation: Advances in data science, trends and applications. Neurocomputing, 410, 237–270.

Gou, W., Ling, C.W., He, Y., Jiang, Z., Fu, Y., Xu, F. and Zheng, J.S. (2021). Interpretable machine learning framework reveals robust gut microbiome features associated with type 2 diabetes. Diabetes Care, 44(2), 358–366.

Gupta, A., Sheth, P. and Xie, P. (2022). Neural architecture search for pneumonia diagnosis from chest X-rays. Scientific reports, 12(1), 11309.

Gupta, R., Srivastava, D., Sahu, M., Tiwari, S., Ambasta, R.K. and Kumar, P. (2021). Artificial intelligence to deep learning: machine intelligence approach for drug discovery. Molecular diversity, 25, 1315–1360.

Habuza, T., Navaz, A.N., Hashim, F., Alnajjar, F., Zaki, N., Serhani, M.A. and Statsenko, Y. (2021). AI applications in robotics, diagnostic image analysis and precision medicine: current limitations, future trends, guidelines on CAD systems for medicine. Informatics in Medicine Unlocked, 24, 100596.

Hamilton, A.J., Strauss, A.T., Martinez, D.A., Hinson, J.S., Levin, S., Lin, G. and Klein, E.Y. (2021). Machine learning and artificial intelligence: Applications in healthcare epidemiology. Antimicrobial Stewardship & Healthcare Epidemiology, 1(1).

Hartmaier, R.J., Albacker, L.A., Chmielecki, J., Bailey, M., He, J., Goldberg, M.E. and Lipson, D. (2017). High-throughput genomic profiling of adult solid tumors reveals novel insights into cancer pathogenesis. Cancer Research, 77(9), 2464–2475.

Hassani, H., Silva, E.S., Unger, S., TajMazinani, M. and Mac Feely, S. et al. (2020). Artificial intelligence (AI) or intelligence augmentation (IA): what is the future? AI, 1(2), 8.

Hassoun, S., Jefferson, F., Shi, X., Stucky, B., Wang, J. and Rosa Jr, E. (2021). Artificial intelligence for biology. Integrative and Comparative Biology, 61(6), 2267–2275.

Higaki, T., Nakamura, Y., Tatsugami, F., Nakaura, T., Awai, K. et al. (2019). Improvement of image quality at CT and MRI using deep learning. Japanese journal of radiology, 37, 73–80.

Hoffmann, T. and Gastreich, M. (2019). The next level in chemical space navigation: going far beyond enumerable compound libraries. Drug discovery today, 24(5), 1148–1156.

Holland, J.H. (1986). Induction: Processes of inference, learning, and discovery. MIT press.

Huang, C., Clayton, E.A., Matyunina, L.V., McDonald, L.D., Benigno, B.B., Vannberg, F. and McDonald, J.F. (2018). Machine learning predicts individual cancer patient responses to therapeutic drugs with high accuracy. Scientific reports, 8(1), 16444.

Irimia, A., Torgerson, C.M., Goh, S.Y.M., Van Horn, J.D. et al. (2015). Statistical estimation of physiological brain age as a descriptor of senescence rate during adulthood. Brain imaging and behavior, 9, 678–689.

Janssens, Y., Nielandt, J., Bronselaer, A., Debunne, N., Verbeke, F., Wynendaele, E. and De Spiegeleer, B. (2018). Disbiome database: linking the microbiome to disease. BMC microbiology, 18(1), 1–6.

Jiang, B., Azad, T.D., Cottrill, E., Zygourakis, C.C., Zhu, A.M., Crawford, N. and Theodore, N. (2019). New spinal robotic technologies. Frontiers of Medicine, 13, 723–729.

Jiang, D., Dou, W., Vosters, L., Xu, X., Sun, Y. and Tan, T. (2018). Denoising of 3D magnetic resonance images with multi-channel residual learning of convolutional neural network. Japanese journal of radiology, 36, 566–574.

Jongeneel, C.V. (2000). Searching the expressed sequence tag (EST) databases: panning for genes. Briefings in Bioinformatics, 1(1), 76–92.

Jorgensen, A.L., Prince, C., Fitzgerald, G., Hanson, A., Downing, J., Reynolds, J. and Pirmohamed, M. (2019). Implementation of genotype-guided dosing of warfarin with point-of-care genetic testing in three UK clinics: a matched cohort study. BMC medicine, 17, 1–11.

Kaur, H., Singh, Y., Singh, S. and Singh, R.B. (2021). Gut microbiome-mediated epigenetic regulation of brain disorder and application of machine learning for multi-omics data analysis. Genome, 64(4), 355–371.

Kayser, K., GÃśrtler, J., Bogovac, M., Bogovac, A., Goldmann, T., Vollmer, E. and Kayser, G. (2009). AI (artificial intelligence) in histopathology–from image analysis to automated diagnosis. Folia Histochemica et Cytobiologica, 47(3), 355–361.

Klambauer, G., Clevert, D.A., Shah, I., Benfenati, E., Tetko, I.V. et al. (2023). Introduction to the Special Issue: AI Meets Toxicology. Chemical Research in Toxicology, 36(8), 1163–1167.

Kumar, A., Gadag, S. and Nayak, U.Y. (2021). The Beginning of a New Era: Artificial Intelligence in Healthcare. Advanced Pharmaceutical Bulletin, 11(3), 414.

Lai, L.S., Redington, A.N., Reinisch, A.J., Unterberger, M.J., Schriefl, A.J. et al. (2016). Computerized automatic diagnosis of innocent and pathologic murmurs in pediatrics: a pilot study. Congenital heart disease, 11(5), 386–395.

Lancellotti, C., Cancian, P., Savevski, V., Kotha, S.R.R., Fraggetta, F., Graziano, P. and Di Tommaso, L. (2021). Artificial intelligence & tissue biomarkers: advantages, risks and perspectives for pathology. Cells, 10(4), 787.

Leinonen, R., Sugawara, H., Shumway, M. and International Nucleotide Sequence Database Collaboration. (2010). The sequence read archive. Nucleic acids research, 39(suppl_1), D19–D21.

Leite, M.L., de Loiola Costa, L.S., Cunha, V.A., Kreniski, V., de Oliveira Braga Filho, M., da Cunha, N.B. and Costa, F.F. (2021). Artificial intelligence and the future of life sciences. Drug Discovery Today, 26(11), 2515–2526.

Lester, C., Byrd, E., Shobair, M. and Yan, G. (2023). Quantifying Analogue Suitability for SAR-Based Read-Across Toxicological Assessment. Chemical Research in Toxicology, 36(2), 230–242.

Levakov, G., Rosenthal, G., Shelef, I., Raviv, T.R. and Avidan, G. (2020). From a deep learning model back to the brain—Identifying regional predictors and their relation to aging. Human brain mapping, 41(12), 3235–3252.

Li, Z. and Chiu, P.W.Y. (2018). Robotic endoscopy. Visceral medicine, 34(1), 45–51.

Liang, N. (2021, April). MRI Image Reconstruction Based on Artificial Intelligence. In Journal of Physics: Conference Series (Vol. 1852, No. 2, p. 022077).

Luger, G.F. (2021). Modern AI and How We Got Here. In: Knowing our World: An Artificial Intelligence Perspective. Springer, Cham. https://doi.org/10.1007/978-3-030-71873-2_3

Marya, N.B., Powers, P.D., Chari, S.T., Gleeson, F.C., Leggett, C.L., Dayyeh, B.K.A. and Levy, M.J. (2021). Utilisation of artificial intelligence for the development of an EUS-convolutional neural network model trained to enhance the diagnosis of autoimmune pancreatitis. Gut, 70(7), 1335–1344.

Mathur, S. and Sutton, J. (2017). Personalized medicine could transform healthcare. Biomedical reports, 7(1), 3–5.

McCarthy, J. (2022). Artificial Intelligence, Logic, and Formalising Common Sense. Machine Learning and the City: Applications in Architecture and Urban Design, 69–90.

Moinul, M., Amin, S.A., Kumar, P., Patil, U.K., Gajbhiye, A., Jha, T. and Gayen, S. (2022). Exploring sodium glucose cotransporter (SGLT2) inhibitors with Machine Learning Approach: A novel hope in anti-diabetes drug discovery. Journal of Molecular Graphics and Modelling, 111, 108106.

Mukherjee, J., Sharma, R., Dutta, P. and Bhunia, B. (2023). Artificial intelligence in healthcare: a mastery. Biotechnology and Genetic Engineering Reviews, 1–50.

Mukhopadhyay, A., Sumner, J., Ling, L.H., Quek, R.H.C., Tan, A.T.H., Teng, G.G. and Motani, M. (2022). Personalised Dosing Using the CURATE. AI Algorithm: Protocol for a Feasibility Study in Patients with Hypertension and Type II Diabetes Mellitus. International Journal of Environmental Research and Public Health, 19(15), 8979.

Murdoch, B. (2021). Privacy and artificial intelligence: challenges for protecting health information in a new era. BMC Medical Ethics, 22(1), 1–5.

National Research Council. (2011). Toward precision medicine: building a knowledge network for biomedical research and a new taxonomy of disease.

Nayor, J., Borges, L.F., Goryachev, S., Gainer, V.S., Saltzman, J.R. et al. (2018). Natural language processing accurately calculates adenoma and sessile serrated polyp detection rates. Digestive diseases and sciences, 63, 1794–1800.

Newton, D.E. (2018). The opioid crisis: A reference handbook. Bloomsbury Publishing USA.

Oliveira, A.L. (2019). Biotechnology, big data and artificial intelligence. Biotechnology journal, 14(8), 1800613.

Paul, D., Sanap, G., Shenoy, S., Kalyane, D., Kalia, K. and Tekade, R.K. (2021). Artificial intelligence in drug discovery and development. Drug discovery today, 26(1), 80.

Pauli, H., Eladawy, M. and Park, J. (2019). Anesthesia for robotic thoracic surgery. Annals of Cardiothoracic Surgery, 8(2), 263.

Peters, B.S., Armijo, P.R., Krause, C., Choudhury, S.A., Oleynikov, D. et al. (2018). Review of emerging surgical robotic technology. Surgical endoscopy, 32, 1636–1655.

Pinaire, J., Azé, J., Bringay, S. and Landais, P. (2017). Patient healthcare trajectory. An essential monitoring tool: a systematic review. Health Information Science and Systems, 5, 1–18.

Preuer, K., Klambauer, G., Rippmann, F., Hochreiter, S., Unterthiner, T. et al. (2019). Interpretable deep learning in drug discovery. Explainable AI: interpreting, explaining and visualizing deep learning, 331–345.

Pun, F.W., Liu, B.H.M., Long, X., Leung, H.W., Leung, G.H.D., Mewborne, Q.T. and Zhavoronkov, A. (2022). Identification of therapeutic targets for amyotrophic lateral sclerosis using PandaOmics–An AI-Enabled Biological Target Discovery Platform. Frontiers in Aging Neuroscience, 14, 638.

Pun, F.W., Ozerov, I.V. and Zhavoronkov, A. (2023). AI-powered therapeutic target discovery. Trends in Pharmacological Sciences.44(9), 561–572

Rassweiler, J.J., Autorino, R., Klein, J., Mottrie, A., Goezen, A.S., Stolzenburg, J.U. and Liatsikos, E. (2017). Future of robotic surgery in urology. BJU international, 120(6), 822–841.

Reiman, D., Farhat, A.M. and Dai, Y. (2021). Predicting host phenotype based on gut microbiome using a convolutional neural network approach. Artificial Neural Networks, 249–266.

Reiman, D., Metwally, A.A., Sun, J. and Dai, Y. (2020). PopPhy-CNN: a phylogenetic tree embedded architecture for convolutional neural networks to predict host phenotype from metagenomic data. IEEE Journal of Biomedical and Health Informatics, 24(10), 2993–3001.

Reiman, D., Metwally, A. and Dai, Y. (2017, July). Using convolutional neural networks to explore the microbiome. In 2017 39th Annual International Conference of the IEEE Engineering in Medicine and Biology Society (EMBC). 4269–4272.

Ren, F., Ding, X., Zheng, M., Korzinkin, M., Cai, X., Zhu, W. and Zhavoronkov, A. (2023). AlphaFold accelerates artificial intelligence powered drug discovery: efficient discovery of a novel CDK20 small molecule inhibitor. Chemical Science, 14(6), 1443–1452.

Routledge, I., Takahashi, S., Epstein, A., Hakim, J., Janson, O., Turcios, K. and Rodríguez-Barraquer, I. (2022). Using sero-epidemiology to monitor disparities in vaccination and infection with SARS-CoV-2. Nature communications, 13(1), 2451.

Savage, N. (2021). Tapping into the drug discovery potential of AI. https://www.nature.com/articles/d43747-021-00045-7. (Accessed on January 12, 2024).
Schüffler, P., Steiger, K. and Weichert, W. (2023). How to use AI in pathology. Genes, Chromosomes and Cancer, 62 (9), 564–567
Schwab, K. (2017). The fourth industrial revolution. Currency.
Sellwood, M.A., Ahmed, M., Segler, M.H. and Brown, N. (2018). Artificial intelligence in drug discovery. Future medicinal chemistry, 10(17), 2025–2028.
SF, A. (1990). Basic local alignment search tool. J Mol Biol, 215, 403–410.
Sheth, K.R. and Koh, C.J. (2019). The future of robotic surgery in pediatric urology: Upcoming technology and evolution within the field. Frontiers in Pediatrics, 7, 259.
Shi, W., Qi, H., Sun, Q., Fan, G., Liu, S., Wang, J. and Ma, J. (2019). gcMeta: a Global Catalogue of Metagenomics platform to support the archiving, standardization and analysis of microbiome data. Nucleic acids research, 47(D1), D637–D648.
Starr, T.K., Allaei, R., Silverstein, K.A., Staggs, R.A., Sarver, A.L., Bergemann, T.L. and Largaespada, D. A. (2009). A transposon-based genetic screen in mice identifies genes altered in colorectal cancer. Science, 323(5922), 1747–1750.
Sy, M.F., Roman, B., Kerrien, S., Mendez, D.M., Genet, H., Wajerowicz, W. and Hill, S.L. (2023). Blue Brain Nexus: An open, secure, scalable system for knowledge graph management and data-driven science. Semantic Web, 14(4), 697–727.
Taylor, R.A., Moore, C.L., Cheung, K.H. and Brandt, C. (2018). Predicting urinary tract infections in the emergency department with machine learning. PloS one, 13(3).
Thai, Q.M., Pham, T.N.H., Hiep, D.M., Pham, M.Q., Tran, P.T., Nguyen, T.H. and Ngo, S.T. (2022). Searching for AChE inhibitors from natural compounds by using machine learning and atomistic simulations. Journal of Molecular Graphics and Modelling, 115, 108230.
Turing, I.B.Y.A.M. (2007). Computing machinery and intelligence-AM Turing. Mind, 59(236), 433.
van IJzendoorn, D.G., Szuhai, K., Briaire-de Bruijn, I.H., Kostine, M., Kuijjer, M.L. and Bovée, J.V. (2019). Machine learning analysis of gene expression data reveals novel diagnostic and prognostic biomarkers and identifies therapeutic targets for soft tissue sarcomas. PLoS computational biology, 15(2).
Walters, W.P. (2018). Virtual chemical libraries: miniperspective. Journal of medicinal chemistry, 62(3), 1116–1124.
Wang, H., Wang, K., Yan, T., Zhou, H., Cao, E., Lu, Y. and Pang, Y. (2023). Endoscopic image classification algorithm based on Poolformer. Frontiers in Neuroscience, 17, 1273686.
Wang, T., Li, H., Pu, T. and Yang, L. (2023). Microsurgery Robots: Applications, Design, and Development. Sensors, 23(20), 8503.
Williams, A.M., Liu, Y., Regner, K.R., Jotterand, F., Liu, P. and Liang, M. (2018). Artificial intelligence, physiological genomics, and precision medicine. Physiological genomics, 50(4), 237–243.
World Health Organization (WHO). Ethics and governance of artificial intelligence for health. 2021 [Accessed on 2024 January 12]; https://www.who.int/publications/i/item/9789240029200.
Yan, J., Bhadra, P., Li, A., Sethiya, P., Qin, L., Tai, H.K. and Siu, S.W. (2020). Deep-AmPEP30: improve short antimicrobial peptides prediction with deep learning. Molecular Therapy-Nucleic Acids, 20, 882–894.
Yang, C., Rathman, J.F., Mostrag, A., Ribeiro, J.V., Hobocienski, B., Magdziarz, T. and Barton-Maclaren, T. (2023). High Throughput Read-Across for Screening a Large Inventory of Related Structures by Balancing Artificial Intelligence/Machine Learning and Human Knowledge. Chemical Research in Toxicology, 36(7), 1081–1106.
Ye, S. (2022). Estimating Epileptic Networks with High Density Electroencephalography and Magnetoencephalography (Doctoral dissertation, Carnegie Mellon University).
Zaouter, C. et al., Autonomous Systems in Anesthesia: Where Do We Stand in 2020? A Narrative Review. Anesth Analg, 2020. 130(5): p. 1120–1132.
Zhang, W., Chien, J., Yong, J., & Kuang, R. (2017). Network-based machine learning and graph theory algorithms for precision oncology. NPJ precision oncology, 1(1), 25.
Zhang, Y., Xi, H. and Juhas, M. (2021). Biosensing detection of the SARS-CoV-2 D614G mutation. Trends in Genetics, 37(4), 299–302.
Zhao, C., Shao, M., Carass, A., Li, H., Dewey, B.E., Ellingsen, L.M. and Prince, J.L. (2019). Applications of a deep learning method for anti-aliasing and super-resolution in MRI. Magnetic resonance imaging, 64, 132–141.
Zhao, Z., Woloszynek, S., Agbavor, F., Mell, J.C., Sokhansanj, B.A. and Rosen, G.L. (2021). Learning, visualizing and exploring 16S rRNA structure using an attention-based deep neural network. PLoS computational biology, 17(9), e1009345.
Ziegelstein, R.C. (2017). Personomics and precision medicine. Transactions of the American Clinical and Climatological Association, 128, 160.
Zou, K.H., Li, J.Z., Imperato, J., Potkar, C.N., Sethi, N., Edwards, J. and Ray, A. (2020). Harnessing real-world data for regulatory use and applying innovative applications. Journal of Multidisciplinary Healthcare, 671–679.

CHAPTER

16

Artificial Intelligence-based Methods for Drug Screening

Rahul Raghavan Thekke Veedu[1*]

1. Introduction

Recent years have seen a large increase in financing and pharma partnerships to utilize the potential of artificial Intelligence tools in drug discovery research. In 2021, German Biotechnology company Evotech in partnership with Exscientia, Oxford UK announced the Phase 1 clinical trial of an anticancer molecule, by applying artificial intelligence in small molecule drug discovery. The A2 receptor antagonist designed to help T helper cells fight the solid tumor was developed within 8 months using Artificial Intelligence-based methods, which would be time consuming by traditional techniques. Recent financing for artificial intelligence-based methods by Companies Schrodinger, Insitro, Insitro, Exscientia, and Recursion Pharmaceuticals Drug discovery points towards the potential of artificial intelligence-based methods in speeding up the drug discovery process(Savage, 2021).

The major steps in the drug discovery process involve screening the small molecule against a target, studying ADMET, and finally conducting clinical trials on humans. These are time-consuming steps; therefore, the classical hit-and-miss approach will be time consuming for preclinical studies. The selected drug can still fail in later stages due to variations in other parameters associated with pharmacokinetics, drug toxicity, and drug-drug interactions. Hence, a successful drug has to pass through all these hurdles and artificial intelligence-based approaches can predict the outcome at various stages of drug development. Artificial intelligence-based methods utilize machine learning or deep learning approaches to handle multidimensional data. Various research studies have shown that machine learning and deep learning approaches can be adopted in drug discovery steps, such as drug screening, to study drug toxicity and drug-drug interactions. One of the primary requirements for the processes of machine learning-based drug discovery methods is molecular fingerprints. The chemical structure of a molecule is represented in machine readable linear formats like SMILES and InChl format, Figure 1 shows the linear representation of Aspirin in SMILE and Chl format.

[1] Hospital for Special Surgery-Weill Cornell Medicine, New York, U.S.A.
* Corresponding author: rrtvactrec@gmail.com

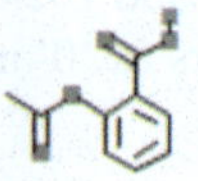

aspirin; ACETYLSALICYLIC ACID; 50-78-2; 2-Acetoxybenzoic acid; 2-(Acetyloxy)benzoic acid; ...

Compound CID: 2244

MF: $C_9H_8O_4$ **MW:** 180.16g/mol

IUPAC Name: 2-acetyloxybenzoic acid

Isomeric SMILES: CC(=O)OC1=CC=CC=C1C(=O)O

InChIKey: BSYNRYMUTXBXSQ-UHFFFAOYSA-N

InChI: InChI=1S/C9H8O4/c1-6(10)13-8-5-3-2-4-7(8)9(11)12/h2-5H,1H3,(H,11,12)

Create Date: 2004-09-16

Fig. 1 Shows Molecular structure of Aspirin represented in SMILES, InChl and InChlKey (*Source:* PubChem)

Molecular fingerprints are a method of encoding molecular structures in binary codes that show the presence and absence of functional groups in the molecule. Molecular fingerprints are utilized to search drug databases for similarity, to predict drug properties, virtual screening, and to map chemical spaces. Molecular fingerprints are generated by various programs and each having its own rules. Therefore, deciding the appropriate molecular fingerprint method in machine learning and Bayesian model studies is very important. Molecular fingerprints are generated through MACCS keys, PubChem fingerprints, ECFP4, and MHFP6 algorithms. MACCS-based fingerprints are generated based on keys. They include a list of pre-generated questions and their answers are represented as true or false corresponding to 0 or 1 in the binary fingerprint (Brown & Martin, 1996). In ECFP4 (Extended connectivity fingerprint)-based algorithm fingerprints are generated by considering the circular environment of each atom up to radius 4; the circular hashing method is used to generate a 1024 binary vector. In this method, the molecule is broken into multiple fragments based on radius

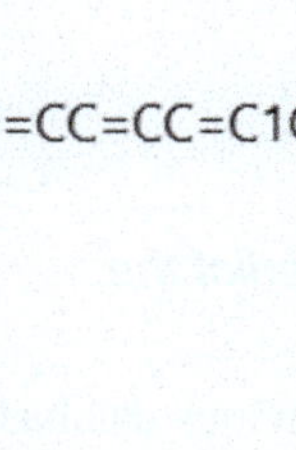

CC(=O)OC1=CC=CC=C1C(=O)O aspirin

>aspirin 29 bits set
00000000 00000008 00000000 00000000 00000000 00000000
00000000 00000000 00000000 00000000 00000000 00000000
00000000 00000080 00000000 00004020 00000000 00000000
00000200 00000000 00000500 00000004 00000000 00000000
00000000 00000000 00000000 00000000 00000000 00000000
00000000 00000000 00000000 00000000 00000000 00000000
00000000 00000000 00000000 00000000 00000000 00000000
00000000 40000800 00000000 00002000 00000000 00000000
00000000 00000000 00400000 00000000 00000000 00000000
00000000 00000000 00000000 00000000 00000000 00000000
00000000 00000000 00000000 20000000 00080000 00000000
00000000 00000000 00000000 80000000 00000000 00000000
00000000 00000000 00000000 08000000 00000000 10008000
00000000 00000400 00000000 00000000 00000000 00000200
00000000 00800008 00000800 00000000 00000000 00000000
00000000 00000000 00000000 00000004 00000000 00000000
00000000 00000020 00000000 00000000 00000000 00000000
00080000 00000000 00000000 00000000 00000000 00000000
00000000 00000000 00000000 00000000 00000000 10000000
00000000 00000000 00004000 00000000 00000000 00000000
00000000 00000000 00000000 00000800 00000000 00000000
00000000 00000000

ECFP4 based finger print for Aspirin

Fig. 2 Shows the ECFP4-based molecular fingerprint for aspirin

and generates a string using the fragments. Each string is hashed onto numbers between 1 and 1024, corresponding bits are represented as '1' based on the presence of a fingerprint and others as '0' in the absence of a finger print.

2. Artificial Intelligence in the Prediction of Drug Targets

In this section, different methods used for high throughput screening of drugs are explored. Drug targets (i.e. inhibitors) are usually identified by experimental high throughput screening. Experimental throughput screening is very costly and time-consuming, and many compound libraries are unsuitable for certain assays. Therefore, virtual screening is a suitable way to speed up drug screening. Traditionally, virtual drug screening can be done by docking studies and ligand-based approaches such as QSAR studies. Artificial intelligence-based methods use machine learning and deep learning models for drug discovery. Artificial intelligence-based approaches require an enormously large database that contains historical knowledge of ligand and target-binding features. With the knowledge of the likelihood of molecular fingerprints appearing in ligands and non-ligand molecules in a database, approaches like Bayesian models can predict the binding capacity of novel compounds.

2.1 *Bayesian Based Machine Learning Approach*

2.1.1 *Profile QSAR Method*

The Profile QSAR is a two-dimensional-based model that can accurately predict the binding of drug candidates to the target protein. This model was used to predict the effect of small molecules against the target kinase protein. This method relies on an enormous database that has historical experimental data information on the target (kinase) protein. It utilizes affinity fingerprinting methods to predict the ability of a new compound to inhibit the kinase activity. Profile QSAR works with data from a family of related enzymes. The database has experimental information such as IC50, IC50 values of a large number of compounds, screened against the specific family of kinases utilized from the database to predict the activity of novel compounds without experimental data. Thus, profile QSAR can be utilized to screen large commercial databases and virtual libraries. Profile QSAR theory assumes that the ligand's binding property to a conserved active site on the kinase family of protein depends upon a linear combination of different features. These features include the number of hydrogen bonds and hydrophobic core region, which determine the magnitude of interaction with the active site. These are represented by the equation.

$$pK_{i(x,\,\mathrm{new})} = \sum_{j=1}^{n} Cj \times pKi_{(x,j)} \quad \text{(Martin et al., 2011)}$$

pK_i - the negative logarithm of compound X against New kinase
Cj - Coefficient for J^{th} basis set kinase
$pKi_{(x,j)}$ - inhibition constant of compound X against kinase j

2.1.2 *Machine Learning and Naive Bayesian Approach to Predict the Inhibitors Against the Target Enzyme*

In a study, structure-derived fingerprints were identified from the previous database and classified into bins based on IC50 and *Pki* values, and the naïve Bayesian method was applied to predict the binding capacity of the novel compound. The method was tested using aminopeptidase inhibitor data extracted from ChEMBL v 20. Four groups of the molecular structure were divided into seven bins based on their activity. The Laplacian corrected naïve Bayesian method based on ECFP6 fingerprints was then applied to determine molecular structures in the test compound and its probability in each bin. The compounds were classified into 4 bins based on the activity of the training set. Following the prediction, 55.4% of test entries were predicted to be in the correct bin. This method successfully predicted and categorized the aminopeptidase N-inhibitor compound based on its known activity.

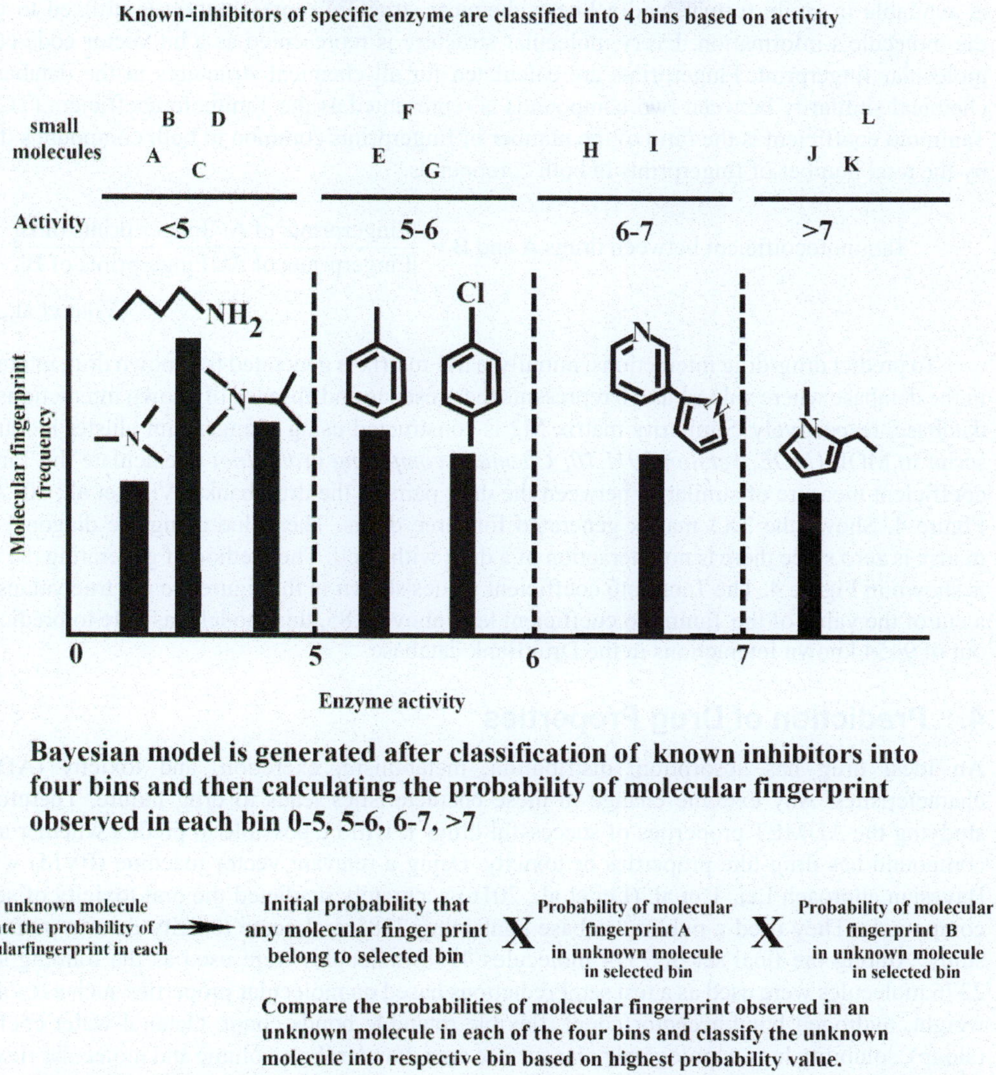

Fig. 3 Shows the naïve Bayesian based approach used to screen the enzyme inhibitors with a training data set.

(Clark et al., 2016). Neel S. Madukar et.al. utilized Bayesian-based machine learning approach (BANDIT) to identify and validate 14 novel microtubule inhibitors and compounds that have inhibitory activity against resistant cancers (Madhukar et al., 2019).

3. Molecular Similarity-based Approach to Predict Drug-drug Interaction

Drug-drug interactions are a serious problem that can cause an adverse condition in patients and cause a drug to be taken off the market. Drugs such as simvastatin are used to reduce cholesterol by inhibiting HMG-CoA reductase. They have been shown to interact with fluconazole and cause the risk for myopathy or rhabdomyolysis. It was suggested that a drug with a similar structure to simvastatin can also interact with fluconazole (Micromedex, 2011; Molden et al., 2008). Clinical trials for identifying drug-drug interactions are expensive therefore, methods for predicting the drug-drug interaction and suggestions on ideal doses are very helpful for physicians. Molecular structure similarity analysis is one way of predicting drug-drug interactions. The molecular structure is available in smile format, and a linear chemical structure representation is utilized to extract the molecule's information. Every molecular structure is represented as a bit vector code called a molecular fingerprint. Fingerprints are calculated for all chemical structures in the database and chemical similarity between two compounds is represented as the Tanimoto coefficient (TC). The Tanimoto coefficient is the ratio of the number of fingerprints common to both compounds divided by the total number of fingerprints in both compounds.

$$\text{Tanimotocoificient between drugs A and B} = \frac{|\text{Fingerprints of A} \cap \text{Finger prints of B}|}{|\text{Fingerprints of A} \cup \text{Fingerprints of B}|}$$

(Vilar et al., 2012)

To predict drug-drug interactions, initially, a M1 matrix is generated for known drug interactions in the database where value 1 and 0 represents the presence and absence of known interactions in the database, respectively. Similarity matrix M2 is constructed using a fingerprint cluster module and script in MOE (*MOE, Version 2011.10; Chemical computing group,Inc*) to calculate the Tanimoto coefficient measure of similarity between the drug pairs in the drug banks (Vilar et al., 2012). The Figure 4. Shows the 3X3 matrix generated for three drugs. The value along the diagonal of the matrix is zero since there is no interaction of a drug with itself. The method of generating the matrix is shown in Figure 4. The Tanimoto coefficient values shown in the figure are not true values. With a cut of the value of the Tanimoto coefficient kept above 0.85, this model was able to predict 4335 out of 9454 known interactions in the DrugBank database.

4. Prediction of Drug Properties

An Ideal drug has absorption, distribution, metabolism, excretion, and toxicity (ADMET) characteristics. Any extreme change in these characteristics leads to drug failure. Therefore, by studying the ADMET properties of successful drugs it will be possible to predict whether a novel compound has drug-like properties or toxicity. Using a relevant vector machine (RVM) with the Bayesian approach Lei, T et al. (Lei et al., 2016) accurately predicted the oral toxicity of selected compounds. They used a public database containing 7385 oral acute toxicity data to perform the study. Among the final set of 7314 molecules 4938 molecules were used as the training set and 2376 molecules were used as a test set. Predictions based on molecular properties such as molecular weight, hydrogen bond acceptor count, flexible-rotatable bonds count, Octanol-water coefficient (slogp), solubility logs, topological polar surface area, Vander Wals volume and molecular flexibility index were unreliable and were not correlated with rat oral cytotoxicity. The prediction based on the QSAR model by MOE descriptors were reliable; the RVM model gave the best prediction for the Test sets. It was found that 9 pubchem FP fragments and sub-FP fragments were responsible for oral acute toxicity. In this method, the number of fragments in each molecule is counted. A compound is considered to have a high contribution to oral toxicity when the number of fragments contributing pLD50 is >3 and the fragments contributing pLD50 is <3 (Lei et al., 2016)

5. Deep Learning Approach

Deep learning models use a neural network. A neural network is a connection of programs; it starts with an input program and goes to the next in a linear or collective fashion. Cai et al. used the deep learning-based method to predict the drug-induced cardiotoxicity (Cai et al., 2019). A basic neural network has three layers (1) Input layer, (2) Hidden layers, and (3) Output layers. The input layer shows the features that are used to make predictions; each feature requires one neuron. The number of neurons in the hidden layer may vary. In case of neural networks that predict bioactivity, molecular fingerprints in the numerical form will be the input value. Connections between the neurons are called synapses and each input value is multiplied by the corresponding weight. Weights can be

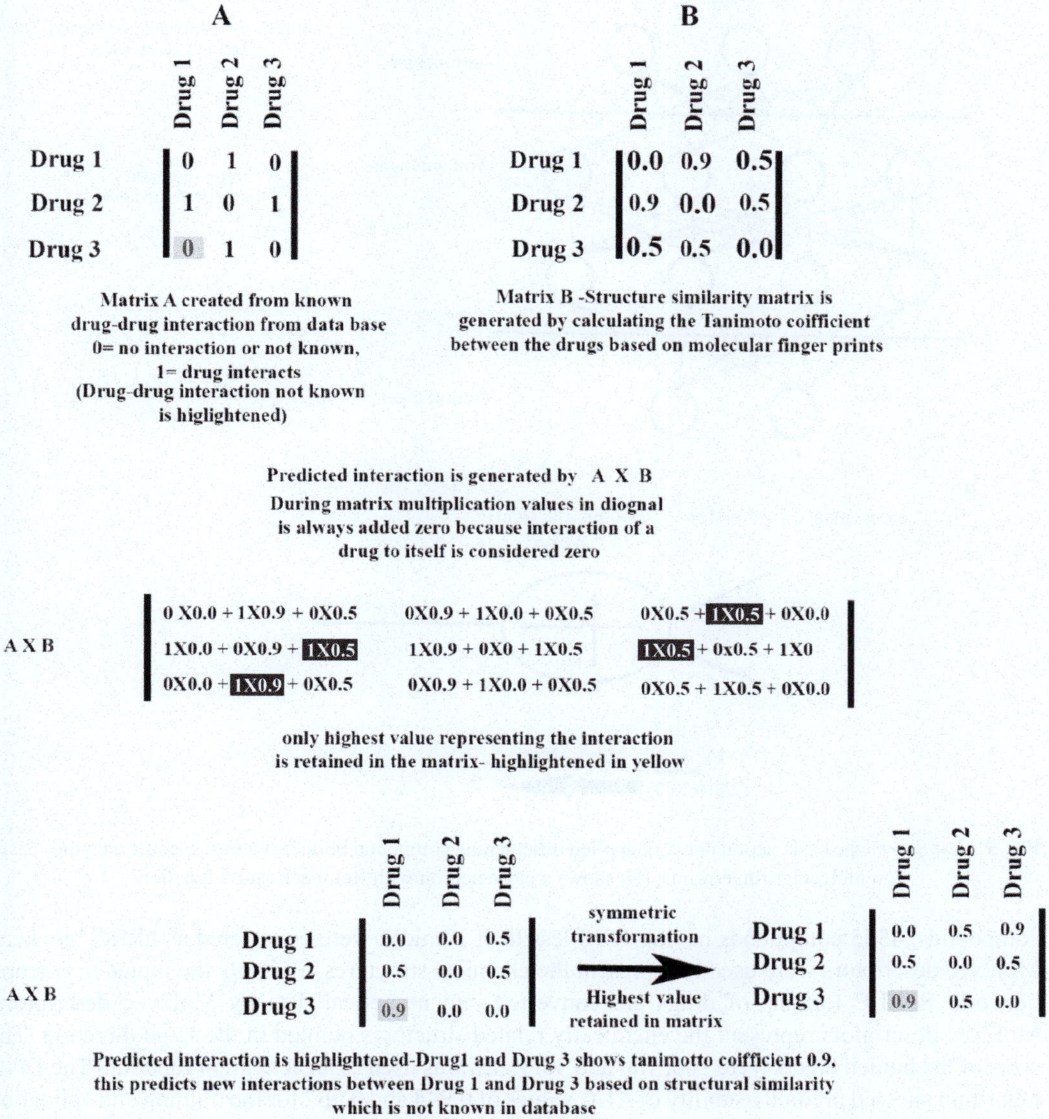

Fig. 4 Shows the identification of drug-drug interaction by molecular structural similarity (*Source:* Vilar et al., 2012)

either positive or negative according to the feature. An activation function determines whether each neuron should be fired to activate the next neuron. The activation function involves a rectified linear unit (ReLU). This function gives off x if x is a positive integer or else it remains 0. The function is represented as f(x) = max (0, x). A linear activation function used is a(x) = x for hidden or output layers (Sharma et al., 2017). In artificial neural networks many of the functions use the sigmoidal function. Sigmoid functions help to squash the values and keep the output between 1 and –1 (Han et al., 2018). The output layer usually makes predictions by regression or binary classification. Other activation functions used in neuronal networks are reviewed in detail by Sharma et al., (Sharma et al., 2017).

Chuipu Cai et al., used deep learning-based methods to predict the drug induced cardiac toxicity. The inhibitors of human ether-a-go-go-related (hERG) gene are seen to induce long-QT syndrome and will lead to fatal cardiac toxicities (Cai et al., 2019). Among the collection of 7889

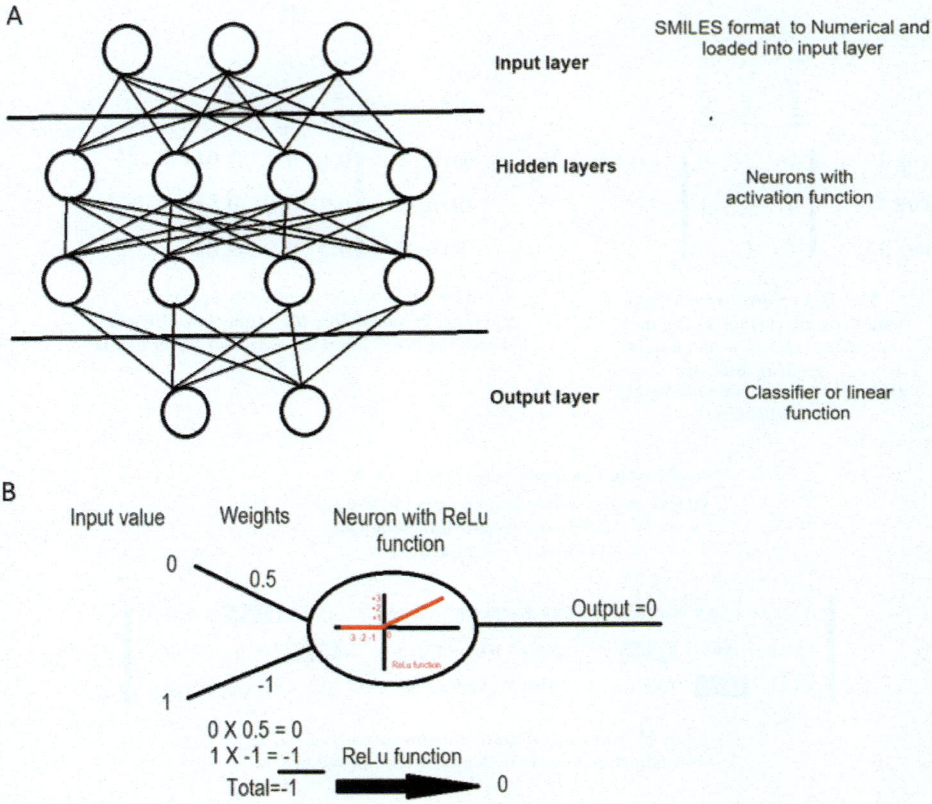

Fig. 5 (A) Shows the basic neural network applied in screening compound libraries against specific enzymes based on molecular fingerprints, (B) shows a basic neuron with ReLu activation function.

compounds, 4355 compounds having IC50 less than 10micM were considered as hERG blockers. Mol2vec descriptors were used to represent the chemical structures. To apply the inputs in a neural network, SMILE formats of drugs are converted into numerical data by Mol2vec descriptors. Mol2vec descriptors represent the chemically related structures pointed in the same direction. The sizes of the hidden layers were 200, 100 and 50. ReLu was used as an activation function. The DNN algorithm showed predictive ability of AUC values of 0.944 and 0.967 for the training and validation sets respectively. Hamza Hentable et al. used a convolutional Neural network-based method to predict bioactivity. Molecules were represented in a new Mol2mat format which is a modified 2D-fingerprint descriptor. A combination of three fingerprinting techniques ECFP4, EPFP4 and ECFC4 was used to get a high predictive ability of 98% AUC compared to naïve Bayesian methods (Hentabli et al., 2022).

Graph based neural networks have been widely used in biology because they can directly use biological network structure as input. They can use various levels of information like expression profile, sequencing and structures to identify drug targets (You et al., 2022). Selvaraj et al., predicted phosphothreonine in MAPK1 as a target in lung adenocarcinoma and identified candidate drugs by evaluating protein-protein and protein-drug interactions (Selvaraj et al., 2018).

6. Future of Artificial Intelligence in Drug Discovery Methods

Deep learning methods have helped in object identification and visual recognition. Application of these methods could screen drugs for binding efficiency and also to predict toxicity. The future of artificial intelligence is promising as more and more machine learning-based and deep learning-based methods are developing. Since a large database of experimental data is available, machine learning-based approaches have been used to predict novel drug-drug interactions, drug properties such as toxicity and screening for enzyme inhibitors. Recent studies show that Bayesian based neural networks predict drug binding more efficiently and a combination of three fingerprinting techniques ECFP4, EPFP4 and ECFC4 and a convolutional Neural Network were shown to better predict the bioactivity (Kim et al., 2021) (Hentabli et al., 2022). New methods such as feature attribution, deep graph networks with molecular counterfactuals, and uncertainty estimations were proposed to increase the acceptance of artificial Intelligence based approaches(Nguyen et al., 2015; Numeroso & Bacciu, 2020; Riniker & Landrum, 2013). Gasten-Mathe concluded that in the future there will be software that can design the drug 100% in silico, and we will only have to test it and it will be good to go (Savage, 2021).

References

Brown, R.D. and Martin, Y.C. (1996). Use of structure–activity data to compare structure-based clustering methods and descriptors for use in compound selection. *Journal of Chemical Information and Computer Sciences*, *36*(3), 572–584.

Cai, C., Guo, P., Zhou, Y., Zhou, J., Wang, Q. et al. (2019). Deep Learning-Based Prediction of Drug-Induced Cardiotoxicity. *J Chem Inf Model*, *59*(3), 1073–1084. https://doi.org/10.1021/acs.jcim.8b00769

Clark, A.M., Dole, K. and Ekins, S. (2016). Open source bayesian models. 3. Composite models for prediction of binned responses. *Journal of chemical information and modeling*, *56*(2), 275–285.

Han, S.-H., Kim, K.W., Kim, S. and Youn, Y.C. (2018). Artificial neural network: understanding the basic concepts without mathematics. *Dementia and Neurocognitive Disorders*, *17*(3), 83–89.

Hentabli, H., Bengherbia, B., Saeed, F., Salim, N., Nafea, I. et al. (2022). Convolutional Neural Network Model Based on 2D Fingerprint for Bioactivity Prediction. *International Journal of Molecular Sciences*, *23*(21), 13230.

Kim, Q., Ko, J.-H., Kim, S., Park, N. and Jhe, W. et al. (2021). Bayesian neural network with pretrained protein embedding enhances prediction accuracy of drug-protein interaction. *Bioinformatics*, *37*(20), 3428–3435.

Lei, T., Li, Y., Song, Y., Li, D., Sun, H et al. (2016). ADMET evaluation in drug discovery: 15. Accurate prediction of rat oral acute toxicity using relevance vector machine and consensus modeling. *Journal of cheminformatics*, *8*, 1–19.

Madhukar, N.S., Khade, P.K., Huang, L., Gayvert, K., Galletti, G. et al. (2019). A Bayesian machine learning approach for drug target identification using diverse data types. *Nature communications*, *10*(1), 5221.

Martin, E., Mukherjee, P., Sullivan, D. and Jansen, J. (2011). Profile-QSAR: a novel meta-QSAR method that combines activities across the kinase family to accurately predict affinity, selectivity, and cellular activity. *Journal of chemical information and modeling*, *51*(8), 1942–1956.

Micromedex. (2011). Healthcare Series [Internet database]. Greenwood Village. In (Vol. 2011). USA: Thomson Reuters (Healthcare) Inc.

MOE, Version 2011.10; Chemical computing group, Inc. http://www.chemcomp.com

Molden, E., Skovlund, E. and Braathen, P. (2008). Risk management of simvastatin or atorvastatin interactions with CYP3A4 inhibitors. *Drug safety*, *31*, 587–596.

Nguyen, A., Yosinski, J. and Clune, J. (2015). Deep neural networks are easily fooled: High confidence predictions for unrecognizable images. Proceedings of the IEEE conference on computer vision and pattern recognition,

Numeroso, D. and Bacciu, D. (2020). Explaining deep graph networks with molecular counterfactuals. *arXiv preprint arXiv:2011.05134*.

Riniker, S. and Landrum, G.A. (2013). Similarity maps-a visualization strategy for molecular fingerprints and machine-learning methods. *Journal of cheminformatics*, *5*, 1–7.

Savage, N. (2021). Tapping into the drug discovery potential of AI. *Biopharma Dealmakers*, *2021*(2021), B37–B39. https://doi.org/https://doi.org/10.1038/d43747-021-00045-7

Selvaraj, G., Kaliamurthi, S., Kaushik, A.C., Khan, A., Wei, Y.-K et al. (2018). Identification of target gene and prognostic evaluation for lung adenocarcinoma using gene expression meta-analysis, network analysis and neural network algorithms. *Journal of biomedical informatics*, *86*, 120–134.

Sharma, S., Sharma, S. and Athaiya, A. (2017). Activation functions in neural networks. *Towards Data Sci*, *6*(12), 310–316.

Vilar, S., Harpaz, R., Uriarte, E., Santana, L., Rabadan, R. et al. (2012). Drug—drug interaction through molecular structure similarity analysis. *Journal of the American Medical Informatics Association*, *19*(6), 1066–1074.

You, Y., Lai, X., Pan, Y., Zheng, H., Vera, J. et al. (2022). Artificial intelligence in cancer target identification and drug discovery. *Signal Transduction and Targeted Therapy*, *7*(1), 156.

CHAPTER

17

Artificial Intelligence in Synthetic Biology

Naorem Bela Devi,[1] *Suparna Sen*[1] and *Kannan Pakshirajan*[1]*

1. Introduction

Over the last two decades, biology has undergone a profound transition, shifting from a primarily descriptive discipline to one characterized by a focus on design and engineering concepts. The identification of DNA as the genetic information repository and the usefulness of recombinant DNA for genetic manipulation initially led to genetic engineering and, subsequently, the field of synthetic biology. Synthetic biology employs engineering concepts to create and assemble biological elements and systems, suitable for various fields of industrial, environmental, agricultural, and pharmaceutical applications (Meng and Ellis 2020). It encompasses a dual purpose. On one hand, it is driven by practical applications, like engineering cells to produce specific chemicals. On the other hand, it has scientific objectives, delving into a deeper understanding of life, such as minimizing cell genome size or demonstrating the assembly of living organisms from non-living molecules. The shift towards industrialized synthetic biology is anticipated to affect a diverse range of human endeavors, from enhancing health to generating sustainable biofuels as a means to address climate change. In this effort, cutting-edge tools empower us to bioengineer cells at an unprecedented pace. CRISPR-based genetic editing has totally transformed our ability to modify DNA within living organisms. DNA synthesis efficiency is increasing at a rate comparable to Moore's law, with transcriptomics data volume of 7 months doubling with time, alongside the growing prominence of high-throughput workflows for proteomics and metabolomics (Carbonell et al. 2019). Consequently, the miniaturization and automation of these approaches via microfluidic chips (Heinemann et al. 2017) imply that the bottleneck in biological research will move from data creation to data processing in the future.

AI (Artificial Intelligence) plays an important role in advancing synthetic biology by speeding up research, optimizing experiments, and enabling more precise design of complex biological

[1] Department of Biosciences and Bioengineering, Indian Institute of Technology Guwahati, Guwahati 781039, Assam, India.

* Corresponding author: pakshi@iitg.ac.in

systems. This synergy between AI and SB has significant advantages and has the potential to revolutionize different applications in biotechnology and beyond. The concept of integrating AI with SB is to expand the exploration of natural cognition, which has traditionally been the domain of AI using computer science and robotics methodologies, into the realm of synthetic biology. In the field of biological research, the rise of synthetic biology (Chiarabelli et al. 2009 and Morange et al. 2009) stands out as a particularly relevant novelty. This growing scientific and technological field's attraction originates from its hybrid nature. Synthetic biology emerged at the turn of the millennium, collaborating biology and engineering to develop and construct biological components or systems that do not naturally exist, with the main goal of reaching practical targets such as the production of chemicals, biofuels, pharmaceuticals, and more. As AI enters into the third wave, with an emphasis on introducing context into models, its capacity to influence the field of synthetic biology grows stronger. For many years, biologists have dedicated extensive efforts to construct and maintain an extensive collection of qualities, encompassing aspects such as association, regulation, rate of change, and functions, to describe this intricate and constantly changing system.

The AI models offer biologists the chance to tackle far more intricate queries concerning the biological system and construct comprehensive and transparent models, thereby expediting the discovery process (Eslami et al. 2022). However, several hurdles must be overcome to make progress in the development of synthetic biology by using AI. The goal of this chapter is to cover topics on dynamic interaction between AI and synthetic biology, genetic circuit design, impact of AI on genomic engineering tools such as CRISPR-Cas9, importance of AI in therapeutic discovery, deep learning approaches to SB, and data integration and analysis (Figure 1). Ethical challenges and considerations in AI-driven research are finally discussed in this chapter.

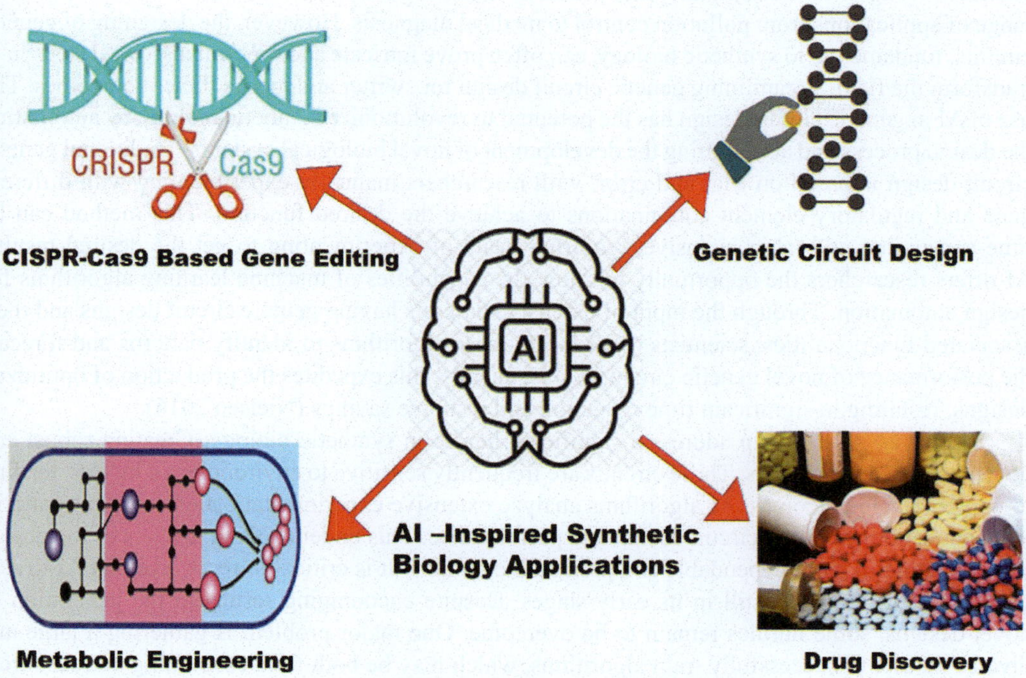

Fig. 1 A graphical representation of the major applications of artificial intelligence (AI) in synthetic biology.

2. Scope and Objectives

In the realm of synthetic biology, AI has emerged as a revolutionary force, providing new prospects for innovation and growth. AI applications in genetic circuit design leverage machine learning and optimization algorithms to create customized genetic circuits tailored for specific biological functions, thereby enhancing the predictability and reliability of engineered biological systems. Details about the role of AI in genetic circuit design are discussed in section 3. Similarly, the integration of AI with the CRISPR/Cas9 system for vaccine development accelerates the identification of vaccine targets and optimizes genome editing, promising more effective and efficient vaccine development (Section 4). Metabolic engineering, a critical aspect of synthetic biology, benefits from AI's machine learning and deep learning capabilities, allowing for the optimization of metabolic pathways to produce biofuels, chemicals, and pharmaceuticals with increased yield and efficiency (Section 5). Furthermore, AI-driven drug discovery platforms leverage AI models to identify potential drug targets, optimize candidates, and predict their biological effects, substantially reducing the time and cost associated with drug development in synthetic biology (Section 6). AI is also pivotal in integrating and analyzing complex biological data sources, enabling a comprehensive understanding of biological systems and enhancing research outcomes in synthetic biology. As AI's role in synthetic biology expands, it is essential to address ethical and regulatory considerations to ensure the responsible and transparent use of AI in research and development, respecting boundaries and maintaining ethical integrity.

3. AI Applications in Genetic Circuit Design

The design and execution of genetic circuits (Brophy and Voigt 2014), which enable cells to perform predetermined functions, is at the heart of synthetic biology. An example can be seen in the construction of progressively intricate Boolean logic circuits using a series of transcriptional regulators (Nielsen et al. 2016). Various organisms, including mammalian cells (Lillacci et al. 2018), bacteria (Wong et al. 2015), and yeast (Chen et al. 2020) are used to access circuits in the wide range of applications from pollution control to medical diagnosis. However, the designing of genetic circuits, fundamental to synthetic biology, can often prove intricate and time-intensive. AI steps in to transform the field, streamlining genetic circuit design for swifter and more efficient outcomes. The use of AI in genetic circuit design has the potential to revolutionize synthetic biology by automating the design process and accelerating the development of novel biological systems. Traditional genetic circuit design is based on trial and error, with researchers manually experimenting with different gene and regulatory element combinations to achieve the desired function. This method can be time-consuming and labor-intensive, requiring years of experimenting to get the desired results. AI offers researchers the opportunity to allow the capabilities of machine learning algorithms for design automation. Through the input of extensive datasets having genetic circuit designs and their associated functionalities, scientists can educate these algorithms to identify patterns and forecast the performance of novel genetic circuits. Consequently, this expedites the production of optimized designs, resulting in significant time reduction and resource savings (Nielsen 2016).

Moreover, AI can help address a significant hurdle in synthetic biology: Creating robust and dependable genetic circuits. These circuits are frequently sensitive to environmental factors, leading to unpredictable outcomes. AI algorithms analyze extensive experimental data to uncover patterns and connections between circuit design and performance. This understanding influences the design step, resulting in more dependable and predictable circuits. It is critical to recognize that AI-driven genetic circuit design is still in its early stages. Despite encouraging results in the generation of novel designs, some hurdles remain to be overcome. One major problem is gathering a large and diverse dataset to successfully train algorithms, which may be both time-consuming and resource-intensive. Moreover, ethical concerns surround the application of AI in synthetic biology. With the increasing sophistication of AI algorithms, responsible and ethical usage becomes highly significant. This encompasses dealing with issues like bias in training data and the possibility of unintended consequences in biological system design.

3.1 Gene Sequence Optimization

Genetic sequence optimization uses AI algorithms, mainly ML and Deep learning models, to study enormous volumes of genetic data and discover the best gene sequences for certain biological functions. This allows researchers to create synthetic genes with enhanced efficiency and stability. Benchling (Capobianco 2023) is a pioneer company in this field, offering a cloud-based platform for life science research. It enables scientists to use machine learning methods to create, edit, and simulate gene sequences. Similarly, DNA2.0, now ATUM^{TM}, uses its unique GeneGPS^{TM} technology to design genes that are optimized for expression in a range of host species. Twist Bioscience^{TM} also uses silicon-based DNA synthesis to create optimized, high-quality genes for a variety of purposes, establishing them as leaders in gene sequence optimization.

4. AI and CRISPR/Cas9 Based Gene Editing

CRISPR/CRISPR-associated protein 9 (Cas9) system was a ground-breaking discovery that significantly transformed the arena of genome editing. It was first found in prokaryotic organisms as a defense mechanism against viruses, but it turned out to be an efficient tool for precise gene editing. Researchers realized it could be applied in genetic research and engineering in an accurate manner, opening up new possibilities in the arena (Bak et al. 2018, Maserat 2022). However, several factors can affect how precise and effective the gene-editing tool functions are based on genetic patterns, how the CRISPR guide RNA (gRNA) is designed, and more. Several features affecting the efficiency of gRNA in CRISPR gene editing have been identified, leading to off-target effects *viz.* protospacer adjacent motifs (PAMs), gRNA sequence motifs, nucleotide usage, position-specific nucleotide composition, secondary structure, and epigenetic features. The design of gRNA is crucial for successful gene editing, and recent efforts focus on refining gRNA design via several algorithms for improved on-target efficacy and minimal off-target effects (Bak et al. 2018). To make CRISPR accurate, researchers are using computer programs and AI to optimize the gRNA design. AI programs predict which gRNAs are more likely to work correctly and cause fewer unintended changes in the DNA. Knowledge discovery, epitope prediction, and modeling-based approaches are examples of AI techniques These prediction tools have contributed largely in the success of CRISPR (Maserat 2022). Machine learning models, using three main methods (regression-based, classification-based, and ensemble-based), are trained on existing datasets to predict on/off-target effects for testing datasets (Moreno-Mateos et al. 2015, Xu et al. 2015, Chari et al. 2017, Wilson et al. 2018, Hiranniramol et al. 2020). Advanced machine learning, including deep learning (DL) with artificial neural networks (ANNs), allows high-precision target predictions by analyzing gRNA-DNA sequences in matrix format (Chuai et al. 2017).

Apart from predicting on-target efficacy, algorithms like FlashFry (Haeussler et al. 2016), dsNickFury (Lin and Wong 2018), CRISPR (Peng et al. 2018), DeepCRISPR (Wong et al. 2015), CNN_std (Kleinstiver et al. 2016), and Elevation (Lin and Wong 2018) help predict crucial off-target effects of CRISPR-Cas9. Despite the advancements, existing models and algorithms face challenges such as data imbalance, heterogeneity, lack of training datasets, and inefficiency across species, requiring refinement for the full integration of the CRISPR system into therapeutics. Ongoing developments in algorithms are expected to enhance the on-target activity and minimize off-target effects, making CRISPR/Cas9 more applicable in clinical and therapeutic settings (Bhat et al. 2022). Companies like Synthego and Inscripta currently use AI to enhance CRISPR. Synthego uses machine learning to analyze and predict the best gRNA designs, reducing off-target effects and improving editing efficiency. Inscripta uses advanced algorithms and machine learning models for high-throughput gene editing, streamlining the process and producing better results (Ahmadi et al. 2020).

5. AI in Metabolic Engineering

A primary objective within the field of metabolic engineering involves utilizing engineered microorganisms to produce valuable chemicals (Fig. 2) (Chae et al. 2017). Nevertheless, the production process is frequently hampered by various factors, including the sensitivity of the biochemical route to environmental conditions, the accumulation of undesired metabolites, and scale-up issues. To meet these challenges, dynamic route engineering, a method in which production strains are outfitted with integrated feedback control systems, has emerged in the last decade. In response to changes in cellular or bioreactor circumstances, these control systems can vary the timing of pathway enzyme expression (Ni et al. 2021). Dynamic pathways comprise two key components (Hartline et al. 2021): A central production pathway and biosensors that control enzyme expression based on metabolite signals. However, building these systems requires integrating various molecular elements, including enzymes, metabolite-sensing proteins, and genetic components. Computational techniques have the ability to improve design iteration by offering powerful tools for *in silico* modeling and simulating system performance in a virtual environment. Until now, the computational tool landscape has been dominated by kinetic models based on ordinary differential equations. However, there has been a recent increase in interest in AI and machine learning approaches (Faulon and Faure 2021). This increased interest is due to their adaptability and ability to recognize nuanced patterns among complicated datasets.

5.1 Integrating Machine Learning in Metabolic Research

Machine learning (ML), a subfield of AI, focuses on the construction of computer programs that autonomously enhance their performance through experience, all without the need for explicit programming. Advances in machine learning research and methodology have been on the increase in recent years, facilitated by the availability of large datasets generated by cutting-edge analytical laboratory tools (Poplin et al. 2018). The primary objective of machine learning is to create algorithms capable of deducing a collection of principles from a pre-established training dataset. Machine learning (ML) applications in metabolic engineering are significant due to the development of high-throughput multi-omics profiling (Mirza et al. 2019). This technology supplies substantial volumes of data required at every regulatory tier (including metabolomic, transcriptomics, genomics and proteomics) (Chen et al. 2019, Fuhrer and Zamboni 2015), enabling dependable and precise forecasts of biological behaviors. However, AI/ML methods operate as black-box systems, depend solely on input data and inherently lack the capacity to establish the mechanistic basis, necessary for explaining complex behaviors and for generating reliable predictions.

Machine learning has the ability to offer predictive capabilities without having a comprehensive understanding of the mechanics involved, as it learns the underlying patterns in experimental data. The training data is used to create a statistical link between a set of inputs and a set of outputs using models with enough flexibility to capture practically any relationship while avoiding limiting biases. Machine learning has also been used to forecast pathway dynamics, improve pathways, detect skin cancer, predict DNA and RNA protein-binding sequences, translational control, and locate tumors in breast tissue (Costello et al. 2018, Esteva et al. 2017, Jervis et al. 2018). Furthermore, machine learning may be used to create synthetic life systems. It can understand the relationship between a phenotype and the genetic components used in genetic circuits, allowing for the development of more reliable circuits.

5.1.1 ML for Pathway Discovery

The metabolic pathway for a target molecule is usually uncertain. Furthermore, the gene(s) involved in its synthesis may need to be transferred to a model organism for easier modification and optimization (García-Granados et al. 2019). Every current modeling approach possesses its own set of limitations. Conversely, when integrating omics data and employing conventional data analysis

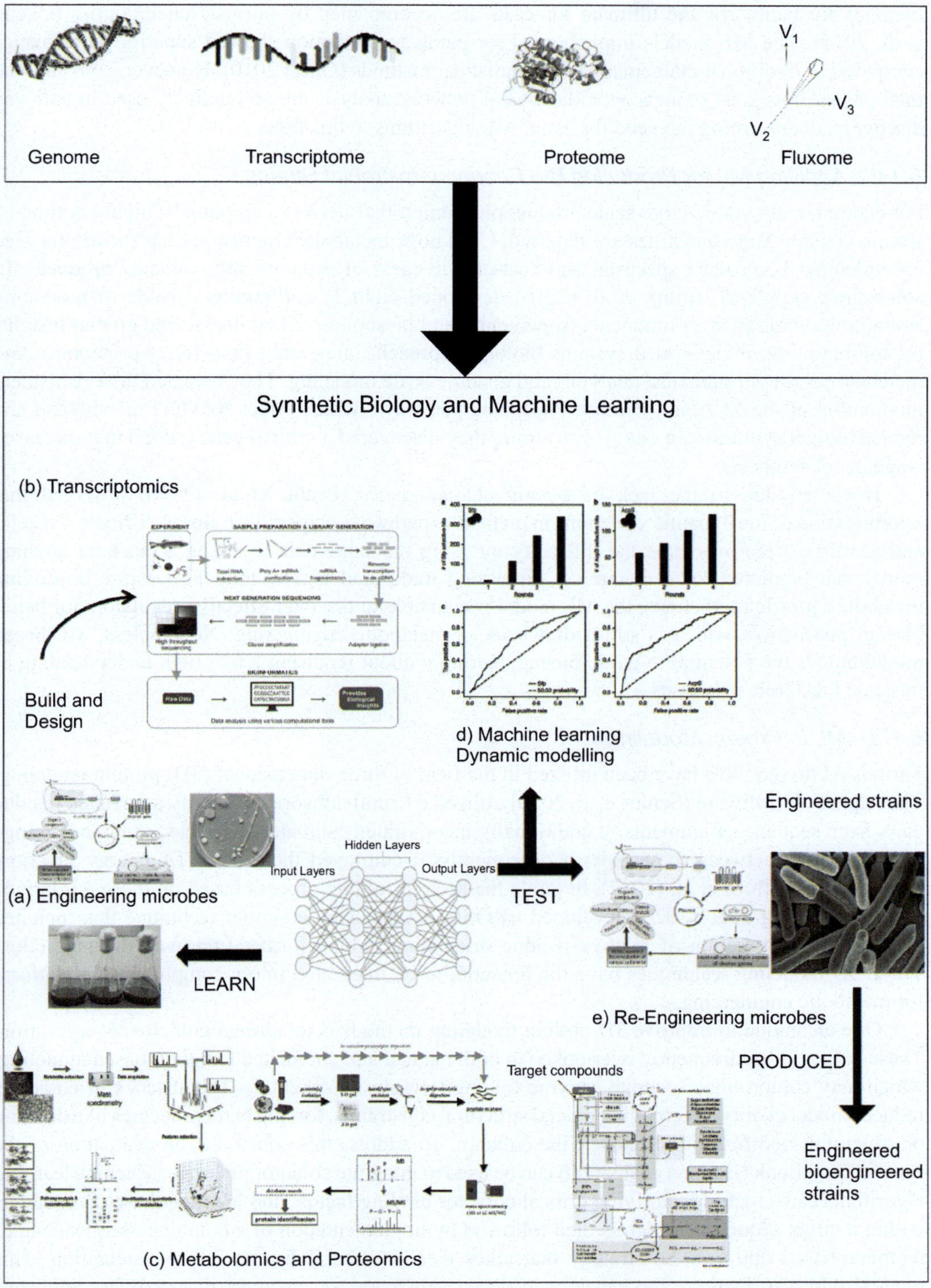

Fig. 2 Incorporation of machine learning and synthetic biology in metabolic engineering studies. Systems biology and machine learning thrive at processing vast amounts of multi-omics data, especially when starting with initial strains and design (a), transcriptomics (b), proteomics and metabolomics data generation (c) generate a large amount of data for data analytics, modeling, and machine learning techniques (d), this aids in identifying targets for redesign or reengineering that require experimental testing (e). This approach produces modified microbes from improved engineering processes suited for modern bio-industry value-added products. [Adapted from Helmy et al. (2020)]

methods for pathways, the ultimate forecasts are accompanied by intrinsic uncertainties (Cheng et al. 2015). The ML models implemented for pathway prediction showed superior performance compared to traditional mathematical and statistical methods (Quest 2010). However, conventional methods such as gene sequence similarity and network analysis are still actively used in pathway discovery, underscoring the need for better ML algorithms in this field.

5.1.2 Applying ML for Producing Bio Economy-microbial Strains

For economically viable large-scale production of microbial-derived compounds, highly optimized strains yielding large quantities are required. Until now, metabolic engineering has mostly focused on widening the product spectrum and cost-effectiveness of economically valuable products. To solve these problems, Brunk et al. (2016) developed eight *E. coli* strains capable of producing essential biofuels such as limonene, isopentenol, and bisabolene. They discovered crucial insights by combining an omics-based systems biology approach, integrating time-series proteomics and metabolomics, with statistical analytics and genome-scale modeling. They revealed time-dependent modulation of the TCA and Pentose-Phosphate pathways, which affect NADPH metabolism and reduce biofuel synthesis. In one *E. coli* strain, they discovered a critical gene (*ydbK*) that increased isopentenol synthesis.

However, due to the lack of kinetic characteristics, Brunk et al. (2016) observed that interpretation of the dynamic variations in metabolic pathways across all strains is difficult. Costello and Martin (2018) overcame the difficulty by using the same data to create a machine learning model that predicted route dynamics, surpassing traditional kinetic models. Despite improving metabolite profile predictions, the ML model was unable to discover effective regulators for better biofuel production, which is still a drawback in metabolic engineering. Nonetheless, ML-based modeling has the potential to guide bioengineering without requiring a thorough understanding of intricate metabolic processes.

5.1.3 ML in Protein Modeling

Various AI driven tools have been utilized in the field of three-dimensional (3D) protein modeling. The AlphaFold software (Senior et al. 2020) utilized a neural network to identify associated residue pairs from sequence alignments. It additionally incorporated estimated distances, spanning a range of 2 to 20 Å, between these pairs. Consequently, it employed the ROSETTA energy function, introduced by Alford et al. (2017), to guide the protein folding process based on these AI-derived constraints. Yang et al. (2020) developed trROSETTA by using a similar technique that included an additional estimation of relative residue orientation, which further improved the prediction. These 3D modeling techniques have the potential to be integrated into a comprehensive platform for metabolic engineering.

One technique to improve 3D protein modeling methods is to address cofactor incorporation. These cofactors are frequently overlooked in both manual and automated template-based modeling, which may compromise accurate enzyme folding (Higgins et al. 2005). The absence of cofactors reduces model quality by removing crucial structural constraints, forcing extra processes like docking or structural modification to reinsert the cofactor. To address this constraint, an evaluation of the Protein Data Bank (Berman et al. 2000) can be used to integrate cofactor presence. Machine learning algorithms can assist in identifying critical cofactor binding factors and locating these components within a target sequence, which is then followed by implementation of a combined sequence-and-template-based optimization strategy that takes these structural factors into consideration. This method will also increase the capacity to detect substrates for enzymes in metabolic pathways, including non-native substrates. This improvement is extremely valuable in the construction of synthetic biological pathways.

5.2 Application of Deep Learning in Synthetic Biology

Deep learning is a subfield of machine learning that focuses on the use of artificial neural networks (ANNs) to model and solve complex problems. It is called "deep" because it typically consists of multiple layers (deep layers) of interconnected artificial neurons or nodes, which are inspired by the structure and functioning of the human brain. It has various applications in the field of synthetic biology (Fig. 3A). A fully trained deep learning network predicts outputs reliably from input data, which is often represented as numerical matrices or vectors. The information input into the model and the appropriate learning methods are determined by the best data representation, which is critical for constructing high-performing and adaptable models.

5.2.1 Design and Modeling of Biological Parts

Researchers have recently achieved notable advancements in employing deep learning to forecast the roles of biological components, including ribosome binding sites (RBSs), promoters, and 5' and 3' untranslated regions (UTRs) (Kotopka and Smolke 2020, Angenent-Mari et al. 2020, Gilliot and Gorochowski 2020, Sample et al. 2019). Due to the limited length of these components, DNA synthesis is used to create large randomized libraries, which are then assessed using parallelized reporter assays and next-generation sequencing. This strategy illustrates how synthetic biology can provide the necessary training data for data-intensive models. Sample et al. (2019) created Optimus 5-Prime, a deep learning algorithm that precisely predicts how the 5' UTR sequence influences ribosome loading. While there are datasets from actual human 5' UTRs that connect sequences to translation efficiency, these natural datasets are inadequate for training the model. This is because sequences having negative impacts are likely to be underrepresented in natural examples, and endogenous transcript data do not fully cover a wide range of expression levels. The authors overcame these problems by synthesizing and analyzing data from a 280,000-member library, each of which contained random 50-nucleotide 5' UTR sequences upstream of the green fluorescent protein coding region. The Optimus 5-Prime model was trained using data from transfected HEK293T cells, with the model input being one-hot encoded 5' UTR sequences and the output being mean ribosome load values. The convolutional Neural Network (CNN)-based model worked significantly, explaining up to 93% of the test dataset containing ribosome loading values.

Similar approaches, involving DNA synthesis, parallel reporter assays, and deep learning, have been utilized for promoter design. In synthetic biology, the traditional practice has been to work with a small number of native promoters in construct design.

5.2.2 Imaging and Computer Vision Applications

Deep learning has fueled remarkable advancements in the field of computer vision (Voulodimos et al. 2018). Imaging applications in synthetic biology include tasks such as automatic feature recognition in images, such as identifying colony development on a plate or analyzing microscope data. Classification (e.g., recognizing the existence of a colony) and segmentation (e.g., outlining the pixel sets representing individual cells in a picture) are two important image analysis tasks. Segmentation, which includes finding the exact location of an object in an image, is a more difficult process, but it is extremely useful for quantification. The introduction of the U-Net algorithm (Ronneberger et al. 2015), a CNN that performs very well in biological data processing. CellPose (Stringer et al. 2021), YeaZ (Dietler et al. 2020), DeepCell (Spahn et al. 2022), MiSiC (Panigrahi et al. 2021), and DeLTA (O'Connor et al. 2022) are some prominent examples of deep learning algorithms used for single cell resolution data.

(A) Various examples of input data and their output prediction using deep learning algorithms, with both computational and biomolecular models' implementations.

(B) Model prediction can be done by giving inputs using deep learning, and it can produce novel designs by giving desired outcomes.

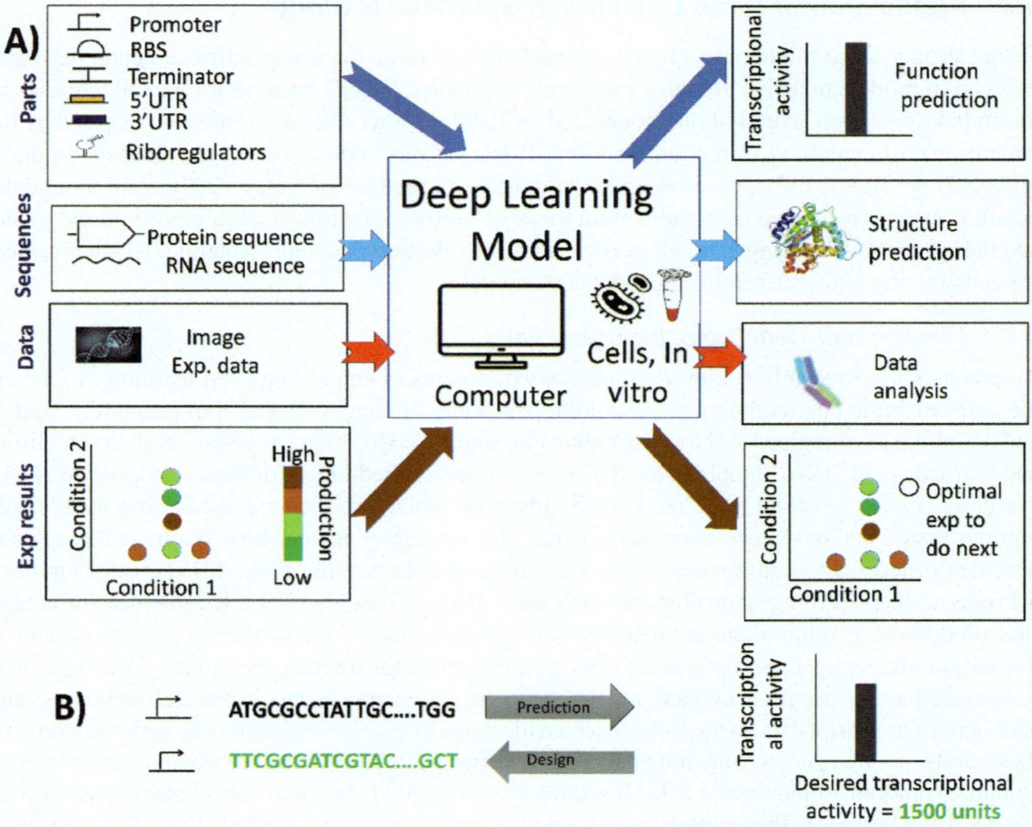

Fig. 3 Applications of deep learning in synthetic biology. [Adapted from Beardall et al. 2022]

Models in synthetic biology are often preferable to possess both predictions, such as from sequence to predicted performance, as well as generative abilities, allowing it to go from the desired performance to a sequence (Fig. 3B).

5.2.3 Optimal Experimental Design

Labeling data for synthetic biology issues is costly compared to other fields. This cost poses a distinct challenge for deep learning models, which rely on significant training data. The application of deep learning methods for optimizing experimental design in synthetic biology is not common at the moment. Nevertheless, the prospects offered by laboratory automation and preliminary results obtained through simulation indicate that this area holds significant potential for future research (Kirsch et al. 2019). Treloar et al. (2020) applied a deep learning approach to control a simulated coculture growing chemostat model in a continuous bioreactor. They showed that a successful control strategy can be acquired within a single 24 hours experiment by simultaneously operating five bioreactors. Deep reinforcement learning proved effective in determining the best sequence of inputs and control actions to maximize the production in a microbial coculture bioprocess within a continuous chemostat.

6. AI-Driven Drug Discovery in Synthetic Biology

A typical drug discovery process comprises multiple stages (Fig. 4). Identification of drug targets is the initiation that exhibits a clear connection to a particular disease, drawn from the vast space

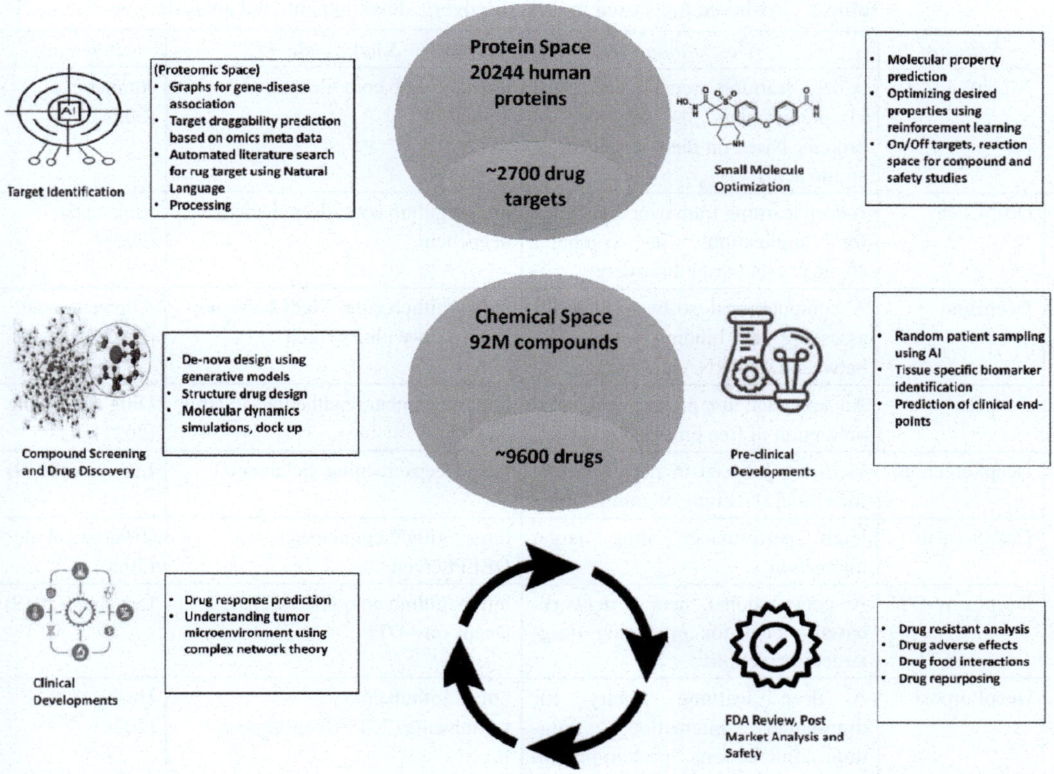

Fig. 4 Applications of AI-based methods at various stages of the drug discovery process. The human body is associated with approximately 2,700 potential drug target proteins, while there are around 9,600 FDA-approved small molecule drugs [Kim et al. 2016, Consortium 2015, Mendez et al. 2019]. Machine learning identifies target proteins, GNNs predict drug-target interactions, and reinforcement learning optimizes molecule properties in drug discovery. Computer vision assesses tumor microenvironment spatial conditions. Generative models aid simulations, inform protein-drug complex properties, new molecule design, NLP extracts information from scientific literature for drug repurposing, post-market analysis, and FDA evaluations. [Adapted from Qureshi et al. 2022]

of proteins within an organism's proteome. This is the first stage in target-based drug discovery (Vamathevan et al. 2019). High-throughput screening of compound libraries against these defined targets yields candidate compounds with potential interactions. Compounds are subsequently refined to improve their appropriateness as drugs, passed through preclinical and clinical testing, and, in the best-case scenario, approved by the FDA. AI could offer advantages across all phases of the drug discovery process (Chen et al. 2018). For instance, Graph Neural Networks (GNNs) can be utilized for forecasting drug-disease associations, drug prediction, and drug repurposing; generative models can aid in the development of novel synthetic molecules (Kadurin et al. 2017); and reinforcement learning (RL) can be employed to enhance specific molecular properties (Khemchandani et al. 2020). Natural Language Processing (NLP) has the potential to speed up drug discovery by extracting insights from scientific literature and to automate other FDA approval steps (Moingeon et al. 2022, Basit et al. 2023).

6.1 AI Based Software Tools for the Development Process

AI tools possess the capacity to revolutionize the field of drug discovery, empowering researchers to swiftly scrutinize large-scale datasets, develop novel molecules, and forecast the effectiveness of potential drug candidates. Table 1 presents some common AI tools that are used in drug discovery.

Table 1 AI-based tools used in drug discovery, development, and analysis.

Software	*Description*	*Source code*	*Reference*
AlphaFold2	A deep learning model is employed to predict the 3D structure of proteins based on their sequences of amino acids	https://github.com /deepmind / alphafold/	Jumper et al. (2021)
DeepChem	A deep learning framework designed for applications in computer chemistry and drug discovery	https://github.com /deepchem / deepchem	Ramsundar (2018)
DeepBind	A computational software tool for assessing the binding interactions between DNA/RNA and proteins	https://github.com /MedChaabane / DeepBind-with-PyTorch	Alipanahi et al. (2015)
DeepBar	An approach for precise and rapid prediction of free binding energy	https://fastmbar.readthedocs.io /en / latest/	Ding and Zhang (2021),
DeepScreening	Web-server based in deep learning for virtual screening of compounds	http://deepscreening.xielab.net/	Liu et al. (2019)
DeepScreen	High performance drug target interaction	https://github.com/cansyl/ DEEPScreen	Rifaioglu et al. (2020)
DeepConv-DTI	A convolutional neural network-based model for predicting drug-target interactions	https://github.com/GIST-CSBL/ DeepConv-DTI	Lee at al. (2019)
DeepPurpose	A deep learning library for drug-target interaction, drug-drug interaction, protein-protein interaction and protein function prediction	https://github.com/ kexinhuang12345/DeepPurpose	Huang et al. (2020)
DeepTox	A deep learning model for toxicity prediction of chemical compounds	http://www.bioinf.jku.at/research/ DeepTox/	Mayr et al. (2016)
TorchDrug	A pytorch based flexible framework for drug discovery models	https://torchdrug.ai/	Zhu et al. (2022)
PathDSP	A deep learning method for predicting drug sensitivity using cancer cell lines	https://github.com/TangYiChing/ PathDSP	Tang et al. (2021)

6.1.1 Graph Neural Network

Graph neural networks (GNNs) are a category of machine learning algorithms designed to handle graphical data, which encode the relationships between entities such as chemical compounds and proteins (Scarselli et al. 2008). GNNs have a wide range of applications in drug discovery (Gaudelet et al. 2021). Directed message-passing GNNs applied to molecular structures, in particular, have provided prospects for drug repurposing as antibiotics (Stokes et al. 2020), and subsequent *in vivo* validation has identified promising candidates with distinct structural features compared to currently available antibiotics. GNN models can generate novel compounds with desirable properties by learning about the connections between atoms and chemical components in a given dataset.

6.1.2 Generative Models

Advanced molecular generative models enable the swift exploration of an extensive chemical space (Atance et al. 2022), encompassing the development of novel structures in combination with pre-existing compounds. GANs (Generative Adversarial Networks) (Goodfellow 2016) are capable of producing synthetic chemicals or molecules with desired characteristics by learning the probability distribution of the training data and then sampling this learned probability distribution to generate novel chemical structures. Nonetheless, determining the uniqueness and, eventually, usefulness of compounds developed by generative models is an unsolved issue.

6.1.3 Reinforcement Learning

Controlling the properties of produced molecules via constant data-driven representations is a challenge (Gómez-Bombarelli et al. 2018). For example, in the setting of generative adversarial networks, development of a molecule with the desired array of physio-chemical properties from a large physico-chemical exploration area is time-consuming and challenging (Kadurin et al. 2017). Reinforcement learning (RL) is a ML algorithm that has been used in drug discovery. It has been mainly used in the development of novel drug molecules by taking dynamic decisions to generate chemical compounds with optimal attributes including bioactivity, solubility, and pharmacokinetics (Olivcrona et al. 2017). This approach involves investigating alternative actions, assessing the statistical association between them and their probable results. Subsequently a strategy aimed at achieving the most favorable attainable outcome is developed. Deep reinforcement learning seeks to determine the ideal sequence of actions within a theoretically infinite action space, a feature that may be used to explore the infinite chemical search space without resorting to extensive computational evaluation of every potential answer. An example of this is ReLeaSE (Popova et al. 2018), a RL model for structural evolution that integrates two deep neural networks—predictive and generative. These networks are trained separately yet work together to build fresh, targeted chemical libraries using deep reinforcement learning approaches.

6.2 Pharmaceutical Start-up Companies Based on AI Technologies

According to research conducted by Emersion Insights, AI-driven start-ups focused on drug development garnered around 2.1 billion USD in funding during the initial half of 2021 (Ding and Zhang, 2021). Many biopharmaceutical companies are already utilizing AI in different phases of the drug discovery process. To illustrate, Pfizer utilizes IBM Watson, a machine learning-based system, in the pursuit of immuno-oncology drug discovery. Meanwhile, Roche Genentech employs GNS Healthcare from Cambridge, Novartis partners with Microsoft for cell and image segmentation research, and AstraZeneca has joined forces with Benevolent AI to advance and commercialize clinical-stage candidates from Janssen (Fleming 2018). Insilico Medicine, Healx, Google, Deep Genomics, and DeepMind are the major companies which have made substantial investments in AI-driven drug discovery applications. The US has emerged as a dominating hub for AI implementation, hosting more than half of the world's AI drug discovery companies. Investors have increased significantly in recent years, with both the US and the European Union witnessing this trend. As a result, these areas, along with the UK, are the most active investors in AI-based drug discovery applications (Qureshi et al. 2023). Recently, China has invested \$5 billion in AI for drug discovery and aims to lead in AI-based drug discovery startups by 2030, with Tianjin investing \$16 billion and Beijing initiating a \$2.12 billion AI development project (Qureshi et al. 2023).

7. Ethical Challenges and Considerations in AI-driven Research

In the near future, economic analysts predicts a significant upswing in the AI health sector. Detailed analysis projects a market growth of more than ten times its current size between 2014 and 2021. But as this expansion occurs, a number of challenges are also expected, highlighting the importance of considering ethical and legal issues when implementing AI in healthcare systems (Gerke et al. 2020). Informed consent for data usage, safety and transparency, algorithmic fairness and biases, and data privacy are the four main ethical issues that must be resolved in order to fully realize the promise of AI in healthcare (Gerke et al. 2020).

7.1 Informed Consent

Privacy protection primarily involves regulating access to an individual's personal information (Winslade and Ross 1985). This access extends to various aspects, including privacy, personal data, property, and control over one's name, image, or likeness (Beauchamp and Childress 2001). In the realm of scientific research, regulating access to protected health data is central to privacy concerns. Data security procedures ensure that precautions are made to prevent unauthorized individuals from obtaining protected health information, hence safeguarding informational privacy. In the context of AI applications, it is crucial for clinical and business entities to protect such data from unethical hacking and to make sure it is stored in secure systems (Brundage et al. 2018). However, the use of extensive databases, a fundamental component of machine learning models, poses a challenge to privacy protection (Ford et al. 2016). Machine learning relies on large datasets to enhance its understanding through valid and generalizable training materials (Safdar et al. 2020). As these training materials are gathered from diverse sources, the risk of tracing data back to patient identities increases, potentially compromising privacy goals (Ford et al. 2016).

The enforcement of the General Data Protection Regulation (GDPR) by the European Union (EU) from May 25, 2018, marked a significant development in global privacy laws, enforcing significant financial penalties in cases of non-compliance (Safdar et al. 2020). The GDPR necessitates transparency from EU companies and those engaged in business partnerships with them regarding the utilization of private data. It mandates a restriction on data collection to only essential information, enforcing temporal limits on data retention. Entities in possession of private information are obligated to inform individuals about the nature of the data, its intended purpose, and, upon request, provide individuals with the right to modify or delete data (Safdar et al. 2020). While the concept of de-identification is often presented as a means to protect privacy, the practicality of this approach is questioned. Traditional manual de-identification approaches for free text are time-consuming and costly, often missing items (Neamatullah et al. 2008). Automated methods, while more efficient, may still struggle to remove all types of protected health information (PHI). In the case of medical imaging data in the DICOM standard, successful de-identification requires addressing metadata and potential pixel-burned-in PHI. Challenges persist in electronic medical records with the risk of reconstruction for re-identification (Rothstein 2010, Milchenko and Marcus 2013). Consequently, GDPR may necessitate companies to inform patients that complete de-identification cannot be guaranteed, usage of data about a patient and the manner it would be used in machine learning applications (Na et al. 2018).

7.2 Ownership

Machine learning has an insatiable appetite for data. Deep learning, in particular, is voracious in its data requirements. Scientists involved in developing deep learning applications often seek large datasets to create and test new tools. The burgeoning demand for creating and marketing new AI applications in the medical field has given rise to a market place for patient-derived data. However, the ownership and usage rights of such data are intricate and vary across jurisdictions, influenced by factors like the degree of de-identification or anonymization (Cartwright-Smith et al. 2016). Determining who owns the genetic data generated in synthetic biology experiments involving AI is crucial. It involves addressing questions about whether patients or participants retain ownership rights over their genetic information. A major challenge lies in the absence of clear legal frameworks for addressing ownership issues in AI-generated synthetic biology outcomes. There is a lack of clarity surrounding companies opting for less-regulated environments with unclear or lenient data ownership regulations. Recent emergence of stricter regulations, such as GDPR (Safdar et al. 2020), emphasizes transparent communication, clearly defined ownership agreements, and ongoing dialogue with all stakeholders involved in AI-driven synthetic biology projects. Furthermore, concerns emerge about the methods European Union (EU) citizens would employ to assert their rights when subjected to jurisdictions like the United States, where waivers of consent for data use may still be permissible. Conversely, there are uncertainties if the GDPR dissuades U.S. companies from collaborating with EU counterparts if aligning with GDPR expectations becomes overly cumbersome. While the prospect of patients successfully claiming ownership of intellectual property resulting from the use of their data seems improbable, these intricate issues are likely to find resolution through both legal proceedings and public opinion. Therefore, addressing ownership in the ethical use of AI in synthetic biology requires careful consideration of data, intellectual property, decision-making autonomy, and fair access are critical aspects of AI-driven research.

7.3 Algorithm Fairness and Bias

One pervasive challenge in developing AI algorithms is the confounding of selection bias in datasets (Buolamwini and Gebru 2018). Machine learning datasets, often derived from majority populations, can lead to algorithms that are more prone to failure when applied to underserved and potentially underrepresented patient groups. If training data used to develop AI models is biased, the algorithms can inherit and perpetuate those biases leading to faulty predictions or recommendations. Additionally, if certain biological entities or experimental conditions are underrepresented in the training data, AI models may not generalize well to these situations, potentially leading to inaccuracies. This has led to raised concerns with the integration of AI in various domains about the potential embedding and deployment of human and social biases at a large scale. However, it is crucial to recognize that the responsibility lies more with the underlying data than the algorithms themselves. AI models can be trained on datasets containing human decisions or reflecting the second-order effects of social and historical inequities. Bias can also be introduced through the collection and utilization of data, with user-generated data creating feedback loops that further amplify bias. Unfortunately, there are currently no established guidelines or standards to report and compare these models, highlighting the need for future efforts to guide researchers and clinicians in addressing these issues (Nelson 2019).

As AI transitions to being an essential component of modern digital systems, there is an increasing need for "responsible AI systems". These systems should prioritize transparency, explainability, and accountability to ensure ethical decision-making and prevent unjust biases. The expanding role of AI in improving patient pathways and surgical outcomes suggests that it may coexist with or even replace existing systems. Embracing AI in healthcare is becoming imperative, and failure to do so might be considered unscientific and unethical (Shah et al. 2020). In commercial applications of AI in healthcare sectors, there is a risk of automation bias, where healthcare providers may be less likely to question erroneous results due to over-reliance on automated systems designed to reduce human error and enhance patient safety. A very recent research underscored the unsettling potential of malware and AI deep learning tools resulting in faulty diagnosis of malignant lung lesions on CT scans (Mirsky et al. 2023). Ensuring diversity and representation in the training data would be instrumental to mitigate biases and ensure that AI models can generalize well across various biological contexts. Continuous monitoring and auditing processes are required to evaluate the performance of AI models over time. This helps in identifying and addressing biases as they emerge. Effectively, addressing algorithm bias in the ethical use of AI in synthetic biology requires a multifaceted approach that includes careful consideration of data, accountability, transparency, and ongoing ethical evaluation. By incorporating these considerations into the development and deployment of AI algorithms, the field can strive for fair, unbiased, and ethically sound applications in synthetic biology.

8. Data Integration and Data Analysis in Synthetic Biology

The synergy of AI in data integration not only streamlines the process but also enables the identification of intricate patterns and correlations within the integrated datasets, empowering researchers to extract meaningful insights and unlock the full potential of synthetic biology advancements. It is widely accepted that a comprehensive understanding of a biological system can only be obtained through a combined analysis of all omics layers. This kind of analysis is frequently called "integration of data" (or "network"). Data integration is a fundamental aspect of harnessing the power of AI in synthetic biology. In this interdisciplinary field, where vast and diverse biological datasets are generated from genomics, transcriptomics, metabolomics, and other sources, AI-driven data integration plays a pivotal role. Automated data integration utilizing AI algorithms facilitates the seamless amalgamation of disparate biological datasets. Through advanced techniques like natural language processing (NLP), AI enhances semantic integration by comprehending and harmonizing data with different formats and semantics. This approach proves particularly valuable when dealing with data originating from various experimental techniques and databases. Data integration builds a joint model that captures all datasets simultaneously by analyzing them collectively. This analysis begins by representing the omics layers using the mathematical concept of networks.

Nodes (also known as vertices) and links (also known as edges) make up a network, or graph. In biological networks, edges indicate the physical, functional, or chemical relationships between pairs of entities, while nodes typically represent discrete biological entities at a molecular (genes, proteins, metabolites, drugs) or phenotypic (diseases) level. In particular, these tools applied to studies of protein–protein interaction (PPI) networks (Pržulj 2011), gene interaction (GI) networks (Tong et al. 2004, Costanzo et al. 2010), metabolic interaction (MI) networks (Ma and Zeng 2003, Tanaka 2005) and gene co-expression (Co-Ex) networks (Prieto-Sánchez et al. 2008) have extracted valuable biological information from these different views of molecular machinery inside a cell. Nonetheless, a collaborative, integrative examination of all these networks is anticipated to yield an improved understanding of a biological system. One of the main challenges in network integration has been building an integrated network representation that best uses all molecular networks to capture all gene-gene associations (Mostafavi et al. 2008). Large amounts of unstructured data can be efficiently stored and processed by AI-driven algorithms, which facilitates quick data extraction-an essential feature for building intelligent computing systems that are capable of complex decision-making. (Sellwood et al. 2018). This technological progression in data generation, storage, and analysis has paved the way for the development of multifaceted products and services across various sectors, including biosciences (Oliveira 2019). While the third industrial revolution, driven by advancements in computing set the stage for AI's remarkable ascent, the advent of Big Data and its analytics has elevated our intellectual capacities to unprecedented levels (Emmert-Streib 2021). AI is now recognized as a pivotal innovation in the fourth industrial revolution (Schwab 2017).

In the field of biology, AI demonstrates diverse applications, particularly in accurately identifying the 3D geometry of biological molecules, such as proteins—an essential undertaking with profound implications for biological research. Furthermore, AI tools and applications are instrumental in automating complex production processes, addressing the escalating demand for medications, industrial chemicals, food products, and other bio-based raw materials. Machine Learning (ML), a subset of AI, contributes by predicting outcomes through extensive analyses of datasets related to drug molecules. ML evaluates numerous permutations and combinations to identify optimal configurations, eliminating the dependence on traditional manual laboratory methods (Dara et al. 2022).

Although conventional model-driven techniques are still valuable for studying biological data, they are not as effective at using large volumes of publicly available data, such as big data, to reveal information, forecast data behavior, and understand intricate data relationships. Big data is becoming increasingly crucial in biotechnology and bioinformatics, and researchers from across the world now have the ability to access and analyze it (Chakraborty et al. 2017). This data, quantified in terms of multi-omics from various biological sources, requires thorough annotation and analysis to understand intricate biological systems. AI, along with deep neural network designs, proves efficient in analyzing genomic data to discern the genetic basis of traits and identify genetic markers associated with specific characteristics (Tong and Nikoloski 2021). The incorporation of AI aids in unraveling complex links within diverse data, providing meaningful insights. Consequently, AI approaches are widely adopted in the field of biological science, and this trend is expected to grow further as the technology matures (Xu et al. 2021).

Drug responses and medical imaging provide complex but essential data that requires effective algorithmic programs for analysis. As a result, AI— especially machine learning (ML) and deep learning (DL)—are gaining prominence due to its capacity to quickly process massive datasets and extract valuable information. In the near future, AI-based virtual drug testing, drug design, and digital image processing could revolutionize medical science (Paul et al. 2021).

8.1 Data Mining using SyBiOnt

Synthetic biology, similar to engineering disciplines, develops complex systems by combining simple, modular components. To ensure predictability, a deep understanding of the parts, modules, and the context of their deployment in biological systems engineering is crucial. However, the scarcity of well-characterized components, coupled with limited biological understanding, hampers the scale and complexity of engineered biological system (Guido et al. 2006, Smolke and Silver 2011). System designers need to gather, integrate, and have access to extensive biological knowledge in order to identify and develop new modular parts, devices, and systems. (Smolke and Silver, 2011). The challenge lies in the scattered nature of information across diverse databases with varying formats and semantics. Synthetic biology faces the task of harmonizing complex, disparate datasets for optimal use in the design process and supporting computationally efficient data mining (Goble and Stevens 2008). Biological data integration remains a significant research challenge, with traditional methods including data warehousing and federated data integration. The lack of consensus on data formats and semantics complicates integration (Shadbolt et al. 2006). Semantic Web technology, leveraging ontologies, has emerged to address this challenge. The identification, integration, and organization of complex data is made easier by ontologies, which standardize data representation and give meaning to data entities and relationships that can be computed (Lin et al. 2011).

Semantic Web technologies, particularly the Resource Description Framework (RDF) format, are widely adopted in the life sciences for modeling, accessing, and exchanging data. In synthetic biology, the Synthetic Biology Open Language (SBOL) has gained traction for electronic information exchange about biological designs. SBOL, with a standardized syntax based on RDF, facilitates the exchange of sequence-based information and has evolved to include various design components (Galdzicki et al. 2014). In order to optimize the design process for synthetic biology, Semantic Web technology, specifically the Web Ontology Language (OWL), is employed to create a semantically well-defined knowledge base called SyBiOnt. This knowledge base integrates data using a warehousing approach and is queried using semantic reasoning to mine information about genetic parts and devices. The resulting novel parts are exported in SBOL format, demonstrating the potential for both manual and automated synthetic biology design (Mısırlı et al. 2016). Misirli et al. (2015) developed the SyBiOnt ontology to model genetic designs in synthetic biology, encompassing key biological parts and their relationships. The ontology allows for the formal representation of information, including annotations for additional knowledge. Using OWL semantics, SyBiOnt enables complex computational queries and automated reasoning across integrated data, contributing to the advancement of synthetic biology design processes.

Future prospects and challenges

The shortage of data poses a major challenge when it comes to implementing synthetic biology and AI. Synthetic biology relies on high-throughput data across various omics levels (metabolomic, genomic, transcriptomic, and proteomic). However, such large data is typically limited to very few numbers of microbes, and notably scarce for food-grade or GRAS (generally recognized as safe) strains. In metabolic engineering research, there is not only a requirement for significant omics data to construct AI models but also a need for an adequate quantity of data from various conditions, such as perturbations, multiple knockouts, and growth conditions to train these models. Training these models also requires downstream gene expression data across multiple knockouts or mutants. However, it becomes more complex when dealing with variations observed at different cultivation scales (Zhang et al. 2015). The behavior microbes often change when moving from small laboratory scale bioreactors to industrial-scale setups due to variations in factors, such as, gas transfer, lighting, bubble dynamics, mixing (Johnson et al. 2018). Therefore, data from lab setups may not be suitable for modeling large-scale cultivation dynamics.

Another challenge is the 'black box problem' which refers to the challenge of understanding how AI techniques produce results, causing uncertainty for end-users and legal concerns (Rieder et al. 2017). For example, in the field of 3D structural modeling and enzyme-substrate identification, new AI-based modeling approaches show potential results. Nevertheless, neural networks' inherent complexity makes it hard to understand precisely what these programs learn about protein folding. In response to the black box problem, AI scientists have introduced a set of techniques known as explainable artificial intelligence (XAI). XAI is designed to render AI results more comprehensible for humans. While it's a relatively new concept, it has the potential to address the issues that hinder the systematic enhancement of AI models (Zednik 2021). AI in drug discovery also faces numerous challenges, including data privacy, model interpretations, limited sample sizes, ethical considerations, label disparities, and learning approaches. For example, properties of proteins and chemicals can change quickly within patients, cell lines, and tissues, possibly causing a shift in their distribution. As a result, developing a dependable system, comprehending the real data representation, and accurately labeling the data are critical barriers to AI success in the field of drug development. Furthermore, owing to the stochastic nature of parameter initialization and optimization, many deep learning systems have repeatability concerns, making them extremely dependent on initial configurations.

The black-box nature of deep learning models is a challenge, as we need to trust predictions without understanding the underlying biology and chemistry. Interpretable models are essential for accurate data utilization, requiring a clear grasp of model accuracy, biological dynamics, and measurement precision (Shockley 2015). Black-box optimization approaches, such as Bayesian optimization, have been used in drug development to find global optima (minima or maxima) of functions. This assists in the identification of small compounds that can improve certain features (Pyzer-Knapp 2018) or in the construction of sequences from starting gene sequences to increase transcription or translation rates. Overall, despite the challenges and constraints associated with AI and ML techniques in handling biological data, there is no better time than the present to fully unlock the potential of these approaches and innovate new approaches to tackle various challenges like the 'black box problem'. Given the inherent complexity of living systems, heuristic approaches and machine learning are essential for advancing systems biology. Improvements in omics data collection will further enhance the integration of systems biology and machine learning for optimizing metabolic engineering.

Conclusions

The intersection of artificial intelligence (AI) and synthetic biology marks a transformative era in the biological sciences. The evolution from a descriptive to an engineering-focused discipline has propelled synthetic biology into the forefront of scientific and technological advancements. This dynamic shift, fuelled by the fusion of biology and engineering principles, has opened up new frontiers in industrial, environmental, agricultural, and pharmaceutical applications. The rapid progress in CRISPR-based genetic editing, increasing efficiency in DNA synthesis, and the automation of processes through microfluidic chips are reshaping the landscape of biological research. These advancements, coupled with the exponential growth of data in transcriptomics, proteomics, and metabolomics, position us at the threshold of a data processing era in biological sciences. AI's role in accelerating research, optimizing experiments, and facilitating precise design of complex biological systems is paramount. As AI enters its third wave, emphasizing context in models, it brings forth a new dimension to the exploration of natural cognition within the realm of synthetic biology.

However, realizing the full potential of this collaboration requires overcoming challenges, both technical and ethical. This chapter has outlined the dynamic interaction between AI and synthetic biology, emphasizing genetic circuit design, the impact of AI on genomic engineering tools like CRISPR-Cas9, therapeutic discovery, deep learning approaches, and data integration. Ethical considerations and regulatory frameworks must be actively addressed to ensure responsible and transparent use of AI in synthetic biology research.

References

Ahmadi, F., A.B. Quach and Shih, S.C. 2020. Is microfluidics the "assembly line" for CRISPR-Cas9 gene-editing? Biomicrofluidics 14: 061301.

Akbari, P., P. Gavidel and M. Gardaneh. 2019. The revolutionizing impact of artificial intelligence on breast cancer management. Archives of Breast Cancer 6: 1–3.

Alford, R.F., A. Leaver-Fay, J.R. Jeliazkov, M.J. O'Meara, F.P. DiMaio, H. Park, M.V. Shapovalov, P.D. Renfrew, V.K. Mulligan, K. Kappel and J.W. Labonte. 2017. The Rosetta all-atom energy function for macromolecular modeling and design. Journal of Chemical Theory and Computation 13: 3031.3048.

Alipanahi, B., A. Delong, M.T. Weirauch and Frey, B.J. 2015. Predicting the sequence specificities of DNA-and RNA-binding proteins by deep learning. Nature Biotechnology 33: 831.838.

Capobianco, J.P. 2023. The Keys to Successful Ventures in the Future, The New Era of Global Services: A Framework for Successful Enterprises in Business Services and IT. Emerald Publishing Limited, Leeds.

Angenent-Mari, N.M., A.S. Garruss, L.R. Soenksen, G. Church and J.J Collins. 2020. A deep learning approach to programmable RNA switches. Nature Communications, 11: 5057.

Atance, S.R., J.V. Diez, O. Engkvist, S. Olsson and R. Mercado. 2022. *De novo* drug design using reinforcement learning with graph-based deep generative models. Journal of Chemical Information and Modeling, 62: 4863–4872.

Bak, R.O., N. Gomez-Ospina and M.H. Porteus. 2018. Gene editing on center stage. Trends in Genetics 34: 600–611.

Basit, S.A., R. Qureshi, S. Musleh, R. Guler, M.S. Rahman, Biswas, K.H. and T. Alam. 2023. COVID-19 Base v3: Update of the knowledgebase for drugs and biomedical entities linked to COVID-19. Frontiers in Public Health 11: 1125917.

Beardall, W.A., G.B. Stan and M.J. Dunlop. 2022. Deep learning concepts and applications for synthetic biology. GEN Biotechnology 1: 360–371.

Beauchamp, T.L. and J.F. Childress. 2001. Principles of biomedical ethics. Oxford University Press, USA.

Berman, H.M., J. Westbrook, Z. Feng, G. Gilliland, T.N. Bhat, H. Weissig, I.N. Shindyalov and Bourne, P.E. 2000. The protein data bank. Nucleic acids research 28: 235–242.

Bhat, A.A., S. Nisar, S. Mukherjee, N. Saha, N. Yarravarapu, S.N. Lone, T. Masoodi, R. Chauhan, S. Maacha, P. Bagga and P. Dhawan. 2022. Integration of CRISPR/Cas9 with artificial intelligence for improved cancer therapeutics. Journal of translational medicine 20: 534.

Bianchini, F. 2018. The problem of prediction in artificial intelligence and synthetic biology. Complex Systems 27: 249–265.

Brophy, J. A. and C. A. Voigt. 2014. Principles of genetic circuit design. Nature Methods, 11: 508–520.

Brundage, M., S. Avin, J. Clark, H. Toner, P. Eckersley, B. Garfinkel, A. Dafoe, P. Scharre, T. Zeitzoff, B. Filar and H. Anderson. 2018. The malicious use of artificial intelligence: Forecasting, prevention, and mitigation. arXiv preprint arXiv:1802.07228.

Brunk, E., K.W. George, J. Alonso-Gutierrez, M. Thompson, Baidoo, E., Wang, G., Petzold, C.J., McCloskey, D., Monk, J., Yang, L. and O'Brien, E.J. 2016. Characterizing strain variation in engineered E. coli using a multi-omics-based workflow. Cell Systems 2: 335–346.

Buolamwini, J. and T. Gebru 2018. Gender shades: Intersectional accuracy disparities in commercial gender classification. The Proceedings of the 1st Conference, Fairness, accountability and transparency, Valladolid, Spain.

Cameron, D.E., C.J. Bashor and J.J. Collins. 2014. A brief history of synthetic biology. *Nature Reviews Microbiology* 12: 381–390.

Capobianco, J.P. 2023. The Keys to Successful Ventures in the Future, The New Era of Global Services: A Framework for Successful Enterprises in Business Services and IT. Emerald Publishing Limited, Leeds.

Carbonell, P., T. Radivojevic and H. Garcia Martin. 2019. Opportunities at the intersection of synthetic biology, machine learning, and automation. ACS Synthetic Biology 8: 1474–1477.

Cartwright-Smith, L., E. Gray and J.H. Thorpe. 2016. Health information ownership: legal theories and policy implications. Vanderbilt journal of entertainment and technology law 19: 207.

Chae, T.U., S.Y. Choi, J.W. Kim, Y.S. Ko and S.Y. Lee. 2017. Recent advances in systems metabolic engineering tools and strategies. Current Opinion in Biotechnology 47: 67–82.

Chakraborty, I., Choudhury, A. and Banerjee, T.S. 2017. Artificial intelligence in biological data. Journal of Information Technology & Software Engineering 7(4): 207.

Chari, R., Yeo, N.C., Chavez, A. and Church, G.M. 2017. sgRNA Scorer 2.0: a species-independent model to predict CRISPR/Cas9 activity. ACS Synthetic Biology 6(5): 902–904.

Chen, H., Engkvist, O., Wang, Y., Olivecrona, M. and Blaschke, T. 2018. The rise of deep learning in drug discovery. Drug Discovery Today 23(6): 1241–1250.

Chen, Y., Guenther, J.M., Gin, J.W., Chan, L.J.G., Costello, Z., Ogorzalek, T.L., Tran, H.M., Blake-Hedges, J.M., Keasling, J.D., Adams, P.D. and García Martín, H., 2019. Automated "cells-to-peptides" sample preparation workflow for high-throughput, quantitative proteomic assays of microbes. Journal of Proteome Research18(10): 3752–3761.

Chen, Y., Zhang, S., Young, E.M., Jones, T.S., Densmore, D. and Voigt, C.A. 2020. Genetic circuit design automation for yeast. *Nature Microbiology*, 5(11), 1349–1360.

Cheng, Y., Liu, J., Zhang, H., Wang, J., Zhao, Y. and Geng, W.2015. Transcriptome analysis and gene expression profiling of abortive and developing ovules during fruit development in hazelnut. PLoS One, 10(4): e0122072.

Chuai, G.-h., Wang, Q.-L. and Liu, Q. 2017. In silico meets *in vivo*: towards computational CRISPR-based sgRNA design. Trends in Biotechnology 35(1):12–21.

Consortium, U. (2015). UniProt: a hub for protein information. *Nucleic Acids Research*, 43(D1), D204–D212.

Costanzo, M., Baryshnikova, A., Bellay, J., Kim, Y., Spear, E.D., Sevier, C.S., Ding H, Koh J.L., Toufighi, K., Mostafavi, S. and Prinz, J. 2010. The genetic landscape of a cell. *Science* 327(5964): 425–431.

Costello, Z. and Martin, H.G. 2018. A machine learning approach to predict metabolic pathway dynamics from time-series multiomics data. NPJ Systems Biology and Applications 4(1): 1–14.

Dalsass, M., Brozzi, A., Medini, D. and Rappuoli, R. 2019. Comparison of open-source reverse vaccinology programs for bacterial vaccine antigen discovery. Frontiers in Immunology10: 113.

Dara, S., Dhamercherla, S., Jadav, S.S., Babu, C.M. and Ahsan, M.J. 2022. Machine learning in drug discovery: a review. Artificial Intelligence Review 55(3): 1947–1999.

Dietler, N., Minder, M., Gligorovski, V., Economou, A.M., Joly, D.A.H.L., Sadeghi, A., Chan, C.H.M., Koziński, M., Weigert, M., Bitbol, A.F. and Rahi, S.J. 2020. A convolutional neural network segments yeast microscopy images with high accuracy. *Nature Communications* 11(1): 5723.

Ding, X. and Zhang, B. 2021. DeepBAR: a fast and exact method for binding free energy computation. *T*he Journal of Physical Chemistry Letters, 12(10), 2509–2515.

Emmert-Streib, F. 2021. From the digital data revolution toward a digital society: Pervasiveness of artificial intelligence. Machine Learning and Knowledge Extraction, 3(1): 284–298.

Eslami, M., Adler, A., Caceres, R.S., Dunn, J.G., Kelley-Loughnane, N., Varaljay, V.A. and Martin, H.G. 2022. Artificial intelligence for synthetic biology. Communications of The ACM 65(5): 88–97.

Esteva, A., Kuprel, B., Novoa, R.A., Ko, J., Swetter, S.M., Blau, H.M. and Thrun, S. 2017. Dermatologist-level classification of skin cancer with deep neural networks. Nature 542(7639): 115–118.

Faulon, J.L. and Faure, L. 2021. *In silico*, *in vitro*, and *in vivo* machine learning in synthetic biology and metabolic engineering. Current Opinion in Chemical Biology 65: 85–92.

Fleming, N. (2018). Computer-calculated compounds. Nature 557(7707): S55–57.

Ford, R.A., Price, W. and Nicholson, I. 2016. Privacy and accountability in black-box medicine. *Michigan Telecommunication & Technology Law Review* 23: 1.

Fuhrer, T. and Zamboni, N. (2015). High-throughput discovery metabolomics. *Current Opinion in Biotechnology*, 31, 73–78.

Galdzicki, M., Clancy, K.P., Oberortner, E., Pocock, M., Quinn, J.Y., Rodriguez, C.A., Roehner, N., Wilson, M.L., Adam, L. anderson, J.C. and Bartley, B.A. 2014. The Synthetic Biology Open Language (SBOL) provides a community standard for communicating designs in synthetic biology. Nature Biotechnology 32(6): 545–550.

García-Granados, R., Lerma-Escalera, J.A. and Morones-Ramírez, J.R. 2019. Metabolic engineering and synthetic biology: synergies, future, and challenges. Frontiers in Bioengineering and Biotechnology 7: 36.

Gaudelet, T., Day, B., Jamasb, A.R., Soman, J., Regep, C., Liu, G., Hayter, J.B., Vickers, R., Roberts, C., Tang, J. and Roblin, D., 2021. Utilizing graph machine learning within drug discovery and development. Briefings in Bioinformatics 22(6):159.

Gerke, S., Minssen, T. and Cohen, G. 2020. Ethical and legal challenges of artificial intelligence-driven healthcare. In Artificial intelligence in healthcare, Academic Press, USA.

Gilliot, P.A. and Gorochowski, T.E. 2020. Sequencing enabling design and learning in synthetic biology. Current Opinion in Chemical Biology 58: 54–62.

Goble, C. and Stevens, R. 2008. State of the nation in data integration for bioinformatics. *Journal of Biomedical Informatics*, 41(5): 687–693.

Gómez-Bombarelli, R., Wei, J.N., Duvenaud, D., Hernández-Lobato, J.M., Sánchez-Lengeling, B., Sheberla, D., Aguilera-Iparraguirre, J., Hirzel, T.D., Adams, R.P. and Aspuru-Guzik, A., 2018. Automatic chemical design using a data-driven continuous representation of molecules. *ACS Central Science* 4(2): 268-276.

Goodfellow, I. 2016. Nips 2016 tutorial: Generative adversarial networks. *arXiv* preprint arXiv:1701.00160.

Guido, N.J., Wang, X., Adalsteinsson, D., McMillen, D., Hasty, J., Cantor, C.R., Timothy, E.C. and Collins, J.J. 2006. A bottom-up approach to gene regulation. Nature 439(7078): 856–860.

Haeussler, M., Schönig, K., Eckert, H., Eschstruth, A., Mianné, J., Renaud, J.B., Schneider-Maunoury, S., Shkumatava, A., Teboul, L., Kent, J. and Joly, J.S., 2016. Evaluation of off-target and on-target scoring algorithms and integration into the guide RNA selection tool CRISPOR. Genome biology 17: 1–12.

Hartline, C.J., Schmitz, A. C., Han, Y. and Zhang, F. 2021. Dynamic control in metabolic engineering: Theories, tools, and applications. Metabolic Engineering 63: 126–140.

Heinemann, J., Deng, K., Shih, S.C., Gao, J., Adams, P.D., Singh, A.K. and Northen, T.R. 2017. On-chip integration of droplet microfluidics and nanostructure-initiator mass spectrometry for enzyme screening. Lab on a Chip, 17(2), 323–331.

Helmy, M., Smith, D. and Selvarajoo, K. 2020. Systems biology approaches integrated with artificial intelligence for optimized metabolic engineering. Metabolic Engineering Communications 11: e00149.

Higgins, C.L., Muralidhara, B.K. and Wittung-Stafshede, P. 2005. How do cofactors modulate protein folding? Protein and Peptide Letters 12(2): 165–170.

Hiranniramol, K., Chen, Y., Liu, W. and Wang, X. 2020. Generalizable sgRNA design for improved CRISPR/Cas9 editing efficiency. Bioinformatics 36(9): 2684–2689.

Huang, K., Fu, T., Glass, L.M., Zitnik, M., Xiao, C. and Sun, J. 2020. DeepPurpose: a deep learning library for drug–target interaction prediction. Bioinformatics 36(22–23): 5545–5547.

Jervis, A.J., Carbonell, P., Vinaixa, M., Dunstan, M.S., Hollywood, K.A., Robinson, C.J., Rattray, N.J., Yan, C., Swainston, N., Currin, A. and Sung, R. 2018. Machine learning of designed translational control allows predictive pathway optimization in *Escherichia coli*. ACS Synthetic Biology 8(1): 127–136.

Johnson, T.J., Katuwal, S. anderson, G.A., Gu, L., Zhou, R. and Gibbons, W.R. 2018. Photobioreactor cultivation strategies for microalgae and cyanobacteria. Biotechnology Progress 34(4): 811–827.

Jumper, J., Evans, R., Pritzel, A., Green, T., Figurnov, M., Ronneberger, O., Tunyasuvunakool, K., Bates, R., Žídek, A., Potapenko, A. and Bridgland, A., 2021. Highly accurate protein structure prediction with AlphaFold. Nature 596(7873): 583–589.

Kadurin, A., Aliper, A., Kazennov, A., Mamoshina, P., Vanhaelen, Q., Khrabrov, K. and Zhavoronkov, A. 2017. The cornucopia of meaningful leads: Applying deep adversarial autoencoders for new molecule development in oncology. Oncotarget 8(7): 10883.

Kadurin, A., Nikolenko, S., Khrabrov, K., Aliper, A. and Zhavoronkov, A. 2017. druGAN: an advanced generative adversarial autoencoder model for de novo generation of new molecules with desired molecular properties *in silico*. Molecular pharmaceutics 14(9): 3098–3104.

Khemchandani, Y., O'Hagan, S., Samanta, S., Swainston, N., Roberts, T.J., Bollegala, D. and Kell, D.B. 2020. DeepGraphMolGen, a multi-objective, computational strategy for generating molecules with desirable properties: a graph convolution and reinforcement learning approach. Journal of Cheminformatics 12(1): 1–17.

Kim, S., Thiessen, P.A., Bolton, E.E., Chen, J., Fu, G., Gindulyte, A., Han, L., He, J., He, S., Shoemaker, B.A. and Wang, J., 2016. PubChem substance and compound databases. Nucleic acids research 44(D1): D1202–D1213.

Kirsch, A., Van Amersfoort, J. and Gal, Y. 2019. Batchbald: Efficient and diverse batch acquisition for deep bayesian active learning. Advances in Neural Information Processing Systems 32.

Kleinstiver, B.P., Pattanayak, V., Prew, M.S., Tsai, S.Q., Nguyen, N.T., Zheng, Z. and Joung J.K. 2016. High-fidelity CRISPR–Cas9 nucleases with no detectable genome-wide off-target effects. Nature 529(7587): 490–495.

Kotopka, B.J. and Smolke, C.D. 2020. Model-driven generation of artificial yeast promoters. Nature Communications 11(1): 2113.

Lee, I., Keum, J. and Nam, H. 2019. DeepConv-DTI: Prediction of drug-target interactions via deep learning with convolution on protein sequences. PLoS Computational Biology 15(6): e1007129.

Lillacci, G., Benenson, Y. and Khammash, M. 2018. Synthetic control systems for high performance gene expression in mammalian cells. Nucleic Acids Research 46(18): 9855–9863.

Lin, J. and Wong, K.-C. 2018. Off-target predictions in CRISPR-Cas9 gene editing using deep learning. Bioinformatics 34(17): i656–i663.

Lin, Y., Xiang, Z. and He, Y. 2011. Towards a Semantic Web Application: Ontology-Driven Ortholog Clustering Analysis, In: International Conference on Biomedical Ontology, Buffalo, NY, USA.

Liu, Z., Du, J., Fang, J., Yin, Y., Xu, G. and Xie, L. 2019. DeepScreening: a deep learning-based screening web server for accelerating drug discovery. *Database*, 2019, baz104.

Ma, H. and Zeng, A.-P. 2003. Reconstruction of metabolic networks from genome data and analysis of their global structure for various organisms. Bioinformatics 19(2): 270–277.

Nielsen A.A., Der B.S., Shin J., Vaidyanathan P., Paralanov V., Strychalski E.A., Ross D., Densmore D. and Voigt C.A. 2016. Genetic circuit design automation. Science 352(6281): aac7341.

Maserat, E. 2022. Integration of Artificial Intelligence and CRISPR/Cas9 System for Vaccine Design. SAGE Publications Sage UK: London, England.

Maserat, E. and Nasiri Hooshmand, M. 2021. Requirements of integrated computational approach for developing personalized cancer vaccines. Human Vaccines and Immunotherapeutics 17(12): 5573–5574.

Mayr, A., Klambauer, G., Unterthiner, T. and Hochreiter, S. 2016. DeepTox: toxicity prediction using deep learning. Frontiers in Environmental Science 3: 80.

Mendez, D., Gaulton, A., Bento, A.P., Chambers, J., De Veij, M., Félix, E., Magariños, M.P., Mosquera, J.F., Mutowo, P., Nowotka, M. and Gordillo-Marañón, M., 2019. ChEMBL: towards direct deposition of bioassay data. Nucleic Acids Research 47(D1): D930–D940.

Meng, F. and Ellis, T. 2020. The second decade of synthetic biology: 2010–2020. Nature Communications 11(1): 5174.

Milchenko, M. and Marcus, D. 2013. Obscuring surface anatomy in volumetric imaging data. Neuroinformatics 11: 65–75.

Mirsky, Y., Demontis, A., Kotak, J., Shankar, R., Gelei, D., Yang, L., et al. 2023. The threat of offensive ai to organizations. Computers and Security 124: 103006.

Mirza, B., Wang, W., Wang, J., Choi, H., Chung, N.C. and Ping, P. 2019. Machine learning and integrative analysis of biomedical big data. Genes 10(2): 87.

Mısırlı, G.k., Hallinan, J., Pocock, M., Lord, P., McLaughlin, J.A., Sauro, H., et al. (2016). Data integration and mining for synthetic biology design. ACS Synthetic Biology 5(10): 1086–1097.

Moingeon, P., Kuenemann, M. and Guedj, M. 2022. Artificial intelligence-enhanced drug design and development: Toward a computational precision medicine. Drug Discovery Today, 27(1): 215–222.

Moreno-Mateos, M.A., Vejnar, C.E., Beaudoin, J.-D., Fernandez, J.P., Mis, E.K., Khokha, M.K. and Giraldez A.J. 2015. CRISPRscan: designing highly efficient sgRNAs for CRISPR-Cas9 targeting *in vivo*. Nature Methods 12(10): 982–988.

Mostafavi, S., Ray, D., Warde-Farley, D., Grouios, C. and Morris, Q. 2008. GeneMANIA: a real-time multiple association network integration algorithm for predicting gene function. Genome Biology 9: 1–15.

Na, L., Yang, C., Lo, C.-C., Zhao, F., Fukuoka, Y. and Aswani, A. 2018. Feasibility of reidentifying individuals in large national physical activity data sets from which protected health information has been removed with use of machine learning. JAMA Network Open 1(8): e186040–e186040.

Nayarisseri, A. 2019. Machine learning, deep learning and artificial intelligence approach for predicting CRISPR for the cancer treatment. In International Conference on Multidisciplinary Sciences. 5th edition session USEDAT-07, USA.

Neamatullah, I., Douglass, M.M., Lehman, L.-W.H., Reisner, A., Villarroel, M., Long, W.J., Szolovits P, Moody G.B., Mark R.G. and Clifford G.D. 2008. Automated de-identification of free-text medical records. BMC Medical Informatics and Decision Making 8(1): 1–17.

Nelson, G.S. 2019. Bias in artificial intelligence. North Carolina Medical Journal 80(4): 220–222.

Ni, C., Dinh, C.V. and Prather, K.L. 2021. Dynamic control of metabolism. Annual Review of Chemical and Biomolecular Engineering 12: 519–541.

Nielsen, A.A., Der, B.S., Shin, J., Vaidyanathan, P., Paralanov, V., Strychalski, E.A., Ross, D., Densmore, D. and Voigt, C.A. (2016). Genetic Circuit Design Automation. *Science 352*(6281): p.aac7341.

O'Connor, O.M., Alnahhas, R.N., Lugagne, J.B. and Dunlop, M.J. 2022. DeLTA 2.0: A deep learning pipeline for quantifying single-cell spatial and temporal dynamics. PLOS Computational Biology 18(1): e1009797.

Olivecrona, M., Blaschke, T., Engkvist, O. and Chen, H. 2017. Molecular de-novo design through deep reinforcement learning. Journal of Cheminformatics 9(1): 1–14.

Oliveira, A.L. 2019. Biotechnology, big data and artificial intelligence. Biotechnology journal 14(8): 1800613.

Panigrahi, Swapnesh, Dorothée Murat, Antoine Le Gall, Eugénie Martineau, Kelly Goldlust, Jean-Bernard Fiche, Sara Rombouts, Marcelo Nöllmann, Leon Espinosa and Tâm Mignot. 2021. Misic, a general deep learning-based method for the high-throughput cell segmentation of complex bacterial communities. Elife 10: e65151.

Paul, D., Sanap, G., Shenoy, S., Kalyane, D., Kalia, K. and Tekade, R.K. 2021. Artificial intelligence in drug discovery and development. Drug Discovery Today 26(1): 80.

Peng, H., Zheng, Y., Zhao, Z., Liu, T. and Li, J. 2018. Recognition of CRISPR/Cas9 off-target sites through ensemble learning of uneven mismatch distributions. Bioinformatics 34(17): i757–i765.

Poplin, R., Chang, P.C., Alexander, D., Schwartz, S., Colthurst, T., Ku, A., Newburger, D., Dijamco, J., Nguyen, N., Afshar, P.T. and Gross, S.S. 2018. A universal SNP and small-indel variant caller using deep neural networks. Nature Biotechnology 36(10): 983–987.

Popova, M., Isayev, O. and Tropsha, A. 2018. Deep reinforcement learning for *de novo* drug design. Science Advances 4(7): eaap7885.

Prieto-Sánchez, C., Risueño, A., Fontanillo, C. and De Las Rivas, J. 2008. Human gene coexpression landscape: Confident network derived from tissue transcriptomic profiles. PloS one 3(12): e3911.

Pržulj, N. 2011. Protein-protein interactions: making sense of networks via graph-theoretic modeling. Bioessays 33(2): 115–123.

Pyzer-Knapp, E.O. 2018. Bayesian optimization for accelerated drug discovery. IBM Journal of Research and Development 62(6): 2–1.

Quest, D.J., Land, M.L., Brettin, T.S. and Cottingham, R.W. 2010. Next generation models for storage and representation of microbial biological annotation. BMC Bioinformatics 11(6): 1–16.

Qureshi, R., Irfan, M., Gondal, T.M., Khan, S., Wu, J., Hadi, M.U., Heymach, J., Le, X., Yan, H. and Alam, T. (2023). AI in Drug Discovery and its Clinical Relevance. *Heliyon*, 9, e17575.

Ramsundar, B. 2018. Molecular Machine Learning with DeepChem. Doctoral dissertation, Stanford University, California, USA.

Rieder, G. and Simon, J. 2017. Big data: A new empiricism and its epistemic and socio-political consequences. Berechenbarkeit der Welt? Philosophie und Wissenschaft im Zeitalter von Big Data: 85–105.

Rifaioglu, A.S., Nalbat, E., Atalay, V., Martin, M.J., Cetin-Atalay, R. and Doğan, T. 2020. DEEPScreen: high performance drug–target interaction prediction with convolutional neural networks using 2-D structural compound representations. Chemical science 11(9): 2531–2557.

Ronneberger, O., Fischer, P. and Brox, T. 2015. U-net: Convolutional networks for biomedical image segmentation. In: *Medical Image Computing and Computer-Assisted Intervention, 18th International Conference,* Munich, Germany.

Rothstein, M.A. 2010. Is deidentification sufficient to protect health privacy in research? The American Journal of Bioethics 10(9): 3–11.

Safdar, N.M., Banja, J.D. and Meltzer, C.C. 2020. Ethical considerations in artificial intelligence. European Journal of Radiology 122: 108768.

Sample, P.J., Wang, B., Reid, D.W., Presnyak, V., McFadyen, I.J., Morris, D.R. and Seelig, G. 2019. Human 5′ UTR design and variant effect prediction from a massively parallel translation assay. Nature Biotechnology 37(7): 803–809.

Scarselli, F., Gori, M., Tsoi, A.C., Hagenbuchner, M. and Monfardini, G. 2008. The graph neural network model. IEEE Transactions On Neural Networks 20(1): 61–80.

Schwab, K. (2017). The fourth industrial revolution. Currency. Crown Publishing Group, New York.

Sellwood, M.A., Ahmed, M., Segler, M.H. and Brown, N. 2018. Artificial intelligence in drug discovery. Future Medicinal Chemistry 10(17): 2025–2028.

Senior, A.W., Evans, R., Jumper, J., Kirkpatrick, J., Sifre, L., Green, T., Qin, C., Žídek, A., Nelson, A.W., Bridgland, A. and Penedones, H., 2020. Improved protein structure prediction using potentials from deep learning. Nature 577(7792): 706–710.

Shadbolt, N., Berners-Lee, T. and Hall, W. 2006. The semantic web revisited. IEEE intelligent systems 21(3): 96–101.

Shah, M., Naik, N., Somani, B.K. and Hameed, B.Z. 2020. Artificial intelligence (AI) in urology-Current use and future directions: An iTRUE study. Turkish Journal of Urology 46: S27–S39.

Shockley, K.R. 2015. Quantitative high-throughput screening data analysis: challenges and recent advances. Drug Discovery Today 20(3): 296–300.

Smolke, C.D. and Silver, P.A. 2011. Informing biological design by integration of systems and synthetic biology. Cell 144(6): 855–859.

Spahn, C., Gómez-de-Mariscal, E., Laine, R.F., Pereira, P.M., von Chamier, L., Conduit, M., Pinho, M.G., Jacquemet, G., Holden, S., Heilemann, M. and Henriques, R. 2022. DeepBacs for multi-task bacterial image analysis using open-source deep learning approaches. Communications Biology 5(1): 688.

Stokes, J.M., Yang, K., Swanson, K., Jin, W., Cubillos-Ruiz, A., Donghia, N.M., MacNair, C.R., French, S., Carfrae, L.A., Bloom-Ackermann, Z. and Tran, V.M. 2020. A deep learning approach to antibiotic discovery. Cell 180(4): 688–702.

Stringer, C., Wang, T., Michaelos, M. and Pachitariu, M. 2021. Cellpose: A generalist algorithm for cellular segmentation. Nature methods 18(1): 100–106.

Tanaka, R. 2005. Scale-rich metabolic networks. Physics Review Letters, 94(16), 168101.

Tang, Y. C. and Gottlieb, A. 2021. Explainable drug sensitivity prediction through cancer pathway enrichment. Scientific reports 11(1): 3128.

Tong, A.H.Y., Lesage, G., Bader, G.D., Ding, H., Xu, H., Xin, X., Young, J., Berriz, G.F., Brost, R.L., Chang M. and Chen Y. 2004. Global mapping of the yeast genetic interaction network. Science 303(5659): 808–813.

Tong, H. and Nikoloski, Z. 2021. Machine learning approaches for crop improvement: Leveraging phenotypic and genotypic big data. J. Plant Physiology 257: 153354.

Treloar, N.J., Fedorec, A.J., Ingalls, B. and Barnes, C.P. 2020. Deep reinforcement learning for the control of microbial co-cultures in bioreactors. PLoS Computational Biology 16(4): e1007783.

Vamathevan, J., Clark, D., Czodrowski, P., Dunham, I., Ferran, E., Lee, G., Li, B., Madabhushi, A., Shah, P., Spitzer, M. and Zhao, S. 2019. Applications of machine learning in drug discovery and development. Nature Reviews Drug discovery 18(6): 463–477.

Voulodimos, A., Doulamis, N., Doulamis, A. and Protopapadakis, E. 2018. Deep learning for computer vision: A brief review. Computational intelligence and neuroscience 2018: 7068349.

Wang, S.W., Gao, C., Zheng, Y.M., Yi, L., Lu, J.C., Huang, X.Y., Cai, J.B., Zhang, P.F., Cui, Y.H. and Ke, A.W. 2022. Current applications and future perspective of CRISPR/Cas9 gene editing in cancer. Molecular Cancer 21(1): 1–27.

Wessels, H.H., Stirn, A., Méndez-Mancilla, A., Kim, E.J., Hart, S.K., Knowles, D.A. and Sanjana, N.E. 2023. Prediction of on-target and off-target activity of CRISPR–Cas13d guide RNAs using deep learning. Nature Biotechnology 2023: 1–10.

Wilson, L.O., Reti, D., O'Brien, A.R., Dunne, R.A. and Bauer, D.C. 2018. High activity target-site identification using phenotypic independent CRISPR-Cas9 core functionality. The CRISPR Journal 1(2): 182–190.

Winslade, W.J. and Ross, J.W. 1985. Privacy, confidentiality and autonomy in psychotherapy. Nebraska Law Review 64: 578.

Wong, A., Wang, H., Poh, C.L. and Kitney, R.I. 2015. Layering genetic circuits to build a single cell, bacterial half adder. BMC Biology 13(1): 1–16.

Wong, N., Liu, W. and Wang, X. 2015. WU-CRISPR: characteristics of functional guide RNAs for the CRISPR/Cas9 system. Genome Biology 16(1): 1–8.

Xu, H., Xiao, T., Chen, C.-H., Li, W., Meyer, C.A., Wu, Q., Wu D., Cong L., Zhang F., Liu J.S. and Brown M. 2015. Sequence determinants of improved CRISPR sgRNA design. Genome Research 25(8): 1147–1157.

Xu, Y., Liu, X., Cao, X., Huang, C., Liu, E., Qian, S., Liu, X., Wu, Y., Dong, F., Qiu, C.W. and Qiu, J. 2021. Artificial intelligence: A powerful paradigm for scientific research. The Innovation 2(4): 100179.

Yang, J., Anishchenko, I., Park, H., Peng, Z., Ovchinnikov, S. and Baker, D. 2020. Improved protein structure prediction using predicted interresidue orientations. Proceedings of the National Academy of Sciences 117(3):1496–1503.

Yeo, H.C. and Selvarajoo, K. 2022. Machine learning alternative to systems biology should not solely depend on data. Briefings in Bioinformatics 23(6): bbac436.

Zednik, C. 2021. Solving the black box problem: A normative framework for explainable artificial intelligence. Philosophy and Technology 34(2): 265–288.

Zhang, D., Dechatiwongse, P., del Rio-Chanona, E.A., Maitland, G.C., Hellgardt, K. and Vassiliadis, V.S. 2015. Dynamic modelling of high biomass density cultivation and biohydrogen production in different scales of flat plate photobioreactors. Biotechnology and Bioengineering 112(12): 2429–2438.

Zhu, Z., Shi, C., Zhang, Z., Liu, S., Xu, M., Yuan, X. and Tang, J. 2022. Torchdrug: A powerful and flexible machine learning platform for drug discovery. arXiv preprint arXiv:2202.08320.

CHAPTER

18

Cognitive Artificial Intelligence

Kavati Pavan Kumar,[1#] *Swarubini PJ*[1#]
and *Nagarajan Ganapathy*[1*]

1. Introduction Foundations of Cognitive AI

1.1 Cognitive Science Principles

Cognitive science principles constitute the foundational framework for understanding the intricate workings of the human mind. Drawing from an interdisciplinary approach that melds insights from psychology, neuroscience, linguistics, philosophy, and artificial intelligence, cognitive science seeks to unravel the complexities of cognition (Zhao et al., 2022). These principles explore how individuals acquire knowledge, process information, and engage in various mental activities such as problem-solving, decision-making, and language comprehension (Simon, 1980). By delving into the underlying mechanisms of memory, attention, and perception, cognitive science provides a holistic perspective on human intelligence (See Fig. 1). It serves as a guiding beacon for research endeavors aimed at unraveling the mysteries of the mind, influencing diverse fields such as education, human-computer interaction, and the development of intelligent systems.

1.2 Neural Networks and Cognitive Models

In the intricate tapestry of artificial intelligence (AI) research, the intersection of neural networks and cognitive models emerges as a fertile ground for exploration. The foundational premise lies in recognizing the complementary strengths of neural networks—inspired by the architecture of the human brain—and cognitive models, which seek to emulate the intricacies of human thought processes (Zhu et al., 2020). It involves a meticulous exploration of the fundamental principles that govern neural networks and cognitive models. It traverses the landscape of machine learning, probing the intricate workings of artificial neural networks and their capacity to mimic aspects of human cognition (Marcus, 2020). From perceptrons to deep learning architectures, the chapters serve as a robust foundation, ensuring readers grasp the intricacies of the artificial side of the equation.

1 Department of Biomedical Engineering, Indian Institute of Technology Hyderabad, India.
# Equal Contribution
* Corresponding author: gnagarajan@bme.iith.ac.in

1.3 Connectionism in Cognitive Systems

Connectionism is a theoretical framework in cognitive science which understands mental processes through the study of interconnected elements, such as neurons in the brain (See Fig. 2). This approach contrasts with other theories like rule-based systems, which focus on individual components and their interactions (Perconti & Plebe, 2020). Connectionism posits that the mind operates through a system of interrelated parts, each contributing to the overall functioning of the whole. The major aspects of connectionism include:

- *Parallel Processing*: Connectionist models simulate the way information is processed simultaneously at different levels of abstraction, allowing for faster problem solving and decision making.
- *Distributed Representations*: Instead of relying on single, centralized representations, connectionist models distribute information across many interconnected elements, enabling them to process complex relationships and patterns.
- *Learning Algorithms*: Connectionist models employ algorithms that enable them to learn from experience, adjusting their internal structures and weights according to feedback received during interaction with the environment.
- *Adaptation*: As connectionist models interact with their surroundings, they continuously update their internal structures and parameters, adapting to new situations and improving their performance over time.

It provides a powerful framework for understanding cognitive systems by focusing on the interplay between interconnected elements, parallel processing, distributed representations, learning algorithms, and adaptation (Knyazev, 2023). This approach has led to significant advancements in fields such as machine learning, artificial intelligence, and neuroscience, offering insights into the workings of the human mind and paving the way for the development of advanced technologies.

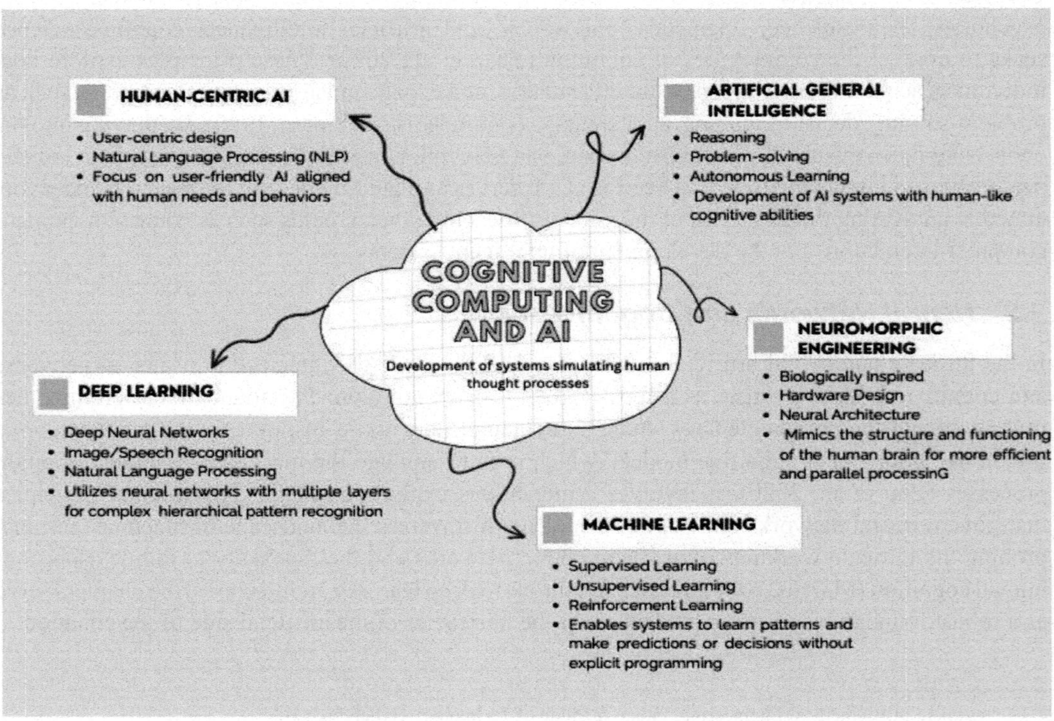

Fig. 1 Overview of Cognitive Computing and AI

2. Cognitive Computing and Machine Learning

2.1 Integration of Cognitive Processes in ML

In the dynamic landscape of artificial intelligence, the fusion of Cognitive Computing and Machine Learning emerges as a frontier reshaping intelligent systems. This partnership amalgamates human-like cognitive capacities with the computational prowess of machine learning algorithms, sparking a revolution in problem-solving and decision-making (Lv et al., 2021). Cognitive Computing mirrors human cognition, empowering machines to comprehend, reason, and learn from complex data in ways similar to human thought patterns, transcending conventional algorithmic computations. Advancements drive the integration of cognitive processes within Machine Learning systems, pushing beyond mere data processing. This integration encompasses perceptual reasoning, contextual decision-making, and adaptive learning mirroring human cognitive capabilities. It propels machines evolve from simple data processors to dynamic entities that interpret, learn, and adapt, echoing human cognitive processes.

This infusion of cognitive elements within Machine Learning signifies a paradigm shift beyond traditional algorithms. Studies such as (Behera et al., 2023; Lv et al., 2021) exemplify this integration, showcasing Cognitive Computing principles in dynamic decision-making scenarios. These initiatives employ granular concept-cognitive systems and coherent radar models, emphasizing an enriched learning paradigm within Machine Learning frameworks. By integrating cognitive processes, these systems exhibit heightened adaptability, efficiency in handling uncertainties, and a remarkable capacity for self-improvement. The interdependence of cognitive principles and machine learning algorithms is evident, serving as a testament to their symbiotic relationship. This integration fosters AI systems that not only process data but also comprehend its underlying essence, heralding a transformative era across industries and domains.

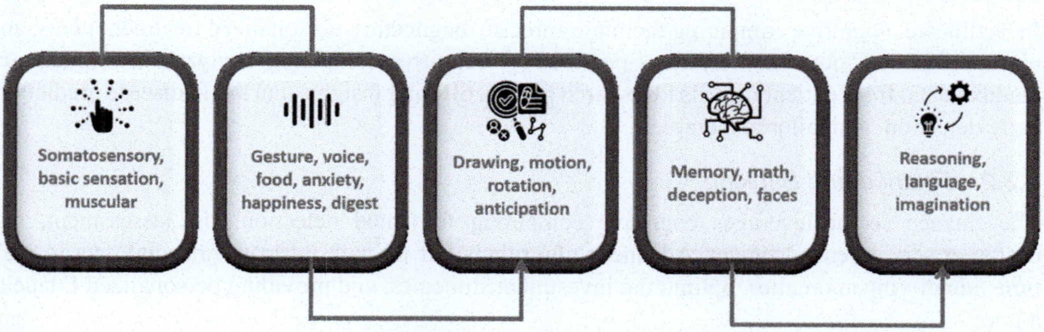

Fig. 2 Objective of cognition pyramid

2.2 Learning Paradigms in Cognitive AI

Cognitive Artificial Intelligence (AI) thrives on a multitude of learning paradigms, with neural networks and deep learning standing as pillars, revolutionizing the understanding and implementation of cognitive computing models. These paradigms delve deep into the intricate workings of human cognition, aiming to replicate and advance its functionalities in machines.

2.2.1 Neural Networks: Mimicking Cognitive Processes

Neural networks, inspired by the human brain's neural structure, form the backbone of cognitive AI. They comprise interconnected nodes, or neurons, capable of processing complex data patterns. Reference studies (Bindu et al., 2007; Parashar et al., 2023) have delved into this paradigm's application in deciphering human behavior and sentiment analysis. Leveraging neural networks, these studies have unveiled profound insights into behavioral prediction and sentiment recognition, mirroring cognitive processes by analyzing intricate patterns within vast datasets.

2.2.2 Deep Learning: Unveiling Complex Patterns

Deep learning, a subset of machine learning, utilizes multiple layers to extract hierarchical representations from data. Studies (Kanumuri et al., 2022; Xiang, 2022) have highlighted the crucial role of deep learning in feature extraction within cognitive computing models. These studies showcase deep neural networks' prowess in extracting nuanced features crucial for understanding human behavior, enabling sentiment sensing, and enhancing cognitive computing processes. The utilization of deep neural networks in these studies signifies the potency of this approach in unraveling complex patterns crucial for various cognitive tasks.

Deep learning, as demonstrated in these studies, emerges as a fundamental tool in cognitive AI, offering powerful capabilities in extracting and understanding complex data structures to drive innovative applications across diverse domains. The integration of neural networks and deep learning within cognitive AI not only enhances the understanding of human-like cognitive processes but also unlocks the potential for diverse applications. These paradigms, as showcased in the referenced studies, play pivotal roles in human behavior prediction, sentiment analysis, and feature extraction, propelling the advancement of cognitive computing models. These learning paradigms pave the way for sophisticated AI systems capable of understanding, interpreting, and responding to human-like patterns and behaviors, heralding a future where cognitive AI aids decision-making, augments human capabilities, and revolutionizes various industries.

2.3 Applications of Cognitive Computing

Cognitive computing has swiftly emerged as a transformative force across various industries, revolutionizing how we process information, make decisions, and interact with technology. Its adaptive and intelligent nature enables a wide range of impactful applications.

2.3.1 Healthcare

In healthcare, cognitive computing facilitates precise diagnostics, personalized treatment plans, and improved patient outcomes. Systems powered by cognitive computing analyze vast amounts of medical data, from patient records to research papers, offering insights that aid in disease prediction, early detection, and tailored therapies.

2.3.2 Finance and Banking

The finance sector leverages cognitive computing for fraud detection, risk assessment, and customer service enhancement. Advanced algorithms sift through intricate financial data in real-time, identifying anomalies, optimizing investment strategies, and providing personalized financial advice.

2.3.3 Education and Learning

Cognitive computing enhances educational experiences through personalized learning paths and adaptive curriculums. Intelligent tutoring systems analyze individual learning patterns, providing customized content and guidance, fostering more effective and engaging learning environments.

3. Artificial General Intelligence (AGI)

3.1 AGI Concepts and Challenges

3.1.1 Introduction to AGI

Artificial General Intelligence (AGI) represents the pinnacle of artificial intelligence research, aiming to create systems with cognitive abilities comparable to human intelligence. Unlike conventional narrow AI systems that excel in specific tasks within defined parameters, AGI seeks to replicate the breadth and depth of human cognition across diverse domains and contexts. Its evolution from philosophical discourse to modern-day advancements highlights a trajectory shaped by theoretical insights, technological breakthroughs, and persistent efforts to emulate human intelligence in machines (Rayhan et al., 2023). AGI stands as a profound pursuit, envisaged as the ultimate quest in AI research, embodying the aspiration to develop intelligent systems capable of flexible learning, adaptation, and problem-solving across complex environments.

3.1.2 Key Characteristics of AGI

Artificial General Intelligence (AGI) embodies a distinctive set of core characteristics essential to emulate human-like cognitive abilities (Salmi, 2023).

Generalization and Adaptability

AGI possesses the ability to generalize knowledge across domains, assimilate new information, and adapt to novel situations without explicit programming. This capability allows AGI systems to learn and extrapolate from previous experiences, applying acquired knowledge to navigate diverse and unstructured environments effectively.

Learning and Reasoning Abilities

AGI is equipped with sophisticated learning mechanisms enabling it to discern complex patterns, extract insights, and reason across multiple domains. These mechanisms facilitate problem-solving, decision-making, and the derivation of actionable insights from vast and varied datasets.

Contextual Awareness and Interpretation

A hallmark of AGI is its capacity for contextual understanding, enabling interpretation of situational cues and adjustments in behavior based on the context. AGI systems possess the ability to comprehend nuances within different contexts, thereby influencing their decision-making and adaptive responses.

Common-Sense Understanding

AGI aims to attain common-sense reasoning, mirroring human-like intuitive judgment and nuanced decision-making. This trait is foundational to AGI's pursuit of replicating human cognitive capabilities, allowing systems to grasp implicit knowledge and make intuitive deductions.

Cognitive Flexibility and Versatility

AGI systems aim for cognitive flexibility and versatility, allowing them to apply learned knowledge across diverse domains and tasks. This trait enables AGI to adapt, generalize, and apply its acquired knowledge to solve a wide range of problems without domain-specific constraints.

3.1.3 Distinctions from Narrow AI

Distinguishing between Artificial General Intelligence (AGI) and Narrow AI reveals fundamental disparities in their scope, adaptability, and cognitive prowess. Narrow AI systems, specialized in executing well-defined tasks within constrained domains, exhibit proficiency in specific functions such as image recognition or natural language processing. Conversely, AGI transcends these limitations, aspiring to embody versatile cognitive abilities akin to human intelligence. While Narrow AI operates within predefined boundaries, excelling in predetermined tasks, AGI endeavors for cross-domain competence, aiming to learn, generalize, and adapt to new, unforeseen situations. The defining characteristic lies in AGI's pursuit of cognitive breadth and flexibility, enabling it to navigate diverse scenarios without task-specific instructions, an attribute that sets it apart from the specialized nature of Narrow AI.

3.1.4 Challenges in AGI Development

The pursuit of Artificial General Intelligence (AGI) confronts multifaceted challenges, spanning technical complexities, ethical considerations, and limitations in data accessibility (*Full Article:*

The Risks Associated with Artificial General Intelligence: A Systematic Review, n.d.). At a technical level, scaling AGI development poses formidable hurdles, encompassing computational constraints, algorithmic intricacies, and the necessity for sophisticated architectures capable of mimicking complex human cognitive processes. Addressing these challenges requires substantial computational resources, innovative algorithmic paradigms, and scalable models that can process vast amounts of data efficiently.

Ethical considerations form another critical dimension, entailing concerns about AGI safety, societal impacts, fairness, and privacy. Ensuring the responsible deployment of AGI systems demands comprehensive ethical frameworks, robust governance, and mitigation strategies for potential biases and unintended consequences. Moreover, data limitations pose a significant impediment, as acquiring diverse, high-quality datasets crucial for training AGI systems remains a formidable challenge. The scarcity of comprehensive datasets spanning multiple domains impedes the development of AGI models capable of comprehensive learning and adaptability. Navigating these challenges necessitates collaborative efforts across interdisciplinary domains, rigorous research, ethical oversight, and the establishment of guidelines ensuring the responsible advancement of AGI technology.

3.2 Cognitive Architectures for AGI

Cognitive Architectures within the sphere of Artificial General Intelligence (AGI) embody foundational frameworks engineered to emulate the intricate nature of human cognition (Slavin, 2023). These architectures comprise comprehensive models that integrate a range of cognitive processes, aiming to replicate versatile human-like intelligence across various domains. Diverging from specialized AI systems tailored to specific tasks, AGI architects aspire to encapsulate perception, learning, reasoning, problem-solving, and decision-making capabilities, mirroring the breadth of human cognitive functionalities. Noteworthy cognitive architectures include ACT-R, Soar, and OpenCog (*Contemporary Cognitive Architectures: A Comparative Study of Structures and Adaptability | International Journal of Artificial Intelligence and Soft Computing*, n.d.), each presenting distinct approaches to simulate cognitive processes. ACT-R leverages declarative and procedural memory systems, employing rules and symbolic representations to model human cognition. Soar emphasizes symbolic reasoning and learning mechanisms akin to human problem-solving strategies. In contrast, OpenCog adopts a broader perspective, amalgamating symbolic reasoning with probabilistic and evolutionary learning algorithms. These architectures serve as conceptual blueprints, striving to emulate the intricacies of human cognition and offering pathways toward AGI systems demonstrating adaptable, multifaceted cognitive capabilities.

However, the pursuit of a unified and scalable cognitive architecture for AGI confronts challenges encompassing scalability, adaptability, and fidelity to human cognition. Crafting a coherent framework capable of encapsulating the entirety of human-like intelligence remains an intricate task (*The Whole Brain Architecture Approach: Accelerating the Development of Artificial General Intelligence by Referring to the Brain - ScienceDirect*, n.d.). Additionally, reconciling the computational intricacies of these architectures with real-world applications while ensuring seamless integration poses a significant concern. The ongoing exploration and refinement of cognitive architectures represent pivotal endeavors in the quest for AGI, laying essential groundwork for the development of systems exhibiting versatile, human-like cognitive abilities adaptable across diverse domains and contexts.

3.3 Ethical Considerations in AGI Development

3.3.1 Societal Alignment and Governance

The pursuit of Artificial General Intelligence (AGI) mandates comprehensive scrutiny of ethical dimensions spanning societal, safety, fairness, privacy, and accountability concerns. AGI's potential to transcend human cognitive capabilities raises profound ethical questions, necessitating careful contemplation and proactive ethical frameworks to steer its development responsibly (*A Survey of Artificial General Intelligence Projects for Ethics, Risk, and Policy by Seth Baum :: SSRN*, n.d.). Ensuring AGI's alignment with societal values and ethical norms stands as a paramount consideration. Safeguarding against adverse societal impacts, including potential job displacement, economic disparities, and ethical dilemmas, requires robust governance and ethical guidelines (*Managing the Risks of Artificial General Intelligence: A Human Factors and Ergonomics Perspective - Salmon - 2023 - Human Factors and Ergonomics in Manufacturing & Service Industries - Wiley Online Library*, n.d.). AGI's deployment demands transparency, accountability, and mechanisms to address unintended consequences or biases ingrained within the systems.

3.3.2 Safety Protocols and Risk Mitigation

Safety emerges as a critical ethical concern, given AGI's potential to operate autonomously. Mitigating risks associated with AGI malfunction or unintended behaviors requires preemptive measures, safety protocols, and fail-safe mechanisms. Fairness and impartiality represent ethical imperatives, necessitating the mitigation of biases ingrained in AGI algorithms, data, and decision-making processes. AGI systems must be calibrated to ensure equitable outcomes and unbiased treatment across diverse demographics and scenarios.

3.3.3 Privacy Protection and Data Governance

Privacy preservation stands as a key ethical tenet, obligating the protection of personal information and data confidentiality within AGI systems. Upholding individuals' privacy rights mandates stringent data protection measures, consent frameworks, and data governance policies. Additionally, ethical considerations extend to AGI's accountability, necessitating mechanisms to attribute responsibility for the actions or decisions made by autonomous systems, ensuring recourse in cases of harm or errors.

3.3.4 Accountability and Responsibility

Navigating these ethical quandaries necessitates interdisciplinary collaboration, ethical oversight, and the development of comprehensive regulatory frameworks. Engaging stakeholders, including ethicists, policymakers, technologists, and the broader society, becomes imperative to address these ethical considerations proactively. The pursuit of AGI mandates a principled approach, ensuring alignment with ethical norms, societal values, and the greater good while fostering responsible and beneficial deployment of advanced AI technologies

4. Neuromorphic Engineering

4.1 Introduction to Neuromorphic Engineering and Neuromorphic Systems

Neuromorphic engineering is a multidisciplinary approach in the field of artificial intelligence, seeking inspiration from the intricate workings of the human brain to design intelligent systems. Neuromorphic systems are implementations or architectures that embody the principles of neuromorphic engineering. The fundamental source of inspiration for neuromorphic systems and models such as spiking neural networks, originates from biological intelligence particularly the neural circuits within the cortical regions of the brain. These systems seek to replicate the efficiency, adaptability, and parallel processing capabilities observed in biological neural networks (Wu et al., 2022)

4.2 Basics of Neuromorphic Systems

4.2.1 Neural Circuits and Parallel Processing

The conventional von Neumann architecture features a central processing unit (CPU) that sequentially fetches and executes instructions stored in a shared memory space. The von Neumann processor architectures, characterized by the separation of processing and memory, stand in contrast to the principles of the brain, which rely on distributed computation (Indiveri & Liu, 2015). In von Neumann architectures, data is encoded as multibit words processed sequentially through instructions coordinated by a global clock. This design, while effective for classical computing tasks, faces challenges such as the von Neumann bottleneck. The brain, on the other hand, employs a fundamentally different approach, co-locating processing and memory with neuron and synapse elements, respectively (Frenkel et al., 2023). Neuromorphic systems are inspired by the intricate structure and operational principles of the human brain. The neural circuits, especially those situated in the cortical regions, serve as a testament to the efficacy of parallel information processing. Neuromorphic systems are designed to emulate the inherent parallelism found in these neural circuits, where billions of neurons collaborate simultaneously to process extensive amounts of information (Sze et al., 2017).

4.2.2 Spiking Dynamics of Neurons

Neurons are considered as the basic building blocks of the nervous system and communicate with each other through electrical signals. When a neuron receives a signal, it can generate a rapid change in voltage known as a spike or action potential. This electrical impulse travels down the neuron's axon facilitating transmission of information to adjacent neurons or specific target cells (Corradi & Indiveri, 2015). The term "spiking dynamics of neurons" refers to the pattern of electrical activity exhibited by neurons in the nervous system, particularly characterized by spikes (See Fig 3). The spiking dynamics represent the temporal patterns of these electrical signals. This unique spiking behavior is harnessed in the design of Spiking Neural Networks (SNNs). Unlike their counterparts in conventional neural networks, SNNs operate with a temporal precision that mirrors the discrete firing patterns observed in biological neurons (Büchel et al., 2021).

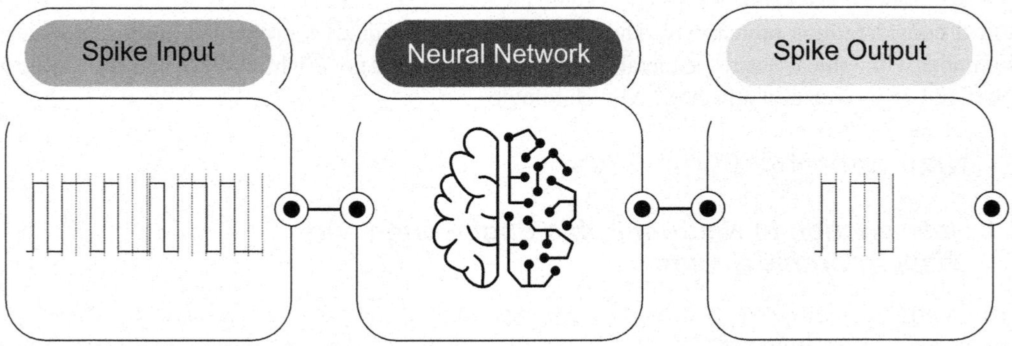

Fig. 3 Representation of neuromorphic architecture

4.2.3 Event Driven Computing

The human brain is exceptionally energy-efficient, capable of performing complex computations with minimal power consumption. The human brain can execute highly complex computations, such as pattern recognition and decision-making, with an energy efficiency that surpasses traditional computing architectures. Neuromorphic systems aim to replicate this energy efficiency by adopting strategies derived from the brain's processing mechanisms (Ielmini, 2018). The event-driven computation model is a fundamental aspect of achieving energy efficiency in neuromorphic systems. In this model, computation is triggered by relevant events or changes in input data, in contrast to the continuous processing of traditional computers. This event-driven approach allows for a more dynamic and efficient utilization of computational resources, as computations are initiated only when there is meaningful information to process (Ansari et al., 2023).

4.3 Neuromorphic Hardware and Software

4.3.1 Neuromorphic Hardware

Neuromorphic hardware employs specialized computing architectures that closely replicate the structure (morphology) of neural networks. This emulation occurs from the foundational level upward: dedicated processing units directly mimic the behavior of neurons within the hardware, and a network of physical interconnections, often in the form of bus systems, enables swift and efficient exchange of information. This method is designed to capture the essence of biological neural processing, emphasizing dedicated hardware components and interconnected systems for more effective and brain-like computation (Xu et al., 2018).

Neuromorphic hardware typically uses elements such as artificial neurons, synapses, and neural circuits, allowing for the implementation of algorithms that simulate synaptic plasticity and other aspects of neural information processing. The goal of neuromorphic hardware is to achieve more efficient and brain-like computation for tasks such as pattern recognition, learning, and cognitive processing (Ghani et al., 2023). Memristors, or memory resistors, exhibit dynamic resistance changes based on the history of applied voltage. This characteristic allows them to emulate the strengthening or weakening of synapses in response to neural activity. Chaeun Kim et. al., stated the potential application of the TiN/ZrOx/Pt structure as a synaptic device within a neuromorphic system. The synaptic connection's strength is subjected to the signals originating from both the anterior and posterior ends of the synapse, which is associated with human learning. Likewise, the resistance of a memristor can be modulated through the application of external pulses (Kim et al., 2023).

Neuromorphic chips are specialized processors designed to replicate the parallel processing capabilities of the brain's neural circuits. These chips are crafted to handle the unique requirements of spiking neural networks (SNNs), the fundamental computational paradigm in neuromorphic systems. They often feature a massively parallel architecture, allowing for the simultaneous processing of multiple information streams, a stark departure from the sequential processing characteristic of traditional von Neumann architecture (Schuman et al., 2022). Tensor-Organized Memory (TOM) introduces a pioneering neuromorphic architecture inspired by the structural and functional principles observed in the human brain. By incorporating spike-timing-dependent plasticity (STDP) and adhering to Hebbian rules, TOM demonstrates cognitive behaviors akin to those exhibited by the human brain. Notably, TOM exhibits resilient performance, particularly in environments with high levels of noise. The potential application of the TOM architecture in neuromorphic chips for intelligent edge devices presents a promising avenue for advancing the field (Khajooei Nejad et al., 2023).

4.3.2 Neuromorphic Software

Neuromorphic software refers to computer programs and algorithms designed to simulate or replicate the principles of neural processing found in biological systems. Neuromorphic software is commonly used in conjunction with neuromorphic hardware to create comprehensive systems that mimic the adaptability, learning capabilities, and parallelism observed in biological neural networks (Mead, 1990). SNNs, inspired by the spiking dynamics of biological neurons, process information based on discrete spikes or pulses. Neuromorphic algorithms are engineered to capture the temporal precision and spike-based encoding observed in biological neural networks. In a supervised learning paradigm, SNNs are trained for temporal classification tasks by emulating a pre-trained dynamical system, leveraging a local learning rule derived from nonlinear control theory. This approach enables the robust deployment of pre-trained networks on mixed-signal neuromorphic hardware, eliminating the need for individual device training or calibration (Büchel et al., 2021).

The utilization of a Continuous Attractor Neural Network (CANN), configured with diverse synaptic dynamics, proves capable of implementing various cognitive functions observed in the human brain. A neuromorphic implementation of the CANN, consisting of spiking neurons on a field-programmable gate array (FPGA), allows for the configuration of synaptic strength in local excitation and global inhibition, facilitating the realization of winner-take-all (WTA) competition. Hardware-based circuit simulations successfully reproduced gradually ramping neural activities observed in decision-making experiments and multiple sustained activities during the delay in working memory experiments (You & Zhao, 2021).

4.4 Applications and Advancements

Autonomous navigation has been explored through the integration of neuromorphic sensors and Edge-AI computing devices (Piñero-Fuentes et al., 2021). A notable advancement in this field is the development of a neuromorphic robotic platform designed for training and testing Spiking Neural Networks (SNNs) to implement trajectory controllers in a robotic arm (Canas-Moreno et al., 2023). An analog circuit design has successfully implemented the neuromorphic prescribed error sensitivity (PES) learning rule using OZ neurons, an analog representation of spiking neurons with demonstrated correspondence to NEF across various parameters. This circuit emulator, in collaboration with a robot simulator, showcased adaptive learning-based control for a six-degree-of-freedom robotic arm (Hazan & Ezra Tsur, 2022).

Proposed by Zuo-Wei Yeh et al., a configurable population-based digital spiking neuromorphic processor, developed in 180nm process technology, features two configurable hierarchy populations. This approach allows for the creation of biomimetic neuromorphic systems and various low-power, low-latency inference processing applications, achieving a normalized energy efficiency of 13.2 pJ/SOP (Yeh et al., 2021). The future of artificial intelligence hardware is envisioned in 3D neuromorphic systems, characterized by enhanced power efficiency and cognitive capabilities compared to conventional 2D circuits. These 3D circuits employ multiple device layers, utilizing unconventional fabrication processes (Lin & Xia, 2021). In the target speed and shape classification domain, Barnes et al. (2022) designed a deep-learning network with high accuracy. Leveraging a p(DVS) neuromorphic sensor, the study demonstrated remarkable computational efficiency and accuracy in classifying different target speeds and shapes. These findings underscore the potential of neuromorphic technologies in advancing artificial intelligence applications. (Barnes et al., 2022).

5. Human-Centric AI

5.1 Introduction to Human-Centric AI

In the landscape of artificial intelligence, the evolution from machine-centric to human-centric approaches marks a profound paradigm shift. Human-Centric AI stands as a pivotal advancement, centering the design and development of intelligent systems around human needs, behaviors, and experiences. At its core, this approach aims to create AI technologies that seamlessly integrate with and augment human capabilities, fostering intuitive and empathetic interactions (Horvatić & Lipic, 2021). By prioritizing user-centric principles, Human-Centric AI seeks to not only enhance the performance and efficiency of AI systems but also to ensure ethical, transparent, and socially responsible deployment. This transformative approach reflects a fundamental commitment to aligning technological advancements with human values and aspirations, ushering in an era where AI serves as an enabler of human potential and well-being.

5.2 User-Centered Design in Cognitive AI

5.2.1 Enhancing User-Centric Cognitive AI Interaction

In the realm of Cognitive AI, enhancing user-centric interaction is a cornerstone for technological advancement. This endeavor encompasses understanding user behavior through robust methodologies aimed at grasping user preferences and behavioral nuances during AI interaction. An equally critical facet lies in the integration of empathetic design principles within Cognitive AI systems (Dodeja et al., 2024). This amalgamation emphasizes the development of AI frameworks that not only excel in performance but also resonate with users on an emotional level. Crucially, the design phase assumes paramount importance. Creating intuitive interfaces becomes the linchpin, ensuring seamless interaction between users and Cognitive AI platforms. These interfaces are meticulously crafted to align with user expectations, fostering effortless and engaging experiences. Additionally, ensuring accessibility remains a foundational element. Cognitive AI systems should embrace inclusivity, accommodating diverse user needs and abilities. The convergence of understanding user behavior, empathetic design, and robust usability considerations forms the bedrock of transformative Cognitive AI experiences, placing user needs and inclusivity at the forefront of technological innovation (Weekes & Eskridge, 2022).

5.2.2 Elevating Cognitive AI Through Transparency and User-Centric Iterations

Enhancing transparency within Cognitive AI systems involves deploying interpretability techniques that demystify complex decision-making processes without compromising performance. Achieving this delicate balance between transparency and high-performance models is a critical challenge (Almufareh et al., 2024). Concurrently, integrating user feedback loops emerges as a cornerstone in Cognitive AI evolution. Agile methodologies rooted in user-centric design principles enable continuous improvements, ensuring adaptability to evolving user needs. This iterative approach, integrating real-time user insights into the development cycle, fosters an AI ecosystem finely attuned to user expectations. This synergy between transparency strategies and iterative, user-centric design delineates a robust framework for Cognitive AI. It not only clarifies AI decisions but also dynamically evolves, driven by real-time user inputs and societal demands. This holistic integration embodies an AI landscape that comprehends, explains, adapts, and evolves harmoniously with users' changing needs and aspirations.

5.3 Explainability and Interpretability

5.3.1 Unraveling Explainability and Interpretability in Human-Centric AI

Within the vast realm of AI explanations, there exists a rich spectrum of methodologies and algorithms specifically designed to furnish insightful explanations within AI systems. These methods encapsulate diverse approaches, ranging from post-hoc interpretation methods to inherently interpretable models that offer transparency into the reasoning behind AI-driven decisions (Endsley, 2023). Evaluating the efficacy of each technique is crucial, requiring meticulous scrutiny to understand how effectively these methods elucidate the inner workings of AI systems (Miller, 2024). This evaluation process involves assessing the clarity, depth, and relevance of the explanations provided, shedding light on the strengths and limitations inherent in each method. Additionally, understanding the trade-offs that accompany the adoption of specific explanation techniques is pivotal, elucidating the nuanced differences and implications associated with employing these methods within AI systems.

5.3.2 Balancing Interpretability and Performance in AI Models

A fundamental challenge entrenched within the development landscape of AI pertains to the intricate equilibrium between achieving model interpretability and upholding high-performance metrics. This juxtaposition necessitates a deep dive into the synergy between interpretability and the accuracy or efficacy of AI models. While striving to render AI models interpretable, there exists a constant need to safeguard and uphold their underlying accuracy and efficiency. This balancing act requires a profound exploration of how interpretability influences user trust and adoption rates, as the seamless fusion of interpretability without compromising accuracy emerges as a cornerstone in the advancement of AI systems. Understanding the profound impact of interpretability on user trust and the adoption of AI-driven systems becomes pivotal in shaping the trajectory of future AI model development.

5.4 Human-AI Collaboration

5.4.1 Enhancing Human-AI Interaction

Strides in the realm of Human-AI Collaboration aim to foster an enriched relationship between individuals and AI systems. These initiatives are centered on refining techniques that promote seamless collaboration and interaction (Kernan Freire, 2023). Efforts are directed toward creating a cohesive environment where AI serves as a supportive adjunct, enhancing human capacities rather than supplanting them. The pursuit involves deploying intuitive interfaces, leveraging natural language processing capabilities, and embracing adaptive learning mechanisms. These strategies facilitate a harmonious coalescence, empowering individuals to leverage AI's capabilities to amplify their own, fostering a symbiotic partnership that drives collective progress.

5.4.2 Ethical Imperatives in Human-AI Collaboration

The landscape of Human-AI collaboration warrants a vigilant and proactive approach toward ethical considerations (Salmon et al., 2023). The pivotal aspect revolves around preemptively addressing ethical concerns and mitigating biases inherent in collaborative AI systems. Anchored in principles of fairness, accountability, and transparency, this approach aims to fortify these systems against inadvertent biases, thereby upholding ethical standards. As collaborative AI systems become increasingly intertwined with societal frameworks, ensuring these systems align with ethical paradigms becomes non-negotiable. This concerted effort propels the trajectory of collaborative AI toward a future where innovation harmonizes with ethical values, forging a path that augments human potential while prioritizing societal well-being.

6. Applications of Cognitive AI

6.1 Cognitive Search Engines

Cognitive search engines represent a significant advancement in information retrieval systems, applying Cognitive AI to understand user intent, and context, and provide more relevant search results. In the domain of decision-making within OFDM-based cognitive radio networks, the cognitive engine employs optimization algorithms. The study conducted an analysis with a specific emphasis on the performance of two modern metaheuristics, namely the gravitational search algorithm (GSA) and the flower pollination algorithm (FPA). The simulation results indicate that, in terms of fitness metrics, the flower pollination algorithm (FPA) exhibits superior performance compared to the gravitational search algorithm (GSA). Conversely, GSA demonstrates superior performance over FPA when considering execution time (Benmammar, 2022).

A fairness framework, adopting a dual perspective, systematically investigates the influence of both system bias and human bias in interactive search, recommendation, and AI-assisted decision-making. This comprehensive analysis, incorporating insights from information retrieval, recommendation systems, AI fairness, and cognitive psychology, introduces two-sided evaluation methods for a more equitable and psychologically realistic approach to evaluating fairness in users' interactions with intelligent information systems (Liu J, 2023).

In the domain of cognitive computing for decision-making in industrial informatics, the current focus is on leveraging deep-learning-based cognitive automation. However, existing algorithms grapple with multiobjective cognitive model challenges, often necessitating significant time investment to continuously adjust various search preference parameters for initiating a new search process. Addressing this concern, E.K. Wang et al. proposed an innovative algorithm named the linear-prefer coevolutionary algorithm. Comparative assessments against both the original user-constrained method and the Pareto-dominant NSGA-II algorithm highlight the linear-prefer coevolutionary algorithm's capability to achieve faster adaptation times and demonstrate a superior quality of adaptation. Simultaneously, the algorithm showed an increased responsiveness to user needs during the reasoning phase. The proposed algorithm outperforms traditional cognitive automation algorithms, particularly in terms of search quality, when tackling multiobjective problems (Wang et al., 2021).

Cognitive search engines transcend conventional keyword matching by employing natural language processing and machine learning techniques to comprehend the context of user queries. This advanced approach results in more precise and contextually relevant search outcomes, ultimately enhancing the overall user experience. In the realm of movie recommendations, a cognitive thinking approach is adopted, taking into account the user's age. Recommendations are tailored based on age group psychology, ensuring a more personalized and thoughtful selection of genres for an enriched user experience. (Ajith et al., 2021; Labde et al., 2023). In exercise recommendations, cognitive search engines use advanced algorithms and machine learning to understand user preferences, health conditions, and fitness goals (Li & Niu, 2023). The application of cognitive search engines in hashtag recommendations on social media platforms like Twitter facilitates more effective communication, content discovery, and community engagement within the platform (Djenouri et al., 2022). Recommender systems provide meaningful suggestions across a diverse range of service offerings within the expansive domain of big data. An effective cognitive 5G communication model using a recommender model for IoMT (Internet of Medical Things) big data processing has been developed. The proposed RSIoMT model executed the recommender model by ranking communication devices that align with the IoMT variables, ultimately enhancing wireless communication quality. This system used the recommender model to estimate direct communication links between IoMT variables within the cognitive radio system (Vekariya et al., 2022).

6.2 Image and Video Analytics

Cognitive AI is used in medical imaging for the diagnosis and interpretation of medical scans. By detecting abnormalities in X-rays, MRIs, and CT scans with the help of image analytics algorithms, healthcare professionals make more accurate and timely diagnoses (Mary Shyni & Chitra, 2022). The diagnostic assessment of COVID-19 infections involved the examination of images obtained from CT scans and X-rays (XR), using various CNN-based models such as ResNet, MobileNet, VGG 16, and EfficientNet. The results indicated that XR images were preferred over CT scan images in accurately distinguishing abnormal COVID-19 patients from normal cases (Jangam et al., 2022). The DL-MOMFO algorithm, introduced by F. Mansour and M. Althobaiti in 2022, incorporates an Adaptive Weighted Mean Filter (AWMF) for noise removal and utilizes contrast-limited adaptive histogram equalization (CLAHE) for contrast enhancement in digital mammograms. This algorithm aims to enhance the quality of digital mammographic images.

In the domain of healthcare applications, specifically within the Internet of Medical Things (IoMT) technology, deep learning (DL) and cognitive computing-based techniques have been integrated. This integration is focused on the detection and classification of breast cancer, showcasing the potential of advanced technologies in improving the diagnosis and management of medical conditions within the e-healthcare framework (Mathapati et al., 2021). The two powerful DL models, ResNet50 and BiGRU, were implemented to analyze the medical images and biomechanical data. This method improves feature representation, contributing to a more enriched framework for sports injury predictions (Xiao et al., 2023).

Dementia and cognitive decline in older adults can be assessed and monitored with the help of DL models by extracting novel drawing metrics from hand-drawn images from such patients (Tasaki et al., 2023). Real-time location systems were used to analyze spatial walking patterns, allowing for the classification of the severity of cognitive impairment (CI) among residents in a memory care unit (Faruk et al., 2023). Data derived from blinks, pupils, and gaze movements can be used for classifying individuals based on these two cognitive functions (Cascone, 2021). Psycho-cognitive computing has garnered significant attention in recent years, especially in advanced human-computer interaction technology. X. He et al. applied the I3D backbone network to extract long video optical flow and original video features, enabling the extraction of effective feature layers. Through downsampling and the incorporation of the BiFPN module with a spatial attention mechanism, multi-scale feature layers were selectively integrated. The resulting classification-regression network successfully distinguished detected expressions and identified the temporal boundaries of these expressions (He X et al., 2022).

In the realm of cognitive AI, deep learning algorithms, natural language processing, and computer vision technologies are employed to understand human emotions derived from both visual and auditory elements in videos. Analyzing emotions in pre-recorded videos aims to simulate the intricate mental processes of humans. For video sentiment analysis, H. Qureshi et al. introduced the Statistical Approach, utilizing machine learning and deep learning algorithms, including Convolutional Neural Networks (CNNs). This approach precisely detects and interprets sentiments within video content by integrating statistical techniques and sophisticated neural network architectures, thereby enhancing the accuracy and effectiveness of sentiment analysis in videos (Qureshi, H., Agrawal, A.P., 2022).

6.3 Motivational Systems for Cognitive Agents

Cognitive agents are sophisticated systems designed to communicate in a manner that is acceptable and understandable to humans. These agents leverage various artificial intelligence methods, including machine learning and deep learning models, natural language processing, text-to-speech, and speech-to-text technologies, among others. By utilizing these advanced technologies, cognitive agents not only interact with humans but also have the capability to learn and adapt to these interactions. This multifaceted approach enables them to engage in meaningful and contextually relevant communication with users.

6.3.1 Design principles for Cognitive Agents

The design of cognitive agents necessitates adherence to certain principles essential for their architecture and functions. One crucial principle involves the integration of data from various sensors, known as multisensory integration, which enhances the functionality of cognitive agents. Dushkin et al. demonstrate the application of this principle in addressing the challenge of symbol grounding, offering insights applicable to the development of artificial cognitive agents at a general level (Dushkin, R.V. and Stepankov, V.Y., 2023). Another critical consideration in cognitive agent design is energy and memory optimization. Incremental learning is a strategy employed by cognitive agents to utilize pertinent data, facilitating knowledge updates and reducing memory usage. Rule-based systems further contribute to memory optimization by selectively using and retrieving information (Lui. H. et al., 2016). Ménager et al. proposed an approach utilizing hybrid event memory, enabling agents to incrementally learn and represent extensive joint probability distributions over states such as Bayesian networks. This method extends the event memory system with a rule-based representation, enhancing scalability for large state estimation problems under various partial observability conditions (Ménager, D.H. and Choi, D, 2023).

Cognitive agents must also possess the ability to interpret information from expressed human emotions and consider human values and preferences. Sz-Ting Tzeng proposed an agent architecture comprising three components: Cognitive architecture, world model, and social mode, accommodating human factors. The cognitive architecture encapsulates a cognitive agent's beliefs, desires, intentions, and personal preferences. The emotion model adopts the OCC model of emotions. The world model represents the contexts in which an agent operates and embodies the agent's general knowledge about the world. The social model includes social values, normative reasoning, and norm fulfillment (Tzeng, S.T. et al., 2022). This comprehensive approach ensures that cognitive agents are equipped to understand, respond to, and incorporate human-centric aspects into their interactions and decision-making processes.

6.3.2 Multi-Agent Systems with Cognitive Abilities

Multi-Agents systems are multiple cognitive agent systems with cognitive abilities that combine and solve a complex problem. These systems can solve problems that cannot be solved by a single cognitive agent system. The DAGHT architecture stands out as an example of a multiagent system comprising five distinct agents, each assigned specific tasks that contribute to distributed intelligence. These agents are interdependent, with each fulfilling its unique role within the system. Motivated by mythological characters beyond Shannon's Theseus, the agents are named "Demiurge" (D), "Argus" (A), "Golem" (G), "Homunculus" (H), and "Theseus" (T) (Huber-Liebl et al., 2022). On a different note, Instance-based Learning Theory (IBLT) has been developed to offer a cognitively plausible explanation for how humans make decisions based on experience and uncertainty in dynamic environments. IBLT has demonstrated its efficacy in accurately representing human choice across various decision-making domains, spanning economic decision-making to complex scenarios such as resource allocation and cybersecurity. Thuy Ngoc Nguyen et al. proposed three variants of multi-agent Instance-based Learning Theory models (MAIBLs). Their study revealed that these MAIBL models exhibit faster learning and better coordination in dynamic coordinated multi-agent object transportation problems, with various settings of stochastic rewards, when compared to current multi-agent deep reinforcement learning models (MADRL) (Nguyen, T.N. et al., 2023). This research highlights the potential of multi-agent IBLT models in outperforming existing approaches in certain dynamic and coordinated tasks.

6.3.3 Real-world Applications of Cognitive Agents

The detection of personality through emotions represents a significant research domain in artificial intelligence. Elmitwally NS et al. introduced a conceptual personality model for cognitive agents capable of determining personality and behavior based on text input. This model utilizes the context subjectivity of the given data and extracts emotions from specific situations or contexts. The proposed work includes the implementation of the Jumbo Chatbot, designed to engage in conversations with humans. During these social interactions, the chatbot predicts human personality by comprehending the emotions and context of the interactive individuals (Elmitwally, N.S. et al., 2022). In the realm of cognitive assessment, automated tools play a crucial role in evaluating cognitive development. A. Kahawanugoda et al. addressed the challenge of limited resources in building automated cognitive ability evaluation tools. Their focus was on developing machine learning models with a restricted amount of data to assess Reasoning IQ, Knowledge IQ, Mental Chronometry, and Attention levels in Sinhala-speaking children aged 7 to 9 years. The study employed open domain speech recognition to evaluate complex Sinhala child verbal responses. Furthermore, limited vocabulary responses were assessed using an end-to-end speech recognition system, achieving 40.1% Word Error Rate (WER) and 97.14% accuracy, respectively (Kahawanugoda, A. et al., 2022). This research demonstrates the feasibility of leveraging machine learning models for cognitive assessments even with limited data, addressing the challenges associated with resource constraints.

6.4 Cognitive Robotics

Cognitive robotics uses robotics, artificial intelligence, and cognitive science to build robotic systems with cognitive capabilities. The core principle of cognitive robotics is to develop robots that can understand, evaluate, gain knowledge, and communicate with their environment and human users in a manner that resembles human cognition. Thus, cognitive robotics brings in deeper and effective human-robot collaboration for varied scenarios (Singh et al., 2022).

6.4.1 Cognitive Approaches in Robotics

To enhance the cognitive capabilities and flexibility of the robotic systems, cognitive robotics derives various methodologies from cognitive science principles. Using ML techniques, such as NN and DL algorithms, robots can learn from their experiences and improvise autonomous decision-making (Piñero-Fuentes et al., 2021). Another cognitive strategy involves the implementation of symbolic reasoning and knowledge representation systems, allowing robots to interpret and manipulate abstract concepts, contributing to higher-level problem-solving capabilities. Furthermore, the incorporation of perception and attention mechanisms, inspired by human cognitive processes, enhances a robot's ability to selectively focus on relevant information in complex environments (Eldardeer et al., 2021).

A robust cognitive architecture serves as the foundation for creating intelligent AI systems that can operate safely and meaningfully in proximity to humans, fostering collaboration. Cognitive architecture involves the design and implementation of structures and processes that mimic human-like cognition, including perception, reasoning, learning, and interaction. One commonly utilized architecture is ACT-R (Adaptive Control of Thought - Rational), where knowledge is organized based on the type of information. ACT-R relies on productions, specifically "IF" - "THEN" rules, to represent cognitive processes. Another cognitive architecture is LIDA (Learning Intelligent Distribution Agent), which operates on cycles with distinct phases. The first phase involves perception, followed by the attention phase, demonstrating a systematic approach to cognitive functioning. For robotic applications, the KnowRob 2.0 architecture is designed to enable robots to perform complex tasks. This architecture leverages ontologies and axioms, providing a structured framework for robotic systems (Franklin, S. et al., 2016). Each of these architectures contributes to the field of cognitive science and artificial intelligence by offering unique approaches to modeling intelligent behavior and facilitating advanced functionalities in different domains.

ICARUS represents a cognitive architecture tailored for physical agents that synthesizes concepts from various traditions, those influenced by results from cognitive psychology. SOAR, another cognitive architecture, centers on problem-solving and decision-making, designed to simulate human cognitive processes. Clarion, yet another cognitive architecture, models human cognition by integrating multiple subsystems that operate in parallel (Zouganeli & Lentzas, 2022). By integrating a robust cognitive architecture, AI systems can transcend basic rule-based functionalities and operate in complex, real-life scenarios. The development of such cognitive architectures not only enhances the intelligence of AI but also contributes to the responsible and ethical deployment of AI technologies in human-centric settings.

6.4.2 Human-Robot Interaction

Human-Robot Interaction (HRI) involves the application of robots to understand, interpret, and interact with humans, aiming to establish a connection between robots and humans. The primary objective of HRI is to develop robots that can comprehend human thoughts, respond to human actions, and perform tasks in alignment with human expectations. Essential to this interaction is the capability of cognitive robots to process human language, a feat made possible through natural language processing technology. Wenpeng Lu et al. introduced a novel 3D CNN-based SSM (3DSSM) method for human-robot language interaction. This method employs a feature cube network architecture to transform a 1D sentence into a multi-dimensional representation called a semantic feature cube. Subsequently, a 3D CNN module is applied to learn a semantic representation for the feature cube, capturing both local features embedded in word representations and sequential information among successive words. The final matching representation is generated by combining the representations of a pair of sentences using another 3D CNN (Lu et al., 2021). Another innovative architecture, the Brain-inspired Active Learning Architecture (BALA), was proposed by Zhang, T. et al., specifically designed for procedural knowledge understanding based on Baxter robot and human interaction (Zhang, T. et al., 2021). Cognitive conflicts may arise during HRI, and S. Aldini et al. demonstrated that among various classification algorithms, CNNs (Convolutional Neural Networks) yielded the best results in classifying different levels of cognitive conflict (Aldini, S. et al., 2023).

6.4.3 Case Studies in Cognitive Robotics

Computer vision plays a pivotal role in real-time 3D object detection within cognitive robotics experiments. Vidal-Soroa et al. implemented and tested an algorithm that combines a deep neural network (YOLOv3 tiny) for processing RGB images, providing object recognition and 2D localization, with a point cloud method. This approach was successfully tested in real-time manipulation experiments with the UR5e robotic arm through ROS (Robot Operating System), utilizing a GPU for efficient execution. The combination of these techniques allowed for efficient real-time learning using cognitive models (Vidal-Soroa, et al., 2023). Socially Assistive Robots (SAR) have shown significant promise in enhancing interventions for children with Autism Spectrum Disorders (ASD). Zhonghao Shi et al. designed and validated personalized models of arousal and valence for children with ASD using a multi-session in-home dataset of SAR interventions. By employing machine learning (ML) algorithms with supervised domain adaptation (s-DA), the personalized models demonstrated the ability to balance the limited individual data with more abundant, less personal data pooled from other study participants (Shi, Z. et al., 2022). This application of SAR and ML showcases the potential for technology to aid in providing tailored and effective interventions for individuals with ASD.

7. Discussion

Cognition has impacted and influenced artificial intelligence in many ways. Cognition involves the understanding, processing, decision making, and problem solving. Some works have drawn inspiration from the cognition processes. Indeed, mimicking biological processes in AI models has driven the development of more sophisticated and intelligent systems. In particular, advances in neuromorphic computing draw inspiration from the intricate structure and function of the human brain. Neuromorphic computing endeavors to replicate the workings of neurons and synapses, with the goal of constructing AI systems that are not only more efficient but also closely emulate the parallel processing capabilities inherent in the human brain. This approach seeks to bridge the gap between traditional computing models and the complex, parallel processing nature of the brain, potentially unlocking new levels of efficiency and capability in artificial intelligence.

AI systems are designed to interpret better and generate human-like text and speech. Cognitive science has influenced the development of learning paradigms within AI. The intersection of cognitive science with other disciplines, such as neuroscience, psychology, and linguistics, has fostered interdisciplinary collaboration. The understanding of information from new experiences, adapting to it, and generalization of knowledge from humans has inspired the creation of machine learning algorithms. This has led to the evolution of AI systems that can dynamically adjust their behavior based on data and feedback. Some researchers have developed cognitive architectures for AI systems based on insights from human cognition. These architectures aim to replicate the hierarchical and parallel processing observed in the human brain, enabling AI systems to perform complex tasks, reason, and make decisions in a manner closer to human thought processes. Conceptual models are essential to facilitate the collaboration between human and artificial cognition in addressing complex tasks. Cognitive principles have contributed to improved problem-solving capabilities in AI. Algorithms inspired by human problem-solving strategies, have been integrated into AI systems. This has enhanced the efficiency and effectiveness of AI in tackling a wide range of problems.

Cognitive science has influenced the design of AI systems with improved memory and attention mechanisms. This has allowed AI models to better retain and recall information, similar to human memory processes. Attention mechanisms, inspired by cognitive attention, enable AI models to focus on relevant information during tasks, enhancing their performance. Integrating emotional intelligence into AI systems, has been greatly influenced by the understanding of emotions and their role in decision making and communication. Cognitive insights into emotional expressions, empathy, and social cues have led to the development of emotionally aware AI, making interactions more human-centric. Cognition has influenced the pursuit of explainable and interpretable AI systems. Cognitive insights have contributed to the development of AI models capable of transfer learning and generalization. AI systems are now designed to transfer learned patterns and information across different domains, leading to more versatile and adaptable models.

Cognition has impacted the development of AI systems with social intelligence, enabling them to understand and navigate social contexts. This is particularly important for AI applications involving human-robot interaction, virtual assistants, and collaborative systems where a nuanced understanding of social cues is essential. In recent years, cognitive computing, driven by advanced machine learning algorithms, has ushered in a transformative era across various industries. This cutting-edge technology empowers systems to emulate complex human-like cognitive processes, enabling them to learn, adapt, and make informed decisions based on discerned patterns within data. The integration of cognitive computing and machine learning has broadened the scope of applications, spanning domains like natural language processing and image recognition. This not only enhances operational efficiency but also facilitates the execution of tasks previously deemed exclusive to human cognition. At the forefront of artificial intelligence, we encounter Artificial General Intelligence (AGI), a formidable aspiration to develop machines with cognitive capacities resembling human intelligence across diverse domains. Unlike specialized AI tasks, AGI exhibits generalized problem-solving skills, necessitating the integration of various cognitive functionalities. This integration enables AGI to adapt to novel situations and learn across diverse domains, positioning it as a holistic and versatile intelligence paradigm. Simultaneously, the Human-Centric AI approach places a strong emphasis on designing AI systems that prioritize the user experience, understanding, and collaboration. This philosophy ensures that AI technologies closely align with human values, needs, and expectations, fostering not only usability but also widespread acceptance in the realm of artificial intelligence. This user-centric perspective strives to create AI systems that seamlessly integrate into human-centric environments, promoting a harmonious relationship between humans and intelligent technologies.

Challenges

Artificial Intelligence (AI) presents a myriad of challenges, each demanding careful consideration. Ensuring transparency in AI decision-making processes is foundational for fostering trust. The pivotal role of high-quality, unbiased input data is accentuated to prevent skewed outcomes. As we advance toward more sophisticated AI, such as Artificial General Intelligence (AGI), grappling with the computational intricacies required for human-like cognitive processes becomes a significant hurdle. Simultaneously, ethical dimensions encompassing societal impact and potential job displacement necessitate diligent management. The concept of transfer learning, where AI leverages knowledge across domains, introduces an additional layer of complexity. In the realm of Human-Centric AI, the challenge extends to designing systems that seamlessly integrate into users' lives, necessitating an understanding of user behavior and adherence to ethical considerations. Balancing technological innovation with ethical responsibility remains an ongoing pursuit, addressing concerns of transparency, unbiased data, computational hurdles, ethical dilemmas, versatile learning, and user-centric design. In this nuanced journey, achieving equilibrium between technological advancement and ethical accountability emerges as a paramount goal.

Conclusion

The combination of cognitive science principles and artificial intelligence has led to distinguished progress, changing the perspective of technology across various sectors. Mimicking the human cognition processes, including learning, knowledge, problem-solving, perception, reasoning, and emotional intelligence has promoted the production of sophisticated AI systems. Neuromorphic computing strives to mimic event driven computing, by effectively utilizing the computational resources that lead to minimal power consumption while mirroring the parallel processing capabilities observed in the human brain.

Inspiring and building AI architecture with the cognitive sciences leads to collaborative efforts with neuroscience, psychology, and linguistics. By this combination of cognitive science and AI systems dynamically adapted in creating versatile and robust models. Complex problems are solved and design making processes are conducted efficiently using the cognitive architectures. These architectures are designed to replicate hierarchical and parallel processing. This strategy aligns the AI system much closer to the human thought process. In the context of explainable and interpretable AI, these architectures can be integrated with emotional intelligence, to improve problem-solving capabilities and to make the process more human- centric. Thereby, the development of emotionally connected AI systems will bring more understanding in human perspective. This leads to the derivation of attention mechanisms from cognitive attention. The fusion of cognitive science principles with artificial intelligence has propelled significant advancements, reshaping the technological landscape and fostering the development of sophisticated AI systems, exemplifying a harmonious convergence of human cognition and computational capabilities.

References

Ajith, T.T., V, A.K.C., J,N., Subramanian, M.S.A. and S, S.R. (2021). Enhanced Movie Recommendation using Knowledge Graph and Particle Filtering. *2021 2nd International Conference on Smart Electronics and Communication (ICOSEC)*, 1139–1144. https://doi.org/10.1109/ICOSEC51865.2021.9591834.

Aldini, S., Singh, A.K., Leong, D., Wang, Y.K., Carmichael, M.G., Liu, D., Lin, C.T. (2023). Detection and Estimation of Cognitive Conflict During Physical Human–Robot Collaboration. IEEE Transactions on Cognitive and Developmental Systems. 5(2):959–68. https://doi.org/10.1109/TCDS.2022.3205168.

Almufareh, M.F., Kausar, S., Humayun, M. and Tehsin, S. (2024). A Conceptual Model for Inclusive Technology: Advancing Disability Inclusion through Artificial Intelligence. *Journal of Disability Research*, *3*, 20230060. https://doi.org/10.57197/JDR-2023-0060.

Ansari, Md. H.R., Kannan, U.M. and El-Atab, N. (2023). Silicon Nanowire Charge Trapping Memory for Energy-Efficient Neuromorphic Computing. *IEEE Transactions on Nanotechnology*, *22*, 409–416. https://doi.org/10.1109/TNANO.2023.3296673.

Barnes, C., Puran, A., Beninati, A., Douard, N., Nowak, M. et al. (2022). Space Situational Awareness (SSA) and Quantum Neuromorphic Computing. *2022 IEEE International Conference on Imaging Systems and Techniques (IST)*, 1–6. https://doi.org/10.1109/IST55454.2022.9827746.

Behera, R.K., Bala, P.K., Panigrahi, P.K. and Dasgupta, S.A. (2023). Adoption of cognitive computing decision support system in the assessment of health-care policymaking. *Journal of Systems and Information Technology*, *25*(4), 395–439. https://doi.org/10.1108/JSIT-10-2021-0221.

Benmammar, B. (2022). Cognitive decision engine design for cognitive radio networks using gravitational search algorithm and flower pollination algorithm. *International Journal of Internet Protocol Technology*, *15*(2), 116. https://doi.org/10.1504/IJIPT.2022.123587.

Bindu, M.H., Gupta, P. and Tiwary, U.S. (2007). Cognitive Model—Based Emotion Recognition From Facial Expressions For Live Human Computer Interaction. *2007 IEEE Symposium on Computational Intelligence in Image and Signal Processing*, 351–356. https://doi.org/10.1109/CIISP.2007.369194.

Büchel, J., Zendrikov, D., Solinas, S., Indiveri, G. and Muir, D.R. et al. (2021). Supervised training of spiking neural networks for robust deployment on mixed-signal neuromorphic processors. *Scientific Reports*, *11*(1), 23376. https://doi.org/10.1038/s41598-021-02779-x.

Canas-Moreno, S., Piñero-Fuentes, E., Rios-Navarro, A., Cascado-Caballero, D., Perez-Peña, F. et al. (2023). Towards neuromorphic FPGA-based infrastructures for a robotic arm. *Autonomous Robots*, *47*(7), 947–961. https://doi.org/10.1007/s10514-023-10111-x.

Cascone, L. (2021). Periocular features as a window on cognitive processing: From memory to understanding the semantic complexity of an image. *2021 4th International Conference on Bio-Engineering for Smart Technologies (BioSMART)*, 1–4. https://doi.org/10.1109/BioSMART54244.2021.9677876.

Corradi, F. and Indiveri, G. (2015). A Neuromorphic Event-Based Neural Recording System for Smart Brain-Machine-Interfaces. *IEEE Transactions on Biomedical Circuits and Systems*, *9*(5), 699–709. https://doi.org/10.1109/TBCAS.2015.2479256.

Djenouri, Y., Belhadi, A., Srivastava, G. and Lin, J. C.-W. (2022). Toward a Cognitive-Inspired Hashtag Recommendation for Twitter Data Analysis. *IEEE Transactions on Computational Social Systems*, *9*(6), 1748–1757. https://doi.org/10.1109/TCSS.2022.3169838.

Dodeja, L., Tambwekar, P., Hedlund-Botti, E. and Gombolay, M. (2024). Towards the design of user-centric strategy recommendation systems for collaborative Human–AI tasks. *International Journal of Human-Computer Studies*, *184*, 103216. https://doi.org/10.1016/j.ijhcs.2023.103216.

Dushkin, R.V., Stepankov, V.Y. (2022). Principles of Solving the Symbol Grounding Problem in the Development of the General Artificial Cognitive Agents. InProceedings of SAI Intelligent Systems Conference (pp. 231–245). Cham: Springer International Publishing. https://doi.org/10.1007/978-3-031-16075-2_15

Eldardeer, O., Gonzalez-Billandon, J., Grasse, L., Tata, M. and Rea, F. et al. (2021). A Biological Inspired Cognitive Framework for Memory-Based Multi-Sensory Joint Attention in Human-Robot Interactive Tasks. *Frontiers in Neurorobotics*, *15*, 648595. https://doi.org/10.3389/fnbot.2021.648595.

Elmitwally, N.S., Kanwal, A., Abbas, S., Khan, M.A., Khan, M.A., Ahmad, M. and Alanazi, S. (2022). Personality Detection Using Context Based Emotions in Cognitive Agents. Computers, Materials & Continua, 70(3).

Endsley, M.R. (2023). Supporting Human-AI Teams:Transparency, explainability, and situation awareness. *Computers in Human Behavior*, *140*, 107574. https://doi.org/10.1016/j.chb.2022.107574.

Franklin, S. et al. (2016). A LIDA cognitive model tutorial. Biologically Inspired Cogn. Architect. 16, 105–130 https://doi.org/10.1016/j.bica.2016.04.003.

Faruk, T., Shum, L.C., Iaboni, A. and Khan, S.S. (2023). Walking path images from real-time location data predict degree of cognitive impairment. *Artificial Intelligence in Medicine*, *144*, 102657. https://doi.org/10.1016/j.artmed.2023.102657.

Frenkel, C., Bol, D. and Indiveri, G. (2023). Bottom-Up and Top-Down Approaches for the Design of Neuromorphic Processing Systems: Tradeoffs and Synergies Between Natural and Artificial Intelligence. *Proceedings of the IEEE*, *111*(6), 623–652. https://doi.org/10.1109/JPROC.2023.3273520.

Full article: The risks associated with Artificial General Intelligence: A systematic review. (n.d.). Retrieved January 13, 2024, from https://www.tandfonline.com/doi/full/10.1080/0952813X.2021.1964003

Ghani, A., Dowrick, T. and McDaid, L.J. (2023). OSPEN: An open source platform for emulating neuromorphic hardware. *International Journal of Reconfigurable and Embedded Systems (IJRES)*, *12*(1), 1. https://doi.org/10.11591/ijres.v12.i1.pp1-8.

Hazan, A. and Ezra Tsur, E. (2022). Neuromorphic Neural Engineering Framework-Inspired Online Continuous Learning with Analog Circuitry. *Applied Sciences*, *12*(9), 4528. https://doi.org/10.3390/app12094528.

He, X., Wu, X., Peng, J., Li, Q., Ma, X., He, Y. (2022). Bidirectional Cross-scale Feature Fusion for Long Video Micro-Expression 3D Spotting Network. In2022 IEEE 21st International Conference on Cognitive Informatics & Cognitive Computing (pp. 110–115). IEEE. https://doi.org/10.1109/ICCICC57084.2022.10101555.

Horvatić, D. and Lipic, T. (2021). Human-Centric AI: The Symbiosis of Human and Artificial Intelligence. *Entropy*, *23*(3), Article 3. https://doi.org/10.3390/e23030332.

Ielmini, D. (2018). Brain-inspired computing with resistive switching memory (RRAM): Devices, synapses and neural networks. *Microelectronic Engineering*, *190*, 44–53. https://doi.org/10.1016/j.mee.2018.01.009.

Indiveri, G. and Liu, S.-C. (2015). Memory and Information Processing in Neuromorphic Systems. *Proceedings of the IEEE*, *103*(8), 1379–1397. https://doi.org/10.1109/JPROC.2015.2444094.

Jangam, E., Barreto, A.A.D. and Annavarapu, C.S.R. (2022). Automatic detection of COVID-19 from chest CT scan and chest X-Rays images using deep learning, transfer learning and stacking. *Applied Intelligence*, *52*(2), 2243–2259. https://doi.org/10.1007/s10489-021-02393-4.

Kahawanugoda, A., Gnanarathna, K., Meegoda, N., Monarawila, R., Samarasinghe, P., Lindamulage, A.G. (2022). Development of low resource machine learning models for child cognitive ability assessments. In2022 4th International Conference on Advancements in Computing (ICAC) (pp. 72–77). IEEE. https://doi.org/10.1109/ICAC57685.2022.10025049.

Kanumuri, C.R., Madhavi, Ch. R. and Chandra, T.R. (2022). Cognitive Computing in diagnosis of COVID-19 from CT scan and radiograph images: A comparative study. *2022 6th International Conference on Computation System and Information Technology for Sustainable Solutions (CSITSS)*, 1–6. https://doi.org/10.1109/CSITSS57437.2022.10026409.

Kernan Freire, S. (2023). The Human Factors of AI-Empowered Knowledge Sharing. *Extended Abstracts of the 2023 CHI Conference on Human Factors in Computing Systems*, 1–5. https://doi.org/10.1145/3544549.3577044.

Khajooei Nejad, A., Jamshidi, M. (Behdad) and B. Shokouhi, S. (2023). Implementing Tensor-Organized Memory for Message Retrieval Purposes in Neuromorphic Chips. *Computers*, *12*(10), 189. https://doi.org/10.3390/computers12100189.

Kim, C., Lee, Y., Kim, S., Kang, M. and Kim, S. et al. (2023). Diverse synaptic weight adjustment of bio-inspired ZrOx-based memristors for neuromorphic system. *Materials Science in Semiconductor Processing*, *157*, 107314. https://doi.org/10.1016/j.mssp.2023.107314.

Knyazev, G.G. (2023). A Paradigm Shift in Cognitive Sciences? *Neuroscience and Behavioral Physiology*, *53*(5), 892–906. https://doi.org/10.1007/s11055-023-01483-9.

Labde, S., Karan, V., Shah, S. and Krishnan, D. (2023). Movie Recommendation System using RNN and Cognitive thinking. *2023 4th International Conference for Emerging Technology (INCET)*, 1–7. https://doi.org/10.1109/INCET57972.2023.10170572.

Li, M. and Niu, L. (2023). Exercise Recommendation with Preferences and Expectations Based on Ability Computation. *Computers, Materials & Continua*, *77*(1), 263–284. https://doi.org/10.32604/cmc.2023.041193.

Lin, P., Xia, Q. (2021). Three-dimensional hybrid circuits: the future of neuromorphic computing hardware. Nano Express. 27;2(3):031003. https://doi.org/10.1088/2632-959X/ac280e

Liu, H., Gegov, A. and Cocea, M. (2016). Rule-based systems: a granular computing perspective. Granul. Comput. 1, 259–274. https://doi.org/10.1007/s41066-016-0021-6

Liu, J. (2023) Toward A Two-Sided Fairness Framework in Search and Recommendation. In Proceedings of the 2023 Conference on Human Information Interaction and Retrieval (pp. 236–246). https://doi.org/10.1145/3576840.3578332

Lu, Z., Qiao, L. and Singh, A.K. (2021). Advanced Machine Learning on Cognitive Computing for Human Behavior Analysis. *IEEE Transactions on Computational Social Systems*, *8*(5), 1194–1202. https://doi.org/10.1109/TCSS.2020.3011158.

Marcus, G. (2020). *The Next Decade in AI: Four Steps Towards Robust Artificial Intelligence* (arXiv:2002.06177). arXiv. https://doi.org/10.48550/arXiv.2002.06177.

Mary Shyni, H. and Chitra, E. (2022). A comparative study of X-ray and CT images in COVID-19 detection using image processing and deep learning techniques. *Computer Methods and Programs in Biomedicine Update*, *2*, 100054. https://doi.org/10.1016/j.cmpbup.2022.100054.

Mathapati, M., Chidambaranathan, S., Nasir, A. W., Vimalarani, G., Sheeba Rani, S et al. (2021). An Intelligent Internet of Medical Things with Deep Learning Based Automated Breast Cancer Detection and Classification Model. In A.E. Hassanien, A. Khamparia, D. Gupta, K. Shankar and A. Slowik (Eds.), *Cognitive Internet of Medical Things for Smart Healthcare* (Vol. 311, pp. 181–193). Springer International Publishing. https://doi.org/10.1007/978-3-030-55833-8_11.

Mead, C. (1990). Neuromorphic electronic systems. *Proceedings of the IEEE*, *78*(10), 1629–1636. https://doi.org/10.1109/5.58356.

Ménager, D.H., Choi, D. (2023). Hybrid Event Memory as a Case Base for State Estimation in Cognitive Agents. InInternational Conference on Case-Based Reasoning (pp. 134–149). Cham: Springer Nature Switzerland. https://doi.org/10.1007/978-3-031-40177-0_9

Miller, S. (2024). *Unveiling the Synergy: Exploring Advances in Applied Artificial Intelligence and Narrow AI* (11651). Article 11651. https://easychair.org/publications/preprint/d1PB.

Nguyen, T.N., Phan, D.N., Gonzalez, C. (2023). Learning in Cooperative Multiagent Systems Using Cognitive and Machine Models. ACM Transactions on Autonomous and Adaptive Systems. 18(4): 1–22. https://doi.org/10.1145/3617835

Parashar, J., Kushwah, V.S. and Rai, M. (2023). Determination Human Behavior Prediction Supported by Cognitive Computing-Based Neural Network. In R. Kumar, A.K. Verma, T.K. Sharma, O.P. Verma and S. Sharma (Eds.), *Soft Computing: Theories and Applications* (pp. 431–441). Springer Nature. https://doi.org/10.1007/978-981-19-9858-4_36.

Perconti, P. and Plebe, A. (2020). Deep learning and cognitive science. *Cognition*, *203*, 104365. https://doi.org/10.1016/j.cognition.2020.104365.

Piñero-Fuentes, E., Canas-Moreno, S., Rios-Navarro, A., Delbruck, T. and Linares-Barranco, A. et al. (2021). Autonomous Driving of a Rover-Like Robot Using Neuromorphic Computing. In I. Rojas, G. Joya and A. Català (Eds.), *Advances in Computational Intelligence* (Vol. 12862, pp. 57–68). Springer International Publishing. https://doi.org/10.1007/978-3-030-85099-9_5.

Qureshi, H., Agrawal, A.P. (2022). Video based sentiment analysis. In2022 Fifth International Conference on Computational Intelligence and Communication Technologies (pp. 388–399). IEEE. https://doi.org/ 10.1109/CCiCT56684.2022.00076.

Rayhan, A., Rayhan, R. and Rayhan, S. (2023). *ARTIFICIAL GENERAL INTELLIGENCE: ROADMAP TO ACHIEVING HUMAN-LEVEL CAPABILITIES.* https://doi.org/10.13140/RG.2.2.33680.79361/1.

Salmi, J. (2023). A democratic way of controlling artificial general intelligence. *AI & SOCIETY*, *38*(4), 1785–1791. https://doi.org/10.1007/s00146-022-01426-x.

Salmon, P.M., Baber, C., Burns, C., Carden, T., Cooke, N. et al. (2023). Managing the risks of artificial general intelligence: A human factors and ergonomics perspective. *Human Factors and Ergonomics in Manufacturing & Service Industries*, *33*(5), 366–378. https://doi.org/10.1002/hfm.20996.

Schuman, C.D., Kulkarni, S.R., Parsa, M., Mitchell, J.P., Date, P. et al. (2022). Opportunities for neuromorphic computing algorithms and applications. *Nature Computational Science*, *2*(1), 10–19. https://doi.org/10.1038/s43588-021-00184-y.

Shi, Z., Groechel, T.R., Jain, S., Chima, K., Rudovic, O., Matarić, M.J. (2022). Toward personalized affect-aware socially assistive robot tutors for long-term interventions with children with autism. ACM Transactions on Human-Robot Interaction (THRI). 11(4):1–28. https://doi.org/10.1145/3526111

Singh, S., Chaudhary, D., Gupta, A.D., Prakash Lohani, B., Kushwaha, P.K. et al. (2022). Artificial Intelligence, Cognitive Robotics and Nature of Consciousness. *2022 3rd International Conference on Intelligent Engineering and Management (ICIEM)*, 447–454. https://doi.org/10.1109/ICIEM54221.2022.9853081.

Sze, V., Chen, Y.-H., Yang, T.-J. and Emer, J.S. (2017). Efficient Processing of Deep Neural Networks: A Tutorial and Survey. *Proceedings of the IEEE*, *105*(12), 2295–2329. https://doi.org/10.1109/JPROC.2017.2761740.

Tasaki, S., Kim, N., Truty, T., Zhang, A., Buchman, A.S. et al. (2023). Explainable deep learning approach for extracting cognitive features from hand-drawn images of intersecting pentagons. *Npj Digital Medicine*, *6*(1), 157. https://doi.org/10.1038/s41746-023-00904-w.

Tzeng, S.T. (2022, May). Engineering Normative and Cognitive Agents with Emotions and Values. In Proceedings of the 21st International Conference on Autonomous Agents and Multiagent Systems (pp. 1878–1880).

Vekariya, Dr. D., Kannan, D.M.J., Gupta, Mr. S., Muthusamy, P., Mahajan et al. (2022). Recommendation Model-Based 5G Network and Cognitive System of Cloud Data with AI Technique in IOMT Applications. *International Journal of Communication Networks and Information Security (IJCNIS)*, *14*(3), 239–256. https://doi.org/10.17762/ijcnis.v14i3.5609.

Vidal-Soroa, D., Furelos, P., Bellas, F., Becerra, J.A. (2023). An Approach to 3D Object Detection in Real-Time for Cognitive Robotics Experiments. In: Tardioli, D., Matellán, V., Heredia, G., Silva, M.F., Marques, L. (eds) ROBOT2022: Fifth Iberian Robotics Conference. ROBOT 2022. Lecture Notes in Networks and Systems, vol 589. Springer, Cham. https://doi.org/10.1007/978-3-031-21065-5_24

Wang, E.K., Xu, S.P., Chen, C.-M. and Kumar, N. (2021). Neural-Architecture-Search-Based Multiobjective Cognitive Automation System. *IEEE Systems Journal*, *15*(2), 2918–2925. https://doi.org/10.1109/JSYST.2020.3002428.

Weekes, T.R. and Eskridge, T.C. (2022). Responsible Human-Centered Artificial Intelligence for the Cognitive Enhancement of Knowledge Workers. *HCI International 2022 – Late Breaking Papers: Interacting with eXtended Reality and Artificial Intelligence: 24th International Conference on Human-Computer Interaction, HCII 2022, Virtual Event, June 26 – July 1, 2022, Proceedings*, 568–582. https://doi.org/10.1007/978-3-031-21707-4_41.

Wu, H., Fang, B., Xiang, C., Cohen, G., Van Schaik, A. et al. (2022). Evolving Neuromorphic Systems on the Ethereum Smart Contract Platform. *2022 IEEE 1st Global Emerging Technology Blockchain Forum: Blockchain & Beyond (iGETblockchain)*, 1–6. https://doi.org/10.1109/iGETblockchain56591.2022.10087086.

Xiang, C. (2022). Research on Real Time Detection of Human Fitness Based on Artificial Intelligence Assisted Cognitive Computing. In B.J. Jansen, H. Liang and J. Ye (Eds.), *International Conference on Cognitive based Information Processing and Applications (CIPA 2021)* (pp. 936–940). Springer. https://doi.org/10.1007/978-981-16-5854-9_129.

Xiao, D., Zhu, F., Jiang, J. and Niu, X. (2023). Leveraging natural cognitive systems in conjunction with ResNet50-BiGRU model and attention mechanism for enhanced medical image analysis and sports injury prediction. *Frontiers in Neuroscience*, *17*, 1273931. https://doi.org/10.3389/fnins.2023.1273931.

Xu, X., Ding, Y., Hu, S.X., Niemier, M., Cong, J. et al (2018). Scaling for edge inference of deep neural networks. *Nature Electronics*, *1*(4), 216–222. https://doi.org/10.1038/s41928-018-0059-3.

Yeh, Z.W., Hsu, C.H., Yeh, C.F., Wu, W.C., Wang, C.T., Lo, C.C. and Tang, K.T. (2021). POPPINS: A population-based digital spiking neuromorphic processor with integer quadratic integrate-and-fire neurons. In2021 IEEE International Symposium on Circuits and Systems (pp. 1–5). IEEE. https://doi.org/10.1109/ISCAS51556.2021.9401426.

You, H. and Zhao, K. (2021). Neuromorphic Implementation of a Continuous Attractor Neural Network With Various Synaptic Dynamics. *IEEE Access*, *9*, 109224–109240. https://doi.org/10.1109/ACCESS.2021.3101290.

Zhang, T., Zeng, Y., Pan, R., Shi, M., Lu, E. (2021). Brain-inspired active learning architecture for procedural knowledge understanding based on human-robot interaction. Cognitive Computation. (2):381–93. https://doi.org/10.1007/s12559-020-09753-1

Zhao, G., Li, Y. and Xu, Q. (2022). From Emotion AI to Cognitive AI. *International Journal of Network Dynamics and Intelligence*, 65–72. https://doi.org/10.53941/ijndi0101006.

Zhu, Y., Gao, T., Fan, L., Huang, S., Edmonds, M. et al. (2020). Dark, Beyond Deep: A Paradigm Shift to Cognitive AI with Humanlike Common Sense. *Engineering*, *6*(3), 310–345. https://doi.org/10.1016/j.eng.2020.01.011.

Zouganeli, E. and Lentzas, A. (2022). Cognitive Robotics—Towards the Development of Next-Generation Robotics and Intelligent Systems. In E. Zouganeli, A. Yazidi, G. Mello and P. Lind (Eds.), *Nordic Artificial Intelligence Research and Development* (Vol. 1650, pp. 16–25). Springer International Publishing. https://doi.org/10.1007/978-3-031-17030-0_2.

CHAPTER

19

Artificial Intelligence in Digital Biology

Kavipriya Murugaiyan,[1#] *Swarubini PJ,*[1#] *Kavati Pavan Kumar,*[1] *Basu Bhattacharjee,*[1] *Aravind Kumar Rengan*[1*] and *Nagarajan Ganapathy*[1*]

1. Introduction

The field of Biology rapidly developed in 19th century where most of the problems were answered. Darwin gave the scientific reason behind evolution. Cell became the basic unit of a living being.. Though the development started, problems aroused based on two different concepts, one was based on understanding the components of a cell and the other was based on holistic problems such as evolution. With the evolving concepts of Physics at that time, everything was addressed in terms of atoms because of which a disparity arose between the atomic and biological worlds. This played a significant role in the origin of 20th century biology (Woese, 2004). 21st century biology aims to unify the concepts of empirical work with computational work. Empirical work is performed in laboratories with hypotheses being tested that can ultimately accept, reject or even suggest a new hypothesis. Computational work supports empirical work through techniques such as simulation, data mining and microarray analysis. Information technology available today has a strong potential to enable data management and capacity to interpret data. It was predicted that computing and information technology will become an integral part of scientific research (Wooley et al., 2005). We can say that in the 21st century, biology has been digitalized to solve the underlying complex problems in biological sciences. Digital biology helps us in gaining insights in real-time problems and also hasten the invention process (Zhydik, 2020). Freddolino and Tavozi predicted the future paths of digital biology depending on the work of Karr *et al.* (Karr et al., 2012), that deals with multiscale modelling of cell-scale simulations. One is the Physicist's perspective that will lead us to the development of in-silico modelling which will enable our understanding of biological systems. The other is the Engineer's perspective which will target the development of computational methods that can replace the wet lab experiments (Sleator, 2012).

---

[1] Department of Biomedical Engineering, Indian Institute of Technology Hyderabad, India.
[#] Equal Contribution
[*] Corresponding author: aravind@bme.iith.ac.in; gnagarajan@bme.iith.ac.in

In the field of biology, Artificial Intelligence (AI) has emanated from a symbolic approach where a set of rules has been used to enable machines to perform operations. Example of symbolic AI is the game of chess. Biological intelligence can make decisions on the spot based on the information available at that time. An example is artificial neural networks (ANN) which are established and influenced by the neural networks of the brain. Hidden Markov model is used for stochastic processes with no repetition of fixed patterns. Machine learning (ML) is employed to identify archetypes between the different data samples. Multiple processing layers are used in Deep learning (DL) focussed on pattern recognition and establishing the relationship between inputs and outputs (Hassoun et al., 2022). Digitalization in healthcare is reducing the timing for detection of underlying diseases/disorders thereby helping in prompt care of the patients with significant reduction of pain and discomfort (Da Fonseca et al., 2021). An example is Digital microfluidics (DMF), an approach based on microchannel based fluidics using a group of electrodes with a hydrophobic insulator. DMF has advantages over microchannels which can perform unique lab-on-chip applications. Recently the DMF technique was used for DNA extraction from whole blood samples to quantify the steroidal hormones from 1mg of breast tissue and 1μl of blood and serum samples (Jebrail & Wheeler, 2010).

This chapter will provide an overview of the common techniques, assays and *invivo* experiments commonly used in biological research and how AI can help in extending their role and improving their functions for successful clinical translation. **Figure 1** shows the role of AI in digitalizing the assays and experiments.

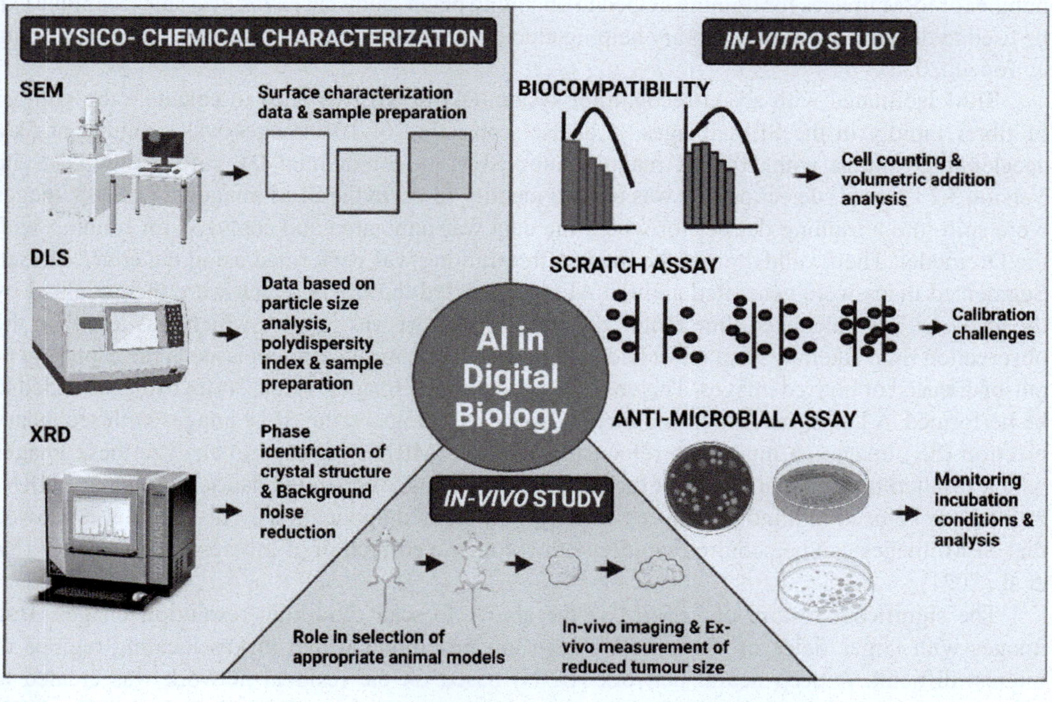

Fig. 1 Role of AI in digitalizing the assays and experiments

2. Physicochemical Characterization of Nanomaterials

2.1 SEM (Scanning Electron Microscope), FESEM and EDX

The Scanning electron microscope (SEM) was devised by Max Knoll in 1935 (Smith et al., 2008). SEM analysis or SEM technique can be used to analyze the size distribution, shape and also the aggregation or dispersed behaviour of nano or micro sized materials (Lin et al., 2014). SEM uses electron sources such as the tungsten filament or solid-state crystals while FESEM uses field emission guns to emit electron beams. The voltages used for SEM are in the range of 0.1 to 30kV (Abd Mutalib et al., 2017); for FESEM they are in the range of 0.5 to 30kV (Mayeen et al., 2018). Energy dispersive X-ray spectroscopy (EDX) is associated with recognition of elements present in the sample. Interaction of samples with the electron beam creates characteristic X-rays and since the emission spectrum for these sample elements differ individually, they can be used to measure their concentration in the samples (Abd Mutalib et al., 2017). The magnification of the SEM ranges from 300,000x to 10,00,000x. The organic and inorganic solid materials used can vary in size ranging from diameters of 200mm and heights of 80mm (Abdullah & Mohammed, 2019). The morphological information obtained is used for further characterization studies. Magnification of FESEM ranges between 10-300,000x (Junoh et al., 2020). Higher resolution images up to 2-3 nm can be achieved in FESEM due to the decrease in interaction volumes.

2.1.1 Limitations and Role of AI in Overcoming the Limitations of SEM, FESEM and EDX

Problems arise in sample preparation for SEM images when the sample needs to be made conductive with metal coatings, heterogeneous sample analysis and drying of samples which causes size reduction (Bootz et al., 2004). A high pressure vacuum chamber is needed to prevent contamination of the solid-state crystals (Abd Mutalib et al., 2017). Analysing the SEM images with AI is a multidisciplinary field. SEM images were trained using supervised and unsupervised ML techniques. The data was then learned and compared. The authors inferred that incrementing the number of SEM images for training helped in obtaining better study based results. This technique can be used to develop a materials library helping elucidate the properties of nanomaterials (DemiRkan & Topcu, 2022).

SEM facilitated with an AI recognition system (AI-SEM) was used to calculate the number of fibres rapidly in the SEM images. A dataset consisting of 108 images was captured at 5kV accelerating voltage with 10000X magnification SEM. A commercial DL software (ViDi Suite Version 3.2 Cognex) detection tool was used to identify fibers in the SEM images. 25 of 108 images were split into a training dataset for which the data was annotated and analyzed for training with the DL model. Then, validation of the model after training was performed using the entire dataset. Segmented maps were generated and the AI tool detected the fibers which were then counted by the analysts. The calculated time required for the trained AI was 5 hours which was less than the observation time taken by the analyst (Iida et al., 2021). The most common issue is the capturing of out-of-focused or blurred images. This results in poor quality images. Hence, refocusing is needed to be performed. A DL-based refocusing method was used for analyzing SEM images with secondary electron (SE) images. A multiscale refocusing network (MRN) was used to analyze these images which resulted in over sharpening of the in-focused regions. Augmented datasets which are MRN-trained can be used in blind-settings essentially for potential applications. This study also showed that SEM images can be acquired rapidly with automatic correction of low resolution images (Na et al., 2021).

The significant feature of FESEM is the ability to scan ultra-high resolution images. But, images with larger fields of view (FOV) have low resolution which makes locating regions of interest difficult. A deep neural network (DNN) based on the residual network was created to enhance the resolution of the FESEM image without changing the physical structure. The DNN consists of an input convolutional layer followed by 32 residual modules, each of which consists of two convolutional layers and a Rectified Linear Unit (ReLU) activation function. The image obtained had a higher resolution along with a large FOV with reduced noise (Gao et al., 2020). Identification of minerals with SEM images is time consuming and labor intensive. Hence, there is a requirement for automated segmentation and classification. ML based techniques for classification have been explored in recent times. Five ML models and a DL model were differentiated based on the performance of mineral segmentation. A random forest (RF) model surpassed other models with a F1 score of 0.92. The U-Net model was better than the RF model for mineral classification in unseen shale samples. Certain elements such as Ca, Si, Al and Mg played a significant role than others in the segmentation of minerals (C. Li et al., 2021).

2.2 STM (Scanning Tunnelling Microscopy)

Scanning tunnelling microscopy was developed by Binning and Rohrer. It involves the use of tips usually of atomic size positioned with piezoelectric actuators. A small voltage is applied between the tip and the electrically conductive sample which leads to the development of a quantum mechanical tunnelling current (Bode, 2003). This technique can provide atomic scale spatial resolution. STM with pulsed electromagnetic radiations has accelerated the examination of atomic and molecular dynamics to an order of attoseconds (Bi et al., 2023). A spatial resolution of 20 nm and time resolution of 2ps was achieved using GaAs probe tip. Tetrahertz pulses coupled to the STM tip offers a resolution as low as 2nm and provides subpicosecond time resolution less than 500 fs (Cocker et al., 2013).

2.2.1 Limitations and Role of AI in Overcoming the Limitations of STM

Samples should be electrically conductive (Bode, 2003). Larger machines are required for a vibration-free environment (Park & Barrett, 1993). A systemic database for computational STM data is still lacking. A STM database is needed to establish the relation between atomic positions and images that can facilitate experimental analysis. Thus, a systematic STM database was created through the density functional theory (DFT) for the analysis of 2D materials. The images in the dataset were augmented and fed into a multi-layer DNN consisting of a single convolution layer activated by ReLU, a fully connected softmax layer with five outputs. The data was divided into trained and validation data (80-20% split). An accuracy of 90% was observed on the validation set (Choudhary et al.).

2.3 TEM (Transmission Electron Microscopy)

Max Knoll and Ernst Ruska innovated the first TEM in 1931(Lewczuk & Szyryńska, 2021). Transmission (TEM) analysis or the TEM technique is used to study crystallinity, size, composition, aggregation or dispersed behaviour of the nanomaterials or nanoclusters (Lin et al., 2014). It has a higher spatial resolution than SEM and helps in the analysis of electronic structures and chemical compositions of nanomaterials (van Benthem et al., 2017). A High energy electron beam of 8o KeV is used (Dorfs et al., 2011). The diffraction images obtained from TEM can be used to study deformities in the perfect crystal (Bendersky & Gayle, 2001). Cryo-TEM can be used to image biological samples such as bacteria, viruses and tissues (Sarabandi et al., 2020). In-situ TEM has been recently designed for use in real-time applications for the study of energy conversion and storage materials (Zheng et al., 2022). Acquiring a high spatial resolution in a microscope is possible with a point resolution of 0.2 nm, while the temporal resolution can range from 10^{-6} to 10^{-15} seconds (Armstrong et al., 2007). Diffraction patterns can be obtained for specific regions in samples which are less than 100 nm (Bendersky & Gayle, 2001). In scanning transmission electron microscopy (STEM), atom positions can be found with high atomic precision of 0.04 to 0.06Å which can be further used to give three dimensional images of the crystal (Raja et al., 2022).

2.3.1 Limitations and Role of AI in Overcoming the Limitations of TEM

TEM offers a limited field of view of samples which is less than 3 mm in diameter. Two dimensional images of three-dimensional objects are obtained. Damage of the samples during sample preparation can occur. They are large instruments and are usually expensive (Abd Mutalib et al., 2017). Difficult sample preparation requires thin samples (Ayache et al., 2010). TEM produces a vast number of datasets. Pre-processing and analyzing this data is arduous through manual techniques and may likely cause human errors. Hence, there is a desire to employ AI methods to enhance autonomous data analysis for TEM images. ML and DL approaches have been utilized for analyzing TEM data and now focuses mainly on image processing to obtain various features such as atomistic, morphological and microstructural features along with emerging applications in automated electron diffraction and spectroscopy analysis (Zheng et al., 2022). One of the challenges in STEM is whether the data stream can be harnessed from the microscope to produce real-time feedback. Big data and ML techniques have been used to extract data from static and dynamic STEM images. This DL analytics and real-time feedback will give a better understanding of electronic, phonon and quantum phenomena of atomically engineered structures (Kalinin et al., 2019).

2.4 Atomic Force Microscopy (AFM)

An AFM is also called a scanning force microscope, which measures the force between the sample and the probe (Yablon, 2013). Measurement of surface texture is important to understand the performance of materials. The components of AFM are cantilever, probe, piezoelectric scanner, laser, data processor and a photodetector (De et al., 2012). The cantilever/probe assembly acts as the heart of an AFM. An AFM is used particularly in industrial research. This is because of its significant characteristics such as high lateral resolution flexibility of operation in variable environments in. The force measured between the sample and the tip is given as a function of the tip-sample distance (Voigtländer, 2019).

2.4.1 Limitations and Role of AI in Overcoming the Limitations for AFM

The area and the height imaged by AFM are of the order of micrometers (Sinha Ray, 2013). It is a time consuming procedure (Chicea et al., 2010). The measurement is non-monotonous since there are two points on the force-distance curve and care needs to be taken to work on any one of these points (Voigtländer, 2019). Artifacts arise due to the probe geometry (Sobola et al., 2017).

2.5 Ultraviolet-Visible Spectrophotometry

This technique is quick and less expensive to determine the release of drugs from the delivery systems (Vigata et al., 2020). It operates on the principle of measuring the attenuated electromagnetic radiation due to an absorbing substance. The spectral energy values is usually in the range of 190-900 nm (L.C. Passos & M.F.S. Saraiva, 2019). Lower concentrations ($<1\mu g/ml$) require high end methods for detection of molecules (Vigata et al., 2020). They can also be used to quantify the presence of certain amino acids (Chutipongtanate et al., 2012).

2.5.1 Limitations and Role of AI in Overcoming the Limitations for Ultraviolet-visible Spectrophotometry

The procedure requires accurate calibration for each sample. Many natural proteins such as heme have strong absorbance in this region (Aitken & Learmonth, n.d.). The challenge with the current spectroscopic techniques was overcome wherein ANN was used to calculate the chemical oxygen demand (COD) values. UV-Visible spectroscopy and ANN have been used to predict water pollution from COD values. COD values from wastewater samples rely on one wavelength to indicate the change. An artificial neural network (ANN) was used to calculate the COD values. UV-Visible spectroscopy and ANN have been used to predict water pollution from COD values. It was found that COD values from the ANN model were closer to the standard values compared to those from a multiple linear regression (MLR) model (Fogelman et al., 2006). The water quality index (WQI) is a criteria that is used extensively for water resource management and planning but it has a drawback whereas generating the chemical waste is a laborious and costly process. This could be addressed by compensating the traditional methods with ultraviolet-visible UV-Vis spectrophotometry along with the use of ANN. The water samples were collected and UV-Vis spectral analysis was performed followed by principal component analysis (PCA). The PCA scores were fed into a three-layer feed-forward NN as inputs. The output layer was described by the WQI values. This technique can be used to create a portable UV-Vis spectrophotometry to detect water quality index (WQI) in places with less infrastructure (Alves et al., 2018).

2.6 XRD (X-ray Diffraction Analysis)

A chemical study using X-ray diffraction patterns was first described by Hull (Delgado, 2017). X-ray diffraction analysis is a non-damaging, elemental analysis technique that gives information on the crystal structure, chemical composition and physical properties of the sample. When light falls on the periodically arranged atoms of the crystal, constructive interference of the scattered light occurs forming a diffraction pattern which provides information about the atoms in the crystal. It can be used to analyze texture and measure the stress of the samples (Raja et al., 2022) (Nasrazadani & Hassani, 2016) (Chauhan, 2014). The international centre for diffraction data performs the function of collecting, distributing and storage of data for identifying new materials (Messick, 2012). Monochromatic synchrotron radiations produce diffractograms with lesser peak overlap. These radiations range from microwaves to hard X-rays (0.05) nm. These radiations produce high signal/noise ratios, enhanced angular resolution, minimal peak overlap and improved peak position. Peak intensity correlates the composition, peak position of unit cell parameters and peak width of lattice formation (Khan et al., 2020).

2.6.1 Limitations and Role of AI in Overcoming the Limitations of XRD

The homogenous phase is preferred for detecting the unknown sample. A standard reference file is required for verification. Detection limit is ~2% for mixed materials. Peak overlay occurs for high angle reflections. Severe errors can occur while filling the samples in the sample holder (Bunaciu et al., 2015). High energy and flux radiation is required to analyze materials in thin films (Thakral et al., 2016). The procedure is time consuming (Hansford, 2013). A well-qualified professional with well-established computational methods might also face difficulty in the identification of unknown components. Hence, a system is needed where even people who lack expertise also can be involved in the identification process. A fully trained CNN model was developed to accurately identify the constituents of multiphase inorganic XRD samples. The trained CNN on simulated XRD data produced results with 100% accuracy for phase identification and 86% for three-step phase fraction. It was also reported that CNN with more layers surpassed those with few layers of phase fraction prediction (Lee et al., 2020).

Spatial knowledge of minerals is essential for interpreting reachable mineral surface areas to study mineral dissolution and weathering reaction rates. Therefore, a new approach for 2D mineral characterization called synchrotron-based ML approach for RasTer (SMART) mineral mapping was developed which uses an ANN to build a classifier which is trained using μXRF elemental intensities and μXRD mineral information. The network was able to derive information between specific elements to a mineral phase and also reduced the time for mineral mapping (Kim et al., 2021).

2.7 Fourier Transform Infrared Spectroscopy (FTIR)

FTIR is a vibrational spectroscopic technique which is a simple, non-damaging technique that gives an assessment of the materials' functional groups, bonding and molecular conformations (Movasaghi et al., 2008). In the mid-infrared region 400-4000 cm^{-1} stretching or bending vibrations occur which helps us in obtaining the spectra (Berthomieu & Hienerwadel, 2009). FTIR records the absorbed energy from the sample which is often particular to that sample and converts it into a special interferogram signal (Mohamed et al., 2017). Small changes in bond length less than 0.2Å are sensitive enough to cause changes in the spectra (Berthomieu & Hienerwadel, 2009).

2.7.1 Limitations and Role of AI in Overcoming the Limitations for FTIR

A strong absorption band in water makes it less favourable for analyzing non-aqueous samples. It requires more sample preparation and cannot do confocal imaging (Talari et al., 2017). Artificially sweetened products are promoted to help avoid obesity by making low calorie foods. In order to ensure consumer safety, it is important to limit the amount of sweeteners in artificially-sweetened products. FTIR combined with computational models was used to recognize and quantify the artificial sweeteners. Among the machine learning techniques, a Support vector machine (SVM) model was selected as a pattern recognition method for categorical assessment of the complex data. Multivariate calibration models based-on partial least squares were developed for quantification of artificial sweeteners and validation by predicting data sets. A rapid tool was developed based on FTIR spectroscopy for identification of artificial sweeteners in beverages (Wang et al., 2020).

Artificial intelligence applications have been used to identify toxic hallucinogenic drugs. Important wavenumbers were identified using genetic algorithms (GA) and the selected database was utilized as an input for building partial least squares regression (PLSR) models and the best model was recognized by two error assessing indices. Using the ATR-FTIR absorptions, the GA-PLSR characterizes the modelled controlled substances.RF was better for class identification of modelled compounds with the correct classification similar to GA-PLSR (Negoita et al., 2019).

2.8 DLS

DLS is a time-resolved technique based on the principle of coherent scattering of incident light by molecules or smaller particles. It calculates the hydrodynamic size of the particles and manipulates the ability of these particles when they are in brownian motion. DLS has the capability to measure a large number of particles and hence provides an accurate size distribution analysis and polydispersity index (PDI) (Bhattacharjee, 2016) (Babick, 2020b). The diffusion quotient can be obtained from the scattered light; it is associated with the hydrodynamic size of the particles. The viscosity of and temperature of the sample play a vital role in precise measures of particle size analysis (Sandhu et al., 2018).

2.8.1 Limitations and Role of AI in Overcoming the Limitations for DLS

The precise measurement of particles remains a challenge due to factors such as scattering intensity which is influenced by instrumental parameters, limited resolution and assumption of optical models that rely on spherical particles (Babick, 2020a). Aggregation of monoclonal antibodies (mAb) is considered as a drawback as it impacts the drug safety and shelf life. Hence, there is a need for an analytical approach to estimate the mAb aggregates. Dynamic light scattering along with machine learning algorithms is used to measure the relative percentage of individual multimers (e.g., monomer, dimer, trimer and tetramer) of a monoclonal antibody. The obtained spectrum and prediction about aggregates from the above approach enables real-time prediction in less than 2 mins which creates opportunities for DLS-NN and high throughput analysis (Shrivastava et al., 2023). An alternative method for DLS was proposed that was collected at equal time intervals termed as DLS time series. ANN was trained to calculate the diameter with spectral entropy of the time series as the input. White noise was added to the big set of simulated time series while the training was carried out. ANN along with computation can be used to fabricate a sensor which is responsive to the particle's size in suspension thereby crucial in identifying microorganisms (Chicea et al., 2022).

2.9 Zeta Potential

The zeta potential is defined as the potential at the interface between the electric double layer and slipping plane of a colloidally dispersed particle (Bhattacharjee, 2016). It determines the electrostatic interactions of particles and also their stability in the dispersions (İlhan et al., 2022). For a colloidal system to be stable the repulsive forces should be greater than the attractive forces. The common electrokinetic effect considered for the determination of Zeta potential is electrophoresis. Electrophoresis is defined as the movement of charged colloidal particles when an external electric field is applied (Olson, 2012). Nanoparticles having Zeta potential of positive/negative values plays a vital role in intracellular delivery of drugs. Particles with extreme positive or negative charges cause repulsion and those with similar charges prevent aggregation. Stabilization of nanoparticles by both electrostatic and steric hindrances requires a minimum zeta potential of ±20mV (Honary & Zahir, 2013).

2.9.1 Limitations and Role of AI in Overcoming the Limitations of Zeta Potential

This requires a dilute sample devoid of settling. Contamination of the samples can occur (Mohammadi-Jam et al., 2022). To be detectable, the charged particles should have a larger charge and smaller size. Though several factors play a role, the upper size limit of the particles for ease of detection is usually between 1-10μm (Olson, 2012). Measuring the zeta potential can be laborious and challenging. Thereby, current research is based on the application of ML tools where the data from various sources was used to create, train and test the ML models. Compared to multiple linear regression (MLR) and support vector machine (SVM) models, the ANN model presented significant correlation coefficients with a high degree of accuracy. The obtained correlation coefficient values significantly matched more than 97% of the actual values (Muneer et al., 2023). Prediction of thermophysical properties of nanofluids is essential without the need of any experimental study. Thermal conductivity and Zeta potential of a Fe_3O_4/nanofluid was accurately predicted by two different ANN models. A multilayer perceptron (MLP) ANN model was created which consisted of input, hidden and output layers. The predicted values were obtained in the output layer (Sahin et al., 2023).

2.10 Rheology

A rheometer helps in studying complex fluid-like materials. The type of solid or liquid response of materials depends on the time or the deformation process. The internal structure of a material can be studied by assessing the relationship between stress, strain or strain rate. Experiments can identify the elastic, viscous or plastic behaviour of materials. A Hookean solid shows an elastic behaviour, because here the stress and strain are in phase with the material. The Newtonian fluids prove to have a plastic behaviour as stress and strain are in or out of phase. In a viscoelastic material which is different from the former materials there occurs a phase lag between stress and strain. The viscoelastic behaviour of materials can be classified into four categories based on its response to the time and magnitude of the applied shear stress. They are classified as pseudoplastic, dilatant, thixotropic and rheopectic. Bingham fluids behave as solids at a low shear force and show Newtonian behaviour after a critical shear force. There is no linear relationship between shear stress and shear strain for non-newtonian fluids and a certain force is required to initiate the flow (Chandran C et al., 2019; Fernanda, 2018).

2.10.1 Limitations and Role of AI in Overcoming the Limitations of Rheology

Instrumental parameters, instrument and sample inertia can occur. Gap errors include parallelism and gap precision. Slips can occur which can affect the rheological characterization (Ewoldt et al., 2015). ML and DL models were developed to find the rheological properties of foods that exhibit complex viscoelastic behaviour (Ahmad & Benjamin, 2023). The materials used in 3D printing need to satisfy definite physical and mechanical criteria. In order to enhance the printability of the biomaterials such as bioactive hydrogels there is a requirement for a system that predicts and interprets the material to be used. Correlation between the rheological factors and printability of the hydrogel was established using ML models. SHAP (Shapley additive explanations) is used for interpreting the expected outcomes from these ML models. A positive SHAP value above zero denotes a prediction that the gel is printable and a value below zero affects the same (Nadernezhad & Groll, 2022).

2.11 Swelling and Degradation Study

Swelling and Degradation studies are carried out mostly in gels and scaffold-like materials for biomedical applications where degradation of the material is expected after achieving its intended purpose (Christy et al., 2020; Singh et al., 2022). The swelling study is usually performed in buffered saline (pH-7.4) and incubated at 37°C (McBath & Shipp, 2010). Degradation is carried out either in saline or in certain enzymes (Shimdov et al., 2020; Donate et al., 2020).

2.11.1 Limitations and Role of AI in Overcoming the Limitations of Swelling and Degradation

Measurement of the exact rate of degradation is difficult (J. Li & Mooney, 2016). Manual handling (Preobrazhenskiy & Putlyaev, 2023) of these experiments may result in error. Swelling behaviour of a novel hydrogel based on acrylamide–2–hydroxyethylmethacrylate and acrylamide-N-isopropylacrylamide is investigated with the development of three-dimensional crosslinked hydrogel networks. Effects of the salt concentrations and crosslinker conditions were tested using the experimental methodology and ANN-assisted optimization algorithms of the hydrogel (Islamkulov et al., 2023). Genipin is used to decrease the degradation rate of gelatin. There are no investigations made in the experimental measurement of the degradation rate of genipin-gelatin scaffolds. ML methods such as ANN and KRR (Kernel ridge regression) are used to compare the degradation results of genipin crosslinked gelatine hydrogels. ANN with a mean square error (MSE) of 2.68% outperformed KRR with a MSE of 4.78% in terms of accuracy (Entekhabi et al., 2020).

3. In vitro Studies

3.1 Biocompatibility

Biocompatibility is a process of safety evaluation and identification of a favourable host response elicited by any synthetic material (compounds, polymers, or nanoparticles). Biocompatibility assessment is also necessary to prevent immunogenic responses that may get elicited later during prolonged exposure. To determine the biocompatibility of any material *in vitro* cell culture is employed. Briefly, the mammalian cells are cultured in 2D using 96 well plates. After incubating the cells with synthetic materials, a colorimetric dye-based assay (MTT, XTT, or Alamar Blue) is used to determine their cytotoxic potential and hence the biocompatibility (Anderson, 2012) (Williams, 2008) (Weimer, 1997).

3.1.1 Limitations of AI on Overcoming the Limitations of the Biocompatibility Study

(a) *Cell counting:* An equal number of cells should be plated in each well, to get similar viability values. Human error, while counting the cells or setting the cell counter brightness during trypan blue staining may occur. This can be taken care of with the help of AI considering the signals can be broken down into binary or Boolean signals.

(b) *Volumetric additions:* While performing the assay, the addition of media or dye or DMSO needs to be precise. In addition, there may be variable pipetting that can be addressed as human errors which can lead to variable outputs with huge standard deviations. This can also be overcome by the intervention of AI, as the coding instructions will make the volumetric pipetting error-free.

(c) *Colorimetric data acquisition:* The spectral analysis of the colored solution of viability assays may get altered slightly by various human errors including sudden jerks, power fluctuations, or various excitation/emission values selected for the same assay. These loopholes in experimental biology can be taken care of by AI intervention.

In order to design biomaterials and evaluate their performance, we generate vast, diverse, complex, and heterogeneous biocompatibility data. To interpret this data effectively, we employ AI algorithms such as natural language processing (NLP), and text mining as suggested in the study (Mateu-Sanz et al., 2023). Integrating AI into volumetric additions during assays (Chen et al., 2023) can provide error-free precision by executing coded instructions. The application of artificial neural networks in modeling MTT assay results on epithelial cancer cell lines of human lung (Vand Aa, 2015) highlights AI's potential to enhance accuracy in biocompatibility studies. Addressing challenges in colorimetric data acquisition (Ghasemi et al., 2021), AI interventions show promise in maintaining consistency and reliability in experimental outcomes. Furthermore, the synergy between in vitro assays and AI approaches (Price & Gesquiere, 2019) emphasizes the transformative impact of AI on refining and advancing biocompatibility assessments.

3.2 ROS Assay

Intracellular ROS (Reactive Oxygen Species) generation is a spontaneous process performed even by healthy cells. However, this ROS generation gets aggravated during the cell death phenomenon, especially in apoptosis and necroptosis. During PTT/PDT mediated anti-cancer therapy, selective amelioration of malignant cells is achieved by apoptosis or necrosis due to ROS generation, which in-turn suppresses tumor microvasculature and stimulates the host immune response. To measure the extent of ROS generation and thereby relative cell death, crucial cell-culture-based assays such as DCFHDA assay or SOS G assay are performed. These wet-lab experiments render us information on ROS-mediated damage to the cells after treatment with drugs or nanoparticle-loaded drugs, in terms of fluorescence intensities or fluorescence images (Frijhoff et al., 2015; Murphy et al., 2022; Villalpando-Rodriguez & Gibson, 2021).

3.2.1 Limitations and AI in Overcoming the Limitation of ROS Assay Study

(a) *Addition of dye and subsequent incubation:* Since the assay is fluorescent dye-based, it is imperative that the dye addition and incubation cycle must be performed with minimal or optimal light exposure such that the quenching of the fluorophore is minimal or else the result interpretation may suffer. Human error pertaining to these reasons is difficult to control.

(b) *Strength and exposure of Laser source:* The fluorescence-based assay either spectroscopy or microscopy depends vitally on the laser source and the exposure time. Mostly, during the initial steps while standardizing the field of view such human errors occur which are beyond the control of humane precision.

AI has emerged as a transformative tool for assessing, categorizing, and predicting oxidative stress using supervised machine learning (Pantic et al., 2022). In particular, the fluorescent dye-based nature of these assays requires careful control over the addition of dye and subsequent incubation to minimize light exposure – a challenge effectively addressed by AI (Li et al., 2023). This emphasizes the potential for AI to revolutionize data interpretation. Furthermore, AI's impact extends to the standardization of the field of view and optimization of laser source exposure during fluorescence-based assays (Idowu & Fatokun, 2021). Collectively, these studies showcase the modern and the future potential of AI in overcoming limitations inherent to reactive oxygen species (ROS) assays, offering promising avenues for advancements in experimental precision and data reliability.

3.3 Live/dead Assay

To ascertain the efficacy of the treatment regimen as an efficient anti-viability/anti-proliferative agent, researchers would over perform the fluorescence-microscopy Or FACS-based Live/dead assay. In the same sample set, the live cells are stained using viability stains such as FDA, CFDA-SE, or Acridine Orange, and the dead cells are stained using stains such as PI or EtBr along with DAPI or Hoescht for nuclear staining (P.S. et al., 2021; Hu et al., 2022; Robertson et al., 2019).

3.3.1 Limitations and AI in Overcoming the Limitation of Live/dead Assay Study

During the whole experimental procedure, two steps can limit the scientific outputs where AI can fruitfully intervene and bring about a positive change in erroneous interpretations. They are:

(a) *Cell counting and plating errors:* To achieve uniform scientific findings in microscopy and threshold events number during FACS analysis, desired number of cells/events may not be there due to human pipetting errors.

(b) *Errors during analysis:* During the acquisition of fluorescent microscopy images or FACS readings, the field of view or population of interest is of utmost importance for getting a proper inference that validates scientific findings such as the cellular population that is manifesting a cancer-hallmark. While selecting all the desired cells from a mixed population, some field of view or population of cells may be ignored due to human error.

As per the above limitations the integration of AI is crucial for addressing challenges in Live/Dead assays, cell counting, and field-of-view selection. Portable AI-enhanced microscopy, such as πM (Jiao et al., 2023), utilizes CNN for real-time target detection and a physical magnification of ×5. Additionally, studies by (Reimann et al., 2020) employ RF, Naive Bayes, and NN for automated binary classification of microalgae, showcasing AI's potential in enhancing assessment accuracy. AI extends its application to label-free cell viability bioassays (Park et al., 2022) and Fluorescence Lifetime Imaging Microscopy (FLIM) using ANN for NAD(P)H imaging. This approach successfully distinguishes microglia from other cell types in the brain (Sagar et al., 2020). Furthermore, CeCILE (Cell Classification and In-vitro Lifecycle Evaluation) DL algorithm (Rudigkeit et al., 2021) addresses limitations in radiobiological studies, providing a comprehensive perspective on AI's diverse applications in cell-based assays, overcoming traditional challenges in Live/Dead assays and fluorescence-based methodologies.

3.4 Scratch Assay

A scratch assay is generally performed to analyze wound healing, anti-metastatic, and anti-migratory potential of any drug candidate or vice-versa. Usually, a confluent 2D-cell culture dish is subjected to an *in vitro* scratch using a 200μL tip or a cell-scraper, and kept in standard cell-culture incubation conditions. The dimensions of the scratches are measured periodically starting from day 1, deciphering the migratory and healing propensity of the cells from one end of the scratch to another. The time taken to completely refill the scratched surface or even the progress in that direction, basically infers the wound-healing potential of the drug candidate (Revi et al., 2022) (Stamm et al., 2016) (Liang et al., 2007).

3.4.1 Limitations and AI in Overcoming the Limitation of the Scratch Assay Study

During the entire experimental process, two steps can limit the scientific output. They are:

(a) *Calibration:* While uniform, the calculation of the width of the scratch, the exact width may not be registered using naked-eye mediated microscopic studies.

(b) *Pressure during scratch-wound generation:* The pressure exerted by the micro-tip during scratch creation on the cultured cells, may rupture certain cells and simultaneously put the remaining population under stress. Additional stress is not desired for certain cell lines (such as stem cells) as it may lead to loss of function or their morphology.

Overcoming critical limitations in scratch-wound generation calibration and pressure is crucial for precise scientific outcomes, and AI emerges as a transformative solution. PyScratch, an open-source Python tool, utilizes AI-based image analysis to automate migration area assessment in scratch assays, addressing calibration challenges and ensuring uniform scratch dimensions (Garcia-Fossa et al., 2020). Additionally, AI-guided techniques optimize pressure during scratch-wound generation, minimizing cell rupture and stress-induced effects. The study employs a multilayer feedforward NN (FNN) model (Bahar & Yoon, 2021) for predicting cancerous cell migration characteristics and introduces an intelligent cell detection (iCD) approach (Oldenburg et al., 2021) for precise endothelial growth prediction. Integrating AI in scratch assays shows promise in enhancing precision, reducing experimental variability, and ensuring reliable interpretations of wound healing and migratory potential in response to drug candidates.

3.5 Photothermal Efficiency

The efficiency of a material or system in transforming absorbed light energy into heat is termed as photothermal efficiency. These materials absorb light, specifically belonging to the near-infrared (NIR) spectrum, where many biological tissues are relatively transparent. Afterwards, the material endures electronic transitions. This transition leads to dissipation of energy in the form of heat (thermal energy). This phenomenon is exploited for cancer treatment specifically amongst various other applications, owing to the localized hyperthermia induction property and selective tumor destruction (Deng et al., 2021; Liu et al., 2013; Nam et al., 2018).

3.5.1 Limitations and AI in Overcoming the Limitation of the Photothermal Efficiency Study

Certain challenges associated with PTE which need to be addressed, are:

(a) *Minimizing off-target effects:* Laser irradiation at the target site can be tampered with due to unavoidable human handling errors. Laser irradiation in adjacent wells for a fraction of a second while setting up the experiment, is inadvertent. But it can surely alter the experimental findings. The off-target or spill-over effects can be tackled adequately by AI interference.

(b) *Ensuring specificity of treatment:* The excitation/emission of the material needs to be tested adequately beforehand to analyze the Photothermal Efficiency (PTE) *in vitro* or *in vivo*. AI can help us hasten the process of checking the specificity of the laser towards the material, hence a better treatment specificity can be achieved.

(c) *Temperature analysis:* To ascertain the PTE of a material, an infrared (IR) thermal camera is employed. There is a time lag between the laser irradiation step and temperature acquisition step due to unavoidable human errors, which may render erroneous findings pertaining to the PTE of any material. This delayed acquisition of temperature rise in the material using thermal cameras can be rectified by AI intervention.

In addressing the limitations of Photothermal Efficiency (PTE) studies, AI emerges as a crucial tool to enhance precision and efficacy. Challenges such as minimizing off-target effects during laser irradiation, ensuring treatment specificity, and analyzing temperature rise can be effectively tackled by AI interventions. Efficient prediction of photothermal conversion efficiency (PCE) in organic photothermal agents (PTAs) for cancer photothermal therapy was achieved through a regression model utilizing ML methods. It demonstrated substantial reliability in early screening of PTAs (Wu, S. et al., 2023). The intricate challenge of overcoming high heterogeneity was addressed by integrating AI's pattern analysis and classification algorithms to find precision cancer medicine in tumor profiles. It offered improved accuracy in patient-specific disease profiling and optimization of nanomaterial properties for enhanced therapeutic outcomes (Adir et al., 2020). AI can address the time lag in temperature acquisition by efficiently synchronizing the laser irradiation step with temperature measurements using thermal cameras (Wang & Zhu, 2023).

3.6 Griess Assay

The Griess assay is performed to measure the presence of nitric oxide (NO) or its derivatives in any sample. During various biological processes (physiological and pathophysiological) such as macrophage activation by LPS or IFN-γ, NO is released as a messenger molecule which can be quantified colorimetrically with the help of sulfanilic acid and N-(1-naphthyl) ethylenediamine dihydrochloride. The end product of the reaction is a pink azo dye, and its concentration is in direct proportion to the biological sample's nitrite concentration (Kumar & Kashyap, 2015) (Green et al., 1982) (Amano & Noda, 1995).

3.6.1 Limitations and AI on Overcoming the Limitation of Griess Assay Study

During the experimental process, there are few steps that can limit the scientific outputs, hence the possibility of erroneous findings increases. They are:

(a) *Sample Preparation and Interference:* During sample preparation or reaction steps, there may be certain chemical contaminants (by products) in biological samples which may interfere with the Griess reaction, resulting in false-positive or false-negative outputs. AI, when used, can consider these errors as standard ones and eliminate for better representable formats.
(b) *Specificity and Sensitivity:* The analyte of Griess assay does not directly measure NO however it measures nitrite, which is a stable breakdown product of nitric oxide (NO) thus, the assay findings can be influenced by other nitrogen-containing compounds or even by low levels of nitrite in biological samples. Here, AI intervention can render valuable inputs regarding appropriate sample preparation techniques or reaction conditions or concentration techniques for uniform and error-free experimental findings.
(c) *Variability and Environmental factors:* Griess assay, being a biochemical assay, can be prone to variability due to unavoidable factors such as as pH, temperature, the presence of light, temperature fluctuations, pipetting errors, and differences in reagent quality or stability. The assay's reproducibility, despite the above-mentioned causes which are beyond human control, can be taken aptly by AI intervention. Hence, cost effectiveness and accurate scientific findings will be achieved using AI tools.

In the context of the Griess assay, addressing critical limitations is essential for obtaining reliable scientific outputs, and AI emerges as a key solution. Challenges such as sample preparation interference, where chemical contaminants may lead to false-positive or false-negative outputs, can be effectively mitigated by AI. By applying ML algorithms, AI can identify and account for common errors in sample preparation, improving the reliability of Griess assay results (Kumar & Kashyap, 2015). Moreover, AI intervention can enhance the specificity and sensitivity of the assay by providing valuable insights into optimizing sample preparation techniques, reaction conditions, and concentration methodologies to minimize the impact of interfering compounds on the assay findings.

3.7 Antimicrobial Studies

Antimicrobial studies are performed to ascertain the growth inhibiting (MIC) or killing (MKC) potential of drug candidates against numerous microbes. These experiments can be performed on planktonic (free cells) or biofilms (adhered cells) or during infection model (invasion) development. The experimental inferences are recorded on the basis of either optical density or CFU/mL counts or dye-based assays. In any of these assays, there may be points stemming from human errors, which can serve as limitations (Wijesundara et al., 2022; Balouiri et al., 2016; Guschin et al., 2015).

3.7.1 Limitations and AI on Overcoming the Limitation of Antimicrobial Study

In any of these assays, there may be points stemming from human errors, which can be prevented by AI intervention. These points can be:

(a) *Incubation conditions:* The incubation conditions in any experimental procedure can deviate due to overrun arising from unattended processes or sudden temperature spikes due to instrument malfunctions. These anomalies can be avoided if the process is monitored by AI. AI ensures that the instrument trips automatically on completion of the incubation period. AI also ensures this when the temperature/incubation parameters deviate beyond a certain threshold.
(b) *Sensitivity of instrument:* The optical density of the microbial culture or distinct colony spots or that of the dye used may not fall under the sensitivity range of the instrument. However,

that doesn't necessarily suggest a failure of the experiment; however, we do consider it as is and replicate the experiment again with the appropriate dilutions. To save time and efforts, chemicals and more importantly the experimental findings AI can be employed to improve the research findings.

(c) *Moisture sensitivity:* During electron microscopy-based morphological analysis of the microbial samples, moisture plays an important role. Without properly air-drying when the microbial samples are put inside the electron microscope, the samples get burnt very easily leading to poor quality images and also the instrument may shut down automatically which also leads to further delays/hustles. Here, with help of AI samples may be pre-scanned for the presence/absence of moisture and thereby the quality of images along with the efficiency of the instrument can be maintained.

In the context of antimicrobial studies, overcoming limitations stemming from human errors becomes paramount, and AI stands out as a potent solution to enhance precision and reliability. Challenges associated with incubation conditions, such as deviations in temperature or unattended processes, can be effectively addressed by AI monitoring systems. In (Tran et al., 2022), it is evident that AI can play a pivotal role in preventing issues related to experiment overrun or instrument malfunction during incubation. Moreover, the sensitivity of instruments, particularly in optical density measurements or dye-based assays, can be optimized through AI algorithms, ensuring accurate readings without the need for repetitive experiments. By leveraging AI in antimicrobial studies, researchers can significantly improve experimental efficiency, data quality, and overall research findings, as highlighted in the provided reference.

3.8 Tumour Models

3D in vitro models serve as an alternative to complex live organisms and 2D in vitro cultures. 3D models are made of either simple cancer cell lines or multiple cancer cell lines (Nyga et al., 2011). 3D in vitro models such as spheroids made of spherical cellular units are developed to mimic the tumours. Methods commonly used to develop spheroids are spontaneous aggregation and other methods include ultra-low attachment plates, hanging drop method and magnetic levitation. These spheroid models provide information regarding the features of the tumour, metabolic profile of the cell, cell-cell interactions and drug discovery (Nayak et al., 2023).

3.8.1 Limitations and AI on Overcoming the Limitation of Tumour Models

Irregular sizes of spheroids can be obtained. The procedure is laborious (Białkowska et al., 2020). There is a lack of standard protocols and difficulty in comparison with other models (Katt et al., 2016). Cell viability in patient-derived tumor models (PDMs) lacks specificity. Hence, ML was employed to rapidly screen the drug susceptibility in gold lactate based PDMs sensor (Zhang et al., 2023). AI can contribute by optimizing spheroid size uniformity through advanced image analysis algorithms, streamlining and automating the experimental procedures to reduce labor intensity, and facilitating the establishment of standardized protocols (K. Z. Wu et al., 2023).

4. In-vivo Experiments

Researchers employ various methods to discover treatments for human illnesses and conditions, encompassing animal experiments (in vivo), computerized simulations, and clinical investigations, as well as in vitro experiments. In vivo models effectively emulate the niche interactions within the Gastro-Intestinal-Tract (GIT) that influences drug uptake and absorption. Animal models can be tailored and categorized based on research requirements into: (a) homologous (mirroring humans), (b) isomorphic (reflecting human disorders), and (c) predictive (enabling human disease and treatment forecasting). Numerous GIT physiological aspects impact oral absorption, including surface area, intestinal transporters, pH levels, retention duration, gastrointestinal fluids, enzyme activity, and gut wall metabolism. Common animal species utilized for studying drug absorption include mice, rats, pigs, and dogs (Shrestha & Preat, 2020; Harloff-Helleberg et al., 2017).

4.1 Limitations and Role of AI in Overcoming the Limitations for in-vivo Experiments

Each of these approaches has its strengths and limitations. While animal models are generally more dependable than in vitro assays, they can still present minor variations in biokinetic parameters essential for drug formulation. There is no single animal species displaying gastrointestinal physiology identical to that of humans, which increases the risk of poor extrapolation of results in human outcomes. Hence, choosing an appropriate animal model requires careful consideration. Additionally, when selecting an animal model species, considerations like age, gender, and health condition are also vital. Moreover, obtaining approval from the relevant ethical committees is mandatory for conducting animal experiments (Morgan et al., 2013; Ghasemi & Dehpour, 2009).

Drug-induced folding is a major challenge that is traceable only by computational techniques. A sequence-to-sequence (Seq2Seq) network architecture is developed that allows language processing and an elementary transformation of the message presenting a molecular change taking place on a fixed time step (Fernandez, A., 2021). AI finds application in drug discovery by designing ligands for proteins that enables easier synthesis and also helps in discovering ligands (Bender, A., 2021).

AI is bound to reshape the landscape of biology by revolutionizing the way we analyze and decipher experimental data for the betterment of healthcare and therapeutic strategies. AI renders better insights into the complex cobweb of biological systems (generated through genomics, proteomics, and other omics technologies) owing to its capacity to handle vast datasets. The algorithms can differentiate between the intricate hallmarks and establish relationships within the datasets. Thereby, fast-tracking the screening of potential drug candidates, understanding complex

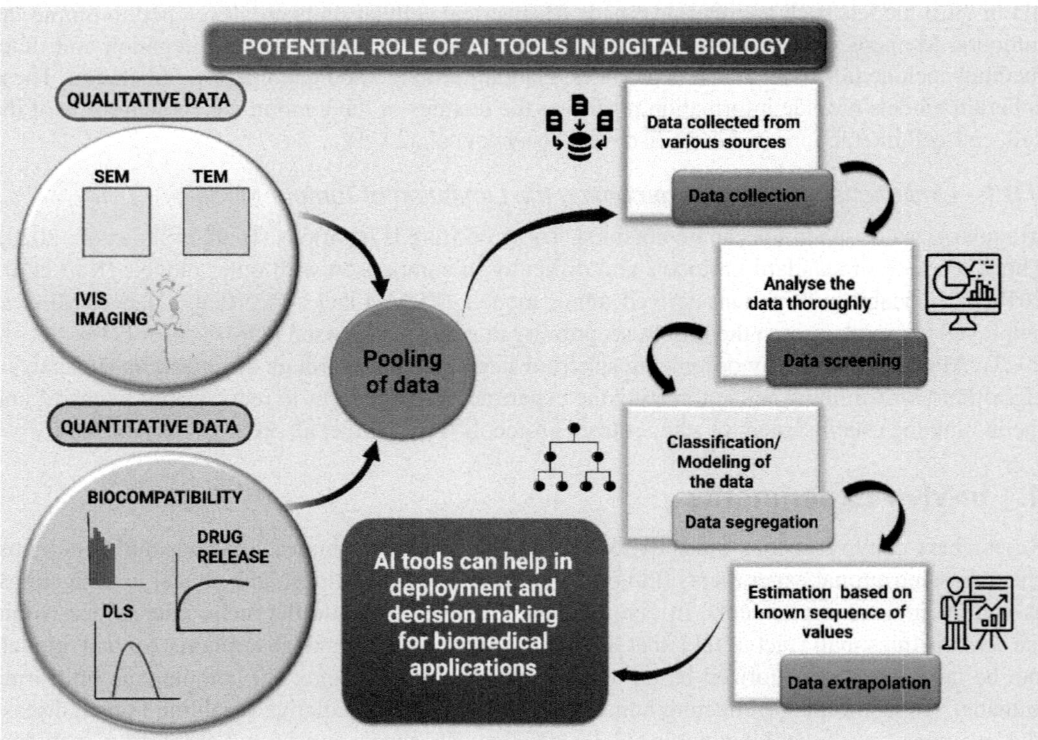

Fig. 2 Potential role of AI tools in digital biology

biological networks, and facilitating personalized medicine by categorizing patient populations based on patterns that are specific to each individual. Furthermore, AI-driven image analysis and natural language processing endows the researchers to extract crucial information from biological images and scientific literature, respectively, augmenting a more wide-ranging understanding of various biological phenomena. With the advancements in AI, its integration with digital biology as shown in **Figure 2**, not only boosts our capability to decipher the complexities of living systems but also envisages transformative breakthroughs in drug discovery, personalized treatment strategies, the way scientists decipher the scientific breakthroughs after rigorous experiments and most importantly healthcare.

Conclusion

The integration of AI and biological sciences has the potential to transform our know-how of living systems and improve healthcare outcomes. There are several ways in which AI can contribute to the field of biology, such as: (a) Large-scale Omics Data analysis and multi-modal integration; b) Virtual Screening and predictive modelling for better drug discovery and development approaches; (c) Patient segregation and predictive analysis for better as well as precise personalized medicine approaches; (d) For modelling complex biological pathways and networks using systems biology; (e) Segregation, feature extraction and computer-aided biological image analysis for better disease diagnosis and research; (f) Facilitating comprehensive data exploration and literature reports extraction through NLP; (g) High-throughput screening and lab workflow automation to enhance overall efficiency and accuracy of experiments by including automation and robotics; and (h) For ensuring the privacy protection and secure handling of large scale biological/healthcare data.

References

Abd Mutalib, M., Rahman, M.A., Othman, M.H.D., Ismail, A.F. and Jaafar, J. et al. (2017). Scanning Electron Microscopy (SEM) and Energy-Dispersive X-Ray (EDX) Spectroscopy In *Membrane Characterization* (pp. 161–179). Elsevier. https://doi.org/10.1016/B978-0-444

Abdullah, A. and Mohammed, A. (2019). *Scanning Electron Microscopy (SEM): A Review*.

Adir, O., Poley, M., Chen, G., Froim, S., Krinsky, N. et al. (2020). Integrating Artificial Intelligence and Nanotechnology for Precision Cancer Medicine. *Advanced Materials (Deerfield Beach, Fla.)*, *32*(13), e1901989. https://doi.org/10.1002/adma.201901989.

Ahmad, I. and Benjamin, T.-A. (2023). Application of artificial intelligence and machine learning to food rheology. In *Advances in Food Rheology and Its Applications* (pp. 201–219). Elsevier. https://doi.org/10.1016/B978-0-12-823983-4.00004-2.

Aitken, A. and Learmonth, M.P. (n.d.). *Protein Determination by UV Absorption*.

Alves, E.M., Rodrigues, R.J., Dos Santos Corrêa, C., Fidemann, T., Rocha, J.C. et al. (2018). Use of ultraviolet–visible spectrophotometry associated with artificial neural networks as an alternative for determining the water quality index. *Environmental Monitoring and Assessment*, *190*(6), 319. https://doi.org/10.1007/s10661-018-6702-7.

Amano, F. and Noda, T. (1995). Improved detection of nitric oxide radical (NO.) production in an activated macrophage culture with a radical scavenger, carboxy PTIO and Griess reagent. *FEBS Letters*, *368*(3), 425–428. https://doi.org/10.1016/0014-5793(95)00700-j.

Anderson, J.M. (2012). 9.19—Biocompatibility. In K. Matyjaszewski and M. Möller (Eds.), *Polymer Science: A Comprehensive Reference* (pp. 363–383). Elsevier. https://doi.org/10.1016/B978-0-444-53349-4.00229-6.

Armstrong, M.R., Boyden, K., Browning, N.D., Campbell, G.H., Colvin, J.D. et al.. (2007). Practical considerations for high spatial and temporal resolution dynamic transmission electron microscopy. *Ultramicroscopy*, *107* (4–5), 356–367. https://doi.org/10.1016/j.ultramic.2006.09.005.

Ayache, J., Beaunier, L., Boumendil, J., Ehret, G. and Laub, D. et al. (2010). *Sample Preparation Handbook for Transmission Electron Microscopy*. Springer New York. https://doi.org/10.1007/978-1-4419-5975-1.

Babick, F. (2020a). Dynamic light scattering (DLS). In V.-D. Hodoroaba, W.E. S. Unger, & A. G. Shard (Eds.), *Characterization of Nanoparticles* (pp. 137–172). Elsevier. https://doi.org/10.1016/B978-0-12-814182-3.00010-9

Babick, F. (2020b). Dynamic light scattering (DLS). In *Characterization of Nanoparticles* (pp. 137–172). Elsevier. https://doi.org/10.1016/B978-0-12-814182-3.00010-9.

Bahar, E. and Yoon, H. (2021). Modeling and Predicting the Cell Migration Properties from Scratch Wound Healing Assay on Cisplatin-Resistant Ovarian Cancer Cell Lines Using Artificial Neural Network. *Healthcare*, *9*(7), Article 7. https://doi.org/10.3390/healthcare9070911.

Bakrania, A., Joshi, N., Zhao, X., Zheng, G. and Bhat, M. et al. (2023). Artificial intelligence in liver cancers: Decoding the impact of machine learning models in clinical diagnosis of primary liver cancers and liver cancer metastases. *Pharmacological Research*, *189*, 106706. https://doi.org/10.1016/j.phrs.2023.106706.

Balouiri, M., Sadiki, M. and Ibnsouda, S.K. (2016). Methods for in vitro evaluating antimicrobial activity: A review. *Journal of Pharmaceutical Analysis*, *6*(2), 71–79. https://doi.org/10.1016/j.jpha.2015.11.005.

Bendersky, L.A. and Gayle, F.W. (2001). Electron diffraction using transmission electron microscopy. *Journal of Research of the National Institute of Standards and Technology*, *106*(6), 997. https://doi.org/10.6028/jres.106.051.

Bender, A. and Cortés-Ciriano, I. (2021). Artificial intelligence in drug discovery: what is realistic, what are illusions? Part 1: Ways to make an impact, and why we are not there yet. Drug discovery today, 26(2), 511–524.

Berthomieu, C. and Hienerwadel, R. (2009). Fourier transform infrared (FTIR) spectroscopy. *Photosynthesis Research*, *101*(2–3), 157–170. https://doi.org/10.1007/s11120-009-9439-x.

Bhattacharjee, S. (2016). DLS and zeta potential – What they are and what they are not? *Journal of Controlled Release*, *235*, 337–351. https://doi.org/10.1016/j.jconrel.2016.06.017.

Bi, L., Liang, K., Czap, G., Wang, H., Yang, K. et al. (2023). Recent progress in probing atomic and molecular quantum coherence with scanning tunneling microscopy. *Progress in Surface Science*, *98*(1), 100696. https://doi.org/10.1016/j.progsurf.2022.100696.

Białkowska, K., Komorowski, P., Bryszewska, M. and Miłowska, K. (2020). Spheroids as a Type of Three-Dimensional Cell Cultures—Examples of Methods of Preparation and the Most Important Application. *International Journal of Molecular Sciences*, *21*(17), 6225. https://doi.org/10.3390/ijms21176225.

Bode, M. (2003). Spin-polarized scanning tunnelling microscopy. *Reports on Progress in Physics*, *66*(4), 523–582. https://doi.org/10.1088/0034-4885/66/4/203.

Bootz, A., Vogel, V., Schubert, D. and Kreuter, J. (2004). Comparison of scanning electron microscopy, dynamic light scattering and analytical ultracentrifugation for the sizing of poly(butyl cyanoacrylate) nanoparticles. *European Journal of Pharmaceutics and Biopharmaceutics*, *57*(2), 369–375. https://doi.org/10.1016/S0939-6411(03)00193-0.

Bunaciu, A.A., Udriştioiu, E.G. and Aboul-Enein, H.Y. (2015). X-Ray Diffraction: Instrumentation and Applications. *Critical Reviews in Analytical Chemistry*, *45*(4), 289–299. https://doi.org/10.1080/10408347.2014.949616.

Chandran C.S., Thomas, S. and Chandran, N. (2019). *Introduction to Rheology* (pp. 1–12). https://doi.org/10.1016/B978-0-12-816957-5.00001-X.

Chauhan, A. (2014). Powder XRD Technique and its Applications in Science and Technology. *Journal of Analytical & Bioanalytical Techniques*, *5*(6). https://doi.org/10.4172/2155-9872.1000212.

Chen, M., Cui, D., Haick, H. and Tang, N. (2023). Artificial Intelligence-Based Medical Sensors for Healthcare System. *Advanced Sensor Research*, 2300009. https://doi.org/10.1002/adsr.202300009.

Chicea, D., Bogdan, N., Chicea, R. and Chicea, L. (2010). The application of AFM for biological samples imaging. *Digest Journal of Nanomaterials and Biostructures*, *5*.

Chicea, D., Doroshkevich, A.S. and Lyubchyk, A. (2022). On the Possibility of Designing an Advanced Sensor with Particle Sizing Using Dynamic Light Scattering Time Series Spectral Entropy and Artificial Neural Network. *Sensors*, *22*(10), Article 10. https://doi.org/10.3390/s22103871.

Choudhary, K., Garrity, K.F., Camp, C., Kalinin, S.V., Vasudevan, R. et al. (n.d.). *Density Functional Theory and Deep-learning to Accelerate Data Analytics in Scanning Tunneling Microscopy*.

Christy, P.N., Basha, S.K., Kumari, V.S., Bashir, A.K.H., Maaza, M. et al. (2020). Biopolymeric nanocomposite scaffolds for bone tissue engineering applications – A review. *Journal of Drug Delivery Science and Technology*, *55*, 101452. https://doi.org/10.1016/j.jddst.2019.101452.

Chutipongtanate, S., Watcharatanyatip, K., Homvises, T., Jaturongkakul, K. and Thongboonkerd, V. et al. (2012). Systematic comparisons of various spectrophotometric and methods to measure concentrations of protein, peptide and amino acid: Detectable limits, linear dynamic ranges, interferences, practicality and unit costs. *Talanta*, *98*, 123–129. https://doi.org/10.1016/j.talanta.2012.06.058.

Cocker, T.L., Jelic, V., Gupta, M., Molesky, S.J., Burgess, J.A.J. et al. (2013). An ultrafast terahertz scanning tunnelling microscope. *Nature Photonics*, *7*(8), 620–625. https://doi.org/10.1038/nphoton.2013.151.

Da Fonseca, M.H., Kovaleski, F., Picinin, C.T., Pedroso, B. and Rubbo, P. et al. (2021). E-Health Practices and Technologies: A Systematic Review from 2014 to 2019. *Healthcare*, *9*(9), 1192. https://doi.org/10.3390/healthcare9091192

De, R.R.L., Albuquerque, D.A.C., Cruz, T.G.S., Yamaji, F.M. and Leite, F.L. et al. (2012). Measurement of the Nanoscale Roughness by Atomic Force Microscopy: Basic Principles and Applications. In V. Bellitto (Ed.),

Atomic Force Microscopy—Imaging, Measuring and Manipulating Surfaces at the Atomic Scale. InTech. https://doi.org/10.5772/37583

Delgado, J.M. (2017). The contributions of Albert W. Hull to X-ray powder diffraction at one hundred years of his landmark publication. *Powder Diffraction*, *32*(1), 2–9. https://doi.org/10.1017/S0885715616000750

Demirkan, A. and Topcu, İ. (2022). Analysis of SEM Images with Artificial Intelligence Methods. *European Journal of Science and Technology*. https://doi.org/10.31590/ejosat.1219252

Deng, X., Shao, Z. and Zhao, Y. (2021). Solutions to the Drawbacks of Photothermal and Photodynamic Cancer Therapy. *Advanced Science (Weinheim, Baden-Wurttemberg, Germany)*, *8*(3), 2002504. https://doi.org/10.1002/advs.202002504

Donate, R., Monzón, M., Alemán-Domínguez, M.E. and Ortega, Z. (2020). Enzymatic degradation study of PLA-based composite scaffolds. *REVIEWS ON ADVANCED MATERIALS SCIENCE*, *59*(1), 170–175. https://doi.org/10.1515/rams-2020-0005

Dorfs, D., Krahne, R., Falqui, A., Manna, L., Giannini, C. et al. (2011). Quantum Dots: Synthesis and Characterization. In *Comprehensive Nanoscience and Nanotechnology* (pp. 17–60). Elsevier. https://doi.org/10.1016/B978-0-12-812295-2.00028-3

Entekhabi, E., Haghbin Nazarpak, M., Sedighi, M. and Kazemzadeh, A. (2020). Predicting degradation rate of genipin cross-linked gelatin scaffolds with machine learning. *Materials Science and Engineering: C*, *107*, 110362. https://doi.org/10.1016/j.msec.2019.110362

Ewoldt, R.H., Johnston, M.T. and Caretta, L.M. (2015). Experimental Challenges of Shear Rheology: How to Avoid Bad Data. In S. E. Spagnolie (Ed.), *Complex Fluids in Biological Systems* (pp. 207–241). Springer New York. https://doi.org/10.1007/978-1-4939-2065-5_6

Fernanda, P. (2018). Methods Used in the Study of the Physical Properties of Fats. In *Structure-Function Analysis of Edible Fats* (pp. 313–385). Elsevier. https://doi.org/10.1016/B978-0-12-814041-3.00011-3

Fernandez, A. (2021). Artificial intelligence deconstructs drug targeting in vivo by leveraging a transformer platform. ACS Medicinal Chemistry Letters, 12(7), 1052–1055.

Fogelan, S., Blumenstein, M. and Zhao, H. (2006). Estimation of chemical oxygen demand by ultraviolet spectroscopic profiling and artificial neural networks. *Neural Computing and Applications*, *15*(3–4), 197–203. https://doi.org/10.1007/s00521-005-0015-9

Frijhoff, J., Winyard, P.G., Zarkovic, N., Davies, S.S., Stocker, R. et al. (2015). Clinical Relevance of Biomarkers of Oxidative Stress. *Antioxidants & Redox Signaling*, *23*(14), 1144–1170. https://doi.org/10.1089/ars.2015.6317

Garcia-Fossa, F., Gaal, V. and de Jesus, M.B. (2020). PyScratch: An ease of use tool for analysis of scratch assays. *Computer Methods and Programs in Biomedicine*, *193*, 105476. https://doi.org/10.1016/j.cmpb.2020.105476

Gao, Z., Ma, W., Huang, S., Hua, P. and Lan, C. et al. (2020). Deep Learning for Super-Resolution in a Field Emission Scanning Electron Microscope. *AI*, *1*(1), Article 1. https://doi.org/10.3390/ai1010001.

Ghasemi, M. and Dehpour, A.R. (2009). Ethical considerations in animal studies. *Journal of Medical Ethics and History of Medicine*, *2*, 12.

Ghasemi, M., Turnbull, T., Sebastian, S. and Kempson, I. (2021). The MTT Assay: Utility, Limitations, Pitfalls, and Interpretation in Bulk and Single-Cell Analysis. *International Journal of Molecular Sciences*, *22*(23), 12827. https://doi.org/10.3390/ijms222312827.

Green, L.C., Wagner, D.A., Glogowski, J., Skipper, P.L., Wishnok, J.S et al. (1982). Analysis of nitrate, nitrite, and [15N]nitrate in biological fluids. *Analytical Biochemistry*, *126*(1), 131–138. https://doi.org/10.1016/0003-2697(82)90118-x.

Guschin, A., Ryzhikh, P., Rumyantseva, T., Gomberg, M. and Unemo, M. et al. (2015). Treatment efficacy, treatment failures and selection of macrolide resistance in patients with high load of Mycoplasma genitalium during treatment of male urethritis with josamycin. *BMC Infectious Diseases*, *15*(1), 40. https://doi.org/10.1186/s12879-015-0781-7.

Hansford, G.M. (2013). X-ray diffraction without sample preparation: Proof-of-principle experiments. *Nuclear Instruments and Methods in Physics Research Section A: Accelerators, Spectrometers, Detectors and Associated Equipment*, *728*, 102–106. https://doi.org/10.1016/j.nima.2013.06.065.

Harloff-Helleberg, S., Nielsen, L.H. and Nielsen, H.M. (2017). Animal models for evaluation of oral delivery of biopharmaceuticals. *Journal of Controlled Release: Official Journal of the Controlled Release Society*, *268*, 57–71. https://doi.org/10.1016/j.jconrel.2017.09.025.

Hassoun, S., Jefferson, F., Shi, X., Stucky, B., Wang, J. et al. (2022). Artificial Intelligence for Biology. *Integrative and Comparative Biology*, *61*(6), 2267–2275. https://doi.org/10.1093/icb/icab188.

Honary, S. and Zahir, F. (2013). Effect of Zeta Potential on the Properties of Nano-Drug Delivery Systems—A Review (Part 2). *Tropical Journal of Pharmaceutical Research*, *12*(2), 265–273. https://doi.org/10.4314/tjpr.v12i2.20.

Hu, C., He, S., Lee, Y.J., He, Y., Kong, E.M. et al. (2022). Live-dead assay on unlabeled cells using phase imaging with computational specificity. *Nature Communications*, *13*(1), 713. https://doi.org/10.1038/s41467-022-28214-x.

Idowu, S.O. and Fatokun, A.A. (2021). Artificial Intelligence (AI) to the Rescue: Deploying Machine Learning to Bridge the Biorelevance Gap in Antioxidant Assays. *SLAS Technology*, *26*(1), 16–25. https://doi.org/10.1177/2472630320962716.

Iida, Y., Watanabe, K., Ominami, Y., Toyoguchi, T., Murayama, T. et al. (2021). Development of rapid and highly accurate method to measure concentration of fibers in atmosphere using artificial intelligence and scanning electron microscopy. *Journal of Occupational Health*, *63*(1), e12238. https://doi.org/10.1002/1348-9585.12238.

İlhan, M., Gültekin, H.E., Rençber, S., Şenyiğit, Z. and Aydın, H.H. et al. (2022). Aquasomes: A novel platform for drug delivery. In *Systems of Nanovesicular Drug Delivery* (pp. 191–206). Elsevier. https://doi.org/10.1016/B978-0-323-91864-0.00020-6.

Islamkulov, M., Karakuş, S. and Özeroğlu, C. (2023). Design artificial intelligence-based optimization and swelling behavior of novel crosslinked polymeric network hydrogels based on acrylamide-2-hydroxyethyl methacrylate and acrylamide-N-isopropylacrylamide. *Colloid and Polymer Science*, *301*(3), 259–272. https://doi.org/10.1007/s00396-023-05064-7.

Jebrail, M.J. and Wheeler, A.R. (2010). Let's get digital: Digitizing chemical biology with microfluidics. *Current Opinion in Chemical Biology*, *14*(5), 574–581. https://doi.org/10.1016/j.cbpa.2010.06.187.

Jiao, Z., Zang, Z., Wang, Q., Chen, Y., Xiao, D. et al. (2023). PAIM (πM): Portable AI-enhanced fluorescence microscope for real-time target detection. *Optics & Laser Technology*, *163*, 109356. https://doi.org/10.1016/j.optlastec.2023.109356.

Junoh, H., Jaafar, J., Nordin, N.A.H. Md., Ismail, A.F., Othman, M.H.D. et al. (2020). Synthetic polymer-based membranes for direct methanol fuel cell (DMFC) applications. In *Synthetic Polymeric Membranes for Advanced Water Treatment, Gas Separation, and Energy Sustainability* (pp. 337–363). Elsevier. https://doi.org/10.1016/B978-0-12-818485-1.00015-0.

Kalinin, S.V., Lupini, A.R., Dyck, O., Jesse, S., Ziatdinov, M. et al. (2019). Lab on a beam—Big data and artificial intelligence in scanning transmission electron microscopy. *MRS Bulletin*, *44*(7), 565–575. https://doi.org/10.1557/mrs.2019.159.

Karr, J.R., Sanghvi, J.C., Macklin, D.N., Gutschow, M.V., Jacobs, J.M. et al. (2012). A Whole-Cell Computational Model Predicts Phenotype from Genotype. *Cell*, *150*(2), 389–401. https://doi.org/10.1016/j.cell.2012.05.044.

Katt, M.E., Placone, A.L., Wong, A.D., Xu, Z.S. and Searson, P.C et al. (2016). In Vitro Tumor Models: Advantages, Disadvantages, Variables, and Selecting the Right Platform. *Frontiers in Bioengineering and Biotechnology*, *4*. https://doi.org/10.3389/fbioe.2016.00012.

Khan, H., Yerramilli, A.S., D'Oliveira, A., Alford, T.L., Boffito, D.C. et al.. (2020). Experimental methods in chemical engineering: X-ray diffraction spectroscopy—XRD. *The Canadian Journal of Chemical Engineering*, *98*(6), 1255–1266. https://doi.org/10.1002/cjce.23747.

Kim, J.J., Ling, F.T., Plattenberger, D.A., Clarens, A.F., Lanzirotti, A. et al. (2021). SMART mineral mapping: Synchrotron-based machine learning approach for 2D characterization with coupled micro XRF-XRD. *Computers and Geosciences*, *156*, 104898. https://doi.org/10.1016/j.cageo.2021.104898.

Kumar, S. and Kashyap, P. (2015). Antiproliferative activity and nitric oxide production of a methanolic extract of <i> Fraxinus micrantha</i> on Michigan Cancer Foundation-7 mammalian breast carcinoma cell line. *Journal of Intercultural Ethnopharmacology*, *4*(2), 109. https://doi.org/10.5455/jice.20150129102013.

L.C. Passos, M. and M.F.S. Saraiva, M.L. (2019). Detection in UV-visible spectrophotometry: Detectors, detection systems, and detection strategies. *Measurement*, *135*, 896–904. https://doi.org/10.1016/j.measurement.2018.12.045.

Li, J., Liang, Y., Zhao, X. and Wu, C. (2023). Integrating machine learning algorithms to systematically assess reactive oxygen species levels to aid prognosis and novel treatments for triple-negative breast cancer patients. *Frontiers in Immunology*, *14*, 1196054. https://doi.org/10.3389/fimmu.2023.1196054.

Lee, J.-W., Park, W.B., Lee, J.H., Singh, S.P. and Sohn, K.-S et al. (2020). A deep-learning technique for phase identification in multiphase inorganic compounds using synthetic XRD powder patterns. *Nature Communications*, *11*(1), 86. https://doi.org/10.1038/s41467-019-13749-3.

Lewczuk, B. and Szyryńska, N. (2021). Field-Emission Scanning Electron Microscope as a Tool for Large-Area and Large-Volume Ultrastructural Studies. *Animals*, *11*(12), 3390. https://doi.org/10.3390/ani11123390.

Li, C., Wang, D. and Kong, L. (2021). Application of Machine Learning Techniques in Mineral Classification for Scanning Electron Microscopy—Energy Dispersive X-Ray Spectroscopy (SEM-EDS) Images. *Journal of Petroleum Science and Engineering*, *200*, 108178. https://doi.org/10.1016/j.petrol.2020.108178.

Li, J. and Mooney, D.J. (2016). Designing hydrogels for controlled drug delivery. *Nature Reviews Materials*, *1*(12), 16071. https://doi.org/10.1038/natrevmats.2016.71.

Liang, C.-C., Park, A.Y. and Guan, J.-L. (2007). In vitro scratch assay: A convenient and inexpensive method for analysis of cell migration in vitro. *Nature Protocols*, *2*(2), 329–333. https://doi.org/10.1038/nprot.2007.30.

Lin, P.-C., Lin, S., Wang, P.C. and Sridhar, R. (2014). Techniques for physicochemical characterization of nanomaterials. *Biotechnology Advances*, *32*(4), 711–726. https://doi.org/10.1016/j.biotechadv.2013.11.006.

Liu, Y., Ai, K., Liu, J., Deng, M., He, Y. et al. (2013). Dopamine-melanin colloidal nanospheres: An efficient near-infrared photothermal therapeutic agent for in vivo cancer therapy. *Advanced Materials (Deerfield Beach, Fla.)*, *25*(9), 1353–1359. https://doi.org/10.1002/adma.201204683.

Mateu-Sanz, M., Fuenteslópez, C.V., Uribe-Gomez, J., Haugen, H.J., Pandit, A. et al., (2023). Redefining biomaterial biocompatibility: Challenges for artificial intelligence and text mining. *Trends in Biotechnology*, S0167779923002895. https://doi.org/10.1016/j.tibtech.2023.09.015.

Mayeen, A., Shaji, L.K., Nair, A.K. and Kalarikkal, N. (2018). Morphological Characterization of Nanomaterials. In *Characterization of Nanomaterials* (pp. 335–364). Elsevier. https://doi.org/10.1016/B978-0-08-101973-3.00012-2.

McBath, R.A. and Shipp, D.A. (2010). Swelling and degradation of hydrogels synthesized with degradable poly(β-amino ester) crosslinkers. *Polymer Chemistry*, *1*(6), 860. https://doi.org/10.1039/c0py00074d.

Messick, J. (2012). The history of the International Centre for Diffraction Data. *Powder Diffraction*, *27*(1), 36–44. https://doi.org/10.1017/S0885715612000085.

Mohamed, M.A., Jaafar, J., Ismail, A.F., Othman, M.H.D. and Rahman, M.A. et al. (2017). Fourier Transform Infrared (FTIR) Spectroscopy. In *Membrane Characterization* (pp. 3–29). Elsevier. https://doi.org/10.1016/B978-0-444-63776-5.00001-2.

Mohammadi-Jam, S., Waters, K.E. and Greenwood, R.W. (2022). A review of zeta potential measurements using electroacoustics. *Advances in Colloid and Interface Science*, *309*, 102778. https://doi.org/10.1016/j.cis.2022.102778.

Morgan, S.J., Elangbam, C.S., Berens, S., Janovitz, E., Vitsky, A. et al. (2013). Use of Animal Models of Human Disease for Nonclinical Safety Assessment of Novel Pharmaceuticals. *Toxicologic Pathology*, *41*(3), 508–518. https://doi.org/10.1177/0192623312457273.

Movasaghi, Z., Rehman, S. and Ur Rehman, Dr. I. (2008). Fourier Transform Infrared (FTIR) Spectroscopy of Biological Tissues. *Applied Spectroscopy Reviews*, *43*(2), 134–179. https://doi.org/10.1080/05704920701829043.

Muneer, R., Hashmet, M.R., Pourafshary, P. and Shakeel, M. (2023). Unlocking the Power of Artificial Intelligence: Accurate Zeta Potential Prediction Using Machine Learning. *Nanomaterials*, *13*(7), Article 7. https://doi.org/10.3390/nano13071209.

Murphy, M.P., Bayir, H., Belousov, V., Chang, C.J., Davies, K.J.A. et al. (2022). Guidelines for measuring reactive oxygen species and oxidative damage in cells and in vivo. *Nature Metabolism*, *4*(6), 651–662. https://doi.org/10.1038/s42255-022-00591-z.

Na, J., Kim, G., Kang, S.-H., Kim, S.-J. and Lee, S. et al. (2021). Deep learning-based discriminative refocusing of scanning electron microscopy images for materials science. *Acta Materialia*, *214*, 116987. https://doi.org/10.1016/j.actamat.2021.116987.

Nadernezhad, A. and Groll, J. (2022). Machine Learning Reveals a General Understanding of Printability in Formulations Based on Rheology Additives. *Advanced Science*, *9*(29), 2202638. https://doi.org/10.1002/advs.202202638.

Nam, J., Son, S., Ochyl, L.J., Kuai, R., Schwendeman, A. et al. (2018). Chemo photothermal therapy combination elicits anti-tumor immunity against advanced metastatic cancer. *Nature Communications*, *9*(1), 1074. https://doi.org/10.1038/s41467-018-03473-9.

Nasrazadani, S. and Hassani, S. (2016). Modern analytical techniques in failure analysis of aerospace, chemical, and oil and gas industries. In *Handbook of Materials Failure Analysis with Case Studies from the Oil and Gas Industry* (pp. 39–54). Elsevier. https://doi.org/10.1016/B978-0-08-100117-2.00010-8.

Nayak, P., Bentivoglio, V., Varani, M. and Signore, A. (2023). Three-Dimensional In Vitro Tumor Spheroid Models for Evaluation of Anticancer Therapy: Recent Updates. *Cancers*, *15*(19), 4846. https://doi.org/10.3390/cancers15194846.

Negoita, C., Praisler, M. and Ion, A. (2019). *Artificial intelligence application designed to screen for new psychoactive drugs based on their ATR-FTIR spectra*. 170026. https://doi.org/10.1063/1.5091391.

Nyga, A., Cheema, U. and Loizidou, M. (2011). 3D tumour models: Novel in vitro approaches to cancer studies. *Journal of Cell Communication and Signaling*, *5*(3), 239–248. https://doi.org/10.1007/s12079-011-0132-4.

Oldenburg, J., Maletzki, L., Strohbach, A., Bellé, P., Siewert, S. et al. (2021). Methodology for comprehensive cell-level analysis of wound healing experiments using deep learning in MATLAB. *BMC Molecular and Cell Biology*, *22*(1), 32. https://doi.org/10.1186/s12860-021-00369-3.

Olson, E.W. (2012). *ZETA POTENTIAL AND COLLOID CHEMISTRY*. https://www.semanticscholar.org/paper/ZETA-POTENTIAL-AND-COLLOID-CHEMISTRY-Olson/6efaeb525f07c330f1428974946f00d455f40d53.

P S, R., Alvi, S.B., Begum, N., Veeresh, B. and Rengan, A.K. (2021). Self-Assembled Fluorosome-Polydopamine Complex for Efficient Tumor Targeting and Commingled Photodynamic/Photothermal Therapy of Triple-Negative Breast Cancer. *Biomacromolecules*, *22*(9), 3926–3940. https://doi.org/10.1021/acs.biomac.1c00744.

Pantic, I., Paunovic, J., Pejic, S., Drakulic, D., Todorovic, A. et al. (2022). Artificial intelligence approaches to the biochemistry of oxidative stress: Current state of the art. *Chemico-Biological Interactions*, *358*, 109888. https://doi.org/10.1016/j.cbi.2022.109888.

Park, S. and Barrett, R.C. (1993). 2. Design Considerations for an STM System. In J.A. Stroscio and W.J. Kaiser (Eds.), *Methods in Experimental Physics* (Vol. 27, pp. 31–76). Academic Press. https://doi.org/10.1016/S0076-695X(08)60007-1.

Park, S., Veluvolu, V., Martin, W.S., Nguyen, T., Park, J. et al. (2022). Label-free, non-invasive, and repeatable cell viability bioassay using dynamic full-field optical coherence microscopy and supervised machine learning. *Biomedical Optics Express*, *13*(6), 3187–3194. https://doi.org/10.1364/BOE.452471.

Preobrazhenskiy, I.I. and Putlyaev, V.I. (2023). The ability to control swelling and degradation processes of hydrogels based on a mixture of PEGMA/PEGDA monomers. *Mendeleev Communications*, *33*(1), 83–85. https://doi.org/10.1016/j.mencom.2023.01.026.

Price, E. and Gesquiere, A.J. (2019). An in vitro assay and artificial intelligence approach to determine rate constants of nanomaterial-cell interactions. Scientific reports, 9(1), 13943.

P S, R., Alvi, S.B., Begum, N., Veeresh, B. and Rengan, A.K. (2021). Self-Assembled Fluorosome-Polydopamine Complex for Efficient Tumor Targeting and Commingled Photodynamic/Photothermal Therapy of Triple-Negative Breast Cancer. Biomacromolecules, 22(9), 3926–3940. https://doi.org/10.1021/acs.biomac.1c00744

Raja, P.B., Munusamy, K.R., Perumal, V. and Ibrahim, M.N.M. (2022). Characterization of nanomaterial used in nanobioremediation. In *Nano-Bioremediation: Fundamentals and Applications* (pp. 57–83). Elsevier. https://doi.org/10.1016/B978-0-12-823962-9.00037-4.

Revi, N., Sankaranarayanan, S.A. and Rengan, A.K. (2022). A study on the role of eugenol encapsulated liposomes in facilitating neuron-microglia mediated wound recovery. *Materialia*, *23*, 101454. https://doi.org/10.1016/j.mtla.2022.101454.

Robertson, J., McGoverin, C., Vanholsbeeck, F. and Swift, S. (2019). Optimisation of the Protocol for the LIVE/DEAD® BacLightTM Bacterial Viability Kit for Rapid Determination of Bacterial Load. *Frontiers in Microbiology*, *10*, 801. https://doi.org/10.3389/fmicb.2019.00801.

Reimann, R., Zeng, B., Jakopec, M., Burdukiewicz, M., Petrick, I. et al. (2020). Classification of dead and living microalgae Chlorella vulgaris by bioimage informatics and machine learning. *Algal Research*, *48*, 101908. https://doi.org/10.1016/j.algal.2020.101908.

Rudigkeit, S., Reindl, J.B., Matejka, N., Ramson, R., Sammer, M. et al. (2021). CeCILE - An Artificial Intelligence Based Cell-Detection for the Evaluation of Radiation Effects in Eucaryotic Cells. *Frontiers in Oncology*, *11*, 688333. https://doi.org/10.3389/fonc.2021.688333.

Sagar, M.A.K., Cheng, K.P., Ouellette, J.N., Williams, J.C., Watters, J.J. et al. (2020). Machine Learning Methods for Fluorescence Lifetime Imaging (FLIM) Based Label-Free Detection of Microglia. *Frontiers in Neuroscience*, *14*, 931. https://doi.org/10.3389/fnins.2020.00931.

Sahin, F., Genc, O., Gökcek, M. and Çolak, A.B. (2023). An experimental and new study on thermal conductivity and zeta potential of Fe3O4/water nanofluid: Machine learning modeling and proposing a new correlation. *Powder Technology*, *420*, 118388. https://doi.org/10.1016/j.powtec.2023.118388.

Sandhu, R., Singh, N., Dhankhar, J., Gandhi, K. and Sharma, R. et al. (2018). *Dynamic light scattering (DLS) technique, principle, theoretical considerations and applications* (pp. 135–137).

Sarabandi, K., Gharehbeglou, P. and Jafari, S.M. (2020). Scanning electron microscopy (SEM) of nanoencapsulated food ingredients. In *Characterization of Nanoencapsulated Food Ingredients* (pp. 83–130). Elsevier. https://doi.org/10.1016/B978-0-12-815667-4.00003-1.

Shmidov, Y., Zhu, Y., Matson, J.B. and Bitton, R. (2020). Effect of Crosslinker Topology on Enzymatic Degradation of Hydrogels. *Biomacromolecules*, *21*(8), 3279–3286. https://doi.org/10.1021/acs.biomac.0c00722.

Shrivastava, A., Mandal, S., Pattanayek, S.K. and Rathore, A.S. (2023). Rapid Estimation of Size-Based Heterogeneity in Monoclonal Antibodies by Machine Learning-Enhanced Dynamic Light Scattering. *Analytical Chemistry*, *95*(21), 8299–8309. https://doi.org/10.1021/acs.analchem.3c00650.

Shrestha N, Préat V. In vivo testing of orally delivered nanoparticles. In Nanotechnology for oral drug delivery 2020 Jan 1 (pp. 459–480). Academic Press. https://doi.org/10.1016/B978-0-12-818038-9.00015-6

Singh, G., Senapati, S., Satpathi, S., Behera, P.K., Das, B. et al. (2022). Establishment of decellularized extracellular matrix scaffold derived from caprine pancreas as a novel alternative template over porcine pancreatic scaffold for prospective biomedical application. *The FASEB Journal*, *36*(10). https://doi.org/10.1096/fj.202200807R.

Sinha Ray, S. (2013). Structure and Morphology Characterization Techniques. In *Clay Containing Polymer Nanocomposites* (pp. 39–66). Elsevier. https://doi.org/10.1016/B978-0444-59437-2.00003-X.

Sleator, R. (2012). Digital biology. *Bioengineered*, *3*, 311–312. https://doi.org/10.4161/bioe.22367.

Smith, K.C.A., Wells, O.C. and McMullan, D. (2008). The fiftieth anniversary of the first applications of the scanning electron microscope in materials research. *Physics Procedia*, *1*(1), 3–12. https://doi.org/10.1016/j.phpro.2008.07.073.

Sobola, D., Ţălu, Ş., Solaymani, S. and Grmela, L. (2017). Influence of scanning rate on quality of AFM image: Study of surface statistical metrics. *Microscopy Research and Technique*, *80*(12), 1328–1336. https://doi.org/10.1002/jemt.22945.

Stamm, A., Reimers, K., Strauß, S., Vogt, P., Scheper, T. et al. and Pepelanova, I. (2016). In vitro wound healing assays – state of the art. *BioNanoMaterials*, *17*(1–2), 79–87. https://doi.org/10.1515/bnm-2016-0002.

Talari, A.C.S., Martinez, M.A.G., Movasaghi, Z., Rehman, S. and Rehman, I.U. et al. (2017). Advances in Fourier transform infrared (FTIR) spectroscopy of biological tissues. *Applied Spectroscopy Reviews*, *52*(5), 456–506. https://doi.org/10.1080/05704928.2016.1230863.

Thakral, S., Terban, M.W., Thakral, N.K. and Suryanarayanan, R. (2016). Recent advances in the characterization of amorphous pharmaceuticals by X-ray diffractometry. *Advanced Drug Delivery Reviews*, *100*, 183–193. https://doi.org/10.1016/j.addr.2015.12.013.

Tran, M.-H., Nguyen, N.Q. and Pham, H.T. (2022). A New Hope in the Fight Against Antimicrobial Resistance with Artificial Intelligence. *Infection and Drug Resistance*, *15*, 2685. https://doi.org/10.2147/IDR.S362356.

van Benthem, K., Pennycook, S. and Rahman, Z. (2017). *Atomic Resolution Characterization of Semiconductor Materials by Aberration-Corrected Transmission Electron Microscopy*. https://doi.org/10.1016/B978-0-12-803581-8.10405-9.

Vand Aa, T.M. (2015). Application of Artificial Neural Network for Modeling and Prediction of MTT Assay on Human Lung Epithelial Cancer Cell Lines. *Journal of Biosensors & Bioelectronics*, *06*(02). https://doi.org/10.4172/2155-6210.1000170.

Vigata, M., Meinert, C., Hutmacher, D.W. and Bock, N. (2020). Hydrogels as Drug Delivery Systems: A Review of Current Characterization and Evaluation Techniques. *Pharmaceutics*, *12*(12), 1188. https://doi.org/10.3390/pharmaceutics12121188.

Villalpando-Rodriguez, G.E. and Gibson, S.B. (2021). Reactive Oxygen Species (ROS) Regulates Different Types of Cell Death by Acting as a Rheostat. *Oxidative Medicine and Cellular Longevity*, *2021*, 9912436. https://doi.org/10.1155/2021/9912436.

Voigtländer, B. (2019). Introduction. In B. Voigtländer, *Atomic Force Microscopy* (pp. 1-13). Springer International Publishing. https://doi.org/10.1007/978-3-030-13654-3_1.

Wang, Y.-T., Li, B., Xu, X.-J., Ren, H.-B., Yin, J.-Y et al. (2020). FTIR spectroscopy coupled with machine learning approaches as a rapid tool for identification and quantification of artificial sweeteners. *Food Chemistry*, *303*, 125404. https://doi.org/10.1016/j.foodchem.2019.125404.

Wang, W.S. and Zhu, L.Q. (2023). Recent advances in neuromorphic transistors for artificial perception applications: FOCUS ISSUE REVIEW. *Science and Technology of Advanced Materials*, *24*(1), 10–41. https://doi.org/10.1080/14686996.2022.2152290.

Weimer, A.W. (Ed.). (1997). *Carbide, nitride and boride materials synthesis and processing* (1. ed). Chapman & Hall.

Wijesundara, N.M., Lee, S.F. and Rupasinghe, H.P.V. (2022). Carvacrol inhibits Streptococcus pyogenes biofilms by suppressing the expression of genes associated with quorum-sensing and reducing cell surface hydrophobicity. *Microbial Pathogenesis*, *169*, 105684. https://doi.org/10.1016/j.micpath.2022.105684.

Williams, D.F. (2008). On the mechanisms of biocompatibility. *Biomaterials*, *29*(20), 2941 2953. https://doi.org/10.1016/j.biomaterials.2008.04.023.

Woese, C.R. (2004). A New Biology for a New Century. *Microbiology and Molecular Biology Reviews*, *68*(2), 173–186. https://doi.org/10.1128/MMBR.68.2.173-186.2004.

Wooley, J.C., Lin, H.S. and Biology, N.R.C. (US) C. on F. at the I. of C. and. (2005). 21st Century Biology. In *Catalyzing Inquiry at the Interface of Computing and Biology*. National Academies Press (US). https://www.ncbi.nlm.nih.gov/books/NBK25458/

Wu, K.Z., Adine, C., Mitriashkin, A., Aw, B.J.J., Iyer, N.G. et al. (2023). Making In Vitro Tumor Models Whole Again. *Advanced Healthcare Materials*, *12*(14), e2202279. https://doi.org/10.1002/adhm.202202279.

Wu, S., Pan, Z., Li, X., Wang, Y., Tang, J. Li, H., Lu, G., Li, J., Feng, Z., He, Y. and Liu, X. (2023). Machine learning assisted photothermal conversion efficiency prediction of anticancer photothermal agents. Chemical Engineering Science, 273, 118619. https://doi.org/10.1016/j.ces.2023.118619

Yablon, D.G. (2013). Overview of Atomic Force Microscopy. In D.G. Yablon (Ed.), *Scanning Probe Microscopy in Industrial Applications* (1st ed., pp. 1–14). Wiley. https://doi.org/10.1002/9781118723111.ch1.

Zhang, J., Hong, Z., Lu, W., Fang, T., Ren, Y. et al. (2023). Assessment of Drug Susceptibility for Patient-Derived Tumor Models through Lactate Biosensing and Machine Learning. *ACS Sensors*, *8*(2), 803–810. https://doi.org/10.1021/acssensors.2c02381.

Zheng, H., Lu, X. and He, K. (2022). In situ transmission electron microscopy and artificial intelligence enabled data analytics for energy materials. *Journal of Energy Chemistry*, *68*, 454–493. https://doi.org/10.1016/j.jechem.2021.12.001.

Zhydik, O. (2020, October 29). *Digital Biology: The New Technological Frontier*. ELEKS: Enterprise Software Development, Technology Consulting. https://eleks.com/blog/digital biology-new-technological-frontier/

CHAPTER

20

Artificial Intelligence in Biological Data Science

Ashtami J,[1] *Megha KB*,[1] *Greeshma N*,[2] *Aswathy V*,[2] *Manjoosha RY*,[2] *Renu John*[2] and *Mohanan PV*[1*]

1. Introduction

The concept of artificial intelligence (AI) was introduced way back in the 1950s (Xu, Y. et al., 2021). Alan Martison Turing introduced the concept of AI and published an article entitled "Can machines talk" (Bellini, V. et al., 2022). Later, he presented the idea of a self-learning machine which works on memory learned from experiences, like human beings (Bhardwaj, A. et al., 2022). The word "artificial intelligence" was coined by John McCarthy at the Dartmouth conference in 1956. Later, he was regarded as the father of AI for his contribution to AI (Andresen, S.L. 2002). The basic concept of AI is to develop self-performing machines capable of carrying out functions at the human intellectual level with minimum use of the actual human workforce. In simple terms, AI is developed to enable machines to perform human actions like studying, thinking, decision-making and crisis-solving (Shabbir, J. and Anwer, T. 2018). AI has gained much popularity over the last few years, especially with the practicable use of self-driven cars, chat GPT, and voice assistant devices like Alexa and Siri (Shabbir, J. and Anwer, T. 2018). Interestingly, AI is finding predominant space in biological science as well. AI has been applied in the fields of medicine, genomics, biotechnology, proteomics, agriculture, animal science, among others (Andresen, S.L. 2002). The key targets in using AI are to process large quantities of biological data, manage health records, make better analyses, make faster decisions and reduce possible human errors (Dlamini, Z. et al., 2020). AI has different approaches to attaining these goals, and they are attained with the help of different mathematical models, algorithms, programming software, data mining, and statistical analysis (Gruson, D. et al., 2009).

AI is broadly classified into weak AI, strong AI, and super AI. The terms weak and strong AI were coined by John Searle. Through weak AI, machines can perform only a single task at a

[1] Toxicology Division, Sree Chitra Tirunal Institute of Medical Science and Technology, Thiruvananthapuram 695 012, Kerala.
[2] Department of Biomedical Engineering, Indian Institute of Technology Hyderabad, Hyderabad, Telangana.
* Corresponding author: mohananpv10@gmail.com

time. Strong AI enables machines to possess actual human intelligence and perform multiple tasks like a real brain. Hence, strong AI is expected to have human-level intelligence (Meskó, B. and Görög, M. 2020). Currently, weak AI is practically in use and strong AI is in the theoretical research stage. Face recognition programs, and voice recognition apps such as Alexa and Siri, work on weak AI. Super AI machines are proposed to have cognitive human power that is yet to be attained practically. Another reported way of classifying AI is based on its functionality (Raghuvanshi, D. 2018). AI in the context of functionalities is classified as (1) reactive (2) limited memory (3) theory of mind and (4) self-awareness. Reactive AI is the primitive form that performs on present input and does not possess stored memory. One example of a primitive AI machine is Deep Blue, the IBM supercomputer capable of playing chess. Deep Blue was reported to have defeated human grandmaster Garry Kasparov in 1997 (Helm, J.M. et al., 2020). Limited memory devices hold stored memory from previous experiences that helps future performance. Both the theory of mind- and self-awareness-based machines are theoretical and require a long way before finding practical application. Mind-based machines are expected to understand human emotions and thoughts while self-awareness machines are expected to be further ahead of others with their super intelligent mind (Raghuvanshi, D. 2018).

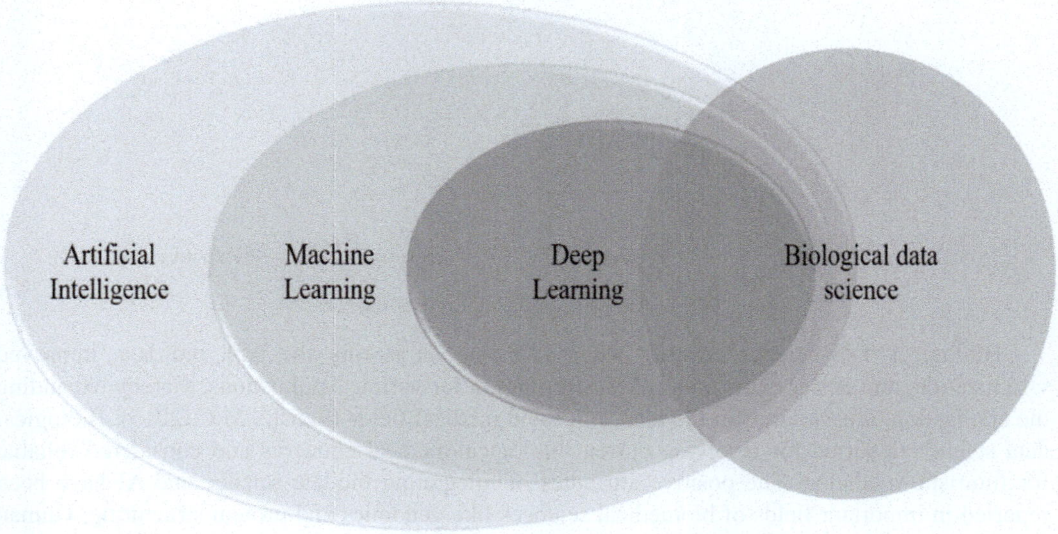

Fig. 1 Artificial intelligence and biological data science

One of the major subsets in AI is machine learning (Figure 1). In machine learning, the system is provided with large sets of inputs and outputs. The algorithm helps the machine learn from the provided data, analyze it, and make predictions / future decisions (Helm, J.M. et al., 2020). Arthur Samuel quoted that " a computer program is said to learn from experience (E) concerning some task (T) and some performance measure (P); if its performance on T, as measured by P, improves with experience E then the program is called a machine learning program (Das, S. et al., 2015).This is considered to be the modern definition of AI. A major subset of machine learning is deep learning. The advancements in AI approaches like machine learning,and deep learning have enabled better feasibility in analyszing biological data.

2. Biological Data Science

Biological systems hold large quantities of data. Furthermore, the high dimensionality, and complex nature of biological data makes it tougher to process, analyze and interpret,making it challenging to predict outcomes (Mahmud, M. et al., 2018). Biological data sciences involve the interdisciplinary integration of biological science, computer science, statistics, and mathematics (Figure 2)(Goh, W. W.B. and Wong, L. 2020). The biological science domain provides data generation corresponding to the particular query that demands resolution. Computer science involves the algorithm required to process a huge amount of data. The mathematical and statistical part involves normalizing and modeling the data (Goh, W.W.B. and Wong, L. 2020).

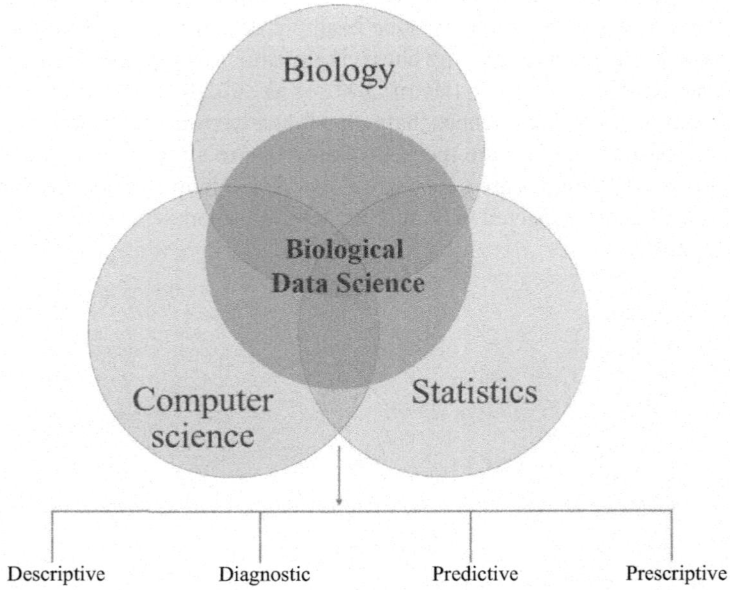

Fig. 2 Domains of biological data science

Biological data science enables better efficiency in storing the bulk bio-data, improved visualization, and easier extraction of bio-information for various applications, thereby expanding the application opportunities in the bioscience and medical fields (Schatz, M.C. 2015). Biological data science is sorted for resolving current biological/medical concerns and considered reliable for futuristic medicine. The positive attributes of integrating biodata science and AI have been reported in important fields of biomedical sciences like genomics and protein structuring. Human genomes are rich in information, and understanding the genetic makeup of humans enables progress in the biomedical field, especially for theragnostics. The capacity to sequence DNA has helped in synthesizing DNA and using the technology for various applications like enzyme synthesis, molecular engineering, gene expression, vaccine development and biotechnology (Ma, S. et al., 2012). Similarly, proteins are important biomolecules. Identifying structural and conformational properties of proteins opens up vast possibilities for understanding protein functions, identifying active sites, determining molecular structure, spotting allosteric sites, mimicking enzyme action, and diagnosing the molecular level and therapy (Wu, G. et al., 2010; Westbrook, J. et al., 2003). Each enzyme is made up of 100s of amino acids and each of which can take different conformations leading to approx. 10^{24} conformational possibilities (Villalobos-Alva, J. et al., 2022). Therefore, the protein data is so huge quantitatively, it is difficult to extract information from it. The application of AI in proteomics helps ease the challenge.

The goals of biological data science are wide-ranging. In a previous report, the goals of biological data science were classified based on the expected outcome as follows: (1) descriptive, (2) diagnostic, (3) predictive, and (4) prescriptive (Goh, W.W.B., and Wong, L. 2020). The descriptive goal is the basic level outcome where the data is evaluated on simple grounds, and basic statistical data normalization is performed to gather information. At the diagnostic level, the target outcome is to identify the cause of a particular event or occurrence. At an advanced level, the predictive level analyzes data and determines the causatives and indicative parameters. Then, it predicts the possibility of the expected causatives. Among the four objectives, the prescriptive goal the most advanced one. It requires a deep understanding of each of the causative factors and their interconnectness instrumental in causing a particular event. Another important terminology is "Big data" which stands for a huge database with a potential utility to improve health and medical care applications (Sagiroglu, S. and Sinanc, D. 2013, May). Big data requires a large storage capacity, and the vast quantity of data demands advanced algorithms to understand and analyze it for future predictions. The advancement in biomedical scenarios and the quantification of healthcare data has indeed expanded big data. Big data stands vital in advancing towards futuristic personalized medicine (Costa, F.F. 2014).

The characteristic feature of Big data is denoted by the three "V"s. The first V stands for Volume of data (i.e., the quantity of big data is high, measured in terabytes and petabytes) (Sagiroglu, S. and Sinanc, D. 2013, May). The traditional data processing methods fail to process such large data, and thus require more advanced approaches like AI to further process the data. The second characteristic feature is Velocity, which refers to the speed of producing and processing data (Luo, J. et al., 2016). The third characteristic feature of big data is Variety: big data is produced from different sources and exists in different forms: (1) structured, (2) semi-structured, and (3) unstructured (Sagiroglu, S. and Sinanc, D. 2013, May). The structured data, as the name suggests, consists of sorted and tagged data collection, while the semi-structured data consists of data tagged but not sorted as in the case of the structured category. The unstructured data is unsorted and untagged and is challenging to analyze and extract further information compared to the other two types.

3. Artificial Intelligences in Biological Data Science

Advancements in biological data science and bulk data generated in different formats like sequences, images, or signals require techniques like AI to process the data and extract useful information for various applications. Research has been extensively carried out to integrate biological data science and AI to facilitate data mining and achieve targeted goals. The important subtypes of AI are: (1) Machine learning, natural language processing, computer vision, and fuzzy Logic (Bellini, V. et al., 2022). Machine learning enables systems to learn from the provided data inputs and use algorithms for analysis and decision-making. Natural language processing uses the input data to make the machines understand human language and perform the required function (Chowdhary, K. and Chowdhary, K.R. 2020). Computer vision helps machines view and recognize or identify image data (Karn, A. et al., 2021). Fuzzy logic uses additional intermediate values for problem solving in contrast to traditional binary systems like yes/no or true-false (Hellmann, M. 2001). Among the different subsets of AI, machine learning is the most important and applied one.

3.1 Machine learning

As discussed in the introduction, machine learning involves using algorithms or computations to identify a given set of data and extract information to perform the specific task, without any explicit programming (Mahesh, B. 2020). Machine learning is classified broadly as supervised and unsupervised learning (Figure 3). Two other subsets of machine learning are semi-supervised and reinforcement learning. In supervised learning, the data sets are well-labelled, and the system is provided with a set of input data for training so that the system draws a relation between the input and output data. The system learns from the input-output pairing pattern from the provided training data set and later predicts the output for test input data (Mahesh, B. 2020). An example of supervised learning is classifying protein sequences to their corresponding secondary structures (Tarca, A.L.

et al., 2007). Supervised learning is again classified based on the nature of output data; the output data is obtained either in a numerical or categorical form (Badillo, S. et al., 2020). Some important supervised learning algorithms are decision trees, neural networks, naïve bayes, and support vector machines. Among these, decision trees, neural networks, and naïve Bayes algorithms give categorical output, whereas support vector machines give numerical outputs.

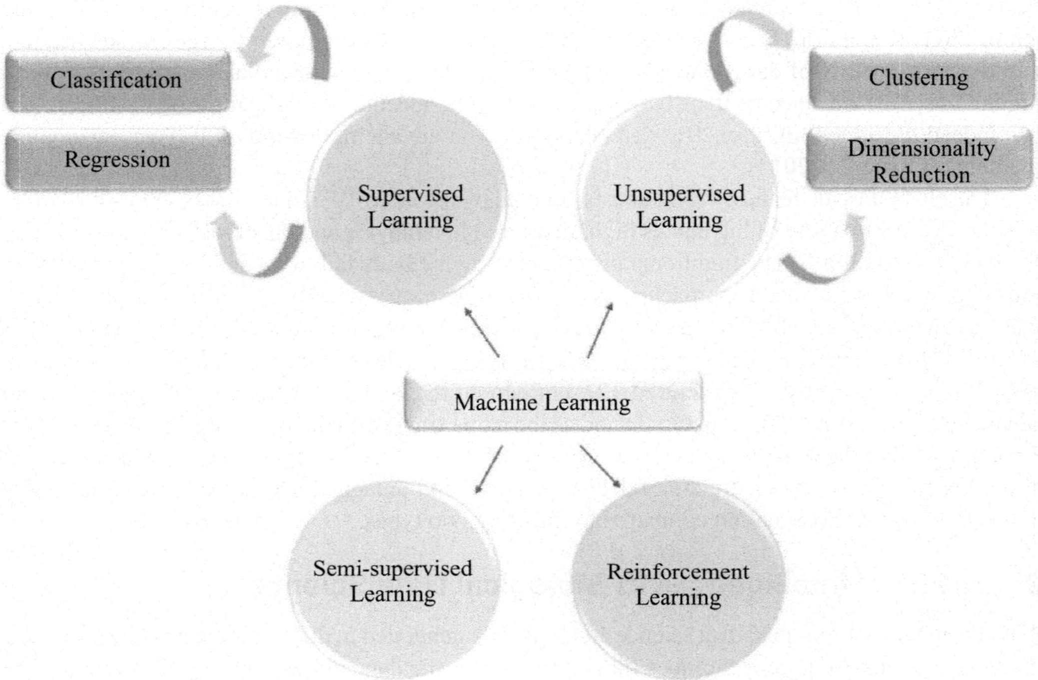

Fig. 3 Machine learning

Unsupervised learning involves unlabelled data. The system is provided with the input data alone and left to figure out the output data on its own (Baştanlar, Y. and Özuysal, M. 2014). Biological data of gene expression, and sorting people into different classes based on social and behavioral patterns are examples of datasets where unsupervised learning is performed. In general, unsupervised learning is usually chosen for clustering and reducing dimensionality (Badillo, S. et al., 2020). Clustering is done in cases of high dimensionality input data as in the case of proteomics. Dimensionality reduction helps remove outliers and is used in omics data sets. It helps ease visualization by removing unwanted data points. A few essential algorithms in unsupervised learning are K-means, and principal component analysis. The K-means algorithm is preferred for clustering, and principal component analysis is deployed in cases of dimensionality reduction. In some cases, the data set may not be completely categorized; such cases are considered semi-supervised learning algorithms. In such cases, a small quantity of labeled data occurs with a larger quantity of unlabelled data. For example, some protein sequences and their corresponding secondary structures are labeled while the remaining protein sequences need to be sorted (Baştanlar, Y. and Özuysal, M. 2014). Another type of machine learning is reinforcement learning. In this subset, an agent is trained to perform actions in a particular environment over time (Sarkar, D. et al., 2018). The training involves constant interaction with the environment and learning from error (Altaf, I. et al., 2023). Reinforcement learning is regarded as behavioral machine learning.

3.2 Deep Learning

Conventional machine learning techniques have limitations in processing raw bulky biological data (or raw data in general), and this is where the deep learning approach comes to the rescue (LeCun, Y. et al., 2015). Deep learning comes as a subset of machine learning. It is developed with the primary aim of mimicking human intelligent behavior. It is developed in neural network patterns like human neurons (Altaf, I. et al., 2023). Deep learning exists as several layers of non-linear processing units or supervised/unsupervised learning on each layer (Cao, C. et al., 2018). The deep learning approach is reported to be successfully applicable in voice/face recognition. Deep learning entails hidden artificial neural networks (Vargas, R. et al., 2017). Artificial neural networks are named after their pattern similarity to natural neuronal systems in human brains (Greener, J.G. et al., 2022). Like neurons in the nervous system, artificial neurons are the basic cell unit of an artificial neural network. This basic cell unit receives input data and applies the trained mathematical function on input data. Then, it provides the output to nearby neurons, serving as input data.

Neural networks are sorted into two varieties: (1) Generative and (2) Discriminative (Shrestha, A. and Mahmood, A. 2019). In the generative neural networks model, the flow of data is from top to bottom with labeled data. The discriminative route follows a bottom-up approach. It is notably done in the classification of supervised data sets. Some important types of neural networks are as follows: (1) Feedforward Neural Network – the data flows unidirectionally from input – to output. (2) Recurrent neural network – data flow occurs in a cycle: output of one-layer acts as input to another layer which in turn helps machines develop memory, (3) Radial basis function neural network – have input layers, hidden layers, and output layers (4) Self-organizing neural network – has two completely connected layers: one input layer based on unsupervised learning and an output layer on supervised learning. (5) Modular neural network – breaks down the sizable neural network into smaller independent units performing a particular assigned task (Shrestha, A. and Mahmood, A. 2019), (6) Deep belief network-uses unsupervised learning for pre-training and supervised learning to develop models, (Chen, X.W. and Lin, X. 2014) (7) Convolutional neural networks-consist of hierarchical layers in an alternating manner having convolutional and sub-sampling layers respectively. Convolutional layers reduce the hyperparameters, while sub-sampling layers are involved in size reduction (C. et al., 2018).

4. Important Applicational Domains using AI in Biological Data

Biological data is primarily large quantitatively, highly variable, and unsorted. The ability to process this data and extract knowledge from biological raw data has immense potential to enhance development in the medical and healthcare fields. AI has been a reliable option to gather information from the input signals/sequences/image data to attain the desired functionality. The earliest known positive feedback on using AI in the healthcare field was reported in 1976 by Gunn when acute abdominal pain diagnosis was done using a computer (Gunn, A.A. 1976). Among the various options, the artificial neural network is the most explored AI option to process biological data and achieve major targets like pattern recognition, image processing, diagnosis, identification of high-risk patterns, and anticipating possible outcomes (Ramesh, A.N. et al., 2004). AI is merged with different fields of biology and contributes positively to various applications like image processing, drug delivery, cancer diagnosis, and disease progression monitoring (Bhardwaj, A. et al., 2022). Interestingly, AI has helped ameliorate even agricultural concerns and provided expert solutions. AI has significantly contributed to agriculture for crop disease control, pest control, crop management, yield predictions, estimating soil quality, suggesting improved water and soil management (Bannerjee, G. et al., 2018).

4.1 AI in Omics Data

Omics data collectively stands for data from fields ending with omics, i.e., genomics, proteomics, transcriptomics, metabolomics (Figure 4). In general, omics data covers the structural and functional information of cells, tissues, and organisms. Among different omics data, genomics data contain information about the structure, function, sequencing, or coding of genetic material. Proteomics data encompass knowledge about the structure, conformational information, amino acid arrangements, and other details of proteins. Metabolomics data covers information on the metabolic activity of cells. Transcriptomics is associated with information related to RNA (Vailati-Riboni, M. et al., 2017).

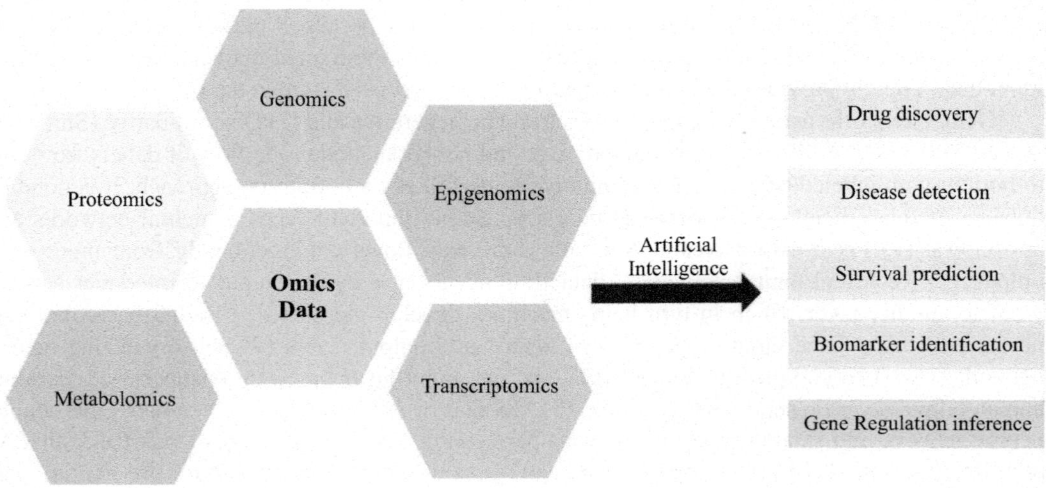

Fig. 4 Application of AI on omics data

Nowadays, multi-omics data is explored to understand the linkage between different constituent factors and get an upper hand in addressing the target issues. Applying a machine learning algorithm on multi-omics data eases data processing and helps decision-making (Ali, A.M., and Mohammed, M.A. 2024). Machine learning aids in minimizing the complexity of bulky multi-omics data and identifying the pattern giving a better understanding of the scenario. The application of AI in multi-omics data has shown promising outputs for accomplishing personalized medicine. Machine learning and deep learning algorithms facilitate genomic sequencing. Genome sequencing provides a wide range of information like mutation sites, drug response, and disease development stage. This information helps achieve disease prediction, diagnosis, personalized treatments, and formulation of precision medicines (Alharbi, W.S., and Rashid, M. 2022).

Antibiotic resistance is a significant hurdle in attaining maximum drug efficacy and curing diseases. The machine learning approach is reliable for estimating antibiotic resistance development based on genomic sequencing information. Ren *et al.* studied efficiency in predicting drug resistance using different machine learning algorithms. The machine learning algorithms used were random forest, logistic regression, convolutional neural network, and support vector machine (SVM). The results showed that machine learning algorithms accurately predicted the drug resistance of four drugs against Escherichia coli bacteria. Also, deep forest and convolutional neural networks showed better efficiency among the tested algorithms than the others (Ren, Y. et al., 2022). Deelder *et al.* and team checked the drug resistance against mycobacterium tuberculosis using a machine learning algorithm on gene sequences (Deelder, W. et al., 2019). Micro-RNA (miRNA) are non-coding RNA sequences responsible for post-transcription modification, gene regulation, and many vital biological events. miRNA is a biomarker for many diseases like cancer, cardiovascular and neuro-related disorders (Parveen, A. et al., 2019). Spotting miRNA from other non-coding and coding RNA sequences is a bit of a challenge. Many machine learning algorithms are better than conventional approaches in precisely working in miRNA identifications (Parveen, A. et al., 2019). Park *et al.* reported a recurrent neural network named "deepMiRGene" for identifying miRNA sequences from a non-coding set (Park, S. et al., 2016). Another important application of AI on omics data is for disease prediction. For example, metastasis, a condition in which tumour cells get detached from their primary site and accumulate at a secondary site, can be detected by applying machine learning and deep learning algorithms on omics data (genomics and epigenomics data) (Albaradei, S. et al., 2021).

4.2 AI in Healthcare

Over the years, AI has found significant applicability in healthcare, especially for patient health record organization, information sorting, patient monitoring, diagnosis, disease monitoring, and drug efficacy prediction (Reddy, S. 2018). Important techniques of AI being explored in healthcare scenarios are machine learning, deep learning, deep neural networks, natural language processing, and computer vision. Many patients' health records have now been computerized with advancements in the medical field. These databases are rich information sources. Proper processing and extraction of this information is beneficial for the right clinical decision-making and for improving patient care. AI methods like machine learning, deep learning, and natural language processing are utilized to extract knowledge from electronic health records and further use it for disease diagnosis (Lin, W.C. et al., 2020). On similar grounds, Liang *et al.* and coworkers reported that using natural language processing based on a deep learning approach to children's electronic health records could diagnose pediatric diseases with high accuracy (Liang, H. et al., 2019). Also, cardiovascular risk predictions were made using the subset of AI called a multilayered feed-forward neural network. The AI neural network algorithm on cardio data showed a risk prediction accuracy of 94.17% (Amma, N.B. 2012, February).

AI is applied to bioimaging for better image processing, pattern recognition, and interpretation. Applying AI to imaging input data is reported to produce output data with better quality, in less time, and more accurate inferences (Selvaraj, J. et al., 2022). The use of AI for pathology image recognition also shows promising results. Khosravi *et al.* reported using deep convolutional neural networks for cancer tissue diagnosis from histograms. The study revealed that deep convolutional neural networks could distinguish cancer cells with 100% accuracy, subsets of lung cancer cells with 92% accuracy, and biomarkers were recognized with 96% accuracy (Khosravi, P. et al., 2018). Among the different bioimaging applications, the remarkable influence of AI is marked in the area of radiology. Deep learning techniques have been effective in pattern recognition, image processing, and image data identification. AI has shown appreciable performance in detecting and monitoring diseases from radiological images (Hosny, A. et al., 2018).

4.3 AI in Biotechnology

AI has been applied to diverse fields of biotechnology such as food processing, agriculture, fermentation, bioprocessing, enzyme production, and microbial biotechnology. One of the AI methods, artificial neural networks, showed successful results in bioprocess monitoring. For instance, applying the artificial neural network approach on near infra-red (NIR) data obtained from a solid-state fermentation system could efficiently predict protein concentration and enzymes (amylase and protease activities) precisely (Barchi, A.C. et al., 2016). Similarly, artificial neural networks are reported to be effective in monitoring food quality and shelf life (Suzuki, K., 2011). Also, machine learning and deep learning are used to monitor food safety, toxicity, and sorting (Amore, A. and Philip, S., n.d). Enzyme production is another primary application of AI in biotechnology. Singh *et al.* used artificial neural networks and genetic algorithms, to predict the production of exocellular glucan-sucrase enzyme. The predicted production was comparable to the experimental yield (Singh, A. et al., 2008).

4.4 AI in Pharmacology

With advancements in data science and technology, AI holds a vital place in pharmacological zones like drug discovery, disease diagnosis, precision medicine, drug response, adverse reaction, ADMET proofing (absorption, distribution, metabolism, excretion, and tolerable toxicity), and drug kinetics (Kumar, M. et al., 2023). Among the subsets of AI, machine learning, more precisely deep learning, is successful in predicting optimized drug molecules exhibiting the targeted activity. The deep learning approach predicts drugs through various approaches like structure-based screening, ligand-based screening, and De Nova designing (Yang, X. et al., 2019). Further toxicity predictions, ADMET profiling and structure-activity relationships are also overlooked (Gupta, R. et al., 2021). AI can also predict the pharmacokinetic pattern in drug delivery studies. Recently Chou *et al.* reported using AI-based structure-activity systems and AI-dependent pharmacokinetics systems to analyze the drug delivery pattern in mice models. The study mentioned that the AI-based system effectively predicts the pharmacokinetics of nanodrugs to the tumour site (Chou, W.C. et al., 2023).

Conclusion

Advancements in AI and integration with multiple areas are anticipated to resolve complex issues. AI techniques like machine learning and deep learning have facilitated better biological data processing. Processing of Big data with the use of conventional methods is tedious and time-consuming beyond feasibility. Storage and extraction of information from Big data have become possible and effective through the utilization of AI algorithms. AI techniques have exhibited appreciable performances in extracting information from genomics data. Other areas of omics data like proteomics, transcriptomics, and metabolomics are explored less with AI compared to genomics but hold potential. Applying deep learning on multi-omics data is promising for achieving effective personalized treatment. Subsets of AI-Machine learning, deep learning, and artificial neural networks have contributed towards improving the quality of healthcare. AI has made it easier to organize patient health records, sort information, monitor patient health, diagnose diseases, track advancement in disease conditions, process images, and predict drug efficacy. AI has also augmented drug discovery, predicting drug response, pharmacokinetics, enzyme production, and agricultural or food technology concerns. In a nutshell, AI is being applied to diverse biological databases to fix different concerns and aid in better decision- making. The current growth phase of AI and biological data science shows promising potential for advancement.

Acknowledgments

The authors wish to express their thanks to the Director Sree Chitra Tirunal Institute for Medical Sciences and Technology (Govt. of India), Trivandrum, Kerala, India, and the Director, Indian Institute of Technology Hyderabad for their support and for providing the infrastructure facilities for carrying out this work.

References

Albaradei, S., Thafar, M., Alsaedi, A., Van Neste, C., Gojobori, T. et al. (2021). Machine learning and deep learning methods that use omics data for metastasis prediction. *Computational and structural biotechnology journal*, *19*, 5008–5018.

Alharbi, W.S. and Rashid, M. (2022). A review of deep learning applications in human genomics using next-generation sequencing data. *Human Genomics*, *16*(1), 1–20.

Ali, A.M. and Mohammed, M.A. (2024). A Comprehensive Review of Artificial Intelligence Approaches in Omics Data Processing: Evaluating Progress and Challenges. *International Journal of Mathematics, Statistics, and Computer Science*, *2*, 114–167.

Altaf, I., Butt, M.A. and Zaman, M. (2023). Artificial Intelligence: For Biological Data. In *Applications of Machine Learning and Deep Learning on Biological Data* (pp. 135–158). Auerbach Publications.

Amma, N.B. (2012, February). Cardiovascular disease prediction system using genetic algorithm and neural network. In *2012 international conference on computing, communication and applications* (pp. 1–5). IEEE.

Amore, A. and Philip, S. Artificial Intelligence in Food Biotechnology: trends and perspectives. *Frontiers in Industrial Microbiology*, *1*, 1255505.

Andresen, S.L. (2002). John McCarthy: father of AI. *IEEE Intelligent Systems*, *17*(5), 84–85.

Badillo, S., Banfai, B., Birzele, F., Davydov, I.I., Hutchinson, L. et al. (2020). An introduction to machine learning. *Clinical pharmacology & therapeutics*, *107*(4), 871–885.

Bannerjee, G., Sarkar, U., Das, S. and Ghosh, I. (2018). Artificial intelligence in agriculture: A literature survey. *International Journal of Scientific Research in Computer Science Applications and Management Studies*, *7*(3), 1–6.

Barchi, A.C., Ito, S., Escaramboni, B., de Oliva Neto, P., Herculano, R.D. et al. (2016). Artificial intelligence approach based on near-infrared spectral data for monitoring of solid-state fermentation. *Process Biochemistry*, *51*(10), 1338–1347.

Baştanlar, Y. and Özuysal, M. (2014). Introduction to machine learning. *miRNomics: MicroRNA biology and computational analysis*, 105–128.

Bellini, V., Cascella, M., Cutugno, F., Russo, M., Lanza, R. et al. (2022). Understanding basic principles of Artificial Intelligence: A practical guide for intensivists. *Acta Bio Medica: Atenei Parmensis*, *93*(5).

Bhardwaj, A., Kishore, S. and Pandey, D.K. (2022). Artificial intelligence in biological sciences. *Life*, *12*(9), 1430.

Cao, C., Liu, F., Tan, H., Song, D., Shu, W. et al. (2018). Deep learning and its applications in biomedicine. *Genomics, proteomics & bioinformatics*, *16*(1), 17–32.

Chen, X.W. and Lin, X. (2014). Big data deep learning: challenges and perspectives. *IEEE access*, *2*, 514–525.

Chou, W.C., Chen, Q., Yuan, L., Cheng, Y.H., He, C. et al. (2023). An artificial intelligence-assisted physiologically-based pharmacokinetic model to predict nanoparticle delivery to tumors in mice. *Journal of Controlled Release*, *361*, 53–63.

Chowdhary, K. and Chowdhary, K.R. (2020). Natural language processing. *Fundamentals of artificial intelligence*, 603–649.

Costa, F.F. (2014). Big data in biomedicine. *Drug discovery today*, *19*(4), 433–440.

Das, S., Dey, A., Pal, A. and Roy, N. (2015). Applications of artificial intelligence in machine learning: review and prospect. *International Journal of Computer Applications*, *115*(9).

Deelder, W., Christakoudi, S., Phelan, J., Benavente, E.D., Campino, S. et al. (2019). Machine learning predicts accurately Mycobacterium tuberculosis drug resistance from whole genome sequencing data. *Frontiers in genetics*, *10*, 922.

Dlamini, Z., Francies, F.Z., Hull, R. and Marima, R. (2020). Artificial intelligence (AI) and big data in cancer and precision oncology. *Computational and structural biotechnology journal*, *18*, 2300–2311.

Goh, W.W.B. and Wong, L. (2020). The birth of Bio-data Science: trends, expectations, and applications. *Genomics, proteomics & bioinformatics*, *18*(1), 5.

Greener, J.G., Kandathil, S.M., Moffat, L. and Jones, D.T. (2022). A guide to machine learning for biologists. *Nature Reviews Molecular Cell Biology*, *23*(1), 40–55.

Gruson, D., Helleputte, T., Rousseau, P. and Gruson, D. (2019). Data science, artificial intelligence, and machine learning: opportunities for laboratory medicine and the value of positive regulation. *Clinical biochemistry*, *69*, 1–7.

Gunn, A.A. (1976). The diagnosis of acute abdominal pain with computer analysis. *Journal of the Royal College of Surgeons of Edinburgh*, *21*(3), 170–172.

Gupta, R., Srivastava, D., Sahu, M., Tiwari, S., Ambasta, R.K. et al. (2021). Artificial intelligence to deep learning: machine intelligence approach for drug discovery. *Molecular diversity*, *25*, 1315–1360.

Hellmann, M. (2001). Fuzzy logic introduction. *Université de Rennes*, *1*(1).

Helm, J.M., Swiergosz, A.M., Haeberle, H.S., Karnuta, J.M., Schaffer, J.L. et al. (2020). Machine learning and artificial intelligence: definitions, applications, and future directions. *Current reviews in musculoskeletal medicine*, *13*, 69–76.

Hosny, A., Parmar, C., Quackenbush, J., Schwartz, L.H. and Aerts, H.J. et al. (2018). Artificial intelligence in radiology. *Nature Reviews Cancer*, *18*(8), 500–510.

Karn, A., Mehta, R., Hiriyanna, G.S., Sayyed Johar, K., Chhabra, A. et al. (2021). Artificial Intelligence in Computer Vision. *International Journal of Engineering Applied Sciences and Technology*, *6*(1), 294–254.

Khosravi, P., Kazemi, E., Imielinski, M., Elemento, O. and Hajirasouliha, I. et al. (2018). Deep convolutional neural networks enable discrimination of heterogeneous digital pathology images. *EBioMedicine*, *27*, 317–328.

Kumar, M., Nguyen, T.N., Kaur, J., Singh, T.G., Soni, D. et al. (2023). Opportunities and challenges in application of artificial intelligence in pharmacology. *Pharmacological Reports*, *75*(1), 3–18.

LeCun, Y., Bengio, Y. and Hinton, G. (2015). Deep learning. *nature*, *521*(7553), 436–444.

Liang, H., Tsui, B.Y., Ni, H., Valentim, C.C., Baxter, S.L. et al. (2019). Evaluation and accurate diagnoses of pediatric diseases using artificial intelligence. *Nature medicine*, *25*(3), 433–438.

Lin, W.C., Chen, J.S., Chiang, M.F. and Hribar, M.R. (2020). Applications of artificial intelligence to electronic health record data in ophthalmology. *Translational vision science & technology*, *9*(2), 13–13.

Luo, J., Wu, M., Gopukumar, D. and Zhao, Y. (2016). Big data application in biomedical research and health care: a literature review. *Biomedical informatics insights*, *8*, BII-S31559.

Ma, S., Tang, N. and Tian, J. (2012). DNA synthesis, assembly and applications in synthetic biology. *Current opinion in chemical biology*, *16*(3-4), 260–267.

Mahesh, B. (2020). Machine learning algorithms-a review. *International Journal of Science and Research (IJSR). [Internet]*, *9*(1), 381–386.

Mahmud, M., Kaiser, M.S., Hussain, A. and Vassanelli, S. (2018). Applications of deep learning and reinforcement learning to biological data. *IEEE transactions on neural networks and learning systems*, *29*(6), 2063–2079.

McCorduck, P. and Cfe, C. (2004). *Machines who think: A personal inquiry into the history and prospects of artificial intelligence*. CRC Press.

Meskó, B. and Görög, M. (2020). A short guide for medical professionals in the era of artificial intelligence. *NPJ digital medicine*, *3*(1), 126.

Park, S., Min, S., Choi, H. and Yoon, S. (2016). deepMiRGene: Deep neural network based precursor microrna prediction. *arXiv preprint arXiv:1605.00017*.

Parveen, A., Mustafa, S.H., Yadav, P. and Kumar, A. (2019). Applications of machine learning in miRNA discovery and target prediction. *Current Genomics*, *20*(8), 537–544.

Raghuvanshi, D. (2018). Artificial Intelligence: Basics and Terminology. *International Journal of Trend in Scientific Research and Development*, *2*(6), 1539–1541.

Ramesh, A.N., Kambhampati, C., Monson, J.R. and Drew, P.J. (2004). Artificial intelligence in medicine. *Annals of the Royal College of Surgeons of England*, *86*(5), 334.

Reddy, S. (2018). Use of artificial intelligence in healthcare delivery. In *eHealth-making health care smarter*. IntechOpen.

Ren, Y., Chakraborty, T., Doijad, S., Falgenhauer, L., Falgenhauer, J. et al. (2022). Prediction of antimicrobial resistance based on whole-genome sequencing and machine learning. *Bioinformatics*, *38*(2), 325–334.

Sagiroglu, S. and Sinanc, D. (2013, May). Big data: A review. In *2013 international conference on collaboration technologies and systems (CTS)* (pp. 42–47). IEEE.

Sarkar, D., Bali, R., Sharma, T., Sarkar, D., Bali, R. et al. (2018). Machine learning basics. *Practical Machine Learning with Python: A Problem-Solver's Guide to Building Real-World Intelligent Systems*, 3–65.

Schatz, M.C. (2015). Biological data sciences in genome research. *Genome research*, *25*(10), 1417–1422.

Selvaraj, J., Prabha, T., Kumar, T.D.A. and Palaniappan, S. (2022). Artificial Intelligence in Biomedical Image Processing. In *Machine Learning and Systems Biology in Genomics and Health* (pp. 147–188). Singapore: Springer Nature Singapore.

Shabbir, J. and Anwer, T. (2018). Artificial intelligence and its role in near future. *arXiv preprint arXiv:1804.01396*.

Shrestha, A. and Mahmood, A. (2019). Review of deep learning algorithms and architectures. *IEEE access*, *7*, 53040–53065.

Singh, A., Majumder, A. and Goyal, A. (2008). Artificial intelligence based optimization of exocellular glucansucrase production from Leuconostoc dextranicum NRRL B-1146. *Bioresource Technology*, *99*(17), 8201–8206.

Suzuki, K. (Ed.). (2011). *Artificial neural networks: industrial and control engineering applications*. BoD–Books on Demand.

Tarca, A.L., Carey, V.J., Chen, X.W., Romero, R. and Drăghici, S. et al. (2007). Machine learning and its applications to biology. *PLoS computational biology*, *3*(6), e116.

Vailati-Riboni, M., Palombo, V. and Loor, J.J. (2017). What are omics sciences? *Periparturient diseases of dairy cows: a systems biology approach*, 1–7.

Vargas, R., Mosavi, A. and Ruiz, R. (2017). Deep learning: a review.

Villalobos-Alva, J., Ochoa-Toledo, L. (2022). Protein Science Meets Artificial Intelligence: A Systematic Review and a Biochemical Meta-Analysis of an Inter-Field. *Frontiers in Bioengineering and Biotechnology*, *10*, 788300.

Westbrook, J., Feng, Z., Chen, L., Yang, H. and Berman, H.M. et al. (2003). The protein data bank and structural genomics. *Nucleic acids research*, *31*(1), 489–491.

Wu, G., Feng, X. and Stein, L. (2010). A human functional protein interaction network and its application to cancer data analysis. *Genome biology*, *11*, 1–23.

Xu, Y., Liu, X., Cao, X., Huang, C., Liu, E. et al. Artificial Intelligence: A Powerful Paradigm for Scientific Research. Innovation 2021, 2, 100179.

Yang, X., Wang, Y., Byrne, R., Schneider, G. and Yang, S. et al. (2019). Concepts of artificial intelligence for computer-assisted drug discovery. *Chemical reviews*, *119*(18), 10520–10594.

CHAPTER

21

Artificial Intelligence in Bioinformatics

Abhishek Subramanian[1*] and ***Nishant Kumar***[1]

1. Introduction

Every living organism demonstrates an organizational hierarchy starting from biomolecules (genes, transcripts, proteins), pathways, cells, tissues and organs which finally constitute an individual. This hierarchy spontaneously emerged during evolution to ensure survival and proliferation of a species. However, this organization corresponds to complexity arising from the behavior of simultaneously functioning components at various biological time and length scales. Atomistic / molecular functions of biomolecules such as DNA, RNA, proteins, and metabolites typically function at a length scale of nanometers (10^{-8} to 10^{-9} meters) and timescale of microseconds (10^{-6} seconds). Pathway function, the next level of organization, takes place at a length scale of micrometers (10^{-6} to 10^{-7} meters) & timescale of a few seconds (10^{2} seconds) whereas entire cells function at length-scales of micro to millimeters & a timescale of minutes (Butcher et al., 2004). Finally, an entire tissue functions within a millimeter length scale and a timescale of hours. The variations in these time and length scales of various biomolecules, pathways, cells, and their interactions give rise to biological diversity in response to environmental changes. Fundamental mechanisms underlying biological functions can be discovered only if the quantitative changes in multiple biological entities (genes, transcripts, proteins, cells, tissues) across time-length scales and environmental conditions can be measured.

1.1 High-throughput Omics Technologies

Fundamentals of biological functions in organisms can be realized accurately only through a collective measurement of qualitative/quantitative changes in multiple biological entities (genes, transcripts, proteins, cells, tissues) and their interactions across time-length scales and environmental conditions. High throughput technologies that measure such qualitative / quantitative changes across the hierarchical organization are revolutionizing biology and opening new avenues to perform confirmatory analytical experiments (Figure 1). For measuring high throughput changes in multiple

[1] Department of Biotechnology, Indian Institute of Technology Hyderabad, Telangana 502284, India.
* Corresponding author: abhisheks@bt.iith.ac.in

molecules in each condition and their interactions, 'omics' technologies have become a standard in modern biology. Depending upon the organization level being characterized, the suffix -ome is attached to annotate that pool of biological objects (e.g. the total DNA nucleotide pool that gives information regarding genes and non-coding regions is together called a genome). Thus, genomics, transcriptomics, proteomics, metabolomics are some popular technologies that characterize the biomolecular pool in question (genes, transcripts, proteins, metabolites) respectively. In genomics and transcriptomics, complete nucleotide pools of DNA or RNA are sequenced to identify – (a) sequence-based patterns that constitute protein-coding and non-coding regions and (b) to quantify the levels of transcripts associated with regions of the genome. Proteomics and metabolomics utilize mass spectrometry technologies to quantify protein and metabolite abundances. On the other hand, ChIP sequencing, ATAC sequencing are for the measurement of transcription-factor DNA binding (Li et al., 2019); yeast-two hybrid, co-immunoprecipitation assays (Tanowitz & Zastrow, 2004) are used to characterize large-scale protein-protein interactions. Similarly, affinity mass spectrometry enables rapid screening of drugs (& other ligands) that can bind to a specific biomolecular target. These techniques are collectively termed as 'interactomics'.

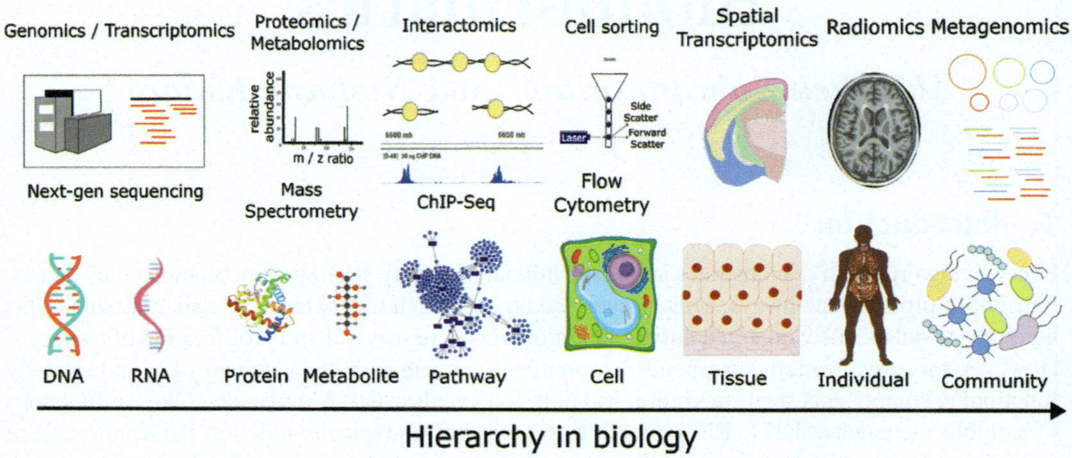

Fig. 1 Hierarchical organization in biology and associated high throughput technologies.

Like molecule-based characterization, technologies like flow cytometry capture cellular information. Cells labeled with fluorescent markers (that establish the identity of the cell) are suspended in a fluid and injected into a flow cytometer such that only one cell is focused on at a time and allowed to flow through a laser beam, where light is scattered to characterize & count that cell type. Recently, flow cytometry has been largely coupled to single-cell omics-based sequencing technologies allowing the characterization of molecular components of every single cell. Similarly, omics technologies are also being coupled with spatial information obtained from histopathological tissue sections and live-cell imaging techniques to characterize tissue-level cellular function based on molecular markers. At an organ-level, a new field of radiological research called radiomics is attempting to leverage routine patient image scans for decoding disease phenotypes (especially in tumors) (Wang et al., 2022). Apart from an individual perspective, omics technologies are gaining traction in understanding species-level interactions as well. Metagenomics and meta-transcriptomics techniques allow the extraction of the genome and transcriptome from a community of organisms to characterize the whole repertoire of micro-organisms surviving in a bulk environmental sample (e.g. human skin, lung, soil, water). The interdisciplinary field of bioinformatics deals with the development of computer models / applications for comprehending such big biological data.

1.2 Big Data in Biology

As opposed to the traditional hypothesis-driven science, also known as day science, where we falsify hypotheses and move onto related hypotheses in small steps, analysis of high throughput biological data has opened up a modern approach called exploration-driven science, also known as night science, which allows the discovery of an unstructured realm of possible hypotheses thereby, helping to create novel hypotheses for testing (Yanai & Lercher, 2019). However, any biological process spans across time-length scales of biological hierarchy and requires the involvement of multiple biological variables working together in unison. Thus, an ideal way to get a holistic perspective of biological functions would be to integrate data from these different hierarchical layers in a logical, interpretable manner (Figure 2A). However, this is a non-trivial and extremely challenging task because of the high dimensionality that exists in biological data. If we look at the numbers associated with every biological data type in a single human individual, the complexity in biology can be realized (Figure 2B).

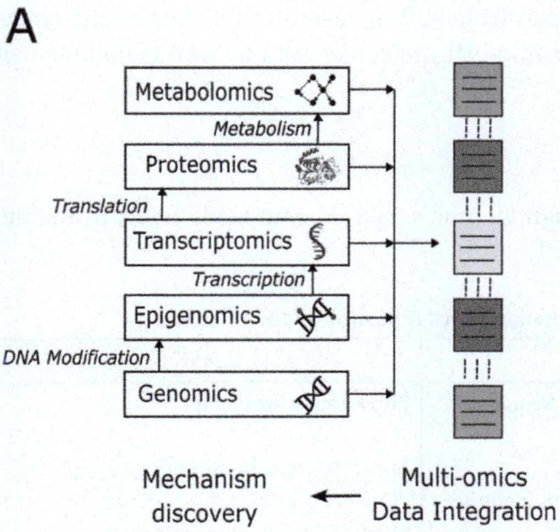

B One Human Individual

# organs	78
# cell types	200
# cells in body	100 trillion cells
# protein-coding genes per cell	~ 20000
# mRNA molecules	~ 300,000
# protein molecules in a cell	~ 10^9
# base pairs in genome per cell	3 billion base pairs
# diseases in humans	~26000

Fig. 2 Biological big data. This figure demonstrates – (A) A schematic of multi-omics data integration where multi-modal omics data is integrated by finding common patterns and (B) The number (#) of different biological variables that occur within a single human individual. This figure demonstrates the multi-modality and high dimensionality of biological big data.

With the advent of high throughput technologies like nucleotide sequencing, whole -omes of human individuals, cell types, various species and strains on earth are rapidly being sampled to discover new mechanisms in the fields of healthcare, agriculture, bioprocess technologies, environmental impact, biodiversity, conservation and evolution. It is presumed that genomics will soon be on par with three other major generators of big data-astronomy, YouTube and Twitter in terms of data acquisition, storage, distribution and analysis (Stephens et al., 2015). This explosion of data, although unprecedented, holds promise in advancing our understanding of biological systems, diseases, development of drugs, diagnostic tests, development of new strains/varieties of crop plants and improving environmental impact.

1.3 Machine Learning Applications

The entire field of bioinformatics and systems biology is being dedicated towards the integration of high-throughput multi-omics data where mathematical and statistical models are employed to identify causal associations between different biological entities. One such popular methodology which enables seamless integration is machine learning (ML). ML is a branch of artificial intelligence which deals in developing algorithms that can learn patterns from data or the underlying data structure without the need for interference from a human programmer. ML techniques outweigh rule-based systems which usually tend to fail outside test scenarios with unseen data (Goodswen et al., 2021). Prediction of molecular biomarkers / gene signatures associated with a disease type using features derived from omics datasets (Sung et al., 2012), cell types in a tissue using marker sets derived from single-cell gene-expression patterns (Nouri et al., 2023), classification of organisms based on genetic/genomic sequences (Kouduka et al., 2007), prediction of essential genes for survival using genotype – phenotype features (Cacheiro et al., 2020), prediction of drug response (e.g. half-maximal inhibitory concentration) using gene expression or protein abundance (Adam et al., 2020), target gene expression using transcription factor expression to infer gene regulatory networks (Zhang et al., 2023), sequence-based evolutionary rates using genotype-phenotype factors (O'Connor & Mundy, 2009), cell-state transitions from image data (Szkalisity et al., 2021) are some popular ML problems in biology. As ML-based techniques derive repeating patterns from high-dimensional data in an efficient manner, machine learning is well suited for multi-omics data integration as well (Feldner-Busztin et al., 2023). Apart from bulk-level molecular data, multi-omics integration techniques have extended also to the single-cell molecular data as well (Subramanian et al., 2022).

2. Omics Repositories

There are multiple repositories on which high throughput omics data is being made freely available. Some of the popular repositories are listed in Table 1.

Table 1 Table listing omics data types, popular omics repositories and their links.

Datatype	*Repository*	*Links*
Genomics	1) INSDC (International Nucleotide Sequence Database Collaboration) 2) Sequence Read Archive 3) National Human Genome Research Institute (NHGRI)	https://www.insdc.org/ https://www.ncbi.nlm.nih.gov/sra/ https://research.nhgri.nih.gov/
Epigenomics	1) IHEC (International Human Epigenome Consortium) Data Portal 2) ENCODE (Encyclopedia of Human Elements)	https://epigenomesportal.ca/ihec/ https://www.encodeproject.org/
Transcriptomics	1) Gene Expression Omnibus 2) Mammalian Transcriptome Database 3) STOmics (Spatial Transcriptomics) Database	https://www.ncbi.nlm.nih.gov/geo/ https://ngdc.cncb.ac.cn/mtd/ https://db.cngb.org/stomics/
Proteomics	1) Peptide Atlas 2) ProteomeXchange	https://peptideatlas.org/ https://www.proteomexchange.org/
Metabolomics	1) Metabolights 2) Metabolomics Workbench 3) Human Metabolome Database	https://www.ebi.ac.uk/metabolights/ https://www.metabolomicsworkbench.org/ https://hmdb.ca/
Radiogenomics	1) The Cancer Imaging Archive (TCIA) 2) Cancer Data Common Repository	https://imaging.cancer.gov/informatics/cancer_imaging_archive.htm https://datascience.cancer.gov/data-commons/repositories
Metagenomics	1) EBI Metagenomics 2) HumanMetagenomeDB	https://www.ebi.ac.uk/metagenomics https://web.app.ufz.de/hmgdb/

3. Nuts and Bolts of Machine Learning Data Analysis

Humans consciously understand the world around them by observing, learning, and reacting to new situations by predicting the action based on the information that was gathered and learnt. Let us consider the simple example of a child learning to ride a bike for the first time. The child does not know anything about the mechanics or physics that drives the movement of a bike; however, by a process of observation, falling many times and getting up (trial-and-error), instinctively predicts possibilities of maneuvering the bike. In other words, the child made a sufficiently accurate mental 'model' after repeating similar actions multiple times and refining the model at every trial-and-error step. Similarly, machine learning (ML) is inspired by the cognitive process of humans to learn from experience. ML does this learning-by-experience exercise by fitting a model on given data iteratively until it achieves sufficient precision and accuracy. Machine learning models have the capability to adapt to new data flexibly and improve. Unlike statistical tests that involve comparisons between finite sets of groups and modest dataset sizes, machine learning makes use of information from vast quantities of data by finding correlated patterns between data points. Machine learning algorithms are broadly classified into supervised, unsupervised, semi-supervised and reinforcement learning algorithms.

Supervised learning algorithms build models on datasets with known quantities or mutually exclusive *classes* (Box1). These models often explain the association of individual data points with other data points (regression task) or with the group (label) that it belongs to (classification task). In other words, typically, the inputs and outputs are known, and the model maps the inputs to their respective outputs. Supervised learning models associated with classification tasks are called classifiers and those with regression tasks are called regressors.

Unsupervised learning algorithms build models on datasets without the requirement of any prior information (no label) regarding the dataset. In other words, input and output combinations are unknown. The relationship between data points is identified automatically based on correlated patterns that exist between data points. Unsupervised algorithms extract generative features, identify meaningful structures in the data, find clusters and exploratory purposes.

Semi-supervised learning algorithms are a hybrid of supervised and unsupervised learning methodologies typically used to find clusters in datasets where there is a lot of unlabelled data and only a few labeled data points for training. Typically, an initial classification model is built on labeled cases and the unlabelled cases are 'labeled' accordingly. The newly labeled data points are

Box 1 Terminologies

Class and class labels: Mutually exclusive, discrete set of values returned by a classifier are called classes. In cases, where these values need not be mutually exclusive, they are called classes.

Training set: refers to a subset of instances belonging to the original dataset that are used to 'train' the model. Training essentially refers to the fitting of linear or non-linear mathematical/statistical models that operate on features.

Parameters: Most of the ML models contain **parameters** that are estimated after fitting the ML model to data.

Hyperparameters: Sometimes new parameters that are not part of the model are added to the existing ML model. Such parameters that are not inherent to the modelling approach but are added to the model during training to avoid model overfitting are called **hyperparameters**. Hyperparameters can be tuned to improve model performance.

Validation set: refers to original dataset minus the training subset that can be used to evaluate model performance, especially to test for overfitting.

Test set: refers to a blind dataset which the model has never seen during training nor validation.

Loss or cost function: Mathematical functions that measure the deviation of the predicted outputs of the machine learning model from the actual values. E.g. mean squared error for regression problems and binary cross entropy for classification problems.

used to perform re-training. Only those newly labeled cases that meet a threshold are used for re-training.

Reinforcement learning enables a software agent (like a PAC-MAN video game) to perform a random trial-and-error procedure scanning a vast number of possible input combinations to achieve an output and grade the performance of every possible path taken to achieve a solution, such that the solution improves every iteration. For building a ML model from scratch, a few mandatory steps need to be followed pre-, during and post-learning'.

3.1 Pre-processing the Dataset

A typical 'input dataset' consists of a I × *F* matrix of *I* data points also called 'instances' and *F* properties of the data points called 'features'. For example, let us consider we are working on a transcriptomics dataset pertaining to a human cell. The given cell contains 20000 protein coding genes, and we are measuring the expression of each gene (transcript levels) across 1000 time points with an aim of classifying genes as disease-causing or having neutral or no effect (matrix of 20000 × 1000). The 20000 genes are instances, and their measurements across 1000 time points signify the gene expression properties. Based on known data, genes are 'labeled' either as disease-causing or neutral genes (Box 1). It is important to know that a minimum dataset size is required to properly train the ML model. It is a thumb rule to consider datasets with instances at least 10 times the features. Typically, the features obtained from the measuring instruments (e.g. sequencer, mass spectrometer) is raw, unprocessed data, which are usually absolute machine measurements which can also be associated with erroneous measurements (technical errors) or missing data (in case of single-cell studies). Similarly, data obtained from annotation databases might have errors as human curators curate annotations from datasets belonging to different laboratories or experimental platforms leading to inconsistencies (human errors) (Prada & Boore, 2019). This cannot be used as it is meant for the modeling process and it will strongly impact the machine learning model performance. Thus, it is the most important step in a ML pipeline. Data pre-processing or data scrubbing can be performed in multiple ways like feature engineering, feature selection and feature extraction.

3.1.1 Feature Engineering

Feature Engineering (FE) is a feature modification methodology that transforms features from raw data. The first step in feature engineering is to clean the data. Data cleaning techniques help to reduce the bias introduced by missing data. If the instances have missing feature values, either the inappropriate instance (row) needs to be removed or the missing value needs to be imputed using mean/median substitution or random values. The second step is to scale the features in relatively comparable ranges using techniques like normalization or standardization. Feature normalization techniques scale the features in the [0, 1] or [−1, 1] interval with respect to the range of the data (min-max normalization), maximum of the data (maximum absolute normalization), mean of the data (mean normalization) or median (median normalization). Feature standardization techniques calculate a Z-score that scales all the features in comparable ranges and preserves the original distribution. Min-max normalization and standardization are the most preferred feature scaling techniques (Eq. 1, Eq. 2).

$$X' = \frac{X - \min(X)}{\max(X) - \min(X)} \qquad ...(1)$$

$$X' = \frac{X - \mu_X}{\sigma_X} \qquad ...(2)$$

where *X* is any given feature, min(*X*) denotes minimum value, max(*X*) denotes maximum value, μ_X is the mean of *X* and σ_X is the standard deviation of *X*.

Tip 1: We recommend a standardization procedure for omics data as all features have comparable means and variances. Standardization is also less sensitive to extreme outliers and preserves the original distribution of the data.

3.1.2 Curse of Dimensionality and Bias-variance Trade-offs

As mentioned above, biological data has a multivariate nature with multiple correlations between instances. In most high throughput biological datasets, feature dimensions are typically so high that there are insufficient instances that cover the entire feature dimension space (volume) due to intrinsic correlations and redundancy between instances. This leads to a sparsity in data and issues where analyzing data in high dimensional spaces becomes a complex problem. This problem is called "the curse of dimensionality". The sparsity in data and the unstructured arrangement in n-dimensional space often leads to models that over- or underfit the data. This leads to prediction errors in the models. There will always be a trade-off between bias and variance to obtain a less accurate but well-balanced model. Bias refers to the difference between the average prediction of the model (predicted value) and the correct value. High bias models pay very little attention to training data thereby oversimplifying predictions leading to high training and testing errors. Variance refers to the variability or degree of scatter or spread of the instances. High variance models pay attention to the training data arrangement but do not generalize well with new, unseen data and hence, have poor testing prediction rates. Models with high variance and low bias are overfitted models whereas models with high bias and low variance are under fitted models. The challenge is to find a well-balanced model with low bias and low variance.

3.1.3 Feature Selection

Feature selection (FS) methodologies help to select the most relevant feature dimensions in the dataset that can optimize model performance. Feature selection avoids overfitting in models by reducing variance in the data. As FS techniques reduce the number of feature columns, the computational effort required to build the model effectively reduces. The ideal way to select features is to generate all possible combinations of features and test the model performance for each combination. In other words, if we consider the dataset has 'n' features and let's say we want to classify instances into two classes. If we use all possible combinations of features to generate the model, around 2^n–1 model evaluations would be required. To avoid this, there are multiple automated feature selection techniques that intelligently pre-select features prior to model building.

Filter methods are computationally inexpensive methods that use intrinsic properties typically univariate statistics to select the best features. Parametric/non-parametric correlation methods like Pearson's correlation coefficient, Chi-square test, Fisher's score are used to detect correlated features, remove them and thereby, select independent features for model building. Information theoretic methods like InformationGain, utilize mutual information to predict features that have the most information and thereby, the least entropy relative to the rest of the features. A skewed distribution has a larger information gain and hence, less entropy as compared to a balanced, equidistant distribution.

Wrapper methods are the easiest to use and do not require prior knowledge in statistics unlike filter methods. Wrapper methods search for possible feature combinations that can optimize model performance. These methods iteratively perform model learning based on individual feature combinations, evaluate the model performance, and thereby choose the combination that demonstrates the best performance. In forward feature selection (FFS) individual features are used to build the model. Based on the performance, all features are ranked from the highest to the lowest performing. Iteratively, feature combinations are generated by adding one feature at a time to the previous feature to check whether model performance increases as compared to the individual features or the previous combination. If the model performance based on the present combination decreases as compared to previous combination, the combination of features in the previous iteration is retained as the best feature set. In backward feature selection (BFS), the sorting is reversed from worst performing to best performing features and again iterative selection of feature combinations is performed as previously entailed. Instead of FFS and BFS, it is also possible to use recursive feature elimination (RFE) where an external estimator is built with each set of features; the weight of the RFE estimator can be used to sort features just like FFS. RFE weight estimation is a faster procedure than FFS.

Embedded methods add the benefits of both the filter and wrapper methods where feature contributions are penalized at every iteration (e.g. LASSO regularization for regression) to find feature combinations that do not overfit the data but still show good performance (well-balanced model). In other words, feature selection is integrated into the learning process itself.

Tip 2: We recommend using Recursive Feature Elimination or embedded methods like regularization to find relevant features as both use mathematical estimators that are derived or added to existing ML algorithms.

3.2 Classic Machine Learning Problems

Machine learning problems can be broadly classified into classification, regression and clustering. Classification implies the prediction of discrete output variables (groups of objects) based on some features/properties. Regression implies the prediction of a continuous output variable (quantities) based on the quantities of the known variables. Cluster analysis or clustering is an unsupervised learning procedure for identifying and grouping data points in large datasets without a specified rule- or prior-information based outcome. These classical techniques have been applied to complex biological problems for a long time (Table 2).

Table 2 Machine learning applications for complex biological problems

Algorithms	*Applications in biology*
A. Classification algorithms	
Naive Bayes	Taxonomic classification of 16S rRNA amplicons (Gao et al., 2017) Assignment of rRNA Sequences into the New Bacterial Taxonomy (Q. Wang et al., 2007)
Logistic regression	Identification of colorectal cancer patients having mutation in DNA repair genes (Barnetson et al., 2006) Identification of non-small lung cancer using circulating plasma DNA as diagnostic biomarker (Paci et al., 2009)
k-Nearest Neighbor	Microarray gene expression analysis (Parry et al., 2010) Identification of disease genes (Xu & Li, 2006)
Support vector machine	Prediction of Sub-cellular location of Protein (Yu et al., 2006) Prediction of metabolic gene essentiality in E. coli using genome, transcriptome and metabolic flux features (Nandi et al 2017)
Decision tree	Identification of Genes in the Vertebrate DNA sequence (Salzberg et al., 1998) Predicting Gene regulatory network (Middendorf et al., 2004)
Random forest	Gene selection and classification of microarray data (Díaz-Uriarte & Alvarez de Andrés, 2006) Detection of Epistatic interactions (Jiang et al., 2009)
ANN	Superfamily Classification of Unknown Proteins (Wu et al., 1992) Prediction of structural and functional features from Protein and DNA sequences (Hirst & Sternberg, 1992)

Contd.

Table 2. *Contd.*

CNN	Identifying the sequence specificity for DNA-Protein binding (Zeng et al., 2016) Identifying Gene regulatory network from scRNA-seq data (Mao et al., 2023)
Autoencoders	Multi-omics analysis of Ovarian Cancer (Hira et al., 2021) Cancer classification using Imbalanced gene expression datasets (Arafa et al., 2023)
B. Regression algorithms	
LASSO regression	Generating Transcriptome Assembly using RNA-seq data (W. Li et al., 2011) Construction of MicroRNA-target regulatory networks (Lu et al., 2011)
Ridge regression	Predicting Peptide-Protein binding affinities (Giguère et al., 2013) Predicting Gene Function (Mostafavi et al., 2008)
C. Clustering techniques	
K-means	Identification of new drugs for lung cancer (J. Lu et al., 2016) Analyzing Protein Sequence motifs representing common structural properties (Zhong et al., 2005)
DBSCAN	Studying Myeloid Heterogeneity in the progression of kidney disease (Conway et al., 2020) Mapping the human Dendritic cells lineage (See et al., 2017)
Hierarchical clustering	Classification of lung cancer cell lines on the basis of DNA methylation markers (Virmani et al., 2002) Detection of Leukemic Cell Population from flow cytometric data (Fišer et al., 2011)
Graph-based clustering	Reconstruction of human transcriptional regulatory modules (Yan et al., 2007) Large scale multiple sequence alignment (Smirnov & Warnow, 2020)
Non-negative matrix factorization	Cancer Class Discovery (Y. Gao & Church, 2005) Identification of disease-related miRNAs (Xiao et al., 2017)

3.2.1 Classification

Classification problems are binary, multi-class or multi-label in nature.

Binary classification refers to the grouping of instances into two classes.

Multi-class classification refers to the grouping of instances into multiple classes (> 2 classes). This is typically useful when there is a range of groups possible in the dataset.

Multi-label classification refers to a specific case, where an instance can be associated with multiple classes. It is a special case of multi-class classification where the classes involved in the problem are hierarchically structured. Many classification algorithms are proposed in ML and data science literature. Algorithms for classification analyses include Naive Bayes, logistic regression, k-nearest neighbors, support vector machines and decision trees. Apart from these classical algorithms, there are ensemble learning algorithms like gradient boosting, adaptive boosting (AdaBoost), that aggregate predictions from multiple weak learning algorithms to improve prediction performance and bagging algorithms like random forest which perform random sampling with replacement to get decision trees of different shapes and sizes whose individual performances are averaged to get a consensus prediction.

3.2.2 Regression

As opposed to classification, regression helps in the prediction of continuous, unknown dependent quantities given known, independent quantities. Simple or multiple linear regression is the simplest and most popular ML technique.

Simple linear regression models the relationships between the dependent quantity or variable with one or more independent variables.

Polynomial regression models the non-linear relationship between dependent and independent variables using a *nth* degree polynomial of the independent variable to predict the dependent variable values.

LASSO and ridge regression are unique and powerful versions of the linear/polynomial regression which are built when there are usually many independent variables as they offer regularization hyperparameters to reduce overfitting and complexity of the regression models. LASSO regularization aims to shrink coefficients of some independent variables to zero thereby selecting the most relevant predictors (independent variables) for regression. Ridge regression on the other hand makes the coefficients small but never zero. Ridge regression performs better when there is a "multi-collinearity" between predictors.

3.2.3 Clustering

Clustering helps to find unknown structures/patterns in the data by grouping similar instances. Clustering techniques can be summarized into partitioning, density-based and hierarchical methods.

Partitioning methods cluster data points into multiple groups based on the feature similarities. The challenge is to pre-set the number of expected clusters that can be seen in the data. K-means and K-medoids are examples of partitioning methods.

Density-based methods on the other hand, identify distinct groups with a differential density of points (either high or low density) in the feature space. These high- or low-density contiguous regions that are isolated from each other are designated as clusters. Points that are not part of the cluster are considered as noise. Such methods struggle to cluster data with similar densities or scatter. DBSCAN is a popular example of density-based clustering.

Hierarchical clustering makes a hierarchical tree structure such that similar data points cluster within every branching in a tree. Hierarchical clustering can be agglomerative or divisive. Agglomerative hierarchical clustering is a bottom-up approach of cluster formation that starts with a pair of similar instances combined as the same cluster and thereby, moves up within a tree. Divisive hierarchical clustering is a top-down approach of cluster formation that starts clustering assuming all points occur within the same cluster and then splits/divisions are performed within the cluster thereby, recursively traversing down the hierarchical tree.

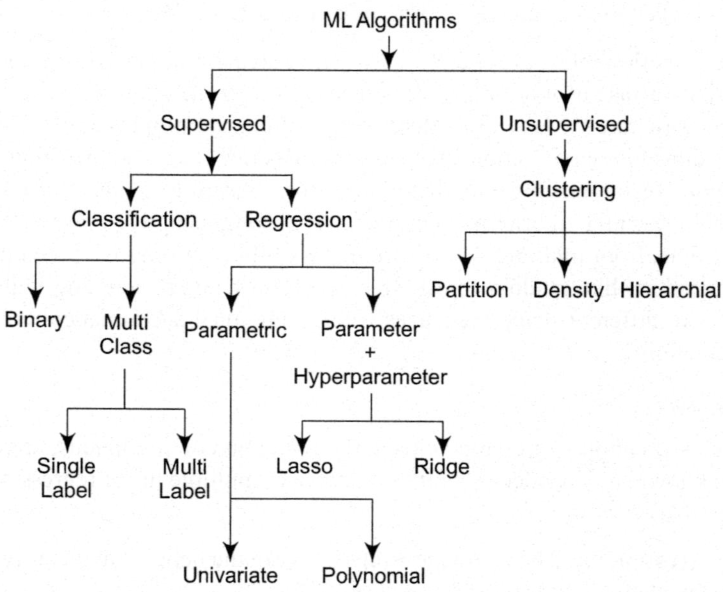

Fig. 3 Types of classic machine learning algorithms

3.2.4 Cross Validation

In supervised learning, the original dataset is split into training, validation and test datasets (Box 1). Usually, the test dataset is kept aside even before model building begins such that it is completely unseen by the ML model. For splitting the dataset into training & validation sets and to test the ability of the built model to predict unseen data (data not used), cross-validation methodologies are used. The aim of cross-validation techniques is to reduce the variance error of the model. Let us consider whether there are *m*-points in a dataset. In ***k*-fold cross validation**, the dataset is split into *k*-folds iteratively such that every fold has a subset of the total *m*-points, out of which, *k*-1 parts are used for training and the remaining one-fold is used for validation in every iteration. In **leave-one-out** cross validation, one point in every iteration is used for validation and the rest of the *m*-1 points are used for training. In **Monte-Carlo cross-validation**, the number of folds (*n*) being used for validation is randomized using repeated random subsampling and the rest of *k-n* folds are used for training. In all cross-validation cases, the prediction error for each train-validation split is averaged-out to get a relatively lower average error.

3.2.5 Imbalanced Data Problem

An **imbalanced dataset** is a dataset in which one class is over-represented as compared to others. Biological/bioinformatics datasets often have a class imbalance. For example, if we want to develop a classifier that can differentiate between essential and non-essential genes, prior experimental literature will typically have more information about essential genes as compared to non-essential genes. In such situations, the unequal ratio in the dataset will create a problem in model training because the classifier would be biased towards the class containing more instances. There are multiple techniques to handle imbalance in the data. Needless to say, the best way is to sample for more instances. However, there might be experimental limitations in certain cases to acquire more data. In such cases, a common strategy is to perform **data class weighting** where different weights are assigned to instances depending on their membership in the majority or minority class. Another method is to perform **under-sampling** where you attempt to downsample the instances in the majority class repeatedly so that it can equate with the number of instances in the minority class and thereby, represent an average prediction behavior of the model.

3.2.6 Evaluation of Model Performance

Model performance refers to the ability of a ML model to predict true outcomes. Many metrics measure the deviation of the predicted outputs of the machine learning model from the actual values (Rainio, O., et al. [2024]).

Regression models:

For regression models, metrics called loss or cost functions that measure the deviation of the predicted outputs of a regression model from the actual values are used. **Mean squared error (MSE)** is a cost function that penalizes even small errors by squaring them, which will magnify errors in the model. Due to the squaring factor, it's fundamentally more prone to outliers than other metrics. **Root Mean Squared Error (RMSE)**, on the other hand, handles the penalization of smaller errors done by MSE by square rooting it. The lower the RMSE the better the fitted model. Error interpretation can be done smoothly since the scale is now the same as the random variable. Since scale factors are essentially normalized, it's less prone to struggle in the case of outliers. **R^2, also called, coefficient of determination or least squared error**, is the proportion of the variance in the dependent variable that is predictable from the independent variable. If the numerator is small, R^2 will be close to 1 (Ideal), the regression was able to capture 100% of the variance in the target variable. Conversely, if the numerator is high, R^2 will be close to 0, the regression won't be able to capture any variance in the target variable.

Classification models

For classification models, the metrics are calculated based on the notion of a **confusion matrix.** The confusion matrix is a table showing the comparison of actual and predicted (positive and negative) values. For understanding the confusion matrix, let us consider a binary classification problem where one class is considered a positive class, and the other class is considered a negative class. The confusion matrix contains two rows and two columns that report the number of true positives, false negatives, false positives, and true negatives. **True positives (TP)**, as the name suggests, refers to the number of actual positive class instances correctly predicted by the model. **True negatives (TN)**, like the true positives, refers to the number of actual negative class instances correctly predicted by the model. **False positives (FP)**, as the name suggests, refers to the number of actual positive class instances that are incorrectly predicted to belong to the negative class. **False negatives (FN)**, as the name suggests, refers to the number of actual negative class instances incorrectly predicted to belong to the positive class. Based on the confusion matrix, several metrics can be calculated for measuring the quality of classification by a classifier. **Accuracy** measures the fraction/percentage of correctly predicted instances from the total set of instances provided for training. **Precision**, also called the positive predictive value (PPV), is the percentage of correctly predicted positive instances from the total set of predicted positive instances. **Sensitivity**, also called the true positive rate (TPR), measures the number of true correctly predicted positive instances from the total set of actual positive instances. **Specificity,** also called the true negative rate (TNR), measures the number of true correctly predicted negative instances from the total set of actual negative instances. **F1 score** is the harmonic mean of precision and sensitivity. F1 score is scaled between 0 to 1, where 1 indicates perfect precision and recall, and 0 indicates either precision or recall is zero.

Receiver-Operating-Characteristic (ROC) curve is a popular and most useful way to visualize the performance of a classifier model. The ROC curve is a graphical plot demonstrating TPR vs. FPR. The TPR and FPR are calculated at different thresholds of parameters and plotted onto the ROC curve. If the ROC curve is towards the upper left quadrant of the plot, the **area under the Receiver-Operating-Characteristic curve (auROC)** is the highest suggesting the model has a high sensitivity and a high specificity (low false positive rate). The auROC ranges from 0 to 1. An auROC = 0 suggests that the model has poor sensitivity & specificity, auROC = 0.5 suggests that the model does not behave better than a random model and auROC = 1 suggests that the model is both sensitive and specific. Matthew's correlation coefficient (MCC) measures the association between the observed and predicted values and returns a value between −1 to +1 indicating the strength of the association. A coefficient of +1 represents a perfect prediction, 0 suggests that the prediction is no better than a random prediction and −1 indicates complete disagreement between the prediction and observation.

4. Machine Learning Breakthroughs in Bioinformatics

Machine learning is being extensively applied for certain high throughput data in modern biology so much so that it has become a routine procedure for data analysis. Machine learning is creating breakthroughs in the field of protein structure prediction (AlQuraishi, 2021), multi-omics data analysis (Reel et al., 2021), single-cell omics technologies (Satija et al. 2015, Aibar et al. 2017) and gene regulatory network predictions (Mochida et al., 2018, Browaeys et al. 2019).

4.1 Graph-based Clustering for Identification of Cell Types Using Single-Cell Transcriptomics

A multicellular organism contains trillions of cells working together to perform biological functions. Different cell types are organized in tissues and tissues form organs. Every tissue has multiple cell types that contribute to overall tissue/organ function. Each cell-type behaves differently depending upon its characteristic property. Moreover, cell progenitors differentiate into cell types with respect to different developmental time points (e.g. zygote to embryos), environmental conditions (e.g. disease, wound healing, tissue damage) and response to chemical treatments. Characterization of such cell-state transitions is a challenging task as transitions can be reversible, irreversible, or highly plastic (Mulas et al., 2021). Cell-type identification has a wide variety of applications such as prediction of cell lineage, ligand receptor interactions, disease diagnosis, tissue engineering and drug designing. Single cell RNA sequencing technologies allow us to characterize cell types and cell states in a high-throughput manner and categorize them based on each single-cell's transcriptomic profile. Such studies provide a snapshot view into the cell types that exist at a given time within the tissue.

There are multiple single-cell data analysis workflows that attempt to group single-cells based on variability in gene expression across single-cells like Seurat, ScanPy, scran (Luecken & Theis, 2019). One such example is the non-parametric, instance-based, graph clustering algorithm implemented in Seurat. A typical single-cell RNA-seq data analysis pipeline starts with a data matrix of genes as features and single-cells as the samples. As introduced above, Seurat pre-processing workflows perform quality control to remove low quality / dying cells, normalize the data, identify highly variable genes by modeling mean-variance relationships (feature selection) and perform dimensionality reduction using PCA to extract new features (feature extraction) to obtain a processed matrix ready for automatic identification of clusters. Graph-based clustering attempts to embed single-cells in a graphical structure and uses community detection algorithms to detect clusters within this graph. Seurat does hierarchical clustering in 2 steps:

(1) Using the PCA – transformed matrix, Seurat constructs a k-nearest neighbors (KNN) graph based on the Euclidean distance between k neighboring cells in PCA space (Figure 4A). The number of principal component dimensions to be chosen can be set by the user. The cells represent the nodes in this graph and edges are drawn between cells that have similarity in gene expression patterns (features) as quantified by the Euclidean distance.

(2) Calculation of a shared nearest neighbor graph within the KNN graph by calculating the shared neighborhood overlap between every cell and its k local neighbors as quantified by the Jaccard similarity distance. In other words, two cells are true local neighbors if they share similar neighbors as well. This represents the edge weight between any two neighbors. After every iteration, the optimization of this edge weight brings similar cells closer to each other and pushes dissimilar cells farther from each other.

(3) Using community detection approaches like the Louvain algorithm on the KNN graph, cells (nodes) are iteratively grouped together. The Louvain method is used to extract non-overlapping communities from large-scale networks. This method optimizes the modularity of a graph. Modules are communities of nodes that share more mutual connections within a community as opposed to fewer connections between communities. Modularity measures the extent of graphs to organize into modules. Modularity has a value between −0.5 (non-modular) to 1 (fully-modular). In the first iteration of the Louvain algorithm, each cell is considered as a community in itself. Each cell is removed from its previous community and inserted into a new community as of another cell and the difference in the modularity of the communities in the previous and current iteration is calculated. If the membership of the cell to a nearby cell community benefits organization (increase in modularity), then it belongs to the new community. This iteration continues until all the cells are assigned to their appropriate communities such that there is no more increase in modularity possible. For the Louvain algorithm, a hyperparameter called "granularity" is optimized where higher granularity suggests a greater number of clusters and vice-versa. This granularity is usually defined by the user depending upon a back-and-forth trial-and-error procedure to get the correct number of clusters according to an expected number of cell types typically known to occur in that given tissue.

(4) Visualization of these communities or clusters using non-linear dimensionality reduction techniques like Uniform Manifold Approximation and Projection (UMAP). UMAP learns the

underlying structure in the dataset such that similar cells are placed together in a lower, reduced dimensional space. Thus, cells grouped together within graph-based clusters thus, co-localize in a better way within these reduced dimension plots.

4.2 Convolutional Neural Network Model for Protein Structure Prediction

Protein folding is a highly complex process, and the structure of any protein is determined by its amino acid sequence. Given the length of an amino acid sequence and the number of possible conformations a protein structure can achieve, it would take an astronomical amount of time for a protein to fold, however, proteins achieve their native structure within tens of microseconds (Dill et al., 2008). Although the general pathways of protein folding are being slowly revealed, the prediction of a native structure of a protein from an amino acid sequence is quite challenging. The neural network model AlphaFold has revolutionized protein structure prediction by generating highly accurate structure predictions of a gold-standard dataset of protein structures, demonstrating the strength of deep learning approaches in solving biological problems (Jumper et al., 2021). Its accurate predictions are opening new possibilities for drug discovery and protein engineering especially in cases where there is a difficulty in solving protein structures experimentally.

Neural networks or artificial neural networks (ANNs) are ML algorithms modeled after the working principle of neurons or neuronal networks in animal brains. In this organization, the input data is converted to output by processing it through a network of interconnected nodes. Depending upon the complexity of the task, each node can be connected to two or more nodes. The nodes are assigned with a weight and bias value. These values are assigned randomly during the initialization of the training phase of the model. When an input data is fed to the node, the data is transformed using weights and transferred onto an activation function. The activation function identifies any non-linear relationship between the input & output and decides whether the node should be activated. The decision to activate the next node depends on whether the calculated value crosses a threshold value or not. The value obtained is calculated by finding the weighted sum. In certain operations, a bias value is added to a node. It is a constant value added to the weighted sum which helps in tuning the model fit to the data. The working architecture of Neural networks is distributed into the **input layer,** which represents the input for the ANN model, **hidden layer**, the main computational component of the model. For a simple ANN, the hidden layer contains a single set of nodes. For deep learning, this hidden layer can contain multiple layers (> 2 layers). The number of layers can be increased further depending upon the complexity of input data and processing tasks. It is at the hidden layer, activation functions are used to establish relationships between inputs and outputs; based on the final decisions, an output is decided at the **output layer.** Deep neural networks are of many types like Convolutional neural networks, recurrent neural networks, autoencoders, perceptrons, generative adversarial networks, reinforcement learning, restricted Boltzmann machine, self-organizing maps, deep belief networks, and more . (Sarker, 2021). AlphaFold uses convolutional neural networks for the prediction of structures.

AlphaFold attempts to combine evolutionary, physical and geometric constraints to create an accurate representation of a native protein structure (Figure 4B). The neural network architecture underlying AlphaFold contains 2 stages:

(1) *Evoformer block:* Evoformer is the building block of the AlphaFold neural network architecture. The Evoformer block jointly embeds a “sequence vs. residue” matrix and a “residue vs. residue” matrix.

   (i) *Sequence vs. residue matrix (Multiple sequence alignment):* AlphaFold workflow starts with an input amino acid sequence of a protein. Using sequence searches, homologs for the input protein are found. These identified homologous sequences are aligned to the input protein sequence using multiple-sequence alignment (MSA). MSA helps to identify conserved blocks of amino acids which occur due to underlying evolutionary constraints (sequence conservation) across species. Every column in a MSA represents a residue within the input protein sequence and every row represents the sequences in which the residue or chemically similar residues are present. This MSA forms the "sequence vs. residue" matrix.

(ii) *Residue vs. residue matrix (Pair representation):* Using attention mechanisms in its neural network architecture, soft weights are assigned to more informative parts of the MSA that can contribute to protein structure prediction. This can be visualized as a graph with edges defined by the proximity of residues. The edge weights represent the distance calculated from the attention weighted MSA. Every block or layer of the Evoformer updates the weights in series and thereby re-calculates pairwise residue distances, ensuring that the spatial relationships are captured by the underlying MSA.

(2) *Structure module:* The embedded MSA - pairwise representation is taken as input for the structure module. The 3D backbone structure is represented as N dependent rotations and translations with respect to a global frame of reference, where N denotes the number of residues represented in the MSA. The rotations and translations are prioritized for the backbone and the location of the side chain is constrained within that global frame of reference. On the other hand, peptide bond geometry is kept unconstrained as breaking the constraint helps in local refinement without solving loop closure issues. The residue gas representation of the input protein sequence is iteratively updated in 2 stages. First, a geometric-aware-attention update is carried out to update the rotation and translation of the local backbone frame of each residue

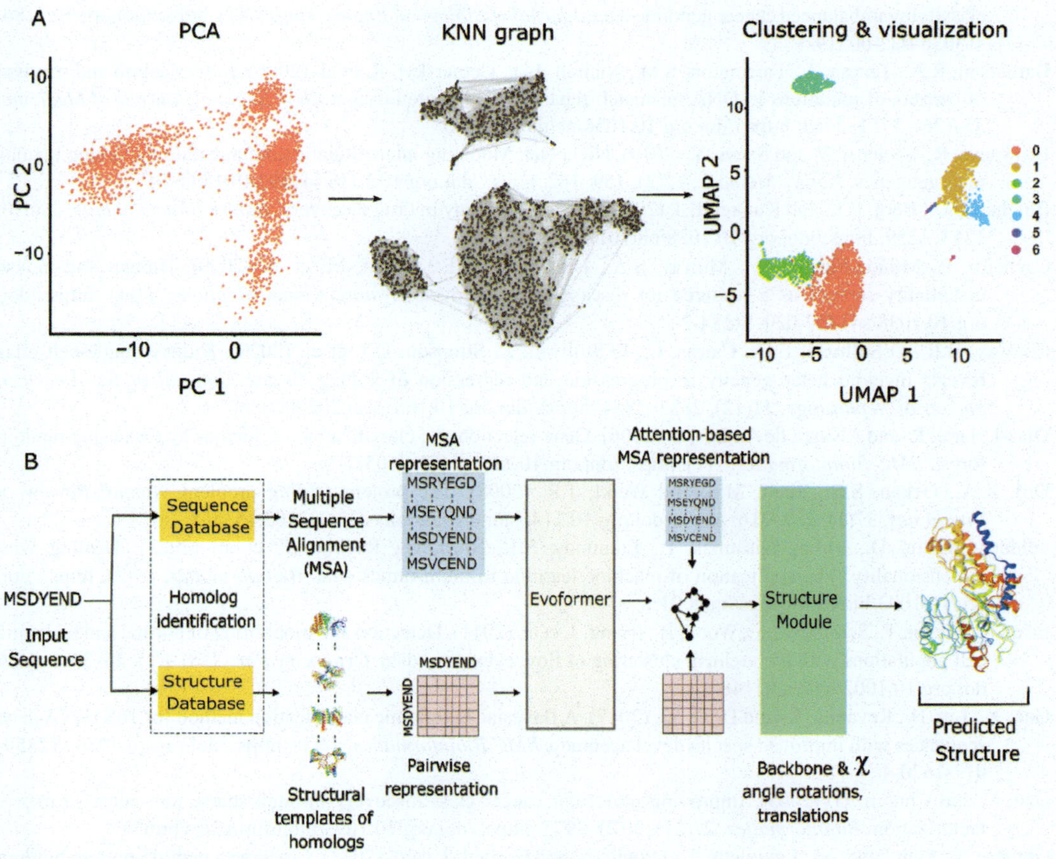

Fig. 4 Modern ML approaches for biological big data analysis. (A) Steps in Seurat workflow for unsupervised graph-based clustering of cell types based on single-cell transcriptomics similarity. (B) Steps in the workflow of the convolutional neural network model AlphaFold for protein structure prediction

using multiple network layers and updation by every layer. Second, side chain χ angles and per-residue accuracy is computed using small per-residue neural networks.

(3) The predicted structure is subjected to relaxation of the structure by gradient descent in the AMBER force field to remove stereochemical violations.

The final loss function, also termed as frame-aligned point error (FAPE) compares the predicted atom positions to the true positions under a variety of MSAs. The FAPE loss function aims to also correct for the atoms with respect to their local frame. The AlphaFold framework was trained on 350,000 diverse sequences from UniClust30. The predicted structures corresponding to these sequences were mixed with experimentally resolved Protein Data Bank (PDB) protein structure data to refine the model predictions.

References

Adam, G., Rampášek, L., Safikhani, Z., Smirnov, P., Haibe-Kains, B. et al. (2020). Machine learning approaches to drug response prediction: Challenges and recent progress. *Npj Precision Oncology*, *4*(1). https://doi.org/10.1038/s41698-020-0122-1

Aibar, S., González-Blas, C.B., Moerman, T., Huynh-Thu, V.A., Imrichova, H. et al. (2017). SCENIC: Single-cell regulatory network inference and clustering. *Nature Methods*, *14*(11), 1083–1086. https://doi.org/10.1038/nmeth.4463

AlQuraishi, M. (2021). Machine learning in protein structure prediction. *Current Opinion in Chemical Biology*, *65*, 1–8. https://doi.org/10.1016/j.cbpa.2021.04.005

Arafa, A., El-Fishawy, N., Badawy, M. and Radad, M. (2023). RN-Autoencoder: Reduced Noise Autoencoder for classifying imbalanced cancer genomic data. *Journal of Biological Engineering*, *17*(1). https://doi.org/10.1186/s13036-022-00319-3

Barnetson, R.A., Tenesa, A., Farrington, S.M., Nicholl, I.D., Cetnarskyj, R. et al. (2006). Identification and survival of carriers of mutations in DNA mismatch-repair genes in colon cancer. *New England Journal of Medicine*, *354*(26), 2751–2763. https://doi.org/10.1056/nejmoa053493

Browaeys, R., Saelens, W. and Saeys, Y. (2019). NicheNet: Modeling intercellular communication by linking ligands to target genes. *Nature Methods*, *17*(2), 159–162. https://doi.org/10.1038/s41592-019-0667-5

Butcher, E.C., Berg, E.L. and Kunkel, E.J. (2004). Systems biology in drug discovery. *Nature Biotechnology*, *22*(10), 1253–1259. https://doi.org/10.1038/nbt1017

Cacheiro, P., Muñoz-Fuentes, V., Murray, S.A., Dickinson, M.E., Bucan, M. et al. (2020). Human and mouse essentiality screens as a resource for disease gene discovery. *Nature Communications*, *11*(1). https://doi.org/10.1038/s41467-020-14284-2

Conway, B.R., O'Sullivan, E.D., Cairns, C., O'Sullivan, J., Simpson, D.J. et al. (2020). Kidney single-cell atlas reveals myeloid heterogeneity in progression and regression of kidney disease. *Journal of the American Society of Nephrology*, *31*(12), 2833–2854. https://doi.org/10.1681/asn.2020060806

Díaz-Uriarte, R. and Alvarez de Andrés, S. (2006). Gene selection and classification of microarray data using random forest. *BMC Bioinformatics*, *7*(1). https://doi.org/10.1186/1471-2105-7-3

Dill, K.A., Ozkan, S.B., Shell, M.S. and Weikl, T.R. (2008). The protein folding problem. *Annual Review of Biophysics*, *37*(1), 289–316. https://doi.org/10.1146/annurev.biophys.37.092707.153558

Feldner-Busztin, D., Firbas Nisantzis, P., Edmunds, S.J., Boza, G., Racimo, F. et al. (2023). Dealing with dimensionality: The application of machine learning to multi-omics data. *Bioinformatics*, *39*(2). https://doi.org/10.1093/bioinformatics/btad021

Fišer, K., Sieger, T., Schumich, A., Wood, B., Irving, J. et al. (2011). Detection and monitoring of normal and leukemic cell populations with hierarchical clustering of flow cytometry data. *Cytometry Part A*, *81A*(1), 25–34. https://doi.org/10.1002/cyto.a.21148

Gao, X., Lin, H., Revanna, K. and Dong, Q. (2017). A Bayesian taxonomic classification method for 16S rRNA gene sequences with improved species-level accuracy. *BMC Bioinformatics*, *18*(1). https://doi.org/10.1186/s12859-017-1670-4

Gao, Y. and Church, G. (2005). Improving molecular cancer class discovery through sparse non-negative matrix factorization. *Bioinformatics*, *21*(21), 3970–3975. https://doi.org/10.1093/bioinformatics/bti653

Giguère, S., Marchand, M., Laviolette, F., Drouin, A. and Corbeil, J. et al. (2013). Learning a peptide-protein binding affinity predictor with kernel ridge regression. *BMC Bioinformatics*, *14*(1). https://doi.org/10.1186/1471-2105-14-82

Goodswen, S.J., Barratt, J.L.N., Kennedy, P.J., Kaufer, A., Calarco, L. et al. (2021). Machine learning and applications in microbiology. *FEMS Microbiology Reviews*, *45*(5). https://doi.org/10.1093/femsre/fuab015

Hira, M.T., Razzaque, M.A., Angione, C., Scrivens, J., Sawan, S. et al. (2021). Integrated multi-omics analysis of ovarian cancer using variational autoencoders. *Scientific Reports*, *11*(1). https://doi.org/10.1038/s41598-021-85285-4

Hirst, J.D. and Sternberg, M.J.E. (1992). Prediction of structural and functional features of protein and nucleic acid sequences by artificial neural networks. *Biochemistry*, *31*(32), 7211–7218. https://doi.org/10.1021/bi00147a001

Jiang, R., Tang, W., Wu, X. and Fu, W. (2009). A random forest approach to the detection of epistatic interactions in case-control studies. *BMC Bioinformatics*, *10*(S1). https://doi.org/10.1186/1471-2105-10-s1-s65

Jumper, J., Evans, R., Pritzel, A., Green, T., Figurnov, M. et al. (2021). Highly accurate protein structure prediction with AlphaFold. *Nature*, *596*(7873), 583–589. https://doi.org/10.1038/s41586-021-03819-2

Kouduka, M., Sato, D., Komori, M., Kikuchi, M., Miyamoto, K. et al. (2007). A solution for universal classification of species based on genomic DNA. *International Journal of Plant Genomics*, *2007*, 1–8. https://doi.org/10.1155/2007/27894

Li, W., Feng, J. and Jiang, T. (2011). IsoLasso: A LASSO regression approach to rna-seq based transcriptome assembly. *Journal of Computational Biology*, *18*(11), 1693–1707. https://doi.org/10.1089/cmb.2011.0171

Li, Z., Schulz, M.H., Look, T., Begemann, M., Zenke, M. and Costa, I.G. (2019). Identification of transcription factor binding sites using ATAC-seq. *Genome Biology*, *20*(1). https://doi.org/10.1186/s13059-019-1642-2

Lu, J., Chen, L., Yin, J., Huang, T., Bi, Y. et al. (2016). Identification of new candidate drugs for lung cancer using chemical–chemical interactions, chemical–protein interactions and a K-means clustering algorithm. *Journal of Biomolecular Structure and Dynamics*, *34*(4), 906–917. https://doi.org/10.1080/07391102.2015.1060161

Lu, Y., Zhou, Y., Qu, W., Deng, M. and Zhang, C. et al. (2011). A Lasso regression model for the construction of microRNA-target regulatory networks. *Bioinformatics*, *27*(17), 2406–2413. https://doi.org/10.1093/bioinformatics/btr410

Luecken, M.D. and Theis, F.J. (2019). Current best practices in single-cell RNA-seq analysis: A tutorial. *Molecular Systems Biology*, *15*(6). https://doi.org/10.15252/msb.20188746

Mao, G., Pang, Z., Zuo, K. and Liu, J. (2023). Gene Regulatory Network Inference Using Convolutional Neural Networks from scRNA-seq Data. *Journal of Computational Biology*, *30*(5), 619–631. https://doi.org/10.1089/cmb.2022.0355

Middendorf, M., Kundaje, A., Wiggins, C., Freund, Y. and Leslie, C. et al. (2004). Predicting genetic regulatory response using classification. *Bioinformatics*, *20*(suppl_1), i232–i240. https://doi.org/10.1093/bioinformatics/bth923

Mochida, K., Koda, S., Inoue, K. and Nishii, R. (2018). Statistical and machine learning approaches to predict gene regulatory networks from transcriptome datasets. *Frontiers in Plant Science*, *9*. https://doi.org/10.3389/fpls.2018.01770

Mostafavi, S., Ray, D., Warde-Farley, D., Grouios, C. and Morris, Q. et al. (2008). GeneMANIA: A real-time multiple association network integration algorithm for predicting gene function. *Genome Biology*, *9*(Suppl 1), S4. https://doi.org/10.1186/gb-2008-9-s1-s4

Mulas, C., Chaigne, A., Smith, A. and Chalut, K.J. (2021). Cell state transitions: Definitions and challenges. *Development*, *148*(20). https://doi.org/10.1242/dev.199950

Nandi, S., Subramanian, A. and Sarkar, R.R. (2017). An integrative machine learning strategy for improved prediction of essential genes in Escherichia coli metabolism using flux-coupled features. *Molecular BioSystems*, *13*(8), 1584–1596. https://doi.org/10.1039/c7mb00234c

Nouri, N., Gaglia, G., Kurlovs, A.H., de Rinaldis, E. and Savova, V. et al. (2023). A marker gene-based method for identifying the cell-type of origin from single-cell RNA sequencing data. *MethodsX*, *10*, 102196. https://doi.org/10.1016/j.mex.2023.102196

O'Connor, T.D. and Mundy, N.I. (2009). Genotype–phenotype associations: Substitution models to detect evolutionary associations between phenotypic variables and genotypic evolutionary rate. *Bioinformatics*, *25*(12), i94–i100. https://doi.org/10.1093/bioinformatics/btp231

Paci, M., Maramotti, S., Bellesia, E., Formisano, D., Albertazzi, L. et al. (2009). Circulating plasma DNA as diagnostic biomarker in non-small cell lung cancer. *Lung Cancer*, *64*(1), 92–97. https://doi.org/10.1016/j.lungcan.2008.07.012

Parry, R.M., Jones, W., Stokes, T.H., Phan, J.H., Moffitt, R.A. et al. (2010). k-Nearest neighbor models for microarray gene expression analysis and clinical outcome prediction. *The Pharmacogenomics Journal*, *10*(4), 292–309. https://doi.org/10.1038/tpj.2010.56

Prada, C.F. and Boore, J.L. (2019). Gene annotation errors are common in the mammalian mitochondrial genomes database. *BMC Genomics*, *20*(1). https://doi.org/10.1186/s12864-019-5447-1

Rainio, O., Teuho, J. and Klén, R. Evaluation metrics and statistical tests for machine learning. Sci Rep 14, 6086 (2024). https://doi.org/10.1038/s41598-024-56706-x

Reel, P.S., Reel, S., Pearson, E., Trucco, E. and Jefferson, E. et al. (2021). Using machine learning approaches for multi-omics data analysis: A review. *Biotechnology Advances*, *49*, 107739. https://doi.org/10.1016/j.biotechadv.2021.107739

Salzberg, S., Delcher, A.L., Fasman, K.H. and Henderson, J. (1998). A decision tree system for finding genes in DNA. *Journal of Computational Biology*, *5*(4), 667–680. https://doi.org/10.1089/cmb.1998.5.667

Sarker, I.H. (2021). Deep learning: A comprehensive overview on techniques, taxonomy, applications and research directions. *SN Computer Science*, *2*(6). https://doi.org/10.1007/s42979-021-00815-1

Satija, R., Farrell, J. A., Gennert, D., Schier, A.F. and Regev, A. et al. (2015). Spatial reconstruction of single-cell gene expression data. *Nature Biotechnology*, *33*(5), 495–502. https://doi.org/10.1038/nbt.3192

See, P., Dutertre, C.-A., Chen, J., Günther, P., McGovern, N. et al. (2017). Mapping the human DC lineage through the integration of high-dimensional techniques. *Science*, *356*(6342). https://doi.org/10.1126/science.aag3009

Smirnov, V. and Warnow, T. (2020). MAGUS: Multiple sequence Alignment using Graph clUStering. *Bioinformatics*, *37*(12), 1666–1672. https://doi.org/10.1093/bioinformatics/btaa992

Stephens, Z.D., Lee, S.Y., Faghri, F., Campbell, R.H., Zhai, C. et al. (2015). Big data: Astronomical or genomical? *PLOS Biology*, *13*(7), e1002195. https://doi.org/10.1371/journal.pbio.1002195

Subramanian, A., Zakeri, P., Mousa, M., Alnaqbi, H., Alshamsi, F.Y. et al. (2022). Angiogenesis goes computational – The future way forward to discover new angiogenic targets? *Computational and Structural Biotechnology Journal*, *20*, 5235–5255. https://doi.org/10.1016/j.csbj.2022.09.019

Sung, J., Wang, Y., Chandrasekaran, S., Witten, D.M. and Price, N.D. et al. (2012). Molecular signatures from omics data: From chaos to consensus. *Biotechnology Journal*, *7*(8), 946–957. https://doi.org/10.1002/biot.201100305

Szkalisity, A., Piccinini, F., Beleon, A., Balassa, T., Varga, I.G. et al. (2021). Regression plane concept for analysing continuous cellular processes with machine learning. *Nature Communications*, *12*(1). https://doi.org/10.1038/s41467-021-22866-x

Tanowitz, M. and Zastrow, M. von. (2004). Identification of protein interactions by yeast two-hybrid screening and coimmunoprecipitation. In *Receptor Signal Transduction Protocols* (pp. 353–370). Humana Press. http://dx.doi.org/10.1385/1-59259-754-8:353

Virmani, A.K., Tsou, J.A., Siegmund, K.D., Shen, L.Y.C., Long, T.I. et al. (2002). Hierarchical Clustering of Lung Cancer Cell Lines Using DNA Methylation Markers. *Cancer Epidemiology, Biomarkers & Prevention*, *11*(3), 291–297.

Wang, Q., Garrity, G.M., Tiedje, J.M. and Cole, J.R. (2007). Naïve Bayesian Classifier for Rapid Assignment of rRNA Sequences into the New Bacterial Taxonomy. *Applied and Environmental Microbiology*, *73*(16), 5261–5267. https://doi.org/10.1128/aem.00062-07

Wang, X., Xie, T., Luo, J., Zhou, Z., Yu, X. et al. (2022). Radiomics predicts the prognosis of patients with locally advanced breast cancer by reflecting the heterogeneity of tumor cells and the tumor microenvironment. *Breast Cancer Research*, *24*(1). https://doi.org/10.1186/s13058-022-01516-0

Wu, C., Whitson, G., Mclarty, J., Ermongkonchai, A. and Chang, T. et al. (1992). Protein classification artificial neural system. *Protein Science*, *1*(5), 667–677. https://doi.org/10.1002/pro.5560010512

Xiao, Q., Luo, J., Liang, C., Cai, J. and Ding, P. et al. (2017). A graph regularized non-negative matrix factorization method for identifying microRNA-disease associations. *Bioinformatics*, *34*(2), 239–248. https://doi.org/10.1093/bioinformatics/btx545

Xu, J. and Li, Y. (2006). Discovering disease-genes by topological features in human protein–protein interaction network. *Bioinformatics*, *22*(22), 2800–2805. https://doi.org/10.1093/bioinformatics/btl467

Yan, X., Mehan, M.R., Huang, Y., Waterman, M.S., Yu, P.S. et al. (2007). A graph-based approach to systematically reconstruct human transcriptional regulatory modules. *Bioinformatics*, *23*(13), i577–i586. https://doi.org/10.1093/bioinformatics/btm227

Yanai, I. and Lercher, M. (2019). Night science. *Genome Biology*, *20*(1). https://doi.org/10.1186/s13059-019-1800-6

Yu, C., Chen, Y., Lu, C. and Hwang, J. (2006). Prediction of protein subcellular localization. *Proteins: Structure, Function, and Bioinformatics*, *64*(3), 643–651. https://doi.org/10.1002/prot.21018

Zeng, H., Edwards, M.D., Liu, G. and Gifford, D.K. (2016). Convolutional neural network architectures for predicting DNA–protein binding. *Bioinformatics*, *32*(12), i121–i127. https://doi.org/10.1093/bioinformatics/btw255

Zhang, S., Pyne, S., Pietrzak, S., Halberg, S., McCalla, S.G. et al. (2023). Inference of cell type-specific gene regulatory networks on cell lineages from single cell omic datasets. *Nature Communications*, *14*(1). https://doi.org/10.1038/s41467-023-38637-9

Zhong, W., Altun, G., Harrison, R., Tai, P.C. and Pan, Y. et al. (2005). Improved K-means clustering algorithm for exploring local protein sequence motifs representing common structural property. *IEEE transactions on nanobioscience*, *4*(3), 255–265. https://doi.org/10.1109/tnb.2005.853667

Ziemski, M., Wisanwanichthan, T., Bokulich, N.A. and Kaehler, B.D. (2021). Beating Naive Bayes at Taxonomic Classification of 16S rRNA Gene Sequences. *Frontiers in Microbiology*, *12*. https://doi.org/10.3389/fmicb.2021.644487

CHAPTER

22

Impact of Artificial Intelligence on Society: A Healthcare Perceptive

Greeshma N,[1] *Aswathy Vijay*,[1] *Manjoosha Ratna Yerrapragada*,[1] *Megha KB*,[2] *Athira SS*,[2] *Mohanan PV*[2] and *Renu John*[1*]

1. Introduction

Artificial intelligence (AI) pertains to computer systems that simulate cognitive functions supported by human intellect, including deep learning, reasoning, participation, adaptation, and sensory comprehension. AI algorithms and applications have enormous diverse uses in various fields. Their ability to be self-reliant and their capability to learn from experience make them strong and easily adaptable. Today, AI technology has permeated diverse sectors including healthcare, finance, education, manufacturing, and marketing. Thus, AI technology is becoming an integral part of our daily lives at a quick pace. The integration of AI technologies expands into transportation, communications, entertainment, and beyond, shaping the very fabric of modern society. It is anticipated that as AI develops further, it will have a greater influence on these sectors, spurring additional innovation and influencing the direction of technology-driven solutions. Historically, from the middle of the twentieth century till now, various landmark works by scientists such as Alan Turing, John McCarthy, Marvin Minsky and Claude Shannon, have led to shaping modern AI. Initial attempts to simulate the human brain using machines later evolved into a practical realization of AI implementation with advances in computational power, large datasets, and algorithms for artificial neural networks. The last decade witnessed a transformative shift with the resurgence of AI, powered by deep learning techniques. Numerous breakthroughs in image and speech recognition, natural language processing, and the advent of powerful GPUs and affordable computing power accelerated the training of deep neural networks (Turing, 1950).

[1] Department of Biomedical Engineering, Indian Institute of Technology Hyderabad, Hyderabad, India.
[2] Toxicology Division, Sree Chitra Tirunal Institute of Medical Science and Technology, Thiruvananthapuram, India.
* Corresponding author: renujohn@bme.iith.ac.in

In the field of healthcare and medicine, AI has shown tremendous potential in improving patient care and empowering individuals by giving them greater control over their health. One of the major advantages of AI in healthcare is the ability to provide personalized healthcare. AI aids in diagnostics and personalized treatment plans. AI can analyze patient data, medical histories, and lifestyle factors to generate personalized health advice. The impact of AI in healthcare has shown an escalation from basic diagnostic decision-making to surgical robots for accurate and precise operations. These algorithms can handle large amounts of data and process them, thereby assisting clinicians in providing better patient care. By adequately training the algorithms based on the data gathered, AI tools have now been developed in domains like automated image processing, diagnostic decision-making, radiology, robotic surgeries, and virtual assistance.

This integration of AI in our practice is not without its challenges. Ethical dilemmas, privacy concerns, and the potential for job displacements have sparked debates about the societal implications of AI. There have been questions pertaining to ethical and moral responsibilities, accountability, transparency, and equitable distribution of benefits when intelligent algorithms share or dominate the cognitive decision-making roles. It is essential that AI is safe and trustworthy to access these tools with confidence, to make human life easy and convenient. This chapter aims to provide a broad overview of the state-of-the-art AI applications, with a specific focus on healthcare and medicine. The objective is to provide a complete perspective of the impacts of AI in the healthcare domain, elaborating on its positive implications on patient care and the expected challenges. We discuss the social and ethical challenges and the possible opportunities for the future development of safer AI practices.

2. Applications of Artificial Intelligence

AI tools and algorithms mimic human intelligence in terms of perception, prediction, problem-solving, cognition, language comprehension and primary learning. When provided with training data, these algorithms can learn, process, and analyze patterns and features from the data made available to the system. Research in AI has led to the development of a rapidly growing discipline of expert systems technology. AI applications are having a significant impact on many facets of society, as expert systems tackle complicated issues in a variety of domains encompassing science, engineering, business, and medicine. The impact of AI in healthcare is transformative, revolutionizing various aspects of the industry to enhance patient care, screening, diagnostics, and overall operational efficiency. Improved diagnostics, faster drug development, and personalized medicine contributing to better patient outcomes and more efficient healthcare delivery. AI can also carry out predictive data analytics in healthcare. A few of the applications in various areas are discussed in this section.

2.1 Healthcare

Artificial Intelligence (AI) is transforming the healthcare industry, presenting novel solutions to longstanding challenges, and fundamentally transforming the way medical professionals approach patient care. The changing demographics, shortage of workforce, increasing morbidity, emergence of novel variants of pathogens, and digitalization of medical records call for an intervention of AI in healthcare (Ramesh et al., 2004). Initially, AI was seen as a tool to automate routine processes and mundane tasks carried out by clinicians and healthcare workers, freeing up human resources for more creative and complex endeavours. This was expected to positively lead to more quality time and effective utilization of clinician time for tasks requiring critical and scientific thinking, innovation, and emotional intelligence. The potential of AI to analyze vast datasets and identify patterns leading to more accurate diagnoses, better treatment plans, and advancements in medical research has revolutionized the state of art diagnosis and treatment scenario. Diagnostic accuracy, precision, and speed has increased dramatically because of the extraordinary speed with which AI can process massive volumes of medical data. Machine learning algorithms can evaluate medical images, such as X-rays, CT scans, and MRI scans with an accuracy that matches or even exceeds that of human experts thus aiding in the detection of abnormalities. This accelerates the diagnostic process and eventually leads to high accuracies aiding healthcare professionals in detecting abnormalities and diseases at an early stage

AI plays a crucial role in predictive analytics in healthcare applications by leveraging advanced algorithms and machine learning techniques to analyze large datasets and forecast potential outcomes. Predictive analytics used by AI tools can detect patterns and trends that might forecast a higher risk of developing specific disorders by evaluating patient data, including medical history, genetic information, and lifestyle preferences (Kaur & Mann, 2018). Thus, a risk stratification approach is possible through AI models based on various factors such as demographics, lifestyle, and medical history. AI can analyze vast datasets to tailor treatment plans based on an individual's unique characteristics. This enables proactive and personalized healthcare interventions, improving patient outcomes and reducing healthcare costs. This approach allows for more targeted and effective medical interventions, minimizing adverse effects. Predictive analytics also helps healthcare organizations in resource allocations and capacity planning. Anticipation of patient volumes, optimization of resource allocations, and plans for addressing crisis like pandemics are feasible through AI tools. This is particularly valuable in managing emergency department loads and planning for surgeries on patient demand. AI models analyze various data sources, including electronic health records and social media, to predict and monitor the spread of infectious diseases. This proactive approach aids in early containment and public health response.

AI analyzes patient data to create personalized treatment plans, considering individual genetic, lifestyle, and medical history factors. AI can also assist in chronic disease management by predicting disease progression and complications for patients. By analyzing patient data over time, healthcare providers can personalize treatment plans and interventions to manage chronic diseases effectively. These AI tools can help in individualized and patient-centered healthcare while minimizing side effects and improving treatment efficacy. Clinicians can select the most appropriate treatment strategy by being aware of the patient's response to therapy (Johnson et al., 2021). Predictive analytics helps identify patients at risk of non-adherence to medication regimens. Healthcare providers can intervene with personalized interventions or reminders to improve medication adherence and overall treatment success. AI solutions can help with remote patient monitoring, allowing medical staff to continuously monitor patients' vital signs and health indicators in real time. Wearable devices and sensors equipped with AI algorithms can detect subtle changes in health conditions, allowing for early intervention and cutting down the need for repeated hospital visits. This is particularly beneficial for managing chronic conditions and ensuring timely responses to evolving health concerns. Existing tele-health platforms could be facilitated through AI tools in efficient management of patients and broadening the patient support. AI plays a crucial role in enhancing the operational efficiency through streamlining administrative tasks, such as appointment scheduling, billing, record-keeping, and resource allocations to patients thus allowing healthcare professionals to focus more on patient care and reducing administrative burdens. AI could assist the administration in hospitals through numerous ways like supply chain management, predictive instrument maintenance systems, workflow and cost optimizations.

Natural Language Processing (NLP) enabled algorithms can understand and interpret human language – a capability that is increasingly applied in healthcare (Ragavi et al., 2021). NLP can analyze electronic health records (EHRs), extracting valuable insights from unstructured data. This enhances clinical decision-making, streamlines documentation processes, and improves communication among healthcare professionals. Similarly, sentiment analysis algorithms can assess patients' written or spoken words, helping in the identification of the indicators of mental well being. Chatbots and virtual mental health assistants powered by AI can provide additional support, offering immediate responses and resources for individuals in need. Surgeons can use AI-powered robotic systems to enhance precision, reduce invasiveness, and improve overall surgical outcomes

(Yoon & Lee, 2019). These technologies offer greater dexterity and accuracy, allowing for complex procedures with minimal impact on patients, shorter recovery times, and reduced postoperative complications. In the field of drug delivery and development, AI has expedited those processes which were formerly slow, laborious, and resource intensive. Biochemical data can be analyzed by ML models to find possible medication/bioactive candidates, forecast their effectiveness, and improve formulations. AI algorithms have surpassed certain limitations like uncertainty in the results of traditional drug testing methods which are usually subject to a high degree of variability. For instance, recently a DL algorithm was trained using a dataset of identified medicinal compounds and the biological activity linked to them. This helps to develop medicines with the fewest possible side effects faster and more precisely when compared to traditional applications which call for manual supervision. (Hansen et al., 2015).

2.2 Healthcare Education and Training

Learning practices and teaching innovations are progressively accelerating due to the implementation of AI applications in medical education (Zhai et al., 2021). AI can enable personalized learning experiences tailored to individual requirements of medical students. ML algorithms can analyze learning styles and preferences, and progress, to create adaptive curricula. This gives students the option to learn at their own pace, covering material they are already familiar with or strengthening subjects they find difficult. Alternatively, AI-based hybrid teaching blends in-person instruction with online learning, by utilizing educational resources available online, while giving equal importance to student initiatives and teacher guidance in the classroom, thus enhancing both the quality of instruction and learning outcomes for students. Applications of AI are also helpful for building immersive virtual reality (VR) and augmented reality (AR) learning environments which will be especially useful for clinicians, surgeons and medical students in enhancing their skills. These technologies offer virtual classrooms, allowing students to explore historical events, scientific phenomena, or procedures at remote locations. Text-to-speech, speech-to-text, and other assistive technologies powered by AI help students with diverse learning needs participate more fully in the educational process (Kumar et al., 2023; Parton, 2006).

2.3 Digital Healthcare Platforms and Communication

Chatbots powered by NLP can handle customer service inquiries, translate languages with near-human fluency, and even engage in basic conversations (Goar et al., 2023). AI algorithms can anticipate the words we are about to type, suggesting completions and correcting typological errors on the fly. This increases accuracy while simultaneously saving time and fluency of our written communication (R. Liu et al., 2023). Intelligent and smart reporting through AI algorithms suggesting content that can match respective cases based on case histories is another example of AI interventions in medical reporting. AI-powered assistants like Siri and Google Assistant can be effectively used in healthcare scenarios for assisting clinicians as well as patients with tasks like setting reminders, scheduling appointments, and managing resources (Natale, 2023). Real time translations, identification, and correction of errors and suggestions for alternative translations are the pivotal tools offered by AI for translators (Kolhar & Alameen, 2021; T. Liu et al., 2020). It allows smooth and error free communication between individuals who do not speak the same language. It generates text in the target language after analyzing the source material and determining translations. Neural Machine Translation is by far the most popular and sophisticated type of AI (NMT) (Stahlberg, 2020).

3. Impact of AI on Society

In the field of medicine and healthcare, AI has already created a new paradigm, by bringing transformations with different degrees of implementation. Opinions on AI-based systems and applications are divided into positive, negative, and controversial based on their implementations. In the case of healthcare, while AI offers patients many advantages there are also threats to patient safety, data security, and health equity. In this section, the social and ethical impacts of AI in healthcare and medicine will be discussed in detail. Table 1 provides a summary of several uses of AI in healthcare and classifies them based on their positive and negative impact on society. This classification was done by calculating a technology availability level (TAL) score, which qualitatively evaluates each application area based on the availability of a technology (Gómez-González et al., 2020).

3.1 Positive Impact

The potential applications of AI include transformative changes in the healthcare industry, impacting various aspects of patient care, diagnosis, treatment, and overall healthcare management (Tai, 2020). For example, the integration of AI in diagnosis and medical imaging has substantially improved the proficiency and accuracy of the diagnostic decision-making process. Some of the ways in which AI tools deliver a positive impact in this field are discussed below.

3.1.1 Speed and Accuracy

AI tools can evaluate massive amounts of data at incredible speeds, surpassing human capabilities. Thus, tasks such as image analysis and decision-making in diagnostics are done instantaneously, thus reducing the wait times associated with conventional diagnostics. Also, when the algorithms are properly trained and validated, AI can achieve high levels of accuracy and precision, especially in the field of healthcare. For example, AI-based robotic surgeries, which are now routinely performed, can perform minimally invasive procedures with a higher degree of precision than manual surgeries. This ultimately improves patient care, as minimally invasive surgeries are associated with lower blood loss, and lower trauma and anxiety levels.

3.1.2 Reduction in Errors

Medicine is a field where human errors are often highly risky and may jeopardize patient care. For an experienced and qualified clinician, the majority of these errors occur due to fatigue, considering their long working hours. However, AI does not suffer from these shortcomings, thereby preventing errors from happening.

3.1.3 Virtual Assistance and Remote Monitoring

AI can power virtual assistants and chatbots that can engage with patients and provide information on-demand. Also, these systems can operate round-the-clock, without any break, thus offering 24/7 customer support and assistance. Certain wearables can also facilitate continuous monitoring, especially for patients with chronic conditions. These continuous and real-time datasets can help in prompt intervention by healthcare providers and preventing complications, thus improving patient care. Telemedicine can help offer remote consultations, making healthcare services more accessible, and is a boon – particularly for patients in rural or resource-limited settings.

3.1.4 Customization and Personalization

Medical care can be tailored to a patient's specific needs, according to the concept of personalized medicine. Conventionally, this idea has limitations since it is difficult to analyze the massive volumes of data required to develop a customized treatment plan. However, AI algorithms are trained in finding patterns and correlations in data, thus helping to create more precise and effective personalized treatment

3.1.5 Efficient Data Management

AI can facilitate seamless analysis and integration of various healthcare-related data from various sources such as imaging data, diagnostic results, and electronic health records. These algorithms

Table 1 AI and AI-mediated technology classification in healthcare and medicine based on social and ethical implications. SW: software, AR: Augmented reality, VR: Virtual reality, IoT: Internet of things (Gómez-González et al., 2020).

AI and AI-mediated technologies	Specific implementations.	Social impact
Algorithms for computer-aided diagnosis	SW for decision support in most of the clinical areas	
Structured reports, eHealth	SW for improved workflow, efficiency	Positive
AR/VR, advanced imaging tools	Tools for information visualization and navigation.	
	Image-guided surgery. Teleoperation	
Digital pathology, 'virtopsy'	SW for automated, extensive analysis	
Personalized, precision medicine	Tailored treatments. Prediction response.	
	'In-silico' modelling and testing. Drug design.	
Apps, chatbots, dashboards, online platforms.	The 'digital doctor' (assistance for professionals and for patients)	
Companion and social robots	For hospitalized person, children and the elderly	
Big Data collection and analysis	Epidemiology, prevention and monitoring of disease outbreaks	
	Fraud detection. Quality control, monitoring of physicians and treatments.	
IoT, wearables, mHealth	Automated clinical/health surveillance in any environment/institution	
	Monitoring, automated drug delivery	
Gene editing	Disease treatment, prevention	
Merging of medical and social data. 'Social' engineering	Prevention of episodes with clinical relevance (e.g. suicide attempts)	Controversial
	Tailored marketing (e.g. related to female cycles)	
Reading and decoding brain signals, interaction with neural processes	Treatment of disease. Restoring damaged functions.	
	Brain-machine inferfaces	
	Control of protheses, exoskeletons. 'Cyborgs'	
	Neurostimulations. Neuromodulation	
	Neuroprostheses (for CNS)	
	Mind reading and manipulation	
Genetic tests. Population screening	Disease tests. Direct-to-consumer tests.	
Personalized, precision medicine	Individual profiling. Personalized molecules (for treatment) at 'impossible' prices.	
Gene editing	'Engineered' humans.	
	Gene- enhanced 'superhumans'	
	Self-experimentation medicine. Biohacking	
Fully autonomous AI systems	The 'digital doctor' 'Robotic surgeon'	
Human-animal embryos	Organs for transplants. Hybrid beings (chimera)	
The quest for immortality	Whole-brain emulsion/ transplant	
The search for artificial life forms	'Living machines' ('biological robots', 'biorobots'	
	Military	
Evil biohacking	Targeting specific individual or groups	
Weaponization	From 'small labs' to military labs	Negative
Bioterrorism	From 'small labs'	

can be trained to perform automated data cleaning and preprocessing, thus streamlining the data analysis procedure. This enables the healthcare providers to obtain complete patient histories and make informed decisions with a holistic approach.

3.1.6 Pandemic Response

AI can play a crucial role in monitoring health data at the population level. AI tools can aid in early detection of potential outbreaks and tracking disease spread. This was particularly relevant during the recent COVID-19 outbreak (Chen et al., 2020). From analyzing the X-ray and CT images to identify patterns of virus infection, to helping in resource allocation like hospital beds and ventilators, to chatbots for disseminating accurate and up-to-date information regarding the outbreak, AI has played a major role in pandemic management (Wakefield, 2020). In addition, AI applications could assist in contact tracing of infected individuals by tracking data from various sources, enabling timely isolation measures to curb the spread. Also, efficient data analytics could provide public health officials with valuable insights, such as mobility patterns and demographic analysis (Fowler, 2020). This paved the way for informed decision making, implementing timely interventions and public health policies.

3.2 Challenges

AI-based applications and devices have an enormous promise to improve healthcare and medicine, but also pose several social and ethical concerns. Though there are several AI-based applications and devices already integrated in daily life, the effects and impact of AI on society and humans is barely studied.

3.2.1 Data Privacy and Security

The principle of privacy and security is the most pertinent concern. Given that developing AI-based technology requires training data, vast amounts of patient data are collected and analyzed. This raises significant concerns about data privacy, security, integrity, and anonymity. In the case of healthcare, unauthorized access to sensitive medical information can pose severe issues, ranging from simple identity theft to the manipulation of treatment plans. In addition to the fact that there are no well-defined limits to the use of data, there is also a possibility for cyber-attacks and data breaches, posing a serious threat to patient confidentiality.

3.2.2 Accountability and Responsibility

Increasing dependence on AI may potentially diminish the importance of human intuition and expertise. Clinicians may be tempted to trust the AI tools without critically assessing the patients' conditions. This can lead to complacency and decreasing development of essential clinical skills, while jeopardizing patient care. With this comes another fundamental issue regarding legal responsibility and accountability – who should be held responsible if something goes wrong? Though rapidly progressing, AI-based systems still have their limitations, and can fail unexpectedly. Most AI applications are not subject to global regulatory requirements. For example, when a clinician is supported by an AI algorithm, they may be held accountable for the improper use of the tool or for justifying the decisions made by the tool. However, in the event of harm caused by improper content or a biased algorithm, then the accountability lies either with those who designed it or those who trained it or those who assured its quality. It is essential that risk assessments are performed, and the human-decision maker is aware of the system's limits.

3.2.3 Algorithmic Bias

Another common question that is posed often is – Are AI-based systems fair and objective with all groups of people (in terms of age, gender, ethnicity), or are they biased? Ultimately, it is the training data that provides information for a bias to be perpetuated. If the training data is not representative, or if the goals are not properly defined, the biases can be amplified and lead to improper decision making. For example, AI algorithms outside of healthcare have been criticized for discriminating based on gender, race and religion. Also, certain chat bots have been misled to propagate hate speeches. A recent study reported that a US-based healthcare algorithm, that refers patients with complex medical needs to certain programmes, has been systematically discriminating against black people (Ledford, 2019). (Ledford, 2019)

3.2.4 Training Data

In the field of healthcare, the major debate is the safe provision of patient care by algorithms. If the AI is poorly programmed, inadequately trained or used in improper situations, it could expose the patient to unforeseen risks. On the other hand, the doctors are expected to provide holistic treatment, taking subtle cues from non-verbal communication, by providing care that is tailored to individual patient requirements, while sharing the decision-making process. The use of AI technology could cause a shift in the fundamental relation between the doctor and the patient, depending on the nature of the interface between AI and the patient. Though AI can be efficient in data processing and predictions, it would still lack the human touch and empathy required in patient care. AI falls short in emotional and psychological factors such as understanding patient concerns, providing comfort and support, and building trust. Ultimately, AI-based tools may lead to dehumanized healthcare experiences, thus impacting the doctor-patient relationships and overall patient satisfaction and social needs. Ultimately, the patient would want to know how well the AI tool works, and whether it is trustworthy, though he/she need not know how it works. One of the most important stages in the development of AI in medicine and healthcare is earning patients trust.

3.2.5 Medical Research

In the field of medical research and innovation, AI algorithms often operate as 'black boxes', which makes it hard for researchers to fully understand how the tool arrived at the specific conclusions. In addition, the research findings are difficult to replicate and validate owing to the complex nature of biological systems. Also, the AI algorithms lack the human creativity and intuition that have historically driven spontaneous discoveries or breakthroughs.

3.2.6 Unethical use of AI Tools

Another ethical concern of using AI-based tools is the possibility of data manipulation resulting in adversarial attacks. For example, hospitals working with insurance providers can tweak the images to do a misdiagnosis while earning larger payouts (Finlayson et al., 2019). On the other hand, certain AI applications related to medicine, especially with the advent of gene editing technology, are potentially prone to evil applications, weaponization or being employed in bioterrorism. With the potential for unanticipated and disastrous outcomes, gene editing may allow humans to experiment on themselves or assist in designing human and human-animal embryos.

3.2.7 Job Displacement and Accessibility Challenges

Other concerns regarding the widespread adoption of AI is the potential for job displacement. As AI takes over traditional human tasks, there is a rising risk of the reducing demand for certain roles, leading to job insecurity and unemployment. For example, the jobs most prone to automation include accountants, auditors, cashiers, budget analysts, etc while surgeons, lawyers, therapists, etc are the least prone to job replacements. Also, developing and implementing AI technology requires substantial economic investment as there is a need for specialized expertise, updated with evolving technology and computational resources. It is expected that the healthcare gap between well-funded institutions and institutions from resource-limited settings may widen leading to an increased risk of wealth inequality.

Therefore, resolving ethical concerns of AI in healthcare and gaining the trust of consumers depends on developing reliable AI algorithms.

4. Opportunities and Suggestions for the Future

AI has led to several advancements in healthcare such as in, radiology – where deep learning algorithms are being utilized in mammography for the detection of breast cancer (Dheeba et al., 2014); CT imaging – for the detection of colon cancer (Summers et al., 2000); MRI – for detecting brain tumours (Perez-Ramirez et al., 2014); identifying neurological disorders such as Alzheimer's disease (Erickson et al., 2017). There has been immense research going on for the past 10 years, publishing with a large number of published research articles yearly (to the tune of 100 to 800 articles), showing a growing interest in developing software for AI (Pesapane et al., 2018). On the one hand, there are many advantages of integrating AI to society, there are also several challenges that must be critically assessed. For example, there is only limited knowledge on how these algorithms might affect humans. One study reported that computer-aided diagnosis for mammography saw an increased usage by the patients, without improving outcomes in comparison to traditional screening techniques (Lehman et al., 2015). Another study reported that excessive warnings and alerts resulted in desensitization and alert fatigue (Phansalkar et al., 2013). It is therefore essential to develop a better understanding of the complex interactions between humans and algorithms. Only through such prospective studies, we can begin to know the true advantage of AI tools.

The major issue with AI applications pertains to human autonomy. The decision-making potential of AI systems can profoundly affect people's lives without their conscious knowledge or consent. People shouldn't be forced or controlled by AI systems, and humans should be able to monitor or interfere with any decision that the software makes. Data security and privacy is another challenge, as AI applications absorb vast amounts of data collected from various sources. Questions on patient consent, confidentiality, data ownership and trust are unavoidable. AI developers should ensure that there exist robust measures for data protection and encryption, while avoiding unethical and unauthorized data usage. This challenge extends to the creation of thorough regulatory systems. Establishing clear regulatory guidelines for the ethical use of data is crucial in overcoming ethical challenges. Concerns about the data quality are important considerations while creating an AI tool for the medical field. Issues such as difficulties in comparing different algorithms, lack of proper gold standards, generalization across different datasets and quality of reporting ML studies must be addressed. Also, AI algorithms suffer because of the training data, which may unintentionally perpetuate bias. Machine learning blind spots can represent the worst aspects of society prejudices, with unintentional and unknown errors in minority data. (Crawford & Calo, 2016). The errors may result from the model bias, model variance or the outcome noise. A greater awareness of the existence of these biases, while empowering clinicians to actively participate in model development can help in overriding them s. The training data should therefore include the global community considering all the population subgroups. A careful performance evaluation and clinical validation should be performed as well.

Advanced AI programs are intricate and challenging, with limitations in understanding their decision-making processes, particularly neural networks. Some AI models are considered as black box, since it is difficult to comprehend their inner workings. If these algorithms are to be utilized in the medical field, they must be applied with care and responsibility. Human-computer interactions must also be considered in order to convey the complex AI decisions to novices since it may be difficult to explain things in a clear and comprehensible manner (Doshi-Velez & Kim, 2017). The future of AI is constantly evolving with the promises of quantum computing, enhanced explainability, and a more profound integration of AI with natural intelligence and other emerging technologies, bringing new frontiers and challenges.

Conclusion

AI has proven to have made considerable progress in all practical aspects of society. It is reshaping healthcare, education, transportation, and more. AI's precision and tireless nature can propel it to excellence in performing tasks like decision-making in medical diagnosis. AI can analyze mountains of medical data, crafting personalized treatment plans (each a potent weapon against specific illnesses). Robotic surgeons, guided by AI's steady hand, operate with unmatched precision, minimizing invasive procedures and scars. Even in our pockets, AI lurks – in smartphone apps monitoring vitals, nudging us towards healthier habits, like a tireless guardian angel. Yet, with great power comes responsibility. Data privacy must be fiercely protected, and algorithms constantly scrutinized for bias to ensure equitable care for all. Several other areas such as education, too, feel the transformative touch of AI. Personalized learning customizes lessons to each student's pace and grasp, nurturing potential like never before. Imagine a virtual tutor, patiently explaining a complex concept until understanding dawns. But beneath the gleaming promise lies a shadow. Automation, the AI-driven engine of efficiency, can render human jobs obsolete, leaving livelihoods precariously balanced. Ethical considerations arise when AI, wielding the power of algorithms, makes decisions that impact lives. Biases, woven into the fabric of data, can skew these decisions, perpetuating inequalities, and injustices. The path forward demands a cautious embrace of AI. We must harness its potential for good, mitigate its risks, and ensure that its development serves humanity without replacing it.

References

Chen, S., Yang, J., Yang, W., Wang, C. and Bärnighausen, T. et al. (2020). COVID-19 control in China during mass population movements at New Year. *The Lancet*, *395*(10226), 764–766. https://doi.org/10.1016/S0140-6736(20)30421-9

Crawford, K. and Calo, R. (2016). There is a blind spot in AI research. *Nature*, *538*(7625), 311–313. https://doi.org/10.1038/538311a

Dheeba, J., Albert Singh, N. and Tamil Selvi, S. (2014). Computer-aided detection of breast cancer on mammograms: A swarm intelligence optimized wavelet neural network approach. *Journal of Biomedical Informatics*, *49*, 45–52. https://doi.org/10.1016/j.jbi.2014.01.010

Doshi-Velez, F. and Kim, B. (2017). *Towards A Rigorous Science of Interpretable Machine Learning*. *Ml*, 1–13.

Erickson, B.J., Korfiatis, P., Akkus, Z. and Kline, T.L. (2017). Machine learning for medical imaging. *Radiographics*, *37*(2), 505–515. https://doi.org/10.1148/rg.2017160130

Finlayson, S.G., Bowers, J.D., Ito, J., Zittrain, J.L., Beam, A.L. et al. (2019). Adversarial attacks on medical machine learning. *Science*, *363*(6433), 1287–1289. https://doi.org/10.1126/science.aaw4399

Fowler, G.A. (2020, March 24). Smartphone data reveal which Americans are social distancing (and not). Retrieved from The Washington Post: https://www.washingtonpost.com/technology/2020/03/24/social-distancing-maps-cellphone-location/

Goar, V., Yadav, N.S. and Yadav, P.S. (2023). Conversational AI for Natural Language Processing: An Review of ChatGPT. *International Journal on Recent and Innovation Trends in Computing and Communication*, *11*(January), 109–117. https://doi.org/10.17762/ijritcc.v11i3s.6161

Gómez-González, E., Gómez, E. and European Commission. Joint Research Centre. (2020). Artificial Intelligence in Medicine and Healthcare: applications, availability and societal impact. In *JRC Science for Policy Report*. https://doi.org/10.2760/047666

Hansen, K., Biegler, F., Ramakrishnan, R., Pronobis, W., Von Lilienfeld, O.A. et al. (2015). Machine learning predictions of molecular properties: Accurate many-body potentials and nonlocality in chemical space. *Journal of Physical Chemistry Letters*, *6*(12), 2326–2331. https://doi.org/10.1021/acs.jpclett.5b00831

Johnson, K.B., Wei, W.Q., Weeraratne, D., Frisse, M.E., Misulis, K. et al. (2021). Precision Medicine, AI, and the Future of Personalized Health Care. *Clinical and Translational Science*, *14*(1), 86–93. https://doi.org/10.1111/cts.12884

Kaur, J. and Mann, K.S. (2018). AI based healthcare platform for real time, predictive and prescriptive analytics. *Communications in Computer and Information Science*, *805*, 138–149. https://doi.org/10.1007/978-981-13-0755-3_11

Kolhar, M. and Alameen, A. (2021). Artificial Intelligence Based Language Translation Platform. *Intelligent Automation & Soft Computing*, *28*(1), 1–9. https://doi.org/10.32604/iasc.2021.014995

Kumar, A., Nayyar, A., Sachan, R.K. and Jain, R. (2023). AI-assisted special education for students with exceptional needs. *AI-Assisted Special Education for Students With Exceptional Needs*, *i*, 1–335. https://doi.org/10.4018/979-8-3693-0378-8

Ledford, H. (2019). Millions of black people are affected by racial bias in health-care algorithms. *Nature*, *574*(7780), 608–609. https://doi.org/10.1038/d41586-019-03228-6

Lehman, C.D., Wellman, R.D., Buist, D.S.M., Kerlikowske, K., Tosteson, A.N.A. et al. (2015). Diagnostic accuracy of digital screening mammography with and witho.ut computer-aided detection. *JAMA Internal Medicine*, *175*(11), 1828–1837. https://doi.org/10.1001/jamainternmed.2015.5231

Liu, R., Gupta, S. and Patel, P. (2023). The Application of the Principles of Responsible AI on Social Media Marketing for Digital Health. *Information Systems Frontiers*, *25*(6), 2275–2299. https://doi.org/10.1007/s10796-021-10191-z

Liu, T., Kim, E., Li, X., Yuizono, T., Nagai, Y. and Lu, Y. et al. (2020). Research and Practice of Hybrid Teaching Based on AI technology for Foreign Language Translation. *Proceedings - 2020 International Conference on Computer Engineering and Application, ICCEA 2020*, 664–668. https://doi.org/10.1109/ICCEA50009.2020.00145

Natale, S. (2023). AI, Human–Machine Communication and Deception. *The Sage Handbook of Human–Machine Communication*, 401–408. https://doi.org/10.4135/9781529782783.n50

Parton, B.S. (2006). Sign language recognition and translation: A multidisciplined approach from the field of artificial intelligence. *Journal of Deaf Studies and Deaf Education*, *11*(1), 94–101. https://doi.org/10.1093/deafed/enj003

Perez-Ramirez, U., Arana, E. and Moratal, D. (2014). Computer-aided detection of brain metastases using a three-dimensional template-based matching algorithm. *2014 36th Annual International Conference of the IEEE Engineering in Medicine and Biology Society, EMBC 2014*, *1*, 2384–2387. https://doi.org/10.1109/EMBC.2014.6944101

Pesapane, F., Volonté, C., Codari, M. and Sardanelli, F. (2018). Artificial intelligence as a medical device in radiology: ethical and regulatory issues in Europe and the United States. *Insights into Imaging*, *9*(5), 745–753. https://doi.org/10.1007/s13244-018-0645-y

Phansalkar, S., van der Sijs, H., Tucker, A.D., Desai, A.A., Bell, D.S. et al. (2013). Drug-drug interactions that should be noninterruptive in order to reduce alert fatigue in electronic health records. *Journal of the American Medical Informatics Association*, *20*(3), 489–493. https://doi.org/10.1136/amiajnl-2012-001089

Ragavi, V., Santha Sheela, A.C. and Narayanan Kannaiyan, G. (2021). Impact of Artificial Intelligence in the field of Health Care. *Journal of Physics: Conference Series*, *1831*(1). https://doi.org/10.1088/1742-6596/1831/1/012006

Ramesh, A.N., Kambhampati, C., Monson, J.R.T. and Drew, P.J. et al. (2004). Artificial intelligence in medicine. *Annals of the Royal College of Surgeons of England*, *86*(5), 334–338. https://doi.org/10.1308/147870804290

Stahlberg, F. (2020). Neural machine translation: A review. *Journal of Artificial Intelligence Research*, *69*, 343–418. https://doi.org/10.1613/JAIR.1.12007

Summers, R.M., Beaulieu, C.F., Pusanik, L.M., Malley, J.D., Jeffrey, R.B. et al. (2000). Automated polyp detector for CT colonography: Feasibility study. *Radiology*, *216*(1), 284–290. https://doi.org/10.1148/radiology.216.1.r00jl43284

Tai, M.C.T. (2020). The impact of artificial intelligence on human society and bioethics. *Tzu Chi Medical Journal*, *32*(4), 339–343. https://doi.org/10.4103/tcmj.tcmj_71_20

Turing, A.M. (1950). Computing machinery and Intelligence. *Mind*, *LIX*(236), 433–460. https://doi.org/10.4324/9780429024689-5

Yoon, S.N. and Lee, D.H. (2019). Artificial intelligence and robots in healthcare: What are the success factors for technology-based service encounters? *International Journal of Healthcare Management*, *12*(3), 218–225. https://doi.org/10.1080/20479700.2018.1498220

Zhai, X., Chu, X., Chai, C.S., Jong, M.S.Y., Istenic, A. et al. (2021). A Review of Artificial Intelligence (AI) in Education from 2010 to 2020. *Complexity*, *2021*. https://doi.org/10.1155/2021/8812542

Wakefield, J. (2020, March 12). Coronavirs: How can AI help fight the pandemic? Retrieved from BBC News: https://www.bbc.com/news/technology-51851292

CHAPTER

23

Application of AI in Digital Pathology

Shanthi PB[1*] and *Ranjini Kudva*[2]

1. Introduction

The advent of novel technology with artificial intelligence paved the way to design an interactive model for pathology investigation that expedites the development of image analysis tools (Jahn et al., 2020). These tools act as necessary aids in research areas, disease diagnosis, and medication through important biological and pathological information extracted from the images. In recent times, analyses based on AI based machine learning and neural network techniques with pathological images attracted greater attention to diagnose patterns. This combination not only reduced costs and time durations for analyses but also mitigated disease incidence and mortality rates through early diagnosis (Lau et al., 2021). Manual screening of pathological smears is considered as one of the most laborious and intensive tasks, especially during the screening of whole population group programs, where there will be an enormous number of pathology samples to be investigated and diagnosed (Barisoni et al., 2020). As manual screening would be time consuming and tedious for pathologists and cyto-technicians, an automated image classification system became an essential tool in medical research that plays a vital role to speed up the process for faster analysis of samples (Baxi et al., 2022).

Clinical pathology includes examination of surgically removed body samples or organs, tissues, and body fluids (Rajkumar 2018). These human samples consist of immunological markers, chemical signatures and gene markers which help the image analysis system to capture the pattern for diagnosis. Pathologists deal with a wide range of diseases, especially cancer diagnosis that utilizes cellular patterns of tissues or body fluids that are observed under a microscope to determine the malignancy of the obtained sample (Gupta et al., 2023). Pathology is the broad area of study of disease, and its specific effects on biological tissues and fluids. Cytopathology or cytology is a branch of pathology that studies and diagnoses disease at cellular level. Cytopathological or

[1] Department of Computer Science and Engg., Manipal Institute of Technology, Manipal Academy of Higher Education, Manipal, India.
[2] Department of Pathology, Kasturba Medical College, Manipal Academy of Higher Education, Manipal, India
* Corresponding author: shanthi.moorkoth@manipal.edu

smear test involves samples that are smeared across slides and stained for subsequent microscopic examination of cells which is considered as one of the main diagnostic tools for cancer diagnosis. Another branch of pathology and histopathology or histology deals with examining the entire section of tissue (Davenport & Kalakota 2019).

Pathologists identify the morphological characteristics of tissues which indicate the presence of diseases like cancer. Automation of timely cancer detection provides a basis for mitigating death and morbidity especially for malignancies such as uterine cervix, lung, leukaemia and skin which is stated as the most preventable disease through early detection (Hou et al., 2022). The changing pathology ecosystem includes image-based acquisition, management, and interpretation of pathology information through some computer assisted image analysis techniques. The overall pathological image analysis system comprises few key steps such as image preprocessing which aid in highlighting the region of interest, image segmentation and feature extraction and finally model construction to classify and analyze the accuracy of the model. Thus, computer assisted disease diagnosis plays a very important role and became a major research subject for disease diagnosis and prognosis (Cui & Zhang 2021). WSI (Whole Slide Imaging) devices transformed the field of pathology through digitization of smear slides. Simultaneous innovations in computational approaches and storage enabled speedy processing of WSI datasets. Some of the potential benefits of WSI include increased efficiency in pathology, remote reviewing and analysis, and digital archiving and accessibility. However, combining WSI with current research setups, is considered as a challenge in digital pathology with respect to managing and analyzing large volumes of data. Whole Slide Imaging has facilitated research organizations such as Pharmaceutics, and other departments to digitize their microscope slides, making them comfortable to store and share with other research professionals. WSI provides high-resolution imaging of collected samples, enabling researchers to detect and analyze molecular transformations within the samples. With the advent of whole slide imaging, developers and researchers are exploring the use of artificial intelligence and machine learning to assist in the analysis of digital images with high resolution (Pantanowitz et al., 2018).

2. Artificial Intelligence (AI) for Digital Pathology

AI is one of the sub parts of information technology that provide support to computers to act intelligently. AI has become an essential part of technology especially in health informatics and clinical pathology (Cui & Zhang 2021). Various AI algorithms are widely used in analyzing and interpreting large volumes of data that massively aid in solving problems in the medical domain. Oncology, Urology, Liver pathology, Cardiology, Orthopaedics, Gynaecology, Neuropathology, Dermatology and Haematology are some of the well-known medical domains, where AI techniques have contributed extremely for the last few decades. Radiological diagnosis, bioinformatics, genome sequencing and drug development are some of the other well-known healthcare sectors, where AI created a significant transformation that enables computers to simulate human intelligence (Niazi et al., 2019).

The Aperio VERSA is a comprehensive digital pathology multiplex whole slide scanner shown in Figure 1 designed and developed to support diverse imaging needs (Dwivedi et al. 2019). Diagnostic analysis in comparison with expert pathologists in histopathology and cytopathology has revealed the accuracy and efficiency of AI models especially in tumour detection and classification (Tizhoosh & Pantanowitz 2018). After being trained in the model using previously acquired data, these AI systems support the clinicians to decide on the diagnosis by anticipating the results. Digital pathology plays a very important role in this as a large amount of diagnoses are based on the pathological tests done by the patients. Development of digital acquisition devices in pathology made it easy to acquire digital image data for computer assisted image analysis. Automated whole slide imaging scanners are used nowadays to produce high resolution images for efficient image analysis and diagnosis (Jiang et al., 2017).

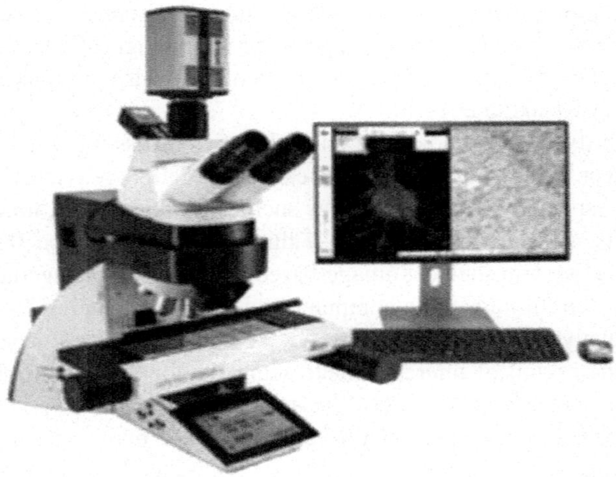

Fig. 1 Multiplex Whole Slide Scanner for Computer Assisted Image Analysis

Whole Slide Imaging is referred to as virtual microscopy used to scan glass slides to produce digital sides of a sample and is called digitization. BLISS (Baccus Laboratories Inc., Slide Scanner), the first commercial scanner, was built by James Bacus in 1994 and it was acquired by the Olympics. Another robotic scanner which assisted integrated imaging techniques that include medical data management is the ultrafast robotic whole slide digitizer. Chromavision introduced their first product ACIS (Automated Cellular Imaging System) that performs quantitative analysis, event detection, object counting and tissue microarray technology. The Multisensor array approach was developed by Aperio which allows simultaneous image acquisition and data processing (Pantanowitz et al., 2018). Figure 2 shows various new scanners in the market which use hybrid imaging techniques and live visualization that helps Pathologists to view the glass slides in real time by remotely controlling features such as magnification and navigation.

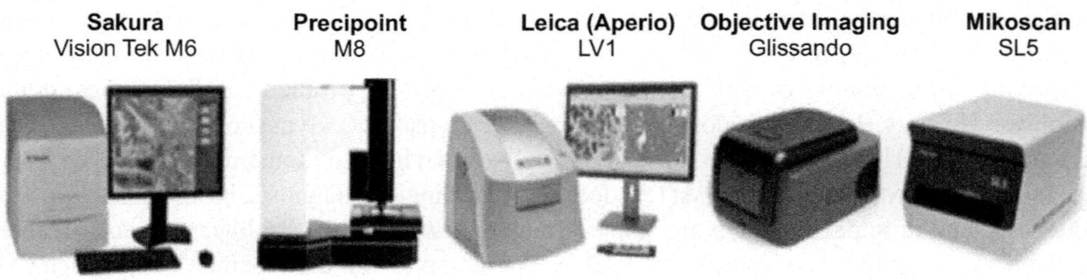

Fig. 2 Various Hybrid Whole Slide Imaging Devices

Many cytologists have incorporated this new technology in their practice especially in the form of automation in cytology in several forms for a few decades. Analyzing specimens in the regions of thyroid, urine, pancreas, lungs, and breasts are some of the cytological areas which are largely confined to research settings. Even though this technology provided benefits over traditional microscopy such as image sharing, workload balancing, and portability of images and demonstrated a good performance especially in cytopathology, the use of WSI in clinical practice is very minimal (Pantanowitz et al., 2018). Table 1 shows the published results of the primary diagnosis conducted in some of the pathology subspecialties using a whole slide scanner in comparison with conventional light microscopy (Pantanowitz et al., 2018). The success of the commercially available automated screening products is limited in their application because of the expensive screening device, cost involved for screening procedures, and difficulty in achieving good sensitivity without getting too many false-negative samples. To achieve good sensitivity (identifying diseased samples correctly) and specificity (identifying non-diseased samples correctly), efficient and reliable classification techniques should be developed for automated systems.

Table 1 Published results of whole-slide imaging validation studies for primary diagnosis

Pathology Subspeciality	*Accuracy*
Surgical	75-98
Breast	90-99
Dermatopathology	94-99
Gastrointestinal	95
Gynaecological	96
Paediatric	90 to 93
Pulmonary	85 to100
Renal	84
Cytopathology	89-97

Improvisation in machine learning algorithms and medical image processing techniques paved the way to screen the samples faster and reliably as best as the visual screening system significantly at a lower cost. The major stages involved in the design of an automated pap smear screening system include image acquisition, noise removal, region of interest segmentation, feature extraction, and classification. In the design process, each stage contributes a major role in the classification and disease diagnosis. False-negative cases are a serious ethical issue in which a patient with the disease is diagnosed as non-disease that leads to women getting cancer instead of treatment. False-positive cases are not a serious ethical issue but will make a patient undergo an expensive treatment process and creates unnecessary anxiety and stress to the patient by the wrong diagnosis. So, an effective, reliable, and low-cost computer-assisted diagnosis system plays a major role in reducing occurrence of cancer and death rates, especially in underdeveloped and developing countries.

Nowadays computational pathology has become an active research area, but usage of these methods in the clinical environment is less. In the forthcoming decades, usage of WSI scanners for diagnosis may increase the usage of digital pathology and enhance the ability of pathologists in patient care. Meaningful implementation of samples in the form of validation, classification and exploration of samples integrated with clinical decision making may improve the quality of pathology informatics. Inclusion of WSI in routine diagnostics is complex and requires expertise and human resources. So, including digital pathology informatics tools in clinical practice is still under research (Jahn et al. 2020).

3. Role of Machine Learning and Deep Learning in Digital Pathology

Various algorithms have been built and verified on pathology images to assist diagnosis based on their morphological patterns (McAlpine & Michelow 2020). So the computational approach has become a part of standard health care that enables clinicians to monitor the progress of complex diseases for better medical care. Machine learning (ML) algorithms automatically detect patterns and perform tasks such as detection and classification. "Linear Regression, Logistic Regression, Naive Bayes, Decision Tree, K Nearest neighbour, Random Forest, Support Vector Machine (SVM) and Neural Network" are some of the widely used traditional machine learning techniques used for classification and disease diagnosis (Shafi & Parwani 2023).

Traditional machine learning algorithms exhibited a poor performance in handling large volumes of data sets and performance of the classifier always depends on the features extracted from the handcrafted segmented images. This resulted in an increase of false negative and false positive cases. Mitigating false negative and false positive scenarios is the major task of a reliable classification algorithm that enables accurate, consistent, and early diagnosis of disease (Shafi & Parwani 2023). By viewing the incompetence of traditional machine learning algorithms, deep learning techniques are introduced which is an extension of classical neural networks with many hidden layers which aid in handling complex data. Deep neural networks can explore complex non-linear patterns as well as handle large volumes of data, their application increases especially in the field of medical research and diagnosis (Rodriguez et al., 2022).

Convolution Neural Network (CNN) is one of the broadly used models in the field of medical application, as it supports high dimensional data and explores complex non-linear patterns in the data (Xie et al., 2021). CNN model learns features of its own at multiple levels of abstraction to learn all possible features from low-level features such as pixels, lines, and edges of an input to high-level features such as objects and shapes without depending on human extracted features. A sample CNN model with all necessary layers is shown in Figure 3.

The convolution layer will compute the output of neurons that are connected to local regions in the input. The Rectified Linear Unit (ReLU) is the most used activation function, returns 0 for negative inputs and for positive values it returns them back. The pooled layer will perform a down sampling operation along the spatial dimensions. FC (fully connected) layer will compute the class scores (Jiang et al. 2017). The model will distinguish dominating and certain low-level features in images over some epochs and classify them using the Soft max classification technique in the model. This aids in normalizing the output of the model to a probability distribution over the predicted output class (Jahn et al. 2020).

Recently, CNN has been successfully implemented in several areas in the medical field to assist clinicians and pathologists to prognosis and diagnose malignancy at its early stages resulting in the reduction of morbidity and mortality rates to a greater extent on a large patient population (Jiang et al. 2017). Congenital cataract disease diagnosis through ocular images, detection of cancer cells from clinical images, detection of diabetic retinopathy through the retinal fundus images, diagnosing heart diseases through cardiac images, grading prostate cancer, identifying biomarkers in early-stage melanoma, identifying morphological features such as nuclear shapes, nuclear orientation, texture, tumour architecture etc. to diagnose cancer are some of the major areas where CNN models demonstrated their performance and assistance in augmenting the ability of pathologists, radiologists and other clinicians to interpret the evidence of cancer and other ailments in their early stages (McAlpine & Michelow 2020). Various metrics are used to evaluate the accuracy of AI algorithms. To achieve good sensitivity (identifying diseased samples correctly) and specificity (identifying non-diseased samples correctly), efficient and reliable classification techniques should be developed for automated systems. Improvisation in machine learning algorithms, deep learning and medical image processing techniques will help pathologists to screen the samples faster and reliably as best as the visual screening system significantly at a lower cost.

4. Automated Screening Systems in Digital Pathology

Substantial efforts have been made in the medical domain, even before the era of artificial intelligence. MYCIN, a computer-based consultation system presented in 1977 was designed to support physicians in the diagnosis and therapy for patients with bacterial infections (Yao et al., 2022). One of the first chatterbots named ELIZA, which was presented in 1966, came with the most famous script DOCTOR, simulating a psychotherapist who treats people for emotional problems and mental illnesses through talk therapy. In 1992, PATHFINDER, an expert system was developed for hematopathology diagnosis. PAPNET, Thin prep imaging system, CYBEST and Cyto-Biologic Electronic Screening System are some of the other expert systems that utilize image analysis

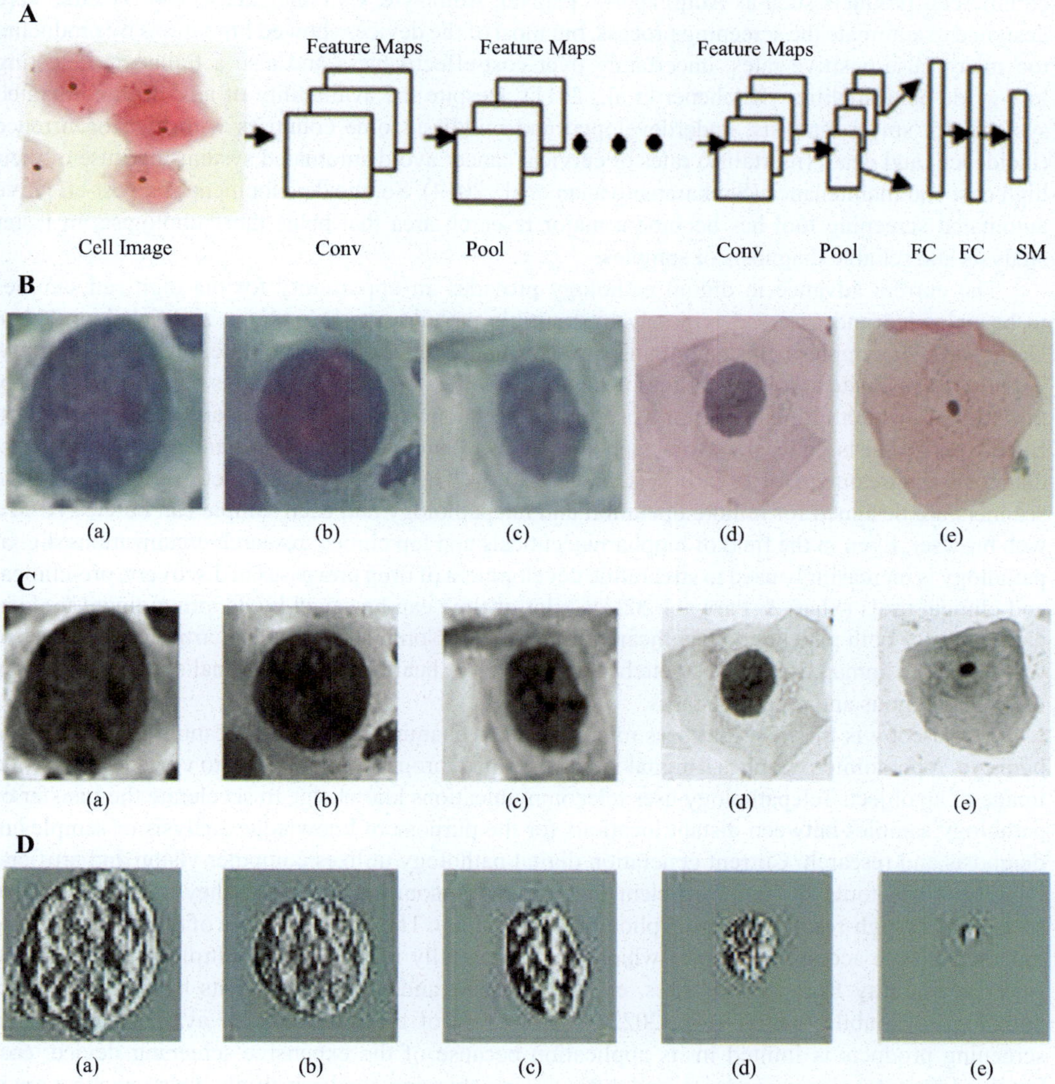

Fig. 3 (A) CNN Model with Convolution, Max Pool, Fully Connected and SoftMax Layers, (B): (a) Carcinoma (b) Severe (c) Moderate (d) Mild (e) Normal Images of Herlev Dataset, (C): Pre-processed Pap Smear Cervical Cell Images of Herlev Dataset (a) Carcinoma (b)Severe (c) Moderate (d) Mild (e) Normal Images, (D): Equalized Gradient Images with Highlighted Nucleus of Pap Smear Cervical Cell Images from Herlev Dataset (a) Carcinoma (b)Severe (c) Moderate (d) Mild (e) Normal Images

technology for the automated pre-screening of cervical cytology specimens. Arterys' medical imaging platform became the first FDA-approved deep learning clinical platform that can help cardiologists to diagnose cardiac diseases (Xie et al. 2021). Manual screening V's automated system has been an open problem for the last few decades, and still remains unsolved. Manual screening of pap smear images is laborious and time-consuming as each sub-image within a single slide has to be studied under the microscope by pathologists for disease diagnosis. During mass screening programs, a huge number of samples must be analyzed, which is more time consuming and requires skilled cytotechnicians. Technology advancement in medical imaging enhanced the quality of medical images leading to better contributions to the early diagnosis of diseases. Computer-assisted commercial products such as AutpPap300, PapNet, AutoCyte, Cytoanalyzer, CYBEST etc., were designed to automate the screening process, but most of the devices showed limitations by producing too many false-positive rates, uncertainty over cost-effectiveness and also a failure in detecting low-grade abnormalities (Kitchener et al., 2011). Despite the availability of advanced automated systems for smear analysis, underdeveloped and middle-income countries with high occurrence (incidence) and death (mortality) rates of cervical cancer avoid automated systems because of their high cost and maintenance (Sankaranarayanan et al., 2014). So, the development of a cost-effective automated screening tool has become a major research area that helps the Pathologists in faster analysis and reliable diagnosis of samples.

The current advance in digital pathology provides an opportunity for the digitized samples to be uploaded and assessed remotely by multiple pathologists (McAlpine & Michelow 2020). Digitization of samples helps pathologists to view and share the images online. So, pathologists do not need to be onsite and sample slides need not be shifted to other places. This eliminates the delay in diagnosis which ultimately improves clinical decision making in cancer and other treatments. Expert pathologists can work from their homes by spending their energy and time for the best diagnosis and treatment at the earliest (Lau et al. 2021). The new generation Ocus series microscope scanners are designed for remote operation and telepathology and each sample can be viewed in a web browser. Even in the field of biopharmaceuticals and top clinical research organizations digital pathology is increasingly used to streamline development of drug processes in discovery, pre-clinical and clinical trials (Shafi & Parwani 2023). A lot of time can be saved by adopting digital tools in pathology for both patient and healthcare professionals. Computer vision and artificial intelligence will not make human experts obsolete but will make the healthcare professional's work simpler by removing tedious and repetitive tasks.

Microscopy is the field that uses microscopes to examine samples that cannot be seen with a bare eye. A photomicrograph is a digital image obtained through a microscope to view the magnified image of an object. Telepathology uses telecommunications knowledge to accelerate the transfer of pathology samples between distant locations for the purpose of knowledge analysis of samples in diagnosis and research. Current generation digital pathology utilizes computer vision and artificial intelligence as tools for more efficient diagnosis (Barisoni et al. 2020). They view samples or specimens as high-resolution digital photographs or video. The notable benefit of digitizing samples includes remote access of samples which can be centrally reviewed by multiple pathologists for better availability from various sites, enabling precise and accurate readouts by reducing inter-pathologist variability (Baxi et al. 2022). The success of the commercially available automated screening products is limited in its application because of the expensive screening device, cost involved for screening procedures, and difficulty in achieving good sensitivity. False-positive cases are not a serious ethical issue but will make a patient undergo an expensive treatment process and creates unnecessary anxiety and stress to the patient by the wrong diagnosis. Thus developing a reliable and economical screening method incorporating AI technologies has become a major research area that aids in early diagnosis of cervical cancer.

5. Role of AI in Pap Smear Cervical Cancer Cell Screening: A Computational Approach

Current AI applications are widely used in the cytopathological field to identify various patterns and morphological changes on cell images. Cervical cancer can be effectively treated and cured if precancerous changes are identified at their early stages. Papanicolaou smear test or Pap test is one of the most cost-effective screening methods used to detect the precancerous and cancerous deformations in the cervix. The cell nucleus is mainly examined during the screening process as its morphological and textural changes contribute a vital role in the identification of abnormalities. Digital pathology paved the way to automate screening facilities which assist pathologists to work on suspicious cells by segregating the normal cells from the huge set of samples for faster analysis and diagnosis. Manual screening of a pap smear is laborious and time-consuming and during mass screening programs, a huge number of samples must be analyzed and diagnosed, which is more tedious and requires skilled cytotechnicians. Advances in medical imaging technology and AI tools have greatly improved the quality of medical images and assessment processes, which contributes much in the reduction of morbidity and mortality rates of cervical cancer. Most of the researchers performed their study on the publicly available Herlev dataset to avoid ethical issues on the smear data. The Herlev dataset consists of 242 normal cells, 182 cells of mild dysplasia, 146 cells of moderate dysplasia, 197 cells of severe dysplasia, and 150 cells of squamous cell carcinoma. The Herlev Database can be downloaded from the url: http://labs.fme.aegean.gr/decision/downloads (Jantzen & Dounias 2006). The images were collected from Herlev University Hospital, Denmark, employing a digital camera and microscope with a resolution of 0.02 micrometer per pixel by skilled cytotechnicians and doctors. Sample Herlev dataset images of all grades are shown in Figure 3B.

As a part of the initial phase of the work, image pre-processing is a mandatory step to remove electronic and other biological noises, which would help and contribute more in segmentation, feature extraction, and classification of images. Image pre-processing was performed using Bi-Histogram Equalization with adaptive sigmoidal function, which highlighted the nucleus of the cell. The size of the nucleus plays a very important role in disease diagnosis and acts as one of the best biomarkers for disease prediction. A preprocessed image sample is shown in Figure 3C.

Gray Level Co-occurrence Matrix (GLCM) is used to quantify the spatial association between pixels within the area of interest. The characteristics of the nuclear region of a cervical cell undergo typical changes because of increased metabolic activity (Mahanta & Bora 2013). The appearance produced by the variation in chromatin content, degree of chromatin dis-aggregation and degeneration results in Hyperchromasia, which thickens the appearance of the nucleus texture, coarse with multiple nucleoli. This variation in chromatin content of abnormal cells diverge very clearly with normal cells and can be used for cancer diagnosis. Equalized gradient images with highlighted nuclei of pap smear cervical cell images from Herlev Dataset are shown in Figure 3D. GLCM features like energy, entropy, contrast, homogeneity, correlation, and autocorrelation were extracted to measure the coarseness in the nuclear region, and the cancer type is addressed numerically.

The purpose of accurate segmentation is to extract textural features from the segmented nuclear region. Textural features extracted from the segmented nucleus had contributed a major role in cell image classification and diagnosis. The explicit changes of the nucleus in terms of its nucleus cytoplasmic ratio, irregular chromatin distribution, and nuclear membrane contributed a lot in the detection of abnormality in pap smear images. Region Based Active Contour algorithm handled Pap smear cervical cell images, which had weak and blur edges and showed better nucleus segmentation when compared with other segmentation techniques. Segmented cell images with the highlighted nucleus of pap smear cervical cell images from Herlev Dataset are shown in Figure 4

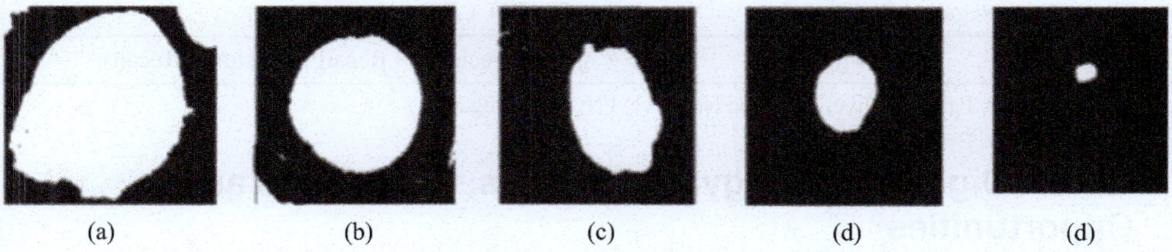

Fig. 4 Segmented Cell Images with Highlighted Nuclei of Pap Smear Cervical Cell Images from Herlev Dataset (a) Carcinoma (b)Severe (c) Moderate (d) Mild (e) Normal Images

After extracting the nucleus morphological and textural features, classification is performed using the extracted features to determine whether the sample has symptoms of the pre-cancerous lesion. This part of the task is considered as the last and the most important stage in the automated cervical cancer screening system. For binary classification, traditional machine learning is performing well and portrays good classification accuracy. But for multi class classification, accuracy slowly diminished. Traditional classification methods depend on hand-engineered features and showed limitations in large, diverse datasets.

So for multi-class classification study is extended to deep learning using convolutional neural networks with augmented datasets. The convolution neural network (CNN) is one of the most popular deep neural network architectures, that automatically learns feature at multiple levels of abstraction and allows a system to learn all possible features from low-level features such as edges and pixel intensity of an image to high-level features such as objects and shapes without depending on extracted human features. Without depending on handcrafted segmented features, the whole image is fed to the proposed CNN network, and it automatically learnt the features and classified the images with acceptable classification accuracy. Performance evaluation of AI algorithms is achieved based on the tool called confusion matrix shown below and represents the accuracy of the classification model. It displays the number of true negative, false negative, true positive and false positive cases and analyzes model performance by identifying misclassification.

		Predicted Condition	
		Normal	Abnormal
Actual Condition	Normal	*TN*	*FP*
	Abnormal	*FN*	*TP*

Confusion Matrix

Assessment of locally developed AI algorithm performance characteristics would include assessing its precision, accuracy, sensitivity, and specificity which is shown in Table 2. The specificity is the probability that a test returns a negative result given that the patient does not have the disease. The sensitivity of a test is the probability that a sample tests positive given that the patient has the disease (Jahn et al. 2020). The F1 score is a metric used in classification models used for both balanced and imbalanced datasets. A high F1 score indicates the strong overall performance of a binary classification model. It signifies that the model can effectively identify positive cases while minimizing false positive and false negative cases (Jahn et al. 2020).

Table 2 Evaluation Metrics

Metric	*Formula to Assess*
Accuracy	TP + TN/TP + FP + FN + TN
Precision	TP/TP + FP
Recall (Sensitivity) or True Positive Rate)	TP/TP + FN
Specificity (True Negative Rate)	TN/TN + FP
F1 Score	2 × (Precision × Recall/Precision + Recall)

TP: True Positive, FP: False Positive, TN: True Negative, FN: False Negative

6. AI and Digital Pathology: Limitations, Challenges and Opportunities

As all technologies have their own merits and demerits, even digital pathology has some clear obstacles that limit easy employment of AI methods. Even though AI technologies have received considerable attention in medical research, their real-life implementation is still facing issues (Tizhoosh & Pantanowitz 2018). In traditional machine learning algorithms, pathologists need to manually delineate the region of interest in all images and handcrafted images always need not to be perfect, resulting in more false negatives and false positives. So besides the time constraint involved, manual annotations create bottlenecks in disease prediction. Low resolution and blurry images are also a cause for bad predictions. Various patterns and minute textural changes make recognizing tissues and cell images by image algorithms exceptionally challenging in digital pathology. Many research papers deal with binary classification problems having just two possible values such as normal and abnormal or benign and malignant. But when it comes to multi-class classification, the accuracy drops drastically as AI requires many training cases for each variation (Baxi et al. 2022).

Deep learning takes machine learning a little progressive and has the capability to handle high dimensional data and provides better accuracy for multi-level classification problems. Current algorithms are using data sets of smaller size and images from a single centre. To increase the data set, images are augmented by methods such as random rotation and flipping. Over-fitting is an unacceptable problem that predicts well for training data and shows poor predictions for new data. This issue can be mitigated by emphasizing the requirement for robust validation of these algorithms using multi-institutional data (Baxi et al. 2022). Differences in the quality of smear slide preparation methods, scanners used for acquisition and the process of digitization between various centres all play a vital role in disease prediction. Some of the technical concerns such as reproducibility, interpretability, the accuracy of competing devices, cost of processing hardware, and regulatory approvals all represent barriers to AI algorithms for clinical adoption (Niazi et al. 2019).

Numerous quality control (QC) techniques should be adopted to overcome issues such as differences in slide preparation, origin, and scanner type. The incapability of deep learning algorithms to demonstrate and interpret the findings have been described as "black boxes". To mitigate interpretability issues some statistical studies can be included to determine the processes and work- flows of the decision made by the models (Jahn et al. 2020). Deep learning models are trained for some specific image recognition tasks, but in radiology that uses imaging technology to diagnose and treat disease, many such narrow detection tasks are adopted to identify all potential findings from the images for disease diagnosis, but AI can perform only a few of those which will not be enough for correct diagnosis. Some of these AI based algorithms are subjected to algorithmic bias where prediction depends on some terms which are not considered as a main causal factor for disease (Davenport & Kalakota 2019). Importance of input data in terms of quality and quantity ensures data fidelity and ensures standardization both in the evaluation and performance of AI algorithms. Accurate prediction and classification mainly depend on the quality and size of data. So, it is essential to have a high specification hardware for processing and analyzing images. with a scalable data storage facility.

Cloud data storage has challenges such as high initial cost, handling bandwidth issues in transmitting images, data security and uninterrupted network. The smear slide digital data can be kept in a central cloud-based space that aids in remote access is utilized for manual review by a pathologist as well as automated review by a data algorithm. Lack of labelled data, dimensionality obstacle of images, computational expenses and lack of transparency and interpretability are some of the highlighted issues to be considered before incorporating AI in clinical practice (Shafi & Parwani 2023). Despite the challenges and limitations of the usage of AI in digital pathology, digitization of slide images paved the way to upload digitized images online and allowing review of images remotely by multiple pathologists. The rapid advances in AI for image analysis made initial investigation much easier for radiologists, pathologists and other medical domains. With a large dataset along with rich information on features, AI algorithms are expected to learn more evidence on the data which is largely related to real time clinical questions that leads to better decision making in disease diagnosis. Recently in 2017, the approval of FDA for using Philip's whole slide scanner in clinical diagnosis marked a way to digitize clinical workflow (Shafi & Parwani 2023). So, the advent of digitization helped the medical domain, especially pathologists to overcome barriers and address associated challenges with the traditional workflow. The improved infrastructure of the network environment, the enhanced computing capacity, and broad integration of informatics have made both computational pathology and collaborative patterns, make data transmission faster enabling the cloud-based central laboratory and data bank to deliver better inexpensive care for patients.

Conclusion

The current advancement in digital pathology incorporating AI technology portrayed a wide improvement in disease diagnosis especially in the field of cancer prediction. Improved accuracy shown in automated morphological image analysis paved a way to expand its contribution to disease severity assessment and prognosis prediction. The facility provided for digital images to be uploaded online and reviewed remotely by multiple pathologists, and the image acquisition and processing of large scale and complex datasets will make next generation pathologists' task easier. Automated whole slide imaging (WSI) scanners provide high resolution images of entire glass slides that paved the way to integrate imaging into all aspects of analysis and diagnosis especially in the fields such as anatomical, clinical, and molecular pathology. Irrespective of the rise in the usage of AI tools in the medical domain, full-fledged integration of these tools into clinical practice is taking time due to issues such as interpretability, validation, regulation, and cost. Image analysis, storage, and throughput can be improved by making use of Graphical Processing Units (GPUs) that have many dedicated parallel processing cores which makes the system more efficient for interpretation. Adoption of AI tools in clinical practice is still facing hurdles and many technical, ethical, and legal questions need to be answered to create trust on usage. Automation of the screening process for early diagnosis of malignancy has been an open problem for over fifty years without attaining any satisfactory results. So, efforts are underway to design a cost-effective automated device by standardizing the process of acquiring, processing, storing, and displaying digital images, especially in the field of cytopathology. The rapid increase of digital image processing in the field of cytopathology resulted in the vast usage of imaging devices such as microscopic digital images, real-time robotic microscopy, and whole slide digital imaging (WSI) to acquire and process images. This effective tool can assist medical doctors effectively in diagnosing the disease in its early stages and allow early treatment for the disease which is completely preventable and curable when detected quickly. By performing further validations through studies with human subjects, acquiring a certain degree of trust and approval from regulatory boards, and creating awareness about the tools among people may significantly promote AI tools in clinical pathology.

References

Barisoni, L., Lafata, K.J., Hewitt, S.M., Madabhushi, A. and Balis, U.G. et al. (2020). Digital pathology and computational image analysis in nephropathology. *Nature Reviews Nephrology*, *16*(11), 669–685.

Baxi, V., Edwards, R., Montalto, M. and Saha, S. (2022). Digital pathology and artificial intelligence in translational medicine and clinical practice, Modern Pathology 35(1): 23–32.

Cui, M. and Zhang, D.Y. (2021). Artificial intelligence and computational pathology, Laboratory Investigation 101(4): 412–422.

Davenport, T. and Kalakota, R. (2019). The potential for artificial intelligence in healthcare, Future healthcare journal 6(2): 94.

Dwivedi, S., Swamy, M., Dubey, A., Verma, Y. (2019). The advent of digital pathology: a depth review, J Entomol Zool Stud 7: 43–9.

Gupta, A., Ranjan, R., Bajaj, A.K. and Pathak, A. (2023). Digital pathology–from slides to screen, Indian Cancer Awareness Journal pp. 1–5.

Hou, X., Shen, G., Zhou, L., Li, Y., Wang, T. et al. (2022). Artificial intelligence in cervical cancer screening and diagnosis, Frontiers in oncology 12: 851367.

Jahn, S.W., Plass, M. and Moinfar, F. (2020). Digital pathology: advantages, limitations and emerging perspectives, Journal of clinical medicine 9(11): 3697.

Jantzen, J. and Dounias, G. (2006, November). Analysis of papsmear image data. In Proceedings of the nature-inspired smart information systems 2nd annual symposium (Vol. 10, pp. 1-11).

Jiang, F., Jiang, Y., Zhi, H., Dong, Y., Li, H. et al. (2017). Artificial intelligence in healthcare: past, present and future, Stroke and vascular neurology 2(4).

Kitchener, H.C., Blanks, R., Cubie, H., Desai, M., Dunn, G. et al. (2011). MAVARIC–a comparison of automation-assisted and manual cervical screening: a randomised controlled trial. Clinical Governance: An International Journal, 16(3).

Lau, R.P., Kim, T.H. and Rao, J. (2021). Advances in imaging modalities, artificial intelligence, and single cell biomarker analysis, and their applications in cytopathology, Frontiers in Medicine 8: 689954.

Mahanta, D., Ahmed, M. and Bora, U. J. (2013). A study of bandwidth management in computer networks. International Journal of Innovative Technology and Exploring Engineering, 2(2), 69–73.

Mahanta, L.B. and Bora, K. (2013). Hyperchromasia and texture as effective features for analysis of malignancy in Pap smear images. International Journal of Signal Processing, Image Processing and Pattern Recognition, 6(4), 451–466.

McAlpine, E.D. and Michelow, P. (2020). The cytopathologist's role in developing and evaluating artificial intelligence in cytopathology practice, Cytopathology 31(5): 385–392.

Niazi, M.K.K., Parwani, A.V. and Gurcan, M.N. (2019). Digital pathology and artificial intelligence, The lancet oncology 20(5): e253–e261.

Pantanowitz, L., Sharma, A., Carter, A.B., Kurc, T., Sussman, A. et al. (2018). Twenty years of digital pathology: an overview of the road travelled, what is on the horizon, and the emergence of vendor-neutral archives. Journal of pathology informatics, 9(1), 40.

Rajkumar, R. (2018). Introductory chapter: cervical cancer-screening, treatment and prevention, Cervical Cancer-Screening, Treatment and Prevention-Universal Protocols for Ultimate Control, Intech Open.

Rodriguez, J.P.M., Rodriguez, R., Silva, V.W.K., Kitamura, F.C., Corradi, G.C.A. et al. (2022). Artificial intelligence as a tool for diagnosis in digital pathology whole slide images: A systematic review, Journal of Pathology Informatics p. 100138.

Shafi, S. and Parwani, A.V. (2023). Artificial intelligence in diagnostic pathology, Diagnostic pathology 18(1): 109.

Tizhoosh, H.R. and Pantanowitz, L. (2018). Artificial intelligence and digital pathology: challenges and opportunities, Journal of pathology informatics 9(1): 38.

Xie, X., Wang, X., Liang, Y., Yang, J., Wu, Y. et al. (2021). Evaluating cancer-related biomarkers based on pathological images: a systematic review, Frontiers in Oncology 11: 763527.

Yao, K., Sadimin, E., Chang, S., Schmolze, D. and Li, Z. et al. (2022). Current applications and challenges of digital pathology in cytopathology, Human Pathology Reports 28: 300634.

CHAPTER

24

Deep Learning and AI in Digital Cytology

Ashwini S Galande,[1] *Vikas Thapa*,[1] *Aswathy Vijay*,[1] *Sandeep Karunakaran*,[1] *Alex C Varghese*,[2] *Megha KB*,[3] *Athira SS*,[3] *Mohanan PV*[3] and *Renu John*[1*]

1. Introduction

Cytology is a broad umbrella of techniques encompassing the study and analysis of individual cells, and their structural and functional characteristics. It aids in diagnosing disease conditions like cancer, inflammation, and several infections. The procedures for extracting cells are safe, quick, simple, and cost-effective. Cytopathology is a diagnostic modality that examines cells present in various parts of the body to determine the cause, nature, progression, and structural and functional characteristics associated with diseases. It involves acquisition, data processing, transfer, sharing, storage, and interpretation on a digital platform. Cytology uses several methodologies: smear, centrifugal, air-dried or fixed (e.g in alcohol, formalin), or liquid-based preparations. The application of digital methods in cytology has accelerated the processes of screening, detection, diagnosis, and classification through quick and accurate interpretations, quality control, research, image analysis, education, training, and proficiency testing from digital images (Weinstein, R.S. et al., 2009; Williams, S. et al., 2010). Moreover, the provision of archiving these images to be used at anytime later enables effective interpretation, learning, training, and research support in analysis using stored databases of images.

It allows imaging in real-time, dynamic microscopic imaging, and whole slide images (WSI). It involves scanning the slide at the microscopic level, and displaying the digital image via the sensor software. It allows the navigation, zooming, and processing of the image, management in

1 Medical Optics and Sensors Laboratory, Department of Biomedical Engineering, Indian Institute of Technology, Hyderabad, India.
2 Fertility Associates, Auckland, New Zealand
3 Division of Toxicology, Sree Chitra Tirunal Institute of Medical Science and Technology, Thiruvananthapuram, Kerala, India.
* Corresponding author: renujohn@bme.iith.ac.in

a system for proper archiving and intelligent retrieval, analysis of the image for quantification of parameters using tools and sharing the information for medical, educational, and research purposes (Giansanti, D. et al., 2015). It allows a single pathologist to assist multiple locations. The specimens in cytological imaging are three-dimensional intrinsically and demand image manipulations to reproduce accurate renditions. In pathology, in an important subsection of cytology the analogue image from a microscope is viewed by a practitioner; digitalization offers the transfer of the images to a digital monitor. It removes the need for physical storage and hassles in transporting the specimen for diagnosis continuously along with cost cutting. The analogue image is converted into an array of pixels to create a well-resolved image without distortion.

The images are transported as static images, as whole-side images, or in real-time. The former is more cost-effective and easier. It delivers only a portion of the entire sample area. In the case of telemedicine, such images can be detrimental. In real-time image transmission, the microscope is located at a remote location, and the user has control over the movement, focusing, and imaging of the sample. Whole-slide imaging is similar to static image transfer. To create whole slide images the high-resolution scanning system takes the image of the specimen in a sequence and stitches the image by digitizing the image across these large arrays. It takes longer scanning time and lacks intrinsic focus. Both the latter techniques demand high memory requirements, and data transfer toward remote destinations over a network requires high bandwidth capacity (Wilbur, D.C. 2011). Z-stacking or multiplanar imaging allows the image capture in multiple planes.

The use of digital cytology was first applied in the 1960s for the transmission of blood film images using black-and-white video microscopy (Weinstein, R.S. 1991). The static images of digital cytology for cervical cells portrayed less accuracy than microscopic images with several cases of false negatives (Raab, S.S. et al., 1996), and a lack of proper focus and resolution (Ziol, M. et al., 2019). During the 1990s the field evolved enough to report only minor discrepancies in the images from digital images and actual consultations from a microscope on evaluating cervical and other nongynecological cells (Allen, E.A. et al., 2001; Lee, E.S. et al., 2003). In the case of telemedicine, rare or limited samples can be effectively used without wastage, stain-related degradation, and slide damage. Static images have been implemented effectively for breast cancer diagnosis (Galvez, J. et al., 1998), pleural effusion (Ayatollahi, H. et al., 2007), cancer studies (Marchevsky, A.M. et al., 2003; Jialdasani, R. et al., 2006). In Japan, tele-cytology has been used as a primary means of diagnosis since 2004 reporting an accuracy of more than 90% (Yamashiro, K. et al., 2004). It has been applied in areas like pulmonology (Kayser, K. et al., 2000), veterinary science (Maiolino, P. et al., 2006), etc. It can assist the doctor or pathologist in fine needle aspiration evaluations with very high accuracy (Shroyer, K.R. and Singh, M., n.d). Real-time evaluation of fine needle aspiration cytology helps in the time-saving diagnosis of breast cancers. The application of rapid on-site evaluation has been beneficial in real-time diagnosis along with affordability and quality. In addition to other modalities of cytology, robotic live image streaming is another option that is widely used in histology. Robotic cytology is yet to be popular but it guarantees a huge potential as it allows the user to control factors like a field of view, focus, and magnification and does not demand human intervention to prepare smears (Singh, N. et al., 2002).

Several curable and common diseases are often unrecognized and not treated properly in resource-deficient and economically backward regions. The clinical implementation of point-of-care digital diagnostics in microscopy has seen significant improvements in this area. Human papillomavirus (HPV) infections must be screened regularly to reduce the incidence and mortality in cases of cervical cancer. Papanicolaou tests aided by such diagnostic methods and artificial intelligence (AI) have proven to be highly effective in detecting abnormal cells (Holmström, O. et al., 2021; Briscoe, D. et al., 2020). AI applications in digital cytology can be used for educational purposes, providing automated feedback and assistance in the training of medical professionals and students. In this chapter, we will give a brief overview of current cytology procedures. The field of cytology has evolved with quantitative phase imaging (QPI) techniques. Furthermore, cytology and disease diagnosis have revolutionized with the evolution of artificial intelligence techniques. We discuss the use of AI techniques for QPI imaging and their applications in cytology. Finally, the challenges of using AI in cytology and the possible solutions are discussed.

1.1 Basic Procedures in Cytology

Cytology slide preparation is a crucial procedure, especially in the case of disease diagnosis. It involves the following steps: (1) Sample collection where methods like fine needle aspiration and swabs are used to collect cells. (2) Fixation of the samples using alcohol or formalin to preserve the cellular structures and prevent degradation. (3) Processing and dehydration using a series of alcohol washes. (4) Cell smearing or preparation on a glass slide by cytocentrifugation and squash method. (5) Staining which is a crucial step to highlight the cellular components and aid in their visualization. Common stains used in cytology include Papanicolaou (Pap) stain and hematoxylin and eosin (H&E) stain. (6) Mounting where a coverslip is placed over the smear. (7) Microscopic examination of the prepared slide by a cytologist or a pathologist to analyze the cell morphology, structure, or other characteristics of the cell under study. (8) Image capture using a digital sensor for documentation, consulting, or educational purposes. (9) Diagnosis and reporting after the examination followed by proper documentation (Gurram, H.P., and John, R. 2018; Raab, Stephen S., et al. 1996). The specific techniques and protocols for cytology slide preparation can vary depending on the type of sample and the intended diagnostic purpose.

There are three methods of sampling-exfoliative, abrasive, and aspirational. In exfoliative cytology, (Kather, Jakob Nikolas, et al. 2019) cells spontaneously shed fluids like urine, sputum, and cerebral spinal fluid from the body among the many that are utilized. Abrasive cytology involves collecting dislodged cells from the body linings and mucosal linings for example pap smears and swabs. In aspirational cytology, cells from a palpable lump or mass in the body are extracted using a needle. In aspirational cytology also known as fine needle aspirational cytology, cells are extracted from a palpable lump or mass in the body using a needle with or without suction and aided by an imaging modality if needed.

2. Artificial Intelligence in Cytology

Artificial intelligence (AI) is applied to automate the analysis of cytological samples. With the advancement of digital scanners, increased computational power, data generation, and storage facilities the field of AI has evolved significantly over time. AI serves as a broad, encompassing concept to mimic human behavior, as shown in Fig. 1. AI models can be rule-based or statistical aiming to perceive, reason, and make decisions (https://synoptek.com/insights/it-blogs/data-insights/ai-ml-dl-and-generative-ai-face-off-a-comparative-analysis). While machine learning (ML) learns patterns from data without the need for explicit programming, these algorithms demonstrate the capacity to improve their performance through exposure to relevant data. Deep learning (DL) leverages deep neural networks for intricate pattern recognition, and Generative AI creates new content. Figure 1(b) shows the general classification process using ML and DL. All these algorithms come under the umbrella of AI hence, we refer to them by the general term 'AI algorithms'.

AI-based systems in digital cytology have demonstrated an exceptional ability to analyze vast amounts of cytological data with high precision. They can detect abnormalities, such as cancerous cells or infectious agents, with remarkable accuracy. AI systems not only improve diagnostic accuracy but also reduce the chances of human error, leading to more reliable results. This efficiency is particularly beneficial when time is of the essence in diagnosing life-threatening diseases. Cytologists often spend a substantial amount of time reviewing slides and images, which can be tedious and time-consuming. AI can automate the initial screening of samples, allowing pathologists to focus on more complex and critical cases. By doing so, AI enhances productivity and enables quicker diagnosis, leading to faster treatment decisions. AI in digital cytology has the potential to extend healthcare services to remote or underserved areas. Tele cytology, enabled by AI, permits the transmission of digital images to specialists for remote consultation. This has the potential to bridge the gap in healthcare access and ensure that patients in remote locations can receive timely diagnosis and treatment recommendations.

One of the applications in pathology involves training the network for making accurate pre-selection, correlations, or judgments for disease diagnosis and therapy response (Ching, T. et al., 2018). Techniques like 'immunoscore' and evaluation of tumor-infiltrating lymphocyte (TIL) assessment can be effectively incorporated into AI algorithms that help oncologists to assess and quantify the therapy effectiveness and prognosis of cancers. AI and machine learning techniques can be applied to various aspects of the analysis, such as image recognition, data interpretation, and prediction modelling (Angell, H. and Galon, J. 2013). Factors like feasibility, reproducibility, simplicity, cost, and robustness must be considered while implementing it in cytological applications. Several CNN-based models have been applied to detect, classify, segment, and grade tumors in the brain, bladder (Harmon, S.A. et al., 2020), lung (Teramoto, A. et al., 2017), stomach (Meier, A. et al., 2018), breast (Ehteshami Bejnordi, B. et al., 2017), and prostate (Litjens, G., 2016). Various deep-learning models have been applied to study CRC tumors, mutation patterns, and prognosis Kather, J.N. et al., 2020; Krisam, J. et al. 2019).

While AI holds great promise in cytology, it is crucial to emphasize the importance of collaboration between AI systems and human experts. Pathologists remain integral to the validation, interpretation, and decision-making processes, ensuring the reliability and ethical use of AI technologies in cytology. Additionally, regulatory and ethical considerations must be considered to ensure patient safety and privacy.

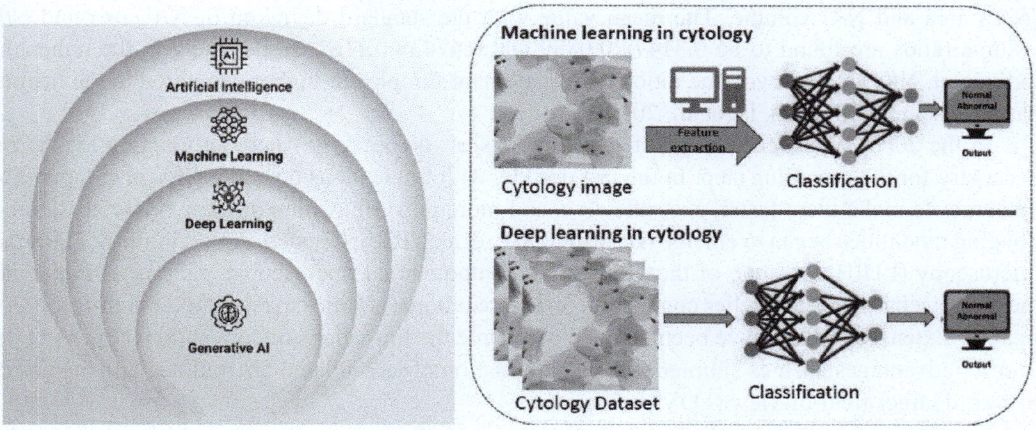

(a) (b)

Fig. 1 (a) Evolution of AI over time (https://synoptek.com/insights/it-blogs/data-insights/ai-ml-dl-and-generative-ai-face-off-a-comparative-analysis), (b) ML and DL for digital cytology

3. Quantitative Phase Imaging (QPI) in Cytology

Cytology is often limited by the availability of suitable stains for labelling and often the sample quantity is also limited. The high cost of consumables in Cytology, the time taken, and the cognitive and physical efforts of the cytologists are critical factors that affect the outcome of traditional cytology procedures. Traditional cytological evaluations often rely on labeling techniques that use fluorescent dyes and specific markers to highlight structures within the cell. However, these methods have limitations, including potential alteration of cell behavior and complexity in the staining process. Label-free digital cytology is a proper solution as it consumes fewer resources and reduced labour for the cytologist. Label-free microscopy techniques can significantly cut down the financial requirements. Most biological samples are transparent to light hence it is difficult to work with a normal bright field microscope unless stained. It allows real-time imaging and preservation of the innate state of the sample. Several methodologies have been adopted for cytological evaluations for label-free non-invasive imaging based on interferometric (Panta, P. et al; 2019; Gurram, H.P.R. et al., 2020; Gerwert, K. (2020); Hermes, M. et al., 2018; Cicchi, R. 2014) and non-interferometric techniques (Poola, P.K. 2016). In this section, we will discuss some of the interferometric and non-interferometric QPI modalities.

3.1 Digital Holographic Microscopy (DHM)

DHM is the well-established interferometric QPI modality for live cell imaging. In DHM, the diffracted wave (object wave) and the reference beam intersect at a specific angle, and their interference is recorded at the detector array, which is referred to as a hologram. The 3D object information is achieved by illuminating the hologram with the conjugate of the reference beam. DHM has proven to be a powerful tool for various biological applications, providing valuable quantitative data and non-invasive imaging of biological samples (Gurram, H.P.R. et al., 2020) (Gurram, H.P. and John, R. (2018); Pandiyan, V.P. and John, R. (2015); Gurram, H.P. et al., 2014; Pandiyan, V.P. and John, R. (2013); Pandiyan, V.P. and John, R. 2016). One such example is the analysis of oral cancer. Oral cells can be imaged using DHM and quantitative parameters can be calculated from the phase image of the cell. Malignancies can be detected and analyzed from the nuclear to cytoplasmic ratio of cells under the imager (Gurram, H. et al., 2020). A preliminary study was carried out for detecting and analyzing precancerous oral cells. Healthy and unhealthy subjects were included in this study where oral cells were extracted from them to record holograms. After reconstructions, quantitative analysis was carried out by calculating the nuclear to cytoplasm ratio (N/C) area and N/C volume. The mean value with the standard deviation of N/C area and N/C volume ratios are found to be 0.3147 ± 0.0889 and 0.4743 ± 0.1167, respectively in the unhealthy individual. N/C area and volume ratios for a malignant sample are higher compared to the healthy patient's sample (Gurram, H. et al., 2020).

In the 2000s, researchers began to explore lensless holography where an imaging lens is not necessary for the recording step. In the last decade, with tremendous developments in computation resources to make them faster, cost-effective, and more powerful, some of these non-conventional imaging modalities began to emerge (Bashshur, R.L. et al., 2020). Lensless digital inline holographic microscopy (LDIHM) is one of those emerging computational microscopies that uses simple and inexpensive hardware and relies completely on computational abilities to reconstruct high-resolution images. Extensive efforts have been made to overcome the limitations of conventional microscopes to offer advantages such as simpler set-ups, elimination of lens-related aberrations, and the ability to record larger field-of-view (FOV) holograms.

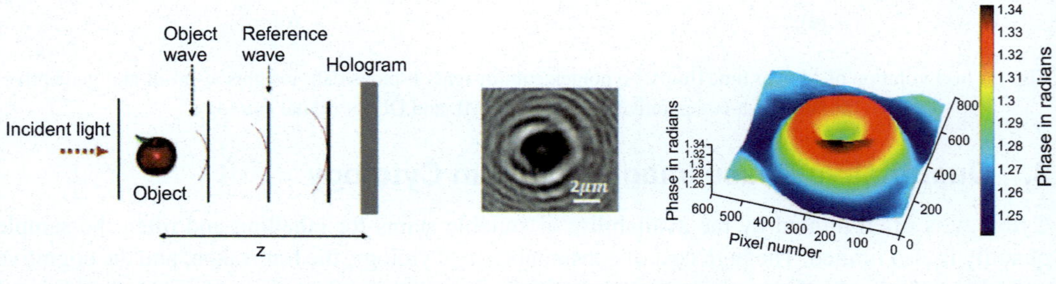

Fig. 2 (a) Schematic of LDIHM setup, (b) hologram recorded of RBC sample, and (c) Phase recovered from the hologram

3.2 Transport of Intensity Equation (TIE)

TIE is a non-interferometric phase imaging modality. The TIE makes use of a well-focused image and two defocused images to extract information about the phase and the object (Hu, C. et al., 2022).

Figure 3(A) shows the experimental set-up of TIE-based microscopy. The oral cells are imaged under the proposed TIE set and the phase is recovered using the TIE algorithm. The requirement of three images limits the live and moving cell imaging. To overcome this problem deep learning algorithms are trained to recover the TIE phase from a single recording. The dataset of input intensity images and corresponding phase images is generated and the generative adversarial network (GAN) is trained to generate a phase image from a single intensity image. Figure 3(B) shows the phase recovery of oral buccal cells using TIE and GAN.

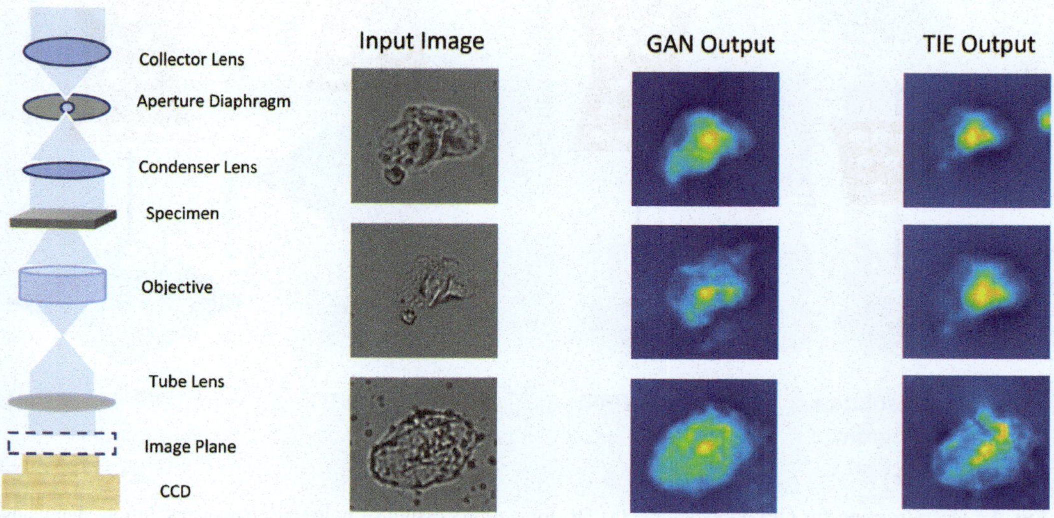

Fig. 3 (A) Experimental setup of TIE, (B): Phase reconstruction of Oral cells using GANs.

4. AI Techniques for QPI

The synergy of QPI and AI techniques (especially ML and DL methods) has improved the conventional cytology procedure for accuracy and faster imaging (as no staining is required). Also, the integration of AI techniques with quantitative phase imaging holds great potential for advancing the capabilities and applications of this imaging modality, allowing (Gurram, H.P.R. et al., 2020) for more accurate and efficient extraction of quantitative information from optical phase data. This can include predicting properties like cell density, refractive index, or other relevant parameters. Different AI methods dominate the QPI image processing and phase recovery with superior performance as shown in Figure 4. In this section, we will discuss the use of deep learning techniques for phase recovery and pre-processing and post-processing applications (Wang, Kaiqiang, et al. 2024 and Harmon, Stephanie A., et al., 2020).

4.1 Deep Learning Methods for Phase Recovery

The deep learning-based reconstruction algorithms leverage trainable neural networks and large datasets to learn better ways of solving imaging inverse problems. These methods have shown great promise over conventional methods in terms of reconstruction speed, improved image quality, and providing better priors for underdetermined problems. Cross-modality training is also demonstrated for efficient reconstruction of holograms to overcome the scarcity of sufficient ground-truth images for the training (Wu, Y. et al., 2019). The use of cross-modality training has enabled phase reconstruction similar to the state-of-the-art phase imaging system with a simple

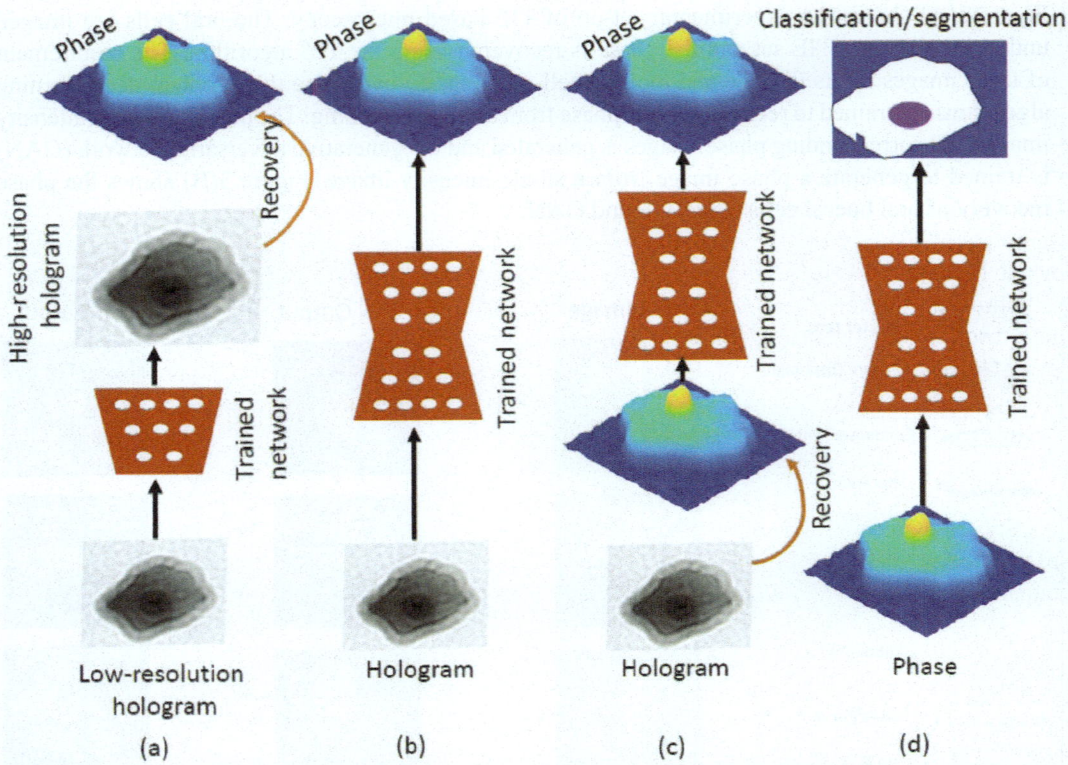

Fig. 4 Deep learning for QPI applications: (a) DL for pre-processing of hologram (super-resolution, denoising, filtering), (b) DL for phase recovery, (c) DL for post-processing (denoising, image enhancement), (d) DL for phase processing (classification, segmentation, feature extraction) (Wang, K. et al., 2024).

and cost-effective imaging setup. However, it is important to note that this method requires many images to be captured at different distances between the sample and the camera. This is essential for providing a reliable training dataset to avoid unknown artifacts in the reconstructed image. In case of unknown artifacts being captured, the model might fail in removing the superimposed phase image. The interconversion of holographic microscopy with other microscopic techniques can overcome the weaknesses of individual modalities by combining their strengths. However, there must be a provision to capture images of the same object by both modalities simultaneously to create a dataset. The interpretability and performance of deep-learning-based reconstructions can be improved significantly by combining domain knowledge of optical physics with deep learning (Galande, A.S. et al., 2023).

4.2 Deep Learning Techniques for Pre-processing and Post-processing

4.2.1 Autofocusing and Depth-Of-Focus (DOF)

Autofocusing is a crucial step in hologram reconstruction as the back-propagation of the hologram to the correct object location determines the quality of the reconstruction. Hence sample-to-sensor distance of different parts of the 3D object needs to be estimated numerically. Deep networks (particularly CNN) are used to obtain the focus distance often formulated as a classification or regression problem (Ren, Z. et al., 2018). These models learn complex features and relationships within the data and can provide highly accurate and efficient autofocusing. However, these methods require in-focus distance as *a priori* knowledge which is calculated based on the autofocusing criteria. Hence, the deviation from the correct focus distance degrades the reconstruction quality outside the system DOF. To overcome this problem, in addition to the in-focus hologram, out-of-focus holograms are also paired with the phase reconstructed from the in-focus hologram during training

(Wu, Y. et al., 2018). This results in an extended DOF for the reconstructed image compared to the previous approaches. Deep learning-based autofocusing methods have demonstrated a significant boost in computation efficiency and increased the robustness of the phase reconstruction against the possible misalignment in the set-up, achieving extended DOF simultaneously.

4.2.2 Super-resolution

The resolution of LDIHM is limited by the pixel size of the sensor (pixelation) and the under-sampling of spatial frequencies at the time of recording. This results in missing object information which then leads to unknown artifacts in the reconstruction. The resolution of the recorded hologram is improved by pixel super-resolution techniques (PSR) (Bishara, W. et al., 2010). In PSR, multiple low-resolution holograms of the same FOV are recorded by shifting the hologram laterally on the grid. However, the recording and computational costs imposed by these techniques are high and time-consuming. Also, this method is not suitable for live and moving cells. Deep CNNs are used to perform super-resolution using low-resolution holograms (Dong, C. et al., 2016). Once trained, the CNN can produce a high-resolution hologram from a single low-resolution hologram. It is worth noting that, even though these deep learning methods produce high-resolution holograms with better quality, compared with conventional image super-resolution methods, they require a massive amount of data for the training. The problem of data requirement is partially solved by using CycleGAN and unpaired datasets (Wang, W., 2022).

4.2.3 Denoising

A hologram is formed by the interference between an object wave and a reference wave, which contains interference-related noise and DC noise. In addition, the use of coherent laser sources introduces speckle noise in the hologram recording. Various filters are proposed in the literature to remove noise from the hologram. However, this results in compromised sharpness and longer recording times. The first feed-forward deep network proposed in (Zhang, K. et al., 2017) for denoising, named DnCNN, embraced developments in the deep learning field of denoising. A conditional generative network was proposed in (Fang, Q. et al., 2022) to remove speckle noise from the wrapped-phase image. Even though the network was trained on simulated data it showed improved results compared to other deep learning-based reconstruction methods. Further, complex noise patterns can be learned by CNNs by learning the relationship between the captured intensity images and the simulation truth images (Chen, L. et al., 2021; Hao, J. et al., 2022). To reduce interference-related noise in LDIHM, different denoisers can be used to plug-and-play along with physics-aware untrained deep networks (Galande, A.S. et al., 2023).

4.2.4 Classification and Segmentation

Volumetric information provided by phase images is a value addition for the classification and segmentation of transparent cells, tissues, and microplastics. Conventional machine learning needs manual extraction of features from the phase image for classification. Support vector machine (SVM) is the most used classification algorithm in cytology (Roitshtain, D. et al., 2017; Mirsky, S.K. et al., 2017; Ozaki, Y. et al., 2019; Bianco, V. et al., 2020; Belashov, A.V., et al., 2020; Lam, V.K., et al., 2020; Nissim, N. et al., 2021). The unsupervised organization of cervical cells using principal component analysis (PCA) is demonstrated in (Mangal, J. et al., 2019). The features were extracted from the cervical cells' nuclei and clustered in an unsupervised manner. The result shows that the additional phase information along with intensity information boosts the performance of PCA as compared to when brightfield-only information is used. In contrast, CNNs extract features from the image automatically and classify them into desired classes, generally achieving higher accuracy as compared to ML algorithms at the cost of a large dataset for training. Feature selection by CNN filters can be viewed using filter visualization. The classification accuracy can be further improved by combining manually extracted features as a prior along with CNNs (Pitkäaho, T., 2018). The detailed classification of various biological cells using different AI algorithms is described in Table 1.

Table 1 Different methods and applications of QPI and AI

Author	*Method*	*Application*	*Remark*
Jo, YoungJu, et al. 2017	Digital Holography Microscopy (DHM) and Deep learning (CNN)	Anthrax spores classification	Point of care diagnosis of pathogens
Park, Han Sang, et al. 2016	DHM and machine learning	*Plasmodium falciparum* classification	Early diagnosis of malaria at trophozoite or schizont stage
Chen, Claire Lifan, et al. 2016	Time stretch QPI (TS-QPI)	Classification of white blood T-cells against colon cancer cells	Novel TS-QPI system which increases accuracy
Mirsky, Simcha K., et al., 2017	Interferometric phase microscopy (IPM) and support vector machine (SVM)	Sperm cell classification	Improvisation in IVF procedure
Yoon, Jonghee, et al., 2018	Optical diffraction tomography (ODT) and machine learning	Lymphocyte subtypes classification	Better 3D morphology classification by ODT
Ryu, DongHun, et al., 2019	ODT and Deep learning	Various bacteria *E. coli, K. pneumoniae, P. aeruginosa* classification	Enhanced visual interpretation
Kim, Geon, et al., 2019	ODT and Deep learning (CNN)	Pathogen classification in the bloodstream	Novel method with improvised Results
Yoon, Jonghee, et al., 2017	ODT and Machine learning (k-NN)	Lymphocytes classification	Novel Comparison of accuracies with different classes
Kim, Sung-Jin, et al. 2018	Lensless inline holography Transfer Learning (VGG19- PCA)	Cancer cells (SkBr3, A431) classification with microbeads	The comparative analysis shown between VGG19-PCA-MLP and CNN
Lam, Van K., et al., 2018	DHM (SVM)	Breast cancer MDA-MB-231 classification	Fast processing rates using DHM and SVM have been observed
Pavillon, Nicolas, et al., 2018	DHM (PCA)	Classification and quantitative information extraction of macrophage-like cells	A novel approach using imaging and spectroscopy to interpret the result.
Feizi, Alborz, et al., 2016	Lensless Holography (SVM)	Classification of yeast cells	Field-portable microscopes are cost effective and SVM shows prominent results.
Nguyen, Tan Huu, et al., 2017	Spatial light interference microscopy (SLIM) Random Forest Tree classification	Prostate cancer classification	Utilizes the merits of QPI to enhance classification
Go, Taesik, et al., 2018	Digital Inline Holography Microscopy (DIHM) (Decision tree classifier)	Erythrocyte classifier	This approach extracted nearly 12 features for classification. 97% accuracy achieved.
Rubin, Moran, et al. 2019	DHM Transfer Learning (DCGAN) CNN	SW 480 (colorectal adenocarcinoma colon cells) Hs 895.T (melanoma cells) classification	This novel approach solved the problem associated with a small training set
Shaked et al. 2019	Interferometric phase microscopy (IPM) PCA and SVM	Cancer cell classification	This method utilizes both PCA and SVM properties to increase accuracy

Image segmentation of the unlabeled cells or tissue samples can be challenging as the contrast of bright field intensity is low. Thus, segmentation on the phase image is efficient with additional volumetric information. The applications reported in the literature are RBC segmentation (Yi, F. et al., 2013), sperm cells' segmentation (Kandel, M.E. et al., 2020), SARS-CoV-2 detection (Goswami, N., 2021), cell live-dead assay (Hu, C. et al., 2022), and cell cycle-stage detection (He, Y.R. et al., 2022). The segmentation accuracy is further improved in densely connected samples with overlapping and multiple-touched cells; the fully convolutional network (FCN) Goswami, N, 2021 phase segmentation is proposed for phase segmentation (Yi, F., Moon, I. and Javidi, B. 2017).

5. Challenges in Applying Deep Learning Techniques in Cytology

The implementation of AI techniques in healthcare environments has revolutionized disease diagnosis in terms of accuracy, cost, and enhanced target treatment. However, AI-based clinical systems pose unique challenges that need to be addressed for accuracy and reliability to be used for commercial purposes (Gedefaw, L. et al., 2023; McAlpine, E.D. et al., 2021; Panayides, A.S. et al., 2020). One of the major challenges is the requirement of large datasets for training AI models. Obtaining accurate training datasets representative of all patient populations along with patient histories (background and clinical condition) is highly desirable for unbiased outcomes. Moreover, annotation of large datasets by domain experts is a time-consuming and laborious task. Hence, sharing and creating benchmark datasets for cytology applications will allow machine learning experts to develop advanced and application-specific algorithms. Then, the AI models should be regularly monitored to ensure accuracy and sensitivity in a continuously changing environment.

The interoperability of various AI models used by different healthcare providers is another issue in ensuring consistency among treatment recommendations. This problem is further elevated by the black box decision-making of AI models. It is difficult to predict the parameters considered by the AI model in decision-making which raises trust issues. Hence, there is a need for standardization of data collection procedures and the implementation of AI models for healthcare applications. With recent developments in explainable AI (XAI), it has become easier to understand and visualize decision-making procedures by AI models (Yang, C.C. 2022). This helps in improving performance, validating the accurate parameter selection, and building trust in the conclusions made by AI models. With the help of XAI, the knowledge gap between cytologists (or domain experts) and machine learning experts can be narrowed significantly.

Another major challenge is data privacy and security. There is a need for strict guidelines, norms, and licenses to protect the confidentiality of patient information and the appropriate use of data in training AI models. Also, regularizing data ownership, storage, and transfer for developing more AI models is highly desirable.

Summary and Conclusion

The field of cytology has significantly improved with the implementation of AI algorithms right from data acquisition to disease diagnosis. The QPI techniques allowed for label-free cytology with additional phase information which will reveal some important quantitative characteristics of the sample. The synergy of QPI and AI techniques has broadened the horizons of live cell imaging in studying flow cytometry and cell dynamics. Further, the complexity of microscope set-ups can be reduced using AI algorithms that will improve the healthcare facilities in low-income countries and rural areas. If some of the challenges in implementing AI algorithms in cytology are overcome properly, then highly accurate, precise, and reliable disease diagnosis is possible in real-life scenarios.

References

Allen, E.A., Ollayos, C.W., Tellado, M.V., Butler, D.R., Buckner, S.B. et al. (2001). Characteristics of a telecytology consultation service. *Human Pathology*, *32*(12), 1323–1326. https://doi.org/10.1053/hupa.2001.29652

Angell, H. and Galon, J. (2013). From the immune contexture to the Immunoscore: The role of prognostic and predictive immune markers in cancer. *Current Opinion in Immunology*, *25*(2), 261–267. https://doi.org/10.1016/j.coi.2013.03.004

Ayatollahi, H., Khoei, A., Mohammadian, N., Sadeghian, M.H., Azari, J.B. et al. (2007). Telemedicine in diagnostic pleural cytology: A feasibility study between universities in Iran and the USA. *Journal of Telemedicine and Telecare*, *13*(7), 363–368. https://doi.org/10.1258/135763307782215343

Bashshur, R.L., Krupinski, E.A., Doarn, C.R., Merrell, R.C., Woolliscroft, J.O. et al. (2020). Telemedicine across time: Integrated health system of the future-A prelude. *Telemedicine Journal and e-Health*, *26*(2), 128–130. https://doi.org/10.1089/tmj.2019.0025

Belashov, A.V., Zhikhoreva, A.A., Belyaeva, T.N., Kornilova, E.S., Salova, A.V. et al. (2020). In vitro monitoring of photoinduced necrosis in HeLa cells using digital holographic microscopy and machine learning. *Journal of the Optical Society of America. A, Optics, Image Science, and Vision*, *37*(2), 346–352. https://doi.org/10.1364/JOSAA.382135

Bianco, V. (2020). 'Microplastic Identification via Holographic Imaging and Machine Learning' adv. *Journal of Intelligent Systems*, *2*, 1900153.

Bishara, W., Su, T.W., Coskun, A.F. and Ozcan, A. (2010). Lensfree on-chip microscopy over a wide field-of-view using pixel super-resolution. *Optics Express*, *18*(11), 11181–11191. https://doi.org/10.1364/OE.18.011181

Briscoe, D., Adair, C.F., Thompson, L.D., Tellado, M.V., Buckner, S.B. et al. (2000). Telecytologic diagnosis of breast fine needle aspiration biopsies. Intraobserver concordance. *Acta Cytologica*, *44*(2), 175–180. https://doi.org/10.1159/000326357

Chen, C.L., Mahjoubfar, A., Tai, L.C., Blaby, I.K., Huang, A. et al. (2016). Deep learning in label-free cell classification. *Scientific Reports*, *6*(1), 21471. https://doi.org/10.1038/srep21471

Chen, L., Chen, X., Cui, H., Long, Y. and Wu, J. et al. (2021). Image enhancement in lensless inline holographic microscope by inter-modality learning with denoising convolutional neural network. *Optics Communications*, *484*, 126682. https://doi.org/10.1016/j.optcom.2020.126682

Ching, T., Himmelstein, D.S., Beaulieu-Jones, B.K., Kalinin, A.A., Do, B.T. et al. (2018). Opportunities and obstacles for deep learning in biology and medicine. *Journal of the Royal Society, Interface*, *15*(141), 20170387. https://doi.org/10.1098/rsif.2017.0387

Cicchi, R. (2014). The new digital pathology: Just say NLO. *Digestive Diseases and Sciences*, *59*(7), 1347–1348. https://doi.org/10.1007/s10620-014-3165-8

Dong, C., Loy, C.C., He, K. and Tang, X. (2016). Image super-resolution using deep convolutional networks. *IEEE Transactions on Pattern Analysis and Machine Intelligence*, *38*(2), 295–307. https://doi.org/10.1109/TPAMI.2015.2439281

Ehteshami Bejnordi, B., Veta, M., Johannes van Diest, P., van Ginneken, B., Karssemeijer, N. et al. (2017). Diagnostic assessment of deep learning algorithms for detection of lymph node metastases in women with breast cancer. *JAMA*, *318*(22), 2199–2210. https://doi.org/10.1001/jama.2017.14585

Fang, Q., Xia, H., Song, Q., Zhang, M., Guo, R. et al. (2022). Speckle denoising based on deep learning via a conditional generative adversarial network in digital holographic interferometry. *Optics Express*, *30*(12), 20666–20683. https://doi.org/10.1364/OE.459213

Feizi, A., Zhang, Y., Greenbaum, A., Guziak, A., Luong, M. et al. (2016). Rapid, portable and cost-effective yeast cell viability and concentration analysis using lensfree on-chip microscopy and machine learning. *Lab on a Chip*, *16*(22), 4350–4358. https://doi.org/10.1039/c6lc00976j

Galande, A.S., Thapa, V., Gurram, H.P.R. and John, R. (2023). Untrained deep network powered with explicit denoiser for phase recovery in inline holography. *Applied Physics Letters*, *122*(13). https://doi.org/10.1063/5.0144795

Galande, A.S., Vijay, A., Gurram, H.P.R. and John, R. (2023, August 14). Physics aware semi-trained deep framework for phase retrieval in digital inline holography. In *Computational optical sensing and imaging* (pp. CTu5B–CTu4). Optica Publishing Group. https://doi.org/10.1364/COSI.2023.CTu5B.4

Galvez, J., Howell, L., Costa, M.J. and Davis, R. (1998). Diagnostic concordance of telecytology and conventional cytology for evaluating breast aspirates. *Acta Cytologica*, *42*(3), 663–667. https://doi.org/10.1159/000331823

Gedefaw, L., Liu, C.F., Ip, R.K.L., Tse, H.F., Yeung, M.H.Y et al. (2023). Artificial intelligence-assisted diagnostic cytology and genomic testing for hematologic disorders. *Cells*, *12*(13). https://doi.org/10.3390/cells12131755

Gerwert, K. (2020). *Label-Free Digital Pathology by Infrared Imaging*, Biomedical Spectroscopy and Imaging, *9*, 5–12.

Giansanti, D., Boschetto, A., Pochini, M., Bottini, L. and Giovagnoli, M.R. et al. (2015). Design of a process for image improvement in digital cytology: A preliminary technology assessement. *Computer Methods in Biomechanics and Biomedical Engineering: Imaging and Visualization*, *3*(1), 13–24. https://doi.org/10.1080/21681163.2014.883940

Go, T., Byeon, H. and Lee, S.J. (2018). Label-free sensor for automatic identification of erythrocytes using digital in-line holographic microscopy and machine learning. *Biosensors and Bioelectronics*, *103*, 12–18. https://doi.org/10.1016/j.bios.2017.12.020

Goswami, N., "Label-free SARS-CoV-2 detection and classification using phase imaging with computational specificity" Light Sci. Appl., *10*, 176. (2021).

Gurram, H.P.R., Galande, A., Thapa, V. and John, R., "Label-free Point-of-care Screening of Cervical Cancer Using Lens-less Holography and Deep Learning". (2019), International conference on optics and Opto electronics ICO-2019. IRDE.

Gurram, H.P.R., Panta, P., Pandiyan, V.P., Ghori, I. and John, R. et al. (2020). Digital holographic microscopy for quantitative and label-free oral cytology evaluation. *Optical Engineering*, *59*(2), 024105. https://doi.org/10.1117/1.OE.59.2.024105

Gurram, H.P. and John, R. (2018). Quantitative phase imaging of low-cost digital in-line holography in comparison with off-axis digital holographic microscopy. *Proceedings of the SPIE* 10677, Unconventional Optical Imaging, 106773C.

Gurram, H.P., Pandiyan, V.P. and John, R. 3D reconstruction of E. coli bacteria using synthetic aperture digital holographic microscopy, International Conference on Fibre Optics and Photonics 2014 (pp. M4A–M1).

Gurram, H. Phani R., Panta, P., Pandiyan, V.P., Ghori, I. et al. (2020). Digital holographic microscopy for quantitative and label-free oral cytology evaluation. *Optical Engineering*, *59*(2), 024105–024105. https://doi.org/10.1117/1.OE.59.2.024105

Hao, J., Lin, X., Chen, R., Lin, Y., Liu, H. et al. (2022). Phase retrieval combined with the deep learning denoising method in holographic data storage. *Optics Continuum*, *1*(1), 51–62. https://doi.org/10.1364/OPTCON.444882

Harmon, S.A., Sanford, T.H., Brown, G.T., Yang, C., Mehralivand, S. et al. (2020). Multiresolution application of artificial intelligence in digital pathology for prediction of positive lymph nodes from primary tumors in bladder cancer. *JCO Clinical Cancer Informatics*, *4*, 367–382. https://doi.org/10.1200/CCI.19.00155

He, Y.R., He, S., Kandel, M.E., Lee, Y.J., Hu, C et al. (2022). Cell cycle stage classification using phase imaging with computational specificity. *ACS Photonics*, *9*(4), 1264–1273. https://doi.org/10.1021/acsphotonics.1c01779

Hermes, M., Morrish, R.B., Huot, L., Meng, L., Junaid, S. et al. (2018). Mid-IR hyperspectral imaging for label-free histopathology and cytology. *Journal of Optics*, *20*(2), 023002. https://doi.org/10.1088/2040-8986/aaa36b

Holmström, O., Linder, N., Kaingu, H., Mbuuko, N., Mbete, J. et al. (2021). Point-of-care digital cytology with artificial intelligence for cervical cancer screening in a resource-limited setting. *JAMA Network Open*, *4*(3), e211740. https://doi.org/10.1001/jamanetworkopen.2021.1740

https://synoptek.com/insights/it-blogs/data-insights/ai-ml-dl-and-generative-ai-face-off-a-comparative-analysis/

Hu, C., He, S., Lee, Y.J., He, Y., Kong, E.M. et al. (2022). Live-dead assay on unlabeled cells using phase imaging with computational specificity. *Nature Communications*, *13*(1), 713. https://doi.org/10.1038/s41467-022-28214-x

Jialdasani, R., Desai, S., Gupta, M., Kothari, A., Deshpande, R. et al. (2006). An analysis of 46 static telecytology cases over a period of two years. *Journal of Telemedicine and Telecare*, *12*(6), 311–314. https://doi.org/10.1258/135763306778558132

Jo, Y., Park, S., Jung, J., Yoon, J., Joo, H. et al. (2017). Holographic deep learning for rapid optical screening of anthrax spores. *Science Advances*, *3*(8), e1700606. https://doi.org/10.1126/sciadv.1700606

Kandel, M.E., Rubessa, M., He, Y.R., Schreiber, S., Meyers, S. et al. (2020). Reproductive outcomes predicted by phase imaging with computational specificity of spermatozoon ultrastructure. *Proceedings of the National Academy of Sciences of the United States of America*, *117*(31), 18302–18309. https://doi.org/10.1073/pnas.2001754117

Kather, J.N., Heij, L.R., Grabsch, H.I., Loeffler, C., Echle, A. et al. (2020). Pan-cancer image-based detection of clinically actionable genetic alterations. *Nature Cancer*, *1*(8), 789–799. https://doi.org/10.1038/s43018-020-0087-6

Kather, J.N., Krisam, J., Charoentong, P., Luedde, T., Herpel, E. et al. (2019) Predicting survival from colorectal cancer histology slides using deep learning: A retrospective multicenter study. *PLoS Medicine*, *16*(1), e1002730. https://doi.org/10.1371/journal.pmed.1002730

Kayser, K., Kayser, G., Becker, H.D. and Herth, F. (2000). Telediagnosis of transbronchial fine needle aspirations – A feasibility study. *Analytical Cellular Pathology*, *21*(3–4), 207–212. https://doi.org/10.1155/2000/103073

Kim, G., Ahn, D., Kang, M., Jo, Y., Ryu, D. et al. (2019). Rapid and label-free identification of individual bacterial pathogens exploiting three-dimensional quantitative phase imaging and deep learning. *bioRxiv*, 596486.

Kim, S.J., Wang, C., Zhao, B., Im, H., Min, J. et al. (2018). Deep transfer learning-based hologram classification for molecular diagnostics. *Scientific Reports*, *8*(1), 1–2.

Lam, V.K., Nguyen, T.C., Chung, B.M., Nehmetallah, G. and Raub, C.B. et al. (2018). Quantitative assessment of cancer cell morphology and motility using telecentric digital holographic microscopy and machine learning. *Cytometry. Part A*, *93*(3), 334–345. https://doi.org/10.1002/cyto.a.23316

Lam, V.K., Nguyen, T., Bui, V., Chung, B.M., Chang, L.C. et al. (2020). Quantitative scoring of epithelial and mesenchymal qualities of cancer cells using machine learning and quantitative phase imaging. *Journal of Biomedical Optics*, *25*(2), 1–17. https://doi.org/10.1117/1.JBO.25.2.026002

Lee, E.S., Kim, I.S., Choi, J.S., Yeom, B.W., Kim, H.K. et al. (2003). Accuracy and reproducibility of telecytology diagnosis of cervical smears: A tool for quality assurance programs. *American Journal of Clinical Pathology*, *119*(3), 356–360. https://doi.org/10.1309/7ytvag4xnr48t75h

Litjens, G., Sánchez, C.I., Timofeeva, N., Hermsen, M., Nagtegaal, I. et al. (2016). Deep learning as a tool for increased accuracy and efficiency of histopathological diagnosis. *Scientific Reports*, *6*, 26286. https://doi.org/10.1038/srep26286

Maiolino, P., Restucci, B., Papparella, S. and De Vico, G. (2006). Evaluation of static telepathology in veterinary diagnostic cytology. *Veterinary Clinical Pathology*, *35*(3), 303–306. https://doi.org/10.1111/j.1939-165x.2006.tb00135.x

Mangal, J., Monga, R., Mathur, S.R., Dinda, A.K., Joseph, J. et al. (2019). Unsupervised organization of cervical cells using bright-field and single-shot digital holographic microscopy. *Journal of Biophotonics*, *12*(8), e201800409. https://doi.org/10.1002/jbio.201800409

Marchevsky, A.M., Nelson, V., Martin, S.E., Greaves, T.S., Raza, A.S. et al. (2003). Telecytology of fine-needle aspiration biopsies of the pancreas: A study of well-differentiated adenocarcinoma and chronic pancreatitis with atypical epithelial repair changes. *Diagnostic Cytopathology*, *28*(3), 147–152. https://doi.org/10.1002/dc.10247

McAlpine, E.D., Pantanowitz, L. and Michelow, P.M. (2021). Challenges developing deep learning algorithms in cytology. *Acta Cytologica*, *65*(4), 301–309. https://doi.org/10.1159/000510991

Meier, A., Nekolla, K., Earle, S., Hewitt, L., Aoyama, T. et al. (2018). End-to-end learning to predict survival in patients with gastric cancer using convolutional neural networks. *Annals of Oncology*, *29*. https://doi.org/10.1093/annonc/mdy269.075

Mirsky, S.K., Barnea, I., Levi, M., Greenspan, H. and Shaked, N.T. et al. (2017). Automated analysis of individual sperm cells using stain-free interferometric phase microscopy and machine learning. *Cytometry. Part A*, *91*(9), 893–900. https://doi.org/10.1002/cyto.a.23189

Mirsky, S.K., Barnea, I., Levi, M., Greenspan, H. and Shaked, N.T. et al. (2017). Automated analysis of individual sperm cells using stain-free interferometric phase microscopy and machine learning. *Cytometry. Part A*, *91*(9), 893–900. https://doi.org/10.1002/cyto.a.23189

Nguyen, T.H., Sridharan, S., Macias, V., Kajdacsy-Balla, A., Melamed, J. et al. (2017). Automatic Gleason grading of prostate cancer using quantitative phase imaging and machine learning. *Journal of Biomedical Optics*, *22*(3), 36015. https://doi.org/10.1117/1.JBO.22.3.036015

Nissim, N., Dudaie, M., Barnea, I. and Shaked, N.T. (2021). Real-time stain-free classification of cancer cells and blood cells using interferometric phase microscopy and machine learning. *Cytometry. Part A*, *99*(5), 511–523. https://doi.org/10.1002/cyto.a.24227

Ozaki, Y., Yamada, H., Kikuchi, H., Hirotsu, A., Murakami, T. et al. (2019). Label-free classification of cells based on supervised machine learning of subcellular structures. *PLOS ONE*, *14*(1), e0211347. https://doi.org/10.1371/journal.pone.0211347

Panayides, A.S., Amini, A., Filipovic, N.D., Sharma, A., Tsaftaris, S.A. et al. (2020). AI in medical imaging informatics: Current challenges and future directions. *IEEE Journal of Biomedical and Health Informatics*, *24*(7), 1837–1857. https://doi.org/10.1109/JBHI.2020.2991043

Pandiyan, V.P. and John, R. (2013). Quantitative phase visualization and aberration correction using digital holographic microscope, Workshop on Recent Advances in Photonics (WRAP) (pp. 1–2).

Pandiyan, V.P. and John, R. (2015). Quantitative phase imaging of cell division in yeast cells and E. coli using digital holographic microscopy. *Biophotonics Japan*, *9792*, 97920Y.

Pandiyan, V.P. and John, R. (2016). Optofluidic bioimaging platform for quantitative phase imaging of lab on a chip devices using digital holographic microscopy. *Applied Optics*, *55*(3), A54–A59. https://doi.org/10.1364/AO.55.000A54

Panta, P., Ram, G. H., Galande, A., Thapa, V. and John, R. et al., "Label-free analysis of Oral Cytology Specimens through Digital Holographic Microscopy and Deep-Neural Networks". (2019, May 19). *InDigital holography and three-dimensional imaging* (pp. Th3A–Th11). Optica Publishing Group.

Park, H.S., Rinehart, M.T., Walzer, K.A., Chi, J.T. and Wax, A. et al. (2016). Automated detection of P, falciparum using machine learning algorithms with quantitative phase images of unstained cells. *PLOS ONE*, *11*(9), e0163045. https://doi.org/10.1371/journal.pone.0163045

Pavillon, N., Hobro, A.J., Akira, S. and Smith, N.I. (2018). Noninvasive detection of macrophage activation with single-cell resolution through machine learning. *Proceedings of the National Academy of Sciences of the United States of America*, *115*(12), E2676–E2685. https://doi.org/10.1073/pnas.1711872115

Pitkäaho, T., Manninen, A. and Naughton, T.J. (2018). Classification of digital holograms with deep learning and handcrafted features. In *Imaging and Applied Optics* (3D, AO, AIO, COSI, DH, IS, LACSEA, LS&C, MATH, pcAOP) DW2F.3 (OSA, 2018).

Poola, P.K. (2016). *'Quantitative label-free sperm imaging by means of transport of intensity', Quantitative Phase Imaging II, International Society for Optics and Photonics, 9718.*

Raab, S.S., Zaleski, M.S., Thomas, P.A., Niemann, T.H., Isacson, C. and Jensen, C.S. (1996). Telecytology: Diagnostic accuracy in cervical-vaginal smears. *American Journal of Clinical Pathology*, *105*(5), 599–603. https://doi.org/10.1093/ajcp/105.5.599

Ren, Z., Xu, Z. and Lam, E.Y. Autofocusing in digital holography using deep learning. In *Three-Dimensional and Multidimensional Microscopy: Image Acquisition and processing XXV vol, 10499* p. 104991V (*International Society for optics and photonics*. 2018).

Roitshtain, D., Wolbromsky, L., Bal, E., Greenspan, H., Satterwhite, L.L. et al. (2017). Quantitative phase microscopy spatial signatures of cancer cells. *Cytometry. Part A*, *91*(5), 482–493. https://doi.org/10.1002/cyto.a.23100

Rubin, M., Stein, O., Turko, N.A., Nygate, Y., Roitshtain, D. et al. (2019). TOP-GAN: Stain-free cancer cell classification using deep learning with a small training set. *Medical Image Analysis*, *57*, 176–185. https://doi.org/10.1016/j.media.2019.06.014

Ryu, D., Jo, Y., Yoo, J., Chang, T., Ahn, D. et al. (2019). Deep learning-based optical field screening for robust optical diffraction tomography. *Scientific Reports*, *9*(1), 15239. https://doi.org/10.1038/s41598-019-51363-x

Shaked, N.T., "Label-Free Quantitative Classification of Cancer Cells Measured by Interferometric Phase Microscopy", *CLEO: Applications and technology*. Optical Society of America. (2019), *AM1* I-3.

Shroyer, K.R. and Singh, M. *Use of telecytology for the immediate assessment of CT guided and endoscopic FNA cytology: Diagnostic accuracy, advantages*.

Singh, N., Akbar, N., Sowter, C., Lea, K.G. and Wells, C.A. et al. (2002). Telepathology in a routine clinical environment: Implementation and accuracy of diagnosis by robotic microscopy in a one-stop breast clinic. *Journal of Pathology*, *196*(3), 351–355. https://doi.org/10.1002/path.1049

Teramoto, A., Tsukamoto, T., Kiriyama, Y. and Fujita, H. (2017). Automated classification of lung cancer types from cytological images using deep convolutional neural networks. *BioMed Research International*, 2017, 4067832. https://doi.org/10.1155/2017/4067832

Thapa, V., Galande, A.S., Gurram, H.P.R. and John, R. (2024). TIE-GANs: Single-Shot Quantitative Phase Imaging using Transport of Intensity equation with integration of GANs. *JBO*.

Wang, K., Song, L., Wang, C., Ren, Z., Zhao, G. et al. (2024). On the use of deep learning for phase recovery. *Light, Science and Applications*, *13*(1), 4. https://doi.org/10.1038/s41377-023-01340-x

Wang, W., Wang, H., Yang, S., Zhang, X., Wang, X. et al. (2022). Resolution enhancement in microscopic imaging based on generative adversarial network with unpaired data. *Optics Communications*, *503*, 127454. https://doi.org/10.1016/j.optcom.2021.127454

Weinstein, R.S. (1991). Telepathology comes of age in Norway. *Human Pathology*, *22*(6), 511–513. https://doi.org/10.1016/0046-8177(91)90225-e

Weinstein, R.S., Graham, A.R., Richter, L.C., Barker, G.P., Krupinski, E.A. et al. (2009). Overview of telepathology, virtual microscopy, and whole slide imaging: Prospects for the future. *Human Pathology*, *40*(8), 1057–1069. https://doi.org/10.1016/j.humpath.2009.04.006

Wilbur, D.C. (2011). Digital cytology: Current state of the art and prospects for the future. *Acta Cytologica*, *55*(3), 227–238. https://doi.org/10.1159/000324734

Williams, S., Henricks, W.H., Becich, M.J., Toscano, M. and Carter, A.B. et al. (2010). Telepathology for patient care: What am I getting myself into? *Advances in Anatomic Pathology*, *17*(2), 130–149. https://doi.org/10.1097/PAP.0b013e3181cfb788

Wu, Y., Luo, Y., Chaudhari, G., Rivenson, Y., Calis, A. et al. (2019). Bright-field Holography: Cross-Modality Deep Learning Enables Snapshot 3d Imaging with Bright-Field Contrast Using a Single Hologram. *Light, Science and Applications*, *8*(1), 25. https://doi.org/10.1038/s41377-019-0139-9

Wu, Y., Rivenson, Y., Zhang, Y., Wei, Z., Günaydin, H. et al. (2018). Extended depth-of-field in holographic imaging using deep-learning-based autofocusing and phase recovery. *Optica*, *5*(6), 704–710. https://doi.org/10.1364/OPTICA.5.000704

Yamashiro, K., Kawamura, N., Matsubayashi, S., Dota, K., Suzuki, H. et al. (2004). Telecytology in Hokkaido Island, Japan: Results of primary telecytodiagnosis of routine cases. *Cytopathology*, *15*(4), 221–227. https://doi.org/10.1111/j.1365-2303.2004.00147.x

Yi, F., Moon, I. and Javidi, B. (2017). Automated red blood cells extraction from holographic images using fully convolutional neural networks. *Biomedical Optics Express*, *8*(10), 4466–4479. https://doi.org/10.1364/BOE.8.004466

Yi, F., Moon, I., Javidi, B., Boss, D. and Marquet, P. et al. (2013). Automated segmentation of multiple red blood cells with digital holographic microscopy. *Journal of Biomedical Optics*, *18*(2), 26006. https://doi.org/10.1117/1.JBO.18.2.026006

Yoon, J., Jo, Y., Kim, M.H., Kim, K., Lee, S. et al. (2017). Identification of non-activated lymphocytes using three-dimensional refractive index tomography and machine learning. *Scientific Reports*, 7(1), 6654. https://doi.org/10.1038/s41598-017-06311-y

Yoon, J., Jo, Y., Kim, Y.S., Yu, Y., Park, J. et al., "Label-Free Identification of Lymphocyte Subtypes Using Three-Dimensional Quantitative Phase Imaging and Machine Learning". (2018). JoVE (journal of visualized experiments), *19*(141), e58305.

Zhang, K., Zuo, W., Chen, Y., Meng, D. and Zhang, L. et al. (2017). Beyond a Gaussian denoiser: Residual learning of deep CNN for image denoising. *IEEE Transactions on Image Processing*, *26*(7), 3142–3155. https://doi.org/10.1109/TIP.2017.2662206

Ziol, M., Vacher-Lavenu, M.C., Heudes, D., Ferrand, J., Mayelo, V. et al. (1999). 'Expert consultation for cervical carcinoma smears', Reliability of selected-field video microscopy. *Analytical and Quantitative Cytology and Histology*, *21*(1), 35–41.

CHAPTER

25

Applications of Artificial Intelligence in RNA-sequencing

Anjana Soorajkumar

1. Introduction

The landscape of sequence-based research has been revolutionized by Next-Generation Sequencing (NGS) technologies, offering unprecedented high throughput, sensitivity, and speed. One of the most significant beneficiaries of this technological evolution is the field of transcriptomics, which has seen remarkable growth, primarily fueled by the emergence of RNA-sequencing (RNA-seq) technologies. RNA-seq, a ubiquitous tool, provides a holistic profile of cellular ribonucleic acid (RNA) species, enabling a deep dive into dynamic biological processes at both the single-cell and population levels. Beyond its conventional role in RNA quantification, RNA-seq is pivotal in exploring the vast transcriptional landscape, revealing that 85% of the human genome is transcribed, with only 3% encoding protein-coding genes (Hangauer et al., 2013). This positions RNA-seq as a crucial method for cataloging diverse transcript species, including long non-coding RNA (lncRNA), microRNA (miRNA), small interfering RNA (siRNA), and other small RNA classes (e.g., snRNA and piRNA) (Robertson et al., 2010; Trapnell et al., 2010).

Recent applications of RNA-seq extend far beyond cataloging, encompassing the precise identification of regulatory elements, allele-specific expression, identification of disease-associated single nucleotide polymorphisms (SNPs) (Jehl et al., 2021), and the detection of gene fusions in cancer. RNA-seq sheds light on endogenous retrotransposon transcription and parasitic repeat elements, influencing neighboring gene transcription and contributing to somatic mosaicism in the brain (Han et al., 2015). The advent of single-cell RNA-seq (scRNA) analysis has further accelerated the exploration of cellular heterogeneity and diversity, notably in stem cell biology and neuroscience (Ofengeim et al., 2017). Figure 1 demonstrates the sequencing workflow, bioinformatic pipeline, and applications of scRNA. With RNA-seq integration into precision medicine trials, the need for efficient data mining becomes paramount. However, the challenges inherent in large-scale data analyses associated with RNA-seq underscore the importance of Artificial Intelligence (AI) as a transformative tool, unlocking the full potential of transcriptomic data (Mao et al., 2023).

1 Dubai Health, Mohammed Bin Rashid University of Medicine and Health Sciences, United Arab Emirates, Dubai.
* Corresponding author: dranjanasooraj@gmail.com

AI, a discipline focused on developing computational algorithms capable of performing tasks requiring human intelligence, emerges as a powerful asset shaping the landscape of precision medicine (C. Zhang & Lu, 2021). It is progressively becoming integral to healthcare and life science research, revolutionizing our scientific understanding and vision, particularly in complex diseases like cancer. The COVID-19 pandemic has accelerated the adoption of AI in biomedical research, extending from remote patient monitoring to predicting the spread of the SARS-CoV-2 coronavirus and developing new drugs (McCall, 2020; Zhou et al., 2020). Figure 2 shows a comparative view of AI, Machine Learning (ML), Deep Learning (DL), and generative AI. Within this context, ML, a crucial subset of AI, focuses on uncovering underlying patterns within datasets. This intricate process enables machines to deduce rules for optimal performance and equips them to adapt seamlessly to dynamic real-world scenarios. DL, positioned as a subset of ML, leverages intricate neural networks to operate. Essentially, it constitutes a cohesive set of techniques akin to other methodological groups, such as Decision Trees (DTs) or Support Vector Machines (SVMs) (Fregoso-Aparicio et al., 2021). This interconnected framework of AI, ML, and DL is pivotal in advancing our capabilities in precision medicine, offering innovative solutions to complex challenges in healthcare and life sciences.

In RNA-seq analysis, AI algorithms are pivotal, enabling faster, more accurate, and user-friendly approaches to cell-type identification. By harnessing the capabilities of AI, transcriptomics research stands to gain unprecedented insights into the intricate landscape of gene expression, paving the way for breakthroughs in personalized medicine and a more profound understanding of human health and disease. This chapter explores the diverse applications of AI in RNA-seq, with a focus on critical areas such as differential expression analysis, gene regulatory network inference, disease biomarker discovery, and drug development.

2. AI in Differential Expression Analysis

Differential expression analysis (DEA) serves as a cornerstone in gene expression research, aiming to identify genes that exhibit significant variations in expression across diverse experimental conditions or sample groups. Integrating AI algorithms is pivotal in this realm, offering a sophisticated means to extract intricate relationships from data and discern subtle yet biologically meaningful differences in gene expression patterns. AI's contribution to DEA extends beyond conventional analytics; it encompasses the modeling and correction of technical biases and confounding factors, thereby enhancing the precision and reliability of DEA outcomes. This advanced approach enables a more nuanced understanding of gene expression dynamics, efficiently managing extensive datasets and considering complex interactions within the transcriptome. Notably, the accuracy of AI algorithms, coupled with their adaptive learning capabilities from diverse datasets, progressively improves performance over time.

The toolkit for implementing AI in DEA includes notable tools such as the versatile Python library scikit-learn (Pedregosa et al., 2011) and MAST in R (Finak et al., 2015). The scikit-learn library offers a comprehensive suite of algorithms for classification and regression, facilitating insightful extractions from gene expression datasets. Meanwhile, MAST specializes in scRNA-seq analysis, addressing unique challenges in single-cell data through hierarchical models and statistical learning. ML models, including Random Forests (RFs), SVMs, and DL architectures like Convolutional Neural Networks (CNNs) and Recurrent Neural Networks (RNNs), excel at deciphering subtle patterns within complex RNA-seq datasets, handling the high dimensionality and non-linearity inherent in gene expression data. Moreover, they have demonstrated remarkable accuracy (Jain et al., 2020). SVMs, known for their versatility in classification tasks, effectively distinguish differentially expressed genes from non-differentially expressed ones. The ensemble learning approach of RFs offers robust and accurate DEA results (Strobl et al., 2007). DL frameworks like TensorFlow (Abadi et al., 2016) and PyTorch (Paszke et al., 2019) serve as instrumental platforms for implementing neural networks in gene expression prediction and dimensionality reduction, providing a nuanced

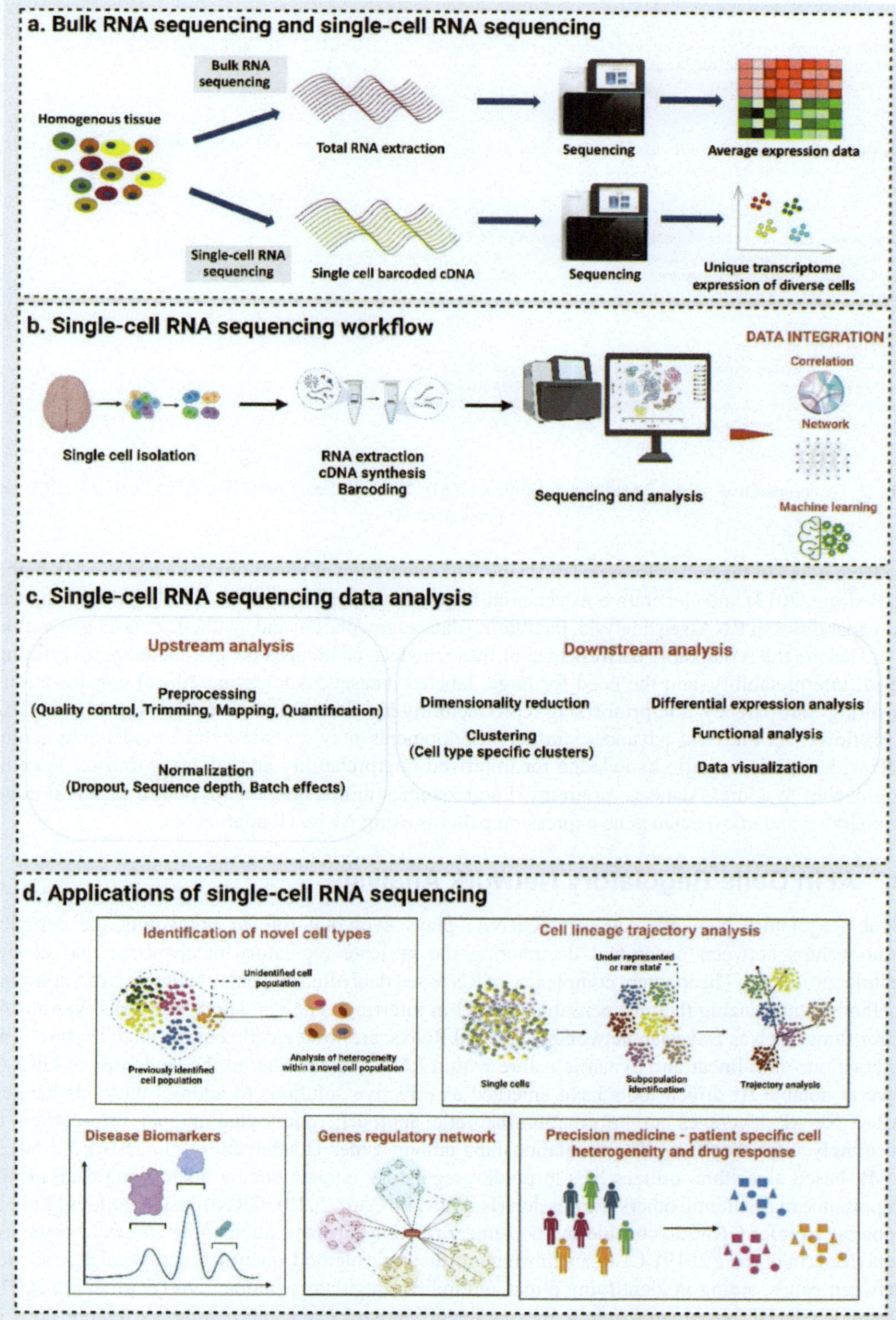

Fig. 1 Workflow, data analysis and applications of single cell RNA-sequencing

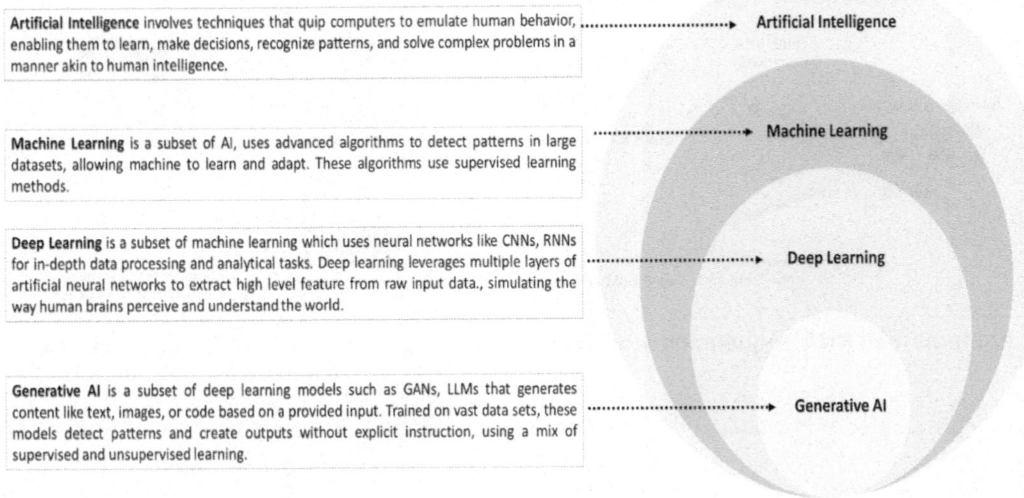

Fig. 2 A comparative view of Artificial Intelligence (AI), Machine Learning (ML), Deep Learning (DL), and generative AI

understanding of the transcriptomic landscape. The integration of Variational Autoencoders (Kingma & Welling, 2013) and Generative Adversarial Networks (Goodfellow et al., 2014) further enhances AI capabilities in RNA-seq analysis, facilitating data augmentation and synthetic dataset generation.

Despite the remarkable performance of these models, challenges persist, including overfitting, result interpretability, and the need for large, labeled datasets. Addressing ethical considerations, ensuring transparency, and prioritizing reproducibility are essential when integrating AI into DEA workflows. As the field advances, further developments may include hybrid models combining ML with domain-specific knowledge for improved interpretability and adopting transfer learning approaches to address dataset variability. Future studies in this domain are poised to reveal novel biomarkers and unexpected gene expression patterns using AI-based approaches.

3. AI in Gene Regulatory Network Analysis

Gene Regulatory Network Analysis (GRNA) plays a pivotal role in unraveling the complex relationships between genes and deciphering the intricate regulatory mechanisms that govern cellular processes. The inherent complexity of RNA-seq data often presents challenges for traditional methods, emphasizing the indispensability of AI in inferring gene regulatory networks. Various AI algorithms, such as Bayesian networks, DTs, and RNNs, are strategically employed to capture gene interactions' non-linear and dynamic nature within RNA-seq datasets. In the landscape of GRNA, several notable AI-driven tools have emerged as effective solutions to address these challenges. ARACNe-AP leverages an information-theoretic approach, employing mutual information to accurately infer direct regulatory relationships among genes (Lachmann et al., 2016). GENIE3, a ML-based algorithm, utilizes RFs to predict regulatory interactions by quantifying each gene's importance in predicting others' expression (Huynh-Thu et al., 2010). GRNBoost2 employs gradient boosting to infer GRNAs, considering non-linear relationships and interactions in gene expression data (Moerman et al., 2019). CLR, an information-theoretic method, quantifies statistical dependence between genes, aiding in identifying direct and indirect regulatory interactions (Faith et al., 2007). DeepGRN, a DL-based approach, harnesses neural networks to capture complex relationships in gene expression data, enhancing the accuracy of regulatory interaction predictions (X. Zhang et al., 2012). Additionally, DREAM Challenges provide a crucial platform for benchmarking various GRN inference methods, fostering community-driven efforts to develop and evaluate AI-based algorithms (Marbach et al., 2012).

These AI-driven tools empower researchers to extract meaningful insights from RNA-seq datasets, shedding light on the complex web of gene regulation within biological systems. AI-driven GRNA takes a holistic approach, considering not only individual gene expression profiles but also temporal aspects, environmental influences, and feedback loops. This comprehensive analysis offers insights into cellular systems' regulatory cascades and hierarchical relationships. Remarkably, the accuracy and scalability of AI algorithms in GRNA are noteworthy, as they can learn complex patterns from large-scale datasets, improving predictive capabilities over time. The scalability ensures the efficient analysis of complex regulatory networks involving numerous genes and interactions. Despite these advantages, challenges in applying AI to GRNA, including the interpretability of complex models, handling noise in the data, and the need for substantial computational resources. Ethical considerations related to potential biases in model predictions also warrant careful attention. The integration of AI into GRNA is rapidly evolving, with future directions potentially involving the development of hybrid models that combine ML with domain-specific knowledge to enhance interpretability. Transfer learning approaches allow models trained on one dataset to be applied to another, addressing challenges related to dataset variability. In conclusion, AI-driven GRNA stands at the forefront of understanding gene interactions, offering profound insights into key regulatory hubs and dynamic gene interactions. Specific studies or applications showcasing the success of AI-driven GRNA in uncovering crucial regulatory mechanisms could further underscore its impact.

4. AI in Disease Biomarker Discovery

Disease Biomarker Discovery (DBD) is critical in the quest for early and accurate disease diagnosis, prognosis, and treatment monitoring. Within this domain, AI has emerged as a transformative force, particularly in harnessing RNA-seq data to identify RNA-based biomarkers (Gadalla et al., 2019). AI-driven DBD transcends conventional biomarker discovery methods by leveraging M models to integrate diverse data dimensions. This encompasses RNA-seq data and incorporates clinical information and genomic data, enabling the identification of robust and context-specific biomarkers. AI's strength lies in its ability to discern intricate patterns and associations within heterogeneous datasets, thereby facilitating the discovery of more reliable and informative biomarkers. Feature selection techniques, including Recursive Feature Elimination (RFE) (Li et al., 2022) and Least Absolute Shrinkage and Selection Operator (LASSO) (Vasquez et al., 2016), play a crucial role in refining high-dimensional datasets. They contribute significantly to pinpointing the most relevant biomarkers associated with disease states, ensuring a focused and informative approach to biomarker discovery while enhancing the efficiency and interpretability of the results.

A critical facet of AI in DBD is its contribution to validating discovered biomarkers. By leveraging large-scale datasets and incorporating diverse sources of information, AI models can systematically assess the reproducibility and generalizability of potential biomarkers across different patient cohorts. This enhances the reliability of identified biomarkers for clinical use and underscores the crucial role of AI in ensuring the robustness of its findings. The accuracy of AI models in DBD is paramount for their clinical application. Well-trained and validated AI models demonstrate high sensitivity and specificity, ensuring the identification of robust biomarkers. The clinical relevance of AI-discovered biomarkers is meticulously assessed through their associations with disease pathology, progression, and response to treatment. Despite the promise of AI in DBD, challenges persist, encompassing the need for extensive and well-annotated datasets, addressing potential biases in training data, and enhancing the interpretability of complex AI models. Ethical considerations, particularly concerning patient data privacy and the responsible use of AI in healthcare, require meticulous attention. The integration of AI in DBD represents an evolving landscape. Future directions may involve the development of explainable AI models to enhance interpretability and foster trust in clinical decision-making. Collaborative efforts between data scientists, clinicians, and domain experts remain pivotal for advancing AI applications in biomarker discovery. As AI continues to shape the landscape of DBD, ongoing research, and innovation hold the promise of uncovering new possibilities and refining the application of AI in the quest for meaningful disease biomarkers.

5. AI in Drug Discovery

The drug discovery process, inherently complex and time-consuming, involves the identification of potential drug candidates for disease treatment. In this intricate landscape, AI employs various algorithms, including ML, DL, and reinforcement learning, to comprehensively analyze RNA-seq data. This encompasses understanding molecular mechanisms, identifying potential drug targets, and predicting the therapeutic effects of candidate compounds. Figure 3 illustrates the AI-powered decision support systems recommending the most effective therapeutic approaches for drug discovery.

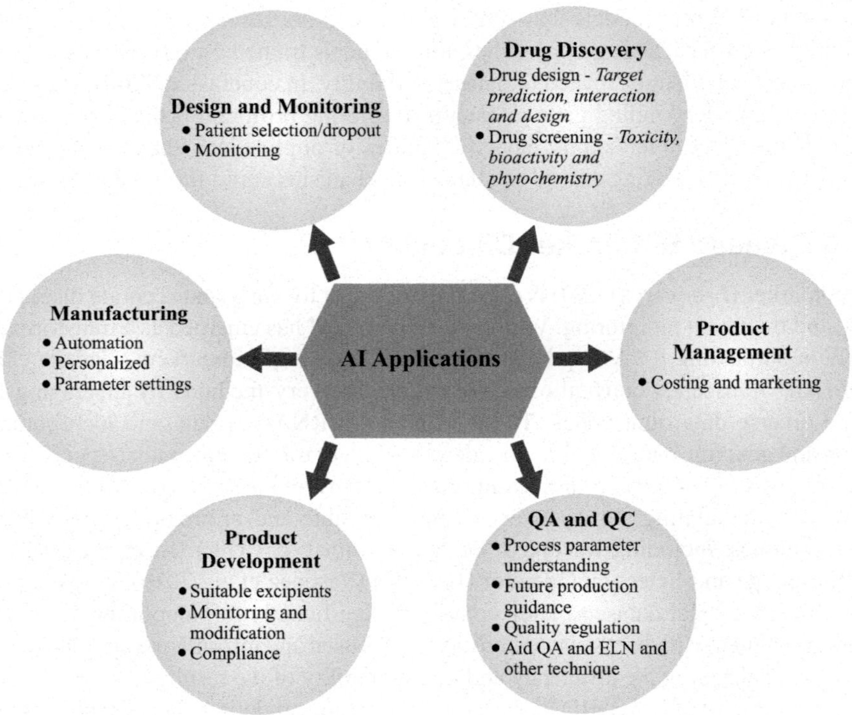

Fig. 3 AI-powered decision support systems for determining the most effective therapeutic approaches in drug discovery

ML models adeptly navigate RNA-seq data, discerning how drugs may influence gene expression profiles and cellular pathways. This predictive modeling facilitates the prioritization of compounds with the highest likelihood of success, thereby reducing the time and resources required for experimental validation. Notably, AI excels in predicting the efficacy of potential drug candidates, leveraging generative models like Generative Adversarial Networks (GANs) to create novel molecular structures with desired pharmacological properties (Kadurin et al., 2017). Collaborative platforms, such as IBM Watson for Drug Discovery, utilize AI to analyze extensive biomedical literature and databases, streamlining the identification of potential drug targets (Martin et al., 2018). Integrating these AI tools not only expedites the identification of promising drug candidates but also optimizes the drug development process, potentially leading to more efficient and targeted therapies. AI plays a pivotal role in drug repurposing efforts, utilizing algorithms to identify existing drugs with untapped potential in treating different conditions by analyzing RNA-seq data from diverse disease contexts. This approach accelerates drug development by capitalizing on existing safety and efficacy data. Furthermore, AI expedites lead optimization by predicting pharmacological properties using RNA-seq data, empowering researchers to make informed decisions about compound prioritization and streamlining the drug development pipeline.

Despite these advancements, challenges persist in AI for drug discovery, including the imperative need for high-quality, diverse datasets, the interpretability of complex models, and ethical considerations related to data privacy and algorithm biases. Overcoming these challenges is crucial for the responsible integration of AI into drug discovery workflows. As AI continues to reshape drug discovery, future directions may involve enhanced collaboration between interdisciplinary teams, the development of explainable AI models, and incorporating additional omics data beyond RNA-seq for a more comprehensive understanding of biological systems. Ongoing research and innovation promise to overcome challenges and unlock new frontiers in the quest for effective and targeted therapeutic interventions.

Conclusion

In conclusion, integrating AI into RNA-seq has inaugurated a transformative era, reshaping our understanding of the transcriptomic landscape. The deployment of AI-driven tools, encompassing ML algorithms and DL frameworks, has exhibited remarkable efficacy in deciphering intricate patterns within high-dimensional RNA-seq data. This synergy has empowered researchers to identify differentially expressed genes, unravel gene regulatory networks, and pinpoint disease biomarkers with unprecedented precision and efficiency. Beyond expediting data analysis, the collaborative power of AI and RNA sequencing enhances the extraction of meaningful biological insights, opening avenues for personalized diagnostics and therapeutics. As the field progresses, the role of AI in RNA-seq stands at the forefront, promising to unravel the complexities of gene expression and contribute significantly to the advancement of precision medicine.

References

Abadi, M., Agarwal, A., Barham, P., Brevdo, E., Chen, Z. et al. (2016). TensorFlow: Large-Scale Machine Learning on Heterogeneous Distributed Systems. *ArXiv*, *76*(1), arXiv:1603.04467. https://doi.org/10.48550/ARXIV.1603.04467

Faith, J.J., Hayete, B., Thaden, J.T., Mogno, I., Wierzbowski, J. et al. (2007). Large-scale mapping and validation of Escherichia coli transcriptional regulation from a compendium of expression profiles. *PLoS Biology*, *5*(1), 0054–0066. https://doi.org/10.1371/JOURNAL.PBIO.0050008

Finak, G., McDavid, A., Yajima, M., Deng, J., Gersuk, V. et al. (2015). MAST: a flexible statistical framework for assessing transcriptional changes and characterizing heterogeneity in single-cell RNA sequencing data. *Genome Biology*, *16*(1). https://doi.org/10.1186/S13059-015-0844-5A

Fregoso-Aparicio, L., Noguez, J., Montesinos, L. and García-García, J.A. (2021). Machine learning and deep learning predictive models for type 2 diabetes: a systematic review. *Diabetology & Metabolic Syndrome 2021 13:1*, *13*(1), 1–22. https://doi.org/10.1186/S13098-021-00767-9

Gadalla, A.A.H., Friberg, I.M., Kift-Morgan, A., Zhang, J., Eberl, M. et al. (2019). Identification of clinical and urine biomarkers for uncomplicated urinary tract infection using machine learning algorithms. *Scientific Reports*, *9*(1). https://doi.org/10.1038/S41598-019-55523-X

Goodfellow, I.J., Pouget-Abadie, J., Mirza, M., Xu, B., Warde-Farley, D. et al. (2014). Generative Adversarial Nets. *Advances in Neural Information Processing Systems*, *27*. http://www.github.com/goodfeli/adversarial

Han, Y., Gao, S., Muegge, K., Zhang, W. and Zhou, B. et al. (2015). Advanced Applications of RNA Sequencing and Challenges. *Bioinformatics and Biology Insights*, *9*(Suppl 1), 29. https://doi.org/10.4137/BBI.S28991

Hangauer, M.J., Vaughn, I.W. and McManus, M.T. (2013). Pervasive transcription of the human genome produces thousands of previously unidentified long intergenic noncoding RNAs. *PLoS Genetics*, *9*(6). https://doi.org/10.1371/JOURNAL.PGEN.1003569

Huynh-Thu, V.A., Irrthum, A., Wehenkel, L. and Geurts, P. (2010). Inferring regulatory networks from expression data using tree-based methods. *PloS One*, *5*(9). https://doi.org/10.1371/JOURNAL.PONE.0012776

Jain, P., Coogan, S.C.P., Subramanian, S.G., Crowley, M., Taylor, S. and Flannigan, M.D. et al. (2020). A review of machine learning applications in wildfire science and management. *Environmental Reviews*, *28*(4), 478–505. https://doi.org/10.1139/ER-2020-0019/ASSET/IMAGES/ER-2020-0019TAB3.GIF

Jehl, F., Degalez, F., Bernard, M., Lecerf, F., Lagoutte, L. et al. (2021). RNA-Seq Data for Reliable SNP Detection and Genotype Calling: Interest for Coding Variant Characterization and Cis-Regulation Analysis by Allele-Specific Expression in Livestock Species. *Frontiers in Genetics*, *12*, 655707. https://doi.org/10.3389/FGENE.2021.655707/BIBTEX

Kadurin, A., Aliper, A., Kazennov, A., Mamoshina, P., Vanhaelen, Q. et al. (2017). The cornucopia of meaningful leads: Applying deep adversarial autoencoders for new molecule development in oncology. *Oncotarget*, *8*(7), 10883. https://doi.org/10.18632/ONCOTARGET.14073

Kingma, D.P. and Welling, M. (2013). Auto-Encoding Variational Bayes. *International Conference on Learning Representations*.

Lachmann, A., Giorgi, F.M., Lopez, G. and Califano, A. (2016). ARACNe-AP: gene network reverse engineering through adaptive partitioning inference of mutual information. *Bioinformatics (Oxford, England)*, *32*(14), 2233–2235. https://doi.org/10.1093/BIOINFORMATICS/BTW216

Li, L., Ching, W.K. and Liu, Z.P. (2022). Robust biomarker screening from gene expression data by stable machine learning-recursive feature elimination methods. *Computational Biology and Chemistry*, *100*. https://doi.org/10.1016/J.COMPBIOLCHEM.2022.107747

Mao, G., Pang, Z., Zuo, K., Wang, Q., Pei, X. et al. (2023). Predicting gene regulatory links from single-cell RNA-seq data using graph neural networks. *Briefings in Bioinformatics*, *24*(6), 1–11. https://doi.org/10.1093/BIB/BBAD414

Marbach, D., Costello, J.C., Küffner, R., Vega, N.M., Prill, R.J. et al. (2012). Wisdom of crowds for robust gene network inference. *Nature Methods 2012 9:8*, *9*(8), 796–804. https://doi.org/10.1038/nmeth.2016

Martin, R.L., Martinez Iraola, D., Louie, E., Pierce, D., Tagtow, B.A. et al. (2018). Hybrid natural language processing for high-performance patent and literature mining in IBM Watson for Drug Discovery. *IBM Journal of Research and Development*, *62*(6). https://doi.org/10.1147/JRD.2018.2888975

McCall, B. (2020). COVID-19 and artificial intelligence: protecting health-care workers and curbing the spread. *The Lancet. Digital Health*, *2*(4), e166–e167. https://doi.org/10.1016/S2589-7500(20)30054-6

Moerman, T., Aibar Santos, S., Bravo González-Blas, C., Simm, J., Moreau, Y. et al. (2019). GRNBoost2 and Arboreto: efficient and scalable inference of gene regulatory networks. *Bioinformatics (Oxford, England)*, *35*(12), 2159–2161. https://doi.org/10.1093/BIOINFORMATICS/BTY916

Ofengeim, D., Giagtzoglou, N., Huh, D., Zou, C. and Yuan, J. et al. (2017). Single-Cell RNA Sequencing: Unraveling the Brain One Cell at a Time. *Trends in Molecular Medicine*, *23*(6), 563. https://doi.org/10.1016/J.MOLMED.2017.04.006

Paszke, A., Gross, S., Massa, F., Lerer, A., Bradbury Google, J. et al. (2019). PyTorch: An Imperative Style, High-Performance Deep Learning Library. *Advances in Neural Information Processing Systems*, *32*.

Pedregosa, F., Varoquaux, G., Gramfort, A., Michel, V., Thirion, B. et al.. (2011). Scikit-learn: Machine Learning in Python. *Journal of Machine Learning Research*, *12*, 2825–2830. https://inria.hal.science/hal-00650905

Robertson, G., Schein, J., Chiu, R., Corbett, R., Field, M. et al. (2010). De novo assembly and analysis of RNA-seq data. *Nature Methods*, *7*(11), 909–912. https://doi.org/10.1038/NMETH.1517

Strobl, C., Boulesteix, A.L., Zeileis, A. and Hothorn, T. (2007). Bias in random forest variable importance measures: illustrations, sources and a solution. *BMC Bioinformatics*, *8*. https://doi.org/10.1186/1471-2105-8-25

Trapnell, C., Williams, B.A., Pertea, G., Mortazavi, A., Kwan, G. et al. (2010). Transcript assembly and quantification by RNA-Seq reveals unannotated transcripts and isoform switching during cell differentiation. *Nature Biotechnology*, *28*(5), 511–515. https://doi.org/10.1038/NBT.1621

Vasquez, M.M., Hu, C., Roe, D.J., Chen, Z., Halonen, M. et al. (2016). Least absolute shrinkage and selection operator type methods for the identification of serum biomarkers of overweight and obesity: simulation and application. *BMC Medical Research Methodology*, *16*(1), 1–19. https://doi.org/10.1186/S12874-016-0254-8/TABLES/13

Zhang, C. and Lu, Y. (2021). Study on artificial intelligence: The state of the art and future prospects. *Journal of Industrial Information Integration*, *23*, 100224. https://doi.org/10.1016/J.JII.2021.100224

Zhang, X., Zhao, X.M., He, K., Lu, L., Cao, Y. et al. (2012). Inferring gene regulatory networks from gene expression data by path consistency algorithm based on conditional mutual information. *Bioinformatics (Oxford, England)*, *28*(1), 98–104. https://doi.org/10.1093/BIOINFORMATICS/BTR626

Zhou, Y., Wang, F., Tang, J., Nussinov, R. and Cheng, F. et al. (2020). Artificial intelligence in COVID-19 drug repurposing. *The Lancet. Digital Health*, *2*(12), e667–e676. https://doi.org/10.1016/S2589-7500(20)30192-8

CHAPTER

26

Artificial Intelligence in Pharmacovigilance and Materiovigilance

Rishi Kumar[1] and *Parveen Kumar Goyal*[2]

1. Introduction

Artificial Intelligence or AI, refers to the simulation of human intelligence in machines, allowing them to perform tasks that typically require human intelligence. These tasks include learning, reasoning, problem-solving, perception, understanding natural language, and even creativity. AI systems are designed to analyze data, extract patterns, and make decisions or predictions based on the information available to them (Atkinson 2023; Sheikh, Prins, and Schrijvers 2023; John McCarthy 2007). Pharmacovigilance (PV) is the science of activities related to the detection, assessment, understanding, and prevention of adverse effects or any other drug-related problems. PV focuses on monitoring the safety of pharmaceutical products (medicines) after they have been approved by competent regulatory agencies/authorities for use and are available in the market. It is evolving immensely after the implementation of AI tools (Ball and Dal Pan 2022). Materiovigilance is almost similar to pharmacovigilance but focuses on the surveillance and monitoring of medical devices, rather than pharmaceutical products. It involves the systematic collection, assessment, and analysis of data related to the safety and performance of medical devices throughout their lifecycle, from development and manufacturing to post-market use (Saifuddin et al. 2022). The integration of AI with PV and materiovigilance has started a new era of development by significantly fostering the extent of accuracy and efficiency. AI tools have made the associated activities much simpler and faster than the earlier ones.

Each country has its own regulatory setup to look into this important part of drug and medical device vigilance. In India, PV and Materiovigilance are being implemented through a Nationwide Central Sponsored Scheme i.e. Pharmacovigilance Programme of India (PvPI). The Indian Pharmacopoeia Commission is working as the National Coordination Centre for the PvPI

1 Indian Pharmacopoeia Commission, Ghaziabad, India.
2 Department of Pharmacy, Panipat Institute of Engineering and Technology (PIET), Samalkha, Panipat, India
* Corresponding author: tawar.rishi@gmail.com

and the Materiovigilance Programme of India (Kumar et al. 2016). In Figure 1, the authors try to illustrate the application of AI in the current workflow of the PvPI. At present the work is being done manually with the help of computer applications like software, mobile applications and email. After the implementation of AI in PV, the ICSR's Processing, quality reviews, signal management, and target identification are quite smooth and effective.

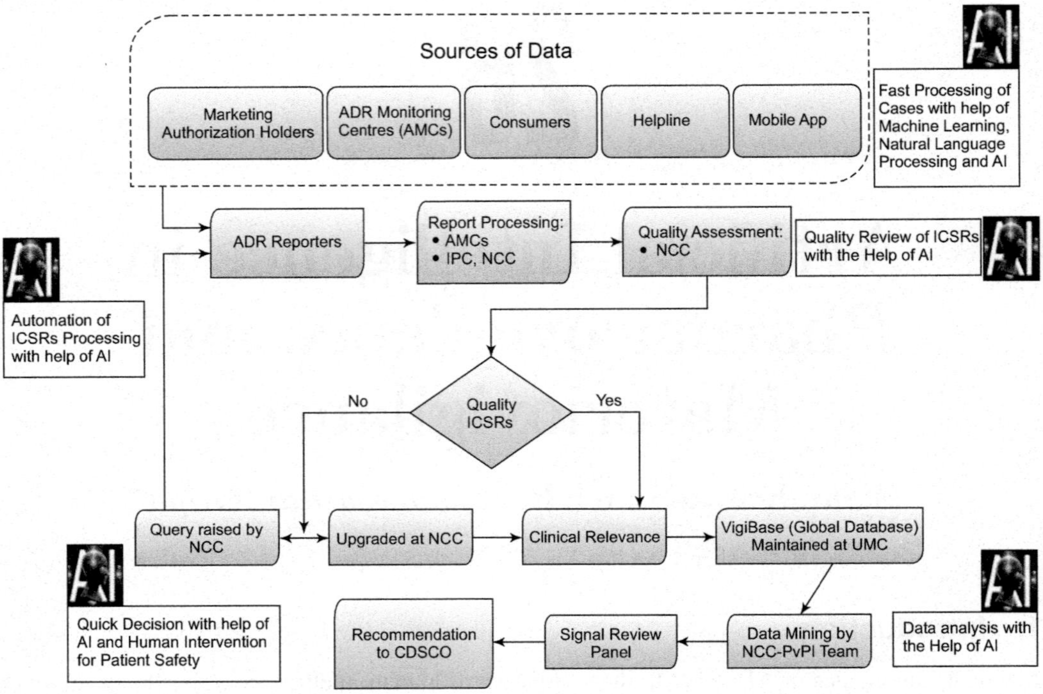

Fig. 1 Proposed Application of AI in the Workflow of Pharmacovigilance Programme of India
AI-Artificial Intelligence, IPC-Indian Pharmacopoeia Commission, NCC-National Coordination Centre, ICSRs-Individual Case Safety Reports, PvPI-Pharmacovigilance Programme of India, UMC-Uppsala Monitoring Centre, CDSCO-Central Drugs Standard Control Organization

In February 2022, the Council for International Organization of Medical Sciences (CIOMS) proposed to form a CIOMS Expert Working Group on AI and PV, with the objective of establishing and promoting principles and guidance for AI use or intelligence augmentation in the field of PV. The Proposal outlines the background, rationale, and scope of the proposed Working Group, and invites interested stakeholders to join the initiative. They also provided some examples of how AI can be used to improve PV tasks, such as data management, signal detection, and causality assessment, as well as the challenges and opportunities involved (Bhardwaj et al., 2022). Ball and Dal Pan 2022 discuss the potential and challenges of applying AI to PV, focusing on the processing and evaluation of ICSRs. They propose a general framework for assessing the readiness of AI for PV, and provide some examples of how AI can be used to improve the quality, efficiency, and consistency of ICSR analysis. They also identify some outstanding scientific and policy issues that need to be addressed before AI can be fully exploited for PV, such as quality assurance, training data, cognitive and sociotechnical frameworks, and best practices. They conclude that AI can be useful for some aspects of ICSR processing and evaluation, but it still requires human oversight and validation.

2. AI in Materiovigilance

AI offers significant potential to enhance Materiovigilance efforts. AI technologies can automate the detection and reporting of adverse events associated with medical devices, analyze data to identify patterns and signals indicative of safety issues, and support post-market surveillance activities by monitoring real-world data. Additionally, AI can assist in predicting and assessing device risks, providing decision support to healthcare professionals, and ensuring compliance with regulatory requirements outlined in the Medical Device Rules, 2018. By leveraging AI capabilities, stakeholders can improve patient safety, facilitate timely regulatory intervention, and enhance overall surveillance and monitoring of medical devices in accordance with CDSCO regulations (Murali et al. 2019); (Liang et al. 2022).

Software as a Medical Device is opening a new use of AI in materiovigilance. It is a powerful and versatile tool for enhancing quality and use of medical devices using softwares. The safety of medical devices stands as a paramount concern. Materiovigilance emerges as a crucial safety net, extending its oversight beyond mere device functionality to encompass the entire journey of a medical device, from creation to patient use. With the advent of Software as a Medical Device, characterized by its software-driven functionality, the landscape of medical device safety has evolved, presenting both opportunities and challenges. Software as a Medical Device, hailed for its transformative potential in healthcare, has witnessed significant growth, valued at approximately $31 billion globally in 2021, with projections of continued expansion. However, alongside its promise, Software as a Medical Device introduces unique safety challenges, including the intricacy of algorithms, rapid technological advancements, interconnectedness within healthcare ecosystems, and the imperative for real-time monitoring and updates. To address these challenges, proactive risk management strategies are essential, encompassing comprehensive risk assessment, iterative risk mitigation, regular compliance audits, and stakeholder collaboration. Moreover, the principles of Materiovigilance seamlessly extend to Software as a Medical Device, emphasizing continuous risk assessment, mitigation, and post-market surveillance to ensure patient safety and regulatory compliance (Ball and Dal Pan 2022;Artificial Intelligence and Machine Learning in Software as a Medical Device | FDA n.d.); Murali et al. 2019).

How the integration of AI with materiovigilance has enhanced safety monitoring and management of medical devices is briefly given as follows:

- **Signal Detection and Surveillance:** AI algorithms can analyze large volumes of data from diverse sources, including electronic health records, medical device registries, post-market surveillance databases, and even social media platforms, to detect signals of potential adverse events or safety issues associated with medical devices. By identifying patterns, trends, and anomalies in real-world data, AI systems can facilitate early detection of safety concerns, enabling proactive intervention and risk mitigation strategies.
- **Adverse Event Detection and Classification:** AI-powered systems can automatically extract and categorize adverse events reported in unstructured data sources such as clinical notes, patient forums, and medical literature. Natural Language Processing (NLP) techniques enable the extraction of relevant information from text data, allowing for efficient adverse event identification and classification. This automation streamlines the adverse event reporting process and improves the accuracy of adverse event analysis.
- **Risk Prediction and Assessment:** AI models can be trained to predict the likelihood of adverse events or device failures based on various factors such as device characteristics, patient demographics, and clinical parameters. By analyzing historical data and identifying risk factors, AI-driven risk prediction models can help prioritize surveillance efforts, guide regulatory decision-making, and inform risk management strategies for medical devices.
- **Post-Market Surveillance and Trend Analysis:** AI technologies enable continuous monitoring and analysis of real-world data streams to identify emerging trends, patterns, or clusters of adverse events associated with specific medical devices. Through advanced analytics and machine learning algorithms, AI systems can uncover hidden relationships and correlations in large-scale data sets, providing insights into device performance, safety profiles, and potential risks.

- **Decision Support Systems:** AI-driven decision support tools can assist healthcare professionals, regulatory authorities, and manufacturers in making informed decisions regarding the safety and regulatory compliance of medical devices. These systems provide actionable insights, risk assessments, and recommendations based on data analysis, helping stakeholders navigate complex materiovigilance challenges and optimize patient safety.
- **Quality Control and Compliance Monitoring:** AI technologies facilitate automated data validation, quality control, and compliance monitoring in materiovigilance processes. By standardizing data entry, ensuring data integrity, and detecting anomalies or inconsistencies, AI systems enhance the accuracy and reliability of materiovigilance data, thereby supporting regulatory compliance and adherence to quality standards.

Overall, the integration of AI into materiovigilance processes holds great promise for enhancing the safety monitoring, risk management, and regulatory oversight of medical devices. By leveraging AI-driven analytics, predictive modeling, and decision support capabilities, materiovigilance stakeholders can improve the efficiency, effectiveness, and responsiveness of safety surveillance efforts, ultimately advancing patient care and public health outcomes (Kalaiselvan, Tripathi, and Prakash 2020).

3. AI in PV

The objectives of implementing AI in PV are multifaceted, aiming to enhance various aspects of safety monitoring, risk assessment, and regulatory compliance in the pharmaceutical domain. Some key objectives include:

- **Fast Data Processing:** AI can help in fast data processing in PV by using machine learning algorithms to learn from large amounts of data and find patterns or features that humans may miss. AI can also automate tasks such as data cleaning, extraction, integration, cataloging, labeling, and reporting, which can improve the speed and efficiency of data analysis. AI can also use natural language processing to enable users to query data using natural language, which can make data more accessible and understandable.
- **Analysis of Large Data of Adverse Drug Reactions:** Analysis of large data of adverse drug reactions (ADRs) with the help of AI is a topic that has gained increasing attention in recent years. AI can help in various aspects of PV, such as data intake, data processing, signal detection, and data analysis. AI can use machine learning algorithms to learn from large amounts of data and find patterns or features that humans may miss.
- **Preventable ADRs and AI:** The exploration of "Preventable ADRs and AI" signifies a pivotal intersection between healthcare and artificial intelligence. This research delves into the potential of AI in mitigating adverse drug reactions that could be avoided with more sophisticated preventative measures. Traditional approaches to preventing these reactions have faced challenges, often relying on human judgment and struggling to process extensive datasets comprehensively. In this context, AI emerges as a promising solution, leveraging machine learning algorithms to analyze diverse patient data and genetic information. By integrating with existing PV systems, AI holds the promise of providing real-time monitoring, enhancing predictive capabilities, and ultimately contributing to more effective prevention strategies. The paper likely discusses both the benefits and limitations of employing AI in this critical healthcare domain, offering valuable insights into the future directions of preventative healthcare practices.

Overall, the objectives of AI in pharmacovigilance encompass leveraging advanced analytics, automation, and predictive modeling capabilities to enhance the efficiency, accuracy, and effectiveness of safety monitoring, risk assessment, and regulatory compliance in the pharmaceutical industry. By harnessing the power of AI-driven technologies, pharmacovigilance stakeholders can proactively identify and mitigate safety risks, ultimately safeguarding patient health and improving public health outcomes (Aronson 2022).

4. Uses of AI in PV

- **Identification of Adverse Drug Events and Reactions:** AI can use machine learning algorithms to analyze large amounts of data from various sources, such as electronic health records, claims databases, social media, and literature, and detect patterns or signals of potential adverse drug events and reactions. AI can also help classify and prioritize the severity and causality of these events and reactions (Bohr and Memarzadeh 2020).
- **Processing of Safety Reports or Clinical Narratives:** AI can use natural language processing to extract relevant information from safety reports or clinical narratives, such as drug names, doses, indications, outcomes, and adverse effects. AI can also help standardize and harmonize the terminology and coding of these reports or narratives across different languages and regions (Ball and Dal Pan 2022).
- **Extraction or Prediction of the Effects of Drug–Drug Interactions:** AI can use machine learning algorithms to learn from existing data on drug–drug interactions and predict the effects of new or unknown combinations of drugs. AI can also help identify the mechanisms and factors that influence these interactions, such as pharmacokinetics, pharmacodynamics, genetics, and environmental factors. There are a number of Signal Detection methods available for Signal Detection, many of them are logarithmic based methods. AI has the Power to identify potential signals by different methods in a fraction of seconds. It helps to make better decisions in terms of patient safety (Murali et al. 2019).
- **Identification of Targeted Population:** AI can also help in identification of targeted populations for focused PV study e.g. like pediatric and geriatric groups who are more prone to adverse events due to the use of medicines and medical devices. During the COVID-19 Pandemic artificial intelligence was used to identify the targeted population for vaccination and to identify adverse events. AI helped in reducing human error during the pandemic in many ways. If we talk specifically about India, The Mobile application Aarogaya Setu and Co-Win Software successfully completed the vaccination drive and also helped in containing the COVID-19 Pandemic (Wang et al. 2023;Vaishya et al. 2020).

5. Limitations of Implementing AI in PV

AI has started a new era of automation and robotics for the better functioning of the healthcare system, but the intervention/role of clinical healthcare professionals is still crucial and inevitable. The implementation of AI in PV, in addition to offering numerous benefits, has several limitations and challenges that need to be considered:

- **Security and Privacy Issues:** AI applications may require access to sensitive data, such as personal health records, social media posts, or genomic information, which may pose risks of data breaches, unauthorized use, or misuse. AI algorithms may also generate outputs that could reveal confidential information about patients, drugs, or adverse events (Hauben and Hartford 2021).
- **Quality and Reliability Issues:** AI applications may produce inaccurate, inconsistent, or biased results, due to errors in data, algorithms, or models. AI algorithms may also lack transparency, explainability, or accountability, making it difficult to understand, validate, or challenge their decisions. AI applications may also malfunction or fail unexpectedly, due to technical glitches, cyberattacks, or human errors (Keogh et al. 2024).
- **Ethical and Regulatory Issues:** AI applications may raise ethical and regulatory questions, such as who is responsible for the outcomes of AI decisions? How to ensure the safety and efficacy of AI tools? How to protect the rights and interests of patients and stakeholders? And how to balance the benefits and risks of AI innovation? AI applications may also have social and cultural implications, such as changing the roles and relationships of humans and machines, or affecting the trust and confidence of the public and professionals (Gruson 2019).

- **Bias and Generalization:** AI models trained on biased or imbalanced datasets may exhibit biased predictions or generalizations, leading to inaccurate or unfair outcomes. In pharmacovigilance, biased data or algorithms could result in the underrepresentation or misclassification of certain patient populations or adverse events, impacting the validity of safety assessments (Norori et al. 2021).
- **Regulatory Compliance and Validation:** Regulatory agencies require transparency, reproducibility, and validation of AI-driven pharmacovigilance systems to ensure compliance with regulatory standards and guidelines. Validating AI algorithms for pharmacovigilance applications involves rigorous testing, validation, and documentation processes, which can be resource-intensive and time-consuming (Huysentruyt et al. 2021).

There are some potential difficulties that new AI techniques may cause in PV. As PV data grows in volume and variety, deep learning and anomaly detection techniques offer new possibilities for improving ADR detection. An AI system that can learn continuously from these diverse data sources could enhance real-time ADR monitoring, as well as assist pharmacotherapy professionals in collecting more information on suspected cases. However, the field of PV is changing fast, and the resources we have mentioned are not enough to solve all the challenges (de Hond et al. 2022; Lavertu et al. 2021).

Pharmacovigilance faces evolving challenges, including emerging safety concerns, novel drug formulations, and changes in regulatory requirements. AI systems must be adaptable and capable of learning from new data inputs to effectively address these challenges and maintain relevance over time. While AI can automate many aspects of pharmacovigilance, human oversight and expertise remain essential for interpreting results, making informed decisions, and addressing complex issues that may arise. Pharmacovigilance professionals need to possess the necessary skills and expertise to validate AI outputs, interpret findings, and ensure the accuracy and relevance of safety assessments (Aronson 2022).

Conclusion

AI has the potential to transform PV by enhancing the detection, analysis, and reporting of adverse events and drug interactions. However, there are also challenges and limitations that need to be addressed, such as data quality, ethical and regulatory issues, and human-AI collaboration. Therefore, PV experts should collaborate with AI experts and stakeholders to ensure the safe and effective use of AI in PV. In terms of regulation of Drugs and Medical Devices AI can help in fast and easy identification of issues of adverse events and targeted areas. By doing so, they can leverage the power of AI to improve patient safety and health outcomes.

Conclusively, the collaborative efforts between healthcare stakeholders, regulatory authorities, and AI developers are crucial for overcoming these challenges and advancing the use of AI in pharmacovigilance and materiovigilance responsibly.

References

Aronson, Jeffrey K., 2022. "Artificial Intelligence in Pharmacovigilance: An Introduction to Terms, Concepts, Applications, and Limitations." *Drug safety* 45(5): 407–18.

"Artificial Intelligence and Machine Learning in Software as a Medical Device | FDA." https://www.fda.gov/medical-devices/software-medical-device-samd/artificial-intelligence-and-machine-learning-software-medical-device (February 21, 2024).

Atkinson, Cameron F., 2023. "Cheap, Quick, and Rigorous: Artificial Intelligence and the Systematic Literature Review." *Social Science Computer Review*: 08944393231196281. https://doi.org/10.1177/08944393231196281.

Ball, Robert and Gerald Dal Pan. 2022. "'Artificial Intelligence' for Pharmacovigilance: Ready for Prime Time?" *Drug Safety* 45(5): 429–38.

Bhardwaj, Kamini, Rabnoor Alam, Ajay Pandeya, and Pankaj Kumar Sharma. 2022. "Artificial Intelligence in Pharmacovigilance and COVID-19." *Current Drug Safety* 18(1): 5–14.

Bohr, Adam, and Kaveh Memarzadeh, 2020. "The Rise of Artificial Intelligence in Healthcare Applications." *Artificial Intelligence in Healthcare*: 25–60.

Gruson, David. 2019. "[Ethics and Artificial Intelligence in Healthcare, towards Positive Regulation]." *Soins; la revue de reference infirmiere* 64(832): 54–57.

Hauben, Manfred, and Craig G. Hartford, 2021. "Artificial Intelligence in Pharmacovigilance: Scoping Points to Consider." *Clinical Therapeutics* 43(2): 372–79.

de Hond, Anne A.H., 2022. "Guidelines and Quality Criteria for Artificial Intelligence-Based Prediction Models in Healthcare: A Scoping Review." *NPJ digital medicine* 5(1): 2. https://europepmc.org/articles/PMC8748878.

Huysentruyt, Kristof, 2021. "Validating Intelligent Automation Systems in Pharmacovigilance: Insights from Good Manufacturing Practices." *Drug safety* 44(3): 261–72.

John McCarthy. 2007. "WHAT IS ARTIFICIAL INTELLIGENCE?" *Computer Science Department.*

Kalaiselvan, V., Santanu Kumar Tripathi and Jai Prakash. 2020. "Materiovigilance Programme of India: A Scheme to Assure Cardiovascular Devices Safety Surveillance." *Indian heart journal* 72(4): 316–18.

Keogh, Rachel J., 2024. "Dealing with Digital Paralysis: Surviving a Cyberattack in a National Cancer Center." *Journal of cancer policy* 39: 100466.

Kumar, Rishi, 2016. "Best Practices for Improving the Quality of Individual Case Safety Reports in Pharmacovigilance." *Therapeutic innovation & regulatory science* 50(4): 464–71.

Lavertu, Adam, 2021. "A New Era in Pharmacovigilance: Toward Real-World Data and Digital Monitoring." *Clinical Pharmacology and Therapeutics* 109(5): 1197–1202.

Liang, Likeng, 2022. "Artificial Intelligence-Based Pharmacovigilance in the Setting of Limited Resources." *Drug Safety* 45(5): 511–19. https://doi.org/10.1007/s40264-022-01170-7.

Murali, Kotni, Sukhmeet Kaur, Ajay Prakash, and Bikash Medhi. 2019. "Artificial Intelligence in Pharmacovigilance: Practical Utility." *Indian journal of pharmacology* 51(6): 373–76.

Norori, Natalia, 2021. "Addressing Bias in Big Data and AI for Health Care: A Call for Open Science." *Patterns (New York, N.Y.)* 2(10): 100347.

Saifuddin, P.K., 2022. "Materiovigilance Programme of India: Current Status and Way Forward." *Indian Journal of Pharmacology* 54(3): 221.

Sheikh, Haroon, Corien Prins, and Erik Schrijvers. 2023. "Artificial Intelligence: Definition and Background." In *Mission AI: The New System Technology*, Cham: Springer International Publishing, 15–41. https://doi.org/10.1007/978-3-031-21448-6_2.

Vaishya, Raju, Mohd Javaid, Ibrahim Haleem Khan, and Abid Haleem et al. 2020. "Artificial Intelligence (AI) Applications for COVID-19 Pandemic." *Diabetes & metabolic syndrome* 14(4): 337–39.

Wang, Jingjing, 2023. "Application of Artificial Intelligence in Geriatric Care: Bibliometric Analysis." *Journal of medical Internet research* 25: e46014.

CHAPTER

27

AI in Imaging: Emphasis on Medical Imaging

Katarzyna Swist-Szulik[1*]

1. Introduction

In recent years, the number of Artificial intelligence (AI) applications in medicine including clinical care and research increased exponentially with radiology playing a major role in the digital transformation of healthcare. AI holds a vast amount of potential to address the growing demands for diagnostic imaging and staff shortages. Most importantly, it can empower radiologists and clinicians to make optimal treatment decisions for their patients. Medical imaging refers to several technologies helping to visualize anatomy, diagnose and treat injuries and disease conditions both acute and chronic. It has been a long and difficult process for medical imaging services to reach the current advanced status but integration of the artificial intelligence (AI)-driven technologies will hopefully take it to another level. During the recent decade of research in medical imaging, the increasing amount of data and burden to medical professionals to process images has grown dramatically. In the post-pandemic world, healthcare systems are still struggling with imaging-related demands, an increase in the number of aging patients, problems with speed and efficiency in imaging analysis, reporting and inaccurate diagnoses due to human error or subjective interpretations. These challenges have laid a perfect foundation for AI to blossom. AI has emerged as a powerful tool, revolutionizing the way medical images are acquired, analyzed, and interpreted. This chapter delves into the remarkable advancements in AI-based medical imaging, exploring its potential to enhance diagnostic accuracy, improve patient outcomes, and reshape the future of healthcare. The application of AI analysis and machine learning in medical imaging could support healthcare professionals to identify problems and details that may be impossible for the human eye to detect.

AI refers to a field of computer science creating systems mimicking human intelligence and able to perform tasks such as learning and problem solving. A subset of AI is Machine learning (ML) that adapts human learning mechanisms and enables computers to learn from data and make predictions and decisions without being explicitly programmed. ML has to be trained with a large amount of data and knowledge base to automatically derive rules and perform tasks including pattern recognition,

[1] Royal Manchester Children's Hospital, UK.
* Corresponding author: Katarzyna.Swist-Szulik@mft.nhs.uk

classification and regression. ML can be further categorized into various subfields such as decision trees, random forests, support vector machines, neural networks or deep learning which have been extensively applied to medical imaging. Among these techniques, deep learning (DL) algorithms, particularly convolutional neural networks (CNNs), have performed remarkably in various imaging tasks, surpassing traditional machine learning approaches. DL has provided several solutions to medical imaging in image processing, detection, and segmentation. CNNs have the ability to use learnable filters on input images to create so-called 'feature maps' to indicate shapes, colours, and structures.

Artificial intelligence (AI) is being increasingly utilized to improve image interpretation and analysis in medical imaging. AI algorithms can be trained to analyze medical images and detect abnormalities or subtle changes that may be difficult for humans to detect, leading to more accurate and efficient diagnosis and treatment of various conditions. AI algorithms can detect early signs of breast cancer on mammograms, lung nodules, tuberculosis or cancer infiltrates on CT and MRI scans, and detect signs of Alzheimer's disease on MRI scans. AI can also be used to analyze echocardiograms to assess heart function and diagnose heart disease. In addition, AI can be used to improve the accuracy and efficiency of radiology reports by automatically generating preliminary information based on the analysis of medical images. This can help reduce the workload of radiologists and improve patient care by providing faster and more accurate reports. Therefore, it is vital to ensure that AI algorithms are developed and used with human healthcare providers to provide the best possible patient care (Figure 1).

There are several factors driving the growth of AI-powered medical imaging technology. The power of Big Data and AI, particularly machine learning and deep learning algorithms is in their ability to analyze large imaging datasets with incredible speed and precision. The most significant applications of AI and data science in medical imaging include their proficiency in identifying patterns and abnormalities, providing a quantitative evaluation of radiographic traits, detecting image modalities at various treatment stages (for example, tumour delineation) and discovering various disease characteristics that are undetectable by the human eye. This helps to make more reliable and timely diagnoses to improve treatment outcomes. AI-powered medical imaging can distinguish pathological changes from health, can identify potential health risks or predict the likelihood of developing certain conditions like signs or early stages of cancer or Alzheimer's disease. Based on this information the healthcare providers and clinicians can make more adequate and informed decisions about treatment options and undertake preventative measures to reduce the risk of certain diseases with significant individual and social impact. However, it is essential to note that AI does not replace radiologists and clinicians. While AI can assist in analyzing medical images, it is ultimately up to the healthcare provider to make the final diagnosis and treatment decisions based on the available information (Figure 2).

AI-powered imaging can provide valuable support in the treatment of various clinical conditions. The most important ailments include cancer and neurological disorders. AI algorithms can assist in the detection, diagnosis, and characterization of various types of cancers, such as lung cancer, breast cancer, or prostate cancer. By analyzing imaging data, AI algorithms can help identify tumours, assess their size, location, and spread microcalcifications. They also provide guidance for treatment planning. AI algorithms can aid in the identification of treatment targets and biomarkers, helping clinicians select the most appropriate therapies, such as targeted therapies or immunotherapies, based on the individual patient's tumour characteristics. AI models provided with MRI-scanned images have achieved an extreme precision and accuracy of 98.56% in brain tumour classification. AI algorithms can monitor treatment response by analyzing changes in tumour size, morphology, or functional characteristics over time. Moreover, AI algorithms can assist in radiation therapy planning by optimizing treatment delivery and minimizing radiation exposure to healthy tissues.

AI-powered medical imaging is also used to successfully diagnose neurological conditions like amyotrophic lateral sclerosis (ALS), Alzheimer's disease, Parkinson's disease, or stroke. AI algorithms

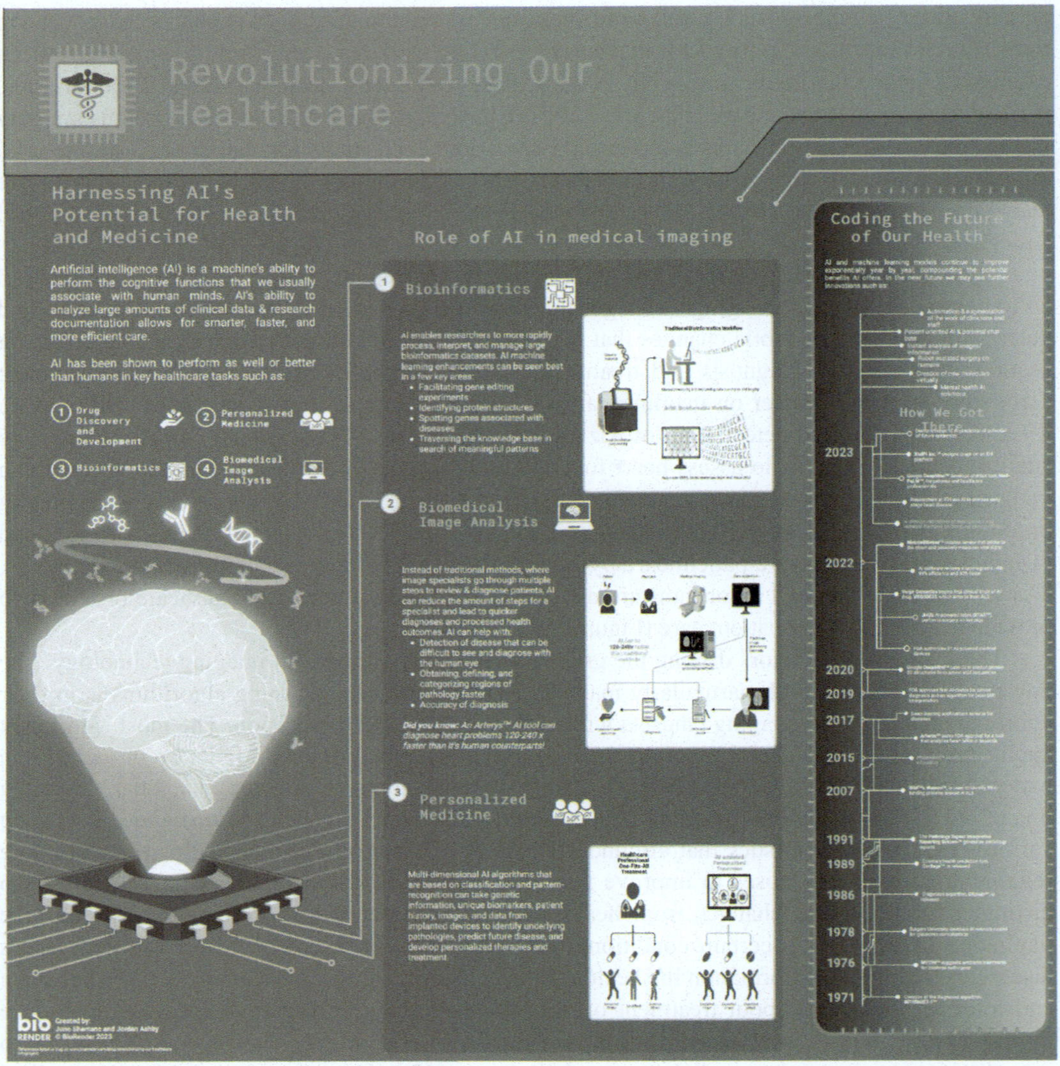

Fig. 1 Role of Artificial Intelligence (AI) in healthcare, especially in medical imaging and personalized medicine highlighting powerful biomedical image analysis and empowerment of a personalized approach to diagnosis and treatment.

can assist in the segmentation and quantification of brain structures, facilitating the localization of lesions or abnormalities and guiding surgical interventions or targeted treatments. AI algorithms can analyze neuroimaging data to predict disease progression and help clinicians develop personalized treatment plans, considering factors such as disease subtype, genetic markers, and response to specific therapies. Cardiovascular abnormalities are often diagnosed by measuring the structure of the patient's heart. Traditionally, chest X-rays are used first to detect potential heart-related abnormalities such as heart enlargement and cardiomegaly. AI-based medical imaging technology can identify cardiac abnormalities such as left atrial enlargement or analyze aortic valve function or diameter of the pulmonary artery. AI algorithms can aid in the identification and quantification of plaque build-up in blood vessels, assisting in the diagnosis and risk stratification of patients with coronary artery disease.

AI algorithms can analyze cardiac imaging data to predict the likelihood of cardiovascular events, such as heart attacks or strokes, and help clinicians develop personalized prevention strategies

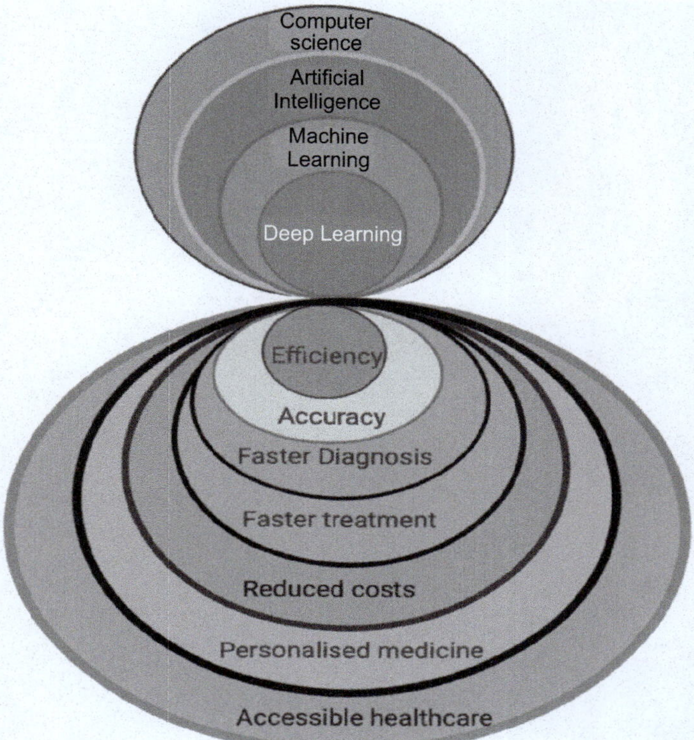

Fig. 2 Structure of computer Science and important benefits of AI-powered medicine in medical imaging.

and treatment plans. AI technology has been widely used to detect and diagnose musculoskeletal disorders such as fractures or dislocations that are hard to detect using standard imaging techniques, various sports' injuries and osteoarthritis. Fractures and musculoskeletal injuries are the common causes of long-term chronic pain and immobility among elderly patients especially. AI-powered tools can detect subtle variations in medical images so that trauma patients can receive the best care with positive outcomes. AI algorithms can aid in the segmentation and quantification of bone structures or joint spaces, facilitating the evaluation of disease progression and response to treatment. AI algorithms can analyze imaging data to assist in surgical planning for orthopaedic procedures, such as joint replacements or spinal surgeries, by optimizing implant positioning and surgical approach. These are just a few examples of how AI-powered imaging can support the treatment of clinical conditions.

The applications of AI in medical imaging are rapidly evolving, and with further advancements, we can expect to witness more utilization as shown in Figure 3.

2. Most Significant Benefits of AI-powered Medical Imaging

2.1 Improving Efficiency

AI algorithms can analyze medical images with a high level of accuracy and precision. By automating certain tasks such as image acquisition, segmentation, and feature extraction, AI allows radiologists to allocate more time to complex cases. This improves workflow leading to faster diagnosis and treatment. AI algorithms can also assist in prioritizing cases based on urgency, helping healthcare providers to manage their workload more effectively.

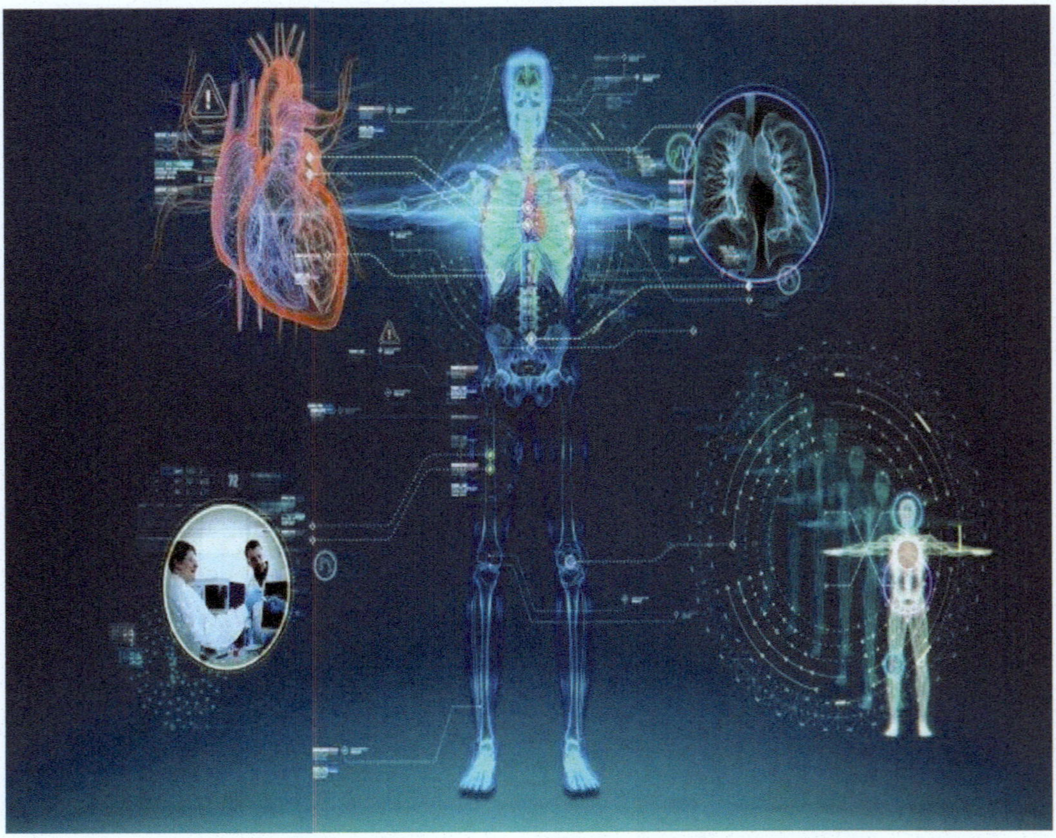

Fig. 3 Source of images-AI in radiology: beyond imaging. European Society of Radiology (ESR).

2.2 Improving Accuracy

AI-based algorithms can detect subtle abnormalities in medical images. By learning from vast datasets, AI models can identify patterns, anomalies, and even predict patient outcomes. This assists radiologists in making more accurate diagnoses, improving early detection of diseases such as cancer, cardiovascular conditions, and neurological disorders. The integration of AI also helps reduce false-positive and false-negative rates, minimizing unnecessary treatments and interventions.

2.3 Faster Diagnosis and Intervention

Radiologists often need valuable time to read and analyze CT or MRI scans. AI-powered medical imaging tools use powerful algorithms in tasks such as image processing, quality control, and data management. AI can automatically segment and label structures in medical images, such as organs and tumors, improving the efficiency of radiologists. AI can also perform quality control checks on medical images, ensuring they meet the necessary standards for diagnosis and treatment and organize large amounts of medical imaging data, making it easier for healthcare providers to access and analyze the data. This can help to provide a comprehensive view of the patient's medical history and imaging data.

2.4 Tracking Patient Care

Traditional imaging techniques face challenges in detecting real-time change in the patient's condition. AI and machine learning technologies in medical imaging are beneficial for tracking the patient's condition and detecting even miniscule changes in large information datasets. This is effective for tracking brain tumours and other cancers, essential for determining the best treatment. For instance, standard medical imaging cannot determine the percentage of tumour cells that are dead or alive.

2.5 Improving Precision Medicine

AI in medical imaging holds great potential for personalized medicine regimes. Personalized treatment planning involves tailoring medical treatments to individual patients based on their unique genetic makeup, medical history, and lifestyle factors. AI algorithms can analyze large datasets of patient information to identify patterns and predict how a patient is likely to respond to different treatments. For example, by integrating imaging data such as CT or MRI scans along with other patient-specific information, AI algorithms can provide better personalized care, treatment plans and treatment response predictions. This enables healthcare providers to deliver more targeted therapies, optimize care and reduce healthcare costs. For example, in radiation therapy, AI algorithms can help determine the optimal radiation dose and target volume for each patient. AI helps with image reconstruction and enhances the quality of images, reduces noise, improves resolution and can reconstruct missing details. These benefits are especially visible in managing cancer. For example, AI and machine learning tools can differentiate between different types of lung cancer. Additionally, they enable more accurate predictions of the survival rate of tumour patients based on the measured grade and stage. AI can also be used in combination with electronic health records (EHRs) containing a wealth of information about a patient's medical history, including past diagnoses, treatments, and medications. This helps to create a patient's unique health profile. Overall, this can help healthcare providers with faster and more accurate diagnosis, support decision-making processes leading to more effective treatment outcomes.

2.6 Reducing Time for Implementing Treatment

AI technology in medical imaging can help detect diseases in early stages when treatment options are more effective and less invasive improving the overall patient outcome. For instance, in critical care scenarios like strokes or cancers, AI tools can save time allowing implementation of appropriate timely treatment. In some cases, the use of AI for stroke cases can reduce the time between CT angiography and intervention by 30-40 minutes. Whether it is through a timelier diagnosis or more precise medical interventions, it can significantly improve the overall patient outcome.

2.7 Reducing Healthcare Costs

AI-powered medical imaging has the potential to reduce healthcare costs in multiple ways. By improving diagnostic accuracy and efficiency, unnecessary tests and procedures can be avoided, leading to cost savings. AI-enabled medical imaging tools can help automate the analysis, thus reducing the overall manual workload. For example, using traditional cancer screening, pathologists had to label and evaluate thousands of images to detect cancer cells. This increased their overall workload, significantly increasing the time to diagnosis. AI can also optimize resource allocation and enhance operational efficiency in healthcare facilities.

2.8 Supporting Continuous Learning and Improvement

AI algorithms can continuously learn from new data, experiences and receive feedback, leading to ongoing improvement in their performance. As more medical images are analyzed, AI systems can refine their algorithms and enhance their diagnostic capabilities. This iterative learning process can lead to ongoing improvement in accurate and effective treatments. Additionally, it has the potential to continually advance the field of medical imaging, benefiting patients and healthcare providers alike.

2.9 Improving Accessibility of Healthcare

AI-powered medical imaging has the potential to improve access to quality healthcare, especially in underserved areas with limited radiology expertise. AI algorithms can be used remotely to analyze medical images, allowing for timely diagnosis and treatment recommendations, even in regions where radiologists are scarce.

3. AI and Robotics in Medical Imaging

AI and robotics have significantly impacted the field of medical imaging. Some of their applications would include image-guided interventions, surgical planning and simulations, and telemedicine. They play a crucial role in image-guided interventions, enabling precise and minimally invasive procedures. Robotic systems can integrate real-time imaging with surgical instruments, allowing surgeons to navigate and perform interventions with high accuracy and control. Surgical planning and simulation is another area where AI and robotics help in surgical planning by providing detailed preoperative imaging analysis and simulation. Surgeons can use virtual reality or augmented reality tools to visualize patient-specific anatomy and plan surgical approaches. This improves surgical precision and reduces complications. AI algorithms can also fuse and register multiple imaging modalities, such as CT, MRI, and ultrasound. This integration enhances the accuracy of image-guided procedures and improves treatment planning by combining complementary information from different modalities. It is important to note that AI algorithms cannot replace human healthcare providers. As highlighted earlier, while they can provide valuable insights and support in diagnosing and treating certain conditions, they do not have the same level of clinical judgement and experience as human healthcare providers. Therefore, AI should be used as a tool to support healthcare providers rather than replace them. The integration of AI and robotics in medical imaging has the potential to improve patient outcomes, enhance workflow efficiency, and optimize resource utilization. However, challenges such as data privacy, regulatory compliance, algorithm transparency, and ethical considerations need to be addressed to ensure responsible and safe implementation. Continued research, clinical validation, and collaboration between engineers, clinicians, and researchers are essential to harness the full potential of AI and robotics in medical imaging.

4. Ethical Considerations

The integration of AI in medical imaging possesses several risks and raises ethical concerns that need to be carefully addressed. AI algorithms often work as black boxes, making it difficult to understand how they arrive at their decisions. In the context of medical imaging, developers should strive to make AI algorithms transparent and explainable, providing clinicians with insights into the decision-making process. As AI plays an increasingly autonomous role in medical imaging, questions arise regarding accountability and liability in case of errors or adverse outcomes. The legal implications of AI systems in healthcare can be relevant when AI technology has the ability to undertake autonomous decisions leading to wrong diagnosis and prognosis and in consequence to inadequate treatment. In current healthcare systems, the ethical and legal responsibility for decision making remains within radiologists and clinicians. However, in situations where the decision-making process is driven by AI algorithms the only advisable solution is a creation of multidisciplinary groups consisting of healthcare providers, medical professionals and AI experts working in the field of medical imaging. It has been already demonstrated that groups of human and AI agents working together make more accurate predictions compared to humans or AI alone. Clear guidelines and regulations need to be established to determine who is responsible for the decisions made by AI algorithms and to ensure appropriate liability allocation.

Other important ethical considerations in the context of AI in medical imaging include privacy and data security including handling, storage, and sharing of sensitive patient information. The use of AI in medical imaging may require patients to provide informed consent for their data to be used in algorithm training or analysis. Patients should be educated about the potential benefits, risks, and limitations of AI technologies to make informed decisions about their participation. Respecting patient autonomy and ensuring their understanding of AI's role is essential. That would require implementing robust data security measures to protect patient confidentiality. AI algorithms should be designed to augment human expertise rather than replace it. The importance of human oversight and involvement in the decision-making process should be emphasized. Clinicians should have the ability to critically evaluate and validate AI-generated results, ensuring that they align with their clinical judgment. AI algorithms in medical imaging should undergo rigorous testing, validation, and ongoing monitoring to ensure their accuracy, safety, and efficacy. Regular audits and assessments of AI systems should be conducted to identify and address any performance or ethical issues that may arise.

Addressing these ethical considerations requires interdisciplinary, transparent, and inclusive collaboration between healthcare professionals, AI developers and ethicists. It is important to develop guidelines, regulations, and best practices that promote the responsible and ethical use of AI in medical imaging.

5. The Place for Radiologists and Clinicians in the Era of AI

In the era of AI-powered medical imaging, radiologists continue to play a crucial role in the interpretation and analysis of imaging studies. While AI algorithms have shown great potential in assisting radiologists and improving efficiency, they are not intended to replace medical professionals. AI technology can provide valuable insights and support in diagnosing and treating certain conditions, however, they do not represent the same level of clinical judgment and experience as human healthcare providers. Radiologists cannot be replaced by machines; their work covers a wider area of expertise beyond the simple interpretation of images. Their assessments and duties also include communication, consideration of patients' values and preferences, medical judgment, interventional procedures, and many other tasks that, so far, cannot be performed by computer programmes alone. AI algorithms are designed to support radiologists by identifying and highlighting abnormalities. However, it is the radiologist's expertise that enables them to interpret these findings and provide meaningful clinical recommendations. Firstly, radiologists have extensive training and expertise in interpreting medical images. They possess in-depth knowledge of anatomy, pathology, and disease patterns, allowing them to make accurate diagnoses and provide valuable clinical insights. They use their clinical judgment to integrate imaging findings with patient history, laboratory results, and other clinical information. Secondly, radiologists are actively involved in the development and validation of AI algorithms. They contribute their clinical expertise and domain knowledge to ensure that the algorithms are accurate, reliable, and clinically relevant. Their expertise helps in ensuring the quality and relevance of the data, thereby improving the performance of the algorithms.

Radiologists are responsible for validating the outputs of AI algorithms and assessing their impact on patient care. They evaluate the algorithm's performance, identify limitations or potential biases, and provide feedback for further refinement. While AI algorithms excel at detecting common abnormalities or patterns, they may struggle with complex or rare cases that require a high level of expertise. Radiologists are trained to handle complex scenarios, subtle findings, and rare diseases that may not be well-represented in training datasets. They can provide valuable insights beyond image interpretation, such as procedural guidance, interventional techniques, or correlation with other diagnostic tests. They offer a holistic approach to patient care that goes beyond the capabilities of AI algorithms. Most importantly, radiologists interact directly with patients, discussing imaging findings, answering questions, and providing reassurance. They play a crucial role in explaining complex medical information in a clear and empathetic manner, helping patients understand their condition and treatment options. They also have the opportunity to collaborate with other medical professionals, including referring physicians and specialists, to provide comprehensive care to patients. They participate in multidisciplinary discussions, contribute to treatment planning, and ensure the effective communication of imaging results within the healthcare team.

While AI-powered medical imaging is transforming the field, radiologists remain indispensable. They bring their clinical expertise, judgment, and patient-focused approach to ensure accurate diagnoses, personalized treatment plans, and optimal patient care. Therefore, AI technology helps to improve the accuracy of diagnosis, the efficiency of treatment, and the overall quality of patient care. However. this technology will not replace radiologists or clinicians who have the ability to integrate patient's clinical and imaging information.

Conclusions and Future Directions

Artificial intelligence has ushered in a new era in medical imaging, transforming the way healthcare professionals diagnose and treat diseases. It will impact radiology much quicker than any other medical field. The future of AI in medical imaging is expected to bring significant advancements in diagnosis, treatment, and personalised care. There are certain areas of medical imaging where AI is expected to have a valuable impact. That would include an ability to quickly analyze large datasets of medical images and identify abnormalities that are difficult to detect using current imaging modalities. This can lead to earlier and more accurate diagnoses and improvement on overall patients' outcomes. AI can help radiologists and clinicians to develop personalized treatment plans according to unique patient characteristics, such as genetics, other comorbidities, and lifestyle factors. This can lead to more effective and efficient treatments tailored to each patient's circumstances. In research, AI can be used to develop new imaging modalities or improve the precision and accuracy of existing ones. AI can also analyze data from past diagnoses, treatments, and medications, to identify potential risks necessary to create adequate personalized treatment plans. Importantly, AI can be also used to analyze medical images taken at home, allowing medical professionals to monitor patients remotely and detect potential health issues in advance.

Advancements in AI algorithms, coupled with the exponential growth in imaging data, will further enhance the accuracy and efficiency of diagnosis and treatment and will help to detect and diagnose abnormalities previously missed using traditional imaging techniques. Overall, the future of AI in medical imaging is expected to bring significant advancements in diagnosis, treatment, and personalized care. With the help of deep learning algorithms, medical imaging technology combined with human expertise, medical practitioners will be able to detect diseases with a higher level of precision leading to improved patient outcomes and better overall healthcare. However, it is important to note that the development and validation of AI algorithms require careful testing and validation to ensure their accuracy and safety. Ethical considerations associated with implementing AI and such extensive data transformations need to be carefully and systematically addressed. Data privacy, algorithm bias, and the need for transparent and interpretable AI models are crucial aspects. Ensuring the integration of AI does not replace human expertise but rather augments it. This is essential in maintaining patient trust and safety.

References

Becker, A.S., Blüthgen, C., Phi van, V.D., Sekaggya-Wiltshire, C., Castelnuovo, B. et al. (2018). Detection of tuberculosis patterns in digital photographs of chest X-ray images using Deep Learning: feasibility study. *The International Journal of Tuberculosis and Lung Disease*, *22*(3), 328–335. https://doi.org/10.5588/ijtld.17.0520

Becker, A.S., Marcon, M., Ghafoor, S., Wurnig, M.C., Frauenfelder, T. et al. (2017). Deep Learning in Mammography: Diagnostic Accuracy of a Multipurpose Image Analysis Software in the Detection of Breast Cancer. *Investigative Radiology*, *52*(7), 434–440. https://doi.org/10.1097/RLI.0000000000000358

Borys, K., Schmitt, Y.A., Nauta, M., Seifert, C., Krämer, N. et al. (2023). Explainable AI in medical imaging: An overview for clinical practitioners – Saliency-based XAI approaches. *European Journal of Radiology*, *162*, 110787. https://doi.org/10.1016/j.ejrad.2023.110787

Chen, H., Zhang, Y., Zhang, W., Liao, P., Li, K. et al. (2017). aLow-dose CT via convolutional neural network. *Biomedical Optics Express*, *8*(2), 679. https://doi.org/10.1364/boe.8.000679

Davnall, F., Yip, C.S.P., Ljungqvist, G., Selmi, M., Ng, F. et al. (2012). Assessment of tumor heterogeneity: an emerging imaging tool for clinical practice? *Insights into Imaging*, *3*(6), 573–589. https://doi.org/10.1007/s13244-012-0196-6

Erickson, B.J., Korfiatis, P., Akkus, Z. and Kline, T.L. (2017). Machine Learning for Medical Imaging. *RadioGraphics*, *37*(2), 505–515. https://doi.org/10.1148/rg.2017160130

Herington, J., McCradden, M.D., Creel, K., Boellaard, R., Jones, E. et al. (2023). Ethical Considerations for Artificial Intelligence in Medical Imaging: Deployment and Governance. *The Journal of Nuclear Medicine*, *64*(10), 1509–1515. https://doi.org/10.2967/jnumed.123.266110

Jha, S. and Topol, E.J. (2016). Adapting to Artificial Intelligence. *JAMA*, *316*(22), 2353. https://doi.org/10.1001/jama.2016.17438

King, A.D., Chow, K.-K., Yu, K.-H., Mo, F.K.F., Yeung, D.K.W. et al. (2013). Head and Neck Squamous Cell Carcinoma: Diagnostic Performance of Diffusion-weighted MR Imaging for the Prediction of Treatment Response. *Radiology*, *266*(2), 531–538. https://doi.org/10.1148/radiol.12120167

Kohli, M., Prevedello, L.M., Filice, R.W. and Geis, J.R. et al. (2017). Implementing Machine Learning in Radiology Practice and Research. *American Journal of Roentgenology*, *208*(4), 754–760. https://doi.org/10.2214/ajr.16.17224

Korfiatis, P., Kline, T.L., Coufalova, L., Lachance, D.H., Parney, I.F. et al. (2016). MRI texture features as biomarkers to predict MGMT methylation status in glioblastomas. *Medical Physics*, *43*(6Part1), 2835–2844. https://doi.org/10.1118/1.4948668

Krittanawong, C. (2018). The rise of artificial intelligence and the uncertain future for physicians. *European Journal of Internal Medicine*, *48*, e13–e14. https://doi.org/10.1016/j.ejim.2017.06.017

Kumar, Y., Koul, A., Singla, R. and Ijaz, M.F. (2022). Artificial Intelligence in Disease diagnosis: a Systematic Literature review, Synthesizing Framework and Future Research Agenda. *Journal of Ambient Intelligence and Humanized Computing*, *14*(7). https://doi.org/10.1007/s12652-021-03612-z

LeCun, Y., Bengio, Y. and Hinton, G. (2015). Deep Learning. *Nature*, *521*(7553), 436–444. https://doi.org/10.1038/nature14539

Lee, J.-G., Jun, S., Cho, Y.-W., Lee, H., Kim, G.B. et al. (2017). Deep Learning in Medical Imaging: General Overview. *Korean Journal of Radiology*, *18*(4), 570. https://doi.org/10.3348/kjr.2017.18.4.570

Liew, C. (2018). The future of radiology augmented with Artificial Intelligence: A strategy for success. *European Journal of Radiology*, *102*, 152–156. https://doi.org/10.1016/j.ejrad.2018.03.019

Obermeyer, Z. and Emanuel, E.J. (2016). Predicting the Future—Big Data, Machine Learning, and Clinical Medicine. *New England Journal of Medicine*, *375*(13), 1216–1219. https://doi.org/10.1056/nejmp1606181

Obrzut, B., Kusy, M., Semczuk, A., Obrzut, M. and Kluska, J. et al. (2017). Prediction of 5–year overall survival in cervical cancer patients treated with radical hysterectomy using computational intelligence methods. *BMC Cancer*, *17*(1). https://doi.org/10.1186/s12885-017-3806-3

Pereira, S., Pinto, A., Alves, V. and Silva, C.A. (2016). Brain Tumor Segmentation Using Convolutional Neural Networks in MRI Images. *IEEE Transactions on Medical Imaging*, *35*(5), 1240–1251. https://doi.org/10.1109/tmi.2016.2538465

Pesapane, F., Codari, M. and Sardanelli, F. (2018). Artificial intelligence in medical imaging: threat or opportunity? Radiologists again at the forefront of innovation in medicine. *European Radiology Experimental*, *2*(1). https://doi.org/10.1186/s41747-018-0061-6

Recht, M. and Bryan, R.N. (2017). Artificial Intelligence: Threat or Boon to Radiologists? *Journal of the American College of Radiology*, *14*(11), 1476–1480. https://doi.org/10.1016/j.jacr.2017.07.007

Samuel, A.L. (1959). Some Studies in Machine Learning Using the Game of Checkers. *IBM Journal of Research and Development*, *3*(3), 210–229. https://doi.org/10.1147/rd.33.0210

Sutton, E.J., Huang, E.P., Drukker, K., Burnside, E.S., Li, H. et al. (2017). Breast MRI radiomics: comparison of computer- and human-extracted imaging phenotypes. *European Radiology Experimental*, *1*(1), 22. https://doi.org/10.1186/s41747-017-0025-2

Tang, X. (2020). The role of artificial intelligence in medical imaging research. *BJR|Open*, *2*(1), 20190031. https://doi.org/10.1259/bjro.20190031

Varoquaux, G. and Cheplygina, V. (2022). Machine learning for medical imaging: methodological failures and recommendations for the future. *Npj Digital Medicine*, *5*(1), 1–8. https://doi.org/10.1038/s41746-022-00592-y

Verghese, A., Shah, N.H. and Harrington, R.A. (2018). What This Computer Needs Is a Physician. *JAMA*, *319*(1), 19. https://doi.org/10.1001/jama.2017.19198

CHAPTER

28

The Decisive Role of Artificial Intelligence in Toxicology: Advancements and Applications

Megha KB,[1] *Athira SS*,[1] *Greeshma N*,[2] *Aswathy V*,[2] *Manjoosha RY*,[2] *Renu J*[2] and *Mohanan PV*[1*]

1. Introduction

Artificial intelligence (AI) is a highly advanced, fast-growing, transformative force that connects various scientific disciplines creating and redefining traditional knowledge and making more relevant options for the betterment of healthcare and industrial developments. The basis of AI is the modern machine-learning (ML) approach including deep neural networks providing an advanced predictive comparison better than the conventional computational methods (Tetko et al., 2022). Machine learning is an area in AI that uses the application of statistical algorithms to learn the system. Computational toxicology is a new approach that includes interconnected methods from both ML and AI to find solutions for the concept of a 3R system to replace, reduce and refine animal testing (Kleinstreuer et al., 2021). AI is a collection of technologies that allows computers to mimic human intelligence and carry out tasks that normally require human intelligence, like learning, logical thinking and making decisions. One such domain that is vastly influenced by AI is toxicology. The branch deals with understanding the adverse effects of physical, chemical and biological agents on living organisms (Goetz & Schork, 2018). The field also deals with the understanding of the mechanism of toxicity, identifying the susceptible group, and creating approaches for risk assessment and management. The unbelievable capability to analyze vast datasets, identify intrinsic patterns and make predictions with maximum accuracy is reshaping toxicological research, regulatory decision making and risk assessment. Traditionally, the safety evaluation of pharmaceuticals, chemicals and environmental pollutants relies on empirical studies, animal testing and epidemiological investigations. These methods provide valuable insights into the toxicological mechanisms and hazards, which are highly time-consuming, costly and associated

[1] Division of Toxicology, Sree Chitra Tirunal Institute of Medical Science and Technology, Thiruvananthapuram, Kerala.
[2] Department of Biomedical Engineering, Indian Institute of Technology, Hyderabad, India.
* Corresponding author: mohananpv10@gmail.com

with ethical concerns. Moreover, the complexity of biological systems and the sheer volume of potential toxicants persist as a very powerful challenge in assessing the risks induced by chemical exposures.

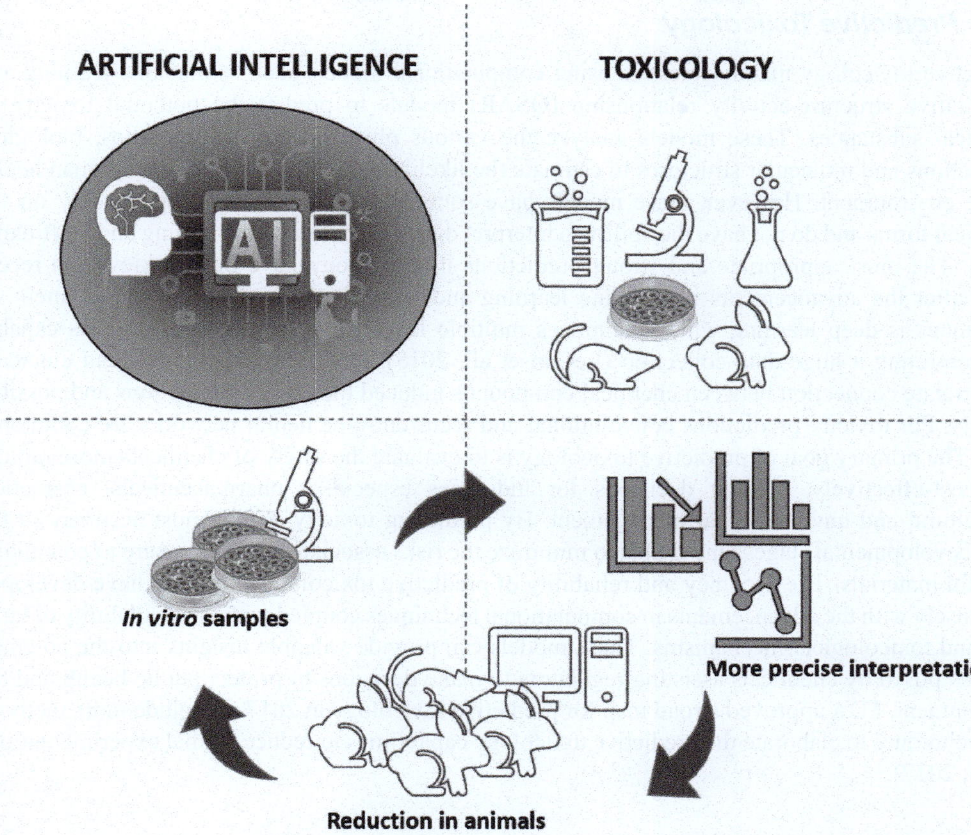

Fig. 1 Impact of artificial intelligence in toxicology

AI can evaluate huge quantities of data and can forecast treatment results or toxicity. From a toxicological and clinical aspect, AI is an application of computational capabilities to trigger human decision-making. The commonly used AI techniques for medical applications include machine learning, artificial neural networks and fuzzy expert systems (Holmes et al., 2004). AI can assist in analyzing the large quantity of data obtained from various toxicological studies, helping in the identification of genetic variations due to drug interactions and enabling people to choose the correct dosage and treatment with the aid of decision-making tools. AI models can help toxicologists to identify the drug or chemical interaction on living organisms. AI-assisted models for recognizing and identifying treatment strategies have been developed (Shen et al., 2021). The advent of AI has made remarkable progress in analyzing the huge data obtained from toxicological analysis and omics applications. Therefore, huge studies that have already been performed repeatedly during drug development can be avoided thus a drop in animal experimentation can be achieved. For this study, the descriptor-based quantitative structure-activity relationship (QSAR) methods are still widely used (Tetko et al., 2022). While considering the voluminous data from various toxicological studies, AI is a better choice for systematically comparing it to provide improved decisions without compromising the time for effective implementation of treatment strategies. This helps in preventing the individual from the adverse effects of a particular drug/chemical exposure and predicting the genomic and molecular level alteration that occurs during the exposure. In this context, the multifaceted impact of AI in toxicology and related areas is explained in detail in this chapter to understand its transformative potential role in safeguarding the health and environment.

1.1 Predictive Toxicology

Predictive toxicology mainly refers to using computational models such as machine learning and quantitative structure-activity relationship (QSAR) models to predict the potential toxicity of chemical substances. These models analyze the various physicochemical properties, biological interactions and molecular structures to estimate the likelihood of adverse effects on human health or the environment. However, these models have some limitations as they depend only on the chemical forms and do not have the ability to interpret diverse types of data (Hartung and Hoffmann, 2009). The most appropriate and strong predictions in toxicology have been achieved in recent years after the advancements in machine learning and AI tools. One of the most prominent AI techniques is deep learning which comprises multiple layers of neural network models capable of interpreting a huge data collection (LeCun et al., 2015). Deep learning models can elucidate the intricate connection between chemical compounds induced biological interactions and possible toxicity. For toxicity predictions convolutional and recurrent-type neural networks are commonly used. The primary goal of predictive toxicology is to evaluate the safety of chemicals more rapidly and cost-effectively, making decisions for industries especially, pharmaceuticals, cosmetics, agricultural and environmental management. By predicting toxicity with utmost accuracy in the early developmental stages, one is able to minimize the risks associated with exposure to potentially harmful materials. The accuracy and reliability of predictive toxicological models have developed immensely with the advancements in computational techniques coupled with the availability of large data and toxicological mechanisms. These models can provide valuable insights into the potential hazards posed by chemicals assisting regulators to make decisions to protect public health and the environment. FDA approved a road map for predictive toxicology in 2017 to include more methods and techniques to elaborate the predictive toxicology capabilities to reduce animal experimentations (FDA, 2017).

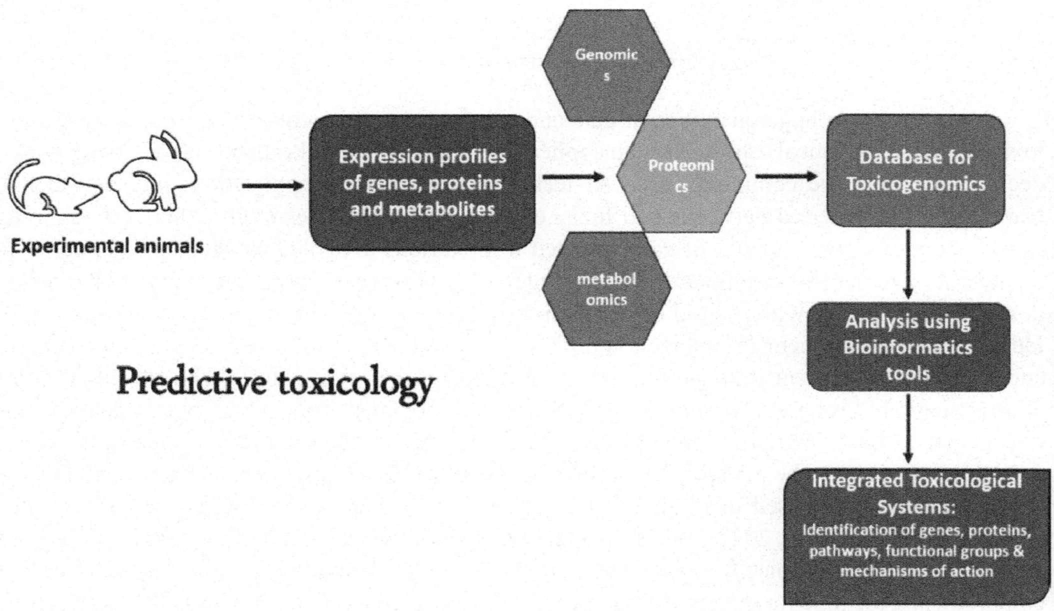

Fig. 2 Schematic representation of basic steps involved in predictive toxicology.

Deep learning models have also made influential advancements in predicting the mutagenic ability of chemicals (Mayr et al., 2016). A better predictive accuracy than the conventional QSAR model was exhibited by deep learning models towards rodent carcinogenicity (Li et al., 2021). The prediction for liver toxicity (Wang et al., 2019), developmental and reproductive toxicity was also accomplished using deep neural networks (Hartung, 2023). Deep learning techniques excel at processing huge data, however, there is a lack of elucidation and susceptibility to biases inherent in the data (Hartung, 2023). Predictive toxicology models are suitable for assessing the safety profiles of potential drug candidates early in the development process. By evaluating the chemical structure and their possible biological interactions these models can predict the potential adverse effects like liver/cardio toxicity making the pharmaceutical companies prioritize safer compounds for further testing (Suter et al., 2004). It assesses the safety of chemicals in cosmetics, skincare products, and personal care items. These models predict skin irritation, sensitization, and other adverse effects for developing safer formulations and complying with regulatory requirements.

Predictive toxicology plays a crucial role in evaluating the potential environmental impact of chemicals including pesticides, industrial chemicals and pollutants. These models can analyze the toxicity of substances to wildlife, aquatic life and ecosystems. The application of these models extends to evaluating the safety of food additives, preservatives and contaminants. Chemical risk assessment associated with exposure to industrial chemicals, household chemicals and environmental pollutants can predict toxicity endpoints such as mutagenicity, carcinogenicity and reproductive toxicity, aiding in risk management and regulatory decision-making. They possess high-throughput screening assays for rapid evaluation of toxicity of many chemical compounds. These models can prioritise compounds for further testing based on the predicted toxicity profiles, accelerated drug discovery, environmental monitoring and chemical safety assessments.

Table 1 List of prediction models used toxicology and toxicogenomic analysis

Tools	*Peculiar features*	*Observations*	*References*
Toxtree	The decision tree approach, which combines statistical analysis with knowledge-based insights, relies on structural and property data for its application	Open source, interpretable, and accessible to non-specialized users,	Yang et al., 2018
Open Tox	Various models for toxicity prediction and risk assessment, machine learning techniques, QSAR modeling and data visualization methods	Open source, enforce language-independent interfaces, facilitates seamless integration across different platforms and environments	O'Boyle et al., 2011
ToxCast	Comprehensive database of *in vitro* toxicity assays predicting the toxicity of chemical mixtures.	Research project, Focuses on exploring the molecular and cellular pathways targeted by chemical interactions	Judson et al., 2014
SafetyPharma	Comprehensive tool for drug safety analysis, featuring data visualization, statistical analysis tools and machine learning capabilities	Commercial	Kolluri et al., 2022
Virtual ToxLab	Forecast the toxic potential, encompass endocrine and metabolic disruption, certain aspects of cardiotoxicity and carcinogenicity for both existing and hypothetical compounds including chemicals, drugs and natural products	Utilizing automated, flexible docking combined with multi-dimensional QSAR, the system stimulates and quantifies the interaction of compounds with a range of proteins implicated in triggering adverse effects.	Vedani et al., 2012

Contd.

Table 1 *Contd.*

Tools	*Peculiar features*	*Observations*	*References*
PreADMET	Statistical tool for predicting genotoxicity and carcinogenicity prediction models	Commercial tools constrained by data primarily derived from mouse models and the Ames test	https://preadmet.webservice.bmdrc.org/
OpenPKG	C++ model focused on the complete body physiology, utilizes data from physicochemical properties and *in vivo* studies to predict the time evolution of drug distribution and clearance at an organ-specific level	Open-source compartmental framework designed for analyzing protein-drug binding, as well as drug metabolism and degradation processes.	McDaniel et al., 2019
DanishQSAR	A repository-based model that provides predictions related to physicochemical characteristics, skin irritation, acute toxicity and environmental toxicity	Limited to a dataset of 600,000 chemicals and 200 QSAR	Rim, 2020
ADRpred	Utilizes machine learning algorithms, the system predicts the probability of adverse drug reactions by analyzing patient data	Open source and employing knowledge embedding to encode drugs for various applications	Zhang et al., 2021

1.2 Risk Assessment using AI in Toxicology

Risk assessment is vital in toxicology for regulatory decision-making. It is mainly performed by checking the evidence over biological levels to evaluate the probability of adverse effects under precise exposure conditions (Maertens et al., 2022). AI can provide risk assessment quantitatively and systemically by using a probabilistic modelling approach (Mak & Pichika, 2019; Perez Santin et al., 2021; Tran et al., 2023) that can enable more quantifiable assessment compared to conventional procedures. The drastic advances in AI techniques have paved the way to recognize the complex mechanisms for the toxicity of chemicals and their associated predictions. The development process of a model is highly complicated with constant assessment and fine-tuning to attain the utmost performance. Translating theoretical knowledge into regulatory practice is the ultimate application in the world. The AI-based risk assessment model will be a valuable tool for faster and more precise resolutions for uncertainties in a more predictive manner. The model can support identifying the risk probabilities by accurately extrapolating the dose or exposure relationship (Figure 3). More active research is carried out to include human and animal studies and also *in vitro* and *in vivo* data to develop probabilistic risk models (Maertens et al., 2022).

2. Role of AI in Precision Medicine and Toxicology

Precision medicine is interconnected with toxicology as it helps to understand the effect of a particular drug on living systems. Precision medicine is providing treatment in a personalized manner by providing appropriate medications in the correct dosage depending on the person's health condition (Gemmati et al., 2019). This makes the treatment more precise and accurate avoiding the unwanted intake of drugs. Precision medicine mainly focuses on extending the implementation zone rather than giving the mistaken dosage of drugs. For the customization of medicine, several parameters like genetic, clinical and molecular data of the patient are used to plan the treatment strategies. Toxicology helps to understand the variability in the pharmacokinetics of a specific medication in that patient. In precision medicine, AI can immensely help to provide more accurate treatment options by analyzing the patient's data, especially molecular and genetic alterations. AI can identify the disease progression and response towards the drugs. Additionally, AI can antedate the treatment response in patients, using genetic data, molecular markers and physiological responses.

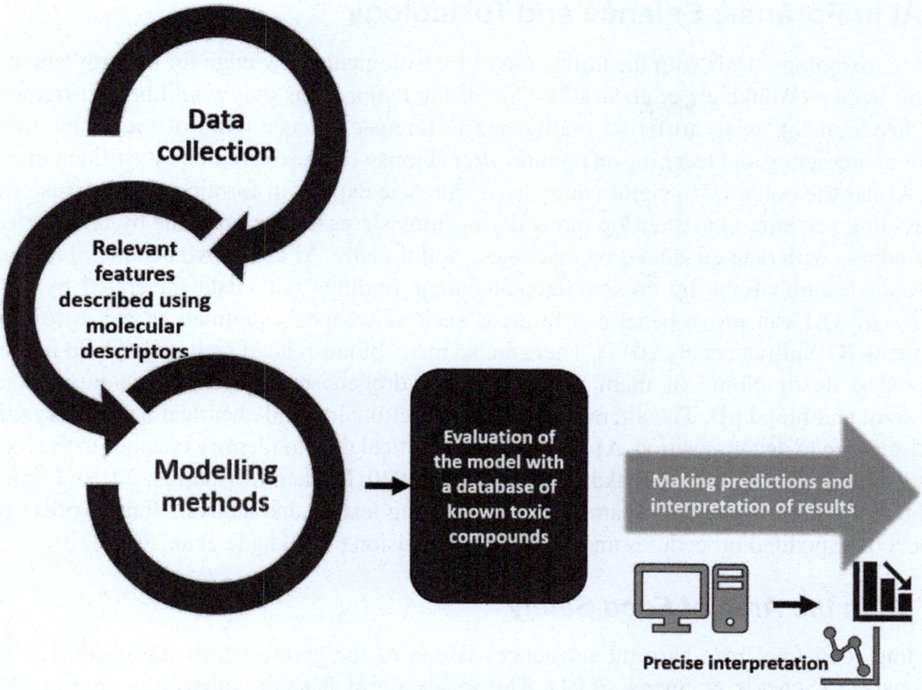

Fig. 3 Steps involved in developing the prediction models.

2.1 AI-assisted Data Analysis in Toxicology

In advanced toxicological research, the analysis and interpretation of huge databases are required from several sources. As mentioned by Hartung these sources can be from scientific literature, animal studies, omics-related measurements and from highly efficient screening assays (Hartung, 2023a). The analysis of such massive data is impracticable. Here AI plays a major role by providing computational resolution for data mining, extraction and interpretation as illustrated in Figure 4. The mining of the data from scientific literature, animal studies and experimental output can be performed with the help of natural language processing methods (Kleinstreuer et al., 2016). That helps in the reliable effective usage of the existing data. Another area on which toxicologists can depend is the graph neural networks that can correlate multi-omics data analysis from transcriptomics, lipidomics, metabolomics (Reel et al., 2021). Another highlight of AI in toxicology is the applicability to identify the pathways that can lead to adverse outcomes is an exceptional contribution (Hemmerich & Ecker, 2020; Lin & Chou, 2022). Development of Tox-GAN based on deep Generative Adversarial Network (GAN) based framework capable of deriving new animal results from existing animal studies without further repeated experimentation. It can also generate high-quality Toxicogenomic profiles without animal experiments (Chen et al., 2022).

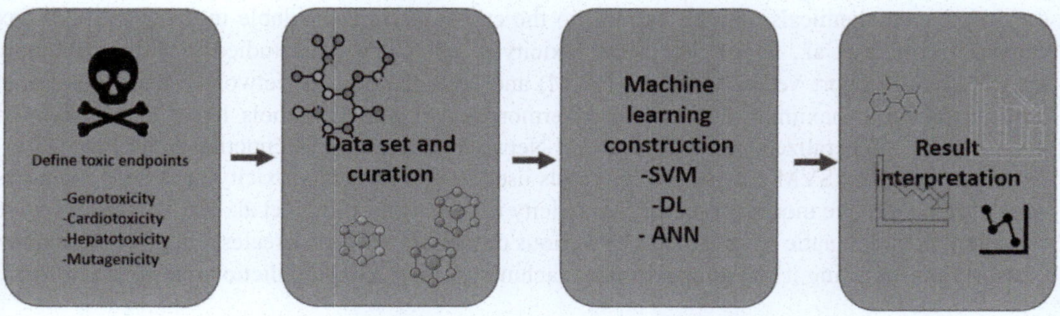

Fig. 4 Schematic representation of data mining and interpretation.

3. AI in Forensic Science and Toxicology

Forensic toxicology deals with the utilization of toxicological knowledge for the implementation of law and justice (Wankhade et al., 2022). One of the major areas that would benefit tremendously from development using artificial intelligence is forensic science. One of the major limitations present in predicting and reaching an opinion after autopsy can be overcome by artificial intelligence tools. AI has the potential to significantly assist forensic experts in forming more precise expertise in providing accurate consistent opinions during forensic case examinations by cross-referencing their findings with data generated by machines. Additionally, AI can prove beneficial in areas such as weapon examinations by cross-referencing their findings with data generated by machines. Additionally, AI can prove beneficial in areas such as weapon examination and estimating time since death (O'Sullivan et al., 2017). There are so many blood-related factors that help in predicting the time of death. Some of them are lactate dehydrogenase, triglyceride, sodium, potassium, cholesterol and blood pH. The alterations in their functionality and chemical nature vary with time after death due to decomposition. AI can analyse statistical data to identify changes in the biomarker levels from the time of death (Khakare & Fatangare, 2020; Hachem & Sharma, 2019). Furthermore, the potential of AI to integrate seamlessly into existing testing and analysis frameworks offers the prospect of expedited procedures and heightened precision (Wankhade et al., 2022).

3.1 AI in the Area of Food Safety

Providing food free from harmful substances is one of the greatest features existing in the food safety aspect (Scheule & Sneed, 2001). The incidence of harmful substances entering food can occur from any source. It can be natural or through pesticides, animal drug residues like hormones, antibiotics, pollutants from the environment, from food processing and packaging. The possible range of chemicals can be identified or analyzed and categorized based on their chances of entry into the food substances. More innovative food safety analytical methods that can provide rapid and reliable identification without affecting the company's economic cost are becoming a prerequisite and can be incorporated with the help of AI. The non-destructive analysis methods and newer technologies like biosensors, e-nose and e-tongue (Gutierrez et al., 2020; Qiu & Wang, 2017; Han et al., 2019) are some of the examples that can be incorporated with AI algorithms to develop new detection systems for food safety monitoring and prediction.

3.2 AI in Monitoring Environmental Toxicity

Great attention has been gained for predicting the toxicity of harmful chemicals that reduce the adverse effects on the environment. AI-based technologies are becoming useful for studying the environmental impact and their possible management approaches (Tebes-Stevens et al., 2018). AI-based technology has been used for decades to forecast the harmful effects of synthetic chemicals that are released into water bodies. This helps in predicting the water quality, estimating the bioaccumulation of chemicals in the species near the aquatic system and foreseeing various biological effects like growth inhibition (Russom et al., 1997). The largest database providing the details of toxic chemicals that are harmful to the environment is available in the EPA's DSSTox database (Grulke et al., 2019). The acute toxicity of chemicals was studied in fathead minnow models using Support Vector Machines (SVM) and Artificial Neural Networks (ANN) to predict the toxicity with maximum accuracy. Furthermore, other AI-based tools like Gene expression programming, Generalized Regression Neural Networks, Radial Basis Function Neural Networks, Decision Trees and SVM are some of the tools used to predict acute toxicity in fathead minnows which is the effective model for predicting toxicity in freshwater (Singh et al., 2013). The extensive contamination of aquatic environments by various chemical pollutants creates a huge accumulation of these toxins making it essential to utilize machine learning-based prediction methods to identify the bioaccumulation of these pollutants in fish and invertebrates. This is crucial for understanding and mitigating potential risks.

Soil is the other major part of the terrestrial system that is getting contaminated with various chemicals, and the adverse effects are finally accumulated in the plants and animals. One of the most toxic pollutants in soil is lead and its derivatives. The ANN was recently advanced for the detection of lead absorption in the soil. A combination of different operating parameters with ANN is used for the removal of various leachates from industrial activity. The soil is contaminated by various sources and ML-based technologies are now prevalent in the prediction of accumulated heavy metals (Sergeev et al., 2019). To understand the soil absorption coefficient (Koc) of pesticides, ML models using physicochemical properties and chemical structure descriptors are developed to monitor the risk assessment in a low-cost and time-consumable manner. The unlimited usage of pesticides and chemicals creates harmful effects on wildlife also (Klein AM et al., 2007; McArt et al., 2017). One of the species that is mostly affected by chemical pollutants is, honey bees (*Apis mellifera*), and there is a great drop in their numbers (Klein AM et al., 2007). Another species that is facing extinction threats is the bumblebee (Cameron et al., 2011). ML-based prediction models can predict probable assessments of environmental hazards induced by chemical pollutants (Lu et al., 2015). AI has created a remarkable influence in identifying the dispersal of toxic materials in the atmosphere that create serious health issues. An ANN-based prediction model was developed to monitor the level of methane (Lauret et al.,2016). A deep learning (DL) based model was used in Seoul to assess the particulate matter in the atmosphere which also warns them when it is about to reach the desired level. Some of the AI-based prediction models for assessing the toxicity induced in environmental science are briefed in Table 2.

Table 2 AI models used in environmental toxicity

Environmental field	*Related study*	*Environmental issues*	*Applied algorithms*	*References*
Water	Quality of water	BOD prediction	SVC	Singh et al., 2011
	Fathead minnow	Chemical categorisation based on toxicity	ML	Singh et al., 2013
		Acute toxicity prediction	SVMs, ANN	Tan et al., 2010
	Gammarus pulex	Predict bioaccumulation of chemicals in freshwater invertebrates	ML	Miller et al., 2019
Soil	Lead and its derivatives	Evaluate biosorption process	ANN	Bingol et al., 2012
	Heavy metals	Predicts accumulation and heavy metal transformation (Cu, Ni, Mn)	ML	Sergeev et al., 2019
	Pesticides	Predicts Soil absorption coefficient	ML	Kobayashi et al., 2020
Wildlife	Bumblebees	Pesticide-associated health alterations in Bumblebees	ML	McArt et al., 2017
	Apis mellifera	Effect of chemical pollutants on honeybees	ML	Li et al., 2017
Atmosphere	Methane	Dispersion rate of methane in the atmosphere	ANN	Lauret et al., 2016
	Particulate matters	Predicts PM in the air for several days used in Seoul	DL	Yang et al., 2020
	Pollutants in Air	Helps in the dispersion and direction of pollutants in the atmosphere	ML	Cervone et al., 2008

3.3 AI-based Emerging Trends in Toxicology

Artificial intelligence has substantially influenced toxicology by improving the understanding of toxicity, prediction of possible risks and development of new treatment strategies. AI is improving prediction capabilities by huge data analysis, identifying the risky groups and imparting interventions. As mentioned earlier, AI can influence understanding of toxicity in several ways such as by finding patterns in huge data from epidemiological studies or high-throughput screening to improve safety regulations and chemical designing. AI can notably predict chemical-induced toxicity for individuals. Thereby reducing the exposure risk and providing interventions to the risky groups. AI assists in developing new treatment methods for poisoned individuals by identifying the treatment methods and diagnostics for new poisons thereby enabling a personalized treatment approach. Reduction in animal testing can be achieved with the benefit of AI by comparing the toxicity studies performed early and with related chemical entities (Hartung, 2023a). AI aids in innovative technologies for cost-effective and rapid improvements in medical treatment. In public health, monitoring and prevention of disease spread can be hastened by incorporating appropriate medical measures. During a sudden outbreak of disease, the etiology and pathogenesis of the causative agents by scientific inquiry can help to arrest the disease spread.

Conclusion

In conclusion, the integration of artificial intelligence in toxicology has reached a new era of innovation and efficiency in assessing chemical safety and risk. Through advanced algorithms and machine learning techniques, AI has enabled the rapid analysis of vast datasets, leading to more accurate toxicity predictions and hazard identification. Moreover, AI-driven models enhance the understanding of toxicological mechanisms, enabling the development of novel therapeutics and safer chemicals. However, AI presents promising opportunities; its application in toxicology poses certain challenges like data quality, results in interpretability and ethical concerns surrounding algorithm biases. Therefore, researchers, regulators and stakeholders must collaborate closely to address these issues and ensure the responsible and transparent deployment of AI in toxicological assessments. Looking ahead, the continued advancement of AI technologies holds immense potential to revolutionize toxicology, allowing more effective risk assessment strategies and ultimately contributing to the protection of human health and the environment. Harnessing the power of AI in conjunction with traditional toxicological approaches, can enable us to strive toward a future where chemical safety is evaluated comprehensively leading to improved public health outcomes worldwide. AI is simply an automated tool that cannot replace human expertise and interaction.

Acknowledgments

The authors wish to express their thanks to the Director Sree Chitra Tirunal Institute for Medical Sciences and Technology (Govt. of India), Trivandrum, India, and the Director, Indian Institute of Technology Hyderabad for their support and for providing the infrastructure facility for carrying out this work

References

Bingöl, D., Hercan, M., Elevli, S. and Kılıç, E. (2012). Comparison of the results of response surface methodology and artificial neural network for the biosorption of lead using black cumin. *Bioresource technology*, *112*, 111–115.

Cameron, S.A., Lozier, J.D., Strange, J.P., Koch, J.B., Cordes, N. et al. (2011). Patterns of widespread decline in North American bumble bees. *Proceedings of the National Academy of Sciences*, *108*(2), 662–667.

Cervone, G., Franzese, P., Ezber, Y. and Boybeyi, Z. (2008). Risk assessment of atmospheric emissions using machine learning. *Natural Hazards and Earth System Sciences*, *8*(5), 991–1000.

Chen, X., Roberts, R., Tong, W. and Liu, Z. (2022). Tox-GAN: an artificial intelligence approach alternative to animal studies—a case study with toxicogenomics. *Toxicological Sciences*, *186*(2), 242–259.

FDA, U. (2017). FDA's predictive toxicology roadmap. *US Food and Drug Administration, Silver Spring, MD.*

Gemmati, D., Varani, K., Bramanti, B., Piva, R., Bonaccorsi, G. et al. (2019). "Bridging the gap" everything that could have been avoided if we had applied gender medicine, pharmacogenetics and personalized medicine in the gender-omics and sex-omics era. *International Journal of Molecular Sciences*, *21*(1), 296.

Goetz, L.H. and Schork, N.J. (2018). Personalized medicine: motivation, challenges, and progress. *Fertility and sterility*, *109*(6), 952–963.

Grulke, C.M., Williams, A.J., Thillanadarajah, I. and Richard, A.M. (2019). EPA's DSSTox database: History of development of a curated chemistry resource supporting computational toxicology research. *Computational Toxicology*, *12*, 100096.

Gutiérrez, P., Godoy, S.E., Torres, S., Oyarzún, P., Sanhueza, I. et al. (2020). Improved antibiotic detection in raw milk using machine learning tools over the absorption spectra of a problem-specific nanobiosensor. *Sensors*, *20*(16), 4552.

Hachem, M. and Sharma, B.K. (2019, February). Artificial Intelligence in Prediction of PostMortem Interval (PMI) Through Blood Biomarkers in Forensic Examination–A Concept. In *2019 Amity International Conference on Artificial Intelligence (AICAI)* (pp. 255–258). IEEE.

Han, F., Huang, X. and Teye, E. (2019). Novel prediction of heavy metal residues in fish using a low-cost optical electronic tongue system based on colorimetric sensors array. *Journal of food process engineering*, *42*(2), e12983.

Hartung, T. (2023). Artificial intelligence as the new frontier in chemical risk assessment. *Frontiers in Artificial Intelligence*, *6*.

Hartung, T. (2023a). ToxAIcology-The evolving role of artificial intelligence in advancing toxicology and modernizing regulatory science. *ALTEX-Alternatives to animal experimentation*, *40*(4), 559–570.

Hartung, T. and Hoffmann, S. (2009). Food for thought... on in silico methods in toxicology. *ALTEX-Alternatives to animal experimentation*, *26*(3), 155–166.

Hemmerich, J. and Ecker, G.F. (2020). In silico toxicology: From structure–activity relationships towards deep learning and adverse outcome pathways. *Wiley Interdisciplinary Reviews: Computational Molecular Science*, *10*(4), e1475.

Holmes, J., Sacchi, L. and Bellazzi, R. (2004). Artificial intelligence in medicine. *Ann R Coll Surg Engl*, *86*, 334–8.

https://preadmet.webservice.bmdrc.org/

Judson, R., Houck, K., Martin, M., Knudsen, T., Thomas, R.S. et al. (2014). In vitro and modelling approaches to risk assessment from the US Environmental Protection Agency ToxCast programme. *Basic & clinical pharmacology & toxicology*, *115*(1), 69–76.

Khakare, M.R. and Fatangare, S. (2020). Survey on Prediction of Post Mortem Interval using Artificial Intelligence in Forensic Examination.

Klein, A.M., Vaissière, B.E., Cane, J.H., Steffan-Dewenter, I., Cunningham, S.A. et al. (2007). Importance of pollinators in changing landscapes for world crops. *Proceedings of the royal society B: biological sciences*, *274*(1608), 303–313.

Kleinstreuer, N.C., Sullivan, K., Allen, D., Edwards, S., Mendrick, D.L. et al. (2016). Adverse outcome pathways: from research to regulation scientific workshop report. *Regulatory toxicology and pharmacology*, *76*, 39–50.

Kleinstreuer, N.C., Tetko, I.V. and Tong, W. (2021). Introduction to special issue: computational toxicology. *Chemical Research in Toxicology*, *34*(2), 171–175.

Kobayashi, Y., Uchida, T. and Yoshida, K. (2020). Prediction of soil adsorption coefficient in pesticides using physicochemical properties and molecular descriptors by machine learning models. *Environmental Toxicology and Chemistry*, *39*(7), 1451–1459.

Kolluri, S., Lin, J., Liu, R., Zhang, Y. and Zhang, W et al. (2022). Machine learning and artificial intelligence in pharmaceutical research and development: a review. *The AAPS journal*, *24*, 1–10.

Lauret, P., Heymes, F., Aprin, L. and Johannet, A. (2016). Atmospheric dispersion modeling using artificial neural network based cellular automata. *Environmental Modelling & Software*, *85*, 5–-69.

LeCun, Y., Bengio, Y. and Hinton, G. (2015). Deep learning. *nature*, *521*(7553), 436–444.

Li, T., Tong, W., Roberts, R., Liu, Z. and Thakkar, S. et al. (2021). DeepCarc: Deep learning-powered carcinogenicity prediction using model-level representation. *Frontiers in artificial intelligence*, *4*, 757780.

Li, X., Zhang, Y., Chen, H., Li, H. and Zhao, Y. et al. (2017). Insights into the molecular basis of the acute contact toxicity of diverse organic chemicals in the honey bee. *Journal of Chemical Information and Modeling*, *57*(12), 2948–2957.

Lin, Z. and Chou, W.C. (2022). Machine learning and artificial intelligence in toxicological sciences. *Toxicological Sciences*, *189*(1), 7–19.

Lu, F., Xu, D., Cheng, Y., Dong, S., Guo, C. et al. (2015). Systematic review and meta-analysis of the adverse health effects of ambient PM2. 5 and PM10 pollution in the Chinese population. *Environmental research*, *136*, 196–204.

Maertens, A., Golden, E., Luechtefeld, T.H., Hoffmann, S., Tsaioun, K. et al.. (2022). Probabilistic risk assessment–the keystone for the future of toxicology. *Altex*, *39*(1), 3.

Mak, K.K. and Pichika, M.R. (2019). Artificial intelligence in drug development: present status and future prospects. *Drug discovery today*, *24*(3), 773–780.

Mayr, A., Klambauer, G., Unterthiner, T. and Hochreiter, S. (2016). DeepTox: toxicity prediction using deep learning. *Frontiers in Environmental Science*, *3*, 80.

McArt, S.H., Urbanowicz, C., McCoshum, S., Irwin, R.E. and Adler, L.S. et al. (2017). Landscape predictors of pathogen prevalence and range contractions in US bumblebees. *Proceedings of the Royal Society B: Biological Sciences*, *284*(1867), 20172181.

McDaniel, M., Carter, J., Keller, J. M., White, S. A. and Baird, A et al. (2019). Open source pharmacokinetic/pharmacodynamic framework: tutorial on the BioGears Engine. *CPT: Pharmacometrics & Systems Pharmacology*, *8*(1), 12–25.

Miller, T.H., Gallidabino, M.D., MacRae, J.I., Owen, S.F., Bury, N.R. et al. (2019). Prediction of bioconcentration factors in fish and invertebrates using machine learning. *Science of the Total Environment*, *648*, 80–89.

O'Sullivan, S., Holzinger, A., Zatloukal, K., Saldiva, P., Sajid, M.I. et al. (2017). Machine learning enhanced virtual autopsy. *Autopsy & case reports*, *7*(4), 3.

O'Boyle, N.M., Banck, M., James, C.A., Morley, C., Vandermeersch, T. et al. (2011). Open Babel: An open chemical toolbox. *Journal of cheminformatics*, *3*, 1–14.

Perez Santin, E., Rodríguez Solana, R., González García, M., García Suárez, M.D.M., Blanco Díaz, G.D. et al. (2021). Toxicity prediction based on artificial intelligence: A multidisciplinary overview. *Wiley Interdisciplinary Reviews: Computational Molecular Science*, *11*(5), e1516.

Qiu, S. and Wang, J. (2017). The prediction of food additives in the fruit juice based on electronic nose with chemometrics. *Food chemistry*, *230*, 208–214.

Reel, P.S., Reel, S., Pearson, E., Trucco, E. and Jefferson, E. et al. (2021). Using machine learning approaches for multi-omics data analysis: A review. *Biotechnology advances*, *49*, 107739.

Rim, K.T. (2020). In silico prediction of toxicity and its applications for chemicals at work. *Toxicology and environmental health sciences*, *12*(3), 191–202.

Russom, C.L., Bradbury, S.P., Broderius, S.J., Hammermeister, D. E. and Drummond, R.A. et al (1997). Predicting modes of toxic action from chemical structure: Acute toxicity in the fathead minnow (Pimephales promelas). *Environmental Toxicology and Chemistry: An International Journal*, *16*(5), 948–967.

Scheule, B. and Sneed, J. (2001). From farm to fork: critical control points for food safety. *Journal of nutrition in recipe & menu development*, *3*(2), 3–23.

Sergeev, A.P., Buevich, A.G., Baglaeva, E.M. and Shichkin, A.V. (2019). Combining spatial autocorrelation with machine learning increases prediction accuracy of soil heavy metals. *Catena*, *174*, 425–435.

Shen, J., Li, W., Deng, S. and Zhang, T. (2021). Supervised and unsupervised learning of directed percolation. *Physical Review E*, *103*(5), 052140.

Singh, K.P., Gupta, S. and Rai, P. (2013). Predicting acute aquatic toxicity of structurally diverse chemicals in fish using artificial intelligence approaches. *Ecotoxicology and environmental safety*, *95*, 221–233.

Singh, Kunwar P., Nikita Basant and Shikha Gupta. "Support vector machines in water quality management." *Analytica chimica acta* 703.2 (2011): 152–162.

Suter, L., Babiss, L.E. and Wheeldon, E.B. (2004). Toxicogenomics in predictive toxicology in drug development. *Chemistry & biology*, *11*(2), 161–171.

Tan, N.X., Li, P., Rao, H.B., Li, Z.R. and Li, X.Y. et al. (2010). Prediction of the acute toxicity of chemical compounds to the fathead minnow by machine learning approaches. *Chemometrics and Intelligent Laboratory Systems*, *100*(1), 66–73.

Tebes-Stevens, C., Patel, J.M., Koopmans, M., Olmstead, J., Hilal, S.H. et al. (2018). Demonstration of a consensus approach for the calculation of physicochemical properties required for environmental fate assessments. *Chemosphere*, *194*, 94–106.

Tetko, I.V., Klambauer, G., Clevert, D.A., Shah, I. and Benfenati, E. et al. (2022). Artificial intelligence meets toxicology. *Chemical research in toxicology*, *35*(8), 1289–1290.

Tran, T.T.V., Surya Wibowo, A., Tayara, H. and Chong, K.T. (2023). Artificial intelligence in drug toxicity prediction: recent advances, challenges, and future perspectives. *Journal of Chemical Information and Modeling*, *63*(9), 2628–2643.

Vedani, A., Dobler, M. and Smieško, M. (2012). VirtualToxLab—a platform for estimating the toxic potential of drugs, chemicals and natural products. *Toxicology and applied pharmacology*, *261*(2), 142–153.

Wang, H., Liu, R., Schyman, P. and Wallqvist, A. (2019). Deep neural network models for predicting chemically induced liver toxicity endpoints from transcriptomic responses. *Frontiers in pharmacology*, *10*, 425033.

Wankhade, T.D., Ingale, S.W., Mohite, P.M., Bankar, N.J., Wankhade, T. et al. (2022). Artificial intelligence in forensic medicine and toxicology: the future of forensic medicine. *Cureus*, *14*(8).

Yang, G., Lee, H. and Lee, G. (2020). A hybrid deep learning model to forecast particulate matter concentration levels in Seoul, South Korea. *Atmosphere*, *11*(4), 348.

Yang, H., Sun, L., Li, W., Liu, G. and Tang, Y. (2018). In silico prediction of chemical toxicity for drug design using machine learning methods and structural alerts. *Frontiers in chemistry*, *6*, 30.

Zhang, F., Sun, B., Diao, X., Zhao, W. and Shu, T. et al. (2021). Prediction of adverse drug reactions based on knowledge graph embedding. *BMC Medical Informatics and Decision Making*, *21*, 1–11.

Index

About the Editor

P.V. Mohanan was a Fellow of National Academy of Science, India and Royal Society of Biologists, UK. He was a JSPS Post doctoral Fellow at the University of Tsukuba, Japan in the field of Neurotoxicity. As a toxicologist, he was associated with all the medical devices/technologies developed at SCTIMST (Government of India). He headed both the Division of Toxicology and Dept. of Applied Biology. Dr Mohanan was a member in the Empowered committee on the 'Rapid Response Regulatory Framework for COVID-19 to deal with applications for development of vaccines, diagnostics, prophylactics and therapeutics and an expert member in the **statutory Committee, RCGM, DBT, New Delhi. Dr Mohanan also served as an expert member at the DST, SERB, DBT, ICMR, CSIR and FSSAI Scientific committees**. He was an Expert member of the Joint Food and Agriculture Organization of the United Nations and the World Health Organization (JECFA), USA. He was a member in the Scientific Advisory Committee of ICMR-NARFBR, Hyderabad and a Visiting Professor and Visiting Researcher at Toyo University, Japan and also a Certified Biological Safety Specialist. Dr Mohanan was an Adjunct Professor at the Indian Institute of Technology, Hyderabad and Jamia Hamdard University, New Delhi, India. He received a lifetime achievement award from the Society of Toxicology India for outstanding contribution in the field of toxicology. He patented an ELISA kit for the measurement of pyrogenicity. Dr Mohanan made significant contributions to the development of medical device regulations in India. He received a certificate of appreciation from the Ministry of Science and Technology, Government of India for his contribution to India getting full adherent status on GLP from OECD. He authored more than 300 publications, edited 10 books, 5 patent filed and 5 design registrations. He was the Secretary General of Society of Toxicology, India and Vice President, Kerala Academy of Sciences, India.

For Product Safety Concerns and Information please contact our EU representative GPSR@taylorandfrancis.com Taylor & Francis Verlag GmbH, Kaufingerstraße 24, 80331 München, Germany

Printed and bound by CPI Group (UK) Ltd, Croydon, CR0 4YY
14/09/2026
02216241-0001